SPECIAL PUBLICATION
of CARNEGIE MUSEUM OF NATURAL HISTORY

CONTRIBUTIONS IN QUATERNARY
VERTEBRATE PALEONTOLOGY:
A VOLUME IN MEMORIAL TO
JOHN E. GUILDAY

edited by

HUGH H. GENOWAYS
Curator of Mammals

MARY R. DAWSON
*Chief Curator, Earth Sciences
and
Curator of Vertebrate Fossils*

NUMBER 8   **PITTSBURGH, 1984**

SPECIAL PUBLICATION OF CARNEGIE MUSEUM OF NATURAL HISTORY
Number 8, pages 1–538

Issued 11 September 1984

Price: $56.00 a copy

Robert M. West, *Director*

ISBN 0-935868-07-0

Library of Congress card number 84-70729

CARNEGIE MUSEUM OF NATURAL HISTORY, 4400 FORBES AVENUE
PITTSBURGH, PENNSYLVANIA 15213

CONTENTS

INTRODUCTION

Mary R. Dawson and Hugh H. Genoways

The loss of a good friend and outstanding colleague, as John Guilday certainly was to us, tends to leave the survivors with mixed feelings of helplessness and "let's do something." Fortunately, following the example set time and again by John himself, our feelings of wanting to take action soon won out and this volume, devoted to vertebrates from the Quaternary of North America, is the result. This broad subject, also set by John, was his consuming scientific interest for the past forty years. His own work, especially on the Quaternary of eastern North America, contributed greatly to understanding of this complex, dynamic interval of time, as can be appreciated fully by reading his bibliography.

Once we had decided on a memorial volume as an appropriate tribute, we wrote to Quaternary specialists in North America and Europe, asking for their participation in the volume. The response to our letters was very encouraging. The people who know John's work best were eager to contribute to a volume in memory of our remarkable friend. Letters of acceptance came in, many accompanied by written tributes to the man we are honoring. Among the expressions of individuals' appreciation for John, are these good examples. "I am honored to be asked to contribute as I admired and valued John greatly." "I greatly appreciate the opportunity to honor John Guilday's memory by writing a chapter for his book." "As you know John and I had been friends for a good many years and have co-authored several papers; I was pleased to learn such a volume honoring his memory is already in the planning stages." "I did, indeed, know that John Guilday had passed away. Sad, not only because of the unparalleled quality of his work on Pleistocene and Holocene North American faunas, but also, on a more personal level, because he was one of my personal heros, and I have very few." Such responses indicated clearly the appropriate nature of our endeavor.

Contributions cover a wide scope of studies on Quaternary faunas, ranging from faunal studies, through paleoecological and taphonomic approaches, to theoretical discussions. The papers are arranged in a more-or-less systematic order, with lower vertebrates heading the list. Theoretical papers bring up the rear, not out of disrespect but because they tend to summarize evidence from all systematic groups. We thank all the contributors for their willingness to participate and their cooperation and promptness in providing us with well prepared manuscripts. We are glad to have been able to assemble this fine collection of papers devoted to the Quaternary. Our only regret is that the volume is unable to contain a paper by John Guilday.

Many of John's accomplishments were recognized by his peers during his lifetime. Bjorn Kurtén and Elaine Anderson, for example, chose to dedicate their "Pleistocene Mammals of North America" to John and Alice. Four species were named for John during his lifetime: one fossil amphibian, *Cryptobranchus guildayi* Holman, 1977; and three fossil rodents, *Leptotomus guildayi* Black, 1971, *Proneofiber guildayi* Hibbard, 1977, and *Microtus guildayi* van der Meulen, 1978. Recognition continues to come to John posthumously. In this volume, we find additions to John's taxonomic spectrum with a new owl, *Otus guildayi,* by Brodkorb and Mourer-Chauvire, a new genus of arvicoline rodent, *Guildayomys,* dedicated by Zakrzewski, and a new subspecies of dire wolf, *Canis dirus guildayi,* provided by Kurtén. Promised by Richard Lund, a long time friend of John, is a new order of Paleozoic fishes.

One gauge of a person's accomplishments is frequency of citation of his work. Here John rates highly. One typical instance is in the chapter, "Terrestrial vertebrate faunas" in the recently published volume 1 of "Late Quaternary Environments of the United States" in which 18 papers with John as senior author are cited.

A further posthumous recognition was the award to John by the Society of American Archaeology of the Fryxell Award for Interdisciplinary Research. The tribute with the award appraised this aspect of his scientific endeavors: "His own research interests, involving especially cave and fissure-deposited fossil assemblages, led him to publish a number of seminal studies on Pleistocene-Holocene faunal changes, which supplied crucial environmental data on the backgrounds of ancient human communities in the Northeast. Additionally, he graciously made his skills available to solving a range of archaeologically generated problems, from the identification of the occasional puzzling specimen to the analysis and interpretation of excavated collections as large as

1

that from Meadowcroft Rockshelter. The quality of his work, whether on his own projects or those of others, set standards for paleontological research and interdisciplinary cooperation that apply continent-wide.

"Guilday contributed significantly to a wide range of scientific problem domains that are shared between archaeology and paleontology: human behavior as manifested in patterns of butchering and food processing; microclimatic information derived from archaeological faunal assemblages; tapho-nomic processes in caves and rockshelters; the discrimination of human from animal contributions to site bone assemblages; relationships between changing faunal assemblages and other environmental indicators such as fossil plant associations; patterns and causes of faunal extinctions; macroclimatic change and its effects on faunal distributions."

So our memorial volume joins other acknowledgements of the life and contributions of a truly remarkable man.

JOHN EDWARD GUILDAY (1925–1982)

Mary R. Dawson

On November 17, 1982, John Guilday, our friend and colleague to whom this volume is dedicated, lost his fight against medical complications following pneumonia. This was one of John's few defeats, for his life was marked by numerous victories, frequently against odds that would have overwhelmed most mortals. John's scientific dedication was to the Quaternary. His contributions to the understanding of this complex, relatively recent interval of earth history were outstanding. But John was more than a dedicated scientist, he was a full, rich human being. Robert J. Gangewere, editor of Carnegie Magazine, expressed this well shortly after John's death, when he wrote, "In a world in which so many people are claimed to be unique, John Guilday truly was. Part of his gift was in being like everyone else while enduring the most disabling effects of polio. At the same time he achieved international esteem as a paleontologist in the Section of Vertebrate Fossils. The fact that he wrote beautifully, in a remarkable, poetic way, was the only fact that the general public may have known about him. But those who knew him personally, knew immediately that he was unforgettable, and that his story was one that would be with them for life."

An active, supportive family and early interests in natural history formed a basic part of the background for John's development. He was born in Pittsburgh, Pennsylvania, on 23 October 1925, first child of John W. and Margaret McKinley Guilday. His sister Marjorie, one and one-half years younger, was his childhood buddy and frequent co-conspirator as well as his lifelong friend. To his second sister, Patricia, sixteen years younger, John was an example and idol. John's bent for natural history expressed itself through collecting moths and similar boyhood activities.

A strong influence on John, as a teen-ager, was his acquaintance with the noted comparative anatomist S. Harmsted Chubb. John, living with his family near New York City in the late 1930s and early 1940s, met Dr. Chubb during a visit to the American Museum of Natural History. As John wrote later, "My first inkling that a bone might be something other than a passive Halloween prop came from the comparative anatomist S. Harmstead Chubb. Dr. Chubb was almost 80 when I first met him, a small, happy man with bright-blue eyes and a most elegant, snow-white, impeccably trimmed Edwardian beard. . . . After we had come to know one another Dr. Chubb, brimming with an enthusiasm that belied his years, would take me from case to case, pointing out small features that only served to emphasize the care and accuracy that went into the making of such skeletal mounts." Thus, John's interest in osteology was fired. He worked as a volunteer preparator in the Osteology section of the American Museum of Natural History in 1941 and 1942. John and the family moved back to Pittsburgh, and John served as a volunteer in the Section of Mammals, Carnegie Museum, in 1942 and 1943. He kept in close contact with Dr. Chubb, who continued to be a strong influence, writing to John in 1943, "Now as we look ahead toward college it seems to me that neither the medical or veterinary course would suit your purpose best. I would rather advise broader scientific study and then specializing in Zoology and Comparative Anatomy."

But college had to wait. In 1944 John entered the United States Army. He had hoped for assignment to the medical corps, but the army chose the infantry for him. He was in the 329th "Buckshot" Infantry Regiment of the 83rd Division. Even during his army service, John's interests in natural history reigned supreme. In March 1944, he wrote to J. Kenneth Doutt, Curator of Mammals at Carnegie Museum, from Camp Wheeler, Georgia: "This afternoon we were having an outdoor lecture on military courtesy. I'm afraid I didn't pay much attention to the speaker though because three gray squirrels were playing tag in a tree right above him. They seem to be very small in comparison with the ones I used to see in New York." To Caroline Heppenstall, then Assistant in the Section of Mammals, Carnegie Museum, he wrote in July, "I'm afraid from the way things are going so far I'll get very little time for collecting mammals, but I'll try my best." John's service included time in Luxembourg and Belgium. His regiment fought through the bitter winter of 1944–1945 in the Battles of the Ardennes and the Bulge. Through all this stress, John kept up his interests in natural history. In field notes sent to Caroline Heppenstall from Europe there is this entry: "Tohogne, Prov. Namur, Belgium, Feb. 1, 1945.

JOHN EDWARD GUILDAY (1925–1982)

While passing through a meadow covered with a snow, during the first warm spell of the season found a dead Microtus (?). Measurements—The circumstances at the time would not permit the taking of any type of measurements. Sex (?) dissection impossible." This masterpiece of understatement described John's preoccupations shortly after he took part in some of the major battles of World War II. In July 1945, after the war in the European theater had ended and John was in Austria, he wrote to Caroline Heppenstall about collecting birds but complained, "Can't make many skins though because a 30 caliber really chews them up . . . ." Such were the difficulties of being an infantryman with interests in collecting specimens. Many of the animals John collected in Europe are in the collections of Carnegie Museum's Section of Mammals.

After discharge from the army, John returned to Pittsburgh and began his undergraduate work in the Zoology Department at the University of Pittsburgh, from which he received his Bachelor of Science degree in 1950. While still an undergraduate, John worked as a biological technician for the Pennsylvania Game Commission, working out of Carnegie Museum between 1947 and 1950 on the Pennsylvania Mammal Survey. In 1950 he worked also as a laboratory assistant in the Section of Vertebrate Fossils. The year 1950 was marked by one of the best things John ever did—he married Alice Mekeel. John met Alice in the museum office, where she worked as secretary to the director. Appropriately enough, John and Alice spent their honeymoon on a fossil collecting expedition to Utah, Montana, and British Columbia with the museum's vertebrate paleontologist, J. LeRoy Kay. In 1951, John began graduate work in zoology at the University of Pittsburgh, and joined the curatorial ranks in the museum as Assistant Curator of Comparative Anatomy. His work included research on the rice rat from archaeological sites in West Virginia and popular articles for Carnegie Museum. In many of his scientific and writing projects, Caroline Heppenstall was his valued mentor.

In the early fall of 1952, when his daughter Kathleen was just ten months old, John was struck down with polio. As John's long time friend and associate Allen McCrady wrote, at this time John began ". . . the most remarkable portion of his career. Although he had little use of his hands and arms, with Alice's unfailing aid and some clever effort-saving devices, John began turning out an incredible amount of quality research from his basement lab." John's di-

rect interests in Quaternary mammals went back at least to 1947 when he took part in a museum expedition to the vicinity of the town of New Paris in Bedford County, Pennsylvania. The nearly complete skeleton of the extinct eastern wapiti, *Cervus canadensis canadensis,* with a flint arrowhead embedded in its neck was excavated from a sinkhole in the Helderberg limestone by John and colleagues from the Pittsburgh Grotto of the National Speleological Society and the Explorers Club of Pittsburgh. So when John's health allowed him to return to scholarly pursuits, it was natural that his interests should become channeled toward animal remains from archaeological sites and thence to the earlier parts of the Quaternary. His studies included careful, reasoned analyses of butchering techniques, with results that have helped many others to determine what are, and are not, signs of human activity related to bones. Very important in John's productivity were the results of a return, in August 1956, to the New Paris region by a group from the museum (Roy Kay) and the Pittsburgh Grotto (including Allen McCrady, Ralph Bossart, John Leppla). Later joined by Harold Hamilton and A. C. Lloyd, these people and a group of enthusiastic volunteers worked on the excavation of Sinkhole No. 4, or Lloyd's Rock Hole, until sterile cave-derived sediments were encountered in March 1963. The paper resulting from this work, co-authored with Paul Martin and Allen McCrady and published in 1964, demonstrated climatic changes from cool taiga parkland through subsequent warming, and gave, by study of plant and animal remains, a clear picture of the late Pleistocene environment of non-glaciated Pennsylvania. The work of John, his co-workers, and dedicated field crews firmly established the nature of late Pleistocene biota and climates in the eastern United States, contributing decisively where there had been almost a void. As John wrote in a review article in 1971, "As late as 1958, Martin stated, 'Rather than in eastern North America, the main full-glacial refugia for tundra mammals and birds lay in unglaciated Alaska.' This was based upon the almost total lack of vertebrate paleontological evidence. Since that time, accelerating research, especially in cave sites, has changed this picture; and tundra forms, as well as those characteristic of all northern ecotopes, have been recovered from at least some part of the Appalachian system." Later in the same review John wrote, "It is frustrating to review the history of the Appalachian mammal fauna from the time of the Wisconsinan maximum to modern times because of the rapidity of new discoveries and the knowledge that many blanks in the paleontological record will be filled in within the next few decades. Research activities within the past 10 years have shown that these deposits exist in such numbers in areas of karst that they can be profitably studied at more than a regional level. Just how refined, geographically, this can be done remains to be seen; but each year brings additional reports of sites as cavers become more aware of the possibilities." Much of the refinement, of course, came from the continuing work of John Guilday. Site after site of Quaternary mammals came under John's scrutiny, and his thoughtful analyses were published. Robinson Cave, Tennessee, Welsh Cave, Kentucky, Clark's Cave, Virginia, Baker Bluff, Tennessee, yielded bones to John's crews and information to the studies of John and his co-workers. Cumberland Cave, Maryland, a pre-Wisconsinan site, was reopened by Allen McCrady and Harold Hamilton, John's two closest "field lieutenants." Remains from the pre-Wisconsinan Hanover Quarry Fissure of Adams County, Pennsylvania, occupied John as one of the last projects he undertook. Throughout his career, John also retained his interest in and activity on archaeological sites. The Buffalo site in West Virginia was a big undertaking, with more than 61,000 bones identified and analyzed; the Meadowcroft Rock Shelter was another nearly Herculean task, with around 300,000 bones and teeth to be identified. During this time of research productivity, John was also an active contributor to other museum activities.

In 1962 he was promoted to Associate Curator of Comparative Anatomy, and in 1964 joined the Section of Vertebrate Fossils. His contributions to the educational and exhibition projects of the museum were always valuable.

John's productivity is known; the full impact of his work is difficult to assess so soon, but it can be anticipated at least partly from a study of his bibliography. Never one to be daunted by investigations avoided by others, John undertook and completed difficult studies of variation. His scientific contributions include analyses of variation in *Blarina* and, surely even more difficult, of dental variation in three species of *Microtus.* Bob Gangewere, though not a scientist himself, had the proper perspective to summarize John's contributions very well: "After surviving some of the bitterest fighting of World War II unharmed, he returned to Pittsburgh to resume his career in the museum and contracted polio. The ravages of this disease left him

permanently disabled, and his periodic use of the iron lung in his home was essential to his health. His visits to the museum were extremely rare, and many of his colleagues never laid eyes on him. Nevertheless, with the constant help of his wife Alice, his scientific work at home prospered, and as a paleontologist his reputation spread around the world. The museum directors under whom John served, and all of his closest associates, knew that John Guilday was one of the research strengths of Carnegie Museum of Natural History."

A glance at John's bibliography shows another aspect of his productivity, its extension to popularizations of science. He contributed frequently to journals such as The Netherworld News, Pennsylvania Game News, Carnegie Magazine, and The Explorer. Enticing titles such as "New Paris Sinks," "Scien-terrific Names," "A Rat is a Rat is a Rat?," "The Sturgeon Surgeon," "The Faunagraph Record" (how I groaned when he told me that one!), "Confessions of an Aging Paleontologist" make it almost impossible not to seek out and read the articles. The popularizations contained good science so well written and interestingly presented that any reader would finish an article by Guilday richer in knowledge as well as understanding.

One of my favorites was John's "Ask the Experts . . . ," published in 1970 in The Explorer. I was involved in the case, as it fell to me to go and look at a very large bone in a public park in Martinsburg, Pennsylvania. John had seen photographs of this enormous specimen, more than six feet across, and began his odyssey about a specimen that he first imagined to be a giant muskox. As Associate Director James Swauger and I set off to see the specimen, John stayed home thinking about it, as he later wrote: "Meanwhile, Guilday is pouring over the pictures with a ten-power magnifying glass, his feet lightly touching the floor now and again. But still no details. Assume that we are looking at the skull from the top, then the horns sweep first down, then out and forward in good ovibovine fashion, in which case, given the relative dimensions of modern members of that subfamily, the beast must have stood no less than eight feet high at the shoulder to accommodate such a tremendous skull! If, on the other hand, the skull is upside down, then the horns would sweep in the opposite manner, somewhat reminiscent of a bison. Now there are bison known from early Pleistocene deposits that were very large indeed. *Bison latifrons* had a horn span of ten feet, but the horns were not as massive and the skull itself much smaller than whatever it is in the park shelter in Martinsburg. Assignment to the *Bison latifrons* group would create additional problems if the skull had originated at Frankstown Cave. The fauna from that cave does not appear to be as old as these fossil bison are known to be. If it were indeed a bison, shoulder height would then be, counting hump and all, truly as high as that of an elephant.

"Guilday spends all night turning restlessly, goaded on all sides by great blue oxen of Bunyanesque proportion, mentally writing and rewriting descriptions of what must surely be a new genus and species. *Preposterobos*! *Preposterobos incredibilous*! First thing to do. Photograph Pittsburgh Pleistocene people standing beside skull, bearing small placard EAT YOUR HEART OUT, CLAYTON! (Dr. Clayton E. Ray, U.S. National Museum, foremost authority on fossil muskoxen). But Walter Mitty thoughts start muddying up this euphoric dream. Suppose a truck hits it between now and tomorrow? What condition is it in? Can we get permission to borrow the specimen and study it? Must get some sleep. But sleep comes fitfully to him whose bed vibrates to the collective hooves of multi-ton muskoxen—especially scientifically undescribed ones—who roll their massive eyeballs as they deftly hook up and carry off one's bedsheets as they thunder through the room.

"The Swauger-Dawson expedition returns from a first-hand assessment of the situation and dull thuds can be heard throughout the paleo section as everyone returns to earth. Everyone except Curator Berman, whose focus is upon Paleozoic reptiles and who is more inclined to excitement over things with scales and forked tongues.

"'It' turns out to be the base of a skull of a large whale donated to the town of Martinsburg by a patriotic citizen to form part of a sailors' memorial in the park. The mighty horn cores are but the remnants of the glenoid fossae—which afford articulation for the lower jaw." John goes on with, "The moral of this story being never to show two paleontologists, one from Michigan and one from western Pennsylvania, any part of a whale skull from high in the Appalachian Mountains. It gets them *too* excited."

John's popular writing was a marvelous combination of science, good style, and a gently satiric poking fun at himself and others. His abilities along this line became recognized not only in Pittsburgh and to his immediate colleagues but also to a wider audience. Bob Gangewere noted, in a postscript to John's "Conjure Bones": "An editor would be re-

miss if he did not point out, at least once, that the writing of Johnny Guilday is very special indeed, and belongs in a class of popular writing about natural history subjects that few authors ever attain. The combination of scientific acumen and poetic insight is remarkable, as more than one reader has noted in the past few years. In fact, a museum administrator from Florida once observed that John's essay defining a natural history museum is the best thing he ever read on the subject, and is required reading for his college students in a museum studies course. That essay is 'The Long Journey'."

In "Caves, Bones and Ice Age Owls" John explained his science in terms full of meaning for scientists and laymen alike: "Science proceeds in no set direction, neither forward toward an unknown future, nor backward to our ultimate beginnings, but radially toward that horizon separating the known from the unknown that stretches unbroken, like the rim of the sky, there in whatever direction we turn. So that research directed toward one point becomes a part of a circuitous whole that expands not by increments, but by ripples from the common center upon which we stand, the whole and its parts inextricably mixed. And, that is why we pick up little bones."

Very little known except to close friends was John's very considerable ability as an artist. Only Alice can explain just how he did it, but John produced Christmas cards, an illustrated Book of Owls for McCrady, Book of Rabbits for Dawson, Book of Birds and Butterflies for Harry and Mary Clench, all aptly captioned by Alice and Kathleen. He made tiles, and his watercolors were masterpieces of oriental-style beauty.

John's poetry ranged from whimsical to the serious. The lighter side of his writing is well illustrated by his *Thoughts on Migration*:

> Oh, little feathered dinosaur,
> Kin to croc and pterosaur,
> Tell us, oh continent-spanning babbler,
> Were you the original Archosaur Traveler?

John Guilday's friends and colleagues gathered together to memorialize him on 19 November 1982. M. Graham Netting, Director Emeritus of Carnegie Museum, under whom John had worked for many years, shared his impression of John—excellent scientist, keen, sensitive observer; a man, if he once had faults had had them burned away by polio. Leonard Krishtalka, fellow Associate Curator in Vertebrate Fossils, named John, "A man for all rea-

Water Shrew. An original painting in color on tile by John E. Guilday. John's feelings about shrews are expressed by the following two quotations from his own writings: "They are nervous, high-strung little beasts with the energy of junior executives." "A shrew's head looks as if it had freshly emerged from a pencil sharpener. In life its rubbery little trunk is constantly bobbing, testing, probing; in death the little snout sticks stiffly forward, the trade mark of a shrew."

sons." Bob Gangewere concluded his tribute to John with a lovely, apt summation of the man: "In many ways John Guilday's life was rich and normal, although circumscribed. Alice and John had reason to be proud of their daughter, Kathleen, and their home was full, full of the everyday business of their lives, and of the scientific work. In the basement John had the evidence, the information, that he needed. To the untutored eye an incredible collection of miniature bones from ages past—to him, a world of meaning about past life, past environments.

"For many years John was a contributor to Carnegie Magazine, where the poetic side of his scientific mind found a natural outlet. Readers of his prose knew that word for word he was one of the best writers about nature they were ever likely to encounter. In 1981 he published his series of observations about the months of the year. His last thought about November, the month in which he died, was, 'It snows again, and deep, and the stars draw closer.'"

BIBLIOGRAPHY

Compiled by Alice Guilday

1948

The fallen monarch. Carnegie Mag., 22:46–49.
Little brown bats copulating in winter. J. Mamm., 29:416–417.

1949

Winter food of Pennsylvania mink. Pennsylvania Game News, 20(9):12, 32.

1950

Winter fetus in the little brown bat, *Myotis lucifugus*. J. Mamm., 31:96–97.

1951

What are fossils? Carnegie Mag., 25:352–354.

Sexual dimorphism in the pelvic girdle of *Microtus pennsylvanicus*. J. Mamm., 32:216–217.

1952

Rhinoceroses—a family tree. Carnegie Mag., 26:388–390.

An occurrence of the rice rat (*Oryzomys*) in West Virginia. J. Mamm., 33:253–255 (with W. J. Mayer-Oakes).

1955

Some biological aspects of archeology. Archeological Newsletter, 10:5–7.

Animal remains from the Globe Hill site 46Hk34-1. Appendix 1. Pp. 27–28, *in* The Globe Hill shell heap (site 46Hk34-1), Hancock County, West Va. (W. J. Mayer-Oakes), West Virginia Archeological Soc., Inc., Publ. Ser., 3:1–34.

Animal remains from an Indian village site, Indiana County, Pennsylvania. Pennsylvania Archaeologist, 25:142–147.

1956

Archeological evidence of the fisher in West Virginia. J. Mamm., 37:287.

[Letter to the editor]. Netherworld News, 4:31 and 4:68. [Reprinted in Speleo Digest 1956, Pittsburgh Grotto Press, Natl. Speleol. Soc., Pittsburgh, PA.]

New Paris sinks. Netherworld News, 4:85–86. [Reprinted in Speleo Digest 1956, Pittsburgh Grotto Press, Natl. Speleol. Soc., Pittsburgh, PA.]

Fossils and New Paris. Netherworld News, 4:113–114. [Reprinted in Speleo Digest 1956, Pittsburgh Grotto Press, Natl. Speleol. Soc., Pittsburgh, PA.]

Pennsylvania mammals—a history. Pennsylvania Game News, 27(12):20–26.

1957

Pennsylvania's biggest game—the whale. Pennsylvania Game News, 28(4):26–28. [Reprinted, pp. 253–255 *in* Pennsylvania Game News Treasury, Pennsylvania Game Commission, Harrisburg, 528 pp., 1979.]

[Reviews of] Mammals of the air, land and waters of the world, by G. G. Goodwin; and Living mammals of the world, by I. T. Sanderson. Carnegie Mag., 31:212–213.

Scien-terrific names. Pennsylvania Game News, 28(8):31–33.

[Field trip log for July] June 18. Netherworld News, 5:82.

[Letter to the editor]. Netherworld News, 5:89.

A bat's wing. Netherworld News, 5:175–176. [Reprinted in Powdermill Nature Reserve Educational Release No. 10, Carnegie Mus., 7 March 1958; and in Speleo Digest 1957, Pittsburgh Grotto Press, Natl. Speleol. Soc., Pittsburgh, PA.]

The masked shrew at Powdermill Nature Reserve. Powdermill Nature Reserve Educational Release No. 9, Carnegie Mus., 1 November 1957.

Individual and geographic variation in *Blarina brevicauda* from Pennsylvania. Ann. Carnegie Mus., 35:41–68.

1958

The mastodon—forest elephant. Pennsylvania Game News, 29(7):15–17.

Price tag on Penn's Woods? Pennsylvania Game News, 29(9):42–43.

The prehistoric distribution of the opossum. J. Mamm., 39:39–43.

Errata. Netherworld News, 6:28.

Report from New Paris. Netherworld News, 6:31. [Reprinted in Speleo Digest 1958, Pittsburgh Grotto Press, Natl. Speleol. Soc., Pittsburgh, PA.]

Educated guessing. Netherworld News, 6:39–42.

[Review of] The geology of the Hidden Valley Boy Scout area, Perry County, Pennsylvania, by J. T. Miller. Netherworld News, 6:91.

A rat is a rat is a rat! Netherworld News, 6:97–99. [Reprinted in Speleo Digest 1958, Pittsburgh Grotto Press, Natl. Speleol. Soc., Pittsburgh, PA.]

So what's a peccary? Netherworld News, 6:165–167. [Reprinted in Speleo Digest 1958, Pittsburgh Grotto Press, Natl. Speleol. Soc., Pittsburgh, PA.]

And some like it cold. Netherworld News, 6:252–254.

Reports from New Paris. Netherworld News, 6:22–23 (with N. D. Richmond).

A Recent fissure deposit in Bedford County, Pennsylvania. Ann. Carnegie Mus., 35:127–138 (with M. S. Bender).

Glacier bear. Carnegie Mag., 32:185–189 (with O. Agathon, senior author).

[Review of] Bones for the archaeologist, by I. W. Cornwall. Amer. Antiquity, 23:441.

Time and Powdermill. Powdermill Nature Reserve Educational Release No. 16, Carnegie Mus., 11 September 1958. [Reprinted in Carnegie Mag., 34:10; and Carnegie Mag., 52:31].

1959

[Review of] The great chain of life, by J. W. Krutch. Carnegie Mag., 33:32–33.

The flintlock forest. Carnegie Mag., 33:263–266, 269.

Box score—New Paris. Netherworld News, 7:55.

Now that you've got 'em—. Netherworld News, 7:96–98.

[Review of] Soils for the archaeologist, by I. W. Cornwall. Netherworld News, 7:140–141.

Slurp-slurp-slurp. Netherworld News, 7:199–200.

Say—A-a-a-a-h. Pennsylvania Game News, 30(12):45–46.

Muskrats at the Reserve. Powdermill Nature Reserve Educational Release No. 23, Carnegie Mus., 28 August 1959.

1960

Mouse bones. Netherworld News, 8:3–5. [Reprinted in Netherworld News, 17:140–143.]

Of microtines and men. Netherworld News, 8:127–129. [Reprinted in Speleo Digest 1960, Pittsburgh Grotto Press, Natl. Speleol. Soc., Pittsburgh, PA.]

Kangaroo rats. All-Pets Mag., 31(1):19–20.

The Fannin pedigree. Carnegie Mag., 34:307–309, 311.

. . . there's science in dump rooting. Pennsylvania Angler, 29(11):14–15.

Late Pleistocene records of the yellow-cheeked vole, *Microtus xanthognathus* Leach. Ann. Carnegie Mus., 35:315–330 (with M. S. Bender).

April and Powdermill. Powdermill Nature Reserve Educational Release No. 28, Carnegie Mus., 13 April 1960. [Reprinted in Carnegie Mag., 41:113–114; and Carnegie Mag., 52:7.]

1961

Vertebrate remains from the Varner Site (36-Gr-1). Pennsylvania Archaeologist, 31:119–124.

Prehistoric record of *Scalopus* from western Pennsylvania. J. Mamm., 42:117–118.

Plecotus from the Pleistocene of Pennsylvania. J. Mamm., 42:402–403.

Abnormal lower 3rd molar in *Odocoileus*. J. Mamm., 42:551–553.

The sky dance. Pennsylvania Game News, 32(4):44–45.

Jaguar—old style. Netherworld News, 9:41–43. [Reprinted in Speleo Digest 1961, Pittsburgh Grotto Press, Natl. Speleol. Soc., Pittsburgh, PA.]

Bones from Field House Cave, Pendleton County, West Virginia. Netherworld News, 9:93.

Bones from Jones Quarry Cave, 1 mile NE of Falling Waters, Berkeley County, West Virginia. Netherworld News, 9:94.

"Because it's there." Netherworld News, 9:105.

New cave organism discovered. Netherworld News, 9:127–128. [Reprinted in Speleo Digest, 1961, Pittsburgh Grotto Press, Natl. Speleol. Soc., Pittsburgh, PA.]

The pterosaur and the lemming. Netherworld News, 9:142–144.

The Ten Commandments of caving. Netherworld News, 9:197. [Reprinted in Speleo Digest 1961, Pittsburgh Grotto Press, Natl. Speleol. Soc., Pittsburgh, PA.]

The rice rat riddle. SPAAC Speaks, 1:23–25.

The collared lemming (*Dicrostonyx*) from the Pennsylvania Pleistocene. Proc. Biol. Soc. Washington, 74:249–250 (with J. K. Doutt).

1962

Notes on Pleistocene vertebrates from Wythe County, Virginia. Ann. Carnegie Mus., 36:77–86.

The Pleistocene local fauna of the Natural Chimneys, Augusta County, Virginia. Ann. Carnegie Mus., 36:87–122.

Bird remains from Pennsylvania archaeological sites. Sandpiper (Meadville, PA), 4:75–80.

The headhunters. Netherworld News, 10:53–54. [Reprinted in Speleo Digest 1962, Pittsburgh Grotto Press, Natl. Speleol. Soc., Pittsburgh, PA.]

The comings and goings of squirrels. Netherworld News, 10:67–70.

The case of the missing possam [sic]. SPAAC Speaks, 2:27–30.

Refuse and the archaeologist. Compost Science, 3(3):20–21.

Supernumerary molars of *Otocyon*. J. Mamm., 43:455–462.

The deer—of 350 years ago. Pennsylvania Game News, 33(12):10–13.

Portraits of the season. Pennsylvania Game News, 33:(1)47; (2)42; (3)58; (4)45; (5)11; (6)6; (7)25; (8)26; (9)11; (10)19; (11)7; (12)9. [Reprinted in Carnegie Mag. See 1981.]

The sturgeon surgeon. Pennsylvania Angler, 31(2):9.

Bone refuse from the Bristol Hills Site, Can 29-3. Manuscript in files of The Rochester Mus. Arts and Sciences, Rochester, NY.

The woodland jumping mouse and its kin. Powdermill Nature Reserve Educational Release No. 42, Carnegie Mus., 9 November 1962.

Aboriginal butchering techniques at the Eschelman Site (36 La 12), Lancaster County, Pennsylvania. Pennsylvania Archaeologist, 32:59–83 (with P. W. Parmalee and D. P. Tanner).

Animal remains from the Quaker State Rockshelter (36 Ve 27), Venango County, Pennsylvania. Pennsylvania Archaeologist, 32:131–137 (with D. P. Tanner).

1963

Bone refuse from the Oakfield Site, Genesee County, New York. Pennsylvania Archaeologist, 33:12–15.

Evidence for buffalo in prehistoric Pennsylvania. Pennsylvania Archaeologist, 33:135–139.

The cup-and-pin game. Pennsylvania Archaeologist, 33:159–163.

Pleistocene zoogeography of the collared lemming (*Dicrostonyx*). Evolution, 17:194–197. [Reprinted (1) pp. 509–512, *in* Readings in mammalogy (J. K. Jones, Jr. and S. Anderson, eds.), Mus. Nat. Hist., Univ. Kansas Monogr., 2:1–586, 1970; and (2) pp. 547–550, *in* Selected readings in mammalogy (J. K. Jones, Jr., S. Anderson, and R. S. Hoffmann, eds.) Mus. Nat. Hist., Univ. Kansas Monogr., 5:1–640, 1976.]

The faunagraph record. Carnegie Mag., 37:233–237.

Rabbits is rabbits. Pennsylvania Game News, 34(5):30–31.

Of rod and gun. Pennsylvania Game News, 34(9):15–16.

The lemming—New Paris. Pp. 2–72–2–73, *in* Speleo Digest 1961 (H. L. Black and A. P. Haarr, eds), Pittsburgh Grotto Press, Natl. Speleol. Soc., Pittsburgh, PA, 409 pp.

Time. Netherworld News, 11:39–40.

Final report—late Pleistocene fauna of New Paris No. 4 (Lloyds Rock Hole). Netherworld News, 11:153–162 (with A. D. McCrady, senior author). [Reprinted in Speleo Digest 1963, Pittsburgh Grotto Press, Natl. Speleol. Soc., Pittsburgh, PA.]

The Clarksville deer—a case history. Amer. Antiquity, 29:109–111 (with A. D. McCrady).

1964

New Paris No. 4: A Pleistocene cave deposit in Bedford County, Pennsylvania. Bull. Natl. Speleol. Soc., 26:121–194 (with P. S. Martin and A. D. McCrady).

Summary of artifacts, Pp. 9–11, *in* The Footer Site (Can 29-3), mimeographed archaeological site report produced by A. J. Parker, 62 Hinkleyville Road, Spencerport, NY, 22 pp.

First record of the least shrew for Powdermill and for Westmoreland County. Powdermill Nature Reserve Educational Release No. 50, Carnegie Mus., 5 October 1964.

The Cumberland caper. Netherworld News, 12:159–162.

1965

Don't tell the kids. Carnegie Mag., 39:191–194.

The Ice Age Heffalump trap. Canadian Audubon, 27:110–114.

Animal remains from the Sheep Rock Shelter (36 Hu 1), Huntingdon County, Pennsylvania. Pennsylvania Archaeologist, 35:34–49 (with P. W. Parmalee).

Quaternary mammals of North America. Pp. 509–525, *in* The Quaternary of the United States (H. E. Wright, Jr. and D. D. Frey, eds.) Princeton Univ. Press, 922 pp. (with C. W. Hibbard, senior author, C. E. Ray, D. E. Savage, and D. W. Taylor).

Vertebrate remains from the Mount Carbon Site, (46-Fa-7), Fayette County, West Virginia. The West Virginia Archeologist, 18:1–14 (with D. P. Tanner).

1966

Rangifer antler from an Ohio bog. J. Mamm., 47:325–326.

The bone breccia of Bootlegger Sink, York County, Pa. Ann. Carnegie Mus., 38:145–163 (with H. W. Hamilton and A. D. McCrady).

Armadillo remains from Tennessee and West Virginia caves. Bull. Natl. Speleol. Soc. 28:183–184 (with A. D. McCrady).

Lower third molar in *Procyon*. J. Mamm., 47:149–151 (with P. W. Parmalee).

Mammals of Pennsylvania. Pennsylvania Game Commission, Harrisburg, 273 pp. (with J. K. Doutt and C. A. Heppenstall, senior authors).

A Recent record of porcupine from Tennessee. J. Tennessee Acad. Sci., 61(3):81–82 (with P. W. Parmalee, senior author).

The vertebrate fauna. Appendix ii, *in* Mound City revisited, by J. A. Brown and R. S. Baby. Report of the Ohio Historical Society to the National Parks Service, June 1966, 103 pp. (with D. P. Tanner).

Animal remains from the Westmoreland-Barber Site (40 Mi-11), Marion County, Tennessee. Appendix A. Pp. 138–145, *in* Westmoreland-Barber Site (40 Mi-11), Nickajack Reservoir, season 11, by C. H. Faulkner and J. B. Graham. Dept. Anthropology, Univ. Tennessee, Knoxville, 150 pp. (with D. P. Tanner).

Powdermill September. Powdermill Nature Reserve Educational Release No. 66, Carnegie Mus., 21 September 1966. [Reprinted in Carnegie Mag., 42:224; and Carnegie Mag., 52:14.]

1967

The climatic significance of the Hosterman's Pit local fauna, Centre County, Pennsylvania. Amer. Antiquity, 32:231–232.

Notes on the Pleistocene big brown bat (*Eptesicus grandis* (Brown)). Ann. Carnegie Mus., 39:105–114.

Differential extinction during late-Pleistocene and Recent times. Pp. 121–140, *in* Pleistocene extinctions—the search for a cause (P. S. Martin and H. E. Wright, Jr., eds.), Yale Univ. Press, New Haven, 453 pp.

Evolution's test tube. [Review of] The life of the cave, by C. E. Mohr and T. L. Poulson. Carnegie Mag. 41:195–197.

Vertebrate remains from the Banshee Hole, Cumberland County, Tennessee. Speleotype, 2:44–45. [Reprinted in Speleo Digest 1967, Natl. Speleol. Soc.]

What skull is that? Pennsylvania Game News, 38(4):9–11.

Trout fishing. Netherworld News, 15:188–192.

Small mammal remains from Jaguar Cave, Lemhi County, Idaho. Tebiwa, J. Idaho State Univ. Mus., 10:26–36 (with E. K. Adam).

Animal remains from Horned Owl Cave, Albany County, Wyoming. Contrib. Geol., Univ. Wyoming, 6:97–99 (with H. W. Hamilton and E. K. Adam).

A new *Peromyscus* (Rodentia: Cricetidae) from the Pleistocene of Maryland. Ann. Carnegie Mus., 39:91–103 (with C. O. Handley, Jr.).

Extinct Florida spectacled bear *Tremarctos floridanus* (Gidley) from central Tennessee. Bull. Natl. Speleol. Soc., 29:149–162 (with D. C. Irving).

A bestiary for Pleistocene biologists. Pp. 1–62, *in* Pleistocene extinctions—the search for a cause (P. S. Martin and H. E. Wright, Jr., eds.), Yale Univ. Press, New Haven, 453 pp. (with P. S. Martin, senior author).

1968

Grizzly bears from eastern North America. Amer. Midland Nat., 79:247–250.

Archaeological evidence of caribou from New York and Massachusetts. J. Mamm., 49:344–345.

Pleistocene zoogeography of the lemming *Dicrostonyx*, a reevaluation. Univ. Colorado Studies Ser. Earth Sci., 6:61–71.

Cope's mouse. Turtox News, 46:333–335.

[Review of] Recent mammals of the world, by S. Anderson and J. K. Jones, Jr. Ecology, 49:190.

Vertebrate remains from the Fairchance Mound (46 Mr 13), Marshall County, West Virginia. West Virginia Archeologist, 21:41–54 (with D. P. Tanner).

The rains of May. Powdermill Nature Reserve Educational Release No. 76, Carnegie Mus., 29 May 1968.

[Review of] Quaternary extinction of large mammals, by D. I. Axelrod. J. Paleontol., 42:1319.

1969

Faunal remains from Dutchess Quarry Cave No. 1. Pp. 17–19, *in* The archeology of Dutchess Quarry Cave, Orange County, New York (R. E. Funk, G. R. Walters, and W. F. Ehlers, Jr.), Pennsylvania Archaeologist, 39:7–22.

Small mammal remains from the Wasden Site (Owl Cave), Bonneville County, Idaho. Tebiwa, J. Idaho State Univ. Mus., 12:47–57.

A possible caribou-Paleo-Indian association from Dutchess Quarry Cave, Orange County, New York. Bull. New York State Archeological Assoc., 45:24–29.

This planet's diversity of life. [Review of] Vertebrates, by C. J. McCoy, Jr. Carnegie Mag., 43:117.

[Review of] Pleistocene mammals of Europe, by B. Kurtén. Ecology, 50:531.

Mastodon and Paleo-Indian in West Virginia. West Virginia Archeologist. 22:1–3 (with D. S Berman).

The Pleistocene vertebrate fauna of Robinson Cave, Overton County, Tennessee. Palaeovertebrata, 2(2):25–75 (with H. Hamilton and A. D. McCrady).

Pleistocene and Recent vertebrate faunas from Crankshaft Cave, Missouri. Illinois State Mus. Repts. Investigations, 14:1–37 (with P. W. Parmalee and R. D. Oesch, senior authors).

1970

Animal remains from archaeological excavations at Fort Ligonier. Ann. Carnegie Mus., 42:177–186. [Reprinted, pp. 121–132, *in* Experimental archeology (D. Ingersoll, J. E. Yellon, and W. McDonald, eds.), Columbia Univ. Press, New York, 423 pp.]

Ask the experts. The Explorer, 12(4):20–21.

Leviathan. [Review of] The year of the whale, by V. Scheffer. Carnegie Mag., 44:59–61.

1971

The Pleistocene history of the Appalachian mammal fauna. Pp. 232–262, *in* The distributional history of the biota of the Southern Appalachians, Part III: Vertebrates (P. C. Holt, ed.), Virginia Polytechnic Inst. and State Univ., Res. Div. Monogr., 4:1–306.

Two letters to Steve Callen. Pp. 43–44, *in* Discovery of a mastodon tooth—Part V. Karst Kaver, Monongahela Grotto, Natl. Speleol. Soc., 5:39–44.

Biological and archeological analysis of bones from a 17th Century Indian village (46 Pu 31), Putnam County, West Virginia. West Virginia Geol. and Econ. Surv., Rept. Archeological Investigations, 4:1–64.

J. LeRoy Kay. Carnegie Mag., 45:296.

Big game of the Pleistocene. Pp. 46–52, in North American big game (R. C. Alberts, ed.), Boone and Crockett Club, Pittsburgh, 403 pp.

Penn wood folk. Pennsylvania Game News, 42:15–16.

The Welsh Cave peccaries (Platygonus) and associated fauna, Kentucky Pleistocene. Ann. Carnegie Mus., 43:249–320 (with H. W. Hamilton and A. D. McCrady).

Thirteen-lined ground squirrel, prairie chicken and other vertebrates from an archeological site in northeastern Arkansas. Amer. Midland Nat., 86:227–229 (with P. W. Parmalee).

1972

[Introduction to] Mice, men and mastodons, by R. S. Mills. The Explorer, 14(2):9–12.

Archaeological evidence of Scalopus aquaticus in the Upper Ohio Valley. J. Mamm., 53:905–907.

Mouse flowers. The Explorer, 14(4):19–21.

Bear Sink—a pitch. Netherworld News, 20:4–6.

Jaguar (Panthera onca) remains from Big Bone Cave, Tennessee and east central North America. Bull. Natl. Speleol. Soc., 34:1–14 (with H. McGinnis).

Quaternary periglacial records of voles of the genus Phenacomys Merriam (Cricetidae: Rodentia). Quat. Res., 2:170–175 (with P. W. Parmalee).

First record of Cervalces scotti Ledekker from the Pleistocene of Ohio. Amer. Midland Nat., 88:255 (with R. S. Mills, senior author).

1973

Confessions of an aging paleontologist. The Explorer, 15:25–26.

The late Pleistocene small mammals of Eagle Cave, Pendleton County, West Virginia. Ann. Carnegie Mus., 44:45–58 (with H. W. Hamilton).

1974

Arctic life in the South (letter to editor). Geotimes, 19:8.

Caribou (Rangifer tarandus L.) from the Pleistocene of Tennessee. J. Tennessee Acad. Sci., 50:109–112 (with H. W. Hamilton and P. W. Parmalee).

[Review of] The children of pride, edited by R. M. Myers. Carnegie Mag., 49:39.

1975

Faunal remains from the Zebree Site. Appendix I. Pp. 228–234, in Report of excavations at the Zebree Site 1969 (D. F. Morse), Research Rept. No. 4, Arkansas Archeological Surv., Fayetteville, 246 pp. (with P. W. Parmalee).

Extinct peccary (Platygonus compressus LeConte) from a central Kentucky cave. Bull. Natl. Speleol. Soc., 37:83–87 (with R. C. Wilson, senior author, and J. A. Branstetter, junior author).

1976

Appalachian bone caves. Pp. 88–103, in Geology and biology of Pennsylvania caves (W. B. White, ed.), Pennsylvania Geol. Surv., General Geol. Rept., 66:1–103.

Owls and Ice Age caves. Netherworld News, 24:86–89.

First records of the giant beaver (Castoroides ohioensis) from eastern Tennessee. J. Tennessee Acad. Sci., 51:87–88 (with P. W. Parmalee and A. E. Bogan, senior authors).

1977

Sabertooth cat, Smilodon floridanus (Leidy), and associated fauna from a Tennessee cave (40 Dv 40), the First American Bank site. J. Tennessee Acad. Sci., 52:84–94.

The long journey. Carnegie Mag., 51(2):76–79.

The pigeons of Meadowcroft. Carnegie Mag., 51(10):33–37.

Mysterious mouse. Rustlings from the Grove. Newsletter, Grove Heritage Assoc. (Glenview, IL), 9:5.

The Clark's Cave bone deposit and the late Pleistocene paleoecology of the central Appalachian Mountains of Virginia. Bull. Carnegie Mus. Nat. Hist., 2:1–87 (with P. W. Parmalee and H. W. Hamilton).

[Selection, pp. 70–77, in] Meadowcroft Rockshelter: retrospect 1976, by J. M. Adovasio, J. D. Gunn, J. Donahue, and R. Stuckenrath. Pennsylvania Archaeologist, 47:(2–3):1–93.

1978

[Review of] A guide to the measurement of animal bones from archaeological sites, by A. von den Driesch. Amer. Anthropologist, 80:180.

Ecological significance of displaced boreal mammals in West Virginia caves. J. Mamm., 59:176–181 (with H. W. Hamilton).

The Baker Bluff cave deposit, Tennessee, and the late Pleistocene faunal gradient. Bull. Carnegie Mus. Nat. Hist., 11:1–67 (with H. W. Hamilton, E. Anderson, and P. W. Parmalee).

The Pleistocene mammalian fauna of Harrodsburg Crevice, Monroe County, Indiana. Bull. Natl. Speleol. Soc., 40:64–75 (with P. W. Parmalee and P. J. Munson, senior authors).

Meadowcroft Rockshelter. Pp. 140–180, in Early man in America (A. L. Bryan, ed.), Occas. Papers, Dept. Anthropology, Univ. Alberta 1:1–327 (with J. M. Adovasio, J. D. Gunn, J. Donahue, R. Stuckenrath, senior authors, and K. Lord, junior author).

1979

Eastern North American Pleistocene Ochotona (Lagomorpha: Mammalia). Ann. Carnegie Mus., 48:435–444.

The star-nosed mole/A Polish connection. Carnegie Mag., 53(1):18–25.

Section of Vertebrate Fossils, Carnegie Museum of Natural History. Carnegie Mag., 53(8):18–27 (with E. Hill).

Pleistocene and Recent vertebrate remains from Savage Cave (15L011), Kentucky. Pp. 5–10, in Western Kentucky Speleol. Surv. Ann. Rept. 1979 (J. E. Mylroie, ed.), 84 pp. (with P. W. Parmalee).

Meadowcroft Rockshelter—retrospect 1979: part 1. North Amer. Archeologist, 1(1):3–44 (with J. M. Adovasio, J. D. Gunn, J. Donahue, R. Stuckenrath, senior authors, and K. Lord, junior author).

1980

Vertebrate faunal remains from Meadowcroft Rockshelter (36WH297), Washington County, Pennsylvania. Manuscript on file, Dept. Anthropology, Univ. Pittsburgh, 140 pp.

(with P. W. Parmalee and R. C. Wilson). [To be published *in* Meadowcroft Rockshelter and the archaeology of the Cross Creek drainage, J. M. Adovasio et al., Univ. Pittsburgh Press.]

Meadowcroft Rockshelter—retrospect 1977: part 2. North Amer. Archaeologist, 1(2):99–137 (with J. M. Adovasio, J. D. Gunn, J. Donahue, R. Stuckenrath, senior authors, K. Lord, and K. Volman, junior authors).

Yes Virginia, it really is that old: a reply to Haynes and Mead. Amer. Antiq., 45:588–595 (with J. M. Adovasio, J. D. Gunn, J. Donahue, R. Stuckenrath, senior authors, and, K. Volman, junior author).

Cavern de Saint-Elzéar-de-Bonaventure, rapport préliminaire sur les fouilles de 1977 et 1978, Ministère de l'Énergie et des Ressources, Direction Générale de la Rechèrche Géolgique et Minérale, Direction de la Géologie, 31 pp. (with P. LaSalle, senior author).

Additional comments on the Pleistocene mammalian fauna of Harrodsburg Crevice, Monroe County, Indiana. Bull. Natl. Speleol. Soc., 42:78–79 (with P. J. Munson and P. W. Parmalee, senior authors).

1981

[Review of] Pleistocene mammals of North America, by B. Kurtén and E. Anderson. J. Vert. Paleontol., 1:241.

Conjure bones. Carnegie Mag., 55(5):4–9.

February. Carnegie Mag., 55(2):43.

March. Carnegie Mag., 55(3):43.

April. Carnegie Mag., 55(4):41.

May. Carnegie Mag., 55(5):43.

June, July, August. Carnegie Mag., 55(6):33.

September. Carnegie Mag., 55(7):40.

October. Carnegie Mag., 55(8):41.

November. Carnegie Mag., 55(9).49.

December and January. Carnegie Mag., 55(10):32.

1982

Dental variation in *Microtus xanthognathus, M. chrotorrhinus,* and *M. pennsylvanicus* (Rodentia, Mammalia). Ann. Carnegie Mus., 51(11):211–230.

Appalachia 11,000–12,000 years ago: a biological review. Archaeology of Eastern North Amer., 10:22–26.

Carnegie's cornice—science. Carnegie Mag., 56(1):16–23.

Caves, bones and Ice Age owls. Carnegie Mag., 56(3):20–29.

[Introduction to] The scientist's notebook: Kayapo, by D. A. Posey. Carnegie Mag., 56(4):18.

[Introduction to] The scientist's notebook: Endangered species and the museum, by C. J. McCoy. Carnegie Mag., 56(5):28.

"by Golley, J. W." Verbatim, 9(2):17–18.

An introduction to the Meadowcroft/Cross Creek archaeological project: 1973–1982. Pp. 1–30, *in* Meadowcroft—collected papers on the archaeology of Meadowcroft Rockshelter and the Cross Creek drainage (R. C. Carlisle and J. M. Adovasio, eds.) (Dept. Anthropol., Univ. Pittsburgh) 47th Ann. Mtg., Soc. Amer. Archaeol., Minneapolis, Minnesota, 14–17 April 1982, 270 pp. (with R. C. Carlisle, J. M. Adovasio, J. Donahue, and P. Wiegman, senior authors).

Vertebrate faunal remains from Meadowcroft Rockshelter, Washington County, Pennsylvania: summary and interpretation. Pp. 163–174, *in* Meadowcroft—collected papers on the archaeology of Meadowcroft Rockshelter and the Cross Creek drainage (R. C. Carlisle and J. M. Adovasio, eds.) (Dept. Anthropol., Univ. Pittsburgh) 47th Ann. Mtg., Soc. Amer. Archaeol., Minneapolis, Minnesota, 14–17 April 1982, 270 pp. (with P. W. Parmalee).

1983

[Introduction to] The scientist's notebook: The birds of Socorro, by K. C. Parkes. Carnegie Mag., 56(7):31.

Terrestrial vertebrate faunas. Pp. 311–353, *in* The Late Pleistocene (S. C. Porter, ed.), Vol. 1 of, Late-Quaternary environments of the United States (H. E. Wright, Jr., ed.), Univ. Minnesota Press, Minneapolis (with E. L. Lundelius, Jr., R. W. Graham, E. Anderson, senior authors, J. A. Holman, D. Steadman, and S. D. Webb, junior authors).

In Press

Pleistocene extinction and environmental change: case study of the Appalachians. *In* Quaternary extinctions: A prehistoric revolution (P. S. Martin and R. G. Klein, eds.). Univ. Arizona Press.

The physiographic provinces of Pennsylvania. *In* Species of special concern in Pennsylvania (H. H. Genoways and F. J. Brenner, eds.).

Paleoenvironmental reconstruction at Meadowcroft Rockshelter, Washington County, Pennsylvania. *In* Environment and extinctions: man in late glacial North America (J. I. Mead and D. J. Meltzer, eds.), Peopling of the Americas Series, Center for the Study of Early Man, Univ. Maine, Orono (with J. M. Adovasio, R. C. Carlisle, K. Cushman, J. Donahue, senior authors, W. C. Johnson, K. Lord, P. W. Parmalee, R. Stuckenrath and P. Weigman, junior authors).

Address: Section of Vertebrate Fossils, Carnegie Museum of Natural History, 4400 Forbes Ave., Pittsburgh, Pennsylvania 15213.

MID-WISCONSINAN AND MID-HOLOCENE HERPETOFAUNAS OF EASTERN NORTH AMERICA: A STUDY IN MINIMAL CONTRAST

Leslie P. Fay

ABSTRACT

Two herpetofaunas (Strait Canyon, Virginia—mid-Wisconsinan, and St. Elzear, Quebec—mid-Holocene) provide the first view of non-glacial Quaternary environments in eastern North America. Six amphibians and one snake were recovered from St. Elzear, nine amphibians and 16 reptiles from Strait Canyon. All herpetological members of these local faunas presently occur at or near the fossil localities, indicating conditions nearly identical to the current climates for western Virginia 30,000 years ago and about 5,000 years ago in eastern Quebec. No Altonian Stadial (Strait Canyon) or Hypsithermal (St. Elzear) effects are evident in the herpetofaunas, and no apparent similarity to the steepened mammalian faunal gradient of eastern North America is revealed. Non-glacial herpetofaunas may have been nearly identical to the present distribution, with only minor range adjustments in response to changing climate. *Ambystoma laterale* is reported from the fossil record for the first time.

INTRODUCTION

Quaternary vertebrates of eastern North America, outside of Florida, are known to us mostly through the work of John E. Guilday and colleagues. Field work by Carnegie Museum of Natural History crews for over two decades has produced a number of important local faunas from the Appalachian region; especially those of late Wisconsinan to early Holocene age. In these faunas, Guilday delineated a mammalian faunal gradient (Guilday et al., 1978) that is steeper than the present interglacial mammal distribution. With the kind permission and encouragement of John Guilday, I have begun to test for any steepened gradient in the herptiles associated with these mammalian local faunas. Two small herpetofaunas, from Caverne de Saint Elzear de Bonaventure, Gaspe Peninsula, Quebec (mid-Holocene), and Strait Canyon Fissure, Highland County, Virginia (mid-Wisconsinan), were examined first because the mammals had not yet been reported in detail and because no interglacial or verified interstadial herpetofaunas have previously been reported from eastern North America.

Caverne de Saint Elzear is located near the village of Saint Elzear de Bonaventure on the south coast of the Gaspe Peninsula, Quebec, in upland boreal forest. The fossils, originally discovered by local residents, were excavated from a talus deposit within the cave by Carnegie Museum of Natural History and Quebec Natural Resources Ministry personnel in 1977, 1978, and 1979 (LaSalle and Guilday, 1980). Age of the local fauna is considered to be mid-Holocene (about 5,000 y.b.p.), based on stratigraphic and faunal evidence. The mammals strongly resemble the late Wisconsinan local faunas of the central Appalachians, providing contrasts useful for regional paleoclimatic analysis (J. E. Guilday, in litt., 29 March 1982).

Strait Canyon Fissure local fauna, Highland County, Virginia (38°28′14″N, 79°30′42″W) was excavated from a fissure fill matrix in 1968 by a field party of the U.S. National Museum. This is the first pre-Woodfordian Wisconsinan fauna from the Appalachians. A carbon-14 date of 29,870 + 1800/ −1400 y.b.p. (GX-7017-A) places the fossils near the Altonian Stadial–Farmdalian Interstadial boundary of Illinois terminology (Frye and Willman, 1960) or within the Plum Point Interstadial of Ontario nomenclature (Dreimanis and Karrow, 1972).

St. Elzear and Strait Canyon fossils are reposited at the Carnegie Museum of Natural History.

ACKNOWLEDGMENTS

John Guilday launched me on this project and I regret not being able to report these findings to him. Alice M. Guilday has cheerfully provided unpublished information and helped me sort out stratigraphic provenance. J. Alan Holman and Thomas C. LaDuke have aided my passage into the maze that is fossil herptile identification. This paper was improved by Holman's thorough review. I thank them all for their guidance and encouragement.

14

FAUNAL LISTS

STRAIT CANYON, VIRGINIA

Class Amphibia
  Order Caudata
    Family Ambystomatidae
      Ambystoma maculatum (Shaw)
    Family Plethodontidae
      Desmognathus ?fuscus Rafinesque
      Plethodon glutinosus (Green)
  Order Salientia
    Family Bufonidae
      Bufo americanus Holbrook
      Bufo woodhousei fowleri Hinckley
    Family Ranidae
      Rana catesbeiana Shaw
      Rana clamitans Latreille
      Rana cf. *R. pipiens* Schreber
      Rana sylvatica LeConte
Class Reptilia
  Order Chelonia
    Family Emydidae
      Pseudemys, sp. indet.
      Terrapene carolina (Linne)
  Order Squamata
    Family Scincidae
      ?Eumeces
    Family Colubridae
      Coluber or *Masticophis*
      Diadophis punctatus (Linne)
      Elaphe guttata (Linne)
      Elaphe obsoleta (Say)
      Lampropeltis calligaster (Harlan)
      Lampropeltis getulus (Linne)
      Lampropeltis triangulum (Lacepede)
      Opheodrys vernalis (Harlan)
      Nerodia sipedon (Linne)
      Storeria ?occipitomaculata (Storer)
      Thamnophis sirtalis (Linne)
    Family Viperidae
      Agkistrodon contortrix (Linne)
      Crotalus horridus Linne

SAINT ELZEAR, QUEBEC

Class Amphibia
  Order Caudata
    Family Ambystomatidae
      Ambystoma laterale Hallowell
      Ambystoma maculatum (Shaw)
  Order Salientia
    Family Bufonidae
      Bufo americanus Holbrook
    Family Hylidae
      Hyla crucifer Wied
    Family Ranidae
      Rana pipiens Schreber
      Rana sylvatica LeConte
Class Reptilia
  Order Squamata
    Family Colubridae
      Thamnophis sirtalis (Linne)

SPECIES ACCOUNTS

Following is an annotated list of herptiles occurring in the Saint Elzear and Strait Canyon local faunas. Conant (1975) and Logier and Toner (1961) were consulted for modern distribution information. Listed taxa presently occupy the area of the fossil localities unless otherwise stated in the remarks.

Class Amphibia
Order Caudata
Family Ambystomatidae

Ambystoma laterale Hallowell
Blue-spotted Salamander

Occurrence.—St. Elzear.
Remarks.—Ambystomatid salamanders may be distinguished by length-width ratios derived from measurements of trunk vertebrae (Tihen, 1958). All St. Elzear ambystomatid salamanders fit the *maculatum*-group, which includes the two most northeasterly distributed members of the family—*A. laterale* and *A. maculatum.* According to the vertebral ratios (centrum length/anterior centrum width), two taxa are present. By means of comparison with modern skeletal material, the shorter vertebrae (<2.65) are assigned to *A. laterale,* the longer to *A. maculatum.* This is the first report of *Ambystoma laterale* from the fossil record.

Ambystoma maculatum (Shaw)
Spotted Salamander

Occurrence.—St. Elzear and Strait Canyon.
Remarks.—The method used to distinguish *A. maculatum* is described above. All Strait Canyon ambystomatid vertebrae fall within the range of this species.

Family Plethodontidae

Desmognathus ?fuscus Rafinesque
Dusky Salamander

Occurrence.—Strait Canyon.
Remarks.—One salamander vertebra has the opisthocoelous condition of *Desmognathus* (Soler,

1950) and very closely resembles *D. fuscus* in form and size. A degree of uncertainty remains because *D. ochrophaeus* vertebrae are quite similar to those of *D. fuscus,* making identification tenuous based on one fossil specimen.

Plethodon glutinosus (Green)
Slimy Salamander

Occurrence.—Strait Canyon.
Remarks.—One plethodontid vertebra is indistinguishable from comparative material of *P. glutinosus.*

Order Salientia
Family Bufonidae

Bufo americanus Holbrook
American Toad

Occurrence.—St. Elzear and Strait Canyon.
Remarks.—Ilia of *B. americanus* may be distinguished from *B. woodhousei* by the wider base and more anterior deflection of the dorsal protuberance in the former (Holman, 1967, 1977; Wilson, 1975).

Bufo woodhousei fowleri Hinckley
Fowler's Toad

Occurrence.—Strait Canyon.
Remarks.—See *B. americanus* remarks for specific determination. In addition to present distribution, subspecific assignment is possible because the dorsal protuberance is relatively lower in ilia of the eastern form, *B. w. fowleri,* than in the western subspecies, *B. w. woodhousei* (Eshelman, 1975; Holman, 1977).

Family Hylidae

Hyla crucifer Wied
Spring Peeper

Occurrence.—St. Elzear.
Remarks.—Ilia of *H. crucifer* may be distinguished from other hylid frogs by a combination of ilial characteristics (Holman, 1967) and by small size.

Family Ranidae

Rana catesbeiana Shaw
Bullfrog

Occurrence.—Strait Canyon.
Remarks.—*R. catesbeiana* and *R. clamitans* ilia have a steeper posterodorsal slope of the shaft as it merges with the dorsal acetabular expansion than do ilia in a group comprised of *R. palustris, R. pip-*

iens, and *R. sylvatica* (Holman, 1967). *R. catesbeiana* is separated from *R. clamitans* by a more rugged (Holman, 1967) and often slightly concave vastus prominence.

Rana clamitans Latreille
Green Frog

Occurrence.—Strait Canyon.
Remarks.—Distinguishing features for *R. clamitans* are discussed with *R. catesbeiana.*

Rana pipiens Schreber
Northern Leopard Frog

Occurrence.—St. Elzear and Strait Canyon.
Remarks.—According to Holman (1967, pp. 158–159): "*Rana pipiens* and *Rana palustris* . . . may be separated from *R. sylvatica* in that the prominence for the origin of the vastus externus head of the triceps femoris muscle (terminology of Holman, 1965) is larger, less produced, and less roughened than in *R. sylvatica. Rana pipiens* has a somewhat steeper slope of the posterodorsal border of the ilial shaft into the dorsal acetabular expansion than in . . . *R. palustris.*" In the Strait Canyon faunal list, tentative identification as *R. pipiens* is made because no method of separating the ilia of the leopard frog complex (*R. berlandieri, R. blairi, R. pipiens,* and *R. utricularia*) has been devised (Holman, 1977). Therefore, I will only make firm specific assignments within the *pipiens*-group when modern distribution warrants. Leopard frogs do not presently occupy the Strait Canyon area, but *R. pipiens* (as a disjunct record) and *R. utricularia* occur in Maryland and eastern Virginia.

Rana sylvatica LeConte
Wood Frog

Occurrence.—St. Elzear and Strait Canyon.
Remarks.—Identification of *R. sylvatica* by Holman's (1967) criteria is discussed with *R. pipiens.*

Class Reptilia
Order Chelonia
Family Emydidae

Pseudemys, sp. indet.

Occurrence.—Strait Canyon.
Remarks.—Several carapace fragments belong to the genus *Pseudemys.* The material is in somewhat poor condition, rendering identification difficult. Also, no specimens of *P. rubiventris,* the species which most closely approaches Strait Canyon today,

were available for comparison. No *Pseudemys* currently occupy the fossil locality, although several forms inhabit the Piedmont and coastal plain.

Terrapene carolina (Linne)
Eastern Box Turtle

Occurrence.—Strait Canyon.
Remarks.—A left epiplastral is identical to comparative material of *T. carolina*.

Order Squamata
Family Scincidae

?*Eumeces*
Skink

Occurrence.—Strait Canyon.
Remarks.—A small jaw fragment comparable to the genus *Eumeces* was recovered. No further identification was attempted.

Family Colubridae

Coluber or *Masticophis*

Occurrence.—Strait Canyon.
Remarks.—Snakes of the subfamily Colubrinae are best distinguished by a combination of characters, including morphology, size, and ratios of linear measurements of vertebrae (Auffenberg, 1963; Holman, 1981; Meylan, 1982). *Coluber* and *Masticophis* have very long, relatively large vertebrae. The two genera have roughly similar habits and morphology, making them difficult to distinguish. *C. constrictor* lives in the area today, and thus might be considered the more likely choice.

Diadophis punctatus (Linne)
Ringneck Snake

Occurrence.—Strait Canyon.
Remarks.—This diminutive snake is distinct from other colubrids, except *Carphophis* and *Rhadinaea*, with vertebral lengths less than 2 mm. The haemal keel of *Rhadinaea* is distinctly different from the other two genera, and *Diadophis* can be distinguished from *Carphophis* by a slightly higher neural spine and more vaulted neural arch on the former (Holman, 1977).

Elaphe guttata (Linne)
Corn Snake

Occurrence.—Strait Canyon.
Remarks.—*Elaphe* and *Lampropeltis* specimens are best sorted by measuring ratios of linear dimensions of vertebrae and cross-checking the resulting

groups with comparative material to verify identity of morphological features. Results of this analysis for all Quaternary Appalachian herpetofaunas will be reported elsewhere (Fay, in preparation). Numerical and gross morphological approaches are both used because each method is imperfect when applied to such variable objects as snake vertebrae. Variation within individuals as well as within taxa often exceeds the norms established with comparative material.

Elaphe obsoleta (Say)
Rat Snake

Occurrence.—Strait Canyon.
Remarks.—See *E. guttata* remarks.

Lampropeltis calligaster (Harlan)
Kingsnake

Occurrence.—Strait Canyon.
Remarks.—See *E. guttata* remarks.

Lampropeltis getulus (Linne)
Eastern Kingsnake

Occurrence.—Strait Canyon.
Remarks.—See *E. guttata* remarks.

Lampropeltis triangulum (Lacepede)
Milksnake

Occurrence.—Strait Canyon.
Remarks.—See *E. guttata* remarks.

Opheodrys vernalis (Harlan)
Smooth Green Snake

Occurrence.—Strait Canyon.
Remarks.—*Opheodrys* vertebrae have higher neural spines than do *Diadophis* and *Carphophis* (Holman, 1977) and tend to be slightly larger as well. The fossils appear identical to *O. vernalis*.

Nerodia sipedon (Linne)
Water Snake

Occurrence.—Strait Canyon.
Remarks.—Snakes of the subfamily Natricinae possess laterally flattened hypopophyses on all vertebrae (Holman, 1981). *Nerodia* vertebrae are relatively shorter and wider than *Thamnophis* vertebrae (Holman, 1981). *N. sipedon* may be distinguished from its congeners by relatively low neural spines (Holman, 1972).

Storeria ?occipitomaculata (Storer)
Red-bellied Snake

Occurrence.—Strait Canyon.

Remarks.—Holman (1962) discusses criteria for separating three genera of small natricines: *Storeria, Tropidoclonion,* and *Virginia.* I cannot satisfactorily discern *S. dekayi* from *S. occipitomaculata,* but the fossils more nearly resemble the latter. Fossil *Storeria* are not often identified to species level in the literature; most are listed as *Storeria* cf. *S. dekayi* (Holman, 1981). Both species currently inhabit the Strait Canyon area.

Thamnophis sirtalis (Linne)
Garter Snake

Occurrence.—St. Elzear and Strait Canyon.

Remarks.—Holman (1962) describes distinguishing features for *Thamnophis* vertebrae.

Family Viperidae

Agkistrodon contortrix (Linne)
Copperhead

Occurrence.—Strait Canyon.

Remarks.—North American vipers (subfamily Crotalinae) have distinctively elongate hypopophyses which are rounded in cross-section. *Agkistrodon* vertebrae are distinguished from other North American crotalines on the basis of large, single cotylar foraminae, often recessed in deep pits. *Crotalus* and *Sistrurus* have smaller, often multiple cotylar foraminae (Holman, 1982). *A. contortrix* usually have slightly smaller cotylar foraminae, less vaulted neural arches, and lower neural spines than do *A. piscivorus* (Holman, 1982).

Crotalus horridus Linne
Timber Rattlesnake

Occurrence.—Strait Canyon.

Remarks.—*C. horridus* has the lowest neural spines of any eastern North American crotalid and usually the least vaulted neural arches as well (Holman, 1982).

DISCUSSION

STRAIT CANYON

Although all 25 taxa are not found at Strait Canyon today, this herpetofauna does coexist 150 km to the east on the Piedmont, with all but four of the mammals from the local fauna as well. Strait Canyon, like St. Elzear, has a slightly more boreal mammal and bird complement with herptiles which currently occupy the region. More northerly distributed herptiles do not appear in the Strait Canyon local fauna to mirror the steepened mammalian faunal gradient.

ST. ELZEAR

The seven species comprise a harmonious fauna which can be found at St. Elzear today. *Bufo americanus* is by far the most common herptile (87% of 255 MNI), perhaps signifying a hibernaculum mortality. *Ambystoma laterale* is reported from the fossil record for the first time, possibly due to the fact that St. Elzear is the first fossil locality within *A. laterale* range. The herptiles have a less boreal character than the St. Elzear mammals (total 34 species), which include four taxa that do not range as far

southeast as the Gaspe Peninsula today. This steepened gradient cannot be reflected in the herptiles, as there are no "more boreal" taxa than those which currently occupy the fossil locality.

INTERGLACIAL/INTERSTADIAL HERPETOFAUNAS

These first two herpetofaunas reported from nonglacial intervals in eastern North America suggest near-equivalence with present distributions. No influence of terminal Altonian events (presumably cooler than present) are illustrated by the Virginia fauna and Hypsithermal effects are not noted in the Quebec fauna. At least some herptiles were sensitive to climatic change in Irvingtonian time, as indicated by northern/prairie species (especially *Elaphe vulpina*) in Appalachian local faunas (Holman, 1977, 1982; Fay, unpublished manuscript). Verification of the possibility that non-glacial herpetofaunas have remained the same throughout the RancholaBrean (that is, resuming the "modern" pattern after each glacial pulse) must wait for the discovery of more interglacial and interstadial local faunas.

LITERATURE CITED

Auffenberg, W. 1963. The fossil snakes of Florida. Tulane Studies Zool., 10:131–216.

Conant, R. 1975. A field guide to reptiles and amphibians of eastern North America. Houghton Mifflin Co., Boston, 429 pp.

Dreimanis, A., and P. F. Karrow. 1972. Glacial history of the Great Lakes–St. Lawrence region, the classification of the Wisconsin Stage, and its correlation. Intl. Geol. Congr., Sec. 12: Quat. Geol., pp. 5–15.

Eshelman, R. E. 1975. Geology and paleontology of the early Pleistocene (late Blancan) White Rock fauna from north-central Kansas. Mus. Paleont. Univ. Michigan, Papers Paleont., 13:1–60.

Frye, J. C., and H. B. Willman. 1960. Classification of the Wisconsinan Stage in the Lake Michigan glacial lobe. Illinois State Geol. Surv. Circ. 285:1–16.

Guilday, J. E., H. W. Hamilton, E. Anderson, and P. W. Parmalee. 1978. The Baker Bluff Cave deposit, Tennessee, and the late Pleistocene faunal gradient. Bull. Carnegie Mus. Nat. Hist., 11:1–67.

Holman, J. A. 1962. A Texas Pleistocene herpetofauna. Copeia, 1962(2):255–261.

———. 1965. Early Miocene anurans from Florida. Quart. J. Florida Acad. Sci., 28:68–82.

———. 1967. A Pleistocene herpetofauna from Ladds, Georgia. Bull. Georgia Acad. Sci., 25:154–166.

———. 1972. Amphibians and reptiles. In Early Pleistocene preglacial and glacial rocks and faunas of northcentral Ne-braska (M. F. Skinner, et al.). Bull. American Mus. Nat. Hist., 148:55–71.

———. 1977. The Pleistocene (Kansan) herpetofauna of Cumberland Cave, Maryland. Ann. Carnegie Mus., 46:157–172.

———. 1981. A review of North American Pleistocene snakes. Publ. Mus., Michigan State Univ., Paleont. Ser., 1:263–306.

———. 1982. The Pleistocene (Kansan) herpetofauna of Trout Cave, Maryland. Ann. Carnegie Mus., 51:391–404.

LaSalle, P., and J. E. Guilday. 1980. Caverne de Saint Elzear de Bonaventure: Rapport preliminaire sur les fouilles de 1977 et 1978. Ministere de l'Energie et des Ressources, Gouv. du Quebec, 29 pp.

Logier, E. B. S., and G. C. Toner. 1961. Check list of the amphibians and reptiles of Canada and Alaska. Life Sci. Div. Contr., Royal Ontario Mus., 53:1–92.

Meylan, P. A. 1982. The squamate reptiles of the Inglis IA fauna (Irvingtonian: Citrus County, Florida). Bull. Florida State Mus., Biol. Sci. 27:111–195.

Soler, E. I. 1950. On the status of the family Desmognathidae (Amphibia, Caudata). Univ. Kansas Sci. Bull., 33:459–480.

Tihen, J. A. 1958. Comments on osteology and phylogeny of ambystomatid salamanders. Bull. Florida State Mus., 3:1–50.

Wilson, V. V. 1975. The systematics and paleoecology of two late Pleistocene herpetofaunas from the southeastern United States. Unpublished Ph.D. dissert., Michigan State Univ., East Lansing, 67 pp.

Address: The Museum, Michigan State University, East Lansing, Michigan 48824-1045.

HERPETOFAUNAS OF THE DUCK CREEK AND WILLIAMS LOCAL FAUNAS (PLEISTOCENE: ILLINOIAN) OF KANSAS

J. Alan Holman

ABSTRACT

The Duck Creek and Williams local faunas of west-central and central Kansas are considered to represent the Illinoian stage of the Pleistocene based on previous faunal and stratigraphic studies. The Duck Creek fauna yielded one salamander, three anurans, and one snake. The Williams fauna yielded one salamander, four anurans, four turtles, and 13 snakes, and is the largest Illinoian herpetofauna known. The major habitat indicated by both herpetofaunas is a pond or slow-moving stream. The presence of prairie flats and hillsides as well as woodland is also indicated. None of the amphibians and reptiles in either fauna are extinct. The presence of three extralimital forms (*Rana sylvatica*, *Emydoidea blandingii*, and *Elaphe vulpina*) that occur mainly to the northeast of the areas today may indicate somewhat cooler, moister conditions. But a full glacial climate or a dominant coniferous forest community is not indicated by the rest of the herpetological species, all of which can be found in the area today. The topics of "disharmonious" Pleistocene faunas and "mosaic" Pleistocene communities are addressed.

INTRODUCTION

Herpetofaunas from the Illinoian stage of the Pleistocene are quite rare. In fact, only one large herpetofauna is now known from an unquestionably Illinoian deposit (Sandahl local fauna, McPherson County, Kansas; Holman, 1971). Thus, the recent availability for study of two Illinoian herpetofaunas from west-central and central Kansas offers the opportunity to add considerably to the knowledge of the amphibians and reptiles of this stage of the Pleistocene. These faunas are especially important in the light of the statement (Holman, 1980: 133) that evidence from the study of the Sandahl herpetofauna did not support the classical idea of a cooler, moister climate for the Illinoian age in Kansas.

ACKNOWLEDGMENTS

I should here like to acknowledge Dr. Richard J. Zakrzewski of the Sternberg Memorial Museum of Fort Hays State College for the privilege of studying the Duck Creek fauna fossils and Dr. Gerald Smith of the Museum of Paleontology of the University of Michigan for the privilege of studying the Williams fauna material. Rosemarie Attilio made the drawings.

THE DUCK CREEK LOCAL FAUNA

The Duck Creek local fauna near the Smoky Hill River, Ellis County, Kansas (northeast quarter of section 33, T 15 S, R 16 W) was previously studied by Kolb et al. (1975), mollusks; McMullen (1975 and 1978) and Zakrzewski and Maxfield (1971), mammals. Based on its stratigraphic location in the Pfeifer Terrace, and on a comparison with other Pleistocene plains faunas, an Illinoian age was suggested (McMullen, 1978). Five collecting localities (designated by Arabic numerals 1–5) were defined. Different lithologies in these localities suggest different depositional mechanisms. Localities 1 and 2 were in coarse sand with sandy silt lenses that suggested a stream channel. Locality 3 is a clayey deposit that suggested quiet water such as an oxbow lake. Localities 4 and 5 are silt deposits with sandy silt lenses, suggesting a backwater area where current action was slow and where deposition of fine particles might have taken place.

Class Amphibia Linnaeus, 1758
Order Caudata Oppel, 1811
Family Ambystomatidae Hallowell, 1858

Ambystoma tigrinum (Green, 1825)

Material.—Locality 1, vertebra FHSVP 2824; Locality 3, right femur FHSVP 2903; Locality 4, vertebra FHSVP 2905.

Remarks.—Tihen (1958) and Holman (1969) have discussed the identification of vertebrae of *Ambystoma tigrinum*. The large right femur (FHSVP 2903) is indistinguishable from those in modern skeletons of *A. tigrinum*. Today in Kansas, adult *A. tigrinum* spends much of its time beneath the ground in burrows of other animals (Collins, 1974:26). From a

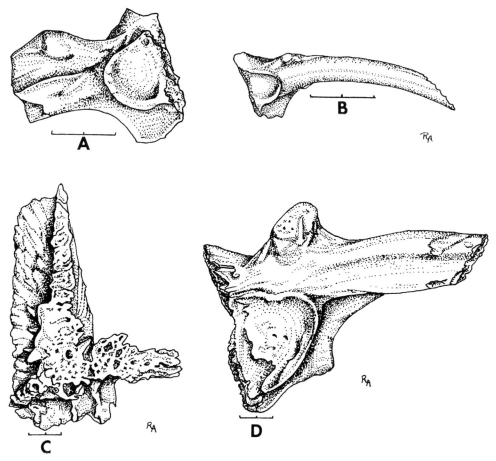

Fig. 1.—A) *Rana sylvatica*, left ilium FHSVP 2859 from Duck Creek local fauna; B) *Acris crepitans,* right ilium UM 81732 from Williams local fauna; C) *Bufo woodhousei woodhousei,* right cranial crest UM 81733 from Williams local fauna; D) *Bufo woodhousei woodhousei,* right ilium UM 81733 from Williams local fauna. Each line equals 2 mm.

taphonomic standpoint it is interesting to note that *A. tigrinum* bones were found in both high energy (Locality 1) and low energy (Localities 3 and 4) situations. This species occurs in Ellis County, Kansas, today (Collins, 1974: map p. 26).

Order Anura Dumeril, 1807
Family Bufonidae Fitzinger, 1826

Bufo woodhousei woodhousei Girard, 1854

Material.—Locality 1, a left and a right ilium FHSVP 2857–2858; Locality 2, right ilium FHSVP 2897; Locality 5, two right ilia FHSVP 2922-2933.

Remarks.—Holman (1971) and Tihen (1962) have shown how elements of *Bufo woodhousei woodhousei* may be identified at the subspecific level. This subspecies has also been reported from the Sandahl Illinoian fauna of McPherson County, Kan-

sas (Holman, 1971). According to Collins (1974:56) *B. w. woodhousei* prefers lowlands and sandy areas and is generally the only toad found on the flood plains of the larger streams and rivers in Kansas today. Taphonomically it is of interest that bones of this species are also found in both high energy (Localities 1 and 2) and low energy (Locality 5) situations. This subspecies occurs in Ellis County, Kansas, today (Collins, 1974: map p. 55).

Family Ranidae Bonaparte, 1831

Rana sylvatica Le Conte, 1825

Material.—Locality 1, left ilium FHSVP 2859 (Fig. 1a).

Remarks.—The ilium of *Rana sylvatica* is quite distinct from those of many other *Rana* species in having a very distinct dorsal protuberance on the anterodorsal part of its lateral surface near the acetabular fossa. In fact, this prominence is rounded and somewhat similar to those of tree toads of the genus *Hyla*. It is noteworthy that several authors have commented on the toad-like gait and short-leggedness of *Rana sylvatica,* especially in northern populations. Moreover, the ilia of *Rana sylvatica,* although of small size, do not show the perforate condition of the acetabular area as do immature individuals of larger species of *Rana. Rana sylvatica* is not found in Kansas today, but mainly occurs to the northeast (Fig. 11). The nearest population today is an isolated one in the Ozark Region of south-west Missouri and north-west Arkansas. Today, the wood frog occurs in wooded areas (Wright and Wright, 1949). The significance of this record will be detailed in the Discussion section.

Rana pipiens complex Schreber, 1782

Material.—Locality 1, left and right ilia FHSVP 2865 and 2860, 2862, sacral vertebra FHSVP 2847; Locality 2, right ilia FHSVP 2898, 2861; Locality 4, left ilia FHSVP 2910, 2911; Locality 5, left ilium FHSVP 2924, sacral vertebra FHSVP 2919.

Remarks.—Holman (1971) discusses characters that distinguish the ilia of the *Rana pipiens* complex from other species of *Rana.* Collins (1974:78) states that *R. pipiens* is found in every aquatic situation in the state today. Taphonomically it is noted that the fossil *R. pipiens* was found in both high energy (Localities 1 and 2) and low energy (Localities 4 and 5) sedimentary situations. Collins (1974: map p. 77) shows that *R. pipiens* occurs in Ellis County, Kansas, today, although he realizes that some workers refer these populations to a distinct species, *Rana blairi.*

Class Reptilia Laurenti, 1768
Order Squamata Oppel, 1811
Family Colubridae Oppel, 1811
Subfamily Colubrinae Cope, 1893

Lampropeltis calligaster (Harlan, 1827)

Material.—Locality 1, trunk vertebra FHSVP 2822.

Remarks.—This species has a very well-defined hemal keel and subcentral ridge, but these processes are not as well-defined as in *Lampropeltis getulus. Lampropeltis calligaster* may be distinguished from *L. triangulum* in that it has a more vaulted neural arch and a more distinct hemal keel. Today, this species is found slightly to the east and south of Ellis County in Barton County, Kansas (Collins, 1974: map p. 183). This snake inhabits a variety of areas in Kansas today from rocky hillsides with open woods to prairie grasslands and often retreats into the burrows of other animals (Collins, 1974: 185).

THE WILLIAMS LOCAL FAUNA

The Williams local fauna of the Great Bend Prairie Area, Rice County, Kansas (north-east corner of south-east ¼ section 21, T 18 S, R 7 W) was previously studied by Hall (1972), mollusks; McMullen (1975), shrews; Lundberg (1975), catfishes; Preston (1979), turtles. It is considered to represent the Illinoian age of the Pleistocene based on its fauna and stratigraphy. The fauna was recovered from a borrow pit by field workers and friends of the late C. W. Hibbard. This is one of the largest fossil herpetofaunas known from the central plains, and is the largest Illinoian herpetofauna known. The mode of accumulation of the fossils in the Williams fauna does not appear to be unlike that of the Duck Creek fauna.

Class Amphibia Linnaeus, 1758
Order Caudata Oppel, 1811
Family Ambystomatidae Hallowell, 1858

Ambystoma tigrinum (Green, 1825)

Material.—Partial right maxilla and distal left humerus UM81731.

Remarks.—This species, also recorded and discussed in the previous fauna, has not been recorded from Rice County, Kansas, today, but according to Collins (1974: map p. 26) should occur there.

Family Hylidae Hallowell, 1857

Acris crepitans Baird, 1854

Material.—One left and two right ilia UM 81732 (Fig. 1b).

Remarks.—These ilia have the dorsal protuberances anterior to the anterior edge of the acetabular cup, and two of them are complete enough to have an ilial shaft ridge. These characters in combination are diagnostic for the genus *Acris.* The above fossils are indistinguishable from those of modern *Acris*

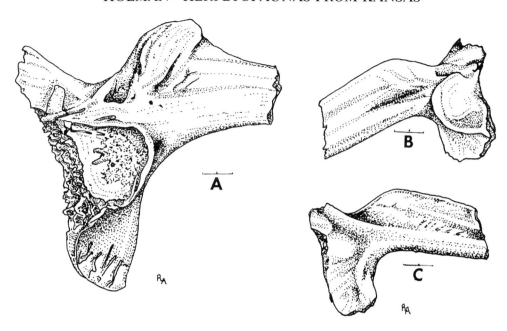

Fig. 2.—A) *Rana catesbeiana*, right ilium UM 81734; B) *Rana pipiens*, left ilium in lateral view UM 81735; C) *Rana pipiens*, left ilium in medial view UM 81735. All from Williams local fauna. Each line equals 2 mm.

crepitans, a species that occurs in Rice County, Kansas, today (Collins, 1974: map p. 57). The preferred habitat is said to be the muddy beach-like edges of small, shallow streams and ponds in Kansas today (Collins, 1974:58).

Family Bufonidae Fitzinger, 1826

Bufo woodhousei woodhousei Girard, 1854

Material.—Right cranial crest and one left and three right ilia UM 81733 (Fig. 1c, d).

Remarks.—The comments on the identification and habitat of this subspecies that were made in the Duck Creek fauna section apply here. *Bufo w. woodhousei* has been recorded from Rice County, Kansas, today (Collins, 1974:55).

Family Ranidae Bonaparte, 1831

Rana catesbeiana Shaw, 1802

Material.—One left and three right ilia UM 81734 (Fig. 2a).

Remarks.—The ilia of *Rana catesbeiana* are distinguishable from those of other species of *Rana* in the middle United States on the basis of (1) the steeper slope of the posterior part of the ilial blade into the dorsal acetabular expansion, (2) its large size, and (3) the very porous area of the bone near the acetabular border. Condition 3 is particularly marked in small individuals. The bullfrog is re-

stricted to permanent lakes, rivers, streams, and swamps today where deep water is available (Collins, 1974:71). This species occurs in Rice County, Kansas, today (Collins, 1974:71).

Rana pipiens complex Schreber, 1782

Material.—Sixteen left and 15 right ilia UM 81735 (Fig. 2b, c).

Remarks.—The comments that were made on the identification and habitat of this form in the Duck Creek section apply here. Collins (1974: map p. 77) has recorded *R. pipiens* from counties surrounding Rice County, Kansas, but there are no actual records from Rice County. Certainly this form occurs in the area today.

Class Reptilia Laurenti, 1768
Order Testudines Batsch, 1788
Family Chelydridae Swainson, 1839

Chelydra serpentina (Linnaeus, 1758)

Material.—Right femur, head of left femur, left ilium, left pubic fragment, nuchal fragment UM 60084.

Remarks.—This material was previously reported by Preston (1979:28). The snapping turtle is not listed as occurring in Rice County, Kansas, today, but according to the distribution shown in Collins (1974: map p. 87) should occur in the area. This species is said to occur in every aquatic situation, but to prefer water with a soft mud bottom, abundant pond-edge vegetation, and numerous sunken logs and branches (Collins, 1974:87).

Family Testudinidae Gray, 1825

Emydoidea blandingii (Holbrook, 1838)

Material.—Right 8th and left 11th peripherals and additional fragments UM 60089.

Remarks.—This species was previously identified and reported from the Williams local fauna by Preston (1979:31). *Emydoidea blandingii* does not occur in Kansas today, the closest records being in south-central Nebraska (Fig. 12). The significance of this occurrence in the Kansas fossil record will be detailed in the Discussion Section of this paper.

Chrysemys picta Schneider, 1783

Material.—Nuchals and epiplastra UM 60085, carapace parts UM 60086, plastral parts UM 60087, and 7th cervical vertebra UM 60577.

Remarks.—This material was previously identified and reported by Preston (1979:37). Although it is not listed as occurring in Rice County, Kansas, today, the subspecies *Chrysemys picta belli* undoubtedly is present there (Collins, 1974: map p. 107). The painted turtle lives in slow-moving, shallow streams and rivers, and shallow ponds and lakes with soft bottoms in Kansas today (Collins, 1974: 107–108).

Pseudemys scripta Schoepff, 1792

Material.—Peripheral fragments and an epiplastron UM 60088.

Remarks.—This material was previously identified and reported by Preston (1979:35). This species is not recorded from Rice County, Kansas, today, but according to Collins (1974: map p. 110) the subspecies *Pseudemys scripta elegans* should occur in the area. This turtle is said to occur in about every permanent body of water throughout its range in Kansas today (Collins, 1974:110).

Order Squamata Oppel, 1811
Family Colubridae Oppel, 1811
Subfamily Xenodontidae Cope, 1893

Heterodon cf. *Heterodon platyrhinos* Latreille, 1802

Material.—Four anterior trunk vertebrae and one fragmental middle trunk vertebra UM 81736.

Remarks.—Most discussion about the identification of *Heterodon* on the basis of vertebral re-

mains has been on the basis of middle trunk vertebrae. Unfortunately, the relatively complete *Heterodon* vertebrae from the Williams fauna are anterior trunk vertebrae and do not appear to be specifically diagnostic. Nevertheless, these vertebrae are quite similar in size to those of a modern *Heterodon platyrhinos* with a total length of 70 cm (27.55 inches), thus I am tentatively referring the vertebrae to this species. Conant (1975) lists the typical ranges in total length as 16–25 inches for *Heterodon nasicus* and 20–33 inches in *H. platyrhinos*. *Heterodon platyrhinos* has not been recorded from Rice County, Kansas, today, but according to Collins (1974: map p. 163) it should occur there. This snake is said to be common along valleys of major rivers in the Great Bend Prairie Area of Kansas today and it is said to prefer sandy areas (Collins, 1974:163).

Subfamily Colubrinae Cope, 1895

Compared to vertebrae of the Subfamily Natricinae in the Williams fauna, vertebrae of the Subfamily Colubrinae were rare (Table 1). Two fragmentary vertebrae UM 81737 are assigned to Colubrinae gen. et sp. indet.

Coluber constrictor Linnaeus, 1758

Material.—Trunk vertebra UM 60256 (Fig. 3).

Remarks.—Several authors (see summary in Holman, 1981) have been unable to distinguish the vertebrae of *Coluber* and *Masticophis*. Nevertheless, based on the smaller size of the specimen and the relatively small size of the neural canal, and based on the modern geographic ranges of the two genera (Collins, 1974: maps pp. 169 and 173) I am cautiously referring this specimen to *Coluber constrictor*. The subspecies *Coluber constrictor flaviventris* occurs in Rice County, Kansas, today (Collins, 1974: map p. 169). This snake occurs in open grassland and prairies in Kansas today (Collins, 1974: map p. 170).

Elaphe vulpina (Baird and Girard, 1853)

Material.—An associated partial skeleton (Fig. 4) consisting of a cervical vertebra, 55 trunk vertebrae, 5 caudal vertebrae, and 5 ribs UM 60231; also 3 trunk vertebrae from another individual or individuals UM 81738.

Remarks.—The associated skeleton was collected by the late C. W. Hibbard and his field party on 8 June 1969. Holman (1982) provided an osteological definition of *Elaphe vulpina*. Diagnostic characters

Table 1.—*Inferred habitat and abundance of herpetological species of the Duck Creek and Williams Local Faunas of the Pleistocene (Illinoian) of Kansas. Habitats based on Collins (1974). Abbreviations: F = frequent, S = seasonal, I = infrequent.*

Species	Habitat					Fossil abundance	
	Prairie pond or slow stream	Pond or stream edge	Prairie flats or hillsides	Woodland-prairie edge	Woodland	Minimum number of individuals	Total number of salamander or snake vertebrae
Duck Creek Fauna							
Ambystoma tigrinum	S	S	F	F	F	1	2
Bufo w. woodhousei	S	F	F			3	
Rana sylvatica					I	1	
Rana pipiens complex	S	F	I			4	
Lampropeltis calligaster		I	F	F		1	1
Williams Fauna							
Ambystoma tigrinum	S	S	F	F	F	1	Cranial only
Acris crepitans	S	F				2	
Bufo w. woodhousei	S	F	F			3	
Rana catesbeiana	F	F				3	
Rana pipiens complex	S	F	I			16	
Chelydra serpentina	F	S				1	
Emydoidea blandingii	F	S				1	
Chrysemys picta	F	S				2	
Pseudemys scripta	F	S				1	
Heterodon platyrhinos		I	F	F		1	4
Colubrinae indet.						1	2
Coluber constrictor			F	F	I	1	1
Elaphe vulpina			I	F		2	63
Lampropeltis triangulum			F	F	F	1	1
Pituophis melanoleucus			F	F		1	1
Sonora episcopa			F			1	1
Tantilla sp.			F			1	3
Natricinae indet.						1	131
Nerodia sipedon	F	F				1	21
Regina grahami	F	F				1	1
Thamnophis radix			F	F		1	21
Thamnophis sp.		F				1	26
Tropidoclonion lineatum			F	F		1	3

of the trunk vertebra are as follow: neural spine usually slightly longer than high, neural arch vaulted; condyle not enlarged; ventral processes of centrum gracile. Holman (1982:40, Table 2) indicates the height of neural spines of the trunk vertebrae of *Elaphe vulpina* compared with related colubrid species.

Elaphe vulpina is well out of its present day range in the Williams fauna. Today, the species gets no closer to Rice County, Kansas, than the northwestern part of Missouri and southeastern part of Nebraska (Fig. 13). Collins (1974:241) states "It is quite possible that the western fox snake will be found in extreme northeast Kansas." The fox snake had a wider distribution in the Pleistocene than it does today (Holman, 1981: map p. 292). The significance of this extralimital species will be detailed in the Discussion section of this paper. The fox snake is a grassland and woodland edge species today (personal observation).

Lampropeltis triangulum (Lacepede, 1788)

Material.—Three trunk vertebrae UM 81739 (Fig. 5).

Remarks.—This is one of the easiest snakes of eastern and central North America to identify on the basis of vertebral characters. It has a low neural spine, a depressed neural arch, and only a moderately distinct hemal keel. The species occurs in Rice County, Kansas, today (Collins, 1974: map p. 188). This snake is said to occur in rocky ledges of prairie canyons today (Collins, 1974:189).

Pituophis melanoleucus (Daudin, 1803)

Material.—Trunk vertebra UM 60255 (Fig. 6).

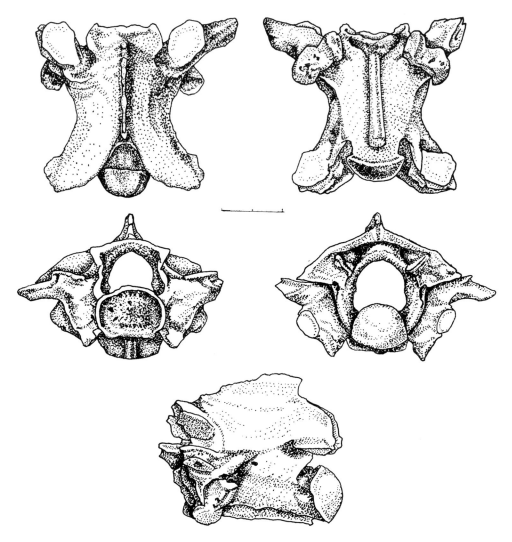

Fig. 3.—*Coluber constrictor,* trunk vertebra UM 60256 from Williams local fauna. Upper left, dorsal, right, ventral; middle left, anterior, right, posterior; lower, lateral. The line equals 2 mm and applies to all views.

Remarks.—Auffenberg (1963:183–184) and Van Devender and Mead (1978:472) give vertebral characters of *Pituophis melanoleucus. Pituophis melanoleucus sayi,* the subspecies found in Rice County, Kansas, today (Collins, 1974: map p. 182) has lower neural spines than the southeastern subspecies *P. m. mugitus.* The Williams fauna vertebra appears identical to *P. m. sayi* in neural spine height (Fig. 6). This snake is said to live in open grassland as well as open woodland and woodland edge today in Kansas (Collins, 1974:182).

Sonora episcopa (Kennicott, 1859)

Material.—Trunk vertebra UM 81740.

Remarks.—The tiny vertebrae of *Sonora* may be distinguished from those of *Diadophis* and *Tantilla* on the basis of their more prominent neural spines, thinner hemal keels, and shorter vertebral form. Van Devender et al. (1977:55–56) state that *S. semiannulata* vertebrae cannot be distinguished from those of *S. episcopa.* The nearest *S. semiannulata* occurs to the area today is in southern New Mexico (Conant, 1975: map 157). Conant (1975: map 158), indicates *S. episcopa* occurs in Rice County, Kansas, today, but the more detailed map of Collins (1974: 163) does not record it from Rice County, although there are records to the north and to the south of Rice County. The subspecies found in Kansas today

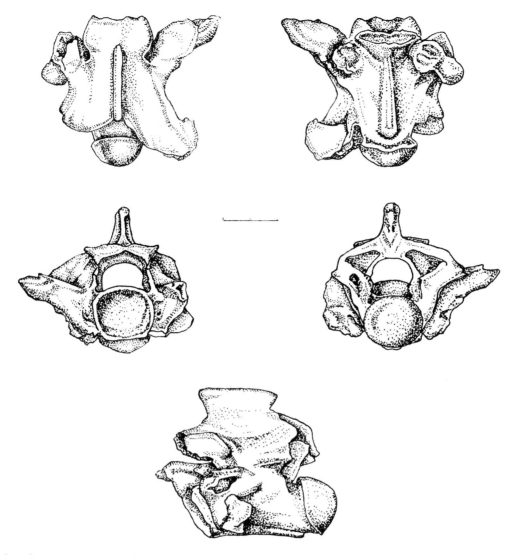

Fig. 4.—*Elaphe vulpina,* trunk vertebra from an associated partial skeleton UM 60231 from Williams local fauna. Upper left, dorsal, right, ventral; middle left, anterior, right, posterior; lower, lateral. The line equals 2 mm and applies to all views.

is *S. e. episcopa.* This snake is said to occupy dry, rocky, prairie hillsides (Collins, 1974:194).

Tantilla sp. indet.

Material.—Three trunk vertebrae UM 81741.

Remarks.—This tiny snake has trunk vertebrae with obsolete neural spines. The species of *Tantilla* appear to be difficult or impossible to distinguish from one another on the basis of vertebral characters. Two species of *Tantilla* occur in Kansas today (Collins, 1974: maps pp. 195 and 197). *Tantilla gracilis* occurs only in the eastern one-third of Kansas, whereas *Tantilla nigriceps nigriceps* occurs rather

extensively in the western three-fourths of the state, although it has never been specifically recorded from Rice County. *Tantilla gracilis* is said to occupy rocky hillsides of open prairies and woodlands in Kansas today, whereas *T. nigriceps* is said to be found on rocky hillsides of grassland prairies (Collins, 1974: 195, 197).

Subfamily Natricinae Bonaparte, 1840

Many more natricine than colubrine vertebrae occur in the Williams fauna (Table 1) and this undoubtedly reflects the fact that these snakes lived nearer the site of deposition than the more terrestrial colubrines. One hundred and thirty-one vertebrae

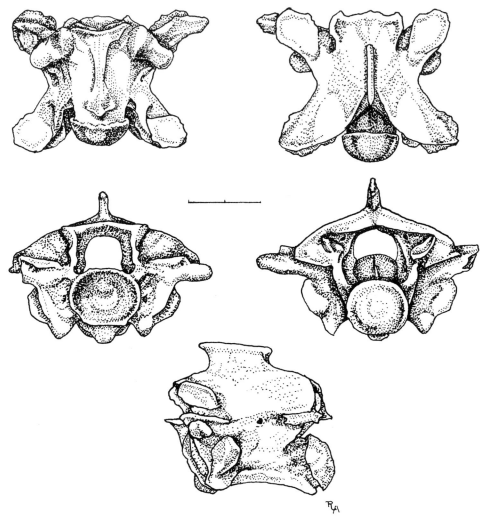

Fig. 5.—*Lampropeltis triangulum,* trunk vertebra UM 81739 from Williams local fauna. Upper left, ventral, right dorsal; middle left, anterior, right, posterior; lower, lateral. The line equals 2 mm and applies to all views.

with diagnostic processes missing and caudal vertebrae are identified as Natricinae gen. et sp. indet. UMV.

Nerodia sipedon (Linnaeus, 1758)

Material.—Twenty-one trunk vertebrae UM 81742 (Fig. 7).

Remarks.—Holman (1967) has given vertebral characters that enable one to distinguish this species from related forms in eastern and central North America. *Nerodia sipedon sipedon* has not been recorded from Rice County, Kansas, today, but according to Collins (1974: map p. 227) should occur there. According to Collins (1974:227) the northern water snake is found in almost every aquatic situation in Kansas today, with both high and low energy bodies of water being frequented.

Regina grahami Baird and Girard, 1853

Material.—A trunk vertebra UM 81743 (Fig. 8).

Remarks.—Holman (1972) gives vertebral characters that separate *R. grahami* from other related species. This species is found in Rice County, Kansas, today (Collins, 1974: map p. 219). This snake lives near ponds and sluggish streams of prairie meadows and river valleys in Kansas today (Collins, 1974:219–220).

Storeria cf. Storeria dekayi (Holbrook, 1842)

Material.—Four trunk vertebrae UM 81744.

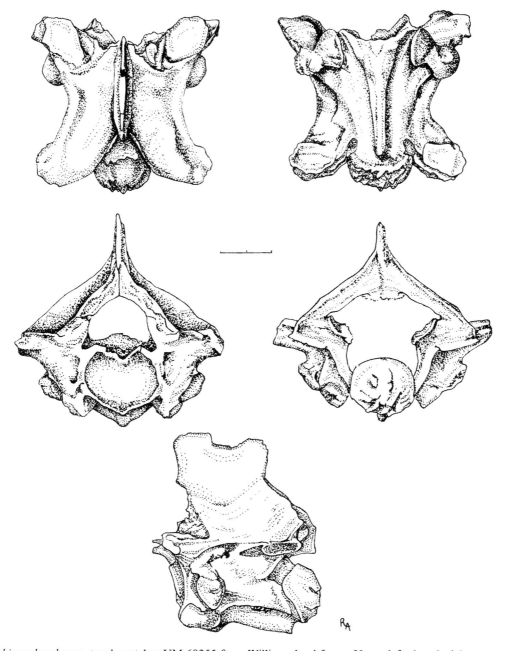

Fig. 6.—*Pituophis melanoleucus,* trunk vertebra UM 60255 from Williams local fauna. Upper left, dorsal, right, ventral; middle left, anterior, right posterior; lower, lateral. The line equals 2 mm and applies to all views.

Remarks.—Auffenberg (1963:192) and Holman (1962:258–259) discuss the identification of *Storeria dekayi* on the basis of vertebrae. *Storeria occipitomaculata* is a woodland species that occurs in the eastern two tiers of counties in Kansas today (Collins, 1974: map p. 217). *Storeria dekayi texana* is not recorded from Rice County, Kansas, today, but has been reported from Ellsworth County just to the north (Collins, 1974: map p. 215). This snake generally lives near moist situations in woodland and along woodland edges (Collins, 1974:215).

Thamnophis radix (Baird and Girard, 1853)

Material.—Twenty-one trunk vertebrae UM 81745 (Fig. 9).

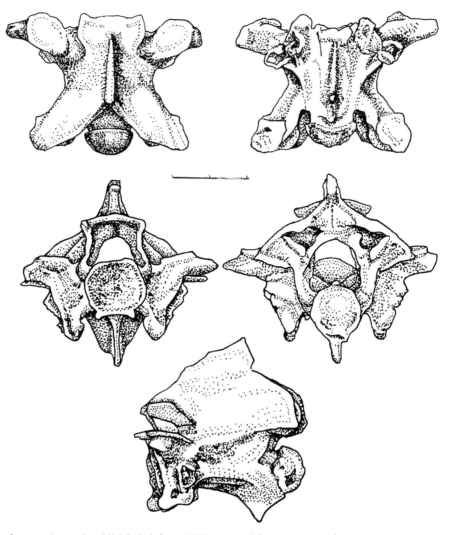

Fig. 7.—*Nerodia sipedon,* trunk vertebra UM 81742 from Williams local fauna. Upper left, dorsal, right, ventral; middle left, anterior, right, posterior; lower, lateral. The line equals 2 mm and applies to all views.

Remarks.—This is only the second published record of this species as a fossil as Rogers (1982) has listed it from an early Kansan site in Kansas. Vertebrae of species of *Thamnophis* are difficult to identify. Nevertheless, trunk vertebrae of modern skeletons of *T. radix* (11) and *T. marcianus* (5) are definitely shorter and also less gracile than those of modern *T. sirtalis* (10) and *T. proximus* (4). Moreover, the neural spines of *T. radix* and *T. marcianus* have more anterior and posterior overhang than in most *T. sirtalis* and *T. proximus.* The neural spines of available modern *T. radix* are somewhat higher than those of *T. marcianus.* The Williams local fauna vertebrae appear identical to those of *T. radix.* The subspecies *T. radix haydeni* is found in Rice

County, Kansas, today (Collins, 1974: map p. 205). These snakes are said to prefer open grassy prairies, particularly along the edges of streams, marshes, and lakes in Kansas today (Collins, 1974:206).

Thamnophis sirtalis (Linnaeus, 1758) or *Thamnophis proximus* (Say, 1823)

Material.—Twenty-six vertebrae, UM 81746 (Fig. 10).

Remarks.—I cannot distinguish whether these more elongate *Thamnophis* vertebrae represent *Thamnophis sirtalis* or *T. proximus. Thamnophis sirtalis parietalis* (Say) is listed as occurring in Rice County, Kansas, today (Collins, 1974: map p. 207), and although *Thamnophis sauritus* occurs in neigh-

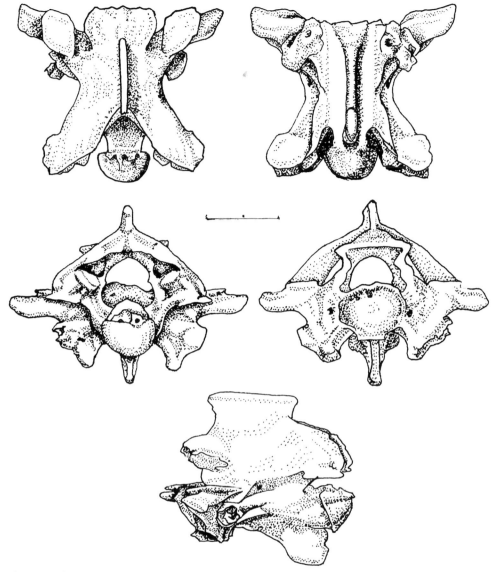

Fig. 8.—*Regina grahami*, trunk vertebra UM 81743 from Williams local fauna. Upper left, dorsal, right, ventral; middle left, posterior, right, anterior; lower, lateral. The line equals 2 mm and applies to all views.

boring Stafford County to the southwest (Collins, 1974: map p. 203) it has not been reported from Rice County, Kansas. *Thamnophis sirtalis* is found in a wide number of habitats such as marshes, wet meadows, pond-margins, woodlands, and woodland edges and floodplains in Kansas today (Collins, 1974:208). It is said to prefer moist situations. *Thamnophis proximus* frequents the edges of swampy marshes, lakes, streams, and rivers (Collins, 1974:204).

Tropidoclonion lineatum (Hallowell, 1856)

Material.—Three trunk vertebrae UM 81747.

Remarks.—This is only the third record of this genus and species as a fossil. It has previously been reported from the Pleistocene of Texas (Holman, 1965) and from the Pleistocene (Illinoian) of central Kansas (Holman, 1971). Holman (1965) gives vertebral characters for this form, which has a distinctive vertebral type. *Tropidoclonion lineatum* is not recorded from Rice County, Kansas, today, but according to Collins (1974: map p. 209) should be expected to occur there. This snake is said to inhabit the hillsides of open prairies and woodland edges in Kansas today (Collins, 1974:209).

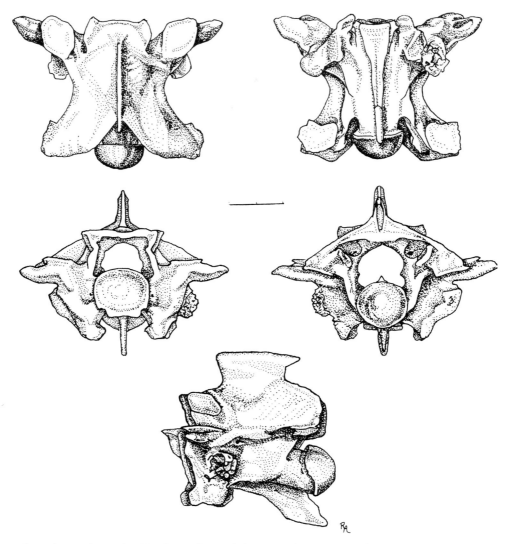

Fig. 9.—*Thamnophis radix,* trunk vertebra UM 81745 from Williams local fauna. Upper left, dorsal, right, ventral; middle left, anterior, right, posterior; lower, lateral. The line equals 2 mm and applies to all views.

DISCUSSION

Vertebrate faunas bear on several compelling Pleistocene problems such as (1) the time of its onset, (2) proper divisional terms, (3) "disharmonious" assemblages, (4) paleoclimates, (5) megafaunal extinction, and recently (6) why no C^{14} ages for extinct Pleistocene genera are younger than 10,000 yrs B.P. (Meltzer and Mead, 1983). The Illinoian herpetofaunas reported herein relate directly to questions 3 and 4 and thus indirectly to questions 5 and 6.

One of the most discussed aspects of North American Pleistocene vertebrate faunas is that many are considered to be "disharmonious" assemblages (Lundelius et al., 1983). The term "disharmonious" is used to describe faunas containing animals that would be "ecologically incompatible" today (term used by Holman, 1976). In such faunas one is unable to find an area on the map where all of the species identified from that fauna could be found living together today ("area of sympatry" of McMullen, 1975, and others). Holman (1980) uses the term "extralimital" to refer to Pleistocene herpetological

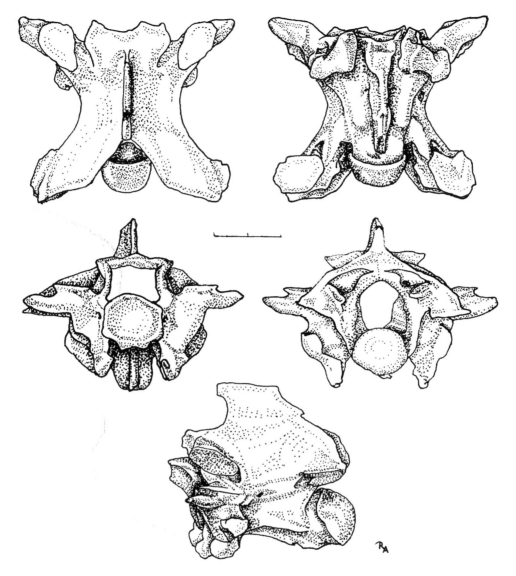

Fig. 10.—*Thamnophis sirtalis* or *Thamnophis proximus*, trunk vertebra UM 81746 from Williams local fauna. Upper left dorsal, right, ventral; middle left, anterior, right, posterior; lower, lateral. The line equals 2 mm and applies to all views.

species that occur outside of their present ranges. The degree of "disharmoniety" in Pleistocene faunas varies; some faunas having only one or two extralimital forms, and other faunas having Boreal and Tropical animals occurring together. Disharmonious faunas may be found in both glacial and interglacial deposits (Holman, 1980).

On the other hand in the British Isles (Stuart, 1982), glacial and interglacial faunas appear to contain more "harmonious" and "ecologically compatible" assemblages, and also to reflect more extreme climatic changes in glacial and interglacial

times than in North America. For instance, interglacial faunas may contain macaque monkeys, spotted hyaenas, African lions, African rhinos, and African hippos, whereas glacial faunas may contain tundra voles, arctic foxes, woolly mammoths, woolly rhinos, reindeer, and musk oxen.

Explanations for disharmonious faunas in the North American Pleistocene vary. A common hypothesis (Hibbard, 1960) repeated by numerous other workers has been called "The Pleistocene Climatic Equability Model" where cooler summers supposedly account for the presence of northern ex-

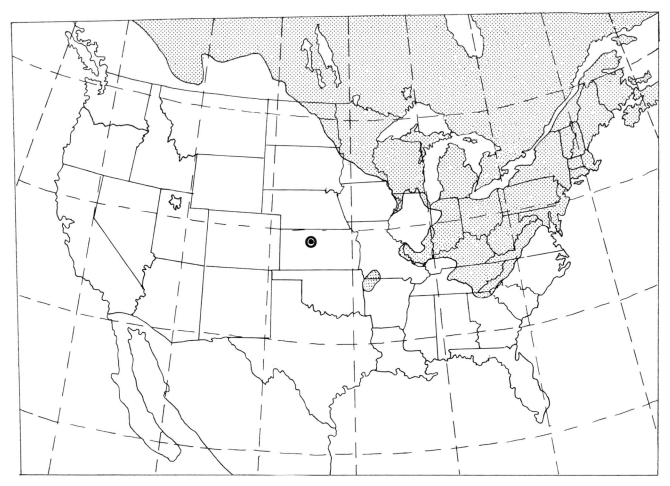

Fig. 11.—Modern range of *Rana sylvatica* (stippling) and record of *Rana sylvatica* from the Duck Creek local fauna, Pleistocene, Illinoian (circled dot).

tralimital species in the fauna, and warmer winters supposedly account for the presence of southern extralimital species in the fauna. Other workers (Lundelius et al., 1983) point out that disharmonious faunas may be associated with mixed vegetational associations. The classic concept that glaciers pushed northern forms down into southern communities may be partially true, but this concept does not explain the presence of southern extralimital forms satisfactorily.

Herpetological species are important in reconstructing fossil communities especially because (1) they are ectothermic and quite sensitive to ecological changes, (2) are probably less vagile than mammalian species, and (3) because most herpetological species of the Pleistocene represent extant ones whose ecological tolerances and habitat preferences are

known. It is noteworthy here that all species of amphibians and reptiles from all Illinoian localities studied in Kansas are indistinguishable from living species.

Duck Creek Local Fauna

Based on the study of the sedimentary environment and on the molluscan fauna, Kolb et al. (1975) state: "The presence of *Pisidium compressum* and *Sphaerium stratinum* within cross-bedded sands indicates a perennial stream with some current action, while the abundance of *Valvata tricarinata* suggests the stream was lake-like in places. The abundance of several strictly woodland species such as *Cionella lubrica* suggests that a continuous strand of trees bordered the stream, while valley slopes were possibly covered with grasses and scattered trees. Like-

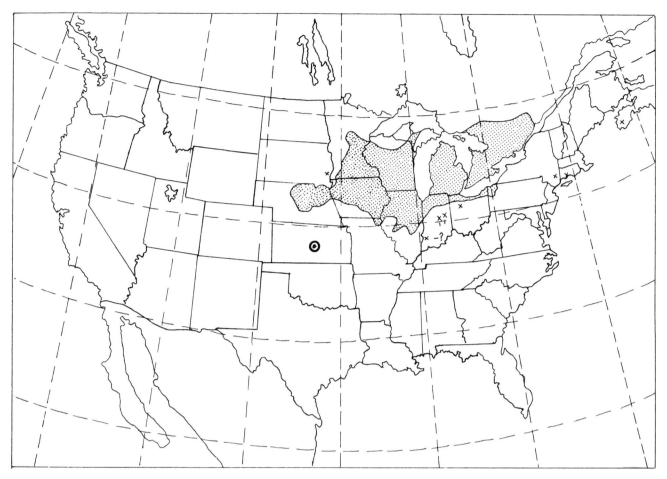

Fig. 12.—Modern range of *Emydoidea blandingii* (stippling, Xs, ?s) and record of *Emydoidea blandingii* from the Williams local fauna, Pleistocene, Illinoian (circled dot).

wise, cooler summers and winters than at present are indicated for Ellis County at the time these taxa lived by the predominance of species with northern distribution."

Based on the mammalian fauna of the Duck Creek locality McMullen (1978) states "In summary, the Duck Creek local fauna lived near a stream which supplied the area with permanent water. Occurring in the moist lowlands and probably bordering the stream in places was a riparian forest. Whether the interspersed trees were coniferous or deciduous is unknown because forest dwelling species in the fauna might have inhabited either situation and palynological data are lacking. The adjacent lowlands area probably was covered with mixed tall and short grasses."

The small herpetofauna of the Duck Creek site appears mainly to fit in with the above interpreta-

tions (Table 1). The only extralimital species is the wood frog, *Rana sylvatica,* which occurs mainly to the north and east of Kansas today (Fig. 11). As is the case with the woodland mammalian species discussed by McMullen (1978), *Rana sylvatica* occurs in both deciduous and coniferous forests. Nevertheless, based on the *R. sylvatica* record, one might suggest the possibility of at least cooler, perhaps moister summers, and at least the presence of some kind of woodland cover near the major aquatic habitat. But it should be pointed out that other herpetological elements of the Duck Creek fauna (*Ambystoma tigrinum, Bufo w. woodhousei, Rana pipiens,* and *Lampropeltis calligaster*) occur in the area today, and do not indicate a climate indicative of northeastern South Dakota as inferred by McMullen's (1975) discussion of the area of sympatry for Duck Creek fauna shrews.

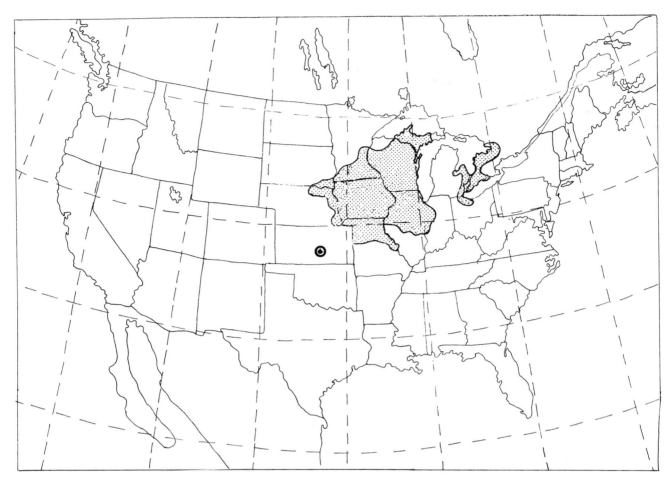

Fig. 13.—Modern range of *Elaphe vulpina* (stippling) and record of *Elaphe vulpina* from the Williams local fauna, Pleistocene, Illinoian (circled dot).

WILLIAMS LOCAL FAUNA

The Williams fauna has not been interpreted as thoroughly as has the Duck Creek fauna. Hall (1972) gave an abstracted paper at the Michigan Academy of Science, Arts, and Letters on the mollusks of the Williams local fauna. McMullen (1975) discussed a new species of shrew that also occurred in the Duck Creek fauna. Lundberg (1975) listed catfishes, and Preston (1979) listed turtles from the Williams fauna. Based on species in these papers, the paleoecology appears to have been generally similar in both the Duck Creek and Williams fauna; that is a permanent body of water with woodland bordering the water, at least in places.

The larger herpetofauna of the Williams fauna indicates (based on habitats of modern Kansas forms, Collins, 1974) some rather specific situations (Table 1). The main habitat appears to have been a pond or the low energy part of a stream and its bank. Thirteen of the 21 species identified from the Williams fauna occur frequently or seasonally near this main habitat. On the other hand, 11 of 21 species are typical of prairie flats or prairie hillside habitats; and six of 21 animals would be found in prairie-woodland edge habitats. One animal is entirely a woodland species.

Two extralimital species, *Emydoidea blandingii* and *Elaphe vulpina,* occur in the Williams fauna. The nearest *Emydoidea blandingii* occurs to Rice County, Kansas, today is southcentral Nebraska (Fig. 12). The nearest *Elaphe vulpina* occurs to Rice County today is in southeastern Nebraska and northeastern Missouri (Fig. 13). These occurrences may indicate somewhat cooler, moister climate for the area in the Pleistocene. On the other hand the rest of the herpetological species in the Williams fauna do not indicate a climate that is different from the area today; in fact all of them are typical animals of the region. Certainly a "full glacial" climate or dominant coniferous forest situation is not indicated by the Williams fauna amphibian and reptile species. Animals of the fauna that occur conspicuously south of coniferous forest situations today include: *Pseudemys scripta, Coluber constrictor, Sonora episcopa, Tantilla* sp., *Regina grahami,* and *Tropidoclonion lineatum.*

COMMENT

The statement made (Holman, 1980) that herpetological evidence from the Sandahl local fauna (Illinoian) did not support the classical idea of a cooler, moister climate for the Illinoian age in Kansas should be somewhat modified based on Illinoian herpetological species in both the Duck Creek and Williams faunas. The Sandahl fauna had two turtles, a salamander, five anurans, three lizards, and six snakes (17 forms), all of which are extant and occur in the area today. But the presence of the wood frog (*Rana sylvatica*) in the Duck Creek fauna and Blanding's turtle (*Emydoidea blandingii*) and the fox snake (*Elaphe vulpina*) in the Williams fauna possibly indicates a somewhat cooler, moister climate,

and the presence of more woodland than occurs in the area today. But, as previously stated, it is emphatically clear that a "full glacial" or Boreal climate, or a dominant coniferous forest association is not indicated by the other herpetological species.

In the past, Pleistocene mammalian faunas have sometimes been interpreted as indicating Boreal or even Tundra situations when the herpetofaunas from the same localities have been similar or identical to those found in the areas today. Certainly a problem of interpretation exists here. Moreover, as Stuart (1979) indicates in his study of Pleistocene occurrences of the European pond turtle (*Emys orbicularis*) in Great Britain, a certain minimum mean summer temperature is needed in order for the turtles to breed. In the future, it would seem that studies on critical temperatures for egg-hatching of turtles and other reptilian species would provide exceedingly important information about Pleistocene summer climates in North America.

Perhaps, rather than indicating the presence of full Boreal or Tundral vegetation and climate, the presence of northern extralimital mammals might best be explained by southward glacial displacement of both mammalian and plant species. This would form a mosaic Pleistocene community of both animals and plants. Such a mosaic community could survive as long as the climate remained equable.

LITERATURE CITED

AUFFENBERG, W. 1963. The fossil snakes of Florida. Tulane Stud. Zool., 10:131–216.

COLLINS, J. T. 1974. Amphibians and reptiles in Kansas. Univ. Kansas Mus. Publ. Nat. Hist., Ed. Ser., 1:1–283.

CONANT, R. 1975. A field guide to reptiles and amphibians of eastern and central North America. Houghton Mifflin Co., Boston, 429 pp.

HALL, S. A. 1972. A new Illinoian molluscan faunule from central Kansas. Abs. Michigan Acad. Sci., Arts, Letts., 76th Ann. Meeting, Program with Abstracts, p. 6.

HIBBARD, C. W. 1960. Pliocene and Pleistocene climates in North America. Ann. Rept. Michigan Acad. Sci., Arts, Letts., 62:5–30.

HOLMAN, J. A. 1962. A Texas Pleistocene herpetofauna. Copeia, 1978:255–261.

———. 1965. Pleistocene snakes from the Seymour Formation of Texas. Copeia, 1965:102–104.

———. 1967. A Pleistocene herpetofauna from Ladds, Georgia. Bull. Georgia Acad. Sci., 25:154–166.

———. 1969. Herpetofauna of the Pleistocene Slaton local fauna of Texas. Southwestern Nat., 14:203–212.

———. 1971. Herpetofauna of the Sandahl local fauna (Pleis-

tocene:Illinoian) of Kansas. Contrib. Mus. Paleont., Univ. Michigan, 23:349–355.

———. 1972. Herpetofauna of the Kanapolis local fauna (Pleistocene: Yarmouth) of Kansas. Michigan Acad., 5:87–98.

———. 1976. Paleoclimatic implications of "ecologically incompatible" herpetological species (late Pleistocene: southeastern United States). Herpetologica, 32:291–295.

———. 1980. Paleoclimatic implications of Pleistocene herpetofaunas of eastern and central North America. Trans. Nebraska Acad. Sci., 8:131–140.

———. 1981. A review of North American Pleistocene snakes. Publ. Mus. Michigan State, Univ. Paleont. Ser., 50:261–306.

———. 1982. A fossil snake (*Elaphe vulpina*) from a Pliocene ash bed in Nebraska. Trans. Nebraska Acad. Sci., 10:37–42.

KOLB, K., M. E. NELSON, and R. J. ZAKRZEWSKI. 1975. The Duck Creek molluscan fauna (Illinoian) from Ellis County, Kansas. Trans. Kansas Acad. Sci., 78:63–74.

LUNDBERG, J. G. 1975. The fossil catfishes of North America. Univ. Michigan Mus. Paleont., Pap. Paleont., 11:1–51.

LUNDELIUS, E. L., R. W. GRAHAM, E. ANDERSON, J. GUILDAY, J. A. HOLMAN, D. STEADMAN, and S. D. WEBB. 1983. Terrestrial vertebrate faunas. *In* Late Quaternary Environments of the United States: the Late Pleistocene, Univ. Minnesota Press, Minneapolis, 407 pp.

MCMULLEN, T. L. 1975. Shrews from the late Pleistocene of Kansas with the description of a new species of *Sorex*. J. Mamm., 56:316–320.

———. 1978. Mammals of the Duck Creek local fauna, late Pleistocene of Kansas. J. Mamm., 59:374–386.

MELTZER, D., and J. I. MEAD. 1983. The timing of the late Pleistocene mammalian extinctions in North American. Quaternary Research, 19:130–135.

PRESTON, R. E. 1979. Late Pleistocene cold-blooded vertebrate faunas from the mid-continental United States. I. Reptilia; Testudines, Crocodilia. Univ. Michigan Mus. Paleont., Pap. Paleont., 19:1–53.

ROGERS, K. L. 1982. Herpetofaunas of the Courland Canal and Hall Ash local faunas (Pleistocene: Early Kansan) of Jewell Co., Kansas. J. Herpetol., 16:174–177.

STUART, A. J. 1979. Pleistocene occurrence of the European pond tortoise (*Emys orbicularis* L.) in Britain. Boreas, 8: 359–371.

———. 1982. Pleistocene vertebrates in the British Isles. Longman, London and New York, 212 pp.

TIHEN, J. A. 1958. Comments on the osteology and phylogeny of ambystomatid salamanders. Bull. Florida State Mus., 3: 1–50.

———. 1962. A review of New World fossil bufonids. Amer. Midland Nat., 68:1–50.

VAN DEVENDER, T. R., and J. I. MEAD. 1978. Early Holocene and late Pleistocene amphibians and reptiles in Sonoran Desert packrat middens. Copeia, 1978:467–475.

VAN DEVENDER, T. R., A. M. PHILLIPS, and J. I. MEAD. 1977. Late Pleistocene reptiles and small mammals from the lower Grand Canyon of Arizona. Southwestern Nat., 22:49–66.

WRIGHT, A. H., and A. A. WRIGHT. 1949. Handbook of frogs and toads of the United States and Canada. Comstock Publ. Co., Ithaca, New York, 640 pp.

ZAKRZEWSKI, R. J., and J. L. MAXFIELD. 1971. Occurrence of *Clethrionomys* in the late Pleistocene of Kansas. J. Mamm., 52:620–621.

Address: The Museum, Michigan State University, East Lansing, Michigan 48824.

PLEISTOCENE BIRDS FROM CUMBERLAND CAVE, MARYLAND

PIERCE BRODKORB AND CÉCILE MOURER-CHAUVIRÉ

ABSTRACT

Seven species of birds are recorded, including a new species of owl, *Otus guildayi*. The avifauna is compatible with the Illinoian age previously postulated on the basis of the mammalian fauna.

The fossil record of *Bonasa umbellus, Ectopistes migratorius,* and *Perisoreus canadensis* is extended to Illinoian time.

INTRODUCTION

Present knowledge of the Appalachian vertebrate fauna older than 20,000 years is limited to fossils from three or four caves—Port Kennedy in Pennsylvania, Cumberland Cave in Maryland, and Trout and perhaps Rapps caves in West Virginia. The fossil vertebrates from these sites include a high percentage of extinct genera and species, some of northern and others of southern affinities. Hay (1923) thought the vertebrates with southern affinities to be of Sangamonian interglacial age, but the entire fauna is now considered to date from the Illinoian glaciation about 100,000 years ago (Guilday, 1971). The only bird fossils previously reported from these caves are turkey bones (*Meleagris gallopavo*) from Port Kennedy Cave (Mercer, 1899) and a bone of the ruffed grouse (*Bonasa umbellus*) from Cumberland Cave (Wetmore, 1927).

Cumberland Cave is located one-half mi south of Corriganville, Alleghany Co., in the Appalachian Mountains of northwestern Maryland, and close to the Pennsylvanian border. Early exploration was carried out by J. W. Gidley of the Smithsonian Institution, who found the grouse bone among the rich non-avian vertebrate fauna (Gidley and Gazin, 1938). In 1965 further excavation was made by Allen D. McCrady and Harold Hamilton, under the direction of John E. Guilday of the Carnegie Museum of Natural History (Guilday and Handley, 1967). The bird remains collected by the Carnegie party form the subject of the present report.

SYSTEMATIC PALEONTOLOGY

The following list describes the bird remains that we have determined to generic or specific level. The collection also includes many indeterminate scraps, mostly tiny fragments of small passerines.

Family Anatidae

Anas crecca Linnaeus
Green-winged or Common Teal

Material.—CM 34027, proximal end of left carpometacarpus.

This specimen agrees with *A. crecca* in having the proximal end of the carpometacarpus wider than in the Blue-winged Teal (*Anas discors* Linnaeus), as shown in Table 1.

The Green-winged Teal is a Holarctic species that breeds in the north and winters as far south as the Equator. Its three subspecies are differentiated on minor plumage differences, and we are unable to separate them on osteological grounds.

In Europe its fossil record extends back to the Mindel glaciation, and possibly even to the Gunz (Mourer-Chauviré, 1975). In America the oldest previous records are of Illinoian or Sangamonian age in Kansas (Galbreath, 1955) and Florida (Brodkorb, 1957, 1959; Ligon, 1965).

Family Accipitridae

Aquila chrysaëtos (Linnaeus)
Golden Eagle

Material.—CM 34018, right claw core with tip and base missing.

In eagles the claw of digit I (the hind toe) is largest, and the claw of digit II (the inner anterior toe) is next largest. The lateral surface of an eagle's claw is flat, and the medial side is rounded in the three anterior toes. The same situation would hold in the hind toe if all four toes were directed forward.

The damaged state of the fossil makes it difficult to determine the digit represented. Nevertheless, the fossil agrees in size with the Golden Eagle and exceeds the largest of our series of the Bald Eagle,

Table 1.—*Measurements (in mm) of carpometacarpus of* Anas crecca *and* A. discors.

Material	Width of trochlea	Depth of internal rim of trochlea
CM 34027	4.2	5.0
A. c. nimia 1 ♂ Aleutians	4.3	4.9
A. c. carolinensis 3 ♂, 3 ♀ North America	3.7–4.2	4.8–5.2
A. c. crecca 1 ♂ Holland	3.8	5.1
A. discors 3 ♂, 2 ♀ North America	3.6–3.8	4.6–5.0

Table 2.—*Measurements (in mm) of largest claw core of eagles.*

Material	Plantar width where broken	Height at same level
CM 34018	7.9	9.7
Aquila chrysaëtos canadensis 1 ♂, 2 ♀ United States (Pennsylvania, Texas, Wyoming)	8.70 (7.3–10.0)	9.0 (8.4–9.7)
Haliaeetus leucocephalus washingtoniensis 2 ♀, Alaska	7.35	9.45
Haliaeetus leucocephalus leucocephalus 2 ♂, 2 ♀ Florida	6.70 (5.9–6.9)	8.40 (7.6–9.1)

Haliaeetus leucocephalus (Linnaeus), as shown in Table 2. It also agrees with *Aquila* in lacking the pronounced distal tapering of the claws in *Haliaeetus*.

The Golden Eagle is a widespread Holarctic species. In the eastern United States it is relatively rare, but it has bred south in the mountains to New York and North Carolina.

In Europe this species first appears as fossil during the Mindel stage of the Pleistocene (Mourer-Chauviré, 1975) and in America during the Illinoian stage (Ritchie, 1980).

Family Phasianidae

Bonasa umbellus (Linnaeus)
Ruffed Grouse

Material.—USNM 11690, distal half of left humerus. We have not seen this specimen reported by Wetmore (1927).

The Ruffed Grouse is a bird of North Woods and similar environments in the mountains. In the Appalachians, it extends southwards into northern Georgia. This is the oldest known occurrence of the species. The site at Arredondo, Florida, formerly also thought to be of Illinoian glacial age (Brodkorb, 1959), is now considered to represent the Sangamonian interglacial age (Webb, 1974).

Meleagris gallopavo Linnaeus
Wild Turkey

Material.—CM 34028, upper part of left coracoid.

Male turkeys are much larger than females, but it is not possible to determine the sex of this isolated specimen with certainty. Its measurements are as follows: head through scapular facet, 28.4; depth of head, 12.2 mm. Comparison with data in the latest revision (Steadman, 1980) shows that both measurements of the Cumberland Cave coracoid fall within the size range of female *M. gallopavo.*

The genus first appears in Blancan times (late Pliocene or early Pleistocene) and *M. gallopavo* is known in several localities of Illinoian age.

Wild turkeys formerly occurred in most of the United States and much of Mexico. Their major requirement seems to be trees for roosting at night.

Family Columbidae

Ectopistes migratorius (Linnaeus)
Passenger Pigeon

Material.—CM 34020, upper end of right coracoid, lacking procoracoid process.

In both the Passenger Pigeon and the feral domestic pigeon (*Columba livia*) the coracoid is much larger than in the Mourning Dove (*Zenaida macroura*). Measurements of the wild species are given in Table 3. The coracoid of *Ectopistes* is shorter, its proximal end is wider, and its shaft is narrower than

Table 3.—*Measurements (in mm) of coracoid of* Ectopistes *and* Zenaida.

Material	Clavicular facet		Humeral facet		Head through procoracoid
	Length	Width	Length	Width	
CM 34020	5.3	3.6	6.1	3.8	9.5+
Ectopistes migratorius (n = 4)	4.3–5.1	3.2–4.0	5.4–6.6	4.3–5.1	9.7–9.9
Zenaida macroura (n = 20)	3.3–4.3	2.4–2.5	4.3–5.1	2.1–2.6	6.5–7.0

in *C. livia.* Other differences between these two species are also apparent. In *Ectopistes* the brachial tuberosity narrows toward the tip, which is rather pointed, whereas in *C. livia* the tuberosity is wider and its tip is more rounded. Additionally, in *Ectopistes* the posterior face of the bone above the sternal facet is much less pneumatic than in *C. livia.*

Exterminated 70 years ago, the Passenger Pigeon formerly bred in the North Woods and thence south to the Midwest and the Ohio River Valley. It wintered in the Southeast.

It is reported from pre-Columbian and Wisconsinan age sites west to Los Angeles (Howard, 1937) and south to southern Florida (Woolfenden, 1959; Weigel, 1962). It is also known from sites now throught to be of Sangamonian age, at Reddick, Haile, and Arredondo in northern Florida (Brodkorb, 1957, 1959; Ligon, 1965).

Family Strigidae

Otus guildayi, new species
Fig. 1

Etymology.—Named in memory of John E. Guilday, late specialist in Pleistocene and Holocene mammals of eastern North America, who directed the excavation of the Cumberland Cave fossils, and who sent us the bird material for identification.

Holotype.—Carnegie Museum of Natural History, Section of Vertebrate Fossils, CM 8040, left tarsometatarsus, lacking proximal end. Collected in 1965 by Allen D. McCrady and Harold Hamilton.

Type locality.—Cumberland Cave, one-half mi south of Corriganville, Alleghany County, in the Appalachian Mountains of northwestern Maryland.

Age.—Middle Pleistocene, Illinoian glacial age.

Diagnosis.—A strigid owl, referable to the genus *Otus* Pennant by having the tarsometatarsus of medium length; its shaft moderately long and slender, with the borders nearly parallel (without marked proximal and distal expansion, and without marked

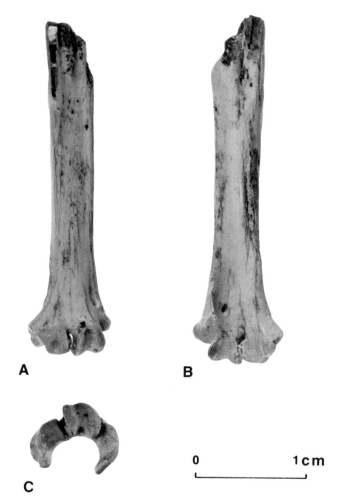

Fig. 1.—*Otus guildayi,* new species, CM 8040, holotype, Cumberland Cave, Maryland. A) anterior, B) posterior, and C) distal views. Scale is 1 cm. Photographs by Ronald G. Wolff.

mid-length constriction); groove of middle trochlea deep.

Similar to Recent and Pleistocene *Otus asio* (Linnaeus), but larger (see Table 4); shaft scarcely con-

Table 4.—*Measurements (in mm) of tarsometatarsus of* Otus.

Material	Total length	Partial length[1]	Distal width	Least width shaft	Trochlear gap[2]	Inner trochlea width	Middle trochlea width
O. guildayi, holotype	38.56[3]	25.4	8.2	4.2	4.2	3.8	3.2
O. a. naevius n = 3 (2 ♀, 1 ?)	33.25 (32.3–34.2)	21.90 (20.6–23.2)	7.17 (6.8–7.4)	3.50 (3.3–3.8)	4.1	3.65 (3.6–3.7)	2.80 (2.5–3.1)
O. a. asio n = 12 (5 ♂, 7 ♀)	30.39 (29.1–31.5)	20.62 (19.1–21.8)	6.29 (5.9–6.8)	2.90 (2.7–3.3)	2.99 (2.3–3.5)	2.85 (2.3–3.3)	2.40 (2.0–2.8)

[1] Length from distal end of Tuberculum musculi tibialis cranialis to distal end of middle trochlea.
[2] Width between tips of outer and inner trochleae.
[3] Estimate; length as preserved is 31.6 mm.

Table 5.—*Measurements (in mm) of tarsometatarsus of* Perisoreus *and* Cyanocitta.

Material	Total length	Proximal width	Proximal width of shaft	Least width of shaft	Width through trochleae
CM 29274	36.7	5.1	2.8	1.8	(3.1)*
C. cristata bromia 3 ♂, 2 ♀ New Brunswick, Maine, Pennsylvania	33.6–37.5 (35.34)	4.4–4.9 (4.68)	2.7–3.2 (2.98)	1.5–1.8 (1.68)	3.2–3.4 (3.32)
P. canadensis canadensis 2 ♂, 4 ♀ Ontario, Maine	33.0–36.0 (35.20)	4.0–4.8 (4.45)	2.3–2.7 (2.50)	1.1–1.6 (1.43)	2.7–3.4 (3.17)
P. c. pacificus 1 ♂ Alaska	36.5	4.3	2.6	1.6	3.0
P. c. capitalis 2 ♂ Rocky Mts	36.8–37.3 (37.05)	4.6	2.4–2.7 (2.55)	1.5–1.6 (1.55)	3.2–3.3 (3.25)
P. infaustus ruthenus 1 adult, Karelia	36.3	4.9	2.8	1.6	3.4

* As preserved. External trochlea is damaged.

stricted in middle, and its sides nearly straight; outer trochlea thrust plantad and mediad toward middle trochlea (in *O. asio* outer trochlea thrust laterad away from middle trochlea, leaving a relatively wider gap between tips of the outer and inner trochleae).

Much larger than other New World species of *Otus.*

Otus asio is reported from many pre-Columbian and Wisconsinan age sites throughout the United States. It also occurs in three Sangamonian age localities, at Reddick, Haile, and Arredondo, Florida (Brodkorb, 1957, 1959; Ligon, 1965).

Family Corvidae

Perisoreus canadensis (Linnaeus)
Canada Jay

Material.—CM 24274, right tarsometatarsus, complete except for damage to hypotarsus; the shaft has been broken in two and glued together.

Tarsometatarsi of *Perisoreus* and the Blue Jay (*Cyanocitta cristata*) are similar in general size, but the element is distally more slender in *Perisoreus*, stouter in *Cyanocitta* (see Table 5).

Perisoreus is a Holarctic genus of the Northern Coniferous Forest. In eastern North America the southern limit of *Perisoreus* is in northern New England and northeastern New York state.

The only previous record of the species is from Natural Chimneys, Virginia (Wetmore, 1962). This locality is from late Wisconsinan to Recent age (Guilday, 1962). The two other members of this genus, the Palearctic *P. infaustus* (Linnaeus) and *P. internigrans* (Thayer and Bangs), are unknown as fossils.

DISCUSSION

CLIMATE OF THE DEPOSIT

Cumberland Cave lies within the modern geographic range of five of the species identified. They are *Anas crecca, Aquila chrysaëtos, Bonasa umbellus, Meleagris gallopavo,* and *Ectopistes migratorius.* Their modern distribution also extends very much farther in all directions. *Otus guildayi* is as yet unknown elsewhere. Its presumed descendant is the screech owl, *Otus asio,* whose modern geographic range is similarly extensive. These six species thus tell us nothing about any difference in climate when the fossils were deposited.

The case of *Perisoreus canadensis* is more helpful. Today its closest occurrence, about 400 mi farther north, is in the Adirondacks Mountains. We may thus conclude that the climate at the time of deposition was cooler than at present, and that the avifauna lived during a glacial age.

AGE OF THE DEPOSIT

The known fossil record of the Cumberland Cave birds gives an insight as to the age of the deposit. *Anas crecca* is known in Europe from the Mindel glaciation and possibly earlier, and its North American record begins with the Illinoian glaciation. *Aquila chrysaëtos* first appears in Europe during the Mindel stage, and during the Illinoian stage in North America. *Meleagris gallopavo* is known from several Illinoian age localities.

Bonasa umbellus and *Ectopistes migratorius* are previously known from localities of Sangamonian interglacial age, as is *Otus asio,* the possible descendant of *Otus guildayi.*

The only other fossil record of *Perisoreus canadensis* is of Wisconsinan glacial age, supposedly the most severe of the American stages.

All the data of the paleoavifauna conform with the hypothesis proposed by Guilday of an Illinoian age for the Cumberland Cave deposit.

RÉSUMÉ

L'avifaune de la Cumberland Cave comporte sept taxons parmi lesquels se trouve une nouvelle espèce de hibou, *Otus guildayi.* Elle est compatible avec l'âge Illinoien indiqué par la faune de mammifères. C'est la plus ancienne apparition connue jusqu'à présent pour les espèces *Bonasa umbellus, Ectopistes migratorius* et *Perisoreus infaustus.*

LITERATURE CITED

BRODKORB, P. 1957. New passerine birds from the Pleistocene of Reddick, Florida. J. Paleont., 31:129–138.

———. 1959. The Pleistocene avifauna of Arredondo, Florida. Bull. Florida State Mus., Biol. Sci., 4:269–291.

GALBREATH, E. C. 1955. An avifauna from the Pleistocene of central Kansas. Wilson Bull., 67:62–63.

GIDLEY, J. W., and C. L. GAZIN. 1938. The Pleistocene vertebrate fauna from Cumberland Cave, Maryland. Bull. U.S. Nat. Mus., 171:1–199.

GUILDAY, J. E. 1962. The Pleistocene local fauna of the Natural Chimneys, Augusta County, Virginia. Ann. Carnegie Mus., 36:87–122.

———. 1971. The Pleistocene history of the Appalachian mammal fauna. Virginia Polytech. Inst. State Univ., Research Div. Monogr., 4:233–262.

GUILDAY, J. E., and C. O. HANDLEY, JR. 1967. A new *Peromyscus* (Rodentia : Cricetidae) from the Pleistocene of Maryland. Ann. Carnegie Mus., 39:91–103.

HAY, O. P. 1923. The Pleistocene of North America and its vertebrated animals from the states east of the Mississippi River and from the Canadian provinces east of longitude 95°. Carnegie Inst. Washington Publ., 322:1–499.

HOWARD, H. 1937. A Pleistocene record of the Passenger Pigeon in California. Condor, 39:12–14.

LIGON, J. D. 1965. A Pleistocene avifauna from Haile, Florida. Bull. Florida State Mus., 10:127–158 (published Jan. 3, 1966)

MERCER, H. C. 1899. The bone cave at Port Kennedy, Pa., and its partial excavation in 1894, 1895, and 1896. J. Acad. Nat. Sci. Philadelphia, 11:269–286.

MOURER-CHAUVIRÉ, C. 1975. Les oiseaux du Pléistocène moyen et supérieur de France. Doc. Lab. Géol. Faculté Sci. Lyon, 64:1–624.

RITCHIE, T. L. 1980. Two mid-Pleistocene avifaunas from Coleman, Florida. Bull. Florida State Mus., Biol. Sci., 26:1–36.

STEADMAN, D. W. 1980. A review of the osteology and paleontology of turkeys (Aves: Meleagridinae). Contrib. Sci., Nat. Hist. Mus. Los Angeles Co., 330:131–207.

WEBB, S. D. 1974. Pleistocene mammals of Florida. Univ. Presses Florida, Gainesville, 270 pp.

WEIGEL, R. D. 1962. Fossil vertebrates of Vero, Florida. Florida Geol. Survey, Special Bull., 10:1–59 pp. Published Jan., 1963.

WETMORE, A. 1927. A record of the ruffed grouse from the Pleistocene of Maryland. Auk, 44:561.

———. 1962. Notes on fossil and subfossil birds. Smithsonian Misc. Coll., 145(2):1–17.

WOOLFENDEN, G. E. 1959. A Pleistocene avifauna from Rock Spring, Florida. Wilson Bull., 71:183–187.

Address (Brodkorb): Department of Zoology, University of Florida, Gainesville, Florida 32611.

Address (Mourer-Chauviré): Centre de Paleontologie stratigraphique et Paleoecologie de l'Universite Claude Bernard, 69622 Villeurbanne Cedex, France.

A VERY LARGE ENIGMATIC OWL (AVES: STRIGIDAE) FROM THE LATE PLEISTOCENE AT LADDS, GEORGIA

STORRS L. OLSON

ABSTRACT

An undescribed species of owl is represented in the late Pleistocene (Rancholabrean) deposits at Ladds, Georgia, by a mandibular symphysis that is larger than in any living species of Strigidae. This is most similar to the mandible in living owls of the genus *Strix*. Paleontologists are alerted to seek more diagnostic specimens so that the species may eventually be described.

INTRODUCTION

One of very few vertebrate faunas known from the late Pleistocene (Rancholabrean) of Georgia was reported from fissure fillings in a limestone quarry at Ladds, Bartow County (Lipps and Ray, 1967; Ray, 1965, 1967). Some of the avian remains from this site were identified by Wetmore (1967), of which the most interesting record was that of a Spruce Grouse (*Dendragapus canadensis*), a species now occurring mainly in boreal coniferous forests of Canada. Whereas this and certain of the mammalian species are characteristic of cooler northern regions today, some of the other mammals have southern affinities (Ray, 1967; Kurtén and Anderson, 1980). Although the deposits at Ladds may have been heterochronic, such "disharmonious" faunas are typical of many late Pleistocene fossil deposits (Lundelius et al., 1983) and are believed to reflect a more equable climate in the Pleistocene.

Among some additional material from the Ladds deposits in the National Museum of Natural History, Smithsonian Institution (USNM), I encountered the anterior portion of a mandible of an owl that appears to have been larger than any of the living species of Strigiformes, although I do not consider the specimen to be sufficiently diagnostic to be named at this time. Nevertheless, it represents a significant addition to the Pleistocene fauna of eastern North America that should be included in faunal surveys and that should be looked for in other Pleistocene deposits.

DESCRIPTION

The specimen (USNM 214769) is a mandibular symphysis with unequal portions of the rami preserved on either side. The length of the symphysis as preserved is 11.5 mm, but as some of the thin, fragile tip is missing, this measurement would have exceeded 12 mm in the intact specimen. The width at the posterior margin of the symphysis is 15.1 mm. The specimen is referable to the Strigidae, the symphysis being broader and more truncate than in the Tytonidae, in which the symphysis is narrow and elongate.

Compared with the three largest North American owls (*Bubo virginianus*, *Nyctea scandiaca*, and *Strix nebulosa*), the fossil is seen to be much larger (Fig. 1). The symphysis is proportionately longer and narrower and the rami less divergent than in either *Bubo* or *Nyctea*. Furthermore, the two large nutrient foramina on the ventral surface of the tip are more widely separated and situated more on the lateral surfaces of the symphysis than in *Bubo* or *Nyctea*. In these and other details the specimen is more similar to *Strix*.

DISCUSSION

Howard (1933) described a presumably extinct owl from the tar pits at Rancho La Brea, California, as *Strix brea*. This was characterized as being larger than either of the living species *S. varia* or *S. occidentalis*, but no size comparisons were made with the panboreal Great Gray Owl, *S. nebulosa*, because at that time the species was regarded as belonging in a monotypic genus, *Scotiaptex*. From Table 1, a rather confusing picture of the size of *Strix brea* emerges. The measurements for the rostrum and

44

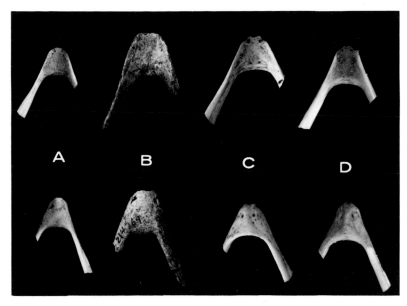

Fig. 1.—Mandibular symphyses of owls in dorsal view (top row) and ventral view (bottow row): A) Great Gray Owl, *Strix nebulosa* (USNM 289980); B) fossil specimen from Ladds Quarry, Georgia (USNM 214769); C) Snowy Owl, *Nyctea scandiaca* (USNM 19901); D) Great Horned Owl, *Bubo virginianus* (USNM 289978). Natural size.

tibiotarsus all lie within the range of variation of a very small sample of *S. nebulosa* and those for the femur and wing elements are either somewhat smaller or in the lower end of the size range of *S. nebulosa*. On the other hand, the holotype and eight referred tarsometatarsi of *Strix brea* are all considerably larger than in *S. nebulosa*. The much greater size of the mandible of the Ladds owl as compared to *S. nebulosa* makes it unlikely that this specimen is referable to *S. brea*, whatever the status of that species may be.

The only extinct strigid owls of immense size belong to the genus *Ornimegalonyx*, originally known from a single species from Quaternary cave deposits in Cuba (see Arredondo, 1976). Kurochkin and Mayo (1973) alluded to the possible existence of additional species of *Ornimegalonyx* and Arredondo (1982) has recently named three as new. Unfortunately, the mandibular symphysis has not been reported for *Ornimegalonyx*. Because *Ornimegalonyx* is believed to be closely related to *Strix* and *Ciccaba* (Kurochkin, personal communication), it is possible that with better material the Ladds owl, which also appears close to *Strix*, may shed some light on the geographical origins of the Cuban birds.

Table 1.—*Measurements of the extinct owl*, Strix brea *(from Howard, 1933) and the mandible from Ladds, Georgia, compared with three specimens of the extant Great Gray Owl*, Strix nebulosa *(from left to right USNM 502543, male; USNM 289980, female; USNM 289429, female).*

Measurements	Strix brea		Strix nebulosa			Ladds owl
Height rostrum	20.7	(n = 1)	—	20.8	20.9	—
Breadth rostrum	21.5	(n = 1)	—	25.0	23.7	—
Length humerus	112.5–121.3	(n = 3)	120.0	134.8	142.0	—
Length carpometacarpus	56.2–59.9	(n = 5)	58.0	66.3	68.0	—
Length femur	75.6–76.6	(n = 4)	78.9	87.2	91.6	—
Length tibiotarsus	112.7–120.0	(n = 10)	106.4	120.0	126.6	—
Length tarsometatarsus	63.5–68.0	(n = 9)	50.0	57.0	59.0	—
Length mandibular symphysis			—	8.1	7.7	12+
Breadth mandibular symphysis			—	9.6	9.9	14.6

ACKNOWLEDGMENTS

Lewis Lipps encouraged me to look through the additional bird material from Ladds quarry and supplied useful information. Kenneth E. Campbell, Jr., John Farrand, Jr., and Clayton E. Ray offered helpful comments on an early draft of the manuscript, and several later drafts were criticized by David W. Steadman. The photographs are by Victor E. Krantz.

LITERATURE CITED

ARREDONDO, O. 1976. The great predatory birds of the Pleistocene of Cuba. *In* Collected Papers in Avian Paleontology Honoring the 90th Birthday of Alexander Wetmore (S. L. Olson, ed.), Smithsonian Contr. Paleobiol., 27:169–187.

———. 1982. Los Strigiformes fósiles del Pleistoceno Cubano. Bol. Soc. Venezolana Cien. Nat., 140:33–55.

HOWARD, H. 1933. A new species of owl from the Pleistocene of Rancho La Brea, California. Condor, 35:66–69.

KUROCHKIN, E. N., and N. MAYO. 1973. Las lechuzas gigantes del Pleistoceno superior de Cuba. Acad. Cien. Cuba Inst. Geol. Resumenes, Comunicaciones, y Notas del V. Consejo Cientifico Actas, 3:56–60.

KURTÉN, B., and E. ANDERSON. 1980. Pleistocene mammals of North America. Columbia Univ. Press, New York, 442 pp.

LIPPS, L., and C. E. RAY. 1967. The Pleistocene fossiliferous deposit at Ladds, Bartow County, Georgia. Bull. Georgia Acad. Sci., 25:113–119.

LUNDELIUS, E. L., JR., R. W. GRAHAM, E. ANDERSON, J. E. GUILDAY, J. A. HOLMAN, D. W. STEADMAN, and S. D. WEBB. 1983. Terrestrial vertebrate faunas. Pp. 311–353, *in* The Late Pleistocene (S. C. Porter, ed.), vol. 1 of, Late Quaternary environments of the United States (H. E. Wright, Jr., ed.), Univ. Minnesota Press, Minneapolis.

RAY, C. E. 1965. A new chipmunk, *Tamias aristus*, from the Pleistocene of Georgia. J. Paleont., 39:1016–1022.

———. 1967. Pleistocene mammals from Ladds, Bartow County, Georgia. Bull. Georgia Acad. Sci., 25:120–150.

WETMORE, A. 1967. Pleistocene Aves from Ladds, Georgia. Bull. Georgia Acad. Sci., 25:151–153.

Address: Department of Vertebrate Zoology, National Museum of Natural History, Smithsonian Institution, Washington, D.C. 20560.

A MIDDLE PLEISTOCENE (LATE IRVINGTONIAN) AVIFAUNA FROM PAYNE CREEK, CENTRAL FLORIDA

DAVID W. STEADMAN

ABSTRACT

Ten taxa of birds are reported from a late Irvingtonian fossil site at Payne Creek, in the Bone Valley region of central Florida. No radiometric dates are available for this alluvial deposit, whose age is based only upon mammalian biochronology. Unlike the extensive marine avifauna of the underlying Bone Valley Formation (Hemphillian), the Payne Creek avifauna represents species characteristic of fresh water and terrestrial habitats. *Tachybaptus dominicus* is reported for the first time as a North American fossil. This species, which no longer occurs in Florida, is another member of the Pleistocene "Gulf Coast Savanna Corridor" fauna. Other living species reported are *Podilymbus podiceps, Colinus virginianus, Rallus* cf. *R. limicola*, and *Zenaida macroura*, each of which still lives in Florida. Less diagnostic fossils are referred to "Anserinae, genus and species indeterminate," *Anas* cf. *A. discors* or *A. cyanoptera*, an extinct but poorly understood species of *Bucephala, Limnodromus* sp., and "Icterinae, genus and species indeterminate." The following described Pliocene or Pleistocene species are regarded as of questionable validity or as being based on inadequate material, or both: *Podilymbus magnus* Shufeldt 1913, *P. wetmorei* Storer 1976, *Podiceps dixi* Brodkorb 1963, *Bucephala fossilis* Howard 1963, *B. ossivallis* Brodkorb 1955, *Colinus suilium* Brodkorb 1959, and *Zenaida prior* Brodkorb 1969.

INTRODUCTION

The Bone Valley region of central Florida is well known for phosphatic deposits that have produced rich faunas of Mio-Pliocene (Hemphillian land mammal age) vertebrates. The fossil birds of Bone Valley were reviewed by Brodkorb (1955), but only part of the hundreds of additional avian fossils that have been collected since that time have ever been mentioned briefly (Brodkorb, 1972), and an updated review of the Hemphillian fauna is needed. This report describes the birds of a much younger vertebrate fossil fauna from the Bone Valley region of Polk County. The site, known as Payne Creek, is an alluvial deposit discovered and collected by John S. Waldrop and Michael K. Frazier. No radiometric dates are available for the Payne Creek local fauna, but the fossil mammals (especially *Ondatra* cf. *O. annectens*) indicate a biostratigraphic age of late Irvingtonian (M. K. Frazier, personal communication). This places the Payne Creek local fauna at approximately 4–6 million years younger than typical Hemphillian vertebrate faunas of the Bone Valley region. Other Irvingtonian mammal sites in Florida are described by Webb (1974).

In Florida and throughout North America, Irvingtonian avifaunas are much rarer than those of the late Pleistocene Rancholabrean land mammal age. Six Irvingtonian avifaunas, including Payne Creek, are known from Florida, whereas fossil birds are known from at least 22 localities from only the latest Rancholabrean (25,000–10,000 years BP) of Florida (Lundelius et al., 1983). The Irvingtonian sites include the Williston local fauna in Levy County (Holman, 1959) and the Coleman IIA local fauna in Sumter County (Ritchie, 1980). Steadman (1980) reviewed the fossil turkeys from these two faunas and from three others in the Florida Irvingtonian— Inglis IA (Citrus County), Haile XVIA (Alachua County), and Santa Fe River IIA (Gilchrist County).

Bird bones are often fragmentary in alluvial sites, and Payne Creek is no exception. The avian fossils from Payne Creek include 27 specimens for which no identification of any sort could be safely made. The remaining 29 specimens are sufficiently diagnostic to allow identification at least to the familial level (Table 1), but I have identified to species only five of the 10 taxa of birds. This conservative treatment is dictated by both the quality of the fossils and the poorly defined nature of species-level systematics in early Pleistocene birds.

Unless noted otherwise, sequence and nomenclature follows Brodkorb's Catalogue of Fossil Birds, Parts 1–5 (1963–1978). Distributions of living species are taken from the American Ornithologists' Union Checklist (1957). All specimens are housed in the collection of the Timberlane Research Organization, Lake Wales, Florida.

47

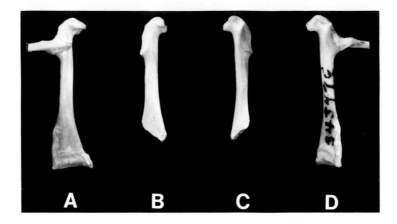

Fig. 1.—Left coracoid of *Tachybaptus dominicus*. A, D) modern specimen (USNM 343476) in dorsal and ventral aspects (scapula attached). B, C) fossil from Payne Creek in dorsal and ventral aspects. All specimens are 2×.

SYSTEMATIC PALEONTOLOGY

Order Podicipediformes
Family Podicipedidae

Tachybaptus dominicus (Linnaeus)
Fig. 1, Table 2

Material.—Coracoid, lacking sternal end.

I follow Storer (1979) in separating the genus *Tachybaptus* from *Podiceps*. This specimen cannot be distinguished qualitatively from the coracoid in five modern specimens of *T. dominicus,* with which it also agrees in size, being much smaller than in other living species of New World grebes or the fossil species reported by Brodkorb (1963a), Murray (1967), and Storer (1976).

Table 1.—*Birds of the Payne Creek local fauna, Polk County, Florida. For each taxon, the minimum number of individuals is one.*

Tachybaptus dominicus—Least Grebe
Podilymbus podiceps—Pied-billed Grebe
Podicipedidae, genus and species indeterminate—grebe
Anserinae, genus and species indeterminate—goose
Anas cf. *A. discors* or *A. cyanoptera*—probable Blue-winged or Cinnamon Teal
Bucephala sp.—goldeneye-type duck
Anatidae, genus and species indeterminate—duck
Colinus virginianus—Bobwhite
Rallus cf. *R. limicola*—probable Virginia Rail
Limnodromus sp.—dowitcher
Zenaida macroura—Mourning Dove
Icterinae, genus and species indeterminate—blackbird

This is the first fossil record of *T. dominicus* from North America. The only other paleontological report of the species is from the apparently Pleistocene site of Lapa da Escrivania, Brazil (Winge, 1888:4). *T. dominicus* is one of many species of vertebrates with tropical or southwestern affinities that inhabited Florida during the late Cenozoic. It is a widespread species today, occurring essentially throughout the lowland Neotropics and much of the West Indies (Storer, 1979). The modern distribution of *T. dominicus* most closely approaches Florida at its northern limits in the Bahamas. On the North American continent, *T. dominicus* occurs no farther north than southern Texas, with accidental records in Louisiana (Eyster, 1978). Unless *T. dominicus* colonized Florida via the West Indies and spread no further, then it seems that its distribution was once continuous across the southern Gulf region of the United States, lending additional support to the

Table 2.—*Measurements (in mm) of the coracoid in modern and fossil* Tachybaptus dominicus.

Specimens	Least width of shaft	Length of glenoid facet
Modern skeletons (N = 5) (4 males, 1 unsexed; USNM 7034, 343476, 344901, 345750, 347811)	1.5–1.8	4.0–4.5
Payne Creek fossil	1.8	4.4

existence of a faunal connection between northern Mexico and Florida during the Pleistocene (the "Gulf Coast savanna corridor"; see Webb, 1974; Steadman, 1980; Pregill and Olson, 1981).

Podilymbus podiceps (Linnaeus)

Material.—Humeral end of scapula, distal end of humerus, distal end of femur, proximal end of tibiotarsus, distal end of tarsometatarsus, anterior one-third of notorium.

Measurements of the Payne Creek fossils (distal width of humerus, 6.5 mm; distal width of femur, 9.9 mm; distal width of tarsometatarsus, 6.0 mm) are within the ranges of late Pleistocene and modern specimens of *P. podiceps* (Murray, 1967; Storer, 1976), with which they also agree qualitatively.

Podilymbus podiceps has been recorded in many Pleistocene localities in Florida and elsewhere in North America (Brodkorb, 1963a), although the only other Irvingtonian record is from the Coleman IIA local fauna (Ritchie, 1980). Three fossil species of *Podilymbus* have been described—*P. majusculus* Murray, 1967, of the Hagerman local fauna, Idaho (Blancan land mammal age); *P. wetmorei* Storer, 1976, of the Reddick local fauna, Florida (Rancholabrean land mammal age); and *P. magnus* Shufeldt, 1913, of Fossil Lake, Oregon (Rancholabrean), which was synonymized with *P. podiceps* by Wetmore (1937). Although Brodkorb (1963a:230) suggested that a subspecies, *P. podiceps magnus,* might be recognizable for certain fossils from Florida and Fossil Lake, this was disproven by Storer (1976), who showed the specimens of *P. magnus* to fall within the normal range of variation of living *P. podiceps*. *P. wetmorei* is known from two tarsometatarsi and two femora that differ from those of *P. podcieps* only in being stouter. Thus the status of *P. wetmorei* as a distinct species is also suspicious.

Podicipedidae, genus and species indeterminate

Material.—Proximal end of humerus, ventral half of cervical vertebra, distal end of pedal phalanx.

These fragmentary specimens are of a size compatible with either *Podilymbus podiceps* or *Podiceps auritus,* each of which occurs commonly in Florida, both living and as fossils. Because these specimens may pertain to *Podilymbus podiceps,* they do not necessarily represent a new taxon for the Payne Creek fauna.

In reviewing the literature on osteology and systematics of Quaternary grebes of North America, the unsatisfactory nature of *Podiceps dixi* Brodkorb, 1963b, became apparent. This name was based on a fragmentary and undiagnostic proximal end of a carpometacarpus from the Reddick local fauna, Florida (Rancholabrean land mammal age). The only difference noted by Brodkorb (1963b) between *Podiceps dixi* and the living *P. auritus* was the slightly greater size of the former. Grebes are known to be sexually dimorphic in size (Storer, 1976), yet neither sex nor sample size were given for the modern specimens of *P. auritus* measured by Brodkorb (1963b). Even if the putative difference in size between *P. dixi* and *P. auritus* were demonstrable from a single fragment, I do not believe that such a specimen can be diagnosed meaningfully as a new, extinct species. In the absence of corroborative fossils, it is preferable to regard *Podiceps dixi* as a synonym of *P. auritus.*

Order Anseriformes
Family Anatidae
Anserinae, genus and species indeterminate

Material.—Proximal end and two distal ends of pedal phalanges.

These undiagnostic specimens belong to some taxon of goose. There is little or nothing in the postcranial osteology of North American geese, to say nothing of that of the pedal phalanges, to distinguish the genera *Chen, Anser,* and *Branta* (Woolfenden, 1961). The only anserine reported from the Pleistocene of Florida is *Branta canadensis* from several sites of Rancholabrean age (Brodkorb, 1964) and the Irvingtonian Coleman IIA local fauna (Ritchie, 1980). The only other Irvingtonian record of any goose is of *B. canadensis* from the type Irvington fauna in California (Wetmore, 1956:26).

Anas cf. *A. discors* or *A. cyanoptera*

Material.—Distal end of humerus, proximal half of carpometacarpus.

These specimens are from a teal the size of *Anas discors* or the closely related and osteologically very similar *A. cyanoptera.* The only other Irvingtonian record of either species is of *A. discors* from the Williston local fauna, Florida (Holman, 1959). *A. discors* is a common Pleistocene species in Florida and elsewhere, whereas the fossil record of *A. cyanoptera* is confined to the late Pleistocene of California and Oregon (Brodkorb, 1964). These specific assignments were doubtless based on geographical probability.

Bucephala sp., ?aff. *B. ossivallis* or *B. fossilis*

Material.—Nearly complete tarsometatarsus, slightly eroded on both ends.

This specimen closely resembles the tarsometatarsi in modern diving ducks *Bucephala albeola* and the larger *B. clangula,* but is intermediate in size and represents an extinct form. It agrees with *Bucephala* and differs from the related genera *Mergus* and *Lophodytes* in all of the characters outlined by Woolfenden (1961) and Alvarez and Olson (1978). Its size suggests affinities to *B. ossivallis* Brodkorb, 1955, from the Bone Valley Formation (Hemphillian land mammal age), or *B. fossilis* Howard, 1963, from the Vallecito Creek local fauna, California (Irvingtonian). *B. ossivallis* is known only from the humeral two-thirds of a coracoid, and *B. fossilis* is based upon the proximal end of a carpometacarpus and a proximal fragment of humerus. Therefore neither is directly comparable to the Payne Creek fossil. Brodkorb (1964) noted that the distal end of a carpometacarpus from the Hagerman local fauna, Idaho (Blancan), listed as *Bucephala* sp. by Brodkorb (1958; 1961), may possibly be referable to *B. fossilis.* However, only the proximal end of the carpometacarpus is known in *B. fossilis.* The only other pre-Rancholabrean record of *Bucephala* is the living *B. albeola* from the Blancan Rexroad local fauna, Kansas (Wetmore, 1944).

In summary, two very poorly known named species, *Bucephala ossivallis* Brodkorb, 1955, and *B. fossilis* Howard, 1963, represent Hemphillian and Irvingtonian forms of *Bucephala* that were intermediate in size between the living *B. clangula* and *B. albeola.* These extinct species are based on specimens inferior to that from Payne Creek, which appears to represent a species of similar size. Because of their fragmentary nature and because none of the elements is comparable, it is impossible to say at this time how many species of *Bucephala* are represented or whether the fossils are from a single species lineage. This is, however, the only certainly extinct species of bird represented in the Payne Creek local fauna.

Anatidae, genus and species indeterminate

Material.—Antero-dorsal fragment of sternum, tiny distal fragment of humerus, two distal ends of carpometacarpi, proximal half of manus digit II, phalanx I.

Each of these specimens represents a small, teal-sized anatid, but otherwise lacks diagnostic features, even at the subfamilial level. These fossils do not necessarily represent a new taxon for the Payne Creek fauna because they could possibly pertain to *Anas discors* or *A. cyanoptera.*

Order Galliformes
Family Phasianidae

Colinus virginianus Linnaeus

Material.—Distal end of ulna, middle one-third of synsacrum, distal end of tibiotarsus.

The distal width of the tibiotarsus (5.2 mm) is the only measurement that I could obtain from these specimens to compare with the extensive series of measurements of living and fossil *Colinus* in Holman (1961). (Unfortunately, Ritchie [1980] provided means but no ranges for his measurements of Pleistocene *Colinus* in Florida.) The measurement 5.2 mm is within the range of living *C. virginianus* as well as the named Pleistocene species *C. suilium* Brodkorb, 1959. Holman (1961) regarded *C. suilium* as a large temporal form of *C. virginianus.* I agree, and because of the lack of consistent qualitative differences and their substantial overlap in size, I regard *C. suilium* as a synonym of *C. virginianus.* Fossils of *Colinus* occur in many Pleistocene sites in Florida (Brodkorb, 1964), including the Irvingtonian sites of Williston (Holman, 1959) and Coleman IIA (Ritchie, 1980). There is a north-south decline in size in living *C. virginianus* of the eastern United States (Holman, 1961), so the larger fossils probably reflect a cooler climate rather than another species.

Order Ralliformes
Family Rallidae

Rallus cf. *R. limicola* Vieillot

Material.—Humeral end of coracoid.

This coracoid resembles that of *R. limicola* and is qualitatively separable from that of the similarly-sized *Porzana carolina.* It is, however, too incomplete for unequivocal identification.

R. limicola occurs in several Pleistocene localities in Florida (Brodkorb, 1967; Olson, 1974). Apparently it was larger in late Pleistocene times than today (Olson, 1974). From other Irvingtonian sites, *R. limicola* is known only from a sternal fragment from Vallecito Creek, California, referred questionably to this species by Howard (1963).

Order Charadriiformes
Family Scolopacidae

Limnodromus sp.

Material.—Distal end of humerus lacking internal condyle.

This fossil agrees with the humerus of *L. griseus* and differs from that of *L. scolopaceus* in that the ectepicondylar spur forms a more acute angle with the shaft. It resembles that of *L. scolopaceus* versus *L. griseus,* however, in having the ectepicondylar spur less sharply pointed. It is within the size range of both living species, and thus could represent either of these or perhaps an extinct form.

L. griseus is unknown in the Pleistocene of Florida, whereas *L. scolopaceus* was recorded from the Rock Spring local fauna of Rancholabrean age (Woolfenden, 1959). Ligon (1965) reported *Limnodromus* sp. from the Rancholabrean Haile local fauna, Florida (listed questionably as *L. scolopaceus* by Brodkorb, 1967). This is the first Irvingtonian record for the genus.

Order Columbiformes
Family Columbidae

Zenaida macroura (Linnaeus)

Material.—Carpometacarpus lacking metacarpal III and distal end of metacarpal II.

This specimen is indistinguishable from the carpometacarpus of *Z. macroura,* a living species that is commonly recorded as a late Pleistocene fossil in Florida and elsewhere (Brodkorb, 1971). The only other record of *Z. macroura* for the Irvingtonian is from the Coleman IIA local fauna, Florida (Ritchie, 1980). A Blancan specimen reported as *Z. macroura* (from Rexroad, Kansas; Wetmore, 1944) was later described by Brodkorb (1969) as a new species, *Z. prior.* Confirmation of the validity and relationships of *Z. prior* must await the discovery of additional specimens.

Order Passeriformes
Family Fringillidae
Icterinae, genus and species indeterminate

Material.—Distal half of ulna.

This fossil is the size of certain species of *Icterus* and *Molothrus,* but otherwise lacks diagnostic features. Other pre-Rancholabrean records of icterines are those of *Euphagus cyanocephalus* from the Blancan age Sanders local fauna, Kansas (Harrell, 1959; listed as Big Springs Ranch local fauna by Brodkorb, 1978), and *Agelaius phoeniceus* and *Pandanaris floridanus* from the Coleman IIA local fauna, Florida (Ritchie, 1980).

DISCUSSION

The small avifauna from Payne Creek provides the first Irvingtonian records for *Tachybaptus dominicus* and *Limnodromus* sp. In addition, it has helped to define systematic problems in Neogene grebes and ducks, although more fossils are needed in both cases to resolve the issues. Grebes, ducks, and semi-aquatic birds (geese, rails, shorebirds) dominate the Payne Creek avifauna. In fact, each of the taxa in the Payne Creek avifauna (Table 1) may be found today at or near some sort of fresh water body, including even the terrestrial taxa such as *Colinus virginianus, Zenaida macroura,* and the indeterminate icterine. The Payne Creek region was composed of a mixture of fresh water and terrestrial habitats at the time of fossil deposition. Thus the Irvingtonian environment at Payne Creek was very different from that reflected by the underlying Bone Valley Formation, where the predominance of cormorants, alcids, loons, and sulids reflects a marine setting, although near-shore because of the lesser numbers of fresh water or terrestrial birds (Brodkorb, 1955; 1972).

ACKNOWLEDGMENTS

The fossils were borrowed through Michael K. Frazier and John S. Waldrop. Storrs L. Olson of the National Museum of Natural History, Smithsonian Institution, and Mary C. McKitrick and G. Scott Mills of the University of Arizona allowed access to skeletal collections under their supervision. Michael K. Frazier supplied information on geology, chronology, and fossil mammals from the sites. Storrs L. Olson and Michael K. Frazier provided beneficial comments on the manuscript and also examined the fossil of *Bucephala.* The photographs are by Victor E. Krantz.

LITERATURE CITED

ALVAREZ, R., and S. L. OLSON. 1978. A new merganser from the Miocene of Virginia (Aves: Anatidae). Proc. Biol. Soc. Washington, 91: 522–532.

AMERICAN ORNITHOLOGISTS' UNION. 1957. Check-list of North American Birds. A.O.U., Baltimore, 5th ed., 691 pp.

BRODKORB, P. 1955. The avifauna of the Bone Valley Formation. Rep. Invest., Florida Geol. Surv., 14:1–57.

———. 1958. Fossil birds from Idaho. Wilson Bull., 79:237–242.

———. 1961. Birds from the Pliocene of Juntura, Oregon. Quart. J. Florida Acad. Sci., 24:169–184.

———. 1963a. Catalogue of fossil birds. Part 1 (Archaeoptergyiformes through Ardeiformes). Bull. Florida St. Mus., Biol. Sci., 7:179–293.

———. 1963b. A new Pleistocene grebe from Florida. Quart. J. Florida Acad. Sci., 26:53–55.

———. 1964. Catalogue of fossil birds. Part 2 (Anseriformes through Galliformes). Bull. Florida St. Mus., Biol. Sci., 8: 195–335.

———. 1967. Catalogue of fossil birds. Part 3 (Ralliformes, Ichthyornithiformes, Charadriiformes). Bull. Florida St. Mus., Biol. Sci., 11:99–220.

———. 1969. An ancestral mourning dove from Rexroad, Kansas. Quart. J. Florida Acad. Sci., 31:173–176.

———. 1971. Catalogue of fossil birds. Part 4 (Columbiformes through Piciformes). Bull. Florida St. Mus., Biol. Sci., 15: 163–266.

———. 1972. New discoveries of pliocene [sic] birds in Florida. Proc. XVth Inter. Ornith. Cong., Abstract, p. 634.

———. 1978. Catalogue of fossil birds. Part 5 (Passeriformes). Bull. Florida St. Mus., Biol. Sci., 23:139–228.

EYSTER, M. B. 1978. Least Grebe in Louisiana. Louisiana Ornith. Soc. News, 81:3.

HARRELL, B. E. 1959. Notes on fossil birds from the Pleistocene of Kansas and Oklahoma. Proc. South Dakota Acad. Sci., 38:103–106.

HOLMAN, J. A. 1959. Birds and mammals from the Pleistocene of Williston, Florida. Bull. Florida St. Mus., Biol. Sci., 5:1–24.

———. 1961. Osteology of living and fossil New World quails (Aves, Galliformes). Bull. Florida St. Mus., Biol. Sci., 6:131–233.

HOWARD, H. 1963. Fossil birds from the Anza-Borrego Desert. Contrib. Sci., Los Angeles Co. Mus., 73:1–33.

LIGON, J. D. 1965. A Pleistocene avifauna from Haile, Florida. Bull. Florida St. Mus., Biol. Sci., 10:127–158.

LUNDELIUS, E. L., JR., R. W. GRAHAM, E. ANDERSON, J. E. GUILDAY, J. A. HOLMAN, D. W. STEADMAN, and S. D. WEBB.

1983. Terrestrial vertebrate faunas. Pp. 311–353, in The late Pleistocene (S. C. Porter, ed.), vol. 1 of, Late Quaternary environments of the United States (H. E. Wright, Jr., ed.), Univ. Minnesota Press, Minneapolis.

MURRAY, B. G., JR. 1967. Grebes from the late Pliocene of North America. Condor, 69:277–288.

OLSON, S. L. 1974. The Pleistocene rails of North America. Condor, 76:169–175.

PREGILL, G. K., and S. L. OLSON. 1981. Zoogeography of West Indian vertebrates in relation to Pleistocene climatic cycles. Ann. Rev. Ecol. Syst., 12:75–98.

RITCHIE, T. L. 1980. Two mid-Pleistocene avifaunas from Coleman, Florida. Bull. Florida St. Mus., Biol. Sci., 26:1–36.

STEADMAN, D. W. 1980. A review of the osteology and paleontology of turkeys (Aves: Meleagridinae). Nat. Hist. Mus. Los Angeles Co., Contrib. Sci., 330:131–207.

STORER, R. W. 1976. The Pleistocene Pied-billed Grebes (Aves: Podicipedidae). Pp. 147–153, in Collected papers in avian paleontology honoring the 90th birthday of Alexander Wetmore (S. L. Olson, ed.), Smithsonian Contrib. Paleobiol., 27: xxvi + 1–211.

———. 1979. Order Podicipediformes. Pp. 149–155 in Checklist of Birds of the World, 2nd ed. (E. Mayr and G. W. Cottrell), Mus. Comp. Zool., Cambridge, Massachusetts, 1:1–547.

WEBB, S. D. (ed.). 1974. Pleistocene mammals of Florida. Univ. Presses Florida, Gainesville, 270 pp.

WETMORE, A. 1937. A record of the fossil grebe, *Colymbus parvus,* from the Pliocene of California, with remarks on other American fossils of this family. Proc. California Acad. Sci., ser. 4, 23:195–201.

———. 1944. Remains of birds from the Rexroad fauna of the upper Pliocene of Kansas. Univ. Kansas Sci. Bull., 30:89–105.

———. 1956. A check-list of the fossil and prehistoric birds of North America and the West Indies. Smithsonian Misc. Coll., 131(5):1–105.

WINGE, O. 1888. Fugle fra Knoglehuler i Brasilien. E Museo Lundii, 1 (art. 2):1–54.

WOOLFENDEN, G. E. 1959. A Pleistocene avifauna from Rock Spring, Florida. Wilson Bull., 71:183–187.

———. 1961. Postcranial osteology of the waterfowl. Bull. Florida St. Mus., Biol. Sci., 6:1–129.

Address: Department of Vertebrate Zoology, National Museum of Natural History, Smithsonian Institution, Washington, D.C. 20560.

AN EVALUATION OF THE FOSSIL CURLEW *PALNUMENIUS VICTIMA* L. MILLER (AVES: SCOLOPACIDAE)

STORRS L. OLSON

ABSTRACT

The holotypical tarsometatarsus of *Palnumenius victima,* from the late Pleistocene (Rancholabrean) of Nuevo Leon, Mexico, is not generically separable from the extant genus *Numenius* and falls within the lower size range of *Numenius americanus,* from which it differs in only a few details. *Numenius victima* is tentatively retained as a problematic taxon that may represent a temporal or geographic form of *N. americanus.*

INTRODUCTION

With re-examination using better comparative material, many supposedly extinct taxa of North American Pleistocene birds have been shown to be synonymous with living forms, some of which, however, have retreated from North America into the tropics (for example, Olson, 1974). The process of re-evaluating nominal fossil taxa is still important to an accurate assessment of the effects of the Pleistocene on North American birds. In this connection I have restudied the supposedly extinct curlew *Palnumenius victima* L. Miller (1942), from late Pleistocene (late Rancholabrean) deposits in San Josecito Cave, Nuevo Leon, Mexico.

Palnumenius victima was founded solely on a complete left tarsometatarsus (LACM (CIT) 2944), and was diagnosed as follows (Miller, 1942:45): "Length about four-fifths that of *Numenius americana* [sic]; shaft almost uniform in transverse diameter throughout; outer cotyla almost the same level as the inner; inner trochlea less elevated." Neither this diagnosis nor the description that followed were organized in a manner that permits one to distinguish generic from specific characters.

RESULTS AND DISCUSSION

Miller quotes a communication from A. Wetmore in which the latter expressed the opinion that *Palnumenius* combined characters of *Numenius* with those of godwits, *Limosa.* Nevertheless, Miller clearly considered *Palnumenius* to be closer to *Numenius* and compared it in particular with the Long-billed Curlew, *Numenius americanus,* an extant species that occurs in Mexico in winter. My examination of the holotype of *P. victima* disclosed no characters linking it with *Limosa,* in which, for example, there is no closed medial hypotarsal canal (clearly shown in Miller's illustration of *P. victima*) and in which the inner trochlea is not as medially flared. I found no characters that will permit *Palnumenius victima* to be separated generically from *Numenius.* The genus *Palnumenius* L. Miller 1942 therefore becomes a junior subjective synonym of *Numenius* Brisson 1760. It thus remains to be determined whether *Numenius victima* can in fact be separated from the extant species of *Numenius.*

The holotype of *N. victima* is much larger than in the smallest of the curlews, *N. borealis* and *N.*

minutus, and decidedly larger and more slender than the tarsometatarsus in *Numenius phaeopus* or *N. tahitiensis.* It is smaller than in *N. arquata* or *N. madagascariensis.* The tarsometatarsus in a single skeleton of the rare Old World species *N. tenuirostris* (AMNH 547) was slightly smaller than the holotype of *N. victima,* with the most internal ridge of the hypotarsus being shorter than in either *N. victima* or *N. americanus.*

In stating that *N. victima* was smaller than *N. americanus,* Miller (1942) clearly did not have adequate comparative material. In curlews, as in many shorebirds, males are smaller than females. There was once a rather heated debate (Oberholser, 1918; Grinnell, 1921) about whether *N. americanus* can be divided into a large southern subspecies and smaller northern one. Although there is overlap between the two forms, the name *N. a. parvus* continues to be applied to the northern populations by some authors (Allen, 1980:7).

The tarsometatarsus in the smallest (USNM 499444, male) of seven skeletons of *N. americanus* that I examined measures 72.9 mm, whereas the holotype of *N. victima* measures 72 mm. Ridgway (1919) records the tarsal length in skins of males of *N. americanus parvus* as ranging from 69.8 to 81.5 mm. In the Smithsonian collections I found skins of two males in which the tarsal length was 71.5 and 72 mm, respectively, when measured along the anterior face from the intercotylar knob to the distal margin of the middle trochlea, as one would measure a skeletal specimen. Thus, the holotype of *N. victima* falls within the lower size range of males of *N. americanus.*

As for the qualitative characters ascribed to *N. victima,* the supposed differences in the relative levels of the cotylae and of the inner trochlea were not apparent to me. The uniform transverse width of the shaft would also seem to occur in *N. americanus* (Fig. 1) and is affected by age in any case, as in juveniles the proximal end is wider than in adults. Perhaps by this, Miller (1942) was attempting to describe the fact that the internal cotyla in *N. victima* projects more abruptly from the shaft than in *N. americanus,* which is the case. Another apparently valid character of *N. victima* mentioned by Miller in his description, but not in the diagnosis, is the larger, more rounded and more distally located distal foramen, with a deeper extensor groove. The significance of these two relatively minor points cannot be evaluated without additional fossil and modern specimens.

The closest relative of *Numenius victima* is undoubtedly *N. americanus,* a species known to breed in central and western North America from southern Canada south to Utah, New Mexico, and Texas, and east to Michigan, Illinois, Iowa, and Kansas, although it is now absent as a breeding bird from the eastern parts of its range. The species is migratory and winters mainly from California to Texas and south to Oaxaca, Mexico, and Guatemala. In the breeding season it inhabits open grasslands, pastures, and shrub steppe. Because San Josecito Cave is presently situated amidst pine and live oak forest at 2,300 m above sea level (Kurtén and Anderson, 1980), the presence of *Numenius* at this site, even if a migrant, would indicate a more open environment in the late Pleistocene, as do certain of the fossil mammals (Kurtén and Anderson, 1980).

It cannot be ascertained whether *Numenius victima* was a migrant or a resident, nor is it possible at this point to say whether it represents a temporal or geographic variant of *N. americanus,* or a distinct, closely related species. Of the two new genera

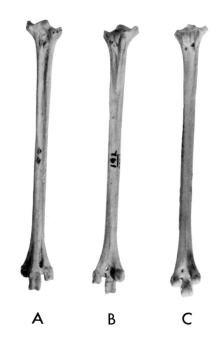

A   B   C

Fig. 1.—Left tarsometatarsi of *Numenius*: A) holotype of *Palnumenius* (=*Numenius*) *victima,* anterior view; B) same, posterior view; C) small individual of *N. americanus* (USNM 499444, male), posterior view. Natural size.

and species of waterbirds proposed by Miller (1942) from San Josecito (the rail *Epirallus natator,* and *Palnumenius victima*) neither genus is valid but neither species can be dismissed unequivocally. Elsewhere I have shown the genus *Epirallus* to be a synonym of *Rallus,* with *natator* being a member of the *Rallus longirostris/elegans* complex (Olson, 1974) that is larger than any of the modern members of that complex. As with *N. victima,* it is uncertain whether its differences are of a specific or subspecific nature.

Thus, of the more than 42 species of birds that Miller (1944) ultimately reported from San Josecito, the only certainly extinct species other than raptorial birds and scavengers are the turkey *Meleagris crassipes* (see Rea, 1980; Steadman, 1980, for documentation of the validity of this species) and the large roadrunner *Geococcyx conklingi* (the specific validity of which has recently been questioned, however [Harris and Crews, 1983]). This supports the idea that most of the Pleistocene extinctions among North American birds involved large species that were for the most part dependent upon the mammalian megafauna (Lundelius et al., 1983; Steadman and Martin, in press).

ACKNOWLEDGMENTS

I thank Robert McKenzie and Armando Solis of the Natural History Museum of Los Angeles County (LACM) for providing a cast and photographs of the holotype of *Palnumenius victima*. The photograph of *Numenius americanus* is by Victor E. Krantz, National Museum of Natural History, Smithsonian Institution (USNM). I am grateful to John Farrand, Jr., for facilitating the loan of a skeleton of *Numenius tenuirostris* from the American Museum of Natural History (AMNH) and for comments on an early draft which was also read by Kenneth E. Campbell. David W. Steadman criticized several subsequent versions of the manuscript.

LITERATURE CITED

ALLEN, J. N. 1980. The ecology and behavior of the Long-billed Curlew in southeastern Washington. Wildlife Monogr., 73: 1–67.

GRINNELL, J. 1921. Concerning the status of the supposed two races of the Long-billed Curlew. Condor, 23:21–27.

HARRIS, A. H., and C. R. CREWS. 1983. Conkling's Roadrunner—a subspecies of California Roadrunner? Southwestern Nat., 28:407–412.

KURTÉN, B., and E. ANDERSON. 1980. Pleistocene mammals of North America. Columbia Univ. Press, New York, 442 pp.

LUNDELIUS, E. L., JR., R. W. GRAHAM, E. ANDERSON, J. E. GUILDAY, J. A. HOLMAN, D. W. STEADMAN, and S. D. WEBB. 1983. Terrestrial vertebrate faunas. Pp. 311–353, *in* The Late Pleistocene (S. C. Porter, ed.), vol. 1 of, Late Quaternary environments of the United States (H. E. Wright, Jr., ed.), Univ. Minnesota Press, Minneapolis.

MILLER, L. 1942. Two new bird genera from the Pleistocene of Mexico. Univ. California Publ. Zool., 47:43–46.

———. 1944. The Pleistocene birds of San Josecito cavern, Mexico. Univ. California Publ. Zool., 47:143–168.

OBERHOLSER, H. C. 1918. Notes on the subspecies of *Numenius americanus* Bechstein. Auk, 35:188–195.

OLSON, S. L. 1974. The Pleistocene rails of North America. Condor, 76:169–175.

REA, A. 1980. Late Pleistocene and Holocene turkeys in the Southwest. *In* Papers in avian paleontology honoring Hildegarde Howard (K. E. Campbell, Jr., ed.), Contr. Sci., Nat. Hist. Mus. Los Angeles Co., 330:209–224.

RIDGWAY, R. 1919. Birds of North and Middle America. Part 8. Bull. U.S. Nat. Mus., 50:xvi + 1–852.

STEADMAN, D. W. 1980. A review of the osteology and paleontology of turkeys (Aves: Meleagridinae). *In* Papers in avian paleontology honoring Hildegarde Howard (K. E. Campbell, Jr., ed.), Contr. Sci., Nat. Hist. Mus. Los Angeles Co., 330:131–207.

STEADMAN, D. W., and P. S. MARTIN. In press. Extinction of birds in the late Pleistocene of North America. *In* Quaternary extinctions (P. S. Martin and R. G. Klein, eds.), Univ. Arizona Press, Tucson.

Address: Department of Vertebrate Zoology, National Museum of Natural History, Smithsonian Institution, Washington, D.C. 20560.

PHYLOGENY AND PALEOBIOGEOGRAPHY OF SHORT-TAILED SHREWS (GENUS *BLARINA*)

CHERI A. JONES, JERRY R. CHOATE, AND HUGH H. GENOWAYS

ABSTRACT

Dental measurements of Pleistocene and Holocene shrews of the genus *Blarina* were analyzed in a multivariate assessment of phylogeny and paleobiogeography of the genus. *Blarina brevicauda,* or an indistinguishable precursor, apparently was the ancestral form. Two semispecies, *brevicauda* and *talpoides,* are recognizable early in the fossil record; these two phena probably appeared as a result of increasing diversity in the environment of the late Irvingtonian. Evidently, repeated fluctuations of climate and the subsequent changes in range have not been sufficient to cause the cessation of gene flow necessary to complete speciation.

A second species, *B. carolinensis,* appeared in the mid-Irvingtonian. Although its original distribution is unclear, the species apparently evolved in temperate conditions in the south. Increased continentality following the Wisconsinan glaciation resulted in the appearance of the species *B. hylophaga,* which is restricted to the southwestern part of the range of the genus.

Sympatry of the two semispecies of *B. brevicauda* and/or sympatry of *B. carolinensis* and *B. hylophaga* detected in some of the paleofaunas are thought to be the result of periods of more equable climate. Post-Wisconsinan fluctuations of climate are the probable cause of the eastward restriction of the genus and the zones of sympatry observed between Holocene phena.

Fossils of *Blarina* have been found in warm-moist and cool-moist faunas. The fossil record (late Blancan–Rancholabrean) and paleoecology of the genus are summarized.

INTRODUCTION

Holocene short-tailed shrews of the genus *Blarina* occur throughout southeastern Canada and the eastern half of the United States (Hall, 1981). In Canada they range westward from the Gaspé Peninsula and Nova Scotia to southeastern Saskatchewan. The northernmost records are from south-central Saskatchewan (Nero, 1960) and western Manitoba (Krivda, 1957). In the United States the distribution of the genus extends from the Atlantic coast westward to eastern Texas, central Oklahoma, northeastern Colorado, central Nebraska, South Dakota, and North Dakota. An isolated population occurs in Aransas County in southern Texas (Jones and Loomis, 1954; Hall, 1981; George et al., 1982; Jones et al., 1982).

Short-tailed shrews inhabit a variety of mesic habitats. In Canada and the northern and eastern United States they have been reported from bogs, swamps, grasslands, and coniferous and deciduous woodland by Sheldon (1936), Blair (1940, 1941), Smith (1959), Buckner (1966), Choate (1972), Kirkland (1978), Handley (1979), and others. Except in Aransas County, their distribution in Texas is limited to the pine-oak and pine forests in the northeastern part of the state (Schmidly and Brown, 1979). In agricultural areas of the Great Plains, short-tailed shrews are confined largely to riparian or otherwise mesic habitats and to grassy roadsides and fence rows in which ground cover and litter are present (Genoways and Choate, 1972; Choate and Fleharty, 1975).

In an analysis of six plant communities on a remnant prairie in western Kansas, a significant preference was shown for the most mesic community in which the deepest litter was found (Choate and Fleharty, 1973). The shrews are less common or absent in grasslands and forests that lack herbaceous ground cover (Smith, 1959; Kirkland, 1978).

Although *Blarina* is ubiquitous over much of its range, local distribution apparently is limited by moisture. Temperature and ground cover are of secondary importance, especially when vegetation is sparse. Temperature and friability of the soil also are important. Short-tailed shrews are not found in areas where the moisture content of the soil is insufficient to keep the air in the soil and litter saturated (Pruitt, 1953, 1959; Getz, 1961).

The first taxonomic revision of the genus *Blarina* was undertaken by Merriam (1895). He recognized three species—*B. brevicauda* from Canada and the northern, central, and eastern United States; *B. carolinensis* from the southeastern United States; and *B. telmalestes,* described on the basis of one specimen from Dismal Swamp, Virginia. Among the several Holocene subspecies of these species that have been recognized by mammalogists are *B. c. peninsulae,* which was described by Merriam (1895) from peninsular Florida, and *B. b. hylophaga,* which was described by Elliot (1899) from southern Oklahoma. The latter was placed in synonymy with *B. b. carolinensis* by Jones and Glass (1960), and all nominal taxa of short-tailed shrews, with the exception of *B. telmalestes,* came to be regarded as subspecies of *B. brevicauda* (Hall and Kelson, 1959).

In 1972 Genoways and Choate concluded, on the basis of multivariate analyses, that the phena *B. b. brevicauda* and what they termed *B. b. carolinensis* [=*B. hylophaga*] behaved as biological species where their ranges were contiguous in southern Nebraska. Subsequent studies of *Blarina* in Illinois (Ellis et al., 1978), Virginia (Tate et al., 1980), and Tennessee (Braun and Kennedy, 1983) likewise revealed no evidence of intergradation between the two phena.

Recently, karyologic data were used to clarify taxonomic relationships among taxa of *Blarina.* Genoways et al. (1977) compared karyotypes of *B. brevicauda* from northern Nebraska and Pennsylvania and what they termed *B. b. carolinensis* [=*B. hylophaga*] from southern Nebraska and Kansas. Their data supported earlier morphometric analyses (Genoways and Choate, 1972) that suggested the two taxa in Nebraska were distinct species. Comparisons of the karyotypes of *B. brevicauda* with those reported for *B. brevicauda talpoides* (by Meylan, 1967) and *B. b. kirtlandi* (by Lee and Zimmerman, 1969) revealed that these northern taxa had the same chromosomal numbers and Robertsonian polymorphism. This chromosomal pattern is one of three that subsequently were interpreted to represent separate species (George et al., 1982).

The three species are *B. brevicauda* in the northern United States and eastern Canada, *B. carolinensis* in the southeastern United States, and *B. hylophaga* in the southwestern part of the range of the genus. A fourth chromosomal pattern from peninsular Florida, for which morphometric data were lacking, tentatively is referred to as *B. c. peninsulae* but might prove to represent a fourth species (George et al., 1981, 1982).

Shrews are poorly known in the fossil record before the Pleistocene. The Tribe Blarinini, which includes the genus *Blarina,* is first represented in the late Miocene of North America by *Adeloblarina,* a genus close to the ancestral stock of *Blarina* (Repenning, 1967). Late Pliocene (Blancan) specimens with *Blarina*-like dentition originally were referred to that genus—*Blarina adamsi* Hibbard (1953a) from the Fox Canyon local fauna and *Blarina gidleyi* Gazin (1933) from the Hagerman local fauna. These two species subsequently were referred to other genera, *adamsi* to *Cryptotis* (Repenning, 1967; Choate, 1970) and *gidleyi* to *Paracryptotis* (Choate, 1970; Hibbard and Bjork, 1971); however, because of the

similarity and presumed close phyletic relationship of those shrews with *Blarina,* the Fox Canyon and Hagerman faunas are included among the faunas summarized herein.

The fossil record of *Blarina* extends from the Blancan through the Rancholabrean land mammal ages. Sites from which specimens of *Blarina* have been recovered are located throughout the eastern half of the United States. Much of their accumulation in caves, fissures, and sinkholes resulted from the activities of owls and other predators. All fossil localities, except those in Texas, Oklahoma, and southwestern Kansas, are within the modern range of the genus. Sites from which fossils of *Blarina* have been reported, including early Holocene sites from the United States and Canada, are listed in Appendix 1.

Because the local distributions of extant species of *Blarina* apparently are limited by moisture, remains of *Blarina* in fossil faunas have been used as paleoecological indicators. The first paleontologist to do so was Hibbard (1956, 1963), who interpreted the presence of *Blarina* in the Mt. Scott local fauna as evidence of "trees and shrubs, with the development of a humus and litter cover" (Hibbard, 1963). "Stratigraphically distinct samples of the Short-tailed Shrew . . . represent different-sized animals that would be ascribed to an evolving lineage in conventional study. Similar differences can be found in the large northern and small southern subspecies of *Blarina brevicauda,* and associated faunas support the interpretation that the observed morphological differences are due to shifts of a cline" (Hibbard et al., 1965). The presence of different sizes of *Blarina* in the same deposit also was interpreted as an indication of changing climate by others, including Guilday et al. (1964) and Oesch (1967).

More detailed analyses of the paleoecology of Pleistocene *Blarina* were conducted by Graham and Semken (1976) and Graham (1976). Graham and Semken (1976) examined the modern distribution of the taxa *B. brevicauda, B. kirtlandi* (regarded herein as a subspecies of *B. brevicauda*), and *B. carolinensis* and concluded that *brevicauda* is separated from *kirtlandi* by moisture extremes, *kirtlandi* from *carolinensis* by temperature extremes, and *brevicauda* from *carolinensis* by temperature and moisture extremes. Fossils that appeared to correspond to the three extant phena seemed sympatric in three Pleistocene deposits examined by Graham and Semken (Cumberland Cave, New Paris No. 4, and Peccary Cave), and two of the phena were thought to be sympatric in the local faunas of Crankshaft Cave, Meyer Cave, Natural Chimneys, and Welsh Cave. Because of the co-occurrence of other currently allopatric species in the fossil deposits, sympatry of different phena of *Blarina* was interpreted as an indication of more equable climate rather than climatic changes as previously suggested by Guilday et al. (1964), Oesch (1967), and others. Subsequent allopatry or parapatry of the three phena of *Blarina* and eastward contraction of the western boundary of distribution since the Pleistocene were attributed to increasing continentality of the climate.

Graham (1976) studied late Wisconsinan environmental gradients by comparing faunal composition, densities of soricid and microtine species, and species distributions of Wisconsinan and Holocene faunas. In 11 of the 12 late Wisconsinan faunas examined, more species of shrews were present during the late Wisconsinan than at present. The greater species density was interpreted as indicating "more equable climate with reduced temperature and moisture gradients." As in the previous paper, Graham thought the presence of more than one phenon of *Blarina* in a fossil fauna indicated sympatric distributions under more equable climatic conditions.

An additional nominal species, *B. ozarkensis,* was recognized by Graham (1972) from the Kansan deposit from Conard Fissure in northern Arkansas. This taxon originally was described by Brown (1908) as a subspecies of *B. brevicauda* on the basis of fossils that were said to be intermediate in size between extant *B. b. brevicauda* and *B. b. kirtlandi.* Additional distinguishing characteristics were given by Graham (1972) and Graham and Semken (1976). The taxonomic status of *ozarkensis* is reviewed herein.

No attempt previously has been made to relate new information on the taxonomy of Holocene *Blarina* to fossils of this genus. Accordingly, the objectives of the present study were 1) to summarize available information regarding the fossil record of *Blarina* and 2) to identify fossils of *Blarina* (most of which previously were assigned to *B. brevicauda*) in an effort to shed light on the phylogeny and paleobiogeography of the species *brevicauda, carolinensis,* and *hylophaga.*

MATERIALS AND METHODS

A total of 672 Holocene specimens was examined. These specimens are housed in the following institutions: Carnegie Museum of Natural History (CM); Museum of Zoology, Louisiana State University (LSUMZ); Museum of Comparative Zoology, Harvard University (MCZ); Museum of the High Plains, Fort Hays State University (MHP); National Museum of Natural Sciences, Canada (NMC); Texas Cooperative Wildlife Collections, Texas A & M University (TCWC); Museum of Natural History, University of Connecticut (UCONN); National Fish and Wildlife Laboratory, National Museum of Natural History (USNM).

Twenty-six dental and dentary measurements, as listed by Graham and Semken (1976), were taken from each Holocene specimen (dental terminology after Choate, 1969, 1970): 1) length of P4–M3; 2) length of upper molar toothrow (M1–M3); 3) labial length of P4; 4) lingual length of P4; 5) anterior width of P4; 6) posterior width of P4; 7) labial length of M1; 8) lingual length of M1; 9) anterior width of M1; 10) posterior width of M1; 11) labial length of M2; 12) lingual length of M2; 13) anterior width of M2; 14) posterior width of M2; 15) length of M3; 16) width of M3; 17) length of u1–m3; 18) length of lower molar toothrow (m1–m3); 19) length of m1; 20) length of trigonid of m1; 21) length of m2; 22) length of trigonid of m2; 23) length of m3; 24) length of trigonid of m3; 25) width of mandibular condyle; 26) depth of body of mandible. Hereafter, these measurements are referred to by the corresponding numbers. Measurements 1, 2, 17, and 18 were taken with dial calipers to the nearest 0.1 mm. Other measurements (14 of the upper teeth and eight of the lower teeth and dentary) were taken with an ocular micrometer at 10×

magnification and rounded to the nearest 0.1 mm. Age classes were assigned according to condition of pelage and degree of wear to teeth (Choate, 1972). Most of the taxa mentioned herein were illustrated by Merriam (1895), Hibbard (1957), Repenning (1967), and Graham and Semken (1976). Therefore, illustrations are not provided herein.

Holocene specimens were assigned to 24 reference samples (Fig. 1) representing all but one (*B. brevicauda hooperi* Bole and Moulthrop) of the 16 nominal subspecies of the three species of *Blarina* currently recognized (Hall, 1981; George et al., 1981, 1982): 1) Autauga, Cullman, Hale, Russell, and Sumter counties, Alabama (*B. carolinensis carolinensis*), 16 specimens; 2) Brevard, Collier, Dade, Highlands, Osceola, Palm Beach, and Polk counties, Florida (*B. carolinensis peninsulae*), 28 specimens; 3) Crawford, Early, Grady, Liberty, Randolph, Richmond, Thomas, and Tift counties, Georgia (*B. carolinensis carolinensis*), 12 specimens; 4) Allen, Newton, and Porter counties, Indiana (*B. brevicauda kirtlandi*), 26 specimens; 5) Hamilton, Henry, Johnson, Marion, and Pottawattamie counties, Iowa (*B. brevicauda brevicauda*), 19 specimens; 6) Ashtabula, Athens, Erie, Fairfield, Geauga, Hamilton, and Meigs counties, Ohio (*B. brevicauda kirtlandi*), 36 specimens; 7) Dukes County, Massachusetts (*B. brevicauda aloga*), 82 specimens; 8) Nantucket County, Massachusetts (*B. brevicauda compacta*), 79 specimens; 9) New Brunswick and Nova Scotia, Canada; Aroostook, Penobscot, and Piscataquis counties, Maine (*B. brevicauda pallida*), 59 specimens; 10) Wake County, North Carolina (*B. carolinensis carolinensis*), 24 specimens; 11) Great Smoky Mountains and Magnetic City

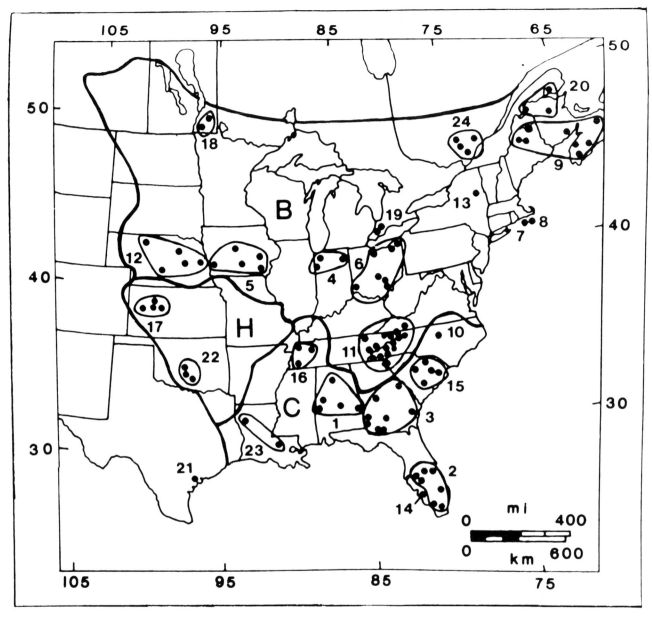

Fig. 1.—Present geographic distribution of *Blarina brevicauda* (B), *B. carolinensis* (C), and *B. hylophaga* (H) (after George et al., 1982). Reference samples are described in text.

(counties uncertain), Buncombe, Cherokee, Haywood, Macon, Mitchell, Transylvania, Watauga, and Yancey counties, North Carolina; Greenville and Oconee counties, South Carolina; Campbell, Carter, Cocke, Johnson, Sevier, and Sullivan counties, Tennessee; and Grayson County, Virginia (*B. brevicauda churchi*), 79 specimens; 12) Antelope, Buffalo, Cherry, Platte, and Washington counties, Nebraska (*B. brevicauda brevicauda*), 10 specimens; 13) Warren County, New York (*B. brevicauda talpoides*), 39 specimens; 14) Lee County, Florida (*B. carolinensis shermani*), 3 specimens; 15) Darlington, Dorchester, Georgetown, Richland, and Williamsburg counties, South Carolina (*B.*

carolinensis carolinensis), 9 specimens; 16) Reelfoot Lake (county uncertain), Benton, and Fayette counties, Tennessee (*B. carolinensis carolinensis*), 5 specimens; 17) Ellis, Rooks, Russell, and Trego counties, Kansas (*B. hylophaga hylophaga*), 44 specimens; 18) Manitoba, Canada (*B. brevicauda manitobensis*), 5 specimens; 19) Ontario, Canada (*B. brevicauda talpoides*), 8 specimens; 20) New Brunswick and Quebec, Canada (*B. brevicauda angusta*), 11 specimens; 21) Aransas County, Texas (*B. hylophaga plumbea*), 7 specimens; 22) Cleveland, Garvin, and Murray counties, Oklahoma (*B. hylophaga hylophaga*), 7 specimens; 23) East Baton Rouge and Red River parishes, Louisiana (*B. caro-*

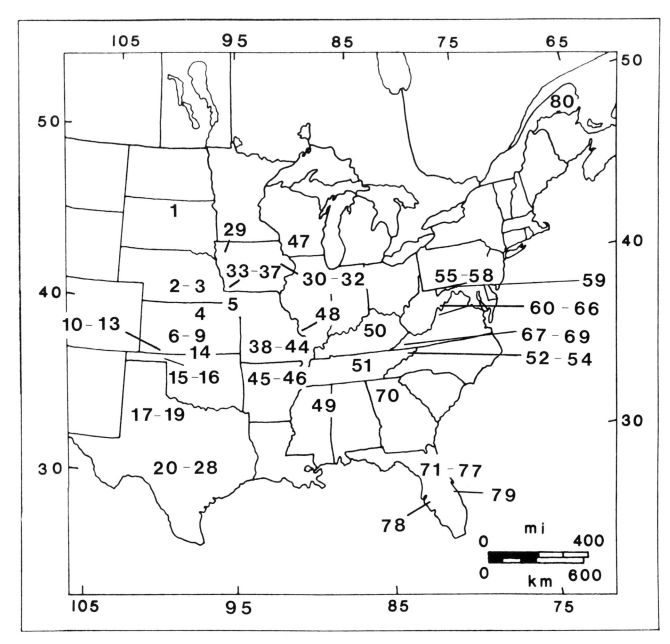

Fig. 2.—Sites from which fossils of *Blarina* were examined for this study. Localities are numbered as in Appendix 1. 1. Java; 2. Bartek Brothers Quarry; 3. Angus; 4. White Rock; 5. Wathena; 6. Rezabek; 7. Kanopolis; 8. Williams; 9. Kentuck; 10. Cudahy; 11. Jinglebob; 12. Mt. Scott; 13. Robert; 14. Dixon; 15. Berends; 16. Doby Springs; 17. Howard Ranch; 18. Easley Ranch; 19. Vera; 20. Miller's Cave; 21. Longhorn Cavern; 22. Felton Cave; 23. Hall's Cave; 24. Klein Cave; 25. Schulze Cave; 26. Friesenhahn Cave; 27. Cave Without a Name; 28. Barton Springs; 29. Cherokee Sewer Site; 30. Willard Cave; 31. Schmitt Cave; 32. Mud Creek; 33. Craigmile; 34. Garrett Farm; 35. Pleasant Ridge; 36. Waubonsie; 37. Thurman; 38, 39. Brynjulfson Cave No. 1 and No. 2; 40. Boney Spring; 41. Trolinger Spring; 42. Crankshaft Cave; 43. Bat Cave; 44. Zoo Cave; 45. Conard Fissure; 46. Peccary Cave; 47. Moscow Fissure; 48. Meyer Cave; 49. Catalpa Creek; 50. Welsh Cave; 51. Robinson Cave; 52. Baker Bluff Cave; 53. Guy Wilson Cave; 54. Riverside Cave; 55. New Paris Sinkhole No. 4; 56. Hanover Quarry Fissure; 57. Bootlegger Sink; 58. Port Kennedy Cave; 59. Cumberland Cave; 60. Eagle Cave; 61. Hoffman School Cave; 62. Trout Cave; 63. Strait Canyon Fissure; 64. Natural Chimneys; 65. Back Creek Cave No. 2; 66. Clark's Cave; 67. Ripplemead Quarry; 68. Jasper Saltpeter Cave; 69. Meadowview Cave; 70. Ladds Quarry; 71. Haile XIB; 72. Arredondo; 73. Reddick IA; 74. Williston IIIA; 75. Waccasassa River; 76. Inglis IA; 77. Coleman IIA; 78. Bradenton 51st Street; 79. Vero; 80. Caverne de Saint-Elzear.

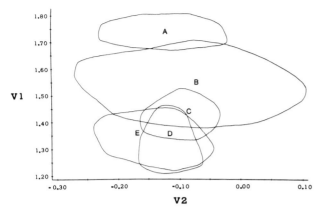

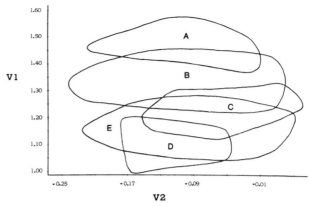

Fig. 3.—Analysis of measurements 3–10 of upper teeth. Range of variation of Holocene specimens of *B. b. brevicauda* (A), other subspecies of *B. brevicauda* (B), *B. hylophaga* (C), *B. c. peninsulae* (D), and *B. carolinensis* (E).

Fig. 4.—Analysis of measurements 19–24 of lower teeth. Range of variation of Holocene specimens of *B. b. brevicauda* (A), other subspecies of *B. brevicauda* (B), *B. hylophaga* (C), *B. c. peninsulae* (D), and *B. carolinensis* (E).

linensis minima), 29 specimens; 24) Quebec, Canada (*B. brevicauda talpoides*), 8 specimens.

A total of 1538 specimens from 82 Pliocene, Pleistocene, and early Holocene faunas was examined (Fig. 2). The fossils are housed in the following collections: Carnegie Museum of Natural History (CM); Vertebrate Paleontology Collection, Central Missouri State University (CMSU); Mississippi Museum of Natural Science (FACM); Illinois State Museum (ISM); Museum of Natural History, University of Kansas (KU); Midwestern University (MWU); Department of Geology, University of Iowa (SUI); Vertebrate Paleontology Laboratory, Balcones Research Center (TMM); Florida State Museum, University of Florida (UF); Museum of Paleontology, University of Michigan (UMMP); Department of Paleobiology, Smithsonian Institution (USNM). Sites from which these specimens were recovered are listed in Appendix 1 and illustrated in Fig. 2.

Because the fossils were fragmented or fragile, length measurements of toothrows generally were not taken. As many as possible of the other dental and dentary measurements used for Holocene specimens were taken. Age classes were assigned according to degree of wear to teeth.

Analyses of mensural data were performed with the Statistical Analysis System, SAS (Helwig and Council, 1979). The Procedure Univariate subroutine was used to calculate univariate statistics (mean, range, standard error, and coefficient of variation) for each character within the reference samples. Age classes and sexes were not run separately because previous studies revealed inconsistent patterns of variation among ages and between sexes (Guilday, 1957; Choate, 1972; Graham and Semken, 1976; Schmidly and Brown, 1979; French, 1981; Moncrief et al., 1982; Braun and Kennedy, 1983). The General Linear Model subroutine (Procedure GLM) was used to test for significant differences among reference samples for each character. Subsequently, a Duncan's Multiple Range Test (Procedure Duncan) was employed to determine maximally non-significant subsets among samples. To assess multivariate relationships among samples, the Manova option of the GLM subroutine was used to compute characteristic roots and vectors. These values then were employed in canonical (discriminant) analyses of variation within

and among reference samples (methodology described by Yates and Schmidly, 1977; Neff and Marcus, 1980).

Because specimens with missing measurements could not be included in multivariate analyses, and because most fossils had missing measurements, we selected suites of measurements from the original 26 for which sizes of fossil samples were greatest. The suites selected were measurements 3 through 10 for upper teeth, and 19 through 22 or 24 for lower teeth. Samples of fossils were analyzed using the one or more of these suites of measurements for which adequate numbers of specimens were available. Accordingly, it was necessary to compute characteristic roots and vectors of the reference samples for these suites of measurements for use in the canonical procedure to compare Holocene and fossil specimens.

Canonical analyses of the appropriate suites of characters for reference samples were used to plot centroids of variation within and among those samples. Initially, centroids were plotted to encompass one standard deviation on either side of the mean for each canonical variate. However, it was judged more informative to use a curved line enclosing peripheral specimens of each cluster so to encompass all known variation within each reference sample even though overlap of clusters resulted. In order to make comparisons with fossils, reference samples were pooled into five clusters (Figs. 3 and 4). Cluster A represented the range of variation among specimens of the largest Holocene subspecies, *B. b. brevicauda* (samples 5 and 12). Remaining samples of this species were not as clearly distinguished. These smaller eastern subspecies were grouped together as "small *brevicauda*" and are referred to as the *talpoides* semispecies. This group (samples 4, 6, 7, 8, 9, 11, 13, 18, 19, 20, and 24) is represented by cluster B. Cluster C represented the range of variation among the individuals of *B. hylophaga* (samples 17, 21, and 22). Specimens of the taxon *plumbea* proved to be more similar to specimens of *hylophaga* than to *carolinensis*, as previously suggested by Schmidly and Brown (1979) and George et al. (1981). Because of the uncertain taxonomic status of *B. carolinensis peninsulae* (George et al., 1982), the range of variation among specimens from sample 2 was represented by a separate cluster (D). Cluster E represented the range of variation among the other three subspecies of *B.*

carolinensis (samples 1, 3, 10, 14, 15, 16, and 23). The relatively large size of the three specimens of *B. c. shermani* available for examination implies that this taxon actually might be a relict population of *B. brevicauda*. Because of this uncertainty, the three specimens were not included in Figs. 3 and 4, nor in comparison with fossils from outside Florida.

Subsequently, fossil samples were analyzed using the corresponding characteristic roots and vectors for the Holocene samples, and were plotted on the illustrations of canonical variation within and among Holocene samples. This procedure enabled identification of fossil specimens, even when fossil samples consisted of two or more species, by comparison with Holocene samples. Of course, this procedure is accurate only if variation within and among taxa has remained essentially uniform throughout their existence; preliminary tests revealed that this assumption is valid. After this procedure was used to sort species in the samples of fossils, univariate analyses of fossil samples were conducted as described for the reference samples.

RESULTS

Univariate statistics and results of Duncan's multiple range tests for the reference samples of Holocene *Blarina* are shown in Table 1. Measurements 3 through 10 of the upper teeth and 19 through 24 of the lower teeth were selected in order to maximize sample sizes for groups of fossils, as discussed in Materials and Methods. Results of Duncan's multiple range tests illustrate the geographic variability and variability within populations that were pointed out previously by George et al. (1981) and Moncrief et al. (1982). The Duncan's multiple range tests consistently showed significant differences between samples of *B. b. brevicauda* from Iowa and Nebraska (samples 5 and 12) and samples of other taxa. Samples of eastern subspecies of *B. brevicauda* were smaller, and differed significantly from samples of *B. hylophaga* and *B. carolinensis* in only one measurement (lingual length of P4); for the other 13 measurements, there was some degree of overlap with samples of *B. hylophaga* and *B. carolinensis*.

Table 1.—*Univariate statistics for Holocene samples for measurements 3–10 of the upper teeth and 19–24 of the lower teeth. Vertical lines represent nonsignificant subsets as indicated by Duncan's multiple range tests. F value for each character is indicated by parentheses. Species represented by each sample are identified as* B. brevicauda *(B for* brevicauda *semispecies, T for* talpoides *semispecies),* B. carolinensis *(C), and* B. hylophaga *(H).*

Table 1.—*Continued.*

Labial length of P4 (F = 82.17)

ID	Sample	Mean	2 SE	Range	CV	N
B	5	2.74	0.04	2.6–2.9	3.46	18
B	12	2.70	0.07	2.5–2.9	4.28	10
T	24	2.59	0.07	2.5–2.7	3.48	7
T	18	2.58	0.08	2.5–2.7	3.24	5
T	13	2.48	0.03	2.3–2.7	3.68	38
T	11	2.47	0.02	2.3–2.6	3.41	78
T	9	2.46	0.02	2.3–2.6	3.23	56
T	4	2.46	0.03	2.3–2.7	3.49	26
T	6	2.43	0.04	2.2–2.6	4.24	34
T	19	2.37	0.04	2.3–2.4	2.19	7
T	20	2.33	0.05	2.2–2.4	3.38	11
T	8	2.31	0.03	2.2–2.5	4.18	52
H	17	2.30	0.03	2.2–2.5	3.98	44
T	7	2.28	0.02	2.1–2.4	3.61	46
C	14	2.27	0.07	2.2–2.3	2.55	3
H	22	2.20	0.05	2.1–2.3	2.88	6
C	1	2.19	0.04	2.0–2.3	4.05	16
C	10	2.18	0.04	2.1–2.4	4.44	24
H	21	2.17	0.04	2.1–2.2	2.25	7
C	15	2.14	0.08	1.9–2.3	5.76	9
C	16	2.14	0.05	2.1–2.2	2.56	5
C	23	2.09	0.03	1.9–2.2	4.04	29
C	3	2.08	0.05	2.0–2.2	3.61	11
C	2	2.08	0.04	1.9–2.3	5.34	27

Lingual length of P4 (F = 68.64)

ID	Sample	Mean	2 SE	Range	CV	N
B	5	2.12	0.04	2.0–2.3	3.61	18
B	12	2.09	0.06	2.0–2.2	4.19	10
T	18	1.88	0.04	1.8–1.9	2.38	5
T	24	1.86	0.11	1.6–2.1	8.14	7
T	20	1.86	0.06	1.7–2.0	5.04	11
T	13	1.85	0.03	1.6–2.0	5.44	38
T	11	1.85	0.02	1.6–2.1	5.57	78
T	19	1.84	0.08	1.6–1.9	2.74	7
T	8	1.84	0.02	1.7–2.0	4.31	52
C	14	1.83	0.07	1.8–1.9	3.15	3
T	4	1.82	0.04	1.7–2.0	4.73	26
T	7	1.82	0.02	1.7–2.0	3.96	46
T	6	1.79	0.03	1.6–2.0	5.11	34
T	9	1.79	0.03	1.6–2.0	5.23	56
H	17	1.71	0.03	1.6–1.9	4.69	44
H	22	1.63	0.07	1.5–1.7	5.00	6
H	21	1.61	0.03	1.6–1.7	2.34	7
C	1	1.59	0.05	1.6–1.9	6.27	16
C	2	1.58	0.03	1.4–1.7	4.98	27
C	10	1.55	0.04	1.4–1.9	6.67	24
C	16	1.54	0.05	1.5–1.6	3.56	5
C	15	1.52	0.06	1.4–1.6	5.48	9
C	23	1.49	0.02	1.4–1.6	4.23	29
C	3	1.46	0.03	1.4–1.5	3.59	11

Table 1.—*Continued.*

Anterior width of P4 (F = 51.79)

ID	Sample	Mean	2 SE	Range	CV	N
B	5	1.61	0.03	1.5–1.7	3.52	18
B	12	1.59	0.07	1.4–1.7	6.92	10
T	18	1.42	0.04	1.4–1.5	3.15	5
T	24	1.41	0.05	1.3–1.5	4.88	7
T	13	1.41	0.03	1.3–1.7	5.59	38
T	11	1.41	0.02	1.2–1.6	5.81	78
T	9	1.38	0.02	1.3–1.6	4.87	56
C	14	1.37	0.07	1.3–1.4	4.23	3
T	6	1.36	0.02	1.2–1.5	5.06	34
T	4	1.36	0.03	1.2–1.5	5.53	26
T	19	1.31	0.07	1.2–1.4	7.12	7
T	8	1.29	0.03	1.1–1.5	8.07	52
T	20	1.25	0.03	1.2–1.3	4.19	11
T	7	1.24	0.02	1.1–1.5	6.06	46
H	17	1.24	0.02	1.1–1.3	4.31	44
C	15	1.23	0.03	1.2–1.3	4.05	9
C	2	1.23	0.03	1.1–1.4	5.49	27
C	3	1.22	0.04	1.1–1.3	4.95	11
C	16	1.20	0.00	1.2	0.00	5
C	1	1.20	0.03	1.1–1.3	4.30	16
H	21	1.20	0.04	1.1–1.3	4.81	7
C	10	1.20	0.03	1.1–1.4	6.29	24
H	22	1.18	0.03	1.1–1.2	3.45	6
C	23	1.13	0.03	1.0–1.3	6.28	29

Posterior width of P4 (F = 29.82)

ID	Sample	Mean	2 SE	Range	CV	N
B	12	2.49	0.05	2.4–2.6	2.96	10
B	5	2.46	0.05	2.3–2.6	4.36	18
T	18	2.40	0.09	2.3–2.5	4.17	5
T	24	2.30	0.11	2.1–2.5	6.15	7
T	4	2.26	0.05	2.0–2.5	5.63	26
T	13	2.25	0.03	2.0–2.4	4.66	38
T	11	2.21	0.03	2.0–2.6	6.14	78
T	9	2.20	0.03	2.0–2.4	5.17	56
T	6	2.18	0.05	1.9–2.6	6.34	34
T	7	2.18	0.03	2.0–2.4	5.06	46
T	8	2.18	0.03	1.9–2.3	4.25	52
T	20	2.17	0.04	2.1–2.3	2.98	11
T	19	2.17	0.04	2.1–2.2	2.25	7
H	17	2.17	0.03	2.0–2.3	3.79	44
C	14	2.17	0.13	2.1–2.3	5.33	3
C	1	2.09	0.06	1.8–2.2	5.77	16
H	22	2.07	0.10	1.9–2.2	5.86	6
H	21	2.06	0.04	2.0–2.1	2.60	7
C	10	2.03	0.05	1.8–2.3	5.98	24
C	3	2.02	0.07	1.9–2.2	5.35	11
C	16	1.98	0.08	1.9–2.1	4.23	5
C	23	1.97	0.03	1.8–2.2	4.52	29
C	15	1.91	0.08	1.7–2.1	6.11	9
C	2	1.91	0.04	1.7–2.1	5.49	27

Table 1.—*Continued.*

Labial length of M1 (F = 58.09)

ID	Sample	Mean	2 SE	Range	CV	N
B	5	2.47	0.03	2.4–2.6	2.78	18
B	12	2.45	0.07	2.2–2.6	4.41	10
T	18	2.28	0.08	2.2–2.4	3.67	5
T	4	2.27	0.04	2.1–2.5	4.79	26
T	13	2.26	0.02	2.2–2.4	2.84	38
T	11	2.25	0.02	2.1–2.4	3.70	78
T	9	2.22	0.02	2.0–2.4	3.63	56
T	19	2.21	0.03	2.2–2.3	1.60	7
T	24	2.21	0.08	2.1–2.3	4.83	7
C	14	2.20	0.00	2.2	0.00	3
T	6	2.20	0.04	2.0–2.4	4.75	34
T	8	2.14	0.03	2.0–2.4	4.34	52
T	20	2.09	0.04	2.0–2.2	3.35	11
T	7	2.09	0.02	2.0–2.2	3.15	46
C	1	2.08	0.05	1.9–2.2	4.38	16
H	17	2.07	0.02	2.0–2.2	3.45	44
C	10	2.02	0.04	1.9–2.2	4.39	24
C	3	2.02	0.04	1.9–2.1	2.99	11
H	22	2.02	0.03	2.0–2.1	2.02	6
C	16	2.00	0.00	2.0	0.00	5
H	21	2.00	0.04	1.9–2.1	2.89	7
C	15	1.99	0.05	1.9–2.1	3.93	9
C	23	1.97	0.02	1.9–2.1	2.88	29
C	2	1.93	0.03	1.8–2.1	4.49	27

Lingual length of M1 (F = 50.59)

ID	Sample	Mean	2 SE	Range	CV	N
B	12	2.24	0.09	2.1–2.6	6.38	10
B	5	2.17	0.03	2.0–2.2	2.74	18
T	18	2.12	0.04	2.1–2.2	2.11	5
T	24	2.11	0.08	2.0–2.2	5.06	7
T	13	2.08	0.02	2.0–2.2	2.59	38
T	4	2.05	0.03	1.9–2.2	4.20	26
T	11	2.04	0.02	1.9–2.2	3.45	78
T	9	2.04	0.02	1.9–2.2	3.17	56
T	19	2.03	0.05	1.9–2.1	3.49	7
T	6	2.00	0.03	1.8–2.2	3.81	34
C	14	2.00	0.00	2.2	0.00	3
T	8	1.99	0.02	1.8–2.1	3.39	52
T	20	1.98	0.05	1.8–2.1	3.79	11
H	17	1.96	0.02	1.8–2.1	2.93	44
T	7	1.94	0.02	1.8–2.0	3.07	46
H	21	1.93	0.06	1.8–2.0	3.92	7
H	22	1.92	0.06	1.8–2.0	3.93	6
C	1	1.91	0.04	1.8–2.0	4.22	16
C	3	1.86	0.05	1.7–2.0	4.42	11
C	10	1.85	0.03	1.7–2.0	3.80	24
C	15	1.84	0.04	1.8–1.9	2.86	9
C	16	1.84	0.05	1.8–1.9	2.98	5
C	23	1.80	0.03	1.6–2.0	3.98	29
C	2	1.76	0.03	1.6–1.9	4.19	27

Measurements of the three samples of *B. hylophaga* and seven samples of *B. carolinensis* overlapped for all characters, although samples of *B. carolinensis* averaged smaller. The sample of *B. c. shermani* (14, N = 3) from Lee County, Florida, was indistinguishable from eastern populations of *B. brevicauda* in all characters.

Statistics for groups of fossils (where N ≥ 4) are

Table 1.—*Continued.*

ID	Sample		Mean	2 SE	Range	CV	N
		Anterior width of M1 (F = 38.18)					
B	12		2.51	0.06	2.4–2.7	3.49	10
B	5		2.45	0.05	2.2–2.6	4.16	18
T	18		2.30	0.09	2.2–2.4	4.35	5
T	24		2.24	0.10	2.1–2.5	5.67	7
T	11		2.23	0.03	1.9–2.6	5.92	78
T	20		2.20	0.05	2.1–2.3	4.07	11
H	17		2.18	0.03	2.0–2.6	4.56	44
T	6		2.18	0.04	2.0–2.4	5.61	34
T	4		2.17	0.05	2.0–2.5	5.34	26
T	8		2.16	0.04	1.9–2.4	5.91	52
T	9		2.14	0.03	1.9–2.4	4.61	56
T	13		2.14	0.03	2.0–2.4	4.73	38
T	19		2.11	0.07	2.0–2.2	4.41	7
C	14		2.10	0.12	2.0–2.2	4.76	3
T	7		2.06	0.02	1.9–2.2	3.55	46
H	22		2.02	0.06	1.9–2.1	3.73	6
H	21		2.00	0.04	1.9–2.1	2.89	7
C	23		1.95	0.03	1.8–2.1	4.21	29
C	16		1.94	0.05	1.9–2.0	2.82	5
C	1		1.93	0.04	1.8–2.1	4.52	16
C	3		1.93	0.07	1.8–2.2	6.18	11
C	15		1.92	0.06	1.8–2.1	4.34	9
C	2		1.92	0.04	1.6–2.1	5.96	27
C	10		1.92	0.04	1.8–2.2	5.08	24
		Posterior width of M1 (F = 41.50)					
B	12		2.66	0.07	2.5–2.9	4.41	10
B	5		2.64	0.07	2.2–2.8	5.55	18
T	18		2.52	0.08	2.4–2.6	3.32	5
T	24		2.50	0.11	2.3–2.7	5.66	7
T	11		2.46	0.03	2.1–2.8	4.79	78
T	4		2.43	0.05	2.2–2.7	5.02	26
T	13		2.42	0.03	2.2–2.6	3.98	38
T	7		2.41	0.04	2.1–2.6	5.32	46
T	8		2.41	0.03	2.2–2.6	3.89	52
T	6		2.40	0.04	2.1–2.6	5.20	34
T	9		2.40	0.03	2.2–2.6	4.57	56
T	20		2.36	0.04	2.2–2.4	2.92	11
T	19		2.34	0.06	2.2–2.4	3.81	7
H	17		2.31	0.02	2.1–2.4	3.03	44
C	14		2.30	0.12	2.2–2.4	4.35	3
H	22		2.22	0.06	2.1–2.3	3.40	6
C	1		2.19	0.05	2.0–2.4	5.88	16
C	3		2.17	0.06	2.0–2.4	4.64	11
C	10		2.16	0.04	2.0–2.4	4.75	24
H	21		2.16	0.04	2.1–2.2	2.48	7
C	23		2.15	0.03	2.0–2.3	4.19	29
C	16		2.14	0.08	2.0–2.2	4.18	5
C	15		2.10	0.08	2.0–2.3	5.32	9
C	2		2.06	0.04	1.8–2.3	5.17	27

Table 1.—*Continued.*

ID	Sample		Mean	2 SE	Range	CV	N
		Length of m1 (F = 85.25)					
B	5		2.48	0.04	2.3–2.6	3.62	19
B	12		2.46	0.05	2.3–2.6	3.47	9
T	18		2.32	0.04	2.3–2.4	1.93	5
T	24		2.29	0.06	2.2–2.4	3.65	8
T	13		2.27	0.02	2.2–2.4	3.09	37
T	11		2.25	0.02	2.0–2.4	3.87	80
T	4		2.25	0.03	2.1–2.4	3.83	24
T	19		2.23	0.03	2.2–2.3	2.08	7
T	6		2.22	0.03	2.0–2.4	3.81	33
T	9		2.19	0.01	2.1–2.3	2.49	56
T	8		2.12	0.03	1.9–2.3	4.92	49
T	20		2.11	0.04	2.0–2.2	2.69	10
C	14		2.10	0.00	2.1	0.00	3
H	17		2.07	0.02	2.0–2.2	2.91	43
H	22		2.03	0.04	2.0–2.1	2.54	6
T	7		2.02	0.03	1.7–2.2	5.39	39
C	16		2.00	0.00	2.0	0.00	4
C	1		1.99	0.03	1.9–2.1	2.52	15
C	3		1.98	0.04	1.9–2.1	3.52	10
H	21		1.97	0.04	1.9–2.0	2.48	7
C	10		1.96	0.03	1.8–2.2	4.01	24
C	15		1.95	0.05	1.8–2.0	3.72	8
C	23		1.91	0.03	1.8–2.0	3.62	28
C	2		1.86	0.03	1.7–2.0	3.92	27
		Length of trigonid of m1 (F = 56.08)					
B	5		1.70	0.04	1.6–1.9	5.55	19
B	12		1.66	0.05	1.6–1.8	4.39	9
T	18		1.62	0.04	1.6–1.7	2.76	5
T	24		1.60	0.05	1.5–1.7	4.73	8
T	13		1.55	0.02	1.5–1.6	3.26	37
T	19		1.54	0.04	1.5–1.6	3.37	7
T	11		1.54	0.02	1.4–1.9	5.32	80
T	4		1.53	0.03	1.4–1.7	5.51	24
T	6		1.51	0.03	1.4–1.7	5.01	33
T	9		1.51	0.02	1.4–1.6	3.85	56
C	14		1.47	0.06	1.4–1.5	3.94	3
T	8		1.44	0.02	1.2–1.6	5.70	49
T	20		1.41	0.04	1.3–1.5	4.03	10
H	17		1.40	0.01	1.3–1.5	2.43	43
T	7		1.37	0.03	1.2–1.6	6.63	39
H	21		1.37	0.04	1.3–1.4	3.56	7
H	22		1.37	0.04	1.3–1.4	3.78	6
C	1		1.37	0.03	1.3–1.4	3.57	15
C	3		1.34	0.03	1.3–1.4	3.88	10
C	10		1.33	0.03	1.2–1.5	5.28	24
C	15		1.31	0.04	1.2–1.4	5.04	8
C	16		1.30	0.00	1.3	0.00	4
C	2		1.29	0.02	1.2–1.4	4.88	27
C	23		1.26	0.03	1.1–1.4	6.43	28

shown in Table 2. Results of Duncan's multiple range tests show variability within and among paleofaunas similar to or greater than that of Holocene populations. The overlapping measurements result in difficulty in identifying fossil taxa, particularly when specimens are incomplete (thus limiting the number

Table 1.—*Continued.*

ID	Sample		Mean	2 SE	Range	CV	N
Length of m2 (F = 48.46)							
B	12		2.06	0.03	1.9–2.2	4.74	9
B	5		2.03	0.03	1.9–2.2	3.23	19
T	24		1.94	0.04	1.9–2.0	2.67	8
T	18		1.92	0.04	1.9–2.0	2.33	5
T	13		1.92	0.02	1.8–2.0	3.14	37
T	11		1.88	0.02	1.7–2.0	3.82	80
T	4		1.88	0.03	1.7–2.1	4.43	24
T	6		1.86	0.03	1.7–2.2	5.43	33
T	19		1.83	0.04	1.8–1.9	2.82	7
T	9		1.83	0.02	1.7–1.9	3.15	56
C	14		1.80	0.00	1.8	0.00	3
T	8		1.78	0.03	1.5–2.0	5.58	49
T	20		1.77	0.04	1.7–1.9	3.81	10
T	7		1.76	0.03	1.5–2.0	6.22	39
H	17		1.75	0.02	1.7–1.8	2.88	43
H	22		1.73	0.04	1.7–1.8	2.98	6
C	1		1.73	0.03	1.6–1.9	3.96	15
C	16		1.70	0.00	1.7	0.00	4
C	10		1.70	0.03	1.6–1.9	3.75	24
H	21		1.67	0.06	1.6–1.8	4.52	7
C	3		1.67	0.04	1.6–1.8	3.87	10
C	15		1.66	0.03	1.6–1.7	3.00	8
C	23		1.64	0.02	1.5–1.7	3.80	28
C	2		1.61	0.02	1.5–1.8	3.90	27
Length of trigonid of m2 (F = 24.70)							
B	12		1.36	0.07	1.2–1.6	8.63	9
B	5		1.34	0.03	1.2–1.4	4.52	19
T	24		1.30	0.05	1.2–1.4	5.82	8
T	13		1.28	0.03	1.1–1.5	6.26	37
T	11		1.25	0.02	1.1–1.5	6.47	80
T	4		1.22	0.03	1.1–1.3	5.39	24
T	6		1.22	0.03	1.1–1.4	6.22	33
T	19		1.21	0.04	1.1–1.3	4.45	7
T	9		1.20	0.01	1.1–1.3	2.44	56
C	14		1.20	0.00	1.2	0.00	3
T	18		1.20	0.06	1.1–1.3	5.89	5
T	8		1.18	0.02	1.0–1.3	5.87	49
T	7		1.18	0.02	1.0–1.3	6.19	39
T	20		1.18	0.03	1.1–1.2	3.57	10
H	17		1.16	0.02	1.1–1.2	4.18	43
C	1		1.16	0.03	1.1–1.3	5.15	15
H	21		1.16	0.04	1.1–1.2	4.62	7
H	22		1.15	0.05	1.1–1.2	4.76	6
C	15		1.14	0.04	1.1–1.2	4.61	8
C	3		1.13	0.05	1.0–1.2	7.12	10
C	16		1.13	0.04	1.1–1.2	3.99	4
C	10		1.12	0.02	1.0–1.2	5.31	24
C	2		1.08	0.02	1.0–1.2	5.27	27
C	23		1.07	0.02	1.0–1.2	5.07	28

Table 1.—*Continued.*

ID	Sample		Mean	2 SE	Range	CV	N
Length of m3 (F = 42.42)							
B	5		1.67	0.03	1.6–1.8	3.49	19
B	12		1.67	0.07	1.5–1.8	6.48	9
T	18		1.58	0.08	1.5–1.7	5.30	5
T	24		1.55	0.05	1.4–1.6	4.88	8
T	4		1.54	0.03	1.4–1.7	4.57	24
T	13		1.54	0.02	1.3–1.6	4.78	37
T	11		1.53	0.02	1.4–1.7	4.70	80
T	6		1.51	0.02	1.4–1.6	4.75	33
T	19		1.50	0.06	1.3–1.6	6.01	7
T	9		1.47	0.02	1.2–1.6	4.93	56
H	17		1.47	0.02	1.4–1.6	4.40	43
T	7		1.44	0.03	1.2–1.6	7.56	39
T	20		1.43	0.03	1.4–1.5	3.38	10
T	8		1.41	0.03	1.2–1.6	7.08	49
C	14		1.37	0.06	1.3–1.4	4.23	3
H	22		1.37	0.04	1.3–1.4	3.78	6
H	21		1.36	0.04	1.3–1.4	3.94	7
C	3		1.33	0.04	1.2–1.4	5.08	10
C	16		1.33	0.05	1.3–1.4	3.77	4
C	10		1.32	0.03	1.2–1.5	5.96	24
C	1		1.32	0.03	1.2–1.4	4.97	15
C	15		1.31	0.03	1.3–1.4	2.69	8
C	23		1.29	0.03	1.1–1.4	5.13	28
C	2		1.27	0.03	1.0–1.4	7.21	27
Length of trigonid of m3 (F = 20.40)							
B	12		1.20	0.06	1.0–1.3	8.78	9
B	5		1.14	0.04	1.0–1.2	6.69	19
T	24		1.11	0.05	1.0–1.2	5.76	8
T	6		1.08	0.02	1.0–1.2	6.49	33
T	11		1.07	0.02	0.9–1.2	6.75	80
T	13		1.07	0.03	1.0–1.2	7.17	37
T	4		1.05	0.03	0.9–1.2	6.84	24
T	19		1.04	0.04	1.0–1.1	5.13	7
T	18		1.04	0.05	1.0–1.1	5.27	5
C	14		1.03	0.06	1.0–1.1	5.59	3
T	9		1.03	0.01	1.0–1.1	4.48	56
H	17		1.01	0.01	0.9–1.1	3.82	43
H	22		1.00	0.00	1.0	0.00	6
T	8		1.00	0.02	0.8–1.2	6.93	49
T	20		0.99	0.02	0.9–1.0	3.19	10
C	3		0.99	0.04	0.9–1.1	5.73	10
C	15		0.99	0.03	0.9–1.0	3.58	8
C	10		0.99	0.02	0.9–1.1	4.97	24
H	21		0.99	0.05	0.9–1.1	7.00	7
C	1		0.98	0.02	0.9–1.0	4.23	15
T	7		0.97	0.03	0.8–1.1	9.02	39
C	2		0.93	0.03	0.8–1.1	7.79	27
C	16		0.93	0.05	0.9–1.0	5.41	4
C	23		0.89	0.03	0.8–1.0	7.96	28

of characters that can be examined) or when sample size is small. Whether this variability reflects the true variation within a contemporaneous popula-

tion or an accident in preservation is difficult to determine. We have tried to indicate (for example, the Ladds Quarry fauna) where conditions of pres-

Table 2.—*Univariate statistics for samples of fossils (where N ≥ 4) for measurements 3–10 of the upper teeth and 19–24 of the lower teeth. Vertical lines represent nonsignificant subsets as indicated by Duncan's multiple range tests. F value for each character is indicated in parentheses. Asterisks indicate local faunas in which more than one phenon is present. Species are identified by ID as in Table 1. Samples are indicated by 6-character acronym (ARREDO = Arredondo, BACKCR = Back Creek Cave No. 2, BAKERB = Baker Bluff, BRYNJU = Brynjulfson Cave No. 2, CAVEWI = Cave Without a Name, CHIMNE = Natural Chimneys, CLARKS = Clark's Cave, CONARD = Conard Fissure, CRANKS = Crankshaft Cave, CUMBER = Cumberland Cave, FRIESE = Friesenhahn Cave, GUYWIL = Guy Wilson Cave, HAILEX = Haile XIB, INDIAN = Vero, JASPER = Jasper Saltpeter Cave, KLEINC = Klein Cave, LADDSQ = Ladds Quarry, MEYERC = Meyer Cave, MOSCOW = Moscow Fissure, NEWPAR = New Paris No. 4, PECCAR = Peccary Cave, REDDIC = Reddick IA, RIVERS = Riverside Cave, ROBINS = Robinson Cave, SCHMIT = Schmitt Cave, SCHULZ = Schulze Cave, STELZE = Caverne de St.-Elzear, STRAIT = Strait Canyon Fissure, TROUTC = Trout Cave, WILLAR = Willard Cave, and ZOO-CAV = Zoo Cave).*

ID	Sample		Mean	2 SE	Range	CV	N
		Labial length of P4 (F = 55.37)					
*B	BACKCR		2.80	0.12	2.7–2.9	3.57	3
*B	NEWPAR		2.80	0.05	2.7–2.9	5.05	2
B	CLARKS		2.79	0.09	2.6–3.0	4.89	9
B	MOSCOW		2.79	0.07	2.7–2.9	3.23	7
*B	CRANKS		2.75	0.04	2.6–2.9	3.44	20
B	ROBINS		2.70	0.08	2.6–2.8	3.02	4
B	CHIMNE		2.70	0.11	2.5–3.0	5.94	8
*B	BAKERB		2.61	0.10	2.5–2.8	5.15	7
T	STELZE		2.61	0.04	2.3–2.9	4.94	34
*T	NEWPAR		2.60	0.18	2.4–2.8	7.02	4
T	CUMBER		2.57	0.08	2.4–2.7	4.02	6
T	RIVERS		2.55	0.06	2.5–2.6	2.26	4
*T	BAKERB		2.53	0.05	2.4–2.6	3.25	10
T	LADDSQ		2.50	0.09	2.4–2.6	4.00	5
T	CONARD		2.45	0.06	2.3–2.7	4.33	15
T	PECCAR		2.43	0.07	2.2–2.8	6.13	20
T	JASPER		2.40	0.16	2.2–2.6	6.80	4
T	WILLAR		2.39	0.05	2.2–2.7	4.97	27
T	BRYNJU		2.39	0.05	2.3–2.5	2.68	8
*T	MEYERC		2.37	0.05	2.2–2.6	4.47	17
H	SCHULZ		2.29	0.06	2.1–2.4	4.34	10
*T	BACKCR		2.25	0.30	2.1–2.4	9.43	2
*T	CRANKS		2.25	0.10	2.2–2.3	3.14	2
H or C	KLEINC		2.20	0.09	2.1–2.3	4.55	5
C	CAVEWI		2.10	0.06	2.0–2.2	3.89	7
*C	MEYERC		2.10	0.03	2.0–2.2	2.92	17
C	ARREDO		1.88	0.05	1.8–1.9	2.67	4
C	HAILEX		1.85	0.03	1.8–2.0	3.37	17
C	INDIAN		1.84	0.10	1.7–2.0	6.20	5

ervation or collection imply the presence of allopatric taxa.

In general, fossils were separated by Duncan's multiple range tests as in the multivariate canonical analyses discussed in Paleontology. Results of the Duncan's tests appear to support our suspicions that more than one taxon was present in several samples (BACKCR, BAKERB, CRANKS, CUMBER, LADDSQ, MEYERC, NEWPAR, and ROBINS). There was a high degree of overlap among almost all samples; only the four groups from Florida (ARREDO, HAILEX, INDIAN, and REDDIC) were significantly different from other samples for most characters.

Table 2.—*Continued.*

ID	Sample		Mean	2 SE	Range	CV	N
		Lingual length of P4 (F = 58.46)					
B	MOSCOW		2.17	0.08	2.0–2.3	5.12	7
*B	NEWPAR		2.15	0.10	2.1–2.2	3.29	2
*B	BACKCR		2.13	0.07	2.1–2.2	2.71	3
B	CLARKS		2.12	0.08	2.0–2.3	5.66	9
*B	CRANKS		2.10	0.03	2.0–2.2	3.46	20
B	ROBINS		2.08	0.05	2.0–2.1	2.41	4
B	CHIMNE		2.03	0.03	2.0–2.1	2.29	8
*T	BAKERB		2.01	0.05	1.9–2.2	3.67	10
*B	BAKERB		1.99	0.10	1.8–2.2	6.77	7
*T	BACKCR		1.95	0.06	1.9–2.0	2.96	2
*T	NEWPAR		1.95	0.10	1.9–2.0	3.63	4
T	STELZE		1.94	0.03	1.7–2.1	4.40	34
T	RIVERS		1.93	0.10	1.8–2.0	4.97	4
T	LADDSQ		1.92	0.12	1.7–2.0	6.79	5
T	PECCAR		1.92	0.04	1.8–2.1	4.57	20
T	CONARD		1.91	0.05	1.8–2.0	4.79	15
T	CUMBER		1.90	0.07	1.8–2.0	4.71	6
T	BRYNJU		1.89	0.05	1.8–2.0	4.42	8
T	WILLAR		1.88	0.04	1.7–2.1	5.11	27
T	JASPER		1.85	0.19	1.7–2.1	10.35	4
*T	MEYERC		1.81	0.03	1.7–1.9	3.31	17
H	SCHULZ		1.72	0.06	1.6–1.8	5.34	10
*T	CRANKS		1.65	0.10	1.6–1.7	4.29	2
H or C	KLEINC		1.62	0.08	1.5–1.7	5.17	5
C	CAVEWI		1.60	0.00	1.6	0.00	7
*C	MEYERC		1.57	0.03	1.5–1.7	4.49	17
C	HAILEX		1.42	0.03	1.3–1.5	3.73	17
C	INDIAN		1.38	0.04	1.3–1.4	3.24	5
C	ARREDO		1.38	0.10	1.3–1.5	6.96	4
		Anterior width of P4 (F = 27.40)					
B	MOSCOW		1.70	0.11	1.5–1.9	8.32	7
T	RIVERS		1.63	0.13	1.5–1.8	7.74	4
B	CLARKS		1.61	0.05	1.5–1.7	4.85	9
B	CHIMNE		1.56	0.07	1.5–1.7	5.86	8
*B	CRANKS		1.54	0.04	1.4–1.7	5.33	20
*B	BACKCR		1.53	0.13	1.4–1.6	7.53	3
*B	BAKERB		1.51	0.11	1.4–1.8	9.67	7
B	ROBINS		1.50	0.08	1.4–1.6	5.44	4
*T	BACKCR		1.50	0.20	1.4–1.6	9.43	2
*B	NEWPAR		1.50	0.20	1.4–1.6	9.43	2
T	LADDSQ		1.48	0.08	1.4–1.6	5.65	5
*T	BAKERB		1.48	0.07	1.3–1.6	6.98	10
*T	NEWPAR		1.48	0.10	1.4–1.6	6.49	4
T	CUMBER		1.47	0.07	1.4–1.6	5.57	6
T	WILLAR		1.45	0.03	1.3–1.7	5.86	27
T	CONARD		1.45	0.06	1.3–1.6	7.34	15
T	PECCAR		1.44	0.04	1.3–1.6	5.66	20
T	STELZE		1.43	0.03	1.3–1.6	6.10	34
H or C	KLEINC		1.38	0.04	1.3–1.4	3.24	5
T	BRYNJU		1.38	0.03	1.3–1.4	3.37	8
T	JASPER		1.35	0.13	1.2–1.5	9.56	4
H	SCHULZ		1.33	0.09	1.1–1.5	10.06	10
*T	MEYERC		1.30	0.02	1.2–1.4	3.85	17
*T	CRANKS		1.20	0.00	1.2	0.00	2
C	CAVEWI		1.19	0.05	1.1–1.3	5.82	7
*C	MEYERC		1.16	0.03	1.1–1.2	4.38	17
C	HAILEX		1.15	0.03	1.1–1.3	5.44	17
C	ARREDO		1.13	0.10	1.0–1.2	8.51	4
C	INDIAN		1.06	0.05	1.0–1.1	5.17	5

Table 2.—*Continued.*

ID	Sample		Mean	2 SE	Range	CV	N
		Posterior width of P4 (F = 34.20)					
*B	NEWPAR		2.70	0.20	2.6–2.8	5.24	2
B	CLARKS		2.67	0.08	2.5–2.9	2.67	9
B	MOSCOW		2.64	0.10	2.5–2.9	4.82	7
*B	BACKCR		2.63	0.07	2.6–2.7	2.19	3
B	ROBINS		2.55	0.06	2.5–2.6	2.26	4
*B	CRANKS		2.50	0.04	2.3–2.6	3.56	20
B	CHIMNE		2.48	0.03	2.4–2.5	1.87	8
T	STELZE		2.43	0.05	2.1–2.7	5.96	34
*T	NEWPAR		2.43	0.15	2.3–2.6	6.19	4
T	LADDSQ		2.40	0.06	2.3–2.5	2.95	5
*T	BACKCR		2.40	0.20	2.3–2.5	5.89	2
*B	BAKERB		2.40	0.04	2.3–2.5	2.41	7
T	CONARD		2.34	0.06	2.1–2.6	5.06	15
T	CUMBER		2.33	0.10	2.2–2.5	5.19	6
T	BRYNJU		2.33	0.05	2.2–2.4	3.04	8
T	PECCAR		2.33	0.07	2.0–2.6	6.96	20
*T	BAKERB		2.30	0.08	2.1–2.5	5.78	10
T	WILLAR		2.27	0.06	2.0–2.6	6.96	27
*T	MEYERC		2.24	0.06	2.0–2.5	5.69	17
T	RIVERS		2.23	0.10	2.1–2.3	4.30	4
T	JASPER		2.15	0.19	1.9–2.3	8.91	4
H	SCHULZ		2.14	0.04	2.0–2.2	3.27	10
*T	CRANKS		2.10	0.20	2.0–2.2	6.73	2
H or C	KLEINC		2.08	0.15	1.9–2.3	7.90	5
C	CAVEWI		2.00	0.00	2.0	0.00	7
*C	MEYERC		1.95	0.05	1.7–2.1	5.16	17
C	INDIAN		1.82	0.10	1.7–2.0	6.02	5
C	HAILEX		1.79	0.06	1.6–2.0	6.96	17
C	ARREDO		1.78	0.10	1.7–1.9	5.39	4
		Labial length of M1 (F = 37.21)					
*B	NEWPAR		2.55	0.10	2.5–2.6	2.77	2
B	CLARKS		2.51	0.13	2.2–2.8	7.57	9
*B	BACKCR		2.43	0.18	2.3–2.6	6.28	3
B	ROBINS		2.43	0.10	2.3–2.5	3.95	4
B	CHIMNE		2.39	0.07	2.3–2.6	4.15	8
*B	CRANKS		2.38	0.05	2.2–2.6	4.71	20
B	MOSCOW		2.37	0.08	2.3–2.6	4.69	7
T	STELZE		2.34	0.04	2.0–2.5	4.93	34
*T	NEWPAR		2.33	0.10	2.2–2.4	4.12	4
T	CUMBER		2.32	0.06	2.2–2.4	3.25	6
*B	BAKERB		2.31	0.09	2.2–2.5	5.25	7
T	RIVERS		2.30	0.08	2.2–2.4	3.55	4
*T	BAKERB		2.27	0.07	2.1–2.5	4.67	10
T	CONARD		2.23	0.04	2.1–2.4	3.66	15
T	PECCAR		2.23	0.04	2.1–2.4	4.62	20
T	LADDSQ		2.22	0.12	2.1–2.4	5.87	5
*T	BACKCR		2.20	0.20	2.1–2.3	6.43	2
T	BRYNJU		2.19	0.05	2.1–2.3	2.93	8
T	WILLAR		2.18	0.04	2.0–2.4	4.82	27
*T	MEYERC		2.15	0.06	2.0–2.4	5.22	17
H	SCHULZ		2.14	0.04	2.0–2.2	3.27	10
T	JASPER		2.13	0.15	2.0–2.3	7.06	4
*T	CRANKS		2.10	0.20	2.0–2.3	6.73	2
H or C	KLEINC		2.06	0.05	2.0–2.1	2.66	5
C	CAVEWI		2.03	0.04	2.0–2.1	2.41	7
*C	MEYERC		1.99	0.04	1.9–2.2	3.93	17
C	ARREDO		1.73	0.05	1.7–1.8	2.90	4
C	HAILEX		1.71	0.07	1.6–1.8	2.83	17
C	INDIAN		1.70	0.06	1.6–1.8	4.16	5

Table 2.—*Continued.*

ID	Sample		Mean	2 SE	Range	CV	N
*B	NEWPAR		2.45	0.10	2.4–2.5	2.89	2
B	CLARKS		2.33	0.09	2.2–2.6	6.06	9
B	MOSCOW		2.27	0.07	2.2–2.4	4.19	7
B	ROBINS		2.23	0.10	2.1–2.3	4.30	4
*B	CRANKS		2.22	0.03	2.1–2.4	3.46	20
*B	BACKCR		2.20	0.20	2.0–2.3	7.87	3
B	CHIMNE		2.19	0.05	2.1–2.3	2.93	8
T	RIVERS		2.15	0.13	2.0–2.3	6.01	4
*T	BAKERB		2.15	0.08	1.9–2.4	5.90	10
T	STELZE		2.15	0.04	1.9–2.5	5.28	34
*B	BAKERB		2.14	0.07	2.0–2.3	4.55	7
T	CUMBER		2.12	0.11	1.9–2.3	6.28	6
*T	BACKCR		2.10	0.20	2.0–2.2	6.73	2
*T	NEWPAR		2.10	0.08	2.0–2.2	3.89	4
T	LADDSQ		2.08	0.10	2.0–2.2	5.27	5
T	CONARD		2.08	0.05	2.0–2.2	4.14	15
T	PECCAR		2.08	0.05	2.0–2.3	4.83	20
T	BRYNJU		2.04	0.05	2.0–2.2	3.65	8
T	WILLAR		2.01	0.03	1.9–2.2	3.89	27
H	SCHULZ		1.99	0.04	1.9–2.1	2.85	10
*T	MEYERC		1.99	0.03	1.9–2.1	3.02	17
T	JASPER		1.98	0.10	1.9–2.1	4.85	4
*T	CRANKS		1.95	0.10	1.9–2.0	3.63	2
H or C	KLEINC		1.94	0.08	1.9–2.1	4.61	5
C	CAVEWI		1.90	0.08	1.8–2.0	5.26	7
*C	MEYERC		1.85	0.03	1.8–2.0	3.38	17
C	ARREDO		1.60	0.08	1.5–1.7	5.10	4
C	INDIAN		1.60	0.06	1.5–1.7	4.42	5
C	HAILEX		1.58	0.03	1.5–1.7	3.57	17

Anterior width of M1 (F = 26.83)

ID	Sample		Mean	2 SE	Range	CV	N
B	CLARKS		2.67	0.11	2.4–2.9	5.93	9
*B	NEWPAR		2.65	0.30	2.5–2.8	8.01	2
B	MOSCOW		2.57	0.13	2.4–2.9	6.63	7
*B	BACKCR		2.50	0.12	2.4–2.6	4.00	3
*T	NEWPAR		2.48	0.13	2.3–2.6	5.08	4
B	ROBINS		2.45	0.06	2.4–2.5	2.36	4
B	CHIMNE		2.45	0.09	2.2–2.6	4.88	8
*B	CRANKS		2.41	0.05	2.2–2.6	4.24	20
T	PECCAR		2.35	0.07	2.1–2.7	6.69	20
T	STELZE		2.34	0.04	2.1–2.6	5.37	34
T	CUMBER		2.33	0.13	2.1–2.5	7.00	6
T	RIVERS		2.33	0.13	2.2–2.5	5.41	4
T	LADDSQ		2.32	0.10	2.2–2.4	4.72	5
T	CONARD		2.32	0.06	2.1–2.5	4.94	15
*B	BAKERB		2.29	0.14	2.0–2.6	8.16	7
*T	BAKERB		2.28	0.07	2.1–2.5	4.98	10
T	WILLAR		2.26	0.05	2.0–2.6	5.68	27
*T	BACKCR		2.25	0.50	2.0–2.5	15.71	2
T	BRYNJU		2.23	0.09	2.1–2.5	5.76	8
*T	MEYERC		2.17	0.06	2.0–2.4	5.82	17
H	SCHULZ		2.17	0.06	2.0–2.3	4.37	10
*T	CRANKS		2.15	0.30	2.0–2.3	9.87	2
H or C	KLEINC		2.12	0.04	2.1–2.2	2.11	5
T	JASPER		2.10	0.22	1.9–2.4	10.29	4
C	CAVEWI		2.01	0.05	1.9–2.1	3.43	7
*C	MEYERC		2.01	0.05	1.9–2.2	4.48	17
C	ARREDO		1.80	0.12	1.7–1.9	6.42	4
C	HAILEX		1.79	0.04	1.7–1.9	4.61	17
C	INDIAN		1.74	0.10	1.6–1.9	6.55	5

Table 2.—*Continued.*

ID	Sample		Mean	2 SE	Range	CV	N
		Posterior width of M1 (F = 32.33)					
*B	NEWPAR		2.85	0.10	2.8–2.9	2.48	2
B	MOSCOW		2.81	0.11	2.6–3.0	5.59	7
B	CLARKS		2.81	0.12	2.5–3.0	6.52	9
*B	BACKCR		2.73	0.07	2.7–2.8	2.11	3
B	ROBINS		2.70	0.12	2.6–2.8	4.28	4
*T	NEWPAR		2.60	0.14	2.5–2.8	5.44	4
B	CHIMNE		2.60	0.05	2.5–2.7	2.91	8
*B	CRANKS		2.60	0.06	2.4–2.9	4.92	20
T	STELZE		2.57	0.05	2.3–2.9	5.25	34
T	BRYNJU		2.54	0.07	2.3–2.6	4.18	8
*B	BAKERB		2.51	0.13	2.2–2.7	6.67	7
T	LADDSQ		2.50	0.06	2.4–2.6	2.83	5
T	PECCAR		2.49	0.04	2.3–2.7	4.19	20
T	CUMBER		2.48	0.11	2.3–2.7	5.35	6
T	RIVERS		2.45	0.13	2.3–2.6	5.27	4
*T	BAKERB		2.45	0.08	2.2–2.6	4.81	10
T	WILLAR		2.44	0.06	2.2–2.8	6.65	27
T	CONARD		2.44	0.08	2.2–2.8	6.35	15
*T	MEYERC		2.37	0.07	2.1–2.7	6.11	17
T	JASPER		2.35	0.24	2.1–2.6	10.13	4
*T	BACKCR		2.35	0.30	2.2–2.5	9.03	2
*T	CRANKS		2.35	0.10	2.3–2.4	3.01	2
H	SCHULZ		2.28	0.08	2.1–2.5	5.77	10
H or C	KLEINC		2.26	0.10	2.1–2.4	5.05	5
*C	MEYERC		2.11	0.04	2.0–2.2	4.06	17
C	CAVEWI		2.10	0.04	2.0–2.2	2.75	7
C	INDIAN		1.90	0.09	1.8–2.0	5.26	5
C	HAILEX		1.90	0.05	1.7–2.1	5.26	17
C	ARREDO		1.85	0.06	1.8–1.9	3.12	4
		Length of m1 (F = 103.66)					
*B	ROBINS		2.49	0.03	2.4–2.7	3.11	22
*B	BAKERB		2.45	0.04	2.3–2.7	3.50	18
*B	CRANKS		2.44	0.02	2.4–2.5	2.03	19
B	CLARKS		2.44	0.02	2.3–2.6	3.08	44
B	CHIMNE		2.42	0.04	2.3–2.6	3.34	18
B	BACKCR		2.42	0.10	2.3–2.6	4.53	5
T	ZOOCAV		2.35	0.09	2.1–2.6	6.10	10
B or T	TROUTC		2.34	0.08	2.2–2.4	3.82	5
*T	LADDSQ		2.30	0.06	2.2–2.4	3.07	5
*T	ROBINS		2.30	0.00	2.3	0.00	2
T	STELZE		2.28	0.02	2.0–2.5	4.03	72
*T	CUMBER		2.28	0.06	2.2–2.4	3.90	8
T	RIVERS		2.26	0.04	2.1–2.3	3.09	10
T	WILLAR		2.25	0.02	2.1–2.5	4.16	59
T	STRAIT		2.25	0.06	2.0–2.4	5.02	13
T	BRYNJU		2.25	0.05	2.2–2.4	3.65	11
*T	CRANKS		2.24	0.02	2.2–2.3	2.25	26
T	SCHMIT		2.22	0.08	2.1–2.3	3.77	5
*T	BAKERB		2.21	0.02	2.0–2.4	4.30	61
T	GUYWIL		2.20	0.06	2.1–2.3	3.21	5
T	PECCAR		2.19	0.03	2.0–2.4	5.42	37
T	CONARD		2.19	0.03	2.1–2.4	3.33	21
*T	MEYERC		2.17	0.02	2.1–2.3	3.12	59
*C	CUMBER		2.06	0.05	1.9–2.1	3.53	9
H	SCHULZ		2.04	0.03	1.8–2.1	3.29	29
*H	CRANKS		2.01	0.02	1.9–2.1	2.22	20
H	KLEINC		1.99	0.05	1.8–2.1	3.93	9
H	FRIESE		1.94	0.05	1.9–2.0	2.82	5
*C	LADDSQ		1.93	0.07	1.9–2.0	2.99	3
*C	MEYERC		1.92	0.02	1.8–2.0	3.38	55
C	REDDIC		1.78	0.05	1.6–1.9	5.70	15
C	HAILEX		1.73	0.03	1.6–1.8	3.40	17
C	INDIAN		1.72	0.04	1.7–1.8	2.60	5
C	ARREDO		1.64	0.12	1.5–1.8	8.18	5

Table 2.—*Continued.*

ID	Sample		Mean	2 SE	Range	CV	N
		Length of trigonid of m1 (F = 75.87)					
*B	ROBINS		1.74	0.03	1.6–1.9	3.83	22
*B	CRANKS		1.70	0.03	1.6–1.8	3.40	19
*B	BAKERB		1.68	0.04	1.5–1.8	5.23	18
B	CLARKS		1.66	0.02	1.5–1.8	3.75	44
B	CHIMNE		1.65	0.03	1.6–1.8	3.75	18
B	BACKCR		1.62	0.04	1.6–1.7	2.76	5
*T	ROBINS		1.60	0.00	1.6	0.00	2
T	ZOOCAV		1.59	0.06	1.4–1.7	5.51	10
*T	LADDSQ		1.58	0.04	1.5–1.6	2.83	5
*T	CUMBER		1.58	0.05	1.5–1.7	4.49	8
*T	CRANKS		1.57	0.02	1.4–1.6	3.39	26
B or T	TROUTC		1.56	0.05	1.5–1.6	3.51	5
T	STELZE		1.55	0.02	1.4–1.7	4.71	72
T	RIVERS		1.54	0.04	1.4–1.6	4.54	10
T	STRAIT		1.54	0.04	1.4–1.6	4.23	13
*T	BAKERB		1.53	0.02	1.4–1.7	5.38	61
T	WILLAR		1.52	0.02	1.3–1.7	5.22	59
T	SCHMIT		1.52	0.08	1.4–1.6	5.50	5
T	BRYNJU		1.50	0.05	1.4–1.6	5.16	11
T	PECCAR		1.50	0.03	1.3–1.6	5.67	37
T	GUYWIL		1.48	0.08	1.4–1.6	5.65	5
*T	MEYERC		1.47	0.02	1.4–1.6	4.77	59
T	CONARD		1.47	0.03	1.4–1.6	3.94	21
*C	CUMBER		1.42	0.04	1.3–1.5	4.69	9
*C	CRANKS		1.38	0.02	1.2–1.4	3.79	20
H	SCHULZ		1.38	0.03	1.2–1.5	5.37	29
H	KLEINC		1.34	0.05	1.2–1.4	5.40	9
H	FRIESE		1.32	0.04	1.3–1.4	3.39	5
*C	LADDSQ		1.27	0.07	1.2–1.3	4.56	3
*C	MEYERC		1.26	0.02	1.2–1.4	5.36	55
C	REDDIC		1.23	0.03	1.1–1.3	5.00	15
C	HAILEX		1.19	0.02	1.1–1.3	3.59	17
C	INDIAN		1.18	0.04	1.1–1.2	3.79	5
C	ARREDO		1.12	0.07	1.0–1.2	7.47	5

TAXONOMY

The three Holocene species recognized in the genus *Blarina* currently consist of a total of 18 subspecies. *Blarina brevicauda* is represented by 13 nominal subspecies arranged herein as representatives of two semispecies, the *brevicauda* semispecies and the *talpoides* semispecies. *Blarina carolinensis* tentatively is represented by four subspecies, whereas *B. hylophaga* is represented by only two subspecies. Synonymies of these taxa precede a discussion of their taxonomic status.

The first citation in synonymies is to the original description. The second is to the first use of the name combination presently employed if it differs from the name originally proposed. Next is a chronological list of synonyms and name combinations applied to the taxon as presently recognized. Extinct taxa are identified (†). References to relevant taxonomic studies and Hall's (1981) "The mammals of North America" are provided when names employed in those publications differed from names presently recognized.

Blarina Gray, 1838

Blarina Gray, Proc. Zool. Soc. London for 1837, p. 124, 1838. Type species *Corsira* (*Blarina*) *talpoides* Gray [=*Sorex talpoides* Gapper], by original designation; elevated to generic rank by Lesson, Nouv. Tableau Mammif., p. 89, 1842 (see also Baird, Mammals, *in* Repts. Expl. Surv. . . . , 8(1):36, 1858).

Brachysorex Duvernoy, Mag. de Zool., 2nd ser., 4:37–41, 1842. Type species *Brachysorex brevicaudus* Duvernoy [=*Sorex brevicaudus* Say], by original designation.

Table 2.—*Continued.*

ID	Sample	Mean	2 SE	Range	CV	N
		Length of m2 (F = 69.57)				
*B	ROBINS	2.06	0.04	1.9–2.2	4.18	22
B	CLARKS	2.05	0.02	1.8–2.1	3.07	44
B	BACKCR	2.04	0.08	2.0–2.2	4.39	5
*B	BAKERB	2.03	0.02	2.0–2.1	2.39	18
B	CHIMNE	2.02	0.04	1.9–2.2	4.66	18
*B	CRANKS	2.02	0.03	1.9–2.1	2.65	19
B or T	TROUTC	1.98	0.08	1.9–2.1	4.23	5
T	ZOOCAV	1.97	0.07	1.7–2.1	5.89	10
*T	LADDSQ	1.96	0.08	1.9–2.1	4.56	5
T	STELZE	1.94	0.02	1.8–2.1	4.04	72
T	RIVERS	1.94	0.03	1.9–2.0	2.66	10
*T	CRANKS	1.94	0.03	1.8–2.0	3.56	26
T	SCHMIT	1.92	0.07	1.8–2.0	4.36	5
T	STRAIT	1.91	0.04	1.8–2.0	3.36	13
*T	CUMBER	1.90	0.06	1.7–2.0	4.87	8
*T	ROBINS	1.90	0.20	1.8–2.0	7.44	2
T	CONARD	1.90	0.03	1.8–2.0	3.33	21
T	BRYNJU	1.88	0.04	1.8–2.0	3.21	11
T	WILLAR	1.88	0.02	1.8–2.0	3.78	59
T	PECCAR	1.86	0.03	1.7–2.0	5.61	37
*T	BAKERB	1.85	0.02	1.7–2.0	4.59	61
*T	MEYERC	1.84	0.02	1.7–2.0	3.47	59
T	GUYWIL	1.82	0.04	1.8–1.9	2.46	5
*C	CUMBER	1.81	0.05	1.7–1.9	4.32	9
H	SCHULZ	1.81	0.02	1.7–1.9	3.29	29
H	KLEINC	1.77	0.05	1.7–1.9	4.00	9
*H	CRANKS	1.75	0.03	1.6–1.9	3.93	20
H	FRIESE	1.68	0.04	1.6–1.7	2.66	5
*C	MEYERC	1.67	0.02	1.6–1.8	3.72	55
*C	LADDSQ	1.67	0.07	1.6–1.7	3.46	3
C	REDDIC	1.53	0.04	1.4–1.7	7.26	15
C	INDIAN	1.52	0.04	1.5–1.6	2.94	5
C	HAILEX	1.48	0.02	1.4–1.5	2.96	17
C	ARREDO	1.44	0.08	1.3–1.5	6.21	5

Talposorex Pomel, Arch. Sci. Phy. Nat. (Geneva), 9:248, 1848.
Type species *Talposorex platyurus* Pomel [=*Sorex brevicaudus* Say], by original designation.

Anotus Wagner, Suppl. Schreber's Saugthiere, 5:550–551, 1855. Type species *Sorex (Anotus) carolinensis* Bachman, by original designation.

Blarina brevicauda (Say, 1823)
brevicauda semispecies

Blarina brevicauda brevicauda (Say, 1823)
  Sorex brevicaudus Say, *in* Long, Account of an exped. . . . to the Rocky Mts., 1:164, 1823.
  Blarina brevicauda: Baird, Mammals, *in* Repts. Expl. Surv. . . . , 8(1):42, 1858.
  Blarina costaricensis J. A. Allen, Bull. Amer. Mus. Nat. Hist., 3:205, 1891; regarded as synonym of *B. brevicauda* by Merriam, N. Amer. Fauna, 10:12–13, 1895.
  †*Blarina fossilis* Hibbard, Univ. Kansas Sci. Bull., 29 (pt. 2): 238, 1943; regarded as subspecies of *B. brevicauda* by Hib-

bard, Trans. Kansas Acad. Sci., 60:333, 1957; regarded as synonym of *B. b. brevicauda* by Graham and Semken, J. Mamm., 57:437, 1976.
Blarina brevicauda manitobensis Anderson, 1947
  Blarina brevicauda manitobensis Anderson, Bull. Natl. Mus. Canada, 102:23, 1947.

talpoides semispecies

Blarina brevicauda aloga Bangs, 1902
  Blarina brevicauda aloga Bangs, Proc. New England Zool. Club, 3:76, 1902.
Blarina brevicauda angusta Anderson, 1943
  Blarina brevicauda angusta Anderson, Ann. Rept. Provancher Soc. Nat. Hist., Quebec, p. 52, 1943.
Blarina brevicauda churchi Bole and Moulthrop, 1942
  Blarina brevicauda churchi Bole and Moulthrop, Sci. Publ., Cleveland Mus. Nat. Hist., 5:109, 1942.
Blarina brevicauda compacta Bangs, 1902

Table 2.—*Continued.*

ID	Sample	Mean	2 SE	Range	CV	N
Length of trigonid of m2 ($F = 53.82$)						
*B	ROBINS	1.39	0.02	1.3–1.5	3.78	22
*B	CRANKS	1.39	0.02	1.3–1.5	3.30	19
B	BACKCR	1.38	0.08	1.3–1.5	6.06	5
B	CLARKS	1.38	0.02	1.3–1.5	3.88	44
B	CHIMNE	1.37	0.03	1.3–1.5	4.35	18
*B	BAKERB	1.36	0.03	1.2–1.4	4.47	18
*T	ROBINS	1.35	0.10	1.3–1.4	5.24	2
*T	CRANKS	1.32	0.03	1.2–1.4	5.46	26
B or T	TROUTC	1.32	0.04	1.3–1.4	3.39	5
*T	LADDSQ	1.32	0.04	1.3–1.4	3.39	5
T	STELZE	1.31	0.02	1.2–1.5	5.84	72
T	ZOOCAV	1.30	0.07	1.1–1.5	8.11	10
*T	CUMBER	1.29	0.05	1.2–1.4	4.98	8
T	RIVERS	1.28	0.03	1.2–1.3	3.29	10
T	STRAIT	1.27	0.04	1.2–1.4	4.97	13
T	CONARD	1.24	0.03	1.2–1.4	4.81	21
*T	BAKERB	1.24	0.02	1.1–1.4	4.98	61
T	GUYWIL	1.24	0.05	1.2–1.3	4.42	5
T	SCHMIT	1.24	0.05	1.2–1.3	4.14	5
T	PECCAR	1.23	0.03	1.1–1.4	6.60	37
T	WILLAR	1.23	0.01	1.1–1.3	4.44	59
*C	CUMBER	1.22	0.03	1.2–1.3	3.61	9
T	BRYNJU	1.21	0.03	1.1–1.3	4.46	11
*T	MEYERC	1.20	0.01	1.1–1.3	2.67	59
H	SCHULZ	1.18	0.02	1.1–1.3	4.56	29
*H	CRANKS	1.17	0.02	1.1–1.2	4.02	20
H	KLEINC	1.16	0.04	1.1–1.2	4.56	9
*C	LADDSQ	1.13	0.07	1.1–1.2	5.09	3
*C	MEYERC	1.10	0.02	1.0–1.2	5.25	55
H	FRIESE	1.08	0.04	1.0–1.1	4.14	5
C	REDDIC	1.07	0.05	1.0–1.2	8.44	15
C	INDIAN	1.00	0.00	1.0	0.00	5
C	HAILEX	1.00	0.02	0.9–1.1	3.54	17
C	ARREDO	0.98	0.04	0.9–1.0	4.56	5

Blarina brevicauda compacta Bangs, Proc. New England Zool. Club, 3:77, 1902.

Blarina brevicauda hooperi Bole and Moulthrop, 1942

   Blarina brevicauda hooperi Bole and Moulthrop, Sci. Publ., Cleveland Mus. Nat. Hist., 5:110, 1942.

Blarina brevicauda kirtlandi Bole and Moulthrop, 1942

   Blarina brevicauda kirtlandi Bole and Moulthrop, Sci. Publ., Cleveland Mus. Nat. Hist., 5:99, 1942; tentatively regarded as distinct species by Graham and Semken, J. Mamm., 57:445, 1976.

†*Blarina brevicauda ozarkensis* Brown, 1908

   Blarina brevicauda ozarkensis Brown, Mem. Amer. Mus. Nat. Hist., 9:170, 1908.

   Blarina ozarkensis: Graham, M.S. thesis, Univ. Iowa, p. 12, 1972; Graham and Semken, J. Mamm., 57:434, 1976.

Blarina brevicauda pallida R. W. Smith, 1940

   Blarina brevicauda pallida R. W. Smith, Amer. Midland Nat., 24:223, 1942.

†*Blarina brevicauda simplicidens* Cope, 1899

Blarina simplicidens Cope, J. Acad. Nat. Sci. Philadelphia, 11:219, 1899.

   Blarina brevicauda simplicidens: Hibbard, Trans. Kansas Acad. Sci., 60:333, 1957.

Blarina brevicauda talpoides (Gapper, 1830)

   Sorex talpoides Gapper, Zool. J., 5:202, 1830; regarded as synonym of *B. brevicauda* by Merriam, N. Amer. Fauna, 10:12, 1895.

   Blarina brevicauda talpoides: Bangs, Proc. New England Zool. Club, 3:75, 1902.

   Sorex dekayi Bachman, J. Acad. Nat. Sci. Philadelphia, 7:362–402, 1837; nomenclatorial history of this name and "*Sorex dekayi* De Kay" were reviewed by Handley and Choate, Proc. Biol. Soc. Washington, 83:195–202, 1970.

   Galemys micrurus Pomel, Arch. Sci. Phy. Nat. (Geneva), 9:249, 1849; nomenclatorial history of this name was reviewed by Handley and Choate, Proc. Biol. Soc. Washington, 83:195–202, 1970.

   Blarina angusticeps Baird, Mammals, *in* Repts. Expl. Surv.

Table 2.—*Continued.*

ID	Sample	Mean	2 SE	Range	CV	N
		Length of m3 (F = 43.85)				
*B	CRANKS	1.67	0.03	1.6–1.8	3.90	19
B	CLARKS	1.67	0.02	1.5–1.9	4.89	44
*B	ROBINS	1.67	0.04	1.5–1.8	5.03	22
B	BACKCR	1.64	0.10	1.5–1.8	6.95	5
T	SCHMIT	1.64	0.05	1.6–1.7	3.34	5
*T	LADDSQ	1.64	0.08	1.6–1.8	5.45	5
B	CHIMNE	1.62	0.04	1.5–1.7	4.51	18
*B	BAKERB	1.62	0.02	1.5–1.8	4.86	18
B or T	TROUTC	1.60	0.06	1.5–1.7	4.42	5
T	CONARD	1.58	0.04	1.4–1.7	5.52	21
*T	CRANKS	1.58	0.04	1.4–1.7	6.46	26
T	ZOOCAV	1.58	0.05	1.4–1.7	4.99	10
T	WILLAR	1.58	0.02	1.4–1.8	5.18	59
T	BRYNJU	1.56	0.04	1.5–1.7	4.31	11
*C	CUMBER	1.56	0.05	1.5–1.7	4.67	9
T	STELZE	1.55	0.02	1.3–1.7	5.19	72
*T	MEYERC	1.55	0.02	1.4–1.7	4.39	59
T	PECCAR	1.53	0.03	1.4–1.7	6.87	37
*T	CUMBER	1.53	0.05	1.4–1.6	4.64	8
T	GUYWIL	1.52	0.04	1.5–1.6	2.94	5
*T	BAKERB	1.51	0.02	1.3–1.8	5.68	61
*T	ROBINS	1.50	0.00	1.5	0.00	2
T	STRAIT	1.50	0.06	1.4–1.7	7.20	13
T	RIVERS	1.48	0.04	1.4–1.6	4.27	10
H	SCHULZ	1.44	0.03	1.3–1.6	5.36	29
*H	CRANKS	1.40	0.03	1.3–1.5	5.44	20
H	KLEINC	1.38	0.04	1.3–1.5	4.84	9
*C	MEYERC	1.37	0.02	1.2–1.7	5.14	55
H	FRIESE	1.34	0.08	1.2–1.4	6.68	5
*C	LADDSQ	1.33	0.07	1.3–1.4	4.33	3
C	REDDIC	1.22	0.03	1.1–1.3	5.54	15
C	INDIAN	1.20	0.06	1.1–1.3	5.89	5
C	HAILEX	1.18	0.03	1.1–1.3	5.65	17
C	ARREDO	1.16	0.05	1.1–1.2	4.72	5

. . . , 8(1):34, 1858; based on "deformed skull" according to Merriam, N. Amer. Fauna, 10:10, 1895 (see also Bole and Moulthrop, Sci. Publ., Cleveland Mus. Nat. Hist., 5:111, 1942).

Blarina brevicauda telmalestes Merriam, 1895
  Blarina telmalestes Merriam, N. Amer. Fauna, 10:15, 1895; Hall, The mammals of North America, 1:57, 1981.
  Blarina brevicauda telmalestes: Handley, Mammals of the Dismal Swamp: a historical account, *in* The Great Dismal Swamp, p. 308, 1979.

Blarina carolinensis (Bachman, 1837)

Blarina carolinensis carolinensis (Bachman, 1837)
  Sorex carolinensis Bachman, J. Acad. Nat. Sci. Philadelphia, 7:366, 1837.
  [*Blarina carolinensis*] *carolinensis:* Genoways and Choate, Syst. Zool., 21:114, 1972.
  Blarina brevicauda carolinensis: Merriam, N. Amer. Fauna,

10:13, 1895; Hall, The mammals of North America, 1:54, 1981.

Blarina carolinensis minima Lowery, 1943
  Blarina brevicauda minima Lowery, Occas. Papers Mus. Zool., Louisiana State Univ., 13:218, 1943; Hall, The mammals of North America, 1:56, 1981.

Blarina carolinensis peninsulae Merriam, 1895
  Blarina carolinensis peninsulae Merriam, N. Amer. Fauna, 10:14, 1895; George et al., J. Mamm., 63:641, 1982.
  [*Blarina brevicauda*] *peninsulae:* Trouessart, Catalogus Mammalium . . . , fasc. 1, p. 188, 1897.
  Blarina brevicauda peninsulae: Hall, The mammals of North America, 1:56, 1981.

Blarina carolinensis shermani Hamilton, 1955
  Blarina brevicauda shermani Hamilton, Proc. Biol. Soc. Washington, 68:37, 1955; Hall, The mammals of North America, 1:56, 1981.
  B[*larina*]. *carolinensis shermani:* George et al., J. Mamm., 63:643, 1982.

Table 2.—*Continued.*

ID	Sample		Mean	2 SE	Range	CV	N

Length of trigonid of m3 (F = 31.54)

ID	Sample	Mean	2 SE	Range	CV	N
B	CHIMNE	1.16	0.04	1.0–1.3	7.40	18
*B	ROBINS	1.15	0.03	1.1–1.4	6.44	22
B	BACKCR	1.14	0.05	1.1–1.2	4.81	5
B	CLARKS	1.14	0.02	1.0–1.2	5.40	44
*B	CRANKS	1.11	0.03	1.0–1.2	5.12	19
*B	BAKERB	1.11	0.03	1.0–1.2	6.56	18
B or T	TROUTC	1.10	0.06	1.0–1.2	6.43	5
*T	ROBINS	1.10	0.00	1.1	0.00	2
*T	CUMBER	1.09	0.05	1.0–1.2	5.89	8
T	SCHMIT	1.08	0.04	1.0–1.1	4.14	5
*T	LADDSQ	1.08	0.04	1.0–1.1	4.14	5
T	STELZE	1.08	0.01	1.0–1.2	5.51	72
T	STRAIT	1.07	0.06	1.0–1.3	9.65	13
*C	CUMBER	1.06	0.04	1.0–1.1	4.99	9
T	CONARD	1.04	0.02	1.0–1.1	4.86	21
*T	CRANKS	1.04	0.02	0.9–1.1	5.54	26
T	GUYWIL	1.04	0.05	1.0–1.1	5.27	5
T	ZOOCAV	1.04	0.04	0.9–1.1	6.72	10
T	WILLAR	1.03	0.01	1.0–1.2	5.53	59
*T	BAKERB	1.03	0.01	0.9–1.2	5.00	61
H	SCHULZ	1.01	0.02	0.9–1.1	4.05	29
T	PECCAR	1.01	0.02	0.9–1.2	7.16	37
*T	MEYERC	1.00	0.01	0.9–1.1	4.52	59
T	RIVERS	1.00	0.00	1.0	0.00	10
T	BRYNJU	1.00	0.03	0.9–1.1	4.47	11
H	KLEINC	0.99	0.02	0.9–1.0	3.37	9
*C	LADDSQ	0.97	0.07	0.9–1.0	5.97	3
*H	CRANKS	0.97	0.02	0.9–1.0	5.07	20
H	FRIESE	0.94	0.05	0.9–1.0	5.83	5
*C	MEYERC	0.93	0.01	0.8–1.0	5.42	55
C	REDDIC	0.89	0.03	0.8–1.0	7.22	15
C	INDIAN	0.86	0.05	0.8–0.9	6.37	5
C	HAILEX	0.84	0.02	0.8–0.9	6.03	17
C	ARREDO	0.84	0.05	0.8–0.9	6.52	5

Blarina hylophaga Elliot, 1899

Blarina hylophaga hylophaga Elliot, 1899

Blarina brevicauda hulophaga [*sic*] Elliot, Field Columbian Mus., Zool. Ser., 1:287, 1899.

Blarina brevicauda hylophaga Elliot, Field Columbian Mus., Zool. Ser., 6:461, 1905 (correction of previous error in spelling or transliteration).

Blarina hylophaga hylophaga: George et al., Ann. Carnegie Mus., 50:504, 1981.

Blarina brevicauda carolinensis: Jones and Glass, Southwestern Nat., 5:138, 1960; Hall, The mammals of North America, 1:54–55, 1981.

[*Blarina carolinensis*] *carolinensis:* Genoways and Choate, Syst. Zool., 21:114, 1972.

Blarina carolinensis carolinensis: Schmidly and Brown, Southwestern Nat., 24:45, 1979.

Blarina hylophaga plumbea Davis, 1941

Blarina brevicauda plumbea Davis, J. Mamm., 22:317, 1941; Hall, The mammals of North America, 1:56, 1981.

Blarina hylophaga plumbea: George et al., Ann. Carnegie Mus., 50:510, 1981.

Blarina carolinensis plumbea: Schmidly and Brown, Southwestern Nat., 24:45, 1979.

Two additional extinct shrews, *Cryptotis adamsi* and *Paracryptotis gidleyi,* also are mentioned in the accounts that follow because they originally were referred to the genus *Blarina.* Synonymies of these taxa are given below.

†*Cryptotis adamsi* (Hibbard, 1953)

Blarina adamsi Hibbard, J. Paleont., 27:29, 1953.

Cryptotis adamsi: Repenning, U.S. Geol. Surv. Prof. Paper, 565:39, 1967; Choate, Univ. Kansas Publ. Mus. Nat. Hist., 19:289, 1970.

†*Paracryptotis gidleyi* (Gazin, 1933)

Blarina gidleyi Gazin, J. Mamm., 14:142, 1933; Repenning, U.S. Geol. Surv. Prof. Paper, 565:43, 1967.

Paracryptotis gidleyi: Choate, Univ. Kansas Publ., Mus. Nat. Hist., 19:294, 1970; Hibbard and Bjork, Contrib. Mus. Paleont., Univ. Michigan, 23:175, 1971.

Ongoing research on geographic variation in *Blarina brevicauda* almost certainly will reduce the number of subspecies recognized in that species. Justification for recognition of two semispecies within *B. brevicauda* follows.

Mayr (1969) defined semispecies as "populations that have acquired some, but not yet all, attributes of species rank; borderline cases between species and subspecies." Semispecies of a species are allopatric over most of their range, but may exist sympatrically where their ranges abut (Dobzhansky et al., 1977). They often hybridize in such areas, but gene flow between them may be limited (Grant, 1971). Eventual elimination of gene flow between semispecies would complete the process of speciation. We think the relationship between what are now eastern and western subspecies of *B. brevicauda* is explained better by their recognition as semispecies than by possible alternative explanations.

Holocene shrews assigned to the western subspecies *B. b. brevicauda* and *B. b. manitobensis* average appreciably larger than Holocene shrews assigned to more easterly subspecies although the extremes of measurements overlap. They do not differ morphologically except in size. Where the ranges of the larger and smaller phena approach, there is no indication of intergradation; where they occur sympatrically in southern Iowa and northern Missouri, hybridization is suggested by the presence within samples of both large and small shrews and shrews variously intermediate between these extremes (Moncrief et al., 1982). The larger and smaller phena have identical standard karyotypes, and in fact share a balanced polymorphism (Genoways et al., 1977), so standard karyotypes cannot be employed to assess the extent of hybridization. Preliminary analysis of unpublished electrophoretic data (J. C. Patton, personal communication) suggests that allelic polymorphisms are not evenly distributed between the larger and smaller phena, which supports the notion of at least a partial barrier to gene flow. Finally, the larger and smaller phena are recognizable in the fossil record as long ago as the early Pleistocene. The distributions of the phena have changed since that time in response to climatic fluctuations, but the mensural relationship between them has not.

We interpret these observations as follows. Two populations of *B. brevicauda* began to diverge genically and morphometrically soon after the species

evolved, but gene flow between them prevented speciation. All known speciation events in *Blarina* apparently have involved karyotypic rearrangements (George et al., 1982), and no such rearrangements have become fixed in the larger and smaller phena of *B. brevicauda.* Nevertheless, the distinctive populations have behaved essentially as separate species, except when sympatric, throughout the Quaternary Period. The mensural variability they exhibit in and near areas of sympatry, apparently as a result of hybridization, has been responsible for difficulty in identifying fossils of *Blarina* in certain local faunas, and led Graham and Semken (1976) to suggest the taxa *B. b. kirtlandi* and "*B. ozarkensis*" are distinct species. Our data show the phena clearly are not distinct species, but indicate that recognition as semispecies is warranted.

Another possible interpretation of this phenomenon, which we elected not to accept, is that the larger and smaller phena of *B. brevicauda* are separate "evolutionary species." Wiley (1978) defined an evolutionary species as "a single lineage of ancestor-descendant populations which maintains its identity from other such lineages and which has its own evolutionary tendencies and historical fate." It is evident that the *talpoides* and *brevicauda* semispecies have retained their identities since the early Pleistocene; in this sense, they would qualify as distinct evolutionary species. However, the lack of morphological and morphometric divergence of the phena over a long period of time and the identical standard karyotypes in the phena indicate evolutionary stasis (in the sense of Gould and Eldredge, 1977), with the phena continuing to behave like populations of the same species when they occur together. We judge that the *talpoides* and *brevicauda* semispecies do not express separate evolutionary tendencies and thus do not qualify as separate evolutionary species.

Ongoing research on the status of two nominal subspecies from Florida, herein referred to the species *Blarina carolinensis* on geographic grounds, might result in taxonomic readjustments. One subspecies, *B. c. peninsulae,* has a karyotype more like that of *B. brevicauda* than that of *B. carolinensis* (George et al., 1982), and might be referred to the former species or elevated to separate species rank when electrophoretic and morphometric analyses are completed. Another subspecies, *B. c. shermani,* is known from few specimens and until recently was thought to be extinct (J. N. Layne, personal communication). Morphometrically it resembles shrews of the *talpoides* semispecies of *B. brevicauda* as much as it does *B. carolinensis,* and it might be referred to the former species if additional morphometric, karyologic, or electrophoretic data become available for analysis. These taxonomic changes would entail reinterpretation of some of the data presented herein, but doubtfully would alter any major phylogenetic or paleobiogeographic conclusions.

The taxonomic status of subspecies of *Blarina hylophaga* has been reviewed (George et al., 1981), and no further taxonomic changes are expected. At the present time, the range of this species is narrowly sympatric with those of shrews of the *brevicauda* semispecies of *B. brevicauda* in southern Nebraska

and with shrews of both semispecies of *B. brevicauda* in their zone of sympatry in southern Iowa, northern Missouri, and possibly northeastern Kansas (Moncrief et al., 1982). The range of *B. hylophaga* approaches that of *B. carolinensis* in northern Louisiana (George et al., 1981). The present range of *B. carolinensis* apparently is sympatric with that of the *talpoides* semispecies of *B. brevicauda* on the piedmont in northern Georgia and on the coastal plain in Virginia and North Carolina (Tate et al., 1980). At the present time the three species recognized herein do not occur together at any one location, but they might have done so in the past during a more equable climatic regime.

PALEONTOLOGY

Blancan

Hagerman

The Hagerman local fauna (l.f.) was recovered from the Glenns Ferry Formation west of Hagerman in Twin Falls County, Idaho. Fossils were collected from approximately 300 sites in fluvial deposits along the Snake River. The Hagerman is considered one of the classic faunas of the late Pliocene of North America (Zakrzewski, 1969). A potassium-argon date of 3.48 ± 0.27 million years BP was obtained by Evernden et al. (1964). Although once considered to be younger than the Rexroad l.f. of southwestern Kansas, recent work suggests that the section at Hagerman represents a period of time equivalent to that beginning with the Fox Canyon and ending with the Rexroad (Zakrzewski, personal communication).

The Hagerman l.f. consists of the remains of molluscs, fishes, herptiles, birds, and mammals. Skinner et al. (1972) published the most recent list of mammals and compared the mammals from Hagerman with those of Rexroad and other Blancan faunas of Kansas, Texas, and Nebraska.

Evidence suggests that the local habitat was warmer and more humid and had more vegetation than at present. Remains of fishes, frogs, and aquatic birds indicate the presence of fresh water (Brodkorb, 1958; Zakrzewski, 1969; Chantell, 1970). Rodents represent marsh-meadow and valley slope communities; representatives of drier upland habitats are rare but increase toward the top of the section. Warmer winters are suggested by the presence of southern elements such as *Baiomys* (Zakrzewski, 1969). Bjork (1970) discussed the zoogeography and paleoecology of carnivores from the deposit. He described the area as "a broad flood plain with trees and grassland adjacent to the tributary streams of Lake Idaho."

Today the area is included in the Artemisian biotic province of Dice (1943) and is characterized by sagebrush steppe dominated by sagebrush (*Artemisia*) and wheatgrass (*Agropyron*). Mean annual precipitation is approximately 25 cm (NOAA, 1974).

Six species of insectivores have been identified from the Hagerman material, including a "*Blarina*"

(Gazin, 1933). Gazin (1933) named *Blarina gidleyi* on the basis of a fragmentary left ramus (USNM 12650 from the locality at T7S, R13E). The specimen was described as "closer in size and general proportions to species of *Blarina* than to any of the other North American shrews . . . . The lower molars . . . show differences which appear to be beyond the range of variation seen in both *Blarina brevicauda* and *Blarina telmalestes*. The trigonid portion of these teeth is elongate giving the crescentic crest a more obtuse appearance than in the living species and the posterior portion of this crest is not so nearly perpendicular to the inner margin of the tooth, being directed slightly forward externally. The talonid portion of the anterior molars is narrower transversely, and the crest meets the posterior wall of the trigonid at a more inward position somewhat as in *Sorex.* Also, the postero-external wall of the talonid crown in m1 and m2 does not project bucally beyond the cingulum so noticeably as in *Blarina brevicauda* . . . . The reduction of the heel of the last lower molar which is often so marked in *Cryptotis* . . . distinguishes *Blarina simplicidens* Cope . . . from the living species of *Blarina* and from *Blarina gidleyi* . . . . It is possible that were more complete remains known the fossil form would be found to represent an undescribed genus, presumably related closely to *Blarina*" (Gazin, 1933). Measurements were given to compare the specimen of *B. gidleyi* with one of *B. brevicauda* (USNM 83073).

Hibbard (1957) identified additional specimens from the Hagerman l.f. as *B. gidleyi.* He noted that this species was distinguished from *B. simplicidens*

Table 3.—*Univariate statistics for specimens of* Paracryptotis gidleyi *from the Hagerman l.f., Twin Falls County, Idaho, for measurements 3–10 of the upper teeth and 19–24 of the lower teeth.*

Variable	N	Mean	2 SE	Range	CV
3	6	2.42	0.10	2.2–2.5	4.84
4	6	1.65	0.10	1.5–1.8	7.42
5	6	1.27	0.08	1.1–1.4	8.15
6	6	2.25	0.04	2.2–2.3	2.43
7	6	2.13	0.04	2.1–2.2	2.42
8	6	2.03	0.04	2.0–2.1	2.54
9	6	2.17	0.07	2.1–2.3	3.77
10	6	2.27	0.07	2.2–2.4	3.60
19	13	2.12	0.04	2.0–2.2	3.79
20	13	1.48	0.03	1.4–1.6	4.06
21	13	1.77	0.05	1.6–2.0	5.36
22	13	1.23	0.04	1.1–1.3	5.12
23	13	1.34	0.04	1.2–1.4	4.86
24	13	0.93	0.04	0.8–1.0	6.77

by its larger size and more developed cingula on the lower molars and p4. He also noted certain similarities between *B. gidleyi* and *Cryptotis kansensis* from the early Pleistocene of Meade County, Kansas. Repenning (1967) examined a well preserved topotype (UMMP 3304) and listed characteristics of the specimen which he considered "clearly primitive." He suggested the possibility that *gidleyi* was intermediate between extant *Blarina* and *Adeloblarina* (described from late Miocene deposits in Malheur County, Oregon) and that *gidleyi* should not be assigned to *Blarina*.

Hibbard and Bjork (1971) reassigned the holotype of *B. gidleyi,* plus additional material from Hagerman, to the genus *Paracryptotis. Paracryptotis gidleyi* shares many of the characteristics of *P. rex,* as described by Hibbard (1950), but differs in the posterior emargination of P4 and M1 and the more anterior position of the lower articular condyle. *Paracryptotis* was described by Repenning (1967) as less advanced than *Blarina* in certain dental features and in mandibular articulation. Probably *P. gidleyi* represents a lineage that evolved near the origin of the Tribe Blarinini (Choate, 1970).

We examined eight maxillaries (UMMP 49899, 53270–71, 53346, 55058, and 59935) and 17 dentaries (UMMP 33904, 34440, 45258, 49900, 49988, 50163–66, 55059; USNM 12650, 21385, and 21386) from Hagerman. Univariate statistics for the most complete specimens are shown in Table 3.

Dixon

As presently defined, the genus *Blarina* is first known from the late Blancan Dixon and White Rock faunas of Kansas. The Dixon l.f. was recovered from two sites on the Dixon farm southwest of Kingman, Kingman County (T29S, R8W). This fauna is one of several from the Meade formation of Kansas (Hibbard, 1956).

Molluscs, fishes, herptiles, birds, and small mammals compose the Dixon l.f. The mammalian families present are the Soricidae, Sciuridae, Geomyidae, Castoridae, Cricetidae, and Mustelidae. All but six of the 14 mammalian species represented are extinct (Hibbard, 1956, 1970).

The Dixon is the earliest cool fauna recognized in Kansas (Skinner et al., 1972). Deposition occurred during a climatic period of more equable temperature and more effective precipitation (Hibbard, 1958, 1970; Brattstrom, 1967; Graham, 1976). The insectivores and microtines indicate marshy habitat with trees and shrubs (Hibbard, 1956). Kingman County now is included in the Illinoian biotic province described by Dice (1943).

The age of the fauna is pre-Nebraskan (Taylor, 1966; Kurten and Anderson, 1980) or Nebraskan (Hibbard, 1956, 1958, 1970; Skinner et al., 1972). It is the only fauna of its age known from the Great Plains (Hibbard, 1970).

Part of a right maxillary with M1 and M2 (UMMP 31966) and an edentulous left dentary (UMMP 31967) from Dixon were referred to *Blarina* sp. by Hibbard (1956). Measurements 7 through 14 (2.1, 2.0, 2.2, 2.3, 1.6, 1.5, 2.3, and 2.1, respectively) of the M1 and M2 were taken for the present study. These measurements were compared with those of reference samples of Holocene specimens (Table 1) and with means derived from the Holocene measurements. Measurements of the fossil are most similar to those of modern *B. brevicauda* (*talpoides* semispecies), and the specimen from Dixon tentatively is assigned to that species.

Fox Canyon

The Fox Canyon local fauna (University of Michigan locality UM-K1-47) is located on the XI Ranch, sec. 35, T34S, R30W, in southwestern Meade County, Kansas. It was one of several fossil sites from the Rexroad Formation of Meade County investigated by Claude Hibbard. Initially Fox Canyon was considered a faunule of the Rexroad fauna; later it was recognized as a separate fauna (Hibbard, 1950, 1967; Zakrzewski, 1967). The Fox Canyon l.f. pertains to the Blancan land mammal age. Fox Canyon is considered older than the Rexroad, Benson, and Hagerman faunas (Hibbard, 1967; Zakrzewski, 1967, 1969). It was one of seven local faunas from Kansas from which sediments were sampled for magnetic polarity by Lindsay et al. (1975). The sample showed reversed magnetism and so probably was deposited during the late Gilbert epoch, approximately 3.5 million years ago. Fox Canyon was considered the oldest of the faunas (Lindsay et al., 1975). Dalquest (1978) suggested that the Beck Ranch l.f. from Scurry County, Texas, dates from between the periods represented by Fox Canyon and Rexroad. The Fox Canyon and Layer Cake (California) local faunas have been correlated (Kurtén and Anderson, 1980).

The fauna of Fox Canyon consists of remains of all vertebrate classes taken from fluvial deposits along the sides of the canyon. Hibbard (1950) provided the initial faunal list. Dalquest (1978) included an unpublished list (compiled by Hibbard) of 34 mammals (19 of which are types) from Fox Canyon. Six insectivores, a bat, a rabbit, 18 rodents, seven carnivores, and one deer were identified by Hibbard. Approximately 24 extinct species are present (Kurtén and Anderson, 1980). Dalquest (1978) compared the mammalian faunas from the Fox Canyon, Beck Ranch, Rexroad, Blanco, and Benson local faunas.

Meade County lies within the High Plains subprovince of the Great Plains. The High Plains are characterized as relatively flat and treeless, with elevation ranging from 610 to 1,220 m above sea level (Self, 1978). The natural vegetation of southwestern Meade County consists of mixed (bluestem-grama) and short-grass (grama-buffalograss) prairie except for floodplain and sand-sage vegetation along the Cimarron River (Küchler, 1974). The climate is semiarid, with the annual water loss by evaporation exceeding annual precipitation. Mean annual precipitation is approximately 51 to 56 cm (Self, 1978).

Notiosorex jacksoni was one of five insectivores described by Hibbard (1950, 1953*a*) as new species. Some of the specimens of *N. jacksoni* later were reassigned to another new species, *Blarina adamsi.* The type of *B. adamsi* is a maxillary bearing P2 through M2 (UMMP 27267). Paratypes include a palate (UMMP 27268), a right maxillary (UMMP 27269), the anterior portion of a palate (UMMP 28403), and a second right maxillary (UMMP 27270); each specimen bears partial dentition (Hibbard, 1953*a*). Mandibles identified as *B. adamsi* also were recovered from Keefe Canyon (UMMP 25777) (Hibbard, 1953*a*). A minimum of 30 rami thus was identified.

Blarina adamsi was described by Hibbard (1953*a*) as smaller than what he termed *B. b. carolinensis* [=*B. hylophaga*], approximately the size of *N. jacksoni.* The holotype has a P2 that is "anteroposteriorly flattened and does not possess a small lingual cusp as in Recent forms. Reduced P3 is just visible from labial side of maxillary. This tooth is not as small as P3 of the Recent forms of *B. b. carolinensis* and *B. b. hulophaga* [sic] Elliot. The posterior edge of P4, M1, and M2 not excavated." Length of P2–M2 was 4.5 mm. In UMMP 27268 the location of large anterior palatine foramina in a depression not observed in extant *Blarina* was noted. The mandibles were separated from those of *N. jacksoni* on the basis of the shape of the mental foramen and the articular surfaces of the condyle. Length of m1–m3 of 41 mandibles averaged 3.91 mm; length of c–m3 of eight specimens averaged 5.14 mm (Hibbard, 1953*a*).

Topotypes (UMMP 24352, 27310, and 28411), all mandibles, from the Fox Canyon l.f. were examined by Repenning (1967), who reassigned the species *adamsi* to the genus *Cryptotis.* Although the dental formula of those specimens is the same as

that of extant *Blarina* (and therefore unlike that of *Cryptotis*), mandibular structure and lower and upper teeth were noted as identical to that of living *Cryptotis.* Repenning listed the differences between the fossils and *Blarina* as "(1) greater anteroposterior shortening of the talonid of m1, (2) more posterior placement of the metaconid of m1, relative to the position of the protoconid, (3) greater reduction of the heel of m3, (4) retention of a primitive blarinine mandibular articulation and associated jaw structure, and (5) a rectangular M2." Another mandible (USGS M6511) from the Christmas Valley l.f. (Hemphillian) of Lake County, Oregon, also was identified as *C. adamsi* (Repenning, 1967). Choate (1970) examined the material from Fox Canyon and supported the assignment to *Cryptotis.* He suggested that, during the mid- and late Pliocene, lineages of *Cryptotis* and *Blarina* were at a similar grade of evolution. We agree with assignment of *adamsi* to the genus *Cryptotis.*

Twenty-five specimens (UMMP 24352, 27268–74, 27305–12, 28403–07, and 28410) of *Cryptotis adamsi* were examined for the present study. Univariate statistics for the most complete specimens examined (nine maxillaries and 13 dentaries) are shown in Table 4.

White Rock

The White Rock l.f. was recovered from 12 sites in the Belleville formation of Republic County, Kansas. The primary collecting sites are located along the White Rock Canal southwest of Republic. Deposition occurred in the paleovalley of the ancestral Republican River. The fauna was described by Eshelman (1975).

Table 4.—*Univariate statistics for specimens of* Cryptotis adamsi *from the Fox Canyon l.f., Meade County, Kansas, for measurements 3–10 of the upper teeth and 19–24 of the lower teeth.*

Variable	N	Mean	2 SE	Range	CV
3	9	1.70	0.05	1.6–1.8	4.16
4	9	1.28	0.06	1.2–1.4	6.52
5	9	0.90	0.03	0.8–1.0	5.56
6	9	1.66	0.06	1.5–1.8	5.33
7	9	1.62	0.04	1.5–1.7	4.11
8	9	1.53	0.03	1.5–1.6	3.26
9	8	1.53	0.03	1.5–1.6	3.04
10	9	1.74	0.05	1.6–1.8	4.17
19	13	1.55	0.05	1.4–1.7	5.67
20	13	1.10	0.05	1.0–1.2	7.42
21	13	1.42	0.04	1.3–1.5	4.87
22	13	1.02	0.02	1.0–1.1	4.29
23	13	1.24	0.03	1.2–1.3	4.09
24	13	0.89	0.05	0.6–1.0	10.69

The fauna consists of molluscs, crustaceans, fishes, herptiles, birds, and mammals. Two species of reptiles are extinct. Forty-three species of mammals are represented, of which 31 are extinct. Small mammals (shrews and mice) predominate in the mammalian fauna (Eshelman, 1975). Of the extant species, all have occurred in the area in historic times (Hall, 1981).

Taxa of the White Rock l.f. are derived from several communities. Permanent water, stream- or river bank, lowland meadow-valley, valley slope, and upland prairie habitats are indicated. The taxa suggest deposition in more equable climate (Eshelman, 1975). Presently, Republic County is in the Smoky Hills subprovince of the Great Plains. It is included in the Illinoian biotic province (Dice, 1943). Natural vegetation is bluestem-grama prairie, and mean annual precipitation is 71 cm (Eshelman, 1975; Self, 1978).

The species composition and paleoecology of the White Rock l.f. are similar to those of the Dixon l.f. of Kingman County. The evolutionary grade of the fossils from White Rock suggests an age younger than that of Sand Draw and older than Borchers (Eshelman, 1975). Eshelman (1975) suggested a pre-Nebraskan age. He thought that the fauna "accumulated under the influence of alpine glaciation in the mountains to the west and the early phases of climatic deterioration, which ultimately resulted in the Nebraskan continental glaciation."

The genus *Blarina* is represented by one specimen (UMMP V60594) from locality UM-K1-66 (NW ¼, SE ¼, NE ¼ sec. 3, T2S, R5W). The specimen is a fragment of right dentary with m1. Eshelman (1975) referred the dentary to the species *Blarina* aff. *carolinensis* based on similarities to Holocene specimens from southeastern Kansas [=*B. hylophaga*]. He noted, however, a stubbier ascending ramus and larger posterointernal fossa. For the present study, measurements 19, 20, and 25 (2.3, 1.6, and 4.0) were taken and were compared with means derived from the reference samples of Holocene specimens. The measurements of the molar from White Rock are most similar to the means of samples of modern *B. brevicauda* (*talpoides* semispecies). The fossil tentatively is assigned to that species.

IRVINGTONIAN

Conard Fissure

Fossils of the genus *Blarina* were recovered from two sites in Newton County, Arkansas. Conard Fissure, located approximately 6 km west of Willcockson (SE ¼, NW ¼ sec. 34, T17N, R21W), is the older of the two faunas.

Birds, herptiles, and mammals are present in the Conard Fissure l.f. The first study of the mammalian fauna was that by Brown (1908). He identified 51 species from the site, including 24 extinct species and subspecies. Later revisions of Brown's work revealed that some of his new names were assigned incorrectly and approximately 44 species are now recognized from Conard (Kurtén and Anderson, 1980). Gidley and Gazin (1938) compared the Conard l.f. with those of Cumberland Cave and Port Kennedy. Oesch (1967) and Saunders (1977) compared the fauna with others from the Ozark Plateau. Graham (1972) reviewed *Blarina* and other small mammals from Conard.

Brown (1908) regarded the fauna as late Wisconsinan, but Conard Fissure now is recognized as an Irvingtonian fauna. Hibbard (1958) considered it Illinoian, but Graham (1972) suggested that the fauna is Kansan.

The biostratigraphy of the site was described by Brown (1908) and Graham (1972). The latter considered the fauna one biostratigraphic unit and correlative with Cudahy. Some remains represent inhabitants of the cave. Deposition of other remains probably was due to activities of owls and other predators (Brown, 1908). The fauna suggests more equable climate at the time of deposition; the local topography would have permitted a variety of mesic and xeric habitats in the immediate area (Graham, 1972).

Newton County is part of the Ozark Mountains region of northwestern Arkansas. Dice (1943) included the area in the Carolinian biotic province. The natural vegetation of the Ozarks consists mainly of oak-hickory forest, although red cedar glades, stands of short-leaf pine, beech-maple forest, and upland prairies also are present (Sealander, 1979). Mean annual precipitation in Newton County ranges from 112 to 142 cm (NOAA, 1974).

Brown (1908) assigned the *Blarina* of Conard Fissure to a new subspecies, *B. b. ozarkensis*. Approximately 550 specimens were referred to this taxon. Diagnostic characters listed by Brown were anteroposterior compression of the fourth incisor, reduction of the last unicuspid, reduction of the last lower molar, and absent or reduced angle on the lower border of the mandible. The interorbital region and the posterior palatine foramina were described as wider, and the heel of the last lower molar smaller, than in living *B. brevicauda*. The subspecies differed similarly from "*B. simplicidens*" of Port Kennedy Cave.

Graham (1972) compared material from Conard Fissure (paratypes and additional specimens) with fossils from Cave Without a Name, Jinglebob, Mt. Scott, and Cumberland, Meyer, Peccary, and Willard caves. Fossils also were compared with Holocene specimens from eight states. Based on these comparisons, he elevated *ozarkensis* to specific rank. Diagnostic characters listed by Graham are "proportionally larger and more bulbous cingula of I4, P2, and lower molars and unicuspids, reduced P3, reduced talonid of m3, and the mandibular condyle shape." Size of *B. ozarkensis* was noted as intermediate between that of *B. b. brevicauda* and *B. b. kirtlandi*. Graham (1972) and Graham and Semken (1976) observed extensive variability in specimens from Conard Fissure and mensural overlap with Holocene samples, but the differences between specimens from Conard Fissure and Holocene specimens were thought sufficient to merit specific rank.

For the present study, 20 maxillaries and 31 dentaries (KU 58676–77; SUI 35672, 35674–79, 35681, 35684–85, 35689, and 35691–92) were examined.

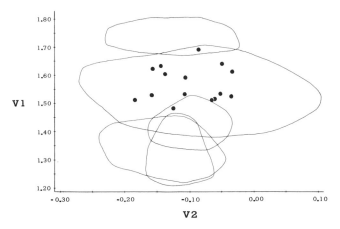

Fig. 5.—Canonical analysis of measurements 3–10 of upper teeth of specimens of *Blarina* from Conard Fissure. Ellipses enclose variation in samples of Holocene taxa of *Blarina* (see Fig. 3).

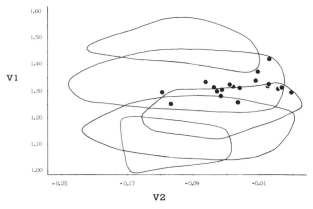

Fig. 6.—Canonical analysis of measurements 19–24 of lower teeth of specimens of *Blarina* from Conard Fissure. Ellipses enclose variation in samples of Holocene taxa of *Blarina* (see Fig. 4).

The sample appeared homogeneous. We observed extensive variability, but the degree of variability seemed comparable to that seen in other Pleistocene faunas. No morphological differences were observed that would distinguish these fossils from modern *B. brevicauda.*

Morphometric features of *ozarkensis* are similar to those observed in Holocene specimens from Iowa and Missouri (Moncrief et al., 1982). Those specimens exhibit intergradation between the nominal subspecies *B. b. brevicauda* and *B. b. kirtlandi,* and it is likely that *ozarkensis* also represents intergradation between larger shrews of the *brevicauda* semispecies and smaller shrews of the *talpoides* semispecies. This would explain the observation by Graham and Semken (1976) that "The size of *Blarina ozarkensis* is intermediate between that of *brevicauda* and *kirtlandi.*" Pending completion of a systematic review of the subspecies of *B. brevicauda,* *ozarkensis* is retained herein as a separate subspecies within the *talpoides* semispecies of *B. brevicauda.*

Results of analyses of 15 maxillaries and 21 dentaries (of which two observations were hidden) are shown in Figs. 5 and 6. Canonical analyses and comparisons with measurements of Holocene samples (Table 1) indicate that fossils from Conard Fissure are most similar to Holocene *B. brevicauda (talpoides* semispecies).

Coleman IIA

Some of the most abundant assemblages of Pleistocene animals have been recovered from Florida. Fossil remains of the genus *Blarina* have been reported from a dozen principal Pleistocene

sites in that state (Appendix 1). Two of these sites are considered Irvingtonian.

Chronological correlation of the Pleistocene faunas of Florida is subject to several major difficulties. Many of the deposits are located in caves and sinkholes formed in limestone, which is soft and soluble. As Auffenberg (1963) pointed out, "Each sinkhole is a unit in itself, containing a fauna which may or may not be related to that in the next sinkhole." Other deposits, in coastal stream beds and marshes, are subject to erosion. Fossils often are discovered through operations such as mining, which usually disturb the deposit, and early field records often were inaccurate (Simpson, 1929b). Many of the fossil assemblages appear to be heterochronic. Simpson (1929b) recognized only four faunas that he considered useful in correlation. To this list, Bader (1957) added the localities at Arredondo and Reddick. Recent advances in stratigraphy, and the discovery of early and middle Pleistocene sites in Florida, have helped clarify some of the difficulties in correlation (Webb, 1974).

The Coleman IIA fauna of central Florida was recovered from a quarry near Coleman in Sumter County (SE ¼, NW ¼ sec. 7, T20S, R23E). The site, which later was destroyed by mining, was a filled sinkhole. Mammalian remains probably represent bats and the prey of owls that roosted at the site and larger species that fell into the open sinkhole (Martin, 1974). Martin (1974) listed thirty-eight species of mammals from Coleman IIA. The list includes 10 extinct species and four extant species no longer present in Florida (*Conepatus* sp., *Felis onca, Lepus alleni,* and *Erethizon dorsatum*).

Martin (1974) and Webb (1974) considered the Coleman IIA fauna to date from the late Illinoian or early Kansan. Deposition probably occurred in less than 1,000 years, but the fossils might not represent sympatric species (Martin, 1974). The presence of *Cryptotis parva, Peromyscus floridanus, Lepus alleni,* and *Platygonus* sp. suggest that a more xeric savanna habitat was present during the period of deposition, although other species (such as *Neofiber alleni, Sciurus carolinensis,* and *Mylohyus* sp.) imply mesic habitat (Martin, 1974; Webb, 1974; Frazier, 1977). Martin (1974) suggested that the Coleman fauna might represent a period of transition from xeric savanna to mesic forest.

Presently, the climate of Florida is humid and subtropical.

Mean annual precipitation north of the everglades ranges from 132 to 142 cm (NOAA, 1974). Dice (1943) included the state in the Austroriparian biotic province. The climax vegetation north of the Everglades is a hardwood forest of pine, oak, magnolia, and sweetgum (Bailey, 1978).

One mandible with i and m1–2 (UF 11626), referred to the species *B. brevicauda,* was reported by Martin (1974). Measurements 19 through 22 and 25 (1.7, 1.2, 1.6, 1.0, and 3.1) of this specimen were taken for the present study. These measurements fall within the range of those reported for modern *B. c. peninsulae* (sample 2 in Table 1), and superficially the fossil resembles a Holocene specimen of that taxon (UF 2532 from Alachua County) with which it was compared. We tentatively assign the specimen to *B. carolinensis* pending completion of a study of the systematic relationships of *peninsulae.*

Inglis IA

Inglis IA is a fissure deposit located on the western coast of Florida in Citrus County. The fauna dates from a dry glacial interval in the very late Blancan or early Irvingtonian (Webb, 1974).

Webb (1974) listed 37 mammalian species from this site. Seventeen extinct species were present, including *Megalonyx jeffersoni, Eremotherium rusconii, Glossotherium chapadmalense, Smilodon gracilis, Chasmaporthetes ossifragus,* and *Equus* sp. Inglis IA is the earliest record of *Hydrochoerus holmesi;* the occurrence of *Hydrochoerus* and that of other members of the fauna suggest the presence of coastal savanna and aquatic habitats and warm winters (Webb, 1974; Kurtén and Anderson, 1980).

Webb (1974) noted the presence of *Blarina brevicauda* in the Inglis IA fauna. For the present study, five dentaries (UF, uncatalogued material) were examined. These specimens are similar in shape and size to modern *B. carolinensis* of Florida. All molars are present in three specimens, and measurements 19 through 26 were taken (1.8, 1.3, 1.6, 1.1, 1.2, 0.8, 3.2, 2.0; 1.8, 1.2, 1.6, 1.0, 1.2, 0.9, 3.2, 2.1; 1.7, 1.2, 1.6, 1.0, 1.2, 0.9, 3.2, 2.1). On the two remaining fossils, measurements 19 through 22 and 26 (1.8, 1.3, 1.6, 1.1, 2.0) and 19, 20, and 25 (1.7, 1.2, 3.2) were taken. The measurements fall within the range of variation of groups of Holocene *B. carolinensis* (Table 1) and are most similar to *B. c. peninsulae* (sample 2). We think the fossils from Inglis pertain to the taxon *peninsulae.*

Cudahy

Two specimens of *Blarina* were identified from two sites from which the Cudahy l.f. has been recovered. The sites are UM-K3-71 (sec. 7, T32S, R28W) and the Sunbrite ash mine (KU loc. 17, sec. 26, T32S, R28W). Both sites are located in Meade County in southwestern Kansas.

The Cudahy l.f. consists of molluscs, fishes, herptiles, birds, and mammals. Thirty-seven species of mammals (largely insectivores and rodents) have been identified, of which 18 are extinct (Paulson, 1961; Hibbard, 1944, 1970).

Remains from Cudahy were deposited immediately below the Pearlette type O ash, suggesting a late Kansan age. The Cudahy was one of five faunas listed by Hibbard (1970) as evidence that local climate was cooler during the Kansan than during the Nebraskan.

Invertebrates and vertebrates of this fauna, particularly the large number of shrews and microtines, indicate the presence of a more boreal fauna at the time of deposition (Leonard, 1950; Blair, 1958; Brattstrom, 1967; Graham, 1976; Hibbard, 1944, 1970). Accumulation occurred in marshy depressions or slow-moving streams. Pollen associated with the ash is poorly preserved (Kapp, 1970).

Presently, Meade County is included in the High Plains physiographic province. Natural vegetation of the area near the Cudahy sites consists of bluestem-grama prairie, with floodplain forest and savanna (*Populus-Salix*) along Crooked Creek. Climate is semiarid; annual water loss through evaporation exceeds annual precipitation. Mean annual precipitation is 41–51 cm (Küchler, 1974; NOAA, 1974; Self, 1978).

A dentary with m1–3 (UMMP V61263) from UM-K3-71 and a broken dentary with m1 and m2 (UMMP 36802) pertain to the genus *Blarina.* Measurements 19 through 24 (2.2, 1.4, 1.9, 1.2, 1.6, and 1.0) and 19 through 22 (2.1, 1.4, 2.0, and 1.3), respectively, were taken and compared with those of reference samples of Holocene specimens (Table 1). The measurements of the fossils are most similar to those of modern specimens of *B. brevicauda* (*talpoides* semispecies), and the two specimens from Cudahy are referred to that species.

Kanopolis

The Kanopolis l.f. was recovered from a gravel pit north of Kanopolis, Ellsworth County, Kansas (SW ¼, NE ¼ sec. 25, T15S, R8W). Kanopolis is the largest Pleistocene fish fauna known from the Great Plains (Neff, 1975). In addition to 15 species of fishes, pelecypods, ostracods, 21 species of herptiles, and 34 species of mammals have been identified (Holman, 1972; Hibbard et al., 1978). Fourteen mammalian species are extinct; all other vertebrates are extant.

The Kanopolis l.f. is considered Yarmouthian. A warm, moist climate is suggested by the fauna. Permanent stream, marsh, and forest communities were present, and savannah and grassland were nearby. Sympatry of extant vertebrates occurs east of the site (Holman, 1972; Neff, 1975; Frazier, 1977; Hibbard et al., 1978).

Ellsworth County presently is part of the Smoky Hills subprovince of the Great Plains. The natural vegetation of the county is bluestem-grama prairie and transitional bluestem-grama and bluestem prairie. Mean annual precipitation is 61 to 71 cm (Küchler, 1974; NOAA, 1974; Self, 1978).

Hibbard et al. (1978) referred four specimens to *Blarina* sp. These specimens are a left dentary with complete dentition (UMMP V61000); two dentaries with i and m1–3 (V60415, V60612); and an isolated M1 (V60412). The possible co-occurrence of more than one species was suggested.

All specimens were examined for the present study. Measurements 7 through 10 (2.3, 2.1, 2.4, and 2.6) of the upper molar were taken and compared with those of Holocene specimens (Table 1). The measurements of the fossil are most similar to those of samples 5, 12, and 18 of *B. brevicauda*; the fossil might represent the subspecies *B. b. brevicauda*. Measurements 19 through 26 of the three remaining specimens (2.2, 1.6, 1.8, 1.2, 1.4, 1.0, 4.0, 2.5; 1.9, 1.4, 1.8, 1.2, 1.5, 1.0, 4.0, 2.4; 2.2, 1.5, 1.9, 1.3, 1.6, 1.0, 4.0, and 2.5) were taken. The means of these measurements are 2.1, 1.5, 1.8, 1.2, 1.5, 1.0, 4.0, and 2.5, respectively. Measurements of the specimens from Kanopolis are most similar to those of Holocene samples of the *talpoides* semispecies of *B. brevicauda* (Table 1). These analyses lend credence to the suggestion by Hibbard et al. (1978) that more than one taxon might be present in the Kanopolis l.f., but all specimens apparently pertain to the species *B. brevicauda*.

Kentuck

Fossils were found in outcrops along Kentuck Creek in McPherson County, Kansas (sec. 13, T18S, R3W). Accumulation occurred in a stream deposit that cut through an older deposit of Pearlette-like ash (Hibbard, 1952).

The Kentuck l.f. consists of fishes, herptiles, birds, and mammals. Birds and snakes were discussed by Galbreath (1955) and Brattstrom (1967). Mammalian faunal lists were published by Hibbard (1952) and Semken (1966). Some of the rodents were examined by Nelson and Semken (1970), Martin (1975), and Van der Meulen (1978). Twenty species of mammals were identified, of which eight are extinct (Hibbard, 1952).

Stream-border and grassland communities are indicated by the fauna. Hibbard (1952) originally designated this group of vertebrates as the Kentuck assemblage because the fossils were thought to represent different communities of Kansan and Yarmouthian age. Semken (1966) and Brattstrom (1967) concurred with this interpretation. Semken and others (Zakrzewski, 1975a) now suggest that the remains are a local fauna deposited during more equable conditions in the Kansan glacial. Van der Meulen (1978) considered the fauna Aftonian.

Most of McPherson County now is part of the Great Bend Prairie physiographic region (Self, 1978). Natural vegetation is mostly bluestem prairie; mean annual precipitation is 61 to 81 cm (Kuchler, 1974; NOAA, 1974).

The genus *Blarina* is represented in this fauna by a lower incisor (UMMP V51248). The species cannot be identified.

Rezabek

Fossils of the Rezabek l.f. were recovered from the Rezabek gravel pit in southwestern Lincoln County, Kansas (sec. 20, T13S,

R10W). The Rezabek l.f. may pertain to the late Irvingtonian or early Rancholabrean. Hibbard (1970) placed it among the early Illinoian faunas of the Great Plains. Hibbard and Dalquest (1973) considered the Rezabek Yarmouthian based on sympatry of *Neofiber* and *Ondatra*. Zakrzewski (1975a) suggested the Rezabek should be retained in the Illinoian, based on the equability model of Pleistocene climate (as discussed by Graham, 1972). The presence of microtines assignable to extant species (except *Neofiber*) suggests the Rezabek l.f. is Rancholabrean (Zakrzewski, personal communication).

Because of the abundant remains of fishes, the fauna was thought to have accumulated in a stream-laid deposit (Hibbard, 1943). Climate was considered temperate. Early Illinoian faunas of the Great Plains suggest a moister climate, with cooler summers, than is present in the region today; a shift toward slightly warmer winters apparently occurred during the late Illinoian (Hibbard, 1970).

Lincoln County is located in the Smoky Hills subprovince of the Great Plains. This subprovince is characterized by rolling hills underlain by Cretaceous rock (Self, 1978). Natural vegetation of the county consisted of bluestem-grama prairie, floodplain vegetation along the Saline River, and transitional vegetation between mixed (bluestem-grama) prairie and bluestem prairie. Climate is of the temperate continental type; normal annual precipitation is 66 to 71 cm (Self, 1978).

The Rezabek l.f. consists of molluscs, fishes, herptiles, birds, and mammals. Fossils of mammals represent the orders Insectivora, Lagomorpha, Rodentia, Perissodactyla, and Artiodactyla. Two nominal species, *Blarina fossilis* and *Neofiber leonardi*, were described and named in the original account of the fauna (Hibbard, 1943).

The genus *Blarina* is represented in the Rezabek l.f. by a fragmentary ramus (KU 6675). The specimen, which bears one molar, is the holotype of *Blarina fossilis*. It was distinguished from *B. simplicidens* Cope and *B. ozarkensis* Brown by its larger talonid on m3. It was differentiated from modern species on the basis of size (Hibbard, 1943). Later, Hibbard (1957) wrote that *B. fossilis*, *B. simplicidens*, and *B. brevicauda* could not be distinguished, and both fossil taxa were relegated to subspecific rank. Dalquest (1965) assigned three specimens from the Howard Ranch l.f. to *B. b. fossilis* because of their large size. Graham and Semken (1976) saw no difference between specimens from Howard Ranch and Rezabek and modern specimens of *B. brevicauda*. They placed *fossilis* in synonymy with *brevicauda*.

Measurements 23 through 25 (1.6, 1.2, and 3.7, respectively) of the specimen from Rezabek were compared with means of those measurements from Holocene samples (Table 1). Measurements 23 and 24 of the fossil were most similar to means of the sample of *B. b. brevicauda* from Nebraska (sample 12); they were also within the range of measurements for the sample of *brevicauda* from Iowa (sample 5). Measurement 25 seems somewhat small for any of the samples of *brevicauda*. Based on the analysis of the first two measurements and the earlier work by Graham and Semken (1976), the Rezabek specimen herein is considered a representative of the species *brevicauda*, and most closely resembles Holocene *B. b. brevicauda*.

Wathena

The Wathena l.f. was recovered south of Wathena (sec. 33, T3S, R22E), Doniphan County, Kansas. Remains were found in lake deposits below the Kansan Nickerson till. Pleistocene deposits in this area were described by Bayne (1969).

The fauna was described in an unpublished thesis (Einsohn, 1971). Molluscs, fishes, herptiles, birds, and mammals were present. Einsohn (1971) listed 11 mammalian species representing the families Soricidae, Sciuridae, Geomyidae, Cricetidae, and Zapodidae. Van der Meulen (1978) discussed *Microtus (Allophaiomys)* sp. of this fauna. Additional remains of unidentified rabbits and rodents are housed at the University of Michigan. The fauna is similar to that of Kentuck l.f. (Zakrzewski, 1975a).

The Wathena l.f. is considered Irvingtonian. Schultz et al. (1978) placed the fauna in the early Irvingtonian (but considered it more recent than the faunas of Dixon and White Rock), based on the presence of "*Allophaiomys*." Van der Meulen (1978) considered the Wathena Aftonian, but Einsohn (1971) thought that the Wathena was more recent than Cudahy.

Doniphan County was glaciated at least twice (Nebraskan and Kansan). Presently the county has a temperate continental climate and receives 81 to 91 cm of precipitation annually. The natural vegetation of the area is oak-hickory forest, with floodplain forest and savanna (*Populus-Salix*) along the Missouri River (Bayne, 1969; Küchler, 1974; NOAA, 1974; Self, 1978). Little can be determined regarding climate during the period of deposition of fossils (Einsohn, 1971).

Shrews are represented by a left dentary fragment with m2–3 (UMMP V60522) and by an isolated M1 (UMMP V61179). The shape and size of both specimens indicate that they pertain to the genus *Blarina*. Measurements 7 through 10 of the M1 (2.0, 1.9, 2.0, and 2.2) resemble those of Holocene *B. carolinensis*. Measurements 21 through 26 (1.5, 1.1, 1.3, 1.0, 3.3, and 2.1) of the lower molars and dentary also correspond to the means of Holocene samples of *B. carolinensis* (Table 1), although careful examination of the dentary revealed some similarity to Holocene specimens of *B. brevicauda* (*talpoides* semispecies). The fragmented condition of the fossils of the Wathena l.f., and the lack of material of comparable age in eastern Kansas, pose major problems in the analysis of the Wathena material. We tentatively assign both specimens to the species *B. carolinensis*.

Cumberland Cave

Cumberland Cave is a fissure located 6.4 km northwest of Cumberland, Allegany County, Maryland, at 39°41½'N and 78°47¼'W (Gidley and Gazin, 1938). The cave was long known to local residents (Nicholas, 1954). Excavation began after exposure of fossils by a railway cut in 1912. Originally there were two entrances. One opening ran 61 m and led into the cave 31 m below the ridgetop. This area was destroyed by quarrying. The second entrance was a sinkhole that led from the top of the ridge through several chambers to the main room of the cave, about 31 m below the ridgetop (Nicholas, 1954). Fossils generally were broken but showed no signs of being worn by water. They were found in unstratified cave clays and breccias (Gidley and Gazin, 1938). Gidley and Gazin suggested that the mode of accumulation was similar to that which occurred at Conard Fissure, as described by Brown (1908); that is, that materials were carried into the cave by predators and that accumulation occurred gradually over a long period of time.

Initial reports describing selected elements of the fauna were produced by Gidley and Gazin; faunal lists were published later (Gidley and Gazin, 1933, 1938). All remains were mammalian except for isolated snake vertebrae and a fragment of the humerus of a ruffed grouse (*Bonasa umbellus*). The latter was identical to the modern form that currently inhabits the eastern United States (Wetmore, 1927). A tooth initially was misidentified as that of a crocodylid (Richmond, 1963). Eight of the 41 genera and 28 of the 46 species identified from the cave fauna are extinct (Gidley and Gazin, 1938).

Gidley and Gazin (1933, 1938) recognized that the fauna of Cumberland Cave included species that presently are allopatric plus species still present in the area. A few boreal species (such as the wolverine, *Gulo luscus*) were present. Western species, including the thirteen-lined ground squirrel (*Spermophilus tridecemlineatus*), also were present. The fauna now is considered autochronic, with deposition having occurred under ecological conditions not known in any one area today (Guilday, 1971).

The cave is located at an altitude of 255 m in a limestone ridge in Wills Creek Valley (Nicholas, 1954). This area was included near the northern boundary of the Carolinian biotic province by Dice (1943), and within the Appalachian physiographic province (Abbe, 1900). Mean annual precipitation in this region is 91 to 101 cm (NOAA, 1974).

The age of the Cumberland Cave fauna is Irvingtonian. A minimum date of 250,000 BP was obtained from racemization of amino acids in a horse phalanx (Zakrzewski, 1975b). Guilday (1971) considered the fauna Illinoian. Van der Meulen (1978) suggested a mid-Irvingtonian age based on his study of the *Microtus* and *Pitymys* of Cumberland Cave. Martin (1974) saw some similarity with the (Irvingtonian) fauna from the Coleman IIA deposit of Florida; he also noted that the Cumberland material had not been reviewed for thirty years and considered additional speculation hazardous. Zakrzewski (1975b) pointed out that "If the Cumberland Cave local fauna had been located on the Great Plains, the association of *Ondatra annectens*, *Neofiber*, and *Atopomys*, and the lack of *Microtus pennsylvanicus* would suggest a pre-Illinoian age. Because of the geographic position of Cumberland and the lack of other pre-Wisconsin faunas in the area with which it can be compared, the determination of the exact age is moot."

Gidley and Gazin (1938) identified the short-tailed shrews from Cumberland Cave as *Blarina brevicauda*. The first specimens to be recovered consisted of a rostral fragment, two maxillary fragments, and

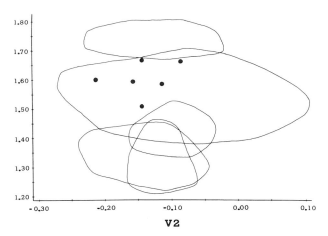

Fig. 7.—Canonical analysis of measurements 3–10 of upper teeth of specimens of *Blarina* from the Cumberland Cave l.f. Centroids enclose variation in samples of Holocene taxa of *Blarina* (see Fig. 3).

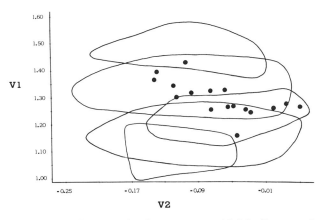

Fig. 8.—Canonical analysis of measurements 19–24 of lower teeth of specimens of *Blarina* from the Cumberland Cave l.f. Centroids enclose variation in samples of Holocene taxa of *Blarina* (see Fig. 4).

eight lower jaws. It was noted that these specimens were "somewhat larger than average" in Holocene specimens and resembled the Conard Fissure specimen described by Brown (1908). Graham and Semken (1976) believed that "three distinct, non-overlapping populations" were represented that could be equated to the modern *brevicauda, kirtlandi,* and *carolinensis.* The coexistence of the three phena was thought to represent a full or early late glacial climate.

We examined 40 maxillary and dentary fragments from Cumberland Cave (CM 8032, 20438, 20441–2, 20445–7, 20449, 20451–3, 20457, 20459, 20462–3, 24218–9, USNM 7779, 12460–2, 12464–7). Of the nine maxillaries, six were sufficiently complete for analysis. These plotted within the range of variation of Holocene specimens of the smaller subspecies (*talpoides* semispecies) of *B. brevicauda* (Fig. 7). Sixteen dentaries fell within the ranges of variation for the smaller subspecies of *B. brevicauda* and for *B. carolinensis* (Fig. 8). Univariate statistics (CUMBER) are shown in Table 2. Specimens that could not be studied by canonical analysis had measurements similar to those for the specimens shown in the two figures. The Cumberland Cave l.f. apparently contains remains of two species of *Blarina.* The largest specimens pertain to *B. brevicauda* (*talpoides* semispecies), whereas the smallest individuals pertain to the species *B. carolinensis.*

Angus

Fossils of the Angus l.f. were recovered from a quarry 2.4 km southwest of Angus, Nuckolls County, Nebraska (SW ¼, NE ¼

sec. 33, T4N, R6W). The fossils were found below the Illinoian Loveland Loess and the Yarmouthian paleosol and were considered late Yarmouthian in age (Schultz and Tanner, 1957; Schultz and Martin, 1970). In the chronology by Schultz et al. (1978), the Angus l.f. was considered younger than the local faunas of Port Kennedy, Cudahy, Conard Fissure, Java, and White Rock, from which specimens of *Blarina* have been reported.

Fossils recovered from the Angus l.f. include remains of fishes, amphibians, reptiles, birds, and mammals (Schultz and Tanner, 1957). Schultz and Martin (1970) published a preliminary list of mammalian species from the site. Remains of extinct species include those of an unidentified sloth, *Cynomys niobrarius, Castoroides* sp., *Ondatra nebracensis, Canis dirus nebrascensis, Mammuthus (Archidiskodon) imperator,* an unidentified mastodont, *Mylohyus browni* [=*nasutus*], *Platygonus* sp., *Camelops kansanus, Capromeryx furcifer,* ?*Stockoceros* sp., *Equus excelsus,* and *E. calobatus.* To this list Kurtén and Anderson (1980) added *Castor accessor* and *Mammuthus meridionalis,* and referred some of the canid material to *Canis armbrusteri.* Extant species reported from the site include *Sorex cinereus, Blarina brevicauda, Scalopus* sp., an unidentified bat, *Spermophilus tridecemlineatus, S. franklinii, S.* cf. *richardsoni, Geomys bursarius, Dipodomys* sp., *Perognathus* sp., *P. hispidus, Castor* sp., *Onychomys* sp., *Reithrodontomys* sp., *Peromyscus* sp., *Neotoma* sp., *Synaptomys* sp., *Clethrionomys gapperi, Microtus ochrogaster, M. pennsylvanicus, Zapus hudsonius, Lepus* sp., *Sylvilagus* sp., *Canis* cf. *latrans, Vulpes velox, Mustela frenata, Taxidea* cf. *taxus,* and *Odocoileus sheridanus* [=*virginianus*] (Schultz and Martin, 1970; Kurtén and Anderson, 1980). Schultz and Martin (1970) noted that most of the smaller mammals of the Angus l.f. have modern descendants in the area today, which suggested that climatic conditions during the period of deposition were no more severe than those of today.

Presently the area is included in the Illinoian biotic province (Dice, 1943). The region receives a mean annual precipitation of 60 to 71 cm (NOAA, 1974).

Two maxillary fragments and three mandibles (KU 33638, 33639, 33658, 33660, and 33664) of

Blarina were examined. Measurements 3 through 6 of two P4's (2.5, 2.0, 1.3, 2.4; 2.4, 1.8, 1.3, 2.2) and measurements 19 through 22 (2.3, 1.5, 2.1, 1.3), 23 through 25 (1.7, 1.1, 4.6), and 21 through 26 (1.9, 1.2, 1.5, 1.0, 4.1, 2.5) of lower molars and dentaries were taken. Measurements of these specimens were compared with means derived for those measurements in Holocene samples (Table 1). Measurements of the fossils were slightly smaller than the means of the modern *B. b. brevicauda* from Nebraska (sample 12), and they were comparable in size to the smaller *talpoides* semispecies (samples 4, 6, 11, 18, 13, 19, and 24). Fossils of *Blarina* from Angus were similar to those of the Cumberland Cave l.f., which also was considered an Irvingtonian fauna. The specimens pertain to the species *B. brevicauda*.

Bartek Brothers Quarry

The Bartek Brothers Fossil Quarry is located approximately 4 km west of Weston (NE ¼ sec. 12, T14N, R5E) in Saunders County, Nebraska. Fossils were taken from the Sappa Formation. The age of the deposit was considered post-Kansan by Schultz and Tanner (1957). Frankel (1963) thought the deposit probably was Yarmouthian. Molluscs and plant material from the site suggest the presence of a pond, probably surrounded by marshy habitat. "The occurrence of fish, reptile, and rodent remains indicates the presence of vertebrates from two distinct environments; one aquatic, the other, terrestrial. If the Microtinae are *Synaptomys*, a marshy environment would be indicated" (Frankel, 1963). The presence of four species of gastropods that no longer occur in the area might indicate a slightly cooler climate at the time of deposition (Frankel, 1963). The date of the type Sappa (1.2 million years BP) and the paleoecological interpretation by Frankel imply a pre-Kansan age for the fauna (Zakrzewski, personal communication).

The region now is included in the Illinoian biotic province by Dice (1943). Mean annual precipitation is approximately 61 to 71 cm (NOAA, 1974).

Seeds, molluscs, and vertebrates were recovered from the quarry. Vertebrates include fishes, snakes, rodents (Microtinae), ground sloth, mastodont, mammoth, horse, camel, llama, wapiti, and bison (Schultz and Tanner, 1957; Frankel, 1963). Preservation of vertebrate remains was poor.

Five specimens (USNM 33094–8) of *Blarina* from the Bartek Brothers l.f. were examined. All were fragments of dentaries with incomplete dentition; isolated teeth also were recovered from the site. Measurements 19 and 20 (2.2, 1.4) of an additional m1 are much smaller; whether they indicate the presence of a second taxon of *B. brevicauda* (representing the *talpoides* semispecies) is uncertain.

Berends

Fossils of the genus *Blarina* have been reported from two sites in Oklahoma. Both sites, in adjacent counties in western Oklahoma, are outside the present range of the genus. The region now is part of the subhumid shortgrass region of the Great Plains (Dice, 1943; Stephens, 1960). Modern climate is described by Stephens (1960) and NOAA (1974).

The Berends l.f. was recovered from deposits in the Berends Sand Draw on the Coy Berends Ranch in Beaver County (secs. 5 and 6, T5N, R28E). This locality is approximately 7.2 km north and 1.6 km west of Gate (Rinker and Hibbard, 1952; Taylor, 1954).

Mengel (1952), Smith (1954), Taylor (1954), Taylor and Hibbard (1955), and Brattstrom (1967) reported on the molluscs, fishes, herptiles, and birds of the Berends l.f. Sixteen species of mammals were reported by Rinker and Hibbard (1952), Starrett (1956), and Hibbard (1963). Six extinct mammals (*Paradipoides stovalli*, *Peromyscus berendsensis* n. sp., *Ondatra nebracensis*, *Equus* sp., *Mammuthus* cf. *M. columbi*, and *Castoroides ohioensis*) were identified (Starrett, 1956; Kurtén and Anderson, 1980).

Both the location of the Berends l.f. above the Pearlette Ash and the composition of the fauna suggest an Illinoian age (Rinker and Hibbard, 1952; Hibbard, 1953b, 1958, 1960, 1970; Taylor, 1954). The Berends l.f. might be older than that of Doby Springs (Jammot, 1972; Kurtén and Anderson, 1980). The fauna and associated pollen suggest deposition during moist conditions, possibly with cooler climate. Sympatry of extant mammals occurs in North Dakota (Starrett, 1956). Lake, marsh, and woodland communities were present (Smith, 1954; Taylor, 1954; Starrett, 1956; Hibbard, 1960, 1970; Kapp, 1970).

Smith (1954) and Starrett (1956) recorded the recovery of *Blarina* from the Berends deposit. The genus is represented by a left M1 (UMMP V60549) and by a fragment of a right ramus with m1–m2 (UMMP 31790). Starrett (1956) tentatively identified the fossils as *B. brevicauda*. Hibbard (1963) initially referred the specimens to the subspecies *B. b. brevicauda*; later (1970) he referred them to *B. b. fossilis*.

Measurements 7 through 10 (2.3, 2.2, 2.5, and 2.7) and 19 through 22 (2.2, 1.4, 1.8, and 1.2) were taken. These measurements were compared with those of the reference samples (Table 1). Measurements of the M1 (V60549) fall within the range of measurements of samples of *B. brevicauda*, and are closest to the means of measurements 7 through 10 recorded for sample 12 (*B. b. brevicauda*). This specimen is referred to *B. b. brevicauda*. Measurements 19 through 22 of the other specimen from Berends are somewhat ambiguous. They fall within the range of measurements of samples of all three Holocene species but are outside the range of measurements of *B. b. brevicauda*. The second specimen is referred tentatively to *B. brevicauda*. Whether two taxa were present in the fauna at the time of deposition cannot be ascertained at this time.

Hanover Quarry Fissure

Remains of *Blarina* were recovered from the Hanover Quarry Fissure in Adams County, southeastern Pennsylvania. The fauna

Table 5.—*Univariate statistics for specimens of* Blarina carolinensis *from Hanover Quarry Fissure, Adams County, Pennsylvania, for measurements 19–22 of the lower teeth.*

Variable	N	Mean	2 SE	Range	CV
19	6	1.80	0.05	1.7–1.9	3.33
20	6	1.22	0.07	1.1–1.3	6.56
21	8	1.55	0.06	1.5–1.7	5.16
22	8	1.01	0.03	1.0–1.1	3.96

from Hanover Quarry is considered Yarmouthian in age and probably older than that of Cumberland Cave (Guilday, in litt., 1982). The *Blarina* from this fissure are unique in that they are smaller than remains from most other Appalachian Pleistocene faunas. Fossils from Hanover are similar in size to some of the smallest specimens from Cumberland Cave. Most of the fossils from Hanover consist of edentulous, fragmented dentaries. Other specimens consist of isolated teeth and dentaries with incomplete dentition.

Measurements of nine specimens (CM 41213, 41406) were taken. One M1 was examined; measurements 7 through 10 (1.9, 1.8, 1.9, 2.0) fall within the range of variation of samples of Holocene *B. carolinensis* (Table 1). Univariate statistics of measurements 19 through 22 of the lower molars of eight dentaries are shown in Table 5. These measurements compare most favorably with those of Holocene *B. carolinensis*, particularly the small individuals from Florida and Louisiana (groups 2 and 23, Table 1). No differences in dental morphology were observed that would merit assignment of the Hanover material to a different species. The appearance of *carolinensis*-sized *Blarina* in this fauna suggests temperate conditions during the period of deposition.

Port Kennedy Cave

Port Kennedy Cave was located in Upper Merion Township, Montgomery County, in southeastern Pennsylvania. The cave was a fissure discovered in 1870 during quarrying of limestone. The fissure later was destroyed by mining. Port Kennedy Cave is one of the few Appalachian faunas considered to be pre-Wisconsinan (Guilday, 1971). Its fauna pertains to the Irvingtonian land mammal age, and generally has been considered Yarmouthian (Hibbard, 1958; Hibbard and Dalquest, 1973) or Illinoian (Guilday, 1971). Kurtén and Anderson (1980), however, tentatively referred the fauna to the Aftonian or early Kansan.

Preliminary investigations of the Port Kennedy l.f. were conducted by Cope (1899). Preservation was poor and the status of many of the taxa described by Cope is uncertain. Brown (1908) listed a total of 56 mammalian species, 40 of which are extinct, from the site. Among the extinct species are ground sloth, Schlosser's wolverine, gracile sabertooth, mastodont, and two species of skunk (Kurtén and Anderson, 1980). Similarities between the local faunas of Port Kennedy, Conard Fissure, and Cumberland

Cave were discussed by Brown (1908), Nicholas (1954), and Kurten and Anderson (1980).

Presently, southeastern Pennsylvania is included in the Carolinian biotic province as described by Dice (1943). The area is part of the Piedmont Plateau, characterized by rolling uplands, low hills, and fertile valleys; mean annual precipitation is approximately 112 cm (NOAA, 1974).

The genus *Blarina* is represented in Port Kennedy Cave by part of a left jaw. The specimen (ANSP 150) is the holotype of *Blarina simplicidens*, described by Cope in 1899. "It is about the size of *B. brevicauda*, and no especial characters can be found to distinguish it from that species excepting the forms of the first premolar and the last true molar . . . . The crown of the first premolar is somewhat worn and has an antero-posteriorly oval section. It does not have the V form as in all the species figured by Merriam [1895], and which seems to be common to all the existing species. The last true molar is also more simple than in the existing species, consisting of a trigon only, and lacking the heel. The heel is present in all the species of this genus, of *Cryptotis*, *Notiosorex*, and *Sorex*, according to Merriam." Length of the molar series, length of m2, and depth of the ramus at m2 were 5, 2.5, and 2 mm, respectively (Cope, 1899).

Hibbard (1957) illustrated and compared the specimen of *B. simplicidens* with *B. gidleyi*, *B. fossilis*, and *B. brevicauda*. He saw no structural differences between *B. simplicidens*, *B. fossilis*, and *B. brevicauda*; the special form of the premolar noted by Cope was attributed to wear. Consequently, *simplicidens* and *fossilis* were relegated to the rank of subspecies of *B. brevicauda* (Hibbard, 1957) and later were placed in synonymy with *B. brevicauda*.

We examined the specimen but did not measure the teeth. No differences were found that would distinguish the specimen from *B. brevicauda* and we concur with that assignment, as a representative of the *talpoides* semispecies.

Java

The Java local fauna was collected in eastern Walworth County, South Dakota, at 45°26′N, 99°52′W (NE ¼ sec 26, T123N, R75W). Fossils were deposited in a paleostream channel 35 km east of the Missouri River trench (Martin, 1973b).

Martin (1973a, 1973b) described the Java l.f., which consists of molluscs, fishes, birds, herptiles, and mammals. Twelve families of mammals are represented. The fauna indicates the presence of woodland or shrubland, grassland, and aquatic communities. Habitat and climate probably were similar to those of today except that climate might have been somewhat cooler (Martin, 1973b). Presently the area is located near the western boundary of the Illinoian biotic province of Dice (1943).

The Java l.f. is one of the northernmost Irvingtonian sites known in North America. A radiocarbon date (>30,000 BP) was obtained from a sample of molluscs (Martin, 1973a). The rodents of the fauna suggest a Kansan age (Martin, 1973b, 1975). The Java l.f. was considered older than those of Conard Fissure, Cudahy, and Cumberland Cave by Martin (1973b), Schultz et al. (1978), and Van der Meulen (1978).

The genus *Blarina* is represented by a single P4 (R. A. Martin, private collection). Measurements 3 through 6 are 1.9, 1.6, 1.1, and 1.7. These measurements are most similar to those of reference samples of *B. carolinensis* (Table 1). The tooth from Java is referred tentatively to that species. Additional speculation is premature because of the small sample size and the age and location of this fauna.

Vera

The Vera l.f. was recovered from three sites in northern Texas. The three sites are UM-T1-56 (SW ¼ sec. 152, Baylor County), UM-T1-57 (N ¼ sec. 101, Knox County), and UM-T1-58 (SW ¼ SE ¼ sec. 110, Knox County) (Getz and Hibbard, 1965). Remains were recovered in the Seymour formation immediately below the Pearlette Ash. Dalquest (1977) estimated the age of the fauna as ca. 600,000 BP. The fauna is similar to that of Cudahy.

Molluscs, fishes, herptiles, birds, and mammals are included in the Vera l.f. Most of the molluscan species occur in sympatry in eastern South Dakota and southwestern Minnesota (Getz and Hibbard, 1965). Thirteen species of mammals were identified, of which five (*Geomys tobinensis*, *Peromyscus cragini*, *Ondatra annectens*, *Microtus paroperarius*, and *M. llanensis*) are extinct (Hibbard and Dalquest, 1966). Of the extant species, *Blarina brevicauda* and *Onychomys* sp. no longer occur in Baylor and Knox counties (Davis, 1978).

The species composition of the Vera l.f. suggests a climate with cooler, moister summers. Streams and wooded floodplains were present (Getz and Hibbard, 1965; Hibbard and Dalquest, 1966; Hibbard, 1970). Presently, the climate of Baylor and Knox counties is semiarid, with an annual precipitation of 58 to 66 cm (Getz and Hibbard, 1965; NOAA, 1974).

Hibbard and Dalquest (1966) identified seven specimens of *Blarina* from UM-T1-57 (UMMP 45834, 45835, 45850–3, and 45855) and two from UM-T1-58 (UMMP 39803 and 45709). These specimens were referred to *Blarina* cf. *B. brevicauda* and a greater range of individual variation than in Holocene faunas was noted. All specimens were examined for the present study. Specimen 45834 consists of two edentulous dentary fragments; all other specimens are isolated teeth. Measurements 3 through 6 (2.3, 1.9, 1.4, 2.3) and 3, 5, and 6 (2.2, 1.3, 2.0) of two P4s, measurements 12 through 14 (1.7, 2.0, 1.9) of a M2, and measurements 19 and 20 (2.0 and 1.4) of a m1 were taken. These measurements (of specimens 45851–3 and 45709) were compared with measurements of reference samples

of Holocene specimens (Table 1). Measurements of the specimens from Vera are most similar to the means of *B. brevicauda*; these fragmentary fossils tentatively are assigned to that species.

Trout Cave

Trout Cave is a limestone cave located along the terrace of the South Fork of the Potomac River, 4.8 km southwest of Franklin, Pendleton County, West Virginia (Kurtén and Anderson, 1980). The upper strata of the deposit appear to be Wisconsinan. The lower strata, separated from the upper levels by a flowstone approximately 1.8 m below the surface, contain many taxa considered equivalent to those found in the Irvingtonian fauna of Cumberland Cave, Maryland (Zakrzewski, 1975b). We examined both Wisconsinan and Irvingtonian specimens.

The fauna of Trout Cave (primarily small mammals) is under study. The presence of *Ondatra annectens*, *Ochotona*, *Phenacomys*, sp., *Neofiber leonardi*, and *Atopomys salvelinus* was reported by Guilday (1971), Guilday and Parmalee (1972), Frazier (1977), and Zakrzewski (1975b), respectively. These five genera were recovered from the lower strata of the deposit.

Fifteen specimens of *Blarina* were examined. Fourteen were dentaries, most with incomplete dentition. Seven specimens (CM 12718) were recovered from the stratum four feet below the surface and three (CM 12752) from five feet below the surface. Both of these strata are above the flowstone, so these specimens are considered Wisconsinan. A maxillary fragment was recovered from the four-foot stratum. All molars were present in the maxillary and measurements 1 through 16 were taken (6.3, 4.4, 2.8, 1.9, 1.6, 2.4, 2.5, 2.2, 2.3, 2.4, 1.7, 1.8, 2.3, 2.2, 0.8, and 1.6, respectively). These measurements compare most favorably with those near the lower end of the range of variation for *B. b. brevicauda* and are similar to those of *B. brevicauda* from Manitoba (sample 18). Of the nine dentaries from these two strata, five were sufficiently complete for analysis (Fig. 9). Four specimens were from the four-foot stratum and one (to the right in the canonical diagram—Fig. 9) was from the five-foot stratum; all fell within the range of variation shown for the *talpoides* semispecies of *B. brevicauda*. Comparisons were made between the measurements of the incomplete dentaries and the means derived for those measurements for extant samples. The fossils all appear to represent *B. brevicauda*. Some are large and correspond most closely to *B. b. brevicauda* from Nebraska and Iowa (samples 12 and 5), whereas others are more similar to the smaller subspecies *churchi* (sample 11), *talpoides* (samples 13, 19, and 24), *kirtlandi* (samples 4 and 6), and the *B. brevicauda* from Manitoba (sample 18).

The five remaining specimens from Trout Cave

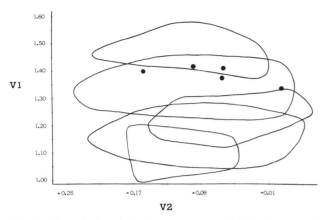

Fig. 9.—Canonical analysis of measurements 19–24 of lower teeth of specimens of *Blarina* from the Trout Cave l.f. Centroids enclose variation in samples of Holocene taxa of *Blarina* (see Fig. 4).

are from eight feet (CM 12822, one dentary), nine to ten feet (CM 20111, one dentary), and ten to eleven feet (CM 12880 and 20109, three dentaries) below the surface. Materials below the flowstone are considered to be Irvingtonian; the trend from younger to older materials is gradual, so the period of deposition cannot be ascertained. Certainly the specimens designated 12880 and 20109 are from the bottom of the excavation and probably are Kansan. These older specimens, when measurements are compared with those of Holocene specimens, are most similar to the smaller *talpoides* semispecies of *B. brevicauda* (samples 19, 4, 6, and 11). There were no discernible differences among the specimens recovered from these strata. Accordingly, all specimens of *Blarina* from Trout Cave pertain to the species *B. brevicauda,* but they apparently represent both the *brevicauda* and *talpoides* semispecies.

RANCHOLABREAN

Peccary Cave

Peccary Cave is located on Ben's Branch of Cave Creek in Newton County, Arkansas (SW ¼, SE ¼ sec. 22, T15N, R19W). The cave consists of more than 300 m of passages. Most excavations have been in a series of numbered trenches near the two entrances of the cave. Preliminary excavations were described by Davis (1969).

Quinn (1972) reported a series of radiocarbon dates from Peccary Cave that range from 2,230 ± 120 to 16,700 ± 250 BP. Generally, the Peccary Cave l.f. is considered Wisconsinan in age.

Remains represent inhabitants of the cave, or were brought in by predators and water (Davis, 1969; Quinn, 1972). Invertebrates, fishes, birds, herptiles, and mammals were found. The invertebrates and herptiles are similar to those of the area today

(Quinn, 1972). The mammals are still under study. Fifty-one species of mammals have been identified, of which eight are extinct (including *Dasypus bellus, Platygonus compressus, Sangamona fugitiva, Mammut,* and *Mammuthus*). Twenty-nine species of small mammals were present (Davis, 1969; Kurtén and Anderson, 1980; Semken, personal communication, 1982).

The fauna indicates the presence of parkland and permanent water at the time of deposition. Climate was cooler, moister, and more equable than that of the present (Quinn, 1972; Graham, 1972, 1976). Semken (personal communication) is conducting a study of the ecology and climate as indicated by the small mammals of trenches 8, 13, and 18. Modern climate and vegetation of the area are summarized in the discussion of Conard Fissure.

Graham and Semken (1976) identified three phena—"*B. b. brevicauda, B. b. kirtlandi,* and *B. b. carolinensis* [=*B. h. hylophaga?*]"—of *Blarina* from Peccary Cave. The three contemporaneous phena at Peccary Cave (and at Cumberland Cave and New Paris No. 4) were thought to represent more equable climate during full or early late glacial time.

For the present study, 13 specimens from Trench 8, 20 specimens from Trench 13, 27 from Trench 15, and 4 from Trench 24 (SUI 38333–36, 38338, 38340, 38342–44, 39341, 49840, and 49843) were measured; only the most complete specimens in the collection were examined. Measurements of the fossils are most similar to those of groups of Holocene *B. brevicauda* (Table 1). The examined sample appeared homogeneous. Results of canonical analyses of 20 maxillaries and 37 dentaries (of which two observations are hidden and one was plotted to the right of Fig. 11) are shown in Figs. 10 and 11. All specimens are shown clustered within or near the centroid showing the range of variation of modern *B. brevicauda* (*talpoides* semispecies). In the examination of these specimens, we saw no indication that more than one phenon, *B. brevicauda,* is present in the Peccary Cave l.f.

Arredondo

The Arredondo fauna was recovered from two pits near Arredondo, Alachua County, Florida (SE ¼ sec. 22, T10S, R19E). Remains were found in fissure fills at an elevation of 25 m (Bader, 1957; Webb, 1974).

Remains of herptiles, birds, and mammals were recovered. Herptiles and birds were studied by Auffenberg (1963, 1967) and Brodkorb (1959). The most recent list of mammals from Arredondo is that by Webb (1974), who listed 33 species from localities IA, IB, and IIA. Eleven species are extinct. The paleofauna indicates the presence of forests, meadows, and freshwater ponds, and winters possibly were warmer than at present (Brodkorb, 1959; Auffenberg, 1963; Webb, 1974).

The remains from localities IA, IB, and IIA are approximately the same age. The fauna is considered Sangamonian (Kurtén, 1965; Webb, 1974; Kurtén and Anderson, 1980).

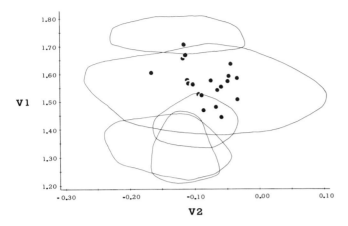

Fig. 10.—Canonical analysis of measurements 3–10 of upper teeth of specimens of *Blarina* from the Peccary Cave l.f. Ellipses enclose variation in samples of Holocene taxa of *Blarina* (see Fig. 3).

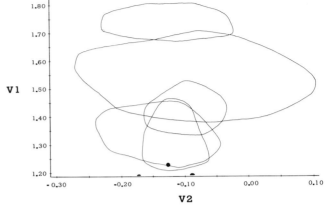

Fig. 12.—Canonical analysis of measurements 3–10 of upper teeth of specimens of *Blarina* from the Arredondo l.f. Centroids enclose variation in samples of Holocene taxa of *Blarina* (see Fig. 3).

Bader (1957) and Webb (1974) reported *B. brevicauda* from Arredondo IB and IIA. We examined six maxillaries and 12 dentaries from Arredondo II (UF 1719, 2098, 3296, 3352, 3559, 12269, 12271, 56073–74, 56076, 56078, 56081, 56545–49, and 56551). Results of the canonical analyses of four maxillaries and five dentaries are shown in Figs. 12 and 13; these specimens are represented within or below the range of variation of Holocene *B. c. peninsulae* and *B. carolinensis*. One maxillary and one dentary are not shown on the figures. The two remaining maxillaries and seven dentaries have mea-

surements comparable to those of the specimens represented in the figures. Univariate statistics for the fossils from Arredondo (ARREDO) are shown in Table 2. We tentatively conclude that the fossils pertain to the species *B. carolinensis*.

Bradenton 51st Street

Fossil vertebrates were recovered from two coastal marsh sites in Bradenton, Manatee County, Florida. One specimen, from the 51st Street locality, was referred to *B. brevicauda* by Webb (1974).

Fourteen species of mammals have been reported from the Bradenton sites. Four extant and six extinct species are known from the 51st Street locality ("*B. brevicauda*," *Sigmodon* sp., *S.*

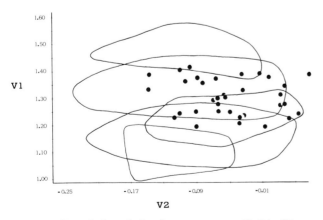

Fig. 11.—Canonical analysis of measurements 19–24 of lower teeth of specimens of *Blarina* from the Peccary Cave l.f. Ellipses enclose variation in samples of Holocene taxa of *Blarina* (see Fig. 4).

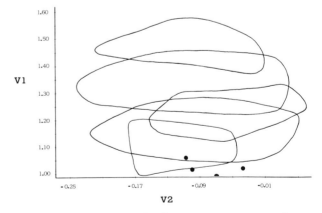

Fig. 13.—Canonical analysis of measurements 19–24 of lower teeth of specimens of *Blarina* from the Arredondo l.f. Centroids enclose variation in samples of Holocene taxa of *Blarina* (see Fig. 4).

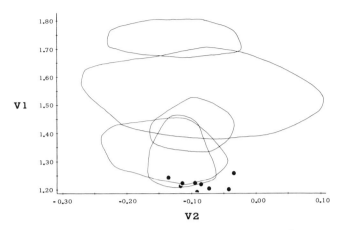

Fig. 14.—Canonical analysis of measurements 3–10 of upper teeth of specimens of *Blarina* from the Haile XIB l.f. Centroids enclose variation in samples of Holocene taxa of *Blarina* (see Fig. 3).

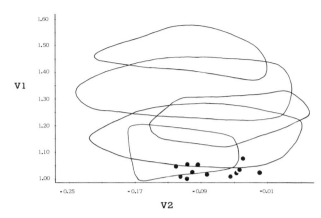

Fig. 15.—Canonical analysis of measurements 19–24 of lower teeth of specimens of *Blarina* from the Haile XIB l.f. Centroids enclose variation in samples of Holocene taxa of *Blarina* (see Fig. 4).

bakeri, Neofiber alleni, Synaptomys australis, Mammuthus columbi, Equus sp., *Paleolama mirifica, Odocoileus virginiana,* and *Bison latifrons*) (Webb, 1974). The Bradenton sites were considered Sangamonian by Webb (1974) and Robertson (1974).

A dentary (UF 2470) with partial dentition was examined. Measurements 19 through 22, 25, and 26 (1.7, 1.2, 1.5, 1.0, 3.4, 1.9) of the two molars and dentary were taken. These measurements are similar to those of the reference samples of Holocene *B. carolinensis* (Table 1), and the fossil apparently pertains to that species (probably to the nominal subspecies *peninsulae*).

Haile XIB

Bones were recovered from a series of quarries near Haile, Alachua County, Florida. The quarries are sinkhole deposits located approximately 24 m above present sea level (Webb, 1974).

Herptiles, birds, and mammals are represented in the Haile deposits. Webb (1974) listed 26 species of mammals from Haile XIB. Two species, *Dasypus bellus* and *Hemiauchenia macrocephala,* are extinct. The other mammals are known from Florida in modern times (Ligon, 1965; Webb, 1974). The presence of *Dasypus bellus* and *Desmodus magnus* implies warmer winters (Ligon, 1965; Webb, 1974). Prairie, scrub, woodland, marsh, and aquatic habitats are suggested (Ligon, 1965). The prevalence of small vertebrates suggests deposition by owls and hawks. The fauna is considered Sangamonian (Kurtén, 1965; Ligon, 1965; Webb, 1974).

Blarina brevicauda is one of three species of shrews reported from Haile XIB by Webb (1974). Seventeen of the most complete maxillaries and seventeen dentaries (UF 4559, 4561, 4563, 4565, 4569, 4571, 4573, 4575–77, 4579, 4581, 4589–91, 4593, 4594,

4603–05, 4608, 4609, 4611–14, 4616–23) were examined for the present study. Results of canonical analyses of these fossils are shown in Figs. 14 and 15. Eight maxillaries and one dentary were plotted outside the range of the diagrams and are not shown. Four dentaries are hidden. The remaining specimens represented in the two figures fell within or near the centroids representing the range of variation of Holocene *B. carolinensis*. The fossils from Haile XIB are approximately the same size as those from Arredondo, and apparently both pertain to the species *B. carolinensis*.

An additional specimen (UF 13094) from Haile XIIIA also was examined. Measurements 3 through 16 (1.9, 1.4, 1.1, 1.8, 1.7, 1.6, 1.8, 1.9, 1.4, 1.3, 1.9, 1.6, 0.7, 1.4) were taken. This fossil is indistinguishable from those from Haile XIB, and we think that it, too, pertains to *B. carolinensis*.

Reddick IA

Vertebrate remains were recovered from a limerock quarry 1.6 km southeast of Reddick, Marion County (SW ¼, NW ¼ sec. 14, T13S, R21E), Florida. Fossils were found in four major deposits (designated A through D) in the quarry. The four sites probably were contemporaneous although there might have been minor temporal differences. Specimens of the short-tailed shrew were recovered from locality A, known as the "Rodent Beds." This extensive deposit contains the remains of many small vertebrates and is thought to be the result of owl-pellet droppings (Gut and Ray, 1963).

A comprehensive list of the vertebrate fauna, which includes herptiles, birds, and mammals, was published by Gut and Ray in 1963. Only the birds and snakes have been studied extensively

(Brodkorb, 1957, 1963; Auffenberg, 1963; Hamon, 1964). Preliminary studies on selected elements of the mammalian fauna were done by Olsen (1958), Gut (1959), and Ray et al. (1963). Initially, the Reddick site was assigned tentatively to the Illinoian. More recently it has been considered Sangamonian (Martin and Webb, 1974; Webb, 1974).

Hypotheses regarding the climate of the area at the time of deposition are somewhat contradictory. The numerous remains of bats, owls, vultures, and swallows suggest that Reddick IA was the site of a Pleistocene cave. Birds of southwestern and northern affinities are present in the A and C deposits. Brodkorb (1957) considered the avifauna "fairly typical of a wet grassland or fresh water marsh" habitat. The presence of birds with northern affinities does not necessarily indicate a cooler climate, because the modern descendants of those birds are migratory. If the climate was slightly cooler, Brodkorb suggested that it would be similar to that of modern Virginia. Hamon (1964) also suggested a slightly cooler climate based on his study of the avifauna. Auffenberg (1963) noted that the reptiles indicated a dry, open forest, similar to that present in the area today. He considered the possibility, however, that the snakes and birds might not have been contemporaries. Olsen (1958) considered the presence of the bog lemming (*Synaptomys australis*) to be further indication of cooler climate; however, Simpson (1928) pointed out that this fossil species often is found in associations that suggest it might not have been boreal. Martin and Webb (1974) thought the high density of mammalian species in Florida during the Rancholabrean indicated tropical or subtropical conditions. Martin (1977) also interpreted the presence of the neotropical bat *Eumops glaucinus* in the Melbourne and Monkey Jungle Hammock faunas of central and southern Florida to indicate warm temperatures during this period.

The short-tailed shrews from Reddick were referred to the species *B. brevicauda* (Gut and Ray, 1963). Fifteen dentaries (KU 14091; UF 8802, 8803, and uncatalogued material; UMMP V60863) were examined for the present study. Univariate statistics are shown in Table 2. Results of canonical analyses of these fossils are shown in Fig. 16. One observation is hidden; another observation fell outside the range of the diagram. All specimens are within or near the range of variation of *B. carolinensis,* which is the shrew present in the southeastern United States today (George et al., 1982), and we refer the fossils to that species.

Vero

The Vero fossil beds are located in Vero Beach, Indian River County, Florida (SE ¼ sec. 35, T32S, R39E). Fossils were discovered in the banks of a canal in three sedimentary beds (Sellards, 1937; Weigel, 1962). The top two layers, beds 3 and 2, contained vegetable debris and vertebrate remains. The first, lowermost bed contained remains of marine organisms only (Weigel, 1962).

The discovery of human remains associated with the Pleistocene fauna stimulated early interest in the Vero site. Sellards (1937) listed most of the early studies of the site. Other vertebrates from Vero include fishes, amphibians, reptiles, birds, and additional mammals. Weigel (1962) listed 45 mammals, including

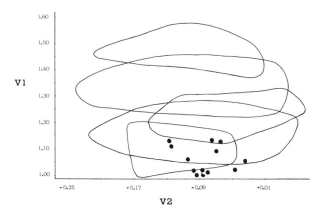

Fig. 16.—Canonical analysis of measurements 19–24 of lower teeth of specimens of *Blarina* from the Reddick l.f. Centroids enclose variation in samples of Holocene taxa of *Blarina* (see Fig. 4).

15 extinct species. Vertebrate remains generally are broken and disarticulated. The fauna suggests the presence of a variety of habitats, including swamps, hammocks, and pine woods (Weigel, 1962).

Simpson (1929a, 1929b) listed Vero as one of four faunas of Florida which he considered contemporaries (early–late Pleistocene). Deposition is thought to have occurred in the Wisconsinan and early Holocene (Bader, 1957; Kurten, 1965; Weigel, 1962). A series of radiocarbon dates from bed 2 suggests deposition from approximately 30,000 to 3,500 BP; bed 3 contains Holocene contaminants and was not dated (Weigel, 1962).

Simpson (1929b) reported "*B. b. peninsulae*" from the Vero deposit. Weigel (1962) reported *B. brevicauda* from beds 2 and 3. For the present study, five maxillaries and 10 dentaries (UF V7413, 7416, 7417, 7423, 7428–29, 7432, 7435–36, 7459, 7464, 7469–70, 7482, 7484), the most complete specimens in the collection, were examined. All specimens were from strata 2 and 3. Five maxillaries and five dentaries were sufficiently complete to be included in statistical (INDIAN, Table 2) and canonical (Fig. 17) analyses. All fossils are most similar to specimens from reference samples of Holocene *B. carolinensis,* and we refer the Vero material to that species.

Waccasassa River

The Waccasassa River sites are fluviatile deposits located in Levy County near the western coast of Florida. The sites are located approximately 6 m above present sea level (Webb, 1974). A detailed description of the fauna has not been published.

Blarina brevicauda was reported from Waccasassa River IIB and III by Webb (1974). Thirty-five additional mammalian species were listed, of which 15 are extinct. The fauna is considered Rancholabrean, probably Wisconsinan, in age (Martin, 1969; Webb, 1974).

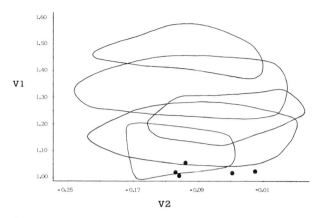

Fig. 17.—Canonical analysis of measurements 19–24 of lower teeth of specimens of *Blarina* from the Vero l.f. Centroids enclose variation in samples of Holocene taxa of *Blarina* (see Fig. 4).

The material tentatively assigned to the genus *Blarina* (UF 16341 and 16362) consists of edentulous jaws, isolated teeth, and a dentary fragment with two molars. Measurements 19 through 22 (1.6, 1.1, 1.4, 1.0) were taken for the present study. These measurements, somewhat smaller than those of reference samples of *B. carolinensis* (Table 1), are similar to measurements of fossils from Arredondo and other Florida sites. We think that these fossils pertain to the species *B. carolinensis*.

Williston IIIA

Fossils were recovered from sink deposits in Williston, Levy County, Florida. Amphibians, reptiles, birds, and mammals from this site were discussed by Holman (1959a, 1959b).

Thirty species of herptiles, six birds, and 19 mammals have been identified. One reptilian species, two birds, and five mammals are extinct; extant forms are present in the area today (Holman, 1959a, 1959b). The composition of the fauna suggests that surrounding habitat consisted of marshy pineland, pine forest, and open sinks during the period of deposition (Holman, 1959b). Present vegetation near the site is mesophytic (Holman, 1959a). Modern vegetation of Florida is summarized in the discussion of the Coleman IIA fauna.

Holman (1959a) considered the Williston fauna Illinoian. More recently, the fauna has been placed in the Sangamonian (Kurtén, 1965; Martin, 1974; Martin and Webb, 1974).

One dentary (UF V-5848) was referred to the species *B. brevicauda* by Holman (1959b). The size of the fossil was thought comparable to that of extant *Blarina* of north-central Florida. For the present study, measurements 19 through 22 (1.6, 1.0, 1.4, 0.9) were taken. We find these measurements somewhat smaller than those of modern *B. c. peninsulae* (Table 1), but no morphological features were ob-

served that would distinguish the fossil from the modern species. We conclude that the Williston specimen is a small *B. carolinensis*.

Ladds Quarry

The first major assemblage of Pleistocene vertebrates reported from Georgia is located in Bartow County. Fossils were recovered from a matrix of red cave earth (in part firmly cemented as a cave breccia) located in small fissures exposed in a limestone quarry (Ray, 1965). The quarry is in the southeastern end of Quarry Mountain (also known as Ladds Mountain), approximately 4 km WSW of Cartersville, 34°09′N, 84°50′W (Lipps and Ray, 1967).

At least 25 species of molluscs, 4 species of birds, 23 species of herptiles, and 48 species of mammals have been identified at Ladds. Preliminary faunal lists were compiled by Holman (1967), LaRocque (1967), Ray (1967), and Wetmore (1967). Mixing of materials in the deposit was recognized early in the investigation of the quarry. "The fossils occur in small fissures exposed in a remnant pinnacle of limestone . . . now isolated from the major part of the mountain by extensive quarrying . . . . The disturbed (by dynamiting) and open nature of the deposit, together with the presence of certain apparently ecologically incompatible species, suggest the possibility of a heterochronic assemblage rather than a unit fauna" (Ray, 1965). A mixed assemblage, at least in part, is suggested by the molluscs, herptiles, and mammals (Holman, 1967; LaRocque, 1967; Ray, 1965, 1967).

A radiocarbon date for this fauna has not been obtained. The presence of some of the extinct forms, particularly *Peromyscus cumberlandensis*, might indicate a pre-Wisconsinan age for at least part of the fauna (Guilday, 1971; Guilday et al., 1978). Watts (1970) compared the results of the analysis of pollen from Ladds by Benninghoff and Stevenson (1967) with pollen spectra obtained from two ponds in Bartow County. In the pollen spectra from the two Georgia ponds, a pollen assemblage (designated as zone Q1) was recognized from which basal dates of 20,100 ± 240 and 22,900 ± 400 years BP were obtained. Watts proposed referral of the assemblage from zone Q1 and a similar assemblage from North Carolina as a southeastern pine-spruce (*Pinus-Picea*) assemblage zone. He raised the possibility that the spectra from Ladds could be referred to this pine-spruce assemblage, which he considered "contemporaneous with the main Wisconsin ice advance." This is, however, a tentative correlation, due to the mixing observed in the deposit at Ladds.

Northern elements in the fauna, including the spruce grouse *Canachites canadensis* and the wood turtle *Clemmys insculpta* (which presently occur no farther south than 44°N and 39°N, respectively), might indicate a cooler environment (Holman, 1967; Ray, 1967; Wetmore, 1967; Guilday et al., 1978). The presence of southern elements in the fauna, including *Neofiber alleni* (presently restricted to Florida and southern Georgia) and *Didelphis virginianus*, demonstrates the mixed nature of the fauna at Ladds (Ray, 1967; Frazier, 1977).

Ray (1967) compared the *Blarina* from the Ladds l.f. with modern samples from Alabama (Cullman County), North Carolina (Transylvania County), and West Virginia (Greenbriar County). Based on the 10 cranial and dental characters used, the fossils were more heterogeneous in size than were the mod-

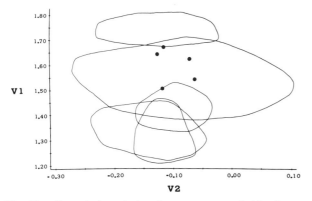

Fig. 18.—Canonical analysis of measurements 3–10 of upper teeth of specimens of *Blarina* from the Ladds Quarry l.f. Centroids enclose variation in samples of Holocene taxa of *Blarina* (see Fig. 3).

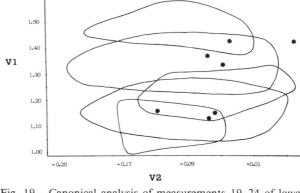

Fig. 19.—Canonical analysis of measurements 19–24 of lower teeth of specimens of *Blarina* from the Ladds Quarry l.f. Centroids enclose variation in samples of Holocene taxa of *Blarina* (see Fig. 4).

ern samples used for comparison, but were most similar to specimens from West Virginia. The fossils were referred to the species *B. brevicauda.*

Twenty-one specimens (USNM 23384, 23385, and uncatalogued material) were examined herein. Five specimens were analyzed for which measurements 3–10 were complete. These specimens clustered within the centroid for the *talpoides* semispecies of *B. brevicauda* (Fig. 18). These fossils are larger than the modern shrews of the southeastern United States, and they are most similar to the samples of *B. b. talpoides* from New York and *B. b. churchi* (samples 13 and 11, respectively). Analyses of measurements of lower teeth were performed for eight specimens, which plotted in two groups (Fig. 19). The larger of the two groups falls within the centroid for the *talpoides* semispecies of *B. brevicauda.* The smaller group consists of three specimens, all of which were uncatalogued but designated by the letters BB, X, and Y. These animals fall within the range of variation for modern *B. c. carolinensis.* Univariate statistics for the Ladds fossils (LADDSQ) are shown in Table 2. Because of the implication that the Ladds l.f. is heterochronic, it is suggested that these are *B. c. carolinensis* that were deposited at a later date (possibly Holocene or sub-Recent). In any event, the fauna evidently consists of two species of *Blarina, B. brevicauda* and *B. carolinensis.*

Meyer Cave

Meyer Cave is located 6.4 km SSW of Columbia, Monroe County, Illinois (NW ¼ sec. 6, T2S, R10W, New Hanover Town-

ship). The opening of the cave is a narrow passageway in a limestone bluff over the Mississippi River. The passageway leads to a small room, the floor of which slopes toward a hole. The bell-shaped fissure located below this room acts as a natural trap for animals (Parmalee, 1967).

Fossils were recovered from the fissure. The invertebrate fauna consists of approximately 27 species of molluscs and insects, whereas the vertebrate fauna consists of approximately 115 species of fishes, herptiles, birds, and mammals. The deposit was not stratified because of burrowing by some of the animals that entered the cave and fell into the fissure. Remains of fishes and some other vertebrates probably were carried in by predators. Six samples of soil subjected to palynological analysis were devoid of pollen (Parmalee, 1967).

No extinct species were found. Fifteen species (including 13 mammals) previously were unknown from Monroe County. Three of the mammals (*Microtus xanthognathus, Clethrionomys gapperi,* and *Erethizon dorsatum*) were not recorded previously from Illinois. The presence of the pigmy shrew (*Microsorex hoyi*) and other species that have modern ranges to the north of Illinois indicates deposition during cool, moist conditions (ca. 9,500–7,500 BC); the presence of the plains pocket gopher (*Geomys bursarius*) and the eastern spotted skunk (*Spilogale putorius*) suggests additional deposition during a warmer, drier period (ca. 3,500–1,500 BC) (Parmalee, 1967).

The area presently receives approximately 91–96 cm of precipitation annually (NOAA, 1974). The fauna and flora are considered part of the Illinoian biotic province as defined by Dice in 1943 (Parmalee, 1967).

Remains representing a minimum of 262 specimens of *Blarina* were removed from Meyer Cave. Two distinct sizes, identified as *B. brevicauda* cf. *brevicauda* and *B. b. carolinensis,* were noted by Parmalee (1967). Graham and Semken (1976) equated the two phena from Meyer Cave with *B. b. kirtlandi* and *B. b. carolinensis.* They regarded the two as contemporaneous phena, and suggested that

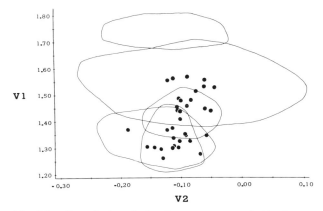

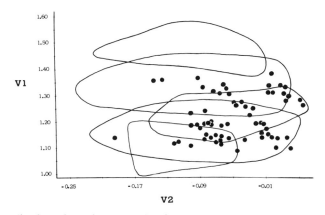

Fig. 20.—Canonical analysis of measurements 3–10 of upper teeth of specimens of *Blarina* from the Meyer Cave l.f. Centroids enclose variation in samples of Holocene taxa of *Blarina* (see Fig. 3).

Fig. 21.—Canonical analysis of measurements 19–24 of lower teeth of specimens of *Blarina* from the Meyer Cave l.f. Centroids enclose variation in samples of Holocene taxa of *Blarina* (see Fig. 4).

sympatry at Meyer Cave (as well as at Crankshaft Cave and Welsh Cave) was indicative of more equable climate at the time of deposition.

For the present study, 148 specimens from Meyer Cave were measured. In the analysis of measurements of upper teeth, 34 specimens were plotted within the centroid for *B. carolinensis* and the centroid for the smaller subspecies (*talpoides* semispecies) of *B. brevicauda* (Fig. 20). In the analysis of the lower tooth measurements, most of 114 fossils again were plotted within the centroids for *B. carolinensis* and the *talpoides* semispecies of *B. brevicauda* (Fig. 21). Six additional fossils were plotted to the right of the range of the diagram shown in Fig. 21; three of these were just outside the range of variation shown for *B. carolinensis* and three were near the centroid for *B. brevicauda*. Forty-five specimens were hidden behind other points within the two centroids. Univariate statistics for the fossils from Meyer Cave are shown in Table 2 (MEYERC). Apparently the species of *Blarina* in the fauna are *B. carolinensis* and *B. brevicauda,* as previously suggested by Graham and Semken (1976); however, it is not possible with the present data to discount the presence of a third species, *B. hylophaga,* which is approximately intermediate in size between *B. brevicauda* and *B. carolinensis*. The three species do not occur together at any locality at the present time, but might have in the past.

Cherokee Sewer Site

The Cherokee Sewer Site is a bison kill site located 3.2 km south of Cherokee, Cherokee County, Iowa (W ½, SW ¼, SE ¼

sec. 4, T91N, R40W). Deposition occurred in an alluvial fan on the west floodplain of the Little Sioux River Valley at an elevation of 360 m (Shutler et al., 1980).

The stratified fan deposits contained three cultural horizons. Radiocarbon dates of ca. 6,350, 7,300, and 8,400 BP were obtained for horizons I, II, and III, respectively (Hoyer, 1980). Seeds, molluscs, and mammals were associated with the archeological remains. Unidentified fish, herptiles, and birds also were present. The site is of special interest because of its location at the fluctuating prairie-forest border. An interdisciplinary study was conducted to reconstruct the paleoecology and cultural history of the area; results were summarized by Shutler et al. (1980).

Remains of canids, bison, cervids, and 23 small mammals were found at the Cherokee site (Pyle, 1980; Semken, 1980). All mammals are extant. Four species (*Microsorex hoyi, Tamiasciurus hudsonicus, Perognathus hispidus,* and *Synaptomys cooperi*) no longer occur in the area. Accumulation of the small mammals probably is due to the presence of bison remains and other garbage; raptorial or human predation also might have contributed to the accumulation (Semken, 1980).

The Cherokee l.f. indicates the presence of forest, arboreal, meadow, prairie, and aquatic communities. The mammals, including the four species no longer present, suggest a local habitat of open gallery forest (Semken, 1980). Sympatry of the species from each horizon show a trend of increasing aridity from approximately 8,400 to 6,350 BP. Climatic conditions of horizon II are the most similar to those of the present (Semken, 1980). Analysis of gastropods from Cherokee and pollen analyses from Lake West Okoboji (80 km north) also indicate a trend of increasing aridity, accompanied by changes in the associated flora and fauna (Baerreis, 1980; Baker and Van Zant, 1980; Wendland, 1980).

Dice (1943) included Iowa in the Illinoian biotic province. Three grassland associations comprised the tall-grass prairie that covered the western and central parts of the state; deciduous forests occurred in eastern Iowa and along waterways and hillsides in the west (Bowles, 1975). The climate, geology, and vegetation of Iowa were summarized by Bowles (1975). In Cherokee County the mean annual precipitation is 72.3 cm (Shutler et al., 1980).

Semken (1980) identified remains of *Blarina* from the Cherokee Sewer Site as *B. b. brevicauda*. Specimens consist of two isolated teeth from horizon I (SUI 38067A, B); a mandible with m1–3, a mandibular fragment with i–m1, a mandibular fragment with m1–2, two mandibular condyles, a mandibular fragment with m2–3, and isolated teeth from II (SUI 38068, 44092A–F); a palate and a mandible from III (SUI 44039A, B). A minimum number of 1, 2, and 2 individuals thus are represented in horizons I, II, and III, respectively.

Six of the most complete specimens were examined for the present study. Measurements 7 through 10 (2.3, 2.2, 2.4, 2.6) and 3 through 16 (2.5, 2.0, 1.4, 2.4, 2.3, 2.1, 2.3, 2.6, 1.7, 1.7, 2.4, 2.0, 0.8, 1.5) of upper teeth and measurements 19 through 22 of two m1s and m2s (2.2, 1.5, 2.0, 1.4; 2.3, 1.6, 2.0, 1.4) and 19 through 26 of a dentary with complete dentition (2.4, 1.7, 2.0, 1.3, 1.6, 1.1, 4.4, 2.8), were compared with those of reference samples (Table 1). Measurements of the specimens from Cherokee were most similar to those of modern *B. b. brevicauda*, and generally fell within the range of variation of the modern sample from Iowa. The specimens from the Cherokee Sewer Site pertain to the subspecies *B. b. brevicauda*.

Craigmile

The Craigmile and Waubonsie local faunas were recovered from the Craigmile farm (SE, NW, NW, NE ¼ sec. 13, T71N, R43W) in the Waubonsie Creek watershed, Mills County, Iowa. The site and the two paleofaunas were described in a dissertation by Rhodes (1982), which serves as the basis for this report.

The two l.f.s were recovered from superposed Wisconsinan alluvial fills. Stratigraphy of the site was described by Rhodes (1982). A minimum age of 23,240 ± 535 RCYBP was obtained from bone residue in the unit from which the Craigmile l.f. was recovered.

The Craigmile l.f. is represented by fishes, herptiles, birds, and mammals. Rhodes (1982) listed 31 mammalian taxa, of which four species (*Dasypus bellus*, *Mammut* sp., *Equus* sp., and *Sangamona fugitiva*) are extinct. Nine species no longer occur in Mills County. Remains of microtines were the most common elements in the deposit. All remains were disarticulated and damaged (Rhodes, 1982).

Twenty-nine mammals indicate grassland, meadow, and brushy edge habitats at Craigmile during the period of deposition. Cooler summer temperatures are suggested. Taxa no longer present in the area are boreal or boreomontane species. Twenty-two species are sympatric in central Wisconsin in an area of oak or pine savannah with patches of dense forest and prairie (Rhodes, 1982).

Pre-settlement vegetation of the Waubonsie watershed probably consisted of tall-grass prairie on the Missouri River floodplain and exposed slopes, and upland riparian forests (Rhodes, 1982). Modern climate and vegetation of the state were summarized by Bowles (1975).

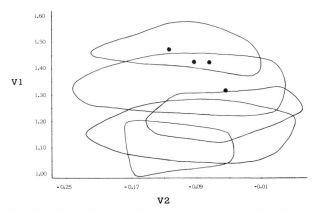

Fig. 22.—Canonical analysis of measurements 19–24 of lower teeth of specimens of *Blarina* from the Craigmile l.f. Centroids enclose variation in samples of Holocene taxa of *Blarina* (see Fig. 4).

Blarina brevicauda was reported from both the Craigmile and Waubonsie faunas. Two mandibles, a maxillary fragment, and isolated teeth (SUI 46405, MNI = 3) were recovered from the Craigmile deposit. The single measurable mandible was referred to the subspecies *B. b. brevicauda* (Rhodes, 1982).

These specimens were examined for the present study. Measurements 7 through 10 of one M1 (2.4, 2.2, 2.5, 2.8) fall within the range of variation of modern *B. brevicauda* (*talpoides* semispecies) and *B. b. brevicauda*; measurements 7 through 10 of a second M1 (2.0, 2.0, 2.1, 2.3) are slightly smaller and are more similar to measurements of the *talpoides* semispecies of *B. brevicauda* (Table 1). Measurements 19 through 26 of a dentary and lower molars (2.5, 1.8, 2.0, 1.3, 1.6, 1.1, 4.4, 3.0) are most similar to measurements of modern *B. b. brevicauda*. This specimen is represented at the top of Fig. 22 within the centroid showing the range of variation of *B. b. brevicauda* from Iowa and Nebraska. These data suggest that Mills County, Iowa, was within the zone of intergradation of the two semispecies of *B. brevicauda* 23,000 years ago just as it is now (Moncrief et al., 1982), and was inhabited by a mixture of large, small, and intermediate-sized shrews. The name *B. b. brevicauda* is appropriate for intergrades recovered from the Craigmile l.f.

Garrett Farm

The Garrett Farm fauna was recovered from alluvial sediments along Waubonsie Creek in Mills County, Iowa (NW ¼, NE ¼, NW ¼, SW ¼ sec. 8, T71N, R42W). A radiocarbon date of 3,400 BP was obtained (Fay, 1980).

Remains of plants, molluscs, and vertebrates were recovered. Fay (1980) listed 21 mammalian taxa from Garrett Farm, of which three (*Neotoma ?micropus, Clethrionomys gapperi,* and what he termed *B. b. kirtlandi*) no longer occur in southwestern Iowa. Mammals associated with steppe habitat dominate the fauna, but boreal and deciduous forest taxa also are present (Fay, 1980). Fay suggested that the allopatric mammals of the Garrett Farm and Pleasant Ridge faunas represent "the last stage of adjustment from Hypsithermal to modern climatic conditions." The modern vegetation and climate of Iowa are summarized in the discussion of the Cherokee Sewer l.f.

Fay (1980) identified *B. b. kirtlandi* (MNI = 2) and *B. carolinensis* (MNI = 1) from Garrett Farm. An isolated P4, M2, and a dentary with M1–3 (SUI 46886) were examined for the present study. Measurements 3 through 6 (2.4, 1.9, 1.3, 2.4), 11 through 14 (1.7, 1.8, 2.4, 2.0), and 19 through 26 (2.2, 1.5, 1.8, 1.2, 1.5, 1.0, 3.6, and 2.5) of these specimens were taken and compared with measurements from the reference samples (Table 1). Measurements of the three specimens from Garrett Farm are most similar to those of the *talpoides* semispecies of *B. brevicauda,* and the specimens possibly pertain to the Holocene subspecies *B. b. kirtlandi.*

Mud Creek

The Mud Creek fauna was recovered from six sites on the banks of Mud Creek, north of Durant, Iowa. The sites were located from the west line of sec. 1 (T79N, R1W) in Cedar County to the east line of sec. 7 (T79N, R1E) in Scott County. The description of the fauna by Kramer (1972) serves as the basis for this report.

The fauna consists of molluscs, fishes, herptiles, and mammals. The mammalian fauna consists of insectivores, rodents, rabbits, deer, and bison. Four species, including what Kramer termed *B. b. kirtlandi,* no longer occur in the area. Sympatry of extant mammals occurs north and east of Iowa (Kramer, 1972).

A radiocarbon date of 6,220 ± 110 BP was reported for the Mud Creek l.f. Members of the fauna indicate the presence of stream and forest habitats. Climate was somewhat cooler and moister, similar to that of modern Michigan, Illinois, Indiana, Ohio, and Pennsylvania (Kramer, 1972). A pronounced meteorological gradient evidently occurred between Mud Creek and the more xeric Cherokee I site (Semken, 1980).

Presently, the mean annual precipitation of Cedar and Scott counties is 81 to 86 cm (NOAA, 1974). Natural vegetation of the area is summarized in the discussion of the Cherokee Sewer Site.

Kramer (1972) referred specimens (MNI = 4) from Mud Creek to *B. brevicauda* and *B. b. kirtlandi.* Measurements 3 through 6 of an isolated P4 (2.3, 1.9, 1.4, 2.0), 7 through 10 of three m1s (2.2, 2.1, 2.4, 2.5; 2.1, 2.1, 2.2, 2.3; and 2.2, 2.1, 2.2, 2.5), 19 and 20 of a m1 (2.2, 1.5), and 21 and 22 of a m2 (1.8, 1.2) were obtained for the present study (SUI 35940A and D). Measurements of these teeth were compared with means obtained for those measurements from samples of Holocene specimens (Table 1). Measurements of the Mud Creek specimens are smaller than those of the *B. b. brevicauda* inhabiting most of Iowa today. Comparisons indicate that the specimens from Mud Creek pertain to the *talpoides* semispecies of *B. brevicauda,* possibly to the subspecies *B. b. kirtlandi.*

Pleasant Ridge

The Pleasant Ridge local biota was recovered from deposits 100 m from the Garrett Farm site in Mills County, Iowa (SE ¼, NW ¼, NW ¼, SW ¼ sec. 8, T71N, R42W). A radiocarbon date of 1,450 BP was reported by Fay (1980).

The Pleasant Ridge fauna consists of molluscs and vertebrates. Plant remains also were recovered. Twenty-four mammalian taxa are present; *Neotoma floridana* is the only species not present in the modern fauna. Mammals of steppe, deciduous forest, and boreal habitats are represented equally in the fossil fauna (Fay, 1980).

Fay (1980) referred remains of *Blarina* (MNI = 5) from Pleasant Ridge to the subspecies *B. b. brevicauda.* For the present study, measurements 3 through 6 (2.5, 2.2, 1.5, 2.6), 11 through 14 (1.8, 1.7, 2.5, 2.1), and 19 and 20 (2.2, 1.5) of a P4, M2, and m1 from Pleasant Ridge were taken (SUI 45973). These measurements were compared with those obtained from modern reference samples (Table 1). This comparison tends to support Fay's original identification of the Pleasant Ridge material as *B. b. brevicauda.*

Schmitt Cave

Schmitt Cave is located in western Dubuque County, Iowa (T88N, R1W, sec. 31), near the town of Farley. The site is a rock shelter and cave located at the base of a cliff, approximately 40 m north of John Creek. Excavation of artifacts was described by Reese (1972). Accumulation of artifacts and faunal remains occurred approximately 1,900 to 1,300 BP (Reese, 1972; Semken, personal communication, 1982). Faunal remains represent inhabitants of the cave and immediate surroundings as well as species hunted by man (Eshelman, 1972).

Eshelman (1972) described the Holocene fauna, which consists of 26 fishes, herptiles, birds, and mammals, which was recovered from the cave. All occur presently in northeastern Iowa except *Erethizon dorsatum.* Sympatry occurs in southwestern Wisconsin and the eastern half of Iowa, indicating little or no climatic changes since deposition. Permanent stream (probably John Creek), stream border, forest, and lowland meadow-prairie communities are represented (Eshelman, 1972). Modern climate and vegetation of Iowa are summarized in the discussion of the Cherokee Sewer Site.

Remains of *Blarina* (SUI 34989) were recovered from levels 1–3, which represent the top 18 inches of the deposit. The nine specimens (MNI = 5) from the site were referred to the species *B. brevicauda*

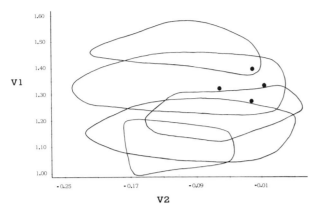

Fig. 23.—Canonical analysis of measurements 19–24 of lower teeth of specimens of *Blarina* from Schmitt Cave. Ellipses enclose variation in samples of Holocene taxa of *Blarina* (see Fig. 4).

by Eshelman (1972). Two maxillaries and five dentaries were examined for the present study. Measurements 3 through 16 of upper teeth of the two maxillaries (2.5, 2.0, 1.6, 2.6, 2.4, 2.2, 2.4, 2.6, 1.7, 1.7, 2.6, 2.2, 0.8, 1.6; 2.4, 1.9, 1.4, 2.3, 2.2, 2.0, 2.3, 2.4, 1.7, 1.7, 2.3, 2.0, 0.8, 1.6) are most similar to those of Holocene *B. b. kirtlandi, B. b. manitobensis,* and *B. b. churchi* (groups 4, 6, 18, and 11 in Table 1). Measurements 19 through 24 of lower teeth of the five dentaries also are comparable to those of modern *B. brevicauda* (*talpoides* semispecies). In canonical analyses these five specimens (one was hidden) are plotted within the centroid showing the range of variation of the *talpoides* semispecies of Holocene *B. brevicauda* (Fig. 23). The seven specimens from Schmitt Cave are assigned to the species *B. brevicauda,* and probably pertain to the subspecies *B. b. kirtlandi.*

Thurman

The Thurman l.f. was recovered near Thurman, Iowa, from the Fremont County Quarry (NW ¼, NW ¼ sec. 23, T70N, R43W). The fauna was described in an unpublished thesis (Jenkins, 1972).

Seeds and remains of gastropods, pelecypods, fishes, amphibians, reptiles, birds, and mammals were recovered from floodplain deposits in the quarry. Jenkins (1972) listed 14 mammalian species from the site, only one of which (*Mammuthus columbi*) is extinct. The plant and animal remains suggest a temperate climate, possibly with more abundant moisture and more equable temperatures than at present (Jenkins, 1972; Graham and Semken, 1976).

A radiocarbon date of 9,800 ± 150 BP was obtained (Graham and Semken, 1976). The presence of mammoths suggests deposition prior to 10,000 BP (Jenkins, 1972).

Four dentaries pertaining to the genus *Blarina* were recovered from the quarry (SUI 35869 and 35888). Graham and Semken (1976) thought the remains represent two taxa, *B. b. brevicauda* and *B. b. carolinensis* [=*B. hylophaga*]. Measurements 19 through 26 of three specimens (2.0, 1.3, 1.7, 1.1, 1.5, 1.0, 3.5, 2.3; 2.0, 1.4, 1.7, 1.2, 1.4, 1.0, 3.4, 2.4; 2.1, 1.4, 1.8, 1.2, 1.4, 1.0, 3.5, 2.2) and 21 through 26 of the remaining specimens (2.1, 1.3, 1.7, 1.1, 4.5, 2.6) were taken for the present study. These measurements were compared with those of Holocene reference samples (Table 1). These comparisons indicate that the three smaller specimens represent the species *B. brevicauda* (*talpoides* semispecies). The larger specimen apparently pertains to *B. b. brevicauda* (of the *brevicauda* semispecies). We assume co-occurrence of these two phena in the Thurman l.f. represents sympatry in southwestern Iowa during the late Wisconsinan/early Holocene period. The two semispecies of *B. brevicauda* remain sympatric in Fremont County, Iowa, today.

Waubonsie

The Waubonsie l.f. was found in a channel fill above the unit from which the Craigmile l.f. was recovered. A radiocarbon date of 14,830 + 1,060, −1,220 RCYBP was obtained from charcoal fragments associated with the vertebrate remains. The Waubonsie l.f. was described by Rhodes (1982).

Remains of fishes, herptiles, birds, and mammals were recovered. Twenty-three mammalian taxa were represented, of which 14 occur presently in Mills County. As at Craigmile, preservation was poor and microtines were the most numerous individuals present. Eight species that have been extirpated locally occur sympatrically in central Alberta. The region of maximum sympatry (17 of 22 taxa considered) is in central Wisconsin and central Minnesota in areas of prairie-forest transition. The area of maximum sympatry in Wisconsin is the same as that of Craigmile (Rhodes, 1982).

The Waubonsie l.f. indicates the presence of conifer and hardwood forests and boreal grasslands (Hallberg et al., 1974; Rhodes, 1982). Temperatures probably were similar to those of Craigmile, but effective precipitation was greater. Rhodes (1982) suggested that the "environmental mosaic" was more strongly developed because of changes in topography and climate.

Rhodes (1982) assigned remains of *Blarina* (SUI 46361, MNI = 7) to the subspecies *B. b. brevicauda.* He noted that four of five measurable mandibles fell within an area of overlap between *B. b. brevicauda* and *B. b. kirtlandi* when using Graham and Semken's (1976) mandibular index.

One complete upper toothrow was examined for the present study. Measurements 3 through 10 (2.4, 2.0, 1.4, 2.4, 2.2, 2.1, 2.3, 2.5) compare most favorably with measurements of reference samples of the *talpoides* semispecies of *B. brevicauda* (Table 1). Three of the mandibles seen are represented in the

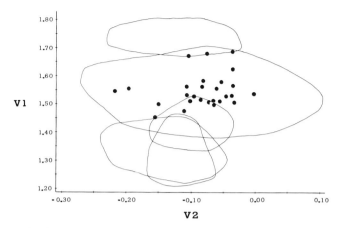

Fig. 24.—Canonical analysis of measurements 3–10 of upper teeth of specimens of *Blarina* from the Willard Cave l.f. Ellipses enclose variation in samples of Holocene taxa of *Blarina* (see Fig. 3).

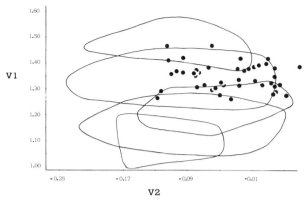

Fig. 25.—Canonical analysis of measurements 19–24 of lower teeth of specimens of *Blarina* from the Willard Cave l.f. Ellipses enclose variation in samples of Holocene taxa of *Blarina* (see Fig. 4).

canonical diagram in Fig. 22. One specimen fell within the centroid representing the range of variation of the *talpoides* semispecies of *B. brevicauda*; the other two plotted within the area of overlap between the *talpoides* and *brevicauda* semispecies of *B. brevicauda*. Measurements of additional fossils with incomplete dentition are similar to those of the specimens included in canonical analyses. We, too, find that these measurements fall into an area of overlap between those of Holocene *B. b. brevicauda* and the smaller *talpoides* semispecies. As with the Craigmile l.f., the Waubonsie l.f. appears to have been deposited in the zone of intergradation between the *brevicauda* and *talpoides* semispecies of *B. brevicauda*. However, fossils of *Blarina* from Waubonsie appear slightly smaller than modern specimens from Iowa, and for that reason we suggest that they be assigned to the *talpoides* semispecies of *B. brevicauda*, subspecies *B. b. kirtlandi*.

Willard Cave

The Willard Cave l.f. was recovered from Delaware County, Iowa (NE ¼, SW ¼, SW ¼ sec. 2, R4W, T90N). The fauna was described in a thesis (Eshelman, 1971) that currently is being prepared for publication (Eshelman, in litt., 1982).

Remains of 37 vertebrate species were recovered from Willard Cave. All species are extant, but eight no longer occur in northeastern Iowa. Thirty-one mammalian species were identified; 74% of the individuals represented pertain to species associated with deciduous forest (Eshelman, 1971). Semken (1980) noted that Willard Cave is the first record of prairie species in eastern Iowa.

The species in the fossil fauna presently are allopatric. Deposition might have occurred during more equable climatic conditions, with a mosaic of habitats in the vicinity (Eshelman, in

litt., 1982). The modern vegetation of the area is summarized in the discussion of the Cherokee Sewer Site.

Radiocarbon dates indicate that accumulation occurred between 3,500 and 1,200 BP (Eshelman, 1971). Eshelman (1971) noted similarities between the Willard Cave l.f. and the faunas of Crankshaft Cave, Meyer Cave, New Paris No. 4, and Robinson Cave. He suggested that the Willard fauna accumulated more recently than these faunas, as evidenced by the absence of extinct taxa.

The genus *Blarina* is represented throughout the deposit by 327 specimens (MNI = 54). These remains (SUI 35185 and 50041) were referred to the subspecies *B. b. brevicauda* by Eshelman (1971) and Graham and Semken (1976).

Twenty-seven maxillaries and 60 dentaries, representing the most complete of all the specimens, were examined for the present study. Analyses of dental measurements of these specimens (Figs. 24 and 25) indicate that the specimens from Willard Cave pertain to the *talpoides* semispecies of *B. brevicauda*.

Jinglebob

Fossils were found in the Jinglebob Pasture of the XI Ranch (UM-K2-47, SW ¼ sec. 32, T33S, R29W) in Meade County, Kansas. The Jinglebob is one of four Rancholabrean sites in Kansas from which specimens of *Blarina* have been recovered.

The fauna consists of molluscs, fishes, herptiles, and mammals. Fragments of avian bones were recovered but have not been studied (Hibbard, 1955). Twenty-three mammalian species have been identified, of which 10 are extant (Hibbard, 1955, 1970). Six vertebrate species, including *Blarina* cf. *brevicauda*, presently occur north or south of Meade County. Taylor and Hibbard (1955) and Hibbard (1963) compared the fauna with others from southwestern Kansas and northwestern Oklahoma.

The fauna was aged as Sangamonian, but more recently it was considered early Wisconsinan (Hibbard, 1970; Kapp, 1970; Zakrzewski, 1975a). The molluscs and vertebrates, particularly the presence of both northern and southern elements in the fauna, indicate more equable climate during the period of deposition (Hibbard, 1955, 1970; Martin, 1968). Grassland with heavy cover and permanent streams were present (Hibbard, 1955; Hibbard and Taylor, 1960). Pine and spruce pollen were abundant in pollen spectra (Kapp, 1970). The modern climate and vegetation of the area were described in the discussion of the Cudahy l.f.

A relatively great diversity of voles and shrews was present in the Jinglebob fauna (Graham, 1976). Eleven specimens of *Blarina* were referred to *B.* cf. *brevicauda* (Hibbard, 1955, 1970). The remains consist of parts of five maxillaries (UMMP 29264, 29265, and 29761) and six dentaries (UMMP 29266–70). All specimens were examined for the present study. The sample appears homogeneous. Measurements of the fossils were compared with those of Holocene specimens (Table 1). Measurements of both upper and lower molars are similar to those of modern *B. brevicauda* and *B. hylophaga*. Measurements 3 through 6 of three P4s (2.3, 1.9, 1.3, 2.2; 2.3, 1.8, 1.4, 2.3; 2.2, 1.8, 1.3, 2.1), 7 through 10 of three M1s (2.1, 1.9, 2.2, 2.3; 2.0, 1.9, 2.1, 2.2; 2.0, 1.8, 2.1, 2.2), and 11 through 14 of two M2s (1.8, 1.6, 2.3, 1.9; 1.7, 1.7, 2.2, 1.8) fall within the range of measurements of modern *B. carolinensis, B. brevicauda* (*talpoides* semispecies), and *B. hylophaga*. Measurements 19 and 20 of four m1s (2.2, 1.5; 2.0, 1.4; 2.1, 1.4; 2.0, 1.4), 21 and 22 of five m2s (1.8, 1.2; 1.8, 1.2; 1.8, 1.2; 1.7, 1.1; 1.7, 1.1), and 23 and 24 of three m3s (1.4, 1.0; 1.3, 1.0; 1.4, 1.0) also are comparable with those of Holocene *B. carolinensis, B. brevicauda,* and *B. hylophaga*. However, measurements of the remains from Jinglebob are most similar to those of *B. brevicauda,* and we think that the fossils pertain to that species.

Mt. Scott

Fossils of the Mt. Scott l.f. were found at three sites on the Big Springs Ranch in Meade County, Kansas. The three sites are UM-K4-53 (SE ¼ sec. 14, T32S, R29W), UM-K2-59 (SE ¼, SE ¼ sec. 18, T32S, R28W), and UM-K1-60 (SW ¼, SW ¼ sec. 13, T32S, R29W). Molluscs only were recovered from a fourth site, UM-K3-60 (NE ¼, SW ¼ sec. 17, T32S, R28W) (Miller, 1966).

In addition to molluscs, remains of fishes, amphibians, reptiles, birds, and mammals were found at the Mt. Scott sites. Six orders of mammals were represented by 25 species (Hibbard, 1963). Mt. Scott was compared to other Pleistocene faunas of Kansas and Oklahoma by Hibbard (1963) and Semken (1966). Six of the extant species, including *Blarina carolinensis,* do not occur presently in Meade County (Hibbard, 1963).

The Mt. Scott l.f. is considered late Illinoian. Deposition occurred during a period of more effective moisture and more equable temperatures than now present (Hibbard, 1963, 1970; Smith, 1963; Miller, 1966). The fossil fauna indicates the presence of a permanent stream, marsh, and shrubs and trees; grassland was

nearby (Etheridge, 1961; Hibbard, 1963; Smith, 1963; Nelson and Semken, 1970). A shift from colder to warmer temperatures occurred after the early Illinoian, accompanied by the appearance of more southern elements as seen in the Mt. Scott fauna (Hibbard, 1970). Pollen from Mt. Scott was poorly preserved, but pollen spectra imply expansion of open plant communities during the period of deposition (Kapp, 1970). Modern climate and vegetation of Meade County are discussed in the section about Cudahy.

Remains identified as *B. b. carolinensis* were recovered from all three sites (Hibbard, 1963). Eighteen dentaries (UMMP 37751, 43804–12, 41233, 44591, and 44600), seven maxillaries (UMMP 43813–15, 44591), and an isolated incisor (UMMP 43816) were found, and all were examined for the present study.

Incomplete measurements were obtained from the seven maxillaries (measurements 7 through 14: 2.1, 2.0, 2.3, 2.4, 1.7, 1.7, 2.3, 1.8; 3 through 14: 2.1, 1.6, 1.2, 2.1, 1.9, 1.8, 2.1, 2.2, 1.5, 1.5, 2.1, 1.7; 3 through 14: 2.1, 1.7, 1.3, 2.1, 1.9, 1.8, 2.1, 2.2, 1.7, 1.6, 2.2, 1.8; 3 through 6: 2.1, 1.7, 1.5, 2.1; 7 through 16: 2.1, 2.0, 1.9, 2.1, 1.7, 1.6, 2.0, 1.7, 0.6, 1.3; 7 through 14: 2.1, 2.0, 2.0, 2.1, 1.7, 1.6, 2.0, 1.7; 11 through 14: 1.7, 1.7, 2.2, 1.8). Comparison with measurements of the reference samples (Table 1) reveals that measurements of these fossils fall within the overlapping ranges of variation of extant *B. carolinensis* and *B. hylophaga*. The same conclusion was reached during analysis of the dentaries from Mt. Scott. Results of canonical analyses of the ten most complete specimens (one observation was hidden) are shown in Fig. 26. We tentatively retain the previous assignment of the Mt. Scott material to the species *B. carolinensis*.

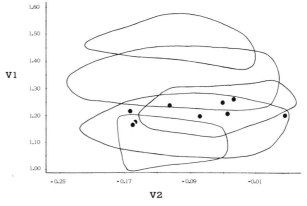

Fig. 26.—Canonical analysis of measurements 19–24 of lower teeth of specimens of *Blarina* from the Mt. Scott l.f. Ellipses enclose variation in samples of Holocene taxa of *Blarina* (see Fig. 4).

Robert

Two specimens of *Blarina* were identified in the Robert l.f. of southwestern Kansas. This Wisconsinan fauna was recovered from SW ¼, SW ¼ sec. 33, T34S, R29W, Meade County (locality UM-K1-57).

The Robert l.f. consists of molluscs, fishes, herptiles, birds, and mammals. No pollen has been found associated with the fauna. All species are extant. Fifteen species of small mammals have been identified; six species (*Sorex cinereus, S. palustris, B. brevicauda, Thomomys* cf. *T. talpoides, Microtus pennsylvanicus,* and *Zapus* cf. *Z. hudsonius*) no longer occur in Meade County (Schultz, 1967, 1969).

Members of the Robert l.f. suggest the presence of marshes and meadows, and nearby woods and upland prairies (Schultz, 1969). A more equable climate, moister and cooler than at present, is indicated (Schultz, 1967, 1969; Miller, 1975). Sympatry of 11 species of the fossil fauna occurs in northeastern South Dakota (Schultz, 1969). The number of shrew and vole species also is correlated with mesic climate (Graham, 1976). A description of the modern climate and vegetation of the area is included in the discussion of the Cudahy l.f.

The Robert is one of the more recent local faunas of southwestern Kansas (Zakrzewski, 1975a). A radiocarbon date of 11,100 ± 390 BP was reported by Schultz (1967, 1969), who thought that the Robert l.f. could not be correlated with any of the other faunas in the vicinity.

The remains of *Blarina* consist of an isolated incisor (UMMP V46004) and a dentary with p4–m1 (UMMP V46022). This material was referred to *B. b. brevicauda* by Schultz (1967). Measurements 19 and 20 (2.3 and 1.6) of the molar indicate that the fossil pertains to the species *B. brevicauda* (Table 1). The two measurements are somewhat smaller than the means of the three samples from Iowa, Nebraska, and Manitoba (5, 12, and 18 in Table 1), although they fall within the range of variation of these three groups. Therefore, the previous assignment of the Robert material to the subspecies *B. b. brevicauda* tentatively is retained.

Williams

The Williams l.f. was recovered from Rice County in central Kansas. A description of the fauna has not yet been published. One new species, *Sorex kansasensis,* was described by McMullen (1975). The stratigraphy and composition of the fauna suggest that deposition occurred during more xeric conditions before the Wisconsinan glacial (Semken, personal communication). Zakrzewski (1975a) considered the fauna Illinoian.

Rice County now is part of the Great Bend Prairie physiographic province (Self, 1978). Natural vegetation consists of bluestem-grama prairie north of the Arkansas, floodplain vegetation along the river, and sand prairie to the south (Küchler, 1974). Mean annual precipitation is 61 to 71 cm (NOAA, 1974).

The genus *Blarina* is represented by a fragmented dentary with m2–3 and an isolated incisor (UMMP V60338). Measurements 21 through 24 of the molars are 2.0, 1.3, 1.6, and 1.1. These measurements are most similar to those of Holocene *B. brevicauda* (Table 1), and the fossil undoubtedly pertains to that species. It might pertain to the subspecies *B. b. brevicauda.*

Welsh Cave

Welsh Cave is located west of the Cumberland Plateau approximately 5.6 km southwest of Troy, Woodford County, Kentucky (84°44′50″W and 37°52′25″N), at an elevation of about 270 m. A preliminary description of the cave and its fauna was provided by Guilday et al. (1971). A subsequent change in ownership of the cave has prevented further research at the site.

Welsh Cave is a "small and relatively featureless" sinkhole that was formed by the collapse of a more extensive cave system. Bones generally were in good condition although there was evidence of washing. The surface of the deposit contained mixed Holocene and fossil material. Materials deeper in the deposit appeared contemporaneous although no stratification was observed. A carbon-14 date of 12,950 ± 550 years BP (obtained from a sample of *Platygonus* bone) was considered applicable to most members of the fauna (Guilday et al., 1971).

Twenty-five mammalian species were identified from the Welsh Cave deposit. Of these, four species now are extinct and thirteen no longer occur in the area. The most common mammal was the thirteen-lined ground squirrel (*Spermophilus tridecemlineatus*), which was a common species in late Pleistocene faunas in the Appalachians but currently is restricted to the prairies of central North America (Guilday et al., 1971). The absence of *Tamias* and the presence of *Spermophilus* and *Geomys bursarius* indicate the existence of grassland communities in the area at the time of deposition (Guilday et al., 1971). A northern element, consisting primarily of species of the Canadian/Hudsonian Zone, also was present in the fossil fauna. Of the 22 extant species represented, the modern ranges of 19 overlap in an area in eastern Minnesota and western Wisconsin. This area of overlap suggests "boreal semi-prairie or parkland" in the Welsh Cave region at the time of deposition (Guilday et al., 1971). In the analysis by Graham (1976), the higher density of vole and shrew species during the late Wisconsinan also was interpreted as evidence of more equable climate at the time of deposition. Presently, the area is surrounded by farmland and is considered typical of the eastern Carolinian biotic province as described by Dice in 1943 (Guilday et al., 1971).

Three genera of soricids were recovered from the deposit. The genus *Blarina* is represented by a right mandible with complete dentition (CM 20144), a mandibular fragment, and an isolated incisor (MNI = 2). The two jaws are of dissimilar size and were assigned tentatively to the taxa *B. b. kirtlandi* and *B. b. brevicauda* (Guilday et al., 1971). Graham and Semken (1976) considered these animals identical with the *B. kirtlandi* (which they assumed was a separate species) and *B. b. brevicauda* from Meyer Cave, Illinois.

The right mandible was examined in the present study. Measurements 17 through 26 (6.0, 4.6, 2.2, 1.5, 2.0, 1.2, 1.6, 1.1, 4.4, and 2.7, respectively) were compared to the means for those measurements for the samples of extant animals. The specimen from Welsh Cave was within the lower end of the range of variation for *B. b. brevicauda* (approximately the same size as *B. b. manitobensis*), and undoubtedly pertains to the species *B. brevicauda*.

Catalpa Creek

One specimen of *Blarina* was recovered from deposits along Catalpa Creek, Lowndes County, Mississippi. Extant and extinct vertebrates have been recovered from the site. The deposits have been reworked, so that the age of the specimen (Pleistocene or Holocene) is impossible to determine (M. Frazier, in litt., 1980). The specimen is housed in the Mississippi Museum of Natural Science (locality cat 1, 47N, 45E).

The specimen is a right mandible with one molar. Measurements 21 and 22 (1.9 and 1.3, respectively) were taken. Comparisons were made with the measurements of samples of modern *Blarina* from the southeastern states (Table 1). Measurements of the specimen from Catalpa Creek fell within the range of variation of the Holocene sample of *B. carolinensis* from Alabama (sample 1) and were slightly larger than the means for the sample of *carolinensis* from Georgia (sample 3). The nearest site that included Pleistocene specimens, Ladds Quarry (Georgia), contained specimens similar in size to that from Catalpa Creek. The specimen from Catalpa Creek therefore is referred to the species *B. carolinensis*.

Bat Cave

Bat Cave is located in Pulaski County, Missouri, in the SE ¼, NW ¼, NE ¼ sec. 4, T36N, R12W (Hawksley et al., 1973). This cave, together with the two subsequent ones, are three of 10 caves in the Ozark Plateau of Missouri that are known to contain remains of Pleistocene faunas (Saunders, 1977). Bat Cave is in Bear Ridge, overlooking the Gasconade River.

Most of the bones were recovered from a crawlway approximately 150 m long (north of the main entrance and main upper passage) known as Bone Passage, and from a room (the Devil's Kitchen) adjacent to Bone Passage. Bones of bats and fragments of *Ursus americanus* were the only remains found outside of Bone Passage and the Devil's Kitchen. Possibly there were at least two modes of deposition of the fossil material. Some of the remains, including material from a short-faced bear, *Arctodus simus,* were found in what seemed to be the original site of a "bear bed." The random distribution of other bones and the nature of the matrix suggest alluvial deposition. The bones might have been carried into the passage or moved within the passage after original deposition by water. The species represented in the fauna and the physical features of the cave indicate that animals entered the cave of their own volition or were carried in by predators. The presence of *Cryptobranchus* suggests the cave was flooded, in which case some of the mammals might have drowned in the cave (Hawksley et al., 1973).

Non-mammalian species represented in the Bat Cave fauna are limited to a gastropod, three genera of amphibians, two snakes, and a passerine. Fourteen families of mammals are represented, of which three species are extinct and nine species currently occur north of the Ozarks. Attempts to analyze pollen from Bone Passage were unsuccessful. Comparisons with the faunas of Boney Spring, Missouri (with radiocarbon dates of 16,500 to 13,500 years BP), Crankshaft Cave, Missouri, and Meyer Cave, Illinois, indicate that the fauna of Bat Cave accumulated in a relatively short period of time (possibly 16,500 to 13,500 years BP) during the late Wisconsinan (Hawksley et al., 1973).

Presently, the bluffs in this area are "covered with second or third growth deciduous forests dominated by oaks and hickories." Average annual precipitation is 91 to 112 cm (Hawksley et al., 1973).

Unlike the faunas of Crankshaft and Zoo caves, only one phenon of *Blarina* was identified from Bat Cave. The three identified remains (MNI = 2) initially were referred to the subspecies *B. b. brevicauda*. The scarcity of material and the presence of only one phenon of *Blarina* were interpreted as indicating deposition during a limited period of relatively stable climate. The lack of a natural trap in this cave probably contributed to the low number of soricids in this fauna (Hawksley et al., 1973).

In the present study, two dentaries (CMSU 388 and 389) with m1-m3 present were examined. Measurements 19–24 for these specimens were 2.3, 2.2; 1.6, 1.5; 2.0, 1.8; 1.4, 1.2; 1.7, 1.6; 1.0, 1.0, respectively. Multivariate analyses were not run on the two specimens; however, comparisons of these measurements with those of Holocene specimens revealed that the fossils are most similar to Holocene shrews from Indiana, Ohio, New York, and Ontario (Table 1, samples 4, 5, 13, and 19). Thus, the fossils are most similar to the *talpoides* semispecies of *B. brevicauda* and likely pertain to that species.

Boney Spring

Boney Spring is located 16 km south of Warsaw in Benton County, Missouri (NW ¼, SW ¼, SW ¼ sec. 29, T39N, R22W) at an elevation of 215 m. The spring is one of eight sites in the Ozark region from which *Blarina* has been recovered (Saunders, 1977).

Boney and Trolinger springs are located in the lower Pomme de Terre River valley. The region is one of transition between oak-hickory forest and western plains (Saunders, 1977). Interdisciplinary research was performed in the area prior to inundation by the Harry S. Truman Reservoir. Saunders (1977) described the historical background, excavation, deposition, and faunas of the two sites.

Remains of ostracods, insects, amphibians, reptiles, and mammals were recovered from Boney Spring. The ostracod and herptile faunas are modern in aspect. Saunders (1977) listed 22 mammalian taxa from the site. Five extinct species (*Paramylodon harlani, Castoroides ohioensis, Mammut americanum, Equus* sp., and *Tapirus* sp.) are present, and *M. americanum* (MNI = 31) is the most abundant species represented in the fossil fauna. *Microtus pennsylvanicus* presently occurs north and northeast of Boney Spring; it has been reported from northern Missouri but not from Benton County (Saunders, 1977; Hall, 1981). *Napaeo-zapus insignis* is found in forested or brushy areas in southeastern Canada, the northeastern United States, and the Appalachians (Saunders, 1977; Hall, 1981). The remaining species of mammals are part of the modern fauna of the study area and are found in prairie, forest border, oak-hickory forest, and bottomlands forest and prairie habitats (Saunders, 1977).

The assemblage from Boney Spring represents at least two communities, one aquatic and one terrestrial (Saunders, 1977). The remains of terrestrial vertebrates and associated pollen suggest the presence of spruce forest and grassland and cool, humid conditions during the period of deposition (Mehringer et al., 1968, 1970; Hallberg et al., 1974; Saunders, 1977). Accumulation of the Boney Spring fauna occurred approximately 16,200–13,600 BP (Saunders, 1977).

Twenty-eight specimens (MNI = 4) of *Blarina* from Boney Spring were referred to the species *B. brevicauda* by Saunders (1977). Fifteen specimens, part of the collection of the Illinois State Museum (E. H. Lindsay field numbers 309–312, 314–317, 474, and BS71), were examined. With the exceptions of one maxillary fragment (EHL 311) and one dentary fragment (EHL BS71), all specimens consist of edentulous fragments and isolated teeth. The sample appears homogeneous. Measurements 3 through 6 of three P4s (2.3, 2.0, 1.4, 2.2; 2.4, 1.8, 1.3, 2.3; 2.3, 2.1, 1.5, 2.4), 7 through 10 of two M1s (2.3, 2.1, 2.2, 2.7; 2.2, 2.2, 2.2, 2.5), 11 through 14 of two M2s (1.6, 1.6, 2.1, 1.9; 1.7, 1.6, 2.3, 2.0), 19 and 20 of four m1s (2.4, 1.6; 2.2, 1.5; 2.2, 1.6; 2.1, 1.5), 21 and 22 of four m2s (2.0, 1.3; 1.8, 1.2; 1.9, 1.2; 1.9, 1.2), and 23 and 24 of two m3s (1.6, 1.1; 1.7, 1.1) were taken. Measurements are similar to those of specimens from Bat and Zoo caves and to larger specimens from Crankshaft Cave (Table 2). Comparisons with the measurements of reference samples (Table 1) suggest referral of the Boney Spring material to the *talpoides* semispecies of the Holocene species *B. brevicauda*.

Brynjulfson Cave No. 1

The Brynjulfson caves are located 5 m above Bonne Femme Creek, approximately 10 km southeast of Columbia, in Boone County, Missouri (SW ¼, NE ¼, SW ¼ sec. 16, T47N, R21W). The faunas of both caves were described in a report by Parmalee and Oesch (1972), on which this summary is based.

Brynjulfson Cave No. 1 contained the older of the two faunas. A radiocarbon date of 9,440 ± 760 BP was obtained from a sample of bone. Gastropods, fishes, herptiles, birds, and mammals were present. Remains of forty-five species of mammals have been identified; nine species (*Brachyprotoma* sp., *Canis dirus*, cf. *Megalonyx*, *Dasypus bellus*, *Tapirus* sp., *Platygonus compressus*, cf. *Sangamona*, *Alces* or *Cervalces*, and *Symbos* sp.) are extinct (Parmalee and Oesch, 1972; Saunders, 1977). Seven species no longer occur in Missouri. Some of the remains probably represent inhabitants of the cave; bones also might have

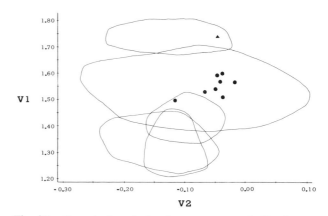

Fig. 27.—Canonical analysis of measurements 3–10 of upper teeth of specimens of *Blarina* from Brynjulfson Cave No. 1 (one observation, represented by a triangle) and Brynjulfson Cave No. 2. Ellipses enclose variation in samples of Holocene taxa of *Blarina* (see Fig. 3).

been carried in by water and by predators (Parmalee and Oesch, 1972).

The presence of *Clethrionomys*, *Alces* or *Cervalces*, *Sangamona*, and *Symbos* suggest cool, moist boreal conditions during the period of deposition. Plentiful remains of beaver indicate the presence of open-water marsh. Grassland habitat also was indicated (Parmalee and Oesch, 1972).

In the vicinity of the caves, forests and meadows are present and swamps occur in the valley bottoms (Parmalee and Oesch, 1972). Mean annual precipitation at Columbia is 94 cm (NOAA, 1974).

Only two soricids were found in the Brynjulfson deposits. At Brynjulfson Cave No. 1, the genus *Blarina* is represented by seven specimens (MNI = 2). Parmalee and Oesch (1972) referred this material to the subspecies *B. b. brevicauda*.

Fragments of two maxillaries and two dentaries (housed at ISM) were examined for the present study. The upper teeth of at least one maxillary (represented by the triangle in Fig. 27) appear large enough to merit assignment to the *brevicauda* semispecies of the species *B. brevicauda*. Measurements 3 through 6 (2.4, 2.0, 1.5, 2.3), measurements 19 through 23 of three lower molars (2.1, 1.4, 1.7, 1.1, 1.6), and measurements 19 through 26 of three lower molars and dentary (2.2, 1.5, 1.9, 1.2, 1.6, 1.0, 4.2, 2.6) are more similar to those of the *talpoides* semispecies of *B. brevicauda*. These three specimens are comparable in size to the fossils from Brynjulfson No. 2.

Brynjulfson Cave No. 2

This site is located approximately 137 m from Brynjulfson Cave No. 1. The methods of accumulation were the same for both.

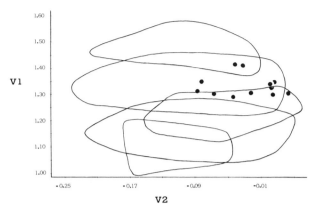

Fig. 28.—Canonical analysis of measurements 19–24 of lower teeth of specimens of *Blarina* from Brynjulfson Cave No. 2. Centroids enclose variation in samples of Holocene taxa of *Blarina* (see Fig. 4).

A radiocarbon date of 2,460 ± 230 BP was obtained for a sample of bone from this fauna, which consists of gastropods, pelecypods, fishes, herptiles, birds, and mammals. Only three (*Dasypus bellus, Canis dirus,* and *Platygonus compressus*) of the 37 species of mammals are extinct (Parmalee and Oesch, 1972; Saunders, 1977). The species *Clethrionomys gapperi, Erethizon dorsatum,* and *Antilocapra americana* no longer occur in Missouri. Fish remains probably indicate periodic flooding. Small mammals are more numerous and boreal species less prevalent than in the fauna of Cave No. 1. Prairie, forest, and marsh habitats are indicated (Parmalee and Oesch, 1972). The presence of species such as the pronghorn and the northern pocket gopher indicate a trend toward increasing aridity after the last glacial retreat (Parmalee and Oesch, 1972).

The genus *Blarina* is represented by 116 remains (MNI = 21) which were referred to the subspecies *B. b. brevicauda* by Parmalee and Oesch (1972). For the present study, nine maxillaries and 11 dentaries were examined. Measurements 3 through 10 of eight of the most complete maxillaries were taken. Results of canonical analyses of these specimens (Fig. 27) demonstrate the similarity of these measurements with those of Holocene samples of the *talpoides* semispecies of the species *B. brevicauda*. Results of analyses of eleven dentaries are shown in Fig. 28; these observations also are plotted within the centroid showing the range of variation of modern *B. brevicauda*. We consider the remains from the deposit at Brynjulfson No. 2 to pertain to that species.

Crankshaft Cave

Crankshaft Cave is located in the SW ¼, SW ¼, SE ¼, NE ¼, NW ¼ sec. 9, T43N, R5E, in Jefferson County, Missouri (Vineyard, 1964, in Oesch, 1967). The entrance to the cave is a small sinkhole that opens into a shaft approximately 1.5 m wide. The floor of the cave lies 20 m below the entrance. The cave acts as a natural trap for animals (Parmalee et al., 1969).

A preliminary list of the fauna of Crankshaft Cave was published by Oesch in 1967; Parmalee et al. published a second account in 1969. Remains of molluscs, herptiles, and birds were identified as genera that inhabit the Ozark region today. Of the mammals, seven (possibly eight) are extinct species typical of late Pleistocene faunas. Remains of 16 Holocene species no longer present in Missouri suggest boreal or short-grass prairie habitats (Parmalee et al., 1969). Attempts were made to date the fauna more precisely using carbon-14 techniques on four samples of bone. The dates obtained seemed too young when compared with those from similar materials from other sites. Possibly the groundwater in Crankshaft Cave altered the composition of the bones and made them useless for dating (Parmalee et al., 1969).

The site presently is surrounded by steep hillsides covered with mixed hardwood forest. Average annual precipitation is approximately 80 cm (Parmalee et al., 1969).

The genus *Blarina* is the most abundant component of the mammalian remains from Crankshaft Cave, with at least 663 individuals represented (Parmalee et al., 1969). Mandibles, teeth, and cranial fragments of *Blarina* initially were referred to what were assumed to be two Holocene subspecies, *B. b. brevicauda* and *B. b. carolinensis* (Oesch, 1967; Parmalee et al., 1969). The implication was that one subspecies had geographically replaced the other as temperature fluctuated with the passage of time. Graham and Semken (1976) subsequently referred these materials (CMSU, ISM uncatalogued material) to the nominal taxa *kirtlandi* and *carolinensis,* which they assumed were contemporaneous species.

In the present study, univariate statistics for the fossils from Crankshaft (CRANKS) are shown in Table 2. Canonical analysis separated 22 maxillae (Fig. 29) into two clusters. Smaller specimens plotted within the centroid for *B. hylophaga,* whereas larger specimens clustered primarily within the centroid for the largest subspecies (*B. b. brevicauda*) of the species *B. brevicauda*. The two species were clearly discriminated by this analysis. Measurements of lower teeth were analyzed twice (Figs. 30 and 31). Most specimens were plotted within the centroids for *B. b. brevicauda,* the smaller subspecies of *B. brevicauda,* and *B. hylophaga*. Four specimens were plotted outside of the range of variation for the Holocene samples of *Blarina* examined in this study. Further analysis (Fig. 32), in which measurements of upper teeth from Crankshaft were compared directly with those of Holocene samples, showed more clearly the similarities between the Crankshaft material (two ellipses numbered 36) and the modern *B. brevicauda* and *B. hylophaga*. Based on these canonical analyses, the modern distribu-

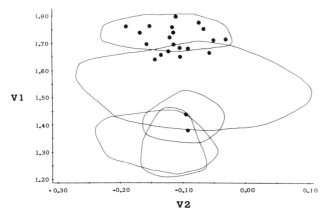

Fig. 29.—Canonical analysis of measurements 3–10 of upper teeth of specimens of *Blarina* from the Crankshaft Cave l.f. Centroids enclose variation in samples of Holocene taxa of *Blarina* (see Fig. 3).

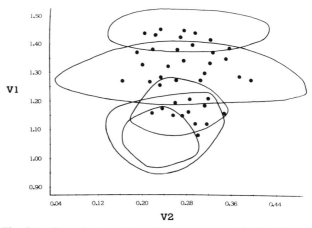

Fig. 31.—Canonical analysis of measurements 19–22 of lower teeth of specimens of *Blarina* from the Crankshaft Cave l.f. Centroids enclose variation in samples of Holocene taxa of *Blarina* (see Fig. 30).

tion of *Blarina* in Missouri, and variation in Holocene samples of *Blarina* from Missouri (Moncrief et al., 1982), the fossils from Crankshaft Cave clearly pertain to the species *B. brevicauda* and *B. hylophaga*.

Trolinger Spring

Trolinger Spring is located 4.5 km southeast of Boney Spring and 19.5 km southeast of Warsaw in Hickory County, Missouri (NE ¼, NW ¼, NE ¼ sec. 9, T38N, R22W) at an elevation of 223 m. Like Boney Spring, Trolinger Spring is in the Pomme de Terre River valley (Saunders, 1977).

The fauna of Trolinger Spring consists of 137 specimens representing seven species of mammals (*B. brevicauda*, *Peromyscus* spp., *Synaptomys* sp., *Mammut americanum*, *Mammuthus* sp., *Equus* sp., and *Symbos* sp.). The most numerous remains in this

deposit are those of the mastodont (MNI = 11). The three extant species occur in the modern fauna of Hickory County (Saunders, 1977).

Accumulation occurred in a peat bog between 29,000 and 34,000 BP (Saunders, 1977). The fossils and associated pollen suggest deposition during the mid-Wisconsinan interstadial when open pine parkland was predominant in the region (Mehringer et al., 1970; Saunders, 1977). The pollen record from Boney and Trolinger springs shows a transition between approximately 32,000 and 13,500 BP from nonarboreal pollen and pine dominance to spruce dominance to spruce with deciduous elements (Mehringer et al., 1970).

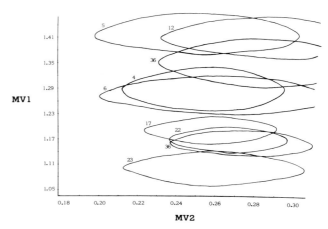

Fig. 32.—Canonical analysis comparing maxillary specimens from the Crankshaft Cave l.f. with specimens from the Holocene reference samples. The Crankshaft specimens are designated by the two ellipses numbered 36. Reference samples are numbered as indicated in the text. Ellipses indicate one standard deviation about the mean.

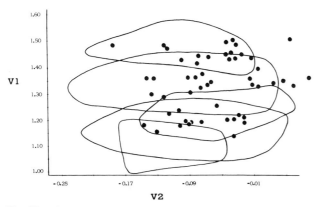

Fig. 30.—Canonical analysis of measurements 19–24 of lower teeth of specimens of *Blarina* from the Crankshaft Cave l.f. Centroids enclose variation in samples of Holocene taxa of *Blarina* (see Fig. 4).

The *Blarina* material, like that of other small mammals from Trolinger Spring, is poorly preserved. Two lower incisors (MNI = 1) were recovered and are housed at the Illinois State Museum (E. H. Lindsay field numbers 150 and 151). The specimens are referred to the species *B. brevicauda*, following Saunders (1977).

Zoo Cave

Zoo Cave is located 1.6 km northeast of Hilda, Taney County, Missouri. The entrance to the cave is low and narrow, and opens into a steeply descending crawlway approximately 6.1 m long. The passage opens into a wide room known as the Armadillo Room. A maze area is south of this room, and in the past there probably were numerous openings from the maze area to the outside. To the north of the Armadillo Room is an area known as the Bone Passage, where most of the bone material has been found. The remains might have been carried into the cave by predators or by water; the fact that no articulated remains have been found suggests the animals did not die in the cave (Hood and Hawksley, 1975).

Fossil material in Zoo Cave was recovered from two deposits. The older of the two deposits contained remains of large vertebrates, including three extinct mammals. The presence of *Dasypus bellus* is considered an indication of deposition during a warm period. Radiocarbon dates of specimens of the three extinct species from other Pleistocene faunas were used to estimate the period of deposition as approximately 9,000 to 13,000 years BP. The younger deposit, in which stratification was present, contained remains of small vertebrates and invertebrates. These animals serve as the best indicators of climate, but several conflicting communities are indicated by the array of gastropods and vertebrates in this deposit. The second deposit probably is more recent than 9,000 BP. Attempts at pollen analysis were unsuccessful (Hood and Hawksley, 1975).

Remains of 59 species were identified in the fauna of Zoo Cave. Non-mammalian representatives include gastropods (aquatic and terrestrial), pelecypods, a crustacean, fishes, and herptiles. All these animals occur in Taney County at present except the fox snake, *Elaphe vulpina* (the modern range of which might include parts of northern Missouri). The remains of fishes and of a hellbender (*Cryptobranchus*) indicate a permanent stream community. The other herptiles include members of deciduous forest, prairie, and streamside communities. Thirty-two species of mammals were present, of which three now are extinct and 10 no longer present in Taney County. One species for which the present distribution is primarily western was present. Seven species (including four shrews) now occur primarily to the north. The area of current sympatry of the eight northern species (seven mammals and the fox snake) was described as an area of mixed conifers, hardwoods, savanna/parkland, and prairie in western Wisconsin and eastern Minnesota (Hood and Hawksley, 1975).

Zoo Cave is located in the Mark Twain National Forest. The site is surrounded by deciduous forest dominated by oaks and hickories with numerous cedar glades (Hood and Hawksley, 1975). Average annual precipitation is approximately 101 to 111 cm (NOAA, 1974).

The soricids of Zoo Cave were neither as diverse nor as numerous as those of Crankshaft Cave (Par-

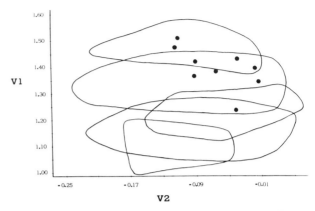

Fig. 33.—Canonical analysis of measurements 19–24 of lower teeth of specimens of *Blarina* from the Zoo Cave l.f. Centroids enclose variation in samples of Holocene taxa of *Blarina* (see Fig. 4).

malee et al., 1969). Remains of the short-tailed shrew were referred to the nominal taxa *B. b. brevicauda* (MNI = 4), *B. b. carolinensis* (MNI = 1), and *B. b. kirtlandi* (MNI = 5). Based on earlier assignment of Holocene shrews in southern Missouri to the taxon *carolinensis*, Hood and Hawksley (1975) surmised that the presence of the more northerly taxa *brevicauda* and *kirtlandi* (as well as *Sorex arcticus* and *S. cinereus*) indicated deposition during a cooler climatic period.

Nine dentaries (CMSU, uncatalogued) with three molars present were analyzed for the present study. Of these, one specimen plotted within the area of overlap of the centroids for *B. carolinensis*, *B. hylophaga*, and the smaller subspecies (*talpoides* semispecies) of *B. brevicauda* (Fig. 33). The remainder were plotted within the centroids for *B. b. brevicauda* and the smaller subspecies of *B. brevicauda*. Based on these data, it cannot be determined unequivocally that these specimens represent more than one species. Tentatively, based on the work by Hood and Hawksley (1975) and Moncrief et al. (1982), the *Blarina* of this fauna is referred to the species *B. brevicauda*.

Doby Springs

The Pleistocene fauna of Doby Springs was collected at five localities approximately 15 km west of Buffalo, Oklahoma (N ¼, SW ¼ sec. 10, T27N, R24W, Harper County). Fossils were found in lake sediments deposited in a collapse basin Illinoian in age. Stephens (1960) described the geology and local fauna.

The Doby Springs l.f. consists of ostracods, molluscs, fishes, herptiles, birds, and mammals. Smith (1958), Etheridge (1961), Brattstrom (1967), and Jammot (1972) discussed the fishes, lizards, snakes, and shrews (*Sorex*) of the fauna. Stephens (1960) and Hibbard (1963) listed 24 species of mammals tentatively identified from the site. Doby Springs is the type locality for *Peromyscus oklahomensis*; seven additional extinct species (*Castoroides* cf. *C. ohioensis, Peromyscus* cf. *P. cochrani, Mammuthus* sp., *Camelops* sp., *Bison* cf. *B. latifrons, Equus* cf. *E. niobrarensis,* and *E.* sp.) were reported.

Stephens (1960) considered the Doby Springs l.f. Illinoian in age on the basis of its stratigraphy and composition of species. The Doby Springs l.f. was compared with the Berends l.f. (Oklahoma) and other Pleistocene faunas in southwestern Kansas, and was found most similar to the Berends l.f. (Stephens, 1960). The Doby Springs and Berends l.f.'s both were considered early Illinoian by Hibbard (1970).

The fauna and associated pollen suggest a cooler climate (with more effective moisture than at present) and the presence of lake, marsh, lowland meadow, and shrub and tree communities during the period of deposition (Smith, 1958; Stephens, 1960; Etheridge, 1961; Brattstrom, 1967; Hibbard, 1963, 1970; Kapp, 1970). Sympatry of extant species of fishes and mammals occurs in the southeastern corner of North Dakota (Stephens, 1960).

One specimen, UMMP 34752, was referred to the species *B. brevicauda* by Stephens (1960). The specimen is a partial left maxillary with P2 and P4; it was illustrated in Fig. 5 by Stephens (1960). This fossil was noted as larger than Holocene material examined at the University of Michigan. Hibbard (1970) referred the fossil to *B. b. fossilis* on the basis of size. For the present study measurements 3 through 6 (2.6, 2.2, 1.5, and 2.4) were taken and were compared with measurements of the reference samples (Table 1). The measurements of the specimen from Doby Springs fall within the range of measurements of samples of *B. brevicauda.* They are most similar to the means of samples 5 and 12 of *B. b. brevicauda,* and the specimen tentatively is assigned to that subspecies.

Bootlegger Sink

Bootlegger Sink is located 0.4 km east of Emigsville, York County, Pennsylvania (40°01′00″N and 76°43′30″W) at an elevation of approximately 107 m. Guilday et al. (1966) described the sinkhole and its fauna.

Remains were recovered from dolomite breccia which occurred as a ledge projecting from the walls of the sinkhole. The ledge is a remnant of surface-derived talus that once filled the sink (Guilday et al., 1966). Attempts were made to date the deposit by fluorine and radiocarbon analyses. Results of the fluorine analysis suggest continuous deposition as the climate changed from boreal to temperate. The carbon-14 date (3,722 ± 200 BP) is the minimum date for brecciation. Pollen analyses were negative. Based on these analyses and the composition of the fauna, accumulation probably occurred from the late Wisconsinan (10,000 to 15,000 BP) through 3,000 or 4,000 years ago (Guilday et al., 1966).

Invertebrate remains from the deposit at Bootlegger Sink were limited to two species of millipede and one snail. Amphibian material consisted of remains of an unidentified salamander and five species of toads and frogs. One species of box turtle, *Ter-*

rapene carolina, was recovered and was considered to imply temperate conditions during the later period of accumulation. Six species of snakes were represented; all occur in the region today except the mole snake, *Lampropeltis calligaster rhombomaculata,* which has a present distribution in the southeastern United States (Guilday et al., 1966).

Remains of 37 species of mammals, representing six orders and a minimum of 97 individuals, were identified from Bootlegger Sink. Eight species no longer occur in York County. The thirteen-lined ground squirrel, *Spermophilus tridecemlineatus,* currently is restricted primarily to the plains of central North America. The long-eared bat, *Plecotus* sp., presently occurs in the southern half of the United States; it and the mole snake were the only species identified from the sink deposit that do not occur as far north as the site today (Guilday et al., 1966). Other species represented that are no longer present in York County—*Sorex arcticus, Glaucomys sabrinus, Synaptomys borealis, Microtus xanthognathus, Napaeozapus insignis,* and *Rangifer* cf. *R. tarandus*—occur to the north of the site (Hall, 1981). Sympatry of the two voles, *M. xanthognathus* and *M. chrotorrhinus,* suggests boreal conditions during at least part of the period of deposition. Bootlegger Sink is one of seven Appalachian sites at which the two species were collected together (Guilday, 1971; Guilday et al., 1977). The remaining species occur in the area today. Such a mixed fauna suggests deposition over a long period of time, as indicated by the fluorine analysis, during varying climatic conditions (Guilday et al., 1966).

Bootlegger Sink is in the Limestone Valley Section of the Piedmont Province. The site is included in the Carolinian biotic province as described by Dice (1943), but the original deciduous forest has been removed and the valley now is farmed. Mean annual precipitation is 101.8 cm (Guilday et al., 1966).

After the preliminary excavation, the genus *Blarina* was represented by three mandibles, a humerus, and isolated teeth (MNI = 2). This material (CM 7924) was referred to the species *B. brevicauda* by Guilday et al. (1966). Bootlegger Sink was among the eleven local faunas examined by Graham (1976) in which greater species density of shrews and voles was interpreted as an indication of more equable climate during the Pleistocene. He also indicated that the Bootlegger material pertained to *B. brevicauda.*

Measurements 3 through 10 of two maxillaries (2.8, 2.0, 1.5, 2.8, 2.4, 2.2, 2.5, 2.9; 2.7, 1.9, 1.7, 2.7, 2.4, 2.3, 2.5, 2.7), 17 through 24 of two dentaries (6.3, 5.1, 2.5, 1.7, 2.0, 1.4, 1.6, 1.1; 6.1, 4.9, 2.5, 1.8, 2.0, 1.4, 1.6, 1.1), and 17 through 23 of a third dentary (5.9, 4.6, 2.1, 1.5, 2.2, 1.4, 1.5) were taken for the present study (CM 7924 and uncatalogued material). Measurements of these specimens were compared with those of Holocene specimens. Although measurements of the fossils vary, they are most similar to means derived from the two samples (5 and 12) of extant *B. brevicauda* from Nebraska and Iowa. The fossil sample appears homogeneous.

New Paris No. 4

New Paris Sinkhole No. 4 is located 2.4 km northeast of New Paris (at 40°5'N and 78°39'W), Bedford County, Pennsylvania. Guilday et al. (1964) published a detailed description of the sinkhole and its fauna, on which this report is based.

Originally the sinkhole was a bell-shaped domepit, formed by circulation of groundwater in the limestone underlying Chestnut Ridge, on which the sinkhole is located. A narrow vertical fissure, 10 m in depth, was located alongside the domepit and connected to it by an underground passage. In the past this fissure was open to the surface, and the fossils recovered from this site were remains of animals that fell down this fissure and were trapped. The large domepit also was opened to the surface by the collapse of its ceiling, but apparently animals were not trapped because of its large opening and rapid rate of fill. Subsequently both the fissure and the domepit were filled by surface-derived material; no stratification was evident. The rate of accumulation of the fill material evidently was faster in the early stages, when the local climate was cooler, than in the final stages. Preservation generally was good; many skeletons were recovered uncrushed and articulated although evidence of disturbance by rodents was noted. A radiocarbon date of 11,300 ± 1,000 BP was obtained from a sample of charcoal (Guilday et al., 1964).

The fauna of the New Paris Sinkhole No. 4 consists of invertebrates (millipedes and snails), amphibians, reptiles, birds, and mammals. The invertebrates all are present in the area today, and it was difficult to separate Pleistocene and Holocene specimens. More than 2,700 vertebrates were recovered. Of the amphibians (salamanders, toads, and frogs) all species occur in the area today with the exception of the northern leopard frog (with modern records from northwestern Pennsylvania) and a tentatively identified subspecies of toad (*Bufo americanus copei*) of the Hudson Bay area (Guilday et al., 1964; Conant, 1975). The snakes represented in the deposit are species currently in the area. The absence of lizards and turtles was due to the boreal climate of the area during the time of deposition (Guilday et al., 1964). Avian remains are scarce (MNI = 9), and all but one of the seven species present search for food on the ground. This material includes remains of the sharp-tailed grouse (*Pedioecetes phasianellus*) and the ruffed grouse (*Bonasa umbellus*). The presence of these birds suggests the forest canopy was closed during the later stages of deposition, although it is possible that the two species were not in the immediate area contemporaneously (Guilday et al., 1964). Remains of these two species also were reported from the Virginian deposits of Back Creek Cave No. 2, Clark's Cave, and Natural Chimneys (Guilday et al., 1977).

The mammalian fauna consists of 40 species of insectivores, bats, rodents, lagomorphs, carnivores, and artiodactyls. The one artiodactyl (*Mylohyus nasutus,* the long-nosed peccary) was the only extinct species present, although an extinct subspecies of squirrel (*Tamiasciurus hudsonicus tenuidens*) also was present. Most mammals still are present in the state but not necessarily at the site. Seven species (*Sorex arcticus, S. tridecemlineatus, Dicrostonyx hudsonius, Phenacomys* cf. *ungava, Synaptomys borealis, Microtus xanthognathus,* and *Mustela nivalis*) no longer occur in the state. The ground squirrel (*S. tridecemlineatus*) presently ranges northward to the edge of the Canadian biotic province, and is an indicator of grassland near the site during the period of deposition. The collared lemming (*D. hudsonius*) indicates the presence of nearby tundra, possibly on Allegheny Mountain or on the crest of Chestnut Ridge (Guilday et al., 1964). The five remaining species presently occur in a variety of boreal habitats in the Hudsonian and Canadian zones (Guilday, 1971; Guilday et al., 1964; Hallberg et al., 1974). The fauna, correlated with analysis of pollen from the deposit, shows a gradual transition from boreal parkland to boreal forest; both vertebrate and pollen remains show more temperate elements above the 6 m level. All mammals from the bottom meter of the excavation occur at least in part in the Hudsonian or Canadian biotic provinces (Guilday et al., 1964).

The sinkhole is located at an altitude of 465 m. The area presently is included in the Ridge and Valley physiographic province. In Pennsylvania, this province is characterized by alternating forested ridges and farmed valleys. The region is not rugged but is sufficiently mountainous that distribution of temperature and precipitation is not uniform (NOAA, 1974). Chestnut Ridge lies within the rain shadow of Allegheny Mountain. The site is relatively arid and jack pine may have been present during the late Pleistocene (Guilday et al., 1964). The biota generally is transitional between the Carolinian and Canadian biotic provinces of Dice (1943), but Guilday et al. (1964) pointed out the local variability in topography which influences biotic distribution.

Guilday et al. (1964) referred specimens of *Blarina* from New Paris No. 4 (MNI = 37, 38.9% of all soricids present) to the taxa *B. b. brevicauda* and *B. b. kirtlandi.* The short-tailed shrews were the most common insectivores in the upper and middle levels of the deposit, but were absent from the lower level. Average size of the remains increased toward the bottom of the deposit, probably due to increasing contamination by modern *B. b. kirtlandi* in the upper levels. The larger *Blarina,* "presumably representing a large northern stock that advanced to the south with mounting glacial conditions," were comparable to the remains found at Robinson Cave, Natural Chimneys, and Clark's Cave (Guilday et al., 1969, 1977). The stratigraphy of the *Blarina* remains was considered supporting evidence for the trend from boreal parkland to boreal forest indicated by pollen and remains of other vertebrates (Guilday et al., 1964). Graham and Semken (1976) detected what they thought were three phena of *Blarina* in the material from New Paris No. 4; they interpreted these as sympatric taxa equivalent to extant *brevicauda, carolinensis,* and *kirtlandi.* The taxon *carolinensis* apparently was based on one specimen, which was not seen in the present study. Graham (1976) considered the greater species density of shrews during the Wisconsin as evidence for "a more equable climate with reduced temperature and moisture gradients," which correlates well with the paleoecological hypotheses of Guilday et al. (1964).

For the present analysis, nine specimens (CM 5843, 5845, 7437, and 7802) were examined. Univariate statistics are shown in Table 2. Six maxil-

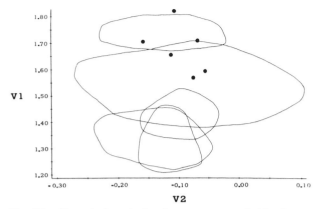

Fig. 34.—Canonical analysis of measurements 3–10 of upper teeth of specimens of *Blarina* from the New Paris No. 4 l.f. Centroids enclose variation in samples of Holocene taxa of *Blarina* (see Fig. 3).

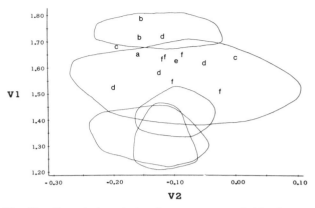

Fig. 35.—Canonical analysis of measurements 3–10 of upper teeth of specimens of *Blarina* from the Baker Bluff Cave l.f. Centroids enclose variation in samples of Holocene taxa of *Blarina* (see Fig. 3).

laries fell within the range of variation for modern samples of *B. brevicauda* (Fig. 34). One maxillary that had incomplete measurements appeared identical to the specimens illustrated in Fig. 34. The two dentaries examined had measurements that compared most favorably with modern specimens of *B. brevicauda* from Nebraska, Iowa, and Manitoba. Accordingly, all specimens from New Paris No. 4 examined for this study are referred to the species *B. brevicauda*, but we cannot say unequivocally that only one species of *Blarina* is represented in the fauna.

Baker Bluff Cave

Remains of *Blarina* have been recovered from several cave sites in Tennessee, three of which are located in Sullivan County (Appendix 1 and Fig. 2). Sullivan County is included in the Valley and Ridge physiographic region, which Miller (1974) described as "characterized by numerous elongate ridges and intervening valleys, all trending in a northeast-southwest direction." The area is in the Carolinian biotic province of Dice (1943). The region is covered by timbered ridges and rolling farmland; annual precipitation at Knoxville is 119 cm (Guilday et al., 1978).

Baker Bluff Cave is located 13 km southeast of Kingsport, at 36°27'30"N and 82°28'W, Boone Dam quadrangle U.S.G.S. 7½' topographic map (Guilday et al., 1978). The cave is a large chamber, approximately 10 m long, located at an elevation of 450 m in a bluff above the South Fork of the Holston River. A detailed description of the cave and its fauna was provided by Guilday et al. (1978).

A 3 m column was excavated, from which remains of 180 taxa of invertebrates and vertebrates were recovered. Stratigraphic levels were designated arbitrarily. The herptiles of the Baker Bluff l.f. are present in the region today. Two species of birds and 10 species of mammals do not occur presently in the area. Six species of mammals are extinct. Thirty-one per cent of the mammalian species have modern ranges in boreal forest, temperate grassland,

or in higher elevations in the southern Appalachians. In the faunal analysis by Guilday et al. (1978), it was noted that the smaller mammals found in cooler areas today were more abundant in the upper levels of the deposit.

Carbon-14 dates were obtained from bone fragments from three stratigraphic levels in the deposit. The basal date was 19,100 ± 850 BP. The dates might be too young due to contamination by plant roots and burrowing by rodents. Based on analysis of distribution of the small mammals in the deposit, Guilday et al. (1978) hypothesized that deposition occurred during a transition from cool-temperate to boreal conditions.

Guilday et al. (1978) identified seven species of shrews, including two phena of *Blarina*, from Baker Bluff Cave. Specimens referred to *B. b. kirtlandi* were found in all stratigraphic levels. The larger *B. b. brevicauda* occurred in the upper levels (3 to 7 feet) and was absent from the lower three feet of the deposit. The latter phenon was considered comparable to *B. b.* cf. *brevicauda* from Clark's Cave, Virginia. As hypothesized by Graham and Semken (1976), Guilday et al. (1978) assumed that coexistence of the two phena in the higher levels of the deposit indicated cooler, more equable climate at the time of deposition.

For the present study, 140 maxillary and dentary fragments (CM 29746–53) were examined. Univariate statistics for these specimens are shown in Table 2. The individuals shown in Figures 35 and 36 were separated stratigraphically: a from the 3'–4' level, b from the 4'–5' level, c from the 5'–6' level, d from the 6'–7' level, e from the 7'–8' level, f from the 8'–9' level, and g from the 9'–10' level. Seventeen maxillary fragments were sufficiently complete for analysis of upper tooth measurements (Fig. 35). Two

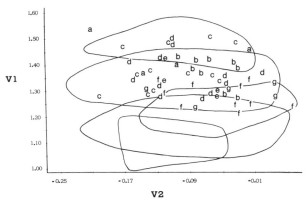

Fig. 36.—Canonical analysis of measurements 19–24 of lower teeth of specimens of *Blarina* from the Baker Bluff Cave l.f. Centroids enclose variation in samples of Holocene taxa of *Blarina* (see Fig. 4).

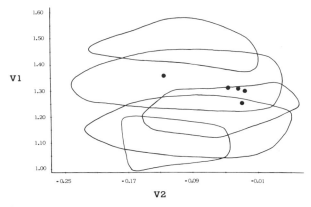

Fig. 37.—Canonical analysis of measurements 19–24 of lower teeth of specimens of *Blarina* from the Guy Wilson Cave l.f. Centroids enclose variation in samples of Holocene taxa of *Blarina* (see Fig. 4).

observations (from strata c and d) were hidden within the centroid for the *talpoides* semispecies of *B. brevicauda*. Although sample size from each stratigraphic level is small, the plotted specimens show the occurrence of large specimens of *B. brevicauda* in strata b and d and smaller specimens of *B. brevicauda* in all strata shown except b. Results of analysis of measurements of lower teeth are shown in Fig. 36. Twenty-three observations (from strata c, d, f, and g) are hidden, and one observation from stratum d fell outside the range of the figure (to the right of the centroid for the *talpoides* semispecies of *B. brevicauda*). The specimens plotted in the centroid for *B. b. brevicauda* in Fig. 36 are from strata a, b, c, and d, which are the uppermost strata of the deposit (3'–7'). Smaller specimens of *B. brevicauda* occurred in all six strata. These results correspond well with those reported by Guilday et al. (1978). There is no doubt that all specimens represented pertain to the species *B. brevicauda*.

Guy Wilson Cave

Guy Wilson Cave is located approximately 4 km south of Bluff City, Sullivan County, Tennessee (Corgan, 1976). The cave is on the south side of the South Fork of the Holston River.

Corgan (1976) reported that at least 25 species of mammals from this locality have been identified by J. E. Guilday. Nine species (of the orders Carnivora, Proboscidea, Perissodactyla, Artiodactyla, Edentata, and Rodentia) have been mentioned in publication. Remains from three caves in Sullivan County (Guy Wilson, Baker Bluff, and Beartown) provide the southernmost record of the caribou *Rangifer tarandus* in eastern North America (Guilday et al., 1975). Guy Wilson, Robinson, Baker Bluff, and Carrier Quarry caves (all in Sullivan County) are noted by Guilday (1971) as sites that indicate the southernmost distribution

of boreal coniferous forest. A late Wisconsinan date (19,700 ± 600 BP) for Guy Wilson Cave was obtained from carbon-14 dating of a sample of peccary bone (Guilday et al., 1975).

We examined 18 specimens (16 dentaries and two maxillaries) of *Blarina* from Guy Wilson Cave. Of these, only five dentaries were sufficiently complete for analysis of measurements 19–24 of the lower teeth. The five specimens fell within the range of *B. brevicauda* (Fig. 37). Measurements of eight of the remaining dentaries fell within the range of variation for those measurements of the samples of extant *B. b. kirtlandi* (samples 4 and 6). Two other dentaries had two measurements slightly larger than those of extant animals, and one specimen appeared smaller than modern *kirtlandi* in five measurements. Measurements of the two maxillaries compared favorably with the means of those measurements of specimens from samples 4 and 6. The species represented is *B. brevicauda*.

Riverside Cave

Riverside Cave is located in Sullivan County, Tennessee. Its fauna has not been described formally. Remains consist largely of isolated teeth and fragmented jaws, probably accumulated from the activities of owls and woodrats. The remains probably include mixed late Wisconsinan and Holocene material (J. E. Guilday, in litt., 1982). Two genera in the fauna, *Sangamona* and *Mylohyus,* now are extinct. The material identified as *Sangamona* (J. E. Guilday, in litt.) might pertain to the fugitive deer, *S. fugitiva,* which has been identified from eight deposits (probably all Wisconsinan) in the central and eastern United States, including the Robinson and Whitesburg local faunas of Tennessee (Kurtén and Anderson, 1980). The long-nosed peccary, *M. nasutus,* has been reported from 35 Irvingtonian and Rancholabrean faunas from the southeastern quarter of the United States

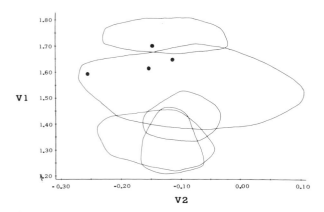

Fig. 38.—Canonical analysis of measurements 3–10 of upper teeth of specimens of *Blarina* from the Riverside Cave l.f. Centroids enclose variation in samples of Holocene taxa of *Blarina* (see Fig. 3).

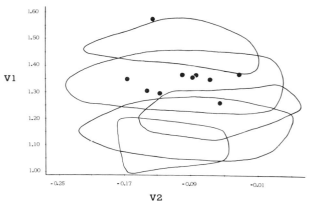

Fig. 39.—Canonical analysis of measurements 19–24 of lower teeth of specimens of *Blarina* from the Riverside Cave l.f. Centroids enclose variation in samples of Holocene taxa of *Blarina* (see Fig. 4).

(Kurtén and Anderson, 1980). In Tennessee this species probably was associated with moist forests and wooded valleys. Remains also were identified as *Tapirus* sp. (J. E. Guilday, in litt., 1982), two species of which have been recorded from eight sites in eastern Tennessee and which may have been associated with deciduous forest (Corgan, 1976). The remaining taxa identified from the deposit are *Sorex cinereus, S. fumeus, Blarina, Parascalops breweri, Myotis* sp., *Eptesicus fuscus, Neotoma floridana, Peromyscus* sp., *Clethrionomys gapperi, Synaptomys cooperi, Microtus (Pitymys)* cf. *pinetorum, Tamias striatus, Sciurus* sp., *Glaucomys* cf. *volans, Zapus hudsonius,* cf. *Sylvilagus* sp., *Ursus americanus, Procyon lotor, Spilogale putorius, Odocoileus virginianus,* and *Cervus elaphus* (J. E. Guilday, in litt., 1982). All of these species have occurred in eastern Tennessee in modern times (Hall, 1981).

Twenty-seven specimens of *Blarina* (CM, uncatalogued material) from the Riverside Cave l.f. were examined. The results of the analysis of maxillary fragments are shown in Fig. 38; one incomplete specimen had measurements similar to those of the four specimens shown. These specimens were plotted within or near the centroid showing the range of variation for modern specimens of the smaller subspecies (*talpoides* semispecies) of *B. brevicauda.* Twenty-two dentaries were examined, nine of which also were plotted within the range of variation of the *talpoides* semispecies of *B. brevicauda* and one near the centroid for *B. b. brevicauda* (Fig. 39). One observation was hidden. With the exception of the one large specimen, the measurements of dentaries were comparable with the means of those measurements for Holocene samples, and were most similar to the Holocene subspecies *manitobensis* and *churchi* (samples 18 and 11). Given the small sample

size, specimens from this deposit appear more similar to fossils from Baker Bluff than to specimens from the Guy Wilson and Robinson caves, and apparently represent only the species *B. brevicauda.*

Robinson Cave

Robinson Cave is located approximately 13 km southwest of Livingston, Overton County, Tennessee (36°17′25″N and 85°22′25″W). It is in the east slope of Maxwell Mountain at an elevation of 366 m (Guilday et al., 1969).

Although bones were found in three sites in the cave, almost all were discovered in the locality known as the Armadillo Pit. The floor of Armadillo Pit consists of talus from an extinct sinkhole. Guilday et al. (1969) thought deposition in Armadillo Pit might have occurred in three stages. The first stage was represented by remains of armadillo and large mammals. A late Sangamonian or early Wisconsinan age was suggested by fluorine dating techniques and the presence of *Dasypus bellus* during this phase of deposition. The second phase was indicated by the deposition of bats and other small vertebrates; the sinkhole might have begun to close so that large animals were not trapped. During the third phase the ceiling was closed, allowing entrance only through the main cave. Remains accumulated during the third period of deposition were uncrushed and articulated. The time of deposition in the second and third stages is thought to be late or post-Wisconsinan, although a precise date has not been obtained.

Most of the animals from Armadillo Pit were bats and woodrats. The other animals entered the cave by their own volition, were trapped in the sinkhole, or were deposited by predators. "Remains of at least 2,615 individual mammals, 14 individual birds, 10 species of reptiles and amphibians, and 461 individual gastropods were recovered from approximately 5 square yards of matrix from the floor of this pit . . ." (Guilday et al., 1969). Six of the vertebrate species are extinct. Forty-nine of the 54 living species are now present in Minnesota and Wisconsin, suggesting a boreal-temperate parkland in the area at the time of deposition (Guilday et al., 1969). Graham (1976) also noted the

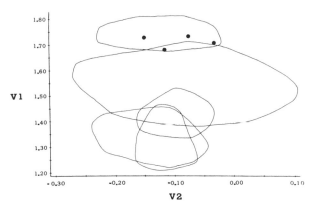

Fig. 40.—Canonical analysis of measurements 3–10 of upper teeth of specimens of *Blarina* from the Robinson Cave l.f. Centroids enclose variation in samples of Holocene taxa of *Blarina* (see Fig. 3).

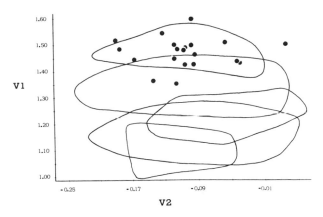

Fig. 41.—Canonical analysis of measurements 19–24 of lower teeth of specimens of *Blarina* from the Robinson Cave l.f. Centroids enclose variation in samples of Holocene taxa of *Blarina* (see Fig. 4).

boreal aspect of the fauna, and suggested a more equable climate at the time of deposition. Guilday (1971) listed Robinson as one of four Appalachian sites that might indicate the southernmost distribution of dense coniferous forests. No mammals that are presently southern in distribution were found.

The site is within the Eastern Highland Rim physiographic region, which was characterized by Miller (1974) as nearly level and underlain by limestone. Guilday et al. (1969) described the immediate vicinity of the cave as an area of broad, cultivated valleys surrounded by uplands "covered with second or third growth deciduous woodland . . . ." Mean annual precipitation is approximately 142 cm (NOAA, 1974).

Specimens of *Blarina* (MNI = 83) from Robinson Cave were identified as *B. brevicauda*. It was noted that they are similar in size to the subspecies *B. b. brevicauda* presently in the Minnesota–Wisconsin area (Guilday et al., 1969). Graham (1976) also considered the animals *B. brevicauda*.

For the present study, 28 specimens of *Blarina* (CM 8133, 8134, 8146) were examined. Measurements of the upper teeth of four specimens (Fig. 40) were analyzed; these specimens fell within the range of variation for Holocene *B. b. brevicauda* from Iowa and Nebraska. In the analysis of lower teeth, two specimens plotted within the centroid for the smaller *talpoides* semispecies of *B. brevicauda*, whereas other specimens (including four hidden observations) plotted near or within the range of variation of Holocene *B. b. brevicauda* (Fig. 41). Univariate statistics for fossils from Robinson Cave (ROBINS) are shown in Table 2. These results are in keeping with the general boreal aspect of the fauna of Robinson Cave, and indicate that only one species, *B. brevicauda* (*brevicauda* semispecies), was represented.

Barton Springs Road Site

The Barton Springs Road Site is located in Travis County, Texas. It is one of several sites on the Edwards Plateau from which late Pleistocene faunas have been recovered. All of the Texan sites from which *Blarina* have been recovered (Appendix 1), except Ben Franklin, Easley Ranch, Howard Ranch, Rex Rodgers, and the Vera faunule, are located on the Edwards Plateau.

The Edwards Plateau was included in the Comanchian biotic province by Dice (1943). He described the vegetation on the plateau as "oaks and junipers . . . often mixed with . . . grass and mesquite, and on steep rocky slopes these trees may form closed stands. It is only rarely that any of them reach a height of over twenty feet. In many places on the plateau the characteristic vegetation is grass, and trees and shrubs are absent or occur only in very open stands." Ford and Van Auken (1982) further described the vegetation on the southeastern part of the plateau. The canyons along the eastern and southern edges are cooler and more humid than the rest of the plateau. They provide ecological diversity today and might have done so in the Wisconsinan (Lundelius, 1967; Dalquest et al., 1969).

In his analysis of the late Wisconsinan faunas and the paleoecology of central Texas, Lundelius (1967) assigned the species of those faunas to one of three groups. The first group consisted of extinct species considered characteristic of the Pleistocene (*Mammut americanum* and *Platygonus,* for example), some of which survived until 7,000 or 6,000 years ago. The second group consisted of 11 extant species that are not present now in central Texas or which occur there as relict populations, including *Sorex cinereus* and what he termed *Blarina brevicauda*. These species presently occur northeast or northwest of the plateau. The third group consists of species present in the area today. Almost all extant species in Texas are represented, with the major exceptions of *Dasypus novemcinctus* and *Tayassu tajacu*. The Pleistocene local faunas of Texas appear "to be a reflection of a temperature gradient in a north–south direction and an east–west change from forest to grassland or savanna" (Lundelius, 1967).

The short-tailed shrew was noted as "abundant in most deposits of Wisconsin age in central Texas"

(Lundelius, 1967). He referred all material from the deposits of central Texas to the subspecies *B. b. carolinensis*, following Hibbard (1963). Lundelius (1967) noted the gradual disappearance of *Blarina* from west to east across the Edwards Plateau; the last western record was from Felton Cave 7,800 years ago. Hall (1982) discussed additional evidence of increasing dryness in the post-Wisconsinan of Texas and Oklahoma.

Lundelius (1967) listed 18 species of mammals recovered from the noncultural unit (3,450 ± 150 BP) and 17 species from the cultural unit (1,015 ± 105 BP) of the Barton Springs site. All species are extant although not all presently occur in the area (Davis, 1978).

The genus *Blarina* is represented in the Barton Springs fauna by a single specimen (TMM 40627-60). Measurements 3 through 16 (2.0, 1.5, 1.3, 1.9, 1.8, 1.7, 1.9, 2.0, 1.5, 1.4, 2.0, 1.6, 0.7, and 1.2, respectively) of this specimen are most similar to those of fossils from Cave Without a Name and Felton Cave. These measurements fall within the lower end of the range of measurements of the reference samples of *B. carolinensis* (Table 1). The specimen from Barton Springs tentatively is referred to the species *B. carolinensis*.

Cave Without a Name

Cave Without a Name is located approximately 18 km northeast of Boerne, Kendall County, Texas. Commercial development of the cave was described by Craun (1948). Lundelius (1967) listed thirty species that have been recovered from this site. Five extinct species and subspecies (*Dasypus bellus*, *Procyon lotor simus*, *Mammut americanum*, an ovibovine, and *Equus* sp.) were recognized from this deposit. The extinct armadillo (*D. bellus*) first appeared in Blancan deposits in Florida, and was identified from Wisconsinan faunas in Missouri, Tennessee, West Virginia, Arkansas, Georgia, and New Mexico, in addition to four other sites in Texas (Kurtén and Anderson, 1980). This species might have been an ecological equivalent of the extant *D. novemcinctus* (Lundelius, 1967). In the Miller's Cave deposit, *D. bellus* was found in the travertine unit and *D. novemcinctus* was found in the more recent clay unit (Patton, 1963b). Remains of a raccoon were referred to an extinct subspecies (*P. l. simus*) first identified from the Pleistocene of northern California (Kurtén and Anderson, 1980). Wisconsinan records of the mastodont (*M. americanum*) extend from Alaska to Florida. Mastodonts were most numerous in eastern forest but also inhabited valleys and lowlands in Texas (Kurtén and Anderson, 1980). Six additional species, including what was termed *B. brevicauda*, were identified that have present distributions north of the site, generally in cooler, wetter climates (Patton, 1963a, Lundelius, 1967). Graham (1976) also suggested a more equable climate at the site during the period of deposition based on higher species density of shrews and voles at the cave during the Wisconsinan.

Kendall County is located along the southeastern escarpment

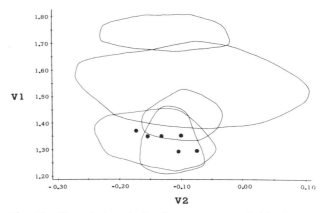

Fig. 42.—Canonical analysis of measurements 3–10 of upper teeth of specimens of *Blarina* from the Cave Without a Name l.f. Centroids enclose variation in samples of Holocene taxa of *Blarina* (see Fig. 3).

of the Edwards Plateau. This region was included in the Comanchian biotic province by Dice (1943); his description of the vegetation of the plateau is included in the account for Barton Springs. Ford and Van Auken (1982) provided additional information regarding the riparian vegetation of Kendall County.

A radiocarbon date of 10,900 ± 190 BP was obtained from a sample of bone from Cave Without a Name. The site has the last record of *Sorex cinereus*, *Microtus pennsylvanicus*, and *Mustela erminea* on the plateau. There was an apparent trend for the species presently distributed farthest north to disappear the earliest from central Texas (Lundelius, 1967).

The short-tailed shrews from Cave Without a Name were referred to the subspecies *B. b. carolinensis* by Lundelius (1967). In his analysis, Graham (1976) also recognized the relatively small size of the remains and referred them to *B. carolinensis*.

Ten specimens of maxillary fragments were examined in the present study (TMM 40450-1254 through 40450-1256, 40450-1260, 40450-1261, and 40450-1263 through 40450-1267). This sample appeared homogeneous. Six specimens were sufficiently complete for analysis (Fig. 42). They clustered within the centroid showing the range of variation for modern *B. carolinensis*, and are referred to that species.

Easley Ranch

Easley Ranch is located in western Foard County, Texas. The faunal remains were recovered from a sinkhole on the Easley Ranch, the type locality of the Good Creek formation (Dalquest, 1962).

The fauna from Easley Ranch has not been dated. Its composition is similar to that of the Jinglebob l.f. (Dalquest, 1962). The Easley Ranch l.f. has been considered Sangamonian (Dalquest, 1962; Slaughter, 1967) and Wisconsinan (Hibbard, 1970) in age.

The mammalian fauna includes 11 extinct species. Of the 19

extant species, 13 presently live in the region (Dalquest, 1962). The Easley Ranch l.f. suggests moderate summers and winters at the time of deposition, with woodland, grassland, and streambank habitats (Dalquest, 1962; Slaughter, 1967).

Blarina cf. *brevicauda* is represented in the fauna by a single incisor (MWU 3119).

Felton Cave

Felton Cave is located in Sutton County, Texas, on the western portion of the Edwards Plateau. The fauna is dated at 7,770 ± 130 BP (Lundelius, 1967). Fourteen species of mammals have been recovered from the site; seven species (*B. brevicauda, Cryptotis parva, Geomys bursarius, ?Peromyscus nasutus, Baiomys taylori, Microtus* sp., and *Bison* sp.) no longer occur in Sutton County (Lundelius, 1967; Davis, 1978).

Felton Cave is the last record of *Blarina* from the western part of the Edwards Plateau. Short-tailed shrews were present in most of the fossil faunas from the western side of the plateau until approximately 1,000 BP, reflecting the trend of increasing warming and drying west to east across the plateau (Lundelius, 1967).

The *Blarina* of Felton Cave were referred to the species *B. b. carolinensis* by Lundelius (1967) and *B. carolinensis* by Graham (1976). In the present study 10 specimens (TMM 41174-19 through 41174-28), consisting of two maxillaries, seven dentaries, and an isolated incisor, were examined. All specimens were fragmented and had incomplete dentition. Measurements were most similar to those of specimens from Barton Springs, Cave Without a Name, and Longhorn Cavern. The specimens are referred to the species *B. carolinensis*.

Friesenhahn Cave

Friesenhahn Cave is located near the southeastern edge of the Edwards Plateau, approximately 34 km north of San Antonio in Bexar County. The cave is a solution cavern developed in the limestone beds underlying the Edwards Plateau. It consists of an underground chamber, approximately 18 by 9 m, which is connected to the surface by a vertical opening ranging from 2 to 3 m in diameter and approximately 9 m deep. A second, inclined entrance was open to the surface during the late Pleistocene, but has filled with debris since that time (Evans, 1961).

Remains were recovered from four distinct units of fill, which were described by Evans (1961). The oldest unit (zone 1) consisted of a mixture of limestone blocks, gravels, and partially cemented red clay; remains of turtle shell and of small mammals were the only vertebrate fossils recovered. Zone 2 consisted mainly of clay that was deposited in an underground pond; vertebrate remains from this unit included articulated specimens. The third zone, in the central and southeastern parts of the chamber, was the most fossiliferous of the four units although preservation was not as good as that in zone 2. The fourth zone was limited to a channel cut by water flowing into the cave; small vertebrates were abundant, and some remains were carried in from the older deposits within the cave (Evans, 1961). In addition, Sellards (1919) and Hay (1920) described the recovery of bones of turtles, proboscideans, saber-toothed cats, and other vertebrates from the surface of the cave floor.

The freshwater pond that once occurred in the main chamber of the cave might have attracted many of the animals represented in the cave deposit. Some animals died within the cave; many others were dragged in by predators. Eventually the entrance was closed by debris, and the cave was sealed until the opening of the sinkhole entrance in modern times (Evans, 1961). Most of the fauna accumulated during the Wisconsinan (zones 2 and 3) during relatively moist conditions. The age of the material from zones 1 and 4 is uncertain. Remains from zone 1 probably date from the early Wisconsinan. Fossils in zone 4, excluding those washed in from older deposits, might date from the late Wisconsinan or early Holocene; they represent a period somewhat more moist than present but drier than the period represented in zones 2 and 3 (Evans, 1961).

Amphibians, reptiles, and mammals are represented in the deposit at Friesenhahn Cave. The herptiles (five species of amphibians and five reptiles) pertain to Holocene species with the exception of a new species of *Testudo* that was described by Milstead (1956). The remaining herps now are associated in the Balconian biotic province, and no extreme faunal shifts in recent times are indicated (Mecham, 1959). Evans (1961) noted the presence of turtle remains throughout the deposit, particularly in zone 3.

Lundelius (1967) listed thirty-two species of mammals from Friesenhahn Cave. Included are eight extinct species—*Canis (Aenocyon) dirus, Arctodus pristinus, Smilodon* sp., *Homotherium (Dinobastis) serus, Mammut americanum, Mammuthus* sp., *Mylohyus* sp., and *Camelops* sp. Two extinct species of *Tapirus* and *Equus* were present also. Remains of the predators (*C. dirus, A. pristinus, Smilodon,* and *H. serus*) were found associated with those of herbivores, probably indicating that the cave regularly served as a den (Evans, 1961). Remains of the mammoth and mastodont were numerous and consisted almost wholly of immature individuals, possibly representing a favorite prey of the cats (Evans, 1961; Meade, 1961). These fossils were found primarily in zones 2 and 3. Bones of *Mylohyus* were recovered from zones 2, 3, and 4, and their association with remains of species such as the tapir and the mastodont suggests the presence of forests near the deposit (Lundelius, 1960). Members of the genus *Camelops* ranged throughout western North America and were extinct by the late Wisconsinan or early Holocene, as were other members of the North American megafauna (Kurtén and Anderson, 1980). *Blarina brevicauda, Peromyscus nasutus, Cynomys ludovicianus,* and possibly *Microtus* were the only extant species listed by Lundelius (1967) that are not present in the region today. Subsequent absence of *P. nasutus* was attributed to increased aridity since the Wisconsinan (Tamsitt, 1957). Additional studies of the small mammals of Friesenhahn Cave were conducted by Kennerly (1956) and Pettus (1956).

Specimens of *Blarina* from Friesenhahn Cave were referred to the subspecies *B. b. carolinensis* by Lundelius (1967). In the present study one maxillary fragment (TMM 933-4008) and seven dentary fragments (TMM 933-4005 through 933-4009 and 933-4011) were examined. Measurements 3 through 14 of the maxillary (2.2, 1.5, 1.2, 1.9, 2.0, 1.8, 2.0, 2.2, 1.6, 1.6, 2.1, and 1.9, respectively) compare most favorably with those of the modern taxa *B. carolinensis* and *B. hylophaga plumbea* (samples 1, 3, 10,

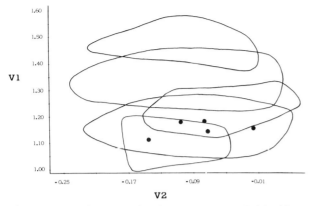

Fig. 43.—Canonical analysis of measurements 19–24 of lower teeth of specimens of *Blarina* from the Friesenhahn Cave l.f. Centroids enclose variation in samples of Holocene taxa of *Blarina* (see Fig. 4).

15, 16, and 21) and are slightly smaller than modern *B. h. hylophaga* (samples 17 and 22). Five of the dentaries that were sufficiently complete for analysis plotted within the ellipse showing the range of variation of modern *B. carolinensis* (Fig. 43). The two specimens with incomplete measurements were comparable to the specimens shown. The fossils tentatively are assigned to *B. carolinensis.*

Hall's Cave

Hall's Cave is located in Kerr County, Texas. Kerr County is part of the southern edge of the Edwards Plateau; modern vegetation of the area was discussed in the account for Barton Springs.

The fauna of Hall's Cave has not been studied. No carbon dates are available, but the fauna is considered Holocene by E. Lundelius, Jr. Bailing wire and barbed wire were found with some of the remains. Artifacts have not been found in the older part of the sequence, but there are no indications of a Pleistocene fauna (E. Lundelius, Jr., in litt., 1983).

One specimen (TMM 41229-452) of *Blarina* from Hall's Cave was examined. The specimen consists of a right dentary with i and m1–m3. Measurements 19 through 26 (1.9, 1.3, 1.6, 1.1, 1.2, 0.8, 3.1, and 2.0) were taken. These measurements fall within the range of variation of Holocene samples of *B. hylophaga* and *B. carolinensis* (Table 1). The measurements are most similar to the means of samples of *B. carolinensis,* however, and the specimen from Hall's Cave is assigned tentatively to that species.

Howard Ranch

The Howard Ranch l.f. was recovered from several deposits along Groesbeck Creek, northwest of the town of Quanah, in Hardeman County, Texas. A list of the mammals now present in the Groesbeck Creek area was included in the description of the Howard Ranch l.f. by Dalquest (1965).

Fossils were recovered from the Groesbeck formation, which consists of sediments deposited during the late Wisconsinan. All of the deposits from which fossils were collected were from the basin of an extinct lake. A carbon-14 date of $16,775 \pm 565$ BP was obtained from a sample of shells taken from below the majority of the vertebrate material. The presence of large extinct mammals in the fauna also suggests a late Wisconsinan age.

The Howard Ranch l.f. consists of the remains of arthropods (crayfishes), molluscs, fishes, amphibians, reptiles, birds, and mammals. Vertebrate remains generally were fragmented and scattered although the quality of preservation varied. Fishes and amphibians were the most common vertebrates recovered and mammals were the least common. Forty species of mammals were identified, seven of which are extinct—*Mammuthus* (*Elephas*) *columbi, Equus* cf. *scotti, E.* cf. *conversidens, Equus* sp., *Camelops* cf. *hesternus, Tanupolama* sp., and *Bison* cf. *antiquus.* Remains of the mammoth, horses, and camels were fairly common. Twelve of the remaining mammalian species are extant but are not present in the Groesbeck Creek area today. Besides *Blarina,* these include two species of *Sorex, Thomomys talpoides, Oryzomys palustris, Reithrodontomys megalotis, Microtus pennsylvanicus, M. ochrogaster* and/or *M. pinetorum, Ondatra zibethicus,* and *Synaptomys cooperi.* These mammals presently are distributed to the north, east, and southeast of the Howard Ranch area. The coexistence of these small mammals during the late Pleistocene might have resulted from a more equable climate. The presence of the Wisconsinan ice sheet would have produced cooler summers and warmer winters as far south as southern Texas (Dalquest, 1965). The slightly younger fauna of the C2 layer of Schulze Cave was considered additional support for this thesis (Dalquest et al., 1969).

The area of the site is located near the northern boundary of the Comanchian biotic province (Dice, 1943). Presently summers are hot and winters short and mild; average annual precipitation is between 50 and 61 cm (NOAA, 1974).

Dalquest (1965) described two maxillaries and a mandible of the short-tailed shrew from deposits at Howard Ranch. He thought that the three specimens were larger than the modern races of *B. brevicauda.* Dalquest referred these specimens to *B. b. fossilis,* following Hibbard (1943).

One maxillary and one dentary (MWU 3083) were examined. Measurements of the fossils were compared to measurements obtained from Holocene specimens (Table 1). The fossils were smaller than the large *B. b. brevicauda* collected from Iowa and Nebraska (samples 5 and 12). They were more similar to the smaller subspecies of *Blarina brevicauda*—*churchi, kirtlandi, manitobensis,* and *talpoides* (samples 11, 4, 6, 18, 13, 19, and 24). They almost certainly pertain to the species *B. brevicauda.*

Klein Cave

Klein Cave is located approximately 19 km WSW of Mountain Home, Kerr County, Texas. The site is a shallow limestone cave located near the southern edge of the Edwards Plateau. The modern vegetation of the area was discussed in the account for Barton Springs. A description of Klein Cave and its mammalian fauna published by Roth (1972) serves as the basis for this account.

The cave is approximately 64 m long, with an entrance ranging in diameter from 4 to 11 m. Roth (1972) identified four layers of sediment in the cave. Bones of extant Holocene rodents, bats, and rabbits were found in the second layer. Fossils were located in a pocket along the south wall of the cave. Few remains of large mammals were discovered. The bones generally were fragmented, and the deposit probably accumulated due to activities of owls and woodrats in the cave. A minimum date of 7,683 ± 643 was obtained from a sample of bone (Roth, 1972).

Remains of amphibians, reptiles, birds, and mammals have been recovered from the cave. Roth (1972) listed 43 mammalian species identified from the deposit. Eight species of bats were included, making Klein Cave the richest bat fauna reported from Texas. No mammals were extinct and all were known from other cave faunas representing the late Pleistocene. Ten species with northern and eastern distributions are no longer present on or near the Edwards Plateau. Four of these (*Tamias striatus, Microtus pennsylvanicus, Synaptomys cooperi,* and *Mustela erminea*) no longer occur in Texas. All of these previously were reported from at least one other fauna on the Edwards Plateau: *T. striatus* from the Schulze Cave fauna (Dalquest et al., 1969), *M. pennsylvanicus* from Cave Without a Name (Lundelius, 1967), *S. cooperi* from Cave Without a Name, Longhorn Cavern, and Miller's Cave (Patton, 1963a), and *M. erminea* from Schulze Cave (Dalquest et al., 1969) and Cave Without a Name (Lundelius, 1967). These four species, whose present ranges extend no closer to the Edwards Plateau than the mountains of New Mexico, are indicative of cooler, moister climate during the period of deposition. More equable climate, with lush grasslands and deciduous forests extending from eastern Texas, was suggested, similar to that indicated by Dalquest et al. (1969) for the fauna of Schulze Cave (Roth, 1972).

Roth (1972) listed six skulls, 5 upper jaws, and 17 lower jaws of what he termed *Blarina brevicauda* from the deposit at Klein Cave. Of this material, six maxillary fragments and 13 mandibles (MWU 9120) were examined. Results of the analysis of the maxillaries are shown in Fig. 44; one specimen had incomplete (but comparable) measurements and is not shown. The five specimens plotted in or near the area of overlap of the ellipses representing *B. carolinensis, B. h. hylophaga,* and the smaller subspecies of *B. brevicauda.* Four dentaries were incomplete but were comparable to the nine specimens included in the analysis and shown in Fig. 45. With the exception of two smaller individuals, these specimens also plotted in the area of overlap between *B. carolinensis, B. hylophaga hylophaga,* and *B. brevicauda.* Comparisons were made between the measurements of the 19 individuals from Klein Cave and the means derived for those measurements for Holocene samples. The specimens from Klein Cave were similar to those of modern groups of *B. carolinensis* (samples 1, 3, 10, 15, and 16) and *B. h.*

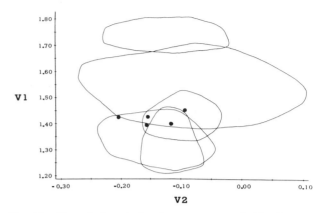

Fig. 44.—Canonical analysis of measurements 3–10 of upper teeth of specimens of *Blarina* from the Klein Cave l.f. Centroids enclose variation in samples of Holocene taxa of *Blarina* (see Fig. 3).

hylophaga (17 and 22). We suspect that the specimens from Klein Cave represent both *B. carolinensis* and *B. hylophaga,* which possibly are sympatric at the present time in northwestern Louisiana and are difficult to distinguish (George et al., 1981), but the present data are inadequate to be certain.

Longhorn Cavern

Longhorn Cavern is located approximately 14 km southwest of Burnet, Burnet County, Texas, on the eastern half of the Edwards Plateau. The cavern was formed by phreatic solution. Fossils were discovered, and many were lost, during commercial development of the cavern (described by Craun, 1948). Geology of Longhorn Cavern was described by Semken (1961) and Matthews (1963). Semken (1961) described the local fauna and stra-

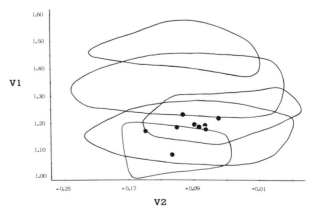

Fig. 45.—Canonical analysis of measurements 19–24 of lower teeth of specimens of *Blarina* from the Klein Cave l.f. Centroids enclose variation in samples of Holocene taxa of *Blarina* (see Fig. 4).

tigraphy; the modern climate and fauna of the area also were discussed.

Vertebrates were collected from three stratigraphic units within Longhorn Cavern (Semken, 1961). The floor of the cavern is overlain by a basal breccia, which was barren of fossils except for a fragment of a horse-tooth. Red fill, the most extensive unit in the cavern, overlies the basal breccia and contained remains of 13 genera of mammals and one extinct species of turtle. Black fill rests in "old stream cuts and in topographic lows" in the red fill (Semken, 1961). Remains of 16 genera of extant mammals and of an unidentified fish were recovered from the black fill. The presence of *Mus musculus* indicates deposition of the black fill within the last 200 years; deposition probably occurred since the 1860's (Semken, 1961). A fourth fill, the Longhorn breccia, was identified in non-commercial areas of the cavern. This breccia, probably derived from the red fill, contained remains of two extinct species of turtles, three extinct species of mammals, and sixteen extant genera of mammals (Semken, 1961). Accumulation of the Longhorn l.f. (remains from the red fill and the Longhorn breccia) and of remains from the black fill probably was due to the activities of carnivores and raptorial birds. Some animals also may have fallen into sinkholes (Semken, 1961).

The Longhorn l.f. was considered late Wisconsinan in age by Semken (1961) and Hibbard (1970). *Pitymys pinetorum,* now restricted in Texas to a relict population near Kerrville and to the eastern portion of the state (Bryant, 1941; Davis, 1978), is the most common member of the l.f. The composition of the Longhorn l.f. suggests the presence of mixed forests and grasslands at the time of deposition (Semken, 1961; Patton, 1963a).

Lundelius (1967) reported the presence of *B. b. carolinensis* in the Longhorn l.f. Two specimens (TMM 40279-163 and 40279-164), both from the red fill, were examined. Measurements 19 through 25 of the first specimen were 1.8, 1.2, 1.6, 1.0, 1.3, 0.9, and 3.4; measurements 19 through 22 of the second specimen were 1.8, 1.2, 1.5, and 1.0. These measurements are somewhat smaller than the range of measurements for samples 17 and 22 of extant *B. hylophaga* (Table 1). The measurements of the fossils from Longhorn Cavern fall within the range of those of extant *B. carolinensis* (particularly the smaller taxa represented in samples 2, 10, 15, and 23) and tentatively are assigned to that species.

Miller's Cave

Miller's Cave is located 21 km southeast of Llano in Llano County, Texas. The site is in the Riley Mountains, at an elevation of 412 m, overlooking Honey Creek. This area is in the central part of the Edwards Plateau. Dice's (1943) description of the vegetation of the plateau is included in the account for Barton Springs. Patton (1963b) published the description of the geology and local fauna of Miller's Cave, on which this report is based.

Miller's Cave consists of two chambers approximately 55 m in combined length. Sediments of the chambers were derived from different sources, so each has a different stratigraphic sequence. The present entrance is a shaft opened by collapse of part of the roof at the junction of the two chambers. Few fossils were discovered in the north chamber. Most of the l.f. was re-

covered from two stratigraphic units within the south chamber. The brown clay unit, consisting of a loose clay-silt conglomerate, contained fossils of bats, armadillo, rabbits, and rodents. A radiocarbon date of 3,008 ± 410 BP was obtained from a sample of charcoal from this unit. A travertine unit located directly below the brown clay unit contained fossils of insectivores, bats, armadillos, rabbits, rodents, carnivores, and deer. Analysis of the composition of the travertine indicated deposition during a cool, moist period. Bone from the travertine unit was dated at 7,200 ± 300 BP. Few remains were found below the travertine (Patton, 1963b).

Unidentified fishes, amphibians, reptiles, and birds were recovered from the brown clay and travertine units. Twenty-two genera and twenty species of mammals were identified tentatively. Remains from the brown clay unit consisted of *Myotis velifer, Dasypus novemcinctus, Sylvilagus* sp., *Geomys bursarius, Perognathus hispidus, Reithrodontomys megalotis, Peromyscus leucopus, Peromyscus* sp., *Neotoma floridana,* and *Microtus ochrogaster.* All of these species, except *D. novemcinctus* and an unidentified bat, also were recovered from the travertine unit. In addition, the travertine yielded remains of what was termed *Blarina brevicauda,* and of *Cryptotis parva, Scalopus aquaticus, Eptesicus fuscus, Lepus californicus, Neotoma floridana, Synaptomys cooperi, Ondatra zibethicus, Canis* sp., *Ursus americanus, Mephitis mephitis, Spilogale putorius,* and *Odocoileus* sp. The only extinct species in the deposit, *Dasypus bellus,* also was recovered from this layer. Patton (1963b) considered this date to mark the minimum terminal date for *D. bellus.* Lundelius (1967) regarded this date (7,300 BP) as probably inaccurate. Some species listed (*B. brevicauda, N. floridana, O. zibethicus, S. cooperi,* and *M. ochrogaster*) have modern distributions north, northeast, and southeast of the Edwards Plateau. Miller's Cave is the last known record of *S. cooperi* (7,300 BP) and *M. ochrogaster* (3,000 BP). They seem to have disappeared from the region after the disappearance of the megafauna and the small mammals that presently are distributed the farthest north from central Texas (Lundelius, 1967). Lundelius considered the order of disappearance due to the warming and drying of the climate to be from west to east across the plateau. Patton (1963b) suggested there was a more equable climate at the time of deposition, with stream and marsh, lowland meadow, shrub and tree, and upland prairie communities in the vicinity. A warming trend was indicated by the mineralogy and the change in number and composition of the fauna of the two units (Patton, 1963b).

The short-tailed shrews of the Miller's Cave deposit were identified as *B. brevicauda* by Patton (1963b). They were noted as falling within the size range of what he termed *B. b. carolinensis* from Kansas and Oklahoma. Ten specimens, all mandibular fragments with incomplete dentition, were examined (TMM 40540-20, 40540-229, 40540-497, 40540-499, 40540-501 through 40540-505, and 40540-507). Measurements were compared with means derived for those measurements from modern samples. The fossils were most similar to the modern samples of *B. carolinensis, B. h. hylophaga,* and *B. h. plumbea* (samples 1, 3, 10, 15, 16, 17, 22, and 21). Probably both *B. hylophaga* and *B. carolinensis* are represented in Miller's Cave l.f.

Schulze Cave

Schulze Cave is located approximately 45 km northeast of Rocksprings, Edwards County, Texas. The cave is a sinkhole near the southern edge of the Edwards Plateau. The Holocene and Pleistocene faunas of the area were described by Dalquest et al. (1969).

Holocene vertebrate material was recovered from underneath the vertical opening. Early Holocene and Pleistocene remains were recovered from a travertine ledge within the shaft (layers B and C) and from debris accumulated at the bottom (E). Radiocarbon dates were obtained from samples of bone from layers C1 (9,680 ± 700 BP) and C2 (9,310 ± 310 BP), indicating deposition from approximately 11,000 to 8,000 years BP (Dalquest et al., 1969). Most of the remains of small mammals probably accumulated as the result of activities of barn owls. Bones of larger mammals probably were from animals that fell into the sinkhole or were washed in or carried in by woodrats (Dalquest et al., 1969).

Remains from Schulze Cave represent 82 species of mammals, 21 of which have not been recorded from the area in modern times. All taxa listed from layer B are present today on the Edwards Plateau, although not necessarily in the immediate vicinity of the sinkhole. Layers C1 and C2 included taxa no longer present in central Texas. Generally layers C1 and C2 contained the same taxa, but species typical of northern, cooler areas were more common in C2, and species typical of warm or semiarid areas were more common in the upper, younger layer C1 (Dalquest et al., 1969). The only extinct species represented in the local fauna are mammoth, horse, and possibly bison. Extant taxa no longer present on the Edwards Plateau included what was termed *Blarina brevicauda*. These taxa presently are distributed in the southeastern United States and southern Rocky Mountains. Dalquest (1965) thought that sympatry of these species in central Texas during the Pleistocene might have resulted from the Wisconsinan ice sheet, which produced cooler summers and warmer winters. More equable climate during the period of deposition also was suggested by Graham (1976) based on his analysis of species density of shrews and voles. Dalquest et al. (1969) assumed that a flora and fauna similar to that of an alpine meadow existed in the area during the late Pleistocene. Deciduous woodland was present in the canyons and in some of the open areas of the plateau, and probably was continuous with the woodland in eastern Texas (Dalquest et al., 1969).

Specimens of *Blarina* from Schulze Cave, identified as *B. brevicauda,* were recovered from layer C. Remains of 79 animals were recovered from layer C1 and 83 from C2, making *Blarina* the most common of the four shrews from the deposit (Dalquest et al., 1969). Dalquest et al. (1969) noted that the average size of the specimens was the same in both layers, but that there was considerable variation in size among the specimens. "The difference in size between the largest and smallest jaws from both layers suggest that two species might be involved, but the measurements do not show a bimodal distribution and indicate a single rather variable population. All but the smallest specimens are too large for the small races found in southeastern Texas to-

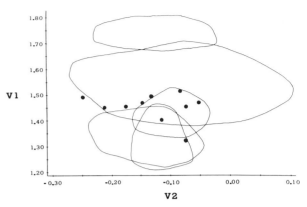

Fig. 46.—Canonical analysis of measurements 3–10 of upper teeth of specimens of *Blarina* from the Schulze Cave l.f. Centroids enclose variation in samples of Holocene taxa of *Blarina* (see Fig. 3).

day; their average size is similar to the average of *B. b. carolinensis* [=*B. hylophaga hylophaga*], from northeastern Texas, Oklahoma, and Kansas" (Dalquest et al., 1969).

Specimens from the C1 layer (MWU 7258) were available for examination in the present study. Eleven maxillaries and 42 jaws were measured. Ten of the maxillary specimens were sufficiently complete for canonical analysis and are shown in Fig. 46; the remaining incomplete specimen is comparable to those shown. Most of these specimens are somewhat larger than the modern samples of *B. carolinensis*. They are shown within or near the two centroids that show the range of variation for modern samples of *B. h. hylophaga* and the smaller subspecies of *B. brevicauda*. Twenty-nine mandibles were analyzed and are plotted in Fig. 47; eight observations are hidden. These specimens fell in or near the centroid showing the range of variation of modern *B. h. hylophaga* from Kansas and Oklahoma (samples 17 and 22). We tentatively conclude that only one species, *B. hylophaga*, is represented in Schulze Cave.

Back Creek Cave No. 2

Fossil *Blarina* have been recovered from several caves in Virginia and West Virginia (Fig. 2). This area is included in the Carolinian biotic province of Dice (1943). However, Guilday (1962b) and Guilday et al. (1977) described how variable the topography and biota are in this mountainous region.

Back Creek Cave, also known as Sheets Cave, is located 24 km west of Clark's Cave in Bath County, Virginia. The site was described as a shallow "rock shelter" by Guilday et al. (1977), and is within the Ridge and Valley physiographic province of western Virginia. The deposit is comparable in age (Ranchola-

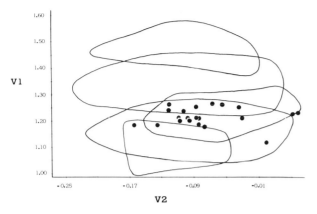

Fig. 47.—Canonical analysis of measurements 19–24 of lower teeth of specimens of *Blarina* from the Schulze Cave l.f. Centroids enclose variation in samples of Holocene taxa of *Blarina* (see Fig. 4).

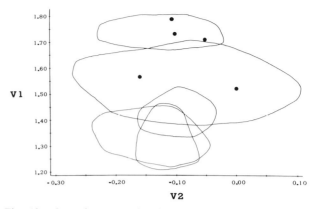

Fig. 48.—Canonical analysis of measurements 3–10 of upper teeth of specimens of *Blarina* from the Back Creek Cave No. 2 l.f. Centroids enclose variation in samples of Holocene taxa of *Blarina* (see Fig. 3).

brean) to the deposits of Natural Chimneys and Clark's Cave (Guilday et al., 1977).

Fossil remains are from a raptor roost deposit with a fauna similar to that of Clark's Cave. Avian remains representing 17 species (MNI = 55) were included in the faunal list of Clark's Cave by Guilday et al. (1977). In several instances identification was tentative because of the small sample size and generally poor preservation of bird bones. Most of the birds, such as bobwhite and woodpeckers, are present in the state today. The spruce grouse (*Canachites canadensis*), ruffed grouse (*Bonasa umbellus*), and sharp-tailed grouse (*Pedioecetes phasianellus*) are among the nine species of birds common to the deposits from Back Creek Cave, Clark's Cave, and Natural Chimneys. These three species presently are distributed in Canada; the grouse are boreal species that currently reside in forests and open glades, whereas the rock ptarmigan, *Lagopus mutus*, indicates open or semi-open tundra vegetation (Guilday et al., 1977). This type of avian fauna implies open coniferous parkland, perhaps "open, tundra-like ridge crests and more heavily wooded intermontane valley parklands or bog-forests" (Guilday et al., 1977).

The mammalian fauna of Back Creek Cave No. 2 is still under study. Five species were mentioned by Guilday et al. (1977). The least chipmunk, *Eutamias minimus,* was identified from the Back Creek Cave and Clark's Cave deposits, which are the first records of this species in the Appalachians. It currently occurs "in open to brushy, boreal, coniferous forest situations, and reaches its greatest abundance in open, sandy, pine, and spruce parklands" (Guilday et al., 1977). The heather vole, *Phenacomys intermedius,* and the yellow-cheeked vole, *Microtus xanthognathus,* presently occur in boreal communities in northern Canada. The northern bog lemming (*Synaptomys borealis*) ranges through boreal forests and taiga, in a variety of habitats, no further south than Minnesota and New Hampshire. The least weasel, *Mustela nivalis,* has a circumboreal distribution and might occur in the mountainous areas of Virginia. The voles, lemming, and weasel also were reported from the deposits of Natural Chimneys and New Paris No. 4 (Guilday et al., 1977).

Fifteen specimens of *Blarina* from Back Creek Cave No. 2 were examined (CM, uncatalogued ma-

terial). A canonical plot of the five maxillaries that were examined is illustrated in Fig. 48; univariate statistics (BACKCR) are shown in Table 2. Three specimens are within the range of variation shown for modern *B. b. brevicauda,* whereas two specimens fall within the range of variation for the *talpoides* semispecies of *B. brevicauda.* Of the ten dentaries examined, five were plotted (Fig. 49) within or near the centroid encompassing the range of variation for *B. b. brevicauda* and certainly represent the species *B. brevicauda.* One specimen appears much smaller, and conceivably represents *B. carolinensis.*

Clark's Cave

Clark's Cave is located 12 km southwest of Williamsville (U.S.G.S. Williamsville quadrangle 15′ series, 38°05′10″N,

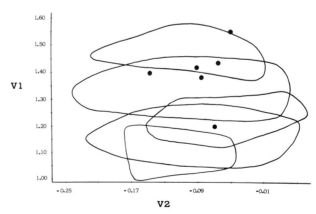

Fig. 49.—Canonical analysis of measurements 19–24 of lower teeth of specimens of *Blarina* from the Back Creek Cave No. 2 l.f. Centroids enclose variation in samples of Holocene taxa of *Blarina* (see Fig. 4).

79°39′25″W) in Bath County, Virginia. The cave is situated in a limestone cliff overlooking the Cowpasture River. In the immediate vicinity of the cave (extending approximately 1.5 km upstream and 1 km downstream) the Cowpasture River valley is gorgelike and talus at the base of the cliff is extensive. Mesic forest covers the slope on which Clark's Cave is located; the present climate is temperate (Guilday et al., 1977).

The cave has six major entrances and more than 2,400 m of passages. Fossils were excavated at the top of loose talus inside the passageway of entrance No. 2. This deposit is thought to be a remnant of nesting and roosting debris accumulated from raptorial birds. Fossils appeared chemically unaltered but fragmented; no articulated remains were found. A detailed description of the cave and analysis of its fauna were published by Guilday et al. (1977), and serve as the basis for this report.

The age of the deposit is considered late Wisconsinan. The large number of boreal birds and mammals indicates deposition during a cooler climatic period. The absence of introduced animals and the size characteristics of several species (including *B. brevicauda*) indicate little or no Holocene accumulation (Guilday et al., 1977). The exposed nature of the deposit has made dating techniques unreliable. All floral remains are believed of Holocene origin. A carbon-14 date (2,260 ± 85 BP) obtained from bone fragments is too young.

Remains of crustaceans (crayfish), insects, molluscs, fishes, amphibians, reptiles, birds, and mammals were recovered from the deposit at Clark's Cave. The two species of crayfish still occur in the Cowpasture River, and the specimens were probably raptor food remains. All insect remains were Holocene. Remains of snails, clams, fishes, and herptiles represent species still present in the area, and were incidental inclusions or food items. The herptiles reflect the selection bias of the raptors. Avian remains, representing 68 species, largely represent raptors and raptor prey. Most are modern residents or migrants through western Virginia. Notable exceptions are the spruce grouse (*Canachites canadensis*), sharp-tailed grouse (*Pediocetes phasianellus*), rock ptarmigan (*Lagopus mutus*), and possibly the gray jay (*Periosoreus canadensis*), all of which presently occur in Canada (Guilday et al., 1977).

The mammalian fauna virtually is identical with that of New Paris No. 4. Fifty-three species of mammals were identified. One species (*Canis dirus*) is extinct, whereas all others are extant. Eight species, all of boreal or western affinities (*Sorex arcticus, Eutamias minimus, Spermophilus tridecemlineatus, Synaptomys borealis, Phenacomys intermedius, Microtus xanthognathus, Martes americana,* and *Mustela erminea*) do not occur in the area today. Eleven species survive at higher elevations and have not been reported from the Cowpasture River valley. Five species, including *Blarina brevicauda*, are of different size than their modern counterparts, and are similar to modern boreal species or to remains from New Paris No. 4 (Guilday et al., 1977). Mammals accounted for 73.5% of the estimated biomass represented by the fossils. Voles (Microtinae) were the most numerous in terms of individuals (45.2%) and body weight (36.6%).

The fauna of the Clark's Cave deposit suggests that the area was covered predominantly by open boreal woodland in which conifers were dominant. No southern birds or mammals were found. The presence of several species (including *P. phasianellus* and *S. tridecemlineatus*) implies semi-prairie or parkland conditions. Remains of ptarmigan at Clark's Cave and Back Creek Cave No. 2 suggest the presence of tundra conditions in the area. The topography of the area probably provided ecological diversity, as it does today (Guilday et al., 1977).

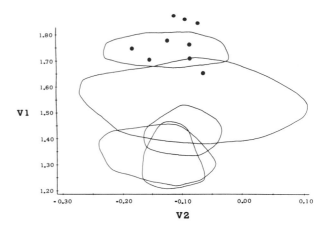

Fig. 50.—Canonical analysis of measurements 3–10 of upper teeth of specimens of *Blarina* from the Clark's Cave l.f. Centroids enclose variation in samples of Holocene taxa of *Blarina* (see Fig. 3).

The fourth most numerous family represented was the Soricidae. Remains identified as *B. brevicauda* were the most abundant, accounting for 42% of the shrews recovered (MNI = 97). Guilday et al. (1977) compared measurements of the specimens from Clark's Cave with measurements reported for extant animals from Minnesota and Pennsylvania (Guilday et al., 1964) and from North Carolina and West Virginia (Ray, 1967). Comparisons also were made with extant specimens from two localities in Bath County and with Pleistocene specimens from the deposits at New Paris No. 4 (Pennsylvania) and Natural Chimneys (Virginia). The specimens from Clark's Cave were comparable to the largest Holocene subspecies (the sample from Minnesota) and to the larger of the two groups from New Paris No. 4.

For the present study, 56 specimens of *Blarina* from Clark's Cave were examined (CM 24541 and 24578). Upper tooth row measurements 3–10 were complete for the nine maxillary specimens examined (Fig. 50). Five of these specimens plotted within the range of variation of the samples of *B. b. brevicauda* from Nebraska and Iowa, the largest Holocene shrews examined for this study. One specimen plotted outside but near this ellipse, in the upper size limit for intermediate-sized animals, whereas the three remaining specimens were larger than the modern *B. b. brevicauda* sampled for this study. Thirty-two of the dentaries are plotted in Fig. 51; 11 additional observations are hidden. These specimens also clustered in or near the centroid showing

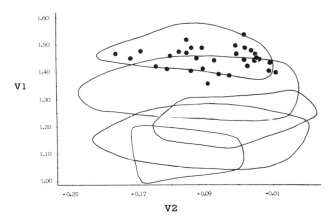

Fig. 51.—Canonical analysis of measurements 19–24 of lower teeth of specimens of *Blarina* from the Clark's Cave l.f. Centroids enclose variation in samples of Holocene taxa of *Blarina* (see Fig. 4).

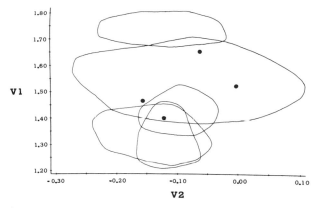

Fig. 52.—Canonical analysis of measurements 3–10 of upper teeth of specimens of *Blarina* from the Jasper Saltpeter Cave l.f. Centroids enclose variation in samples of Holocene taxa of *Blarina* (see Fig. 3).

the range of variation for the two samples of modern *B. b. brevicauda*. The specimens that were not analyzed because of missing measurements appear no different from those shown in the figure. All specimens of *Blarina* from Clark's Cave represent the species *B. brevicauda*.

Jasper Saltpeter Cave

Jasper Saltpeter Cave is located in the southwestern corner of Virginia in Lee County. This area is part of the Appalachian Plateau, a region of sharp ridges and deep valleys, with a mean annual precipitation of 122 cm (NOAA, 1974).

A formal description of Jasper Saltpeter Cave and its fauna has not been published. The fauna consists of the remains of small animals recovered from near the surface in the entrance to the cave. The age of the bone material is unknown, but it is noteworthy that remains of pika (*Ochotona*) were included. This genus was thought to have become extinct in the Appalachians before the Wisconsinan (J. E. Guilday, in litt., 1982). Previous Appalachian records of the pika were from the Irvingtonian faunas of Cumberland Cave, Maryland, and Rapps and Trout caves, West Virginia (Kurtén and Anderson, 1980).

Five maxillaries and nine dentaries of *Blarina* were examined (CM 30244). Results of canonical analysis of the most complete dentaries are shown in Fig. 52. The specimens from the cave are shown to be within the range of variation of the modern *talpoides* semispecies of *B. brevicauda*. The fifth specimen had measurements most similar to those of the larger specimens at the right of the diagram. Comparisons were made between measurements of the fossils (maxillaries and dentaries) and means obtained for those measurements for the modern reference samples. The specimens from the Jasper

Saltpeter Cave deposit again appear most similar to modern samples of *B. brevicauda* (*talpoides* semispecies), and doubtless pertain to that species.

Meadowview Cave

Meadowview Cave is located in Washington County, Virginia. A description of the cave and its fauna has not been published, although the presence of *Phenacomys* in the deposit was reported by Guilday and Parmalee (1972). Also present are fragmentary remains of *Blarina brevicauda, Neotoma, Clethrionomys gapperi, Synaptomys* sp., *Microtus* sp., *Microtus* (*Pitymys*) sp., *Ursus americanus, Dasypus bellus*, cervid sp., and sciurid sp. (J. E. Guilday, in litt., 1982). The extinct armadillo (*D. bellus*) has been reported from Blancan, Irvingtonian, and Wisconsinan localities in Florida, Georgia, Arkansas, Tennessee, West Virginia, Missouri, Texas, and New Mexico (Kurtén and Anderson, 1980). The genus *Phenacomys* currently has a Canadian-Hudsonian distribution (Guilday and Parmalee, 1972). The remaining genera have been recorded from Virginia in Holocene times (Hall, 1981). The age of the deposit is late Wisconsinan or early Holocene (Guilday and Parmalee, 1972).

One dentary (CM 24300) from this deposit was examined. All molars were present and measurements 18 through 24 and 26 (4.6, 2.3, 1.5, 1.9, 1.3, 1.5, 1.0, 2.5) were taken. When compared with the means of modern samples for these measurements (Table 1), the dentary is most similar to the smaller subspecies of *B. brevicauda* (*talpoides* semispecies) and doubtless pertains to this species.

Natural Chimneys

The Natural Chimneys are located 1.6 km north of Mt. Solon, Augusta County, Virginia at 30°22′N and 70°5′W. The site is on the east bank of the North River valley at an elevation of 414 m. The geology and the fauna of the Natural Chimneys were described by Guilday (1962*b*).

The Natural Chimneys are dolomite pillars, which are eroded remnants of the walls of sinkholes that formerly were open on the hilltop. Brown's Cave and the Cave of the Wooden Steps are located at the base of the pillars; these are probably remnants of subterranean connections between the sinkholes (Guilday, 1962b). The bulk of the bone deposit was recovered from cave fill inside the mouth of Brown's Cave, although some specimens also were found in the side passages of Brown's Cave and in the Cave of the Wooden Steps. No stratification was evident. Most of the bones accumulated from the debris of raptorial birds. Few remains of larger mammals were present, and these probably resulted from denning activities in the cave or were carried in by woodrats. Except for the larger bones, the remains were well preserved (Guilday, 1962b).

The vertebrate fauna (especially the lack of southern mammals and introduced species) and the nature of the deposit suggest an early post-Wisconsin age for this cave fauna (Guilday, 1962b). The deposit is considered comparable in age to the Bath County deposits at Back Creek Cave No. 2 and Clark's Cave (Guilday et al., 1977). The fauna also was correlated with that of New Paris No. 4 of Pennsylvania (Guilday, 1962b).

The invertebrates of the Natural Chimneys l.f. consist of a millipede and snails. Fishes are common, but all remains are small and probably were deposited by predators or by occasional flooding of the cave. Of the 22 genera of herptiles, only two species (the coachwhip, *Masticophis flagellum,* and the diamond-backed rattlesnake, *Crotalus adamanteus*) are not found in Virginia today. These also were the only southern animals found in the fauna, and it is unknown whether they were contemporaneous with the remainder of the fauna. Forty species of birds, most of which are found presently in the area as residents or migrants, were identified from Natural Chimneys. Remains of nine of the species also were found at Back Creek Cave No. 2 and Clark's Cave, including the spruce grouse (*Canachites canadensis*) and sharp-tailed grouse (*Pedioecetes phasianellus*), which currently have northern boreal distributions (Guilday et al., 1977). The gray jay (*Perisoreus canadensis*), recovered from Clark's Cave and Natural Chimneys, also is an indicator of boreal conditions. The record of the magpie (*Pica pica*) at Natural Chimneys is the first for the eastern United States. Its presence implies semi-prairie or parkland conditions and may be an indicator of increased continentality (Guilday, 1962b, 1971; Guilday et al., 1977).

Remains of at least 878 mammals were identified, representing 55 species of insectivores, bats, rodents, lagomorphs, and artiodactyls. Four species are extinct. The other species are present in the area today or retreated to higher elevations or latitudes during post-glacial warming (Guilday, 1962b). Several species with modern distributions in the Canadian zone, such as the arctic shrew (*Sorex arcticus*) and the spruce vole (*Phenacomys*), indicate deposition during a period of glacial advance (Guilday, 1962b). Coexistence of the presently allopatric voles *Microtus xanthognathus* and *M. chrotorrhinus* implies taiga-like conditions (Guilday, 1971). In short, the avian and mammalian remains of Natural Chimneys suggest that coniferous forest covered the area during deposition (ca. 10,000–15,000 years ago), and that streams, swampy grasslands, and parkland were in the vicinity (Guilday, 1962b, 1971).

Soricids are represented in the deposit by seven species and a minimum of 133 individuals. A minimum of 56 individuals was referred to the species *B. brevicauda* (Guilday, 1962b; Guilday et al., 1969).

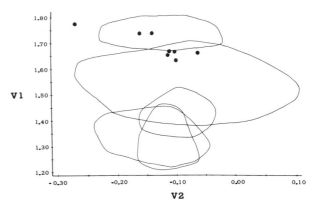

Fig. 53.—Canonical analysis of measurements 3–10 of upper teeth of specimens of *Blarina* from the Natural Chimneys l.f. Centroids enclose variation in samples of Holocene taxa of *Blarina* (see Fig. 3).

Guilday (1962b) noted that the average size of specimens of *Blarina* from Natural Chimneys is larger than that of modern specimens from Pennsylvania, and that "some specimens are distinctly larger and more rugged than any modern *B. brevicauda kirtlandi,* and compare favorably with the large late Pleistocene *B. brevicauda* from the New Paris No. 4 local fauna, Pennsylvania." Graham (1976) asserted that both *brevicauda* and *kirtlandi* were present in the cave fauna, and that vole and shrew species density was higher during the Wisconsinan than in the Holocene. Both of these interpretations were considered evidence of more equable climate, with reduced temperature and moisture gradients, during the period of deposition.

Thirty of the most complete specimens (CM 7544–7545) were selected for examination in this study. Nine maxillary fragments were measured, of which eight were sufficiently complete for canonical analysis. These specimens, shown in Fig. 53, were plotted in or near the centroid showing the range of variation of the two modern samples of *B. b. brevicauda* from Nebraska and Iowa. The ninth specimen had measurements comparable to those of the specimens shown. Of the dentaries examined, eighteen are shown in Fig. 54. These specimens also are within or near the centroid for modern *B. b. brevicauda*. The three remaining dentaries appear no different than those shown. Taking into account the differences in sample sizes, the two diagrams are comparable to those for Clark's Cave (Figs. 50 and 51). Based on the analysis of these specimens, the *Blarina* of the Natural Chimneys fauna appear to

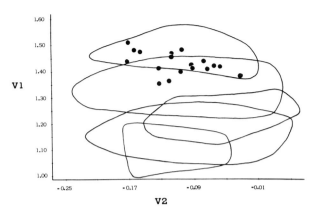

Fig. 54.—Canonical analysis of measurements 19–24 of lower teeth of specimens of Blarina from the Natural Chimneys l.f. Centroids enclose variation in samples of Holocene taxa of *Blarina* (see Fig. 4).

be a relatively homogeneous population of *B. brevicauda* comparable to the extant subspecies *B. b. brevicauda.*

Ripplemead Quarry

Bones were recovered from three fissures in the wall of an abandoned limestone quarry near Ripplemead, Giles County, Virginia. The quarry is on a bluff overlooking the New River. Shells of terrestrial snails and remains of one fish, one bird, 5 species of herptiles, and 26 species of mammals were recovered (Weems, 1972; Weems and Higgins, 1977). Accumulation probably occurred as bones were washed into the fissures from surrounding slopes (Weems and Higgins, 1977). The most numerous remains are those of snakes and rodents. Larger mammals present include horse, peccary, deer, and bear (Weems and Higgins, 1977; J. E. Guilday, in litt., 1982). All of the small mammals still live in the area (J. E. Guilday, in litt., 1982).

Six specimens (CM, uncatalogued material), consisting of fragments of one maxillary and five dentaries, were examined. The sample appeared homogeneous. Measurements of these specimens were compared with those of Holocene specimens (Table 1). Measurements 3 through 14 of the maxillary (2.7, 2.1, 1.7, 2.5, 2.3, 2.2, 2.6, 2.5, 1.8, 1.8, 2.5, 2.1) are most similar to the means derived for measurements 3 through 14 of the sample of *B. b. brevicauda* from Iowa (sample 5). Measurements 19 through 22 (2.5, 1.7, 1.9, 1.3; 2.4, 1.6, 2.0, 1.3) of two dentaries and measurements 18 through 24 (5.0, 2.3, 1.5, 1.9, 1.3, 1.6, 1.1; 4.9, 2.4, 1.7, 2.0, 1.4, 1.6, 1.1; 4.8, 2.5, 1.7, 2.0, 1.3, 1.5, 1.0) of three dentaries also were taken. The dentaries compare favorably with modern samples of the subspecies *brevicauda, manitobensis,* and *churchi* (samples 5, 12, 18, and

11) and are referred to the species *B. brevicauda.* The specimens from Ripplemead Quarry appear somewhat larger than those from Jasper Saltpeter Cave (Lee County). They are similar to some of the specimens from Back Creek Cave No. 2 and Clark's Cave (Bath County), Strait Canyon (Highland County), and Meadowview Cave (Washington County).

Remains from Ripplemead Quarry have not been dated. The presence of the tapir, horse, and peccary, and the absence of domestic animals and of large mammals typical of the Wisconsinan imply deposition approximately 7,000 to 9,000 BP (Weems and Higgins, 1977). Possibly some mixing with more recent material occurred; if the remains are contemporaneous, the presence of hardwood forest and climate similar to that of today is indicated (Weems and Higgins, 1977).

Strait Canyon Fissure

Strait Canyon Fissure is located in Highland County, within the Ridge and Valley physiographic province of western Virginia. A description of the fissure and its fauna is in preparation (J. E. Guilday, in litt., 1982). Specimens are housed at the Carnegie Museum of Natural History. The deposit includes remains of fishes, amphibians, reptiles, birds, and mammals. Remains are fragmentary, and no stratification is evident. The mammalian fauna includes four taxa now extinct: *Mammut, Tapirus, Mylohyus, Equus,* and *Neofiber* cf. *N. leonardi.* The dominant microtine is *Microtus pennsylvanicus* (J. E. Guilday, in litt., 1982). Ray (1965) reported the presence of *Tamias.* The age of the deposit is considered early Wisconsinan, and remains of boreal species (*Phenacomys* and *Synaptomys borealis*) are rare (J. E. Guilday, in litt., 1982).

Thirty-six specimens (four maxillaries and 32 dentaries) collected from talus and from eight stratigraphic levels (arbitrarily designated) were examined. Measurements were compared with the means derived from modern samples. Although the measurements of the fossils were variable, the specimens from the cave are most similar to the modern samples of *B. brevicauda.* Thirteen of the dentaries were sufficiently complete for analysis; results are shown in Fig. 55 (one observation was hidden). The specimens shown were recovered from talus (T) and from levels two feet below the talus (b), nine feet above the talus (i), eleven feet above the talus (k), and sixteen feet above the talus (p). All but one of the specimens are plotted in the centroid that shows the range of variation for the *talpoides* semispecies of *B. brevicauda.* Clearcut differences among specimens from the different levels are not evident. All specimens of *Blarina* from the Strait Canyon Fissure l.f. tentatively are referred to *B. brevicauda.*

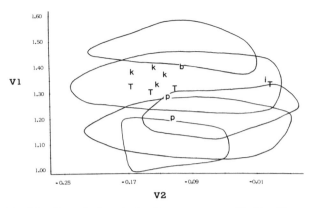

Fig. 55.—Canonical analysis of measurements 19–24 of lower teeth of specimens of *Blarina* from the Strait Canyon l.f. Centroids enclose variation in samples of Holocene taxa of *Blarina* (see Fig. 4).

Eagle Cave

Eagle (Eagle Rock) Cave is one of four caves in Pendleton County, West Virginia, from which remains of *Blarina* have been reported (Appendix 1). Pendleton County is included in the Ridge and Valley physiographic province. The original vegetation of the area was a chestnut-oak forest, which has been replaced by a mixed hardwood forest in which oak predominates. Spruce is found at elevations above 765 m. Grass-sedge meadows occur occasionally in mountain valleys, and may have persisted since the late Pleistocene (Guilday and Hamilton, 1973, 1978). Guilday and Hamilton (1978) reported the average precipitation as 76 to 101 cm annually.

Eagle Cave is located 21.6 km north of Franklin, West Virginia, at 38°50'N, 79°17'W, 762 m (Onego Quad., U.S.G.S. 15' series) (Guilday and Hamilton, 1978). The deposit is in a limestone cave located in Cave Mountain, overlooking the South Branch of the Potomac River. The floor of the cave has been greatly disturbed by human traffic, and no stratification was observed. Fossils were fragmentary. A description of Eagle Cave and its fauna was published by Guilday and Hamilton (1973). Additional notes regarding the fauna were published by Guilday (1971), Guilday and Parmalee (1972), Graham (1976), and Guilday and Hamilton (1978).

The fauna of Eagle Cave consists of the remains of fishes, frogs, snakes, and mammals. Most of the remains accumulated in owl pellets although a few species, such as bats, probably were residents of the cave. Boreal, continental, and temperate species were represented. Species that now have northern distributions are *Microtus xanthognathus, Synaptomys borealis, Sorex arcticus,* and *Phenacomys intermedius. Spermophilus tridecemlineatus,* which currently occupies the prairies of central North America, also was present. A fossil assemblage with these mixed components is thought to be typical of mid-Appalachian late Pleistocene faunas (Guilday and Hamilton, 1978).

Four species of shrews were identified from the cave deposit. Remains of *Blarina* (MNI = 7) were assigned to the species *B. brevicauda* (Guilday and Hamilton, 1973). Graham (1976) also assigned the

fossils to *B. brevicauda*; he considered members of the modern fauna to be *B. kirtlandi.*

Seven dentaries (CM 24397–24399), most with incomplete dentition, were examined. Measurements 19 through 26 of one dentary (2.5, 1.7, 2.0, 1.4, 1.6, 1.1, 4.4, 2.7), 19 through 22 of three dentaries (2.4, 1.6, 2.1, 1.4; 2.4, 1.6, 2.0, 1.3; 2.6, 1.8, 2.0, 1.4), 19 through 24 and 26 of one dentary (2.3, 1.5, 1.9, 1.3, 1.6, 1.1, 2.8), and 21 through 24 of one dentary (2.1, 1.4, 1.6, 1.1) were taken. Measurements 19 and 20 of one molar (2.5, 1.7) also were taken. Measurements were compared with means for those measurements obtained from extant groups (Table 1). The fossils from Eagle Cave seem referable to either *B. b. brevicauda* or *B. b. churchi* although there is more variation among the measurements for the fossils than among their modern counterparts. Generally, the fossils were larger than the two modern samples of *B. b. kirtlandi* (samples 4 and 6). The fossils obviously represent the species *B. brevicauda.*

Hoffman School Cave

Hoffman School Cave is located 8 km south of Franklin, Pendleton County, West Virginia, at 38°34'38"N, 79°21'49"W (Circleville Quad., U.S.G.S. 15' series) at an elevation of 660 m (Guilday and Hamilton, 1978). This limestone cave is located on Thorn Creek, a tributary of the South Branch of the Potomac River. The cave is approximately 30 km south of Eagle Cave.

The bone deposit in Hoffman School Cave, like that in Eagle Cave, consists mainly of remains of small vertebrates that accumulated as a result of roosting birds. Cave rats (*Neotoma floridana*) also inhabited the cave, and probably were responsible for accumulating the larger mammalian teeth found in the deposit. Three components that presently are allopatric—boreal, continental, and eastern temperate species—are represented in the deposit (Guilday and Hamilton, 1978). Species no longer present in the area include *Spermophilus tridecemlineatus, Glaucomys sabrinus, Phenacomys intermedius, Microtus chrotorrhinus, Erethizon dorsatum,* and *Martes americana.* The absence of *M. xanthognathus,* common in other late-Wisconsinan sites in the Appalachians, might indicate that the deposit was formed in the period of climatic change after the Wisconsinan glaciation, during which time other boreal forms persisted in the area (Guilday and Hamilton, 1978). Otherwise, the assemblage is similar to that of Eagle Cave in that both are "typical 'late-boreal' woodland faunas characteristic of the central Appalachians" (Guilday and Hamilton, 1978).

The soricids recovered from this deposit were identified as *Sorex* sp. (MNI = 2) and *Blarina brevicauda* (MNI = 8) by Guilday and Hamilton (1978). Seven specimens of *Blarina* (CM 30020) were examined for this study. Measurements 19 and 20 of four m1s (2.5, 1.8; 2.4, 1.6; 2.4, 1.7; 2.2, 1.5), 21 and 22 of three m2s (2.0, 1.3; 2.1, 1.4; 2.0, 1.3), and

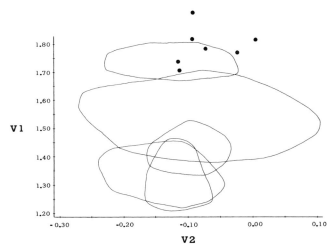

Fig. 56.—Canonical analysis of measurements 3–10 of upper teeth of specimens of *Blarina* from the Moscow Fissure l.f. Centroids enclose variation in samples of Holocene taxa of *Blarina* (see Fig. 3).

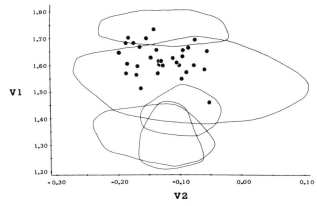

Fig. 57.—Canonical analysis of measurements 3–10 of upper teeth of specimens of *Blarina* from the Caverne De Saint-Elzear-de-Bonaventure l.f. Centroids enclose variation in samples of Holocene taxa of *Blarina* (see Fig. 3).

23 and 24 of two m3s (1.6, 1.1; 1.7, 1.1) were taken. All were dentaries with incomplete dentition; consequently, the measurements were not useful for canonical analysis. Comparisons were made between measurements of these fossils and means obtained from Holocene samples for those measurements. The fossils were most similar to the sample of *B. b. brevicauda* from Iowa and Nebraska (Table 1, samples 5 and 12). One fossil specimen appeared slightly smaller than the others but obviously pertains to the species *B. brevicauda*. Generally the fossils of Hoffman School Cave are similar in size to those of Eagle Cave, which perhaps is to be expected given the overall similarity of the two faunas.

Moscow Fissure

A Rancholabrean fauna was recovered in the driftless area of southwestern Wisconsin. The Moscow Fissure l.f. was recovered approximately 3 km northeast of Blanchardville, Iowa County, in a deposit along the Blue Mounds Branch (SW ¼, NE ¼, SW ¼ sec. 12, T4N, R5E). The fauna was described in a thesis by Robert L. Foley of the University of Iowa (Semken, personal communication, 1982).

A radiocarbon date of 17,050 ± 1,500 BP places the Moscow Fissure l.f. near the late Wisconsin glacial maximum. The fauna is disharmonius but appears predominantly boreal. Nine species, including *Thomomys talpoides*, *Synaptomys borealis*, *Phenacomys intermedius*, *Dicrostonyx torquatus*, *Microtus xanthognathus*, and *Zapus princeps*, no longer occur in the area. The fauna indicates the presence of marsh, steppe, and spruce parkland. The l.f. of Moscow Fissure compares most favorably with that of New Paris No. 4 (Semken, personal communication, 1982). All specimens, as yet uncatalogued, are housed at the Department of Geology, University of Iowa (SUI).

Southwestern Wisconsin was included in the Illinoian biotic province (Dice, 1943). Mean annual precipitation is 76 to 81 cm (NOAA, 1974).

Foley considered the *Blarina* of this fauna equivalent to *B. b. brevicauda* and *B. b. manitobensis* [= the *brevicauda* semispecies of *B. brevicauda*] (Semken, personal communication, 1982). Sixteen of the most complete specimens (eight maxillaries and eight dentaries) were examined for the present study. Results of the canonical analysis of seven maxillaries are shown in Fig. 56. The observations are plotted within and near the centroid showing the range of variation of Holocene *B. b. brevicauda* from Iowa and Nebraska. Measurements 19 through 22 of eight m1s and m2s (means = 2.46, 1.69, 2.06, 1.40) also are most similar to those of Holocene *B. b. brevicauda*. The fossils from Moscow Fissure undoubtedly pertain to that subspecies.

Caverne de St. Elzear

Remains of *Blarina* were recovered from Caverne de Saint-Elzear-de-Bonaventure, Bonaventure County, Quebec. Bonaventure County is located on the southern coast of the Gaspe Peninsula. The area was included in the Canadian biotic province by Dice (1943).

The deposit of this cave is of Holocene origin. A date of 5,110 ± 150 BP was reported by LaSalle and Guilday (1980). Microtines from the deposit include *Microtus chrotorrhinus*, *M. pennsylvanicus*, *M. xanthognathus*, *Dicrostonyx*, *Phenacomys*, and *Synaptomys borealis* (Zakrzewski, personal communication). *Phenacomys* and *Dicrostonyx* do not occur in the area today (Hall, 1981).

One hundred and thirty-five specimens (CM 37906–37910 and 37912–37919) of *Blarina* from

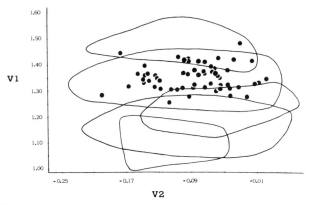

Fig. 58.—Canonical analysis of measurements 19–24 of lower teeth of specimens of *Blarina* from the Caverne De Saint-Elzear-de-Bonaventure l.f. Centroids enclose variation in samples of Holocene taxa of *Blarina* (see Fig. 4).

St. Elzear were examined. Univariate statistics are shown in Table 2. The specimens were recovered from 13 stratigraphic levels, and were separated by level for canonical analysis. Thirty-two maxillaries plotted within the centroids for *B. b. brevicauda* and the smaller subspecies of *brevicauda*; one observation was hidden (Fig. 57). Seventy-two mandibular specimens were examined and are illustrated in Fig. 58 (13 observations are hidden). Again, the specimens were plotted within the two centroids representing subspecies of *B. brevicauda*. No differences were observed among specimens from different levels. Measurements of the St. Elzear material are most similar to those of the subspecies *talpoides, angusta,* and *pallida* (samples 13, 19, 24, 20, and 9).

PHYLOGENY AND PALEOBIOGEOGRAPHY

We attempted to depict the paleobiogeography of *Blarina* in a series of maps (Figs. 59–64). Information was gleaned from studies of fossil pollen, in addition to the paleontological studies previously discussed. Because of the difficulties in the interpretation and correlation of past climates and paleofaunas, the maps were designed only to show general trends in the evolution of *Blarina*; the faunas depicted in each figure are not intended to indicate strict synchrony.

The fossil record of *Blarina* implies evolution of the two species *brevicauda* and *carolinensis* in the early Pleistocene. The earliest specimens of the genus were recovered from late Blancan (early Pleistocene) faunas in Kansas. These specimens, from Dixon and White Rock, are the size of the modern *talpoides* semispecies of *B. brevicauda*. Fossils of both faunas indicate cool, moist climatic conditions. The presence of *Blarina* on the Great Plains in the early Irvingtonian prior to the period of maximum Nebraskan glaciation is implied. The extent of the range of the genus at this time is unknown because few Nebraskan faunas have been found.

We examined fossils from 17 Irvingtonian sites (from the central plains, Florida, Arkansas, West Virginia, Maryland, and Pennsylvania) which comprise the Irvingtonian record of *Blarina* as we know it. The relative ages of these faunas is uncertain (Appendix 1), due to the poor preservation of some of the fossils, the relative scarcity of Irvingtonian faunas, and the heterochronic nature of some of the

deposits, all of which make comparisons among faunas difficult.

The earliest Irvingtonian fossils are thought to be those of the Inglis IA fauna (Nebraskan—Aftonian) of western Florida; these are the first known specimens of *B. carolinensis*. The fauna indicates coastal savanna habitat. Thus far, remains of *Blarina* have not been discovered in any other Aftonian faunas (unless the Port Kennedy l.f. of Pennsylvania is found to date from the Aftonian, as tentatively suggested by Kurtén and Anderson, 1980). Semiarid conditions on the Great Plains during the Aftonian (Hibbard, 1960, 1970; Eshelman and Hibbard, 1981) probably drove *Blarina* into relatively moister areas to the south- and northeast (see Dorf, 1959, for maps of generalized climatic zones of a composite of interglacial and glacial stages).

The genus *Blarina* reappears in the fossil record during the Kansan glacial stage in faunas in South Dakota, Kansas, Texas, West Virginia, and Arkansas (Fig. 59); two of these sites are outside the present range of the genus. Three additional specimens of what appears to be *B. carolinensis* were recovered from Wathena (Kansas) and Java (South Dakota). Identification of these fossils is tentative due to small sample size. Comparative material of similar provenience is lacking and both faunas have undergone preliminary study only. If the fossils from Wathena and Java prove to be *B. carolinensis* of Aftonian-Kansan age, an early, widespread distribution (possibly associated with grassland) of the species would

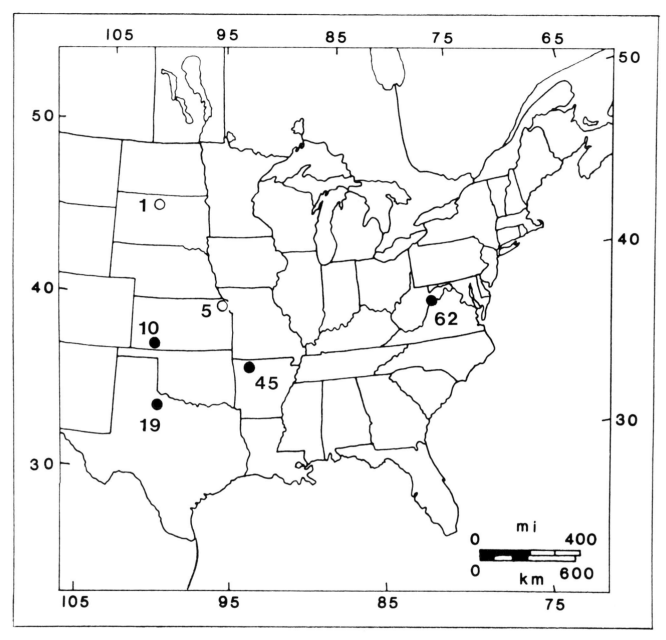

Fig. 59.—Possible distribution of *B. brevicauda* (closed circles) and *B. carolinensis* (open circles) in the middle Irvingtonian, as suggested by Aftonian-Kansan localities; 1. Java; 5. Wathena; 10. Cudahy; 19. Vera; 45. Conard Fissure; 62. Trout Cave. Specimens of *B. b. brevicauda* were found at 62.

be argued. Fossils from other sites of this general period of time (Cudahy, Vera, and Conard) pertain to the species *B. brevicauda*. Specimens from Trout Cave might represent two phena, the *brevicauda* and *talpoides* semispecies of *B. brevicauda*. The climate on the plains at this time continued to be cool but with more effective moisture (Hibbard and Dalquest, 1966; Semken, 1966; Hibbard, 1970; Martin, 1973*b*). The Conard Fissure l.f. (Arkansas) also suggests a climate with cooler and moister summers. Climate at Trout Cave probably was similar to that of today, and the smaller fossils from that cave are similar in size to *B. b. kirtlandi,* the subspecies presently in the northeast.

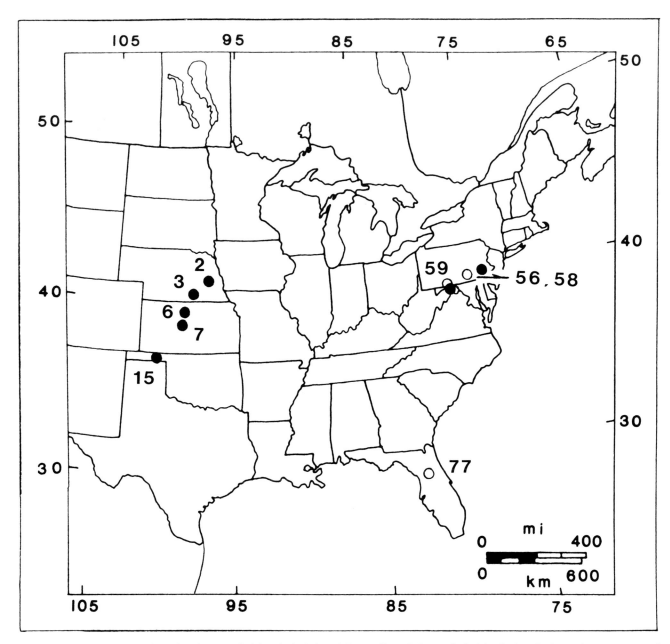

Fig. 60.—Possible distribution of *B. brevicauda* (closed circles) and *B. carolinensis* (open circles) in the late Irvingtonian, as suggested by Yarmouthian and early Illinoian localities: 2. Bartek Brothers Quarry; 3. Angus; 6. Rezabek; 7. Kanopolis; 15. Berends; 56. Hanover Quarry Fissure; 58. Port Kennedy; 59. Cumberland Cave; 77. Coleman IIA. Specimens of *B. b. brevicauda* were found at 2, 6, 7, and 15.

Fossils of *Blarina* are known from five faunas of the Yarmouth interglacial. Two of these (Port Kennedy and Bartek Brothers) might be older than other Yarmouthian faunas. The Angus (Nebraska) and Kanopolis (Kansas) local faunas suggest climatic conditions similar to or moister and warmer than those of today (Schultz and Martin, 1970; Hibbard

et al., 1978). The difference in size of specimens from Angus and Bartek Brothers Quarry suggests there was a zone of contact between two subspecies in southeastern Nebraska or the periods of deposition were not the same. Based on current knowledge of intergradation between Holocene *B. b. brevicauda* and *B. b. kirtlandi* in Iowa and Missouri

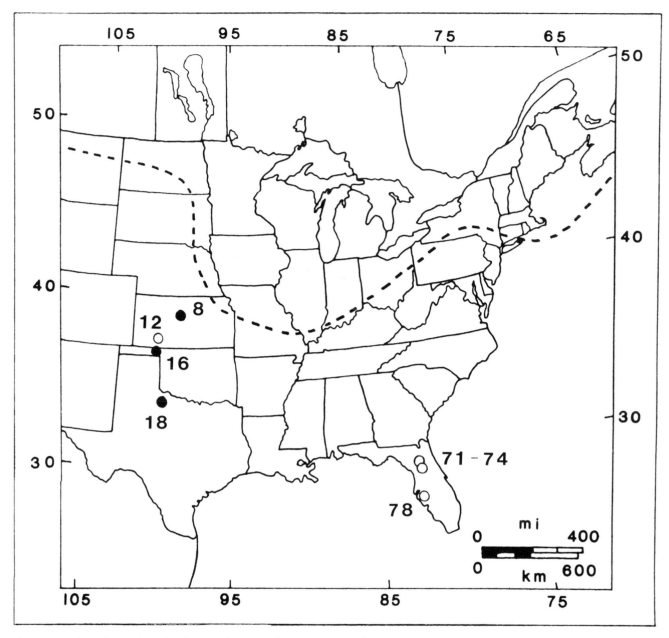

Fig. 61.—Possible distribution of *B. brevicauda* (closed circles) and *B. carolinensis* (open circles) in the early Rancholabrean, as suggested by late Illinoian and Sangamonian localities: 8. Williams; 12. Mt. Scott; 16. Doby Springs; 18. Easley Ranch; 71. Haile XIB; 72. Arredondo; 73. Reddick IA; 74. Williston IIIA; 78. Bradenton 51st Street. Dotted line indicates extent of Illinoian glaciation. Specimens of *B. b. brevicauda* were found at 16.

(Moncrief et al., 1982), the latter alternative seems more likely. The age of the Sappa formation, from which the Bartek Brothers l.f. was recovered, suggests that fauna is older (possibly pre-Kansan) than the Angus fauna. A trend from cool to warm climate is implied, perhaps accompanied by decreasing moisure; this might account for the subsequent rein-

vasion of the smaller subspecies observed at Angus. The size of specimens from Port Kennedy (if, indeed, that fauna is Yarmouthian) indicates climatic conditions during the period of deposition were similar to those of today, and *B. brevicauda* extended across the midwestern and eastern United States (Fig. 60). Temperate conditions in the east are sug-

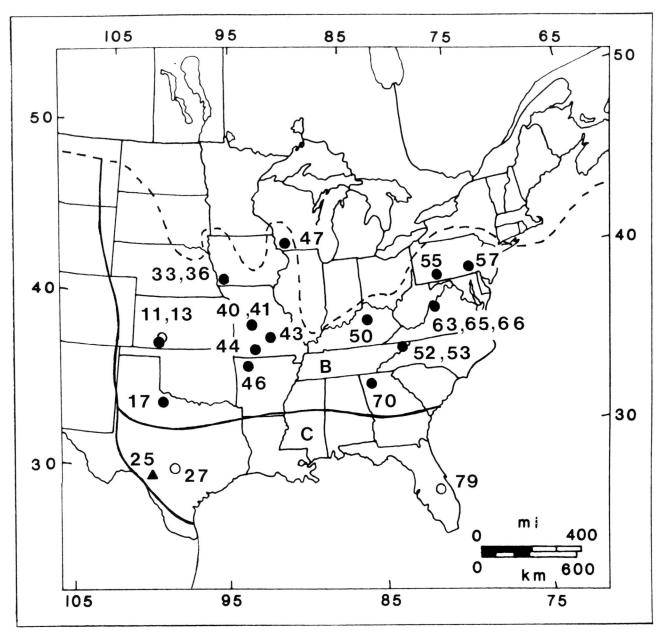

Fig. 62.—Possible distribution of *B. brevicauda* (closed circles), *B. carolinensis* (open circles), and *B. hylophaga* (closed triangles) as suggested by early Wisconsinan localities thought to represent full glacial and late glacial (until approximately 10,500 BP) faunas: 11. Jinglebob; 13. Robert; 17. Howard Ranch; 25. Schulze Cave; 27. Cave Without a Name; 33. Craigmile; 36. Waubonsie; 40. Boney Spring; 41. Trolinger Spring; 43. Bat Cave; 44. Zoo Cave; 46. Peccary Cave; 47. Moscow Fissure; 50. Welsh Cave; 52. Baker Bluff; 53. Guy Wilson Cave; 55. New Paris Sinkhole No. 4; 57. Bootlegger Sink; 63. Strait Canyon Fissure; 65. Back Creek Cave No. 2; 66. Clark's Cave; 70. Ladds Quarry; 79. Vero. Specimens of *B. b. brevicauda* were found at 13, 33, 47, 50, 52, 55, 57, 65, and 66. Dotted line indicates maximum extent of Wisconsinan glaciation.

gested by the fauna of Hanover Quarry Fissure (*B. carolinensis*). The presence of *B. carolinensis* at Hanover and Cumberland might indicate a zone of sympatry of the two species, probably prior to the

maximum Illinoian glaciation. Cumberland Cave is the only Irvingtonian site at which both species are clearly present. The scarcity of *Blarina* from the Yarmouthian probably reflects the sparse fossil rec-

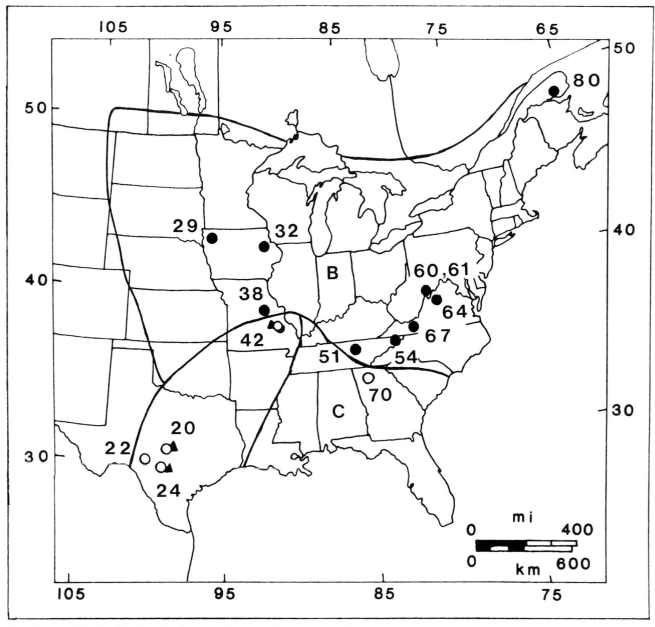

Fig. 63.—Possible distribution of *B. brevicauda* (closed circles), *B. carolinensis* (open circles), and *B. hylophaga* (closed triangles) during the pre-boreal and boreal periods (approximately 10,500 to 5,000 BP) as suggested by late Wisconsinan–early Holocene localities: 20. Miller's Cave; 22. Felton Cave; 24. Klein Cave; 29. Cherokee; 32. Mud Creek; 38. Brynjulfson Cave No. 1; 42. Crankshaft Cave; 51. Robinson Cave; 54. Riverside Cave; 60. Eagle Cave; 61. Hoffman School Cave; 64. Natural Chimneys; 67. Ripplemead Quarry; 70. Ladds Quarry; 80. Caverne de Saint-Elzear. Specimens of *B. b. brevicauda* were found at 29, 42, 51, 61, and 64.

ord on the plains and elsewhere from that interglacial (Hibbard, 1970) and might not be a true indication of the range of *Blarina* at that time.

Early Illinoian (late Irvingtonian) records of *Blarina* consist of the Rezabek (Kansas), Berends (Okla-

homa), and Coleman IIA (Florida) local faunas. The Rezabek fauna, and others from the Great Plains, suggest a climate moister and more equable than now present (Hibbard, 1970; Zakrzewski, 1975*a*). The specimen of *Blarina* that we examined was sim-

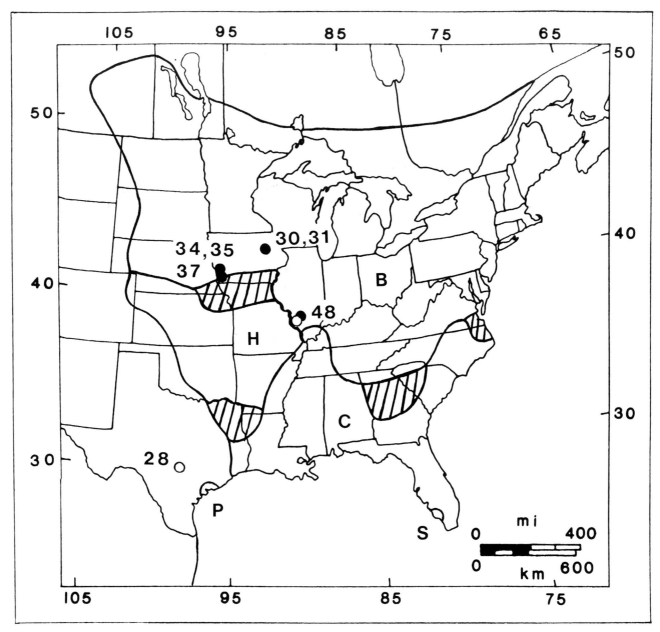

Fig. 64.—Present distribution of *B. brevicauda* (B), *B. carolinensis* (C), and *B. hylophaga* (H); the restricted distributions of the taxa *B. h. plumbea* (P) and "*B. c. shermani*" (S) also are shown (George et al., 1981, 1982; Moncrief et al., 1982). Shaded areas represent sympatry of species. Localities are indicated as in Fig. 63. Early Holocene (<5,000 BP) localities of *Blarina* are included: 28. Barton Springs; 30. Willard Cave; 31. Schmitt Cave; 34. Garrett Farm; 35. Pleasant Ridge; 37. Thurman; 39. Brynjulfson Cave No. 2; 48. Meyer Cave. Specimens of *B. b. brevicauda* were found at 35 and 37.

ilar in size to *B. b. brevicauda*. The material from Berends also indicated moister climate and included specimens that might represent both semispecies of *B. brevicauda*. *B. carolinensis* was present in the Coleman IIA fauna. The Cumberland Cave (Mary-

land) l.f. also might pertain to the early Illinoian (Fig. 60 and Appendix 1).

The earliest Rancholabrean records are probably those from Doby Springs (Oklahoma), Mt. Scott, and Williams (Kansas). On the plains a transition from moist, equable climate to warmer conditions by the late Illinoian was accompanied by shifts in the flora and fauna (Hibbard, 1970; Kapp, 1970). Whether the *B. carolinensis* of Mt. Scott (Fig. 61) reflects this shift, as we suspect, or marks an area of sympatry between the two species during this period cannot be ascertained until additional specimens are found. Sangamonian records of *Blarina* are limited to several sites in Florida and one (Easley Ranch) in Texas. Webb (1974) and Delcourt and Delcourt (1977) cited evidence of warm temperate conditions in the southeast during at least part of the Sangamonian. Dalquest (1962), Hibbard (1970), and Kapp (1970) suggested that the climate on the central plains during the Sangamonian was warmer than during the Illinoian but still moist. However, Graham (1972) suggested, based on the Cragin Quarry l.f. (Kansas), that the Sangamonian climate on the plains was semiarid—sufficiently dry to restrict *Blarina* to the east. The scarcity of *Blarina* remains from this period seems to support Graham's thesis (Fig. 61).

The Wisconsinan fossil record of *Blarina* is more extensive than that of earlier periods. Fossil sites are scattered throughout the southeastern quarter of the United States, and Wisconsinan biogeography of the three species of *Blarina* is complex (Figs. 62 and 63). The history of changes in vegetation during this period also is complicated, and the composition of plant communities was not analogous to that of extant communities (Davis, 1976; Wright, 1981). An overview of the ecology of the Pleistocene was provided by Wright (1976).

During the full glacial period (until approximately 13,000 BP), climate in the central and eastern United States was characterized by cooler summers, milder winters, and more effective moisture than at present (Slaughter, 1967; Hoffmann and Jones, 1970; Guilday, 1971; Wright, 1981). Floral and faunal analyses suggest that vegetation during this period consisted of boreal spruce forest, perhaps with patches of deciduous trees (Mehringer et al., 1970; Ross, 1970; Watts, 1970; Wells, 1970; Wright, 1971). Spruce forests extended from southern Canada to Kansas and Missouri, possibly to the western edge of the modern plains. In the east, forests of spruce and jack pine might have reached the Atlantic coast and extended southward as far as the Coastal Plain of Georgia (Wright, 1968, 1970; Watts, 1970, 1980; Davis, 1976). The extent of tundra and isolated glaciation south of the spruce forest is debatable (Guilday et al., 1977; Wright, 1981). Prairie and decid-

uous forest probably were limited to the southwest and southeast (Hoffmann and Jones, 1970; Ross, 1970; Wright, 1981). In the late glacial period (approximately 13,000 to 10,500 BP) the formerly widespread boreal forest deteriorated, although it might have remained in much of the Great Plains (Hoffmann and Jones, 1970; Wright, 1968, 1970, 1971).

During the moister conditions of the full glacial period, the geographic range of the genus *Blarina* was extended again. Remains from this period have been recovered from throughout the central and eastern United States, including areas in Texas and southwestern Kansas in which *Blarina* are not now present (Fig. 62). The faunas and radiocarbon dates (when available) of many of the Wisconsinan deposits from which *Blarina* were recovered (those of Peccary, Back Creek, Clark's, Baker Bluff, Bootlegger, and New Paris caves, Howard Ranch, part of Ladds, and possibly Bat, Zoo, and Welsh caves) imply deposition during the full glacial and late glacial periods (for example, Dalquest, 1965; Guilday et al., 1977, 1978). The shrews of most of the glacial faunas were identified as *B. brevicauda* (*B. b. brevicauda* from northern and Appalachian sites), indicating a widespread distribution for that species during Wisconsinan boreal conditions. Fossils from Jinglebob fall into an area of overlap between *B. brevicauda*, *B. carolinensis*, and *B. hylophaga*. Otherwise, *B. carolinensis* was restricted to the south during this period. The presence of *B. carolinensis* in these faunas evidently reflects the presence of warmer prairie or forests south of the boreal forest. Remains from Schulze Cave (late glacial) apparently represent the species *B. hylophaga*. Two phena of *Blarina* were detected at Baker Bluff, New Paris, and Back Creek. Sympatry of more than one phenon of *Blarina* and of other presently allopatric species is interpreted as reflecting the more equable climate of full glacial and late glacial periods (Dalquest, 1965; Graham, 1972, 1976, 1979; Graham and Semken, 1976; Klippel and Parmalee, 1982).

During the pre-boreal and boreal periods of the late Pleistocene and early Holocene (approximately 10,500 to 8,450 BP), climate became increasingly continental. Spruce forests were replaced by pine or oak forests in the north, northeast, and in the Ozarks, and grassland spread northward from the southwest (Hoffmann and Jones, 1970; Mehringer et al., 1970; Watts, 1980; Webb, 1981; Wright, 1970). Although post-glacial climates have fluctuated, the overall trend for the last 7,000 years continues to be one of

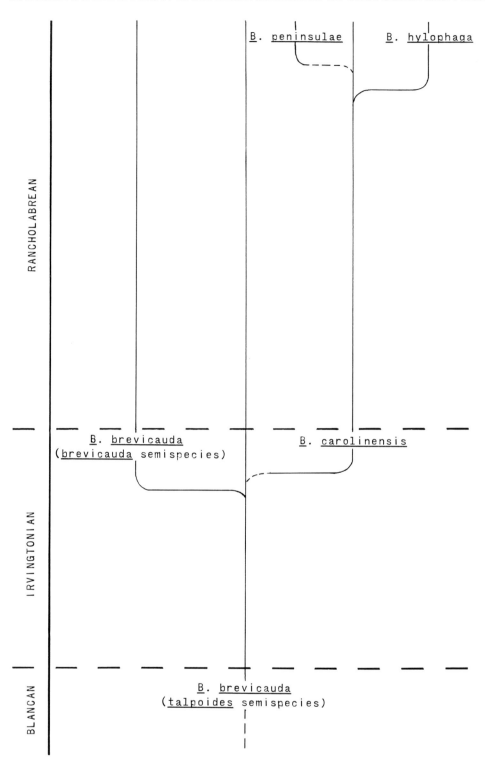

Fig. 65.—Dendrogram showing phylogeny of *Blarina* as indicated by analyses of fossils.

increased continentality (Hoffmann and Jones, 1970; Webb, 1981).

The fossil record of *Blarina* suggests that the species became segregated as a result of increased continentality (Graham and Semken, 1976), as can be seen in the faunas of the late Wisconsinan and early Holocene (Fig. 63). Specimens of *B. brevicauda* from these faunas became increasingly indistinguishable from modern specimens. The large *B. b. brevicauda* was still present in faunas in the north and northeast (Cherokee, Robinson, Riverside, Hoffman School Cave, and Natural Chimneys). Climatic fluctuations since the glacial and late glacial periods have resulted in fluctuating areas of parapatry or sympatry, such as the post-Wisconsinan sympatry of three phena at Cheek Bend Cave (Klippel and Parmalee, 1982) and Crankshaft Cave, and the intergradation and overlapping observed in Holocene distribution (French, 1981; Moncrief et al., 1982; and others). *Blarina hylophaga* appeared in the fossil record in Missouri and Texas, and was sympatric with *B. carolinensis* in some of the Texas faunas. Its relatively recent separation from *carolinensis* is suggested by the past geographic distributions and present similarities of these two species (George et al., 1981). As aridity on the central plains increased, the range of *Blarina* retracted eastward across Texas (Lundelius, 1967; Hall, 1982) and Kansas (Fig. 64). Holocene *B. h. plumbea* is considered a remnant of this formerly widespread population.

Based on the results of this study, we propose that *B. brevicauda,* or an ancestral species similar to *B. brevicauda,* arose from the blarinine stem in the middle or late Pliocene (Fig. 65). The earliest remains of *B. brevicauda* (late Blancan) are the size of the *talpoides* semispecies. This semispecies seems to have had a long association with cool, moist climate. The large semispecies *brevicauda* might have appeared as early as the Kansan glacial (at Trout Cave). It apparently has always occurred near the northern boundary of the distribution of the genus. Its distribution is related presumably to its affinity to the subhumid microthermal regime (as suggested by Graham, 1972).

Blarina brevicauda likely became divided into two populations during the Irvingtonian. Differentiation and chromosomal rearrangement during isolation resulted in the smaller southern species, *B. carolinensis*. The early distribution of *B. carolinensis* is unclear. It became at least narrowly sympatric with *B. brevicauda* during parts of subsequent glacial and interglacial periods. *B. carolinensis* became divided into eastern and western populations shortly after the Wisconsinan glaciation, and subsequent speciation (again involving both morphological and chromosomal changes) resulted in the species *B. hylophaga* in the southwestern part of the range of the genus. All three species possibly were sympatric at various times in the past.

SPECIMENS EXAMINED

MASSACHUSETTS. *Dukes Co.:* Martha's Vineyard, 75 (69 UCONN, 6 USNM); West Tisbury, 7 (MCZ).

Blarina brevicauda angusta

NEW BRUNSWICK. Bald Peak, 7 (NMC); Lac Baker, 3 (NMC). QUEBEC. Bonaventure, 1 (NMC).

Blarina brevicauda brevicauda

IOWA. *Hamilton Co.:* Jewell, 5 (USNM). *Henry Co.:* Hillsboro, 2 (USNM). *Johnson Co.:* Iowa City, 3 (USNM). *Marion Co.:* Knoxville, 2 (USNM). *Pottawattamie Co.:* Council Bluffs, 7 (USNM).
NEBRASKA. *Antelope Co.:* Neligh, 1 (USNM). *Buffalo Co.:* Kearney, 2 (USNM). *Cherry Co.:* Valentine, 3 (USNM). *Platte Co.:* Columbus, 3 (USNM). *Washington Co.:* Blair, 1 (USNM).

Blarina brevicauda churchi

NORTH CAROLINA. Great Smoky Mtns. (county uncertain), 1 (USNM). Magnetic City (county uncertain), 5 (USNM). *Buncombe Co.:* Asheville, 1 (USNM). *Cherokee Co.:* Murphy, 1 (USNM). *Haywood Co.:* Waynesville, 1 (USNM). *Macon Co.:* Highlands, 2 (USNM). *Mitchell Co.:* Roan Mountain, 16 (USNM). *Transylvania Co.:* Pisgah Forest, 11 (USNM). *Watauga Co.:* Boone, 1 (USNM); Grandfather Mountain, 1 (USNM). *Yancey Co.:* Mt. Mitchell, 1 (USNM).
SOUTH CAROLINA. *Greenville Co.:* Caesars Head, 2 (USNM). *Oconee Co.:* Walhalla, 1 (USNM).
TENNESSEE. *Campbell Co.:* La Follette, 1 (USNM). *Carter Co.:* Roan Mountain, 12 (USNM). *Cocke Co.:* 4½ mi SE Cosby, 2 (USNM); Mt. Guyot, 1 (USNM). *Johnson Co.:* Shady Valley, 8 (USNM). *Sevier Co.:* Great Smoky Mountain National Park, 4 (USNM). *Sullivan Co.:* 17 mi SE Bristol, 2 (USNM).
VIRGINIA. *Grayson Co.:* Whitetop Mountain, 5 (USNM).

Blarina brevicauda compacta

MASSACHUSETTS. *Nantucket Co.:* Nantucket Island, 46 (6 MCZ, 35 UCONN, 5 USNM); Siasconset, 4 (UCONN); Quidnet, 29 (UCONN).

Blarina brevicauda kirtlandi

INDIANA. *Allen Co.:* Fort Wayne, 1 (USNM). *Newton Co.:* Rose Lawn, 1 (USNM). *Porter Co.:* no particular locality, 24 (USNM).
OHIO. *Ashtabula Co.:* 5 mi N Geneva, 7 (CM). *Athens Co.:* Athens, 6 (USNM). *Erie Co.:* Sandusky, 1 (USNM). *Fairfield Co.:* no particular locality, 10 (USNM). *Geauga Co.:* Garrettsville, 10 (USNM). *Hamilton Co.:* California, 1 (CM). *Meigs Co.:* Carpenter, 1 (USNM).

Blarina brevicauda manitobensis

MANITOBA. Red Rock Lake, 1 (NMC); Rennie, 4 (NMC).

Blarina brevicauda pallida

NEW BRUNSWICK. Hampton, 5 (USNM).
NOVA SCOTIA. Digby, 4 (USNM); Halifax, 1 (USNM); James River, 6 (USNM); Kedgemakooge Lake, 13 (USNM).
MAINE. *Aroostook Co.:* Presque Isle, 3 (USNM). *Penobscot Co.:* East Branch Penobscot River, 14 (USNM). *Piscataquis Co.:* 20 mi E Mount Katahdin, 3 (USNM); Basin Pond, Mount Katahdin, 1 (USNM); Chimney Pond, Mount Katahdin, 7 (USNM); Tableland Mountain, Mount Katahdin, 2 (USNM).

Blarina brevicauda talpoides

ONTARIO. Chatham, 7 (CM); Komoka, 1 (USNM).
QUEBEC. Montreal, 1 (USNM); Mt. Tremblant Park, 2 (USNM); St. Roche de Mekinac, St. Maurice, 1 (USNM); St. Rose, 4 (USNM).
NEW YORK. *Warren Co.:* Lake George, 39 (USNM).

Blarina carolinensis carolinensis

ALABAMA. *Autauga Co.:* Autaugaville, 4 (USNM). *Cullman Co.:* Ardell, 8 (USNM). *Hale Co.:* Greensboro, 1 (USNM). *Russell Co.:* Seale, 1 (USNM). *Sumter Co.:* York, 2 (USNM).
GEORGIA. *Crawford Co.:* Roberta, 1 (USNM). *Early Co.:* Blakeley, 1 (USNM). *Grady Co.:* Beachton, 4 (USNM). *Liberty Co.:* LeConte Plantation, Riceboro, 1 (USNM). *Randolph Co.:* Cuthbert, 1 (USNM). *Richmond Co.:* Augusta, 2 (USNM). *Thomas Co.:* Boston, 1 (USNM). *Tift Co.:* Tifton, 1 (USNM).
NORTH CAROLINA. *Wake Co.:* Raleigh, 24 (USNM).
SOUTH CAROLINA. *Darlington Co.:* Society Hill, 1 (USNM). *Dorchester Co.:* Saint George, 1 (USNM). *Georgetown Co.:* Georgetown, 1 (USNM); Plantersville, 2 (USNM). *Richland Co.:* Columbia, 3 (USNM). *Williamsburg Co.:* Lane, 1 (USNM).
TENNESSEE. "Reelfoot Lake," (county uncertain), 1 (USNM). *Benton Co.:* Big Sandy, 1 (USNM). *Fayette Co.:* Hickory Withe, 3 (USNM).

Blarina carolinensis minima

LOUISIANA. *East Baton Rouge Par.:* 2 mi S Baker, 1 (LSU); 2 mi NNE Baton Rouge, 1 (LSU); 3 mi NSE Baton Rouge, 1 (LSU);

College, Baton Rouge, 1 (LSU); Baton Rouge, 2 (LSU); ⅓ mi E on Greenville Springs Road, Baton Rouge, 1 (LSU); 300 yds. WNW Plank Rd. and Airline Hwy. Baton Rouge, 1 (LSU); 3 mi SE Baton Rouge, 1 (LSU); 4 mi SE Baton Rouge, 1 (LSU); 7 mi SE Baton Rouge, 1 (LSU); 3 mi S, 2 mi E Baton Rouge, 1 (LSU); 5 mi N University, 1 (LSU); 3 mi NE University, 1 (LSU); ¾ mi NE Indian Mound, University, 1 (LSU); Indian Mound, University, 2 (LSU); 5 mi E University, 1 (LSU); 2 mi ESE University, 2 (LSU); 2 mi S LSU, Baton Rouge, 1 (LSU); 3 mi S University, 1 (LSU); 4.5 mi S University, 1 (LSU); 10 mi S LSU campus, 1 (LSU); 1 mi SE University, 2 (LSU); 13 mi S University Station, 1 (LSU); 7 mi SW Zachary, 1 (LSU). *Red River Par.:* 5 mi N Coushatta, 1 (LSU).

Blarina carolinensis peninsulae

FLORIDA. *Brevard Co.:* Georgiana, 1 (USNM). *Collier Co.:* Deep Lake, 3 (USNM). *Dade Co.:* 1 mi W Chekika State Recreation Area (T.55S, R.37E, Sec. 26), 16 (13 CM, 3 MHP); Miami River, 1 (USNM). *Highlands Co.:* 6 mi S Lake Placid, 1 (MHP) *Osceola Co.:* Kissimmee, 1 (USNM). *Palm Beach Co.:* Ritta, 1 (USNM). *Polk Co.:* Winter Haven, 4 (CM).

Blarina carolinensis shermani

FLORIDA. *Lee Co.:* 2 mi N Ft. Myers, 1 (USNM); Ft. Myers, 2 (USNM).

Blarina hylophaga hylophaga

KANSAS. *Ellis Co.:* 1 mi S, 6½ mi W Antonino (T.15S, R.20W, Sec. 2), 14 (MHP); 1 mi S, 6 mi W Antonino, 3 (MHP); 16 mi N, 1 mi W Hays, 3 (MHP); 9 mi N, 4 mi W Hays (T.12S, R.19W, NE ¼ Sec. 14), 3 (MHP). *Rooks Co.:* 3 mi S, 3 mi W Stockton, 5 (MHP). *Russell Co.:* 5 mi N, 1 mi E Dorrance (T.13S, R.11W, NE ¼ Sec. 17), 2 (MHP); 4 mi N, 4 mi E Dorrance (T.13S, R.11W, SW ¼ Sec. 23), 2 (MHP); 3 mi N, 4 mi E Dorrance (T.13S, R.11W, NW ¼ Sec. 26), 3 (MHP); 6½ mi S, ½ mi E Lucas (T.12S, R.11W, NE ¼ Sec. 34), 1 (MHP); 8½ mi S, ½ mi E Lucas (T.13S, R.11W, NE ¼ Sec. 10), 1 (MHP). *Trego Co.:* ½ mi N, 3¼ mi W Ellis (*in* Ellis Co.) (T.13S, R.21W, NE ¼ Sec. 11), 2 (MHP); ½ mi N, 3 mi W Ellis (*in* Ellis Co.) (T.13S, R.21W, NE ¼ Sec. 11), 2 (MHP); 2½ mi W Ellis (*in* Ellis Co.), 1 (MHP); 11 mi S, 4 mi W Ellis (*in* Ellis Co.) (T.15S, R.21W, NE ¼ Sec. 2), 1 (MHP); 14½ mi S, 5½ mi E Ogallah, 1 (MHP).
OKLAHOMA. *Cleveland Co.:* 2 mi S, 1 mi W Norman, 1 (MHP). *Garvin Co.:* 3 mi N, 2 mi W Davis (by road), 1 (MHP). *Murray Co.:* 1.3 mi S, 2.2 mi W Davis (by road), 2 (MHP); 1.3 mi S, 2 mi W Davis (by road), 2 (MHP); 5.7 mi S, 2.3 mi W Davis (by road), 1 (MHP).

Blarina hylophaga plumbea

TEXAS. *Aransas Co.:* Aransas Wildlife Refuge, 7 (TCWC).

ACKNOWLEDGMENTS

Permission to examine specimens used in this study was granted by L.G. Vostreys (ANSP), J. E. Guilday and A. D. McCrady (CM), O. Hawksley (CMSU), M. F. Frazier (FACM), R. W. Graham and J. J. Saunders (ISM), L. D. Martin (KU), J. A. W. Kirsch (MCZ), W. W. Dalquest (MWU), C. Van Zyll de Jong (NMC), H. A. Semken, Jr. (SUI), D. J. Schmidly (TCWC), E. L. Lundelius, Jr. (TMM), S. D. Webb (UF), R. M. Wetzel (UCONN), P. D. Gingerich (UMMP), M. R. Voorhies (UNSM), T. Bown (USGS), and R. W. Purdy and others (USNM). R. E. Eshelman (Calvert Marine Museum), R. S. Rhodes and H. A. Semken, Jr. (SUI), G. E. Schultz (West Texas State University), E. L. Lundelius, Jr. (TMM), and J. E. Guilday generously shared unpublished data. The senior author also thanks Dr. and Mrs. H. H. Genoways, Mr. and Mrs. Bruce George, and Mr. and Mrs. J. E. Guilday for their hospitality while she was working in Pittsburgh and Washington. Clerical assistance was provided by B. Lange, M. A. Schmidt, and J. Unrine. S. B. George, J. Golden, C. J. Jones, D. C. Lovell, N. D. Moncrief, G. S. Morgan, K. W. Navo, D. K. Tolliver, and J. D. Wieland provided assistance during various stages of the study. Help with the computer was given by D. Eining and N. D. Moncrief. The first author submitted an earlier draft of this manuscript as a thesis in partial fulfillment of the requirements for the M.S. degree in Biology at Fort Hays State University. Members of Jones' thesis committee (C. A. Ely, E. D. Fleharty, G. K. Hulett, and R. J. Zakrzewski) made careful and constructive recommendations regarding the original manuscript. Special thanks are due N. J. Hildreth and C. B. Jones for their generous help with the manuscript; J. E. Guilday and R. J. Zakrzewski for their invaluable comments regarding the Pleistocene of North America; and C. and C. B. Jones for their assistance throughout the course of this study. This study was partially supported by two grants from the National Science Foundation (DEB77-13120 to J. R. Choate and DEB77-12283 to H. H. Genoways).

LITERATURE CITED

ABBE, C., JR. 1900. The physiography of Allegany County. Pp. 27–55, *in* Maryland Geological Survey Allegany County, Johns Hopkins Press, Baltimore, 323 pp.

AUFFENBERG, W. 1963. The fossil snakes of Florida. Tulane Studies Zool., 10:131–216.

———. 1967. Further notes on fossil box turtles of Florida. Copeia, 1967:319–325.

BADER, R. S. 1957. Two Pleistocene mammalian faunas from Alachua County, Florida. Bull. Florida State Mus., 2:53–75.

BAERREIS, D. A. 1980. Habitat and climatic interpretation from terrestrial gastropods at the Cherokee site. Pp. 101–122, *in* The Cherokee excavations: Holocene ecology and human adaptations in northwestern Iowa (D. C. Anderson and H. A. Semken, Jr., eds.), Academic Press, New York, xvi + 277 pp.

BAILEY, R. G. 1978. Description of the ecoregions of the United States. USDA Forest Service, Intermtn. Reg. Ogden, Utah, 77 pp.

BAKER, R. G., and K. L. VAN ZANT. 1980. Holocene vegetational reconstruction in northwestern Iowa. Pp. 123–138, *in* The Cherokee excavations: Holocene ecology and human adaptations in northwestern Iowa (D. C. Anderson and H. A. Semken, Jr., eds.), Academic Press, New York, xvi + 277 pp.

BAYNE, C. K. 1969. Evidence of multiple stades in the lower Pleistocene of northeastern Kansas. Trans. Kansas Acad. Sci., 71:340–349.

BENNINGHOFF, W. S., and A. L. STEVENSON. 1967. Pollen analysis of cave breccia from Ladds Locality, Bartow County, Georgia. Bull. Georgia Acad. Sci., 25:188–191.

BJORK, P. R. 1970. The Carnivora of the Hagerman local fauna (late Pliocene) of southwestern Idaho. Trans. Amer. Phil. Soc., 60(7):1–54.

BLAIR, W. F. 1940. Notes on home ranges and populations of the short-tailed shrew. Ecology, 21:284–288.

———. 1941. Some data on the home ranges and general life history of the short-tailed shrew, red-backed vole, and woodland jumping mouse in northern Michigan. Amer. Midland Nat., 25:681–685.

———. 1958. Distributional patterns of vertebrates in the southern United States in relation to past and present environments. Pp. 433–468, *in* Zoogeography (C. L. Hubbs, ed.), Amer. Assoc. Sci. Publ., 51:x + 1–509.

BOWLES, J. B. 1975. Distribution and biogeography of mammals of Iowa. Spec. Publ. Mus., Texas Tech Univ., 9:1–184.

BRATTSTROM, B. H. 1967. A succession of Pliocene and Pleistocene snake faunas from the high plains of the United States. Copeia, 1967:188–202.

BRAUN, J. K., and M. L. KENNEDY. 1983. Systematics of the genus *Blarina* in Tennessee and adjacent areas. J. Mamm., 64:414–425.

BRODKORB, P. 1957. New passerine birds from the Pleistocene of Reddick, Florida. J. Paleo., 31:129–138.

———. 1958. Fossil birds from Idaho. Wilson Bull., 70:237–242.

———. 1959. The Pleistocene avifauna of Arredondo, Florida. Bull. Florida State Mus., 4:269–291.

———. 1963. A giant flightless bird from the Pleistocene of Florida. Auk, 80:111–115.

BROWN, B. 1908. The Conard Fissure, a Pleistocene bone deposit in northern Arkansas: with a description of two new genera and twenty new species of mammals. Mem. Amer. Mus. Nat. Hist., 9:157–208.

BRYANT, M. D. 1941. A far southwestern occurrence of Pitymys in Texas. J. Mamm., 22:202.

BUCKNER, C. H. 1966. Populations and ecological relationships of shrews in tamarack bogs of southeastern Manitoba. J. Mamm., 47:181–194.

CHANTELL, C. J. 1970. Upper Pliocene frogs from Idaho. Copeia, 1970:654–664.

CHOATE, J. R. 1969. Taxonomic status of the shrew, *Notiosorex* (*Xenosorex*) *phillipsii* Schaldach, 1966 (Mammalia: Insectivora). Proc. Biol. Soc. Washington, 82:469–476.

———. 1970. Systematics and zoogeography of middle American shrews of the genus *Cryptotis*. Univ. Kansas Publ., Mus. Nat. Hist., 19:195–317.

———. 1972. Variation within and among populations of the short-tailed shrew in Connecticut. J. Mamm., 53:116–127.

CHOATE, J. R., and E. D. FLEHARTY. 1973. Habitat preference and spatial relations of shrews in a mixed grassland in Kansas. Southwestern Nat., 18:93–114.

———. 1975. Synopsis of native, Recent mammals of Ellis County, Kansas. Occas. Papers Mus., Texas Tech Univ., 37:1–80.

CHURCHER, C. S., and M. B. FENTON. 1968. Vertebrate remains from the Dickson Limestone Quarry, Halton County, Ontario, Canada. Bull. Natl. Speleol. Soc., 30:11–16.

CONANT, R. 1975. A field guide to reptiles and amphibians of eastern and central North America. Houghton Mifflin Co., Boston, 429 pp.

COPE, E. D. 1899. Vertebrate remains from Port Kennedy bone deposit. J. Acad. Nat. Sci. Philadelphia, 11:193–267.

CORGAN, J. X. 1976. Vertebrate fossils of Tennessee. Tennessee Div. Geol. Bull., 77, vi + 100 pp.

CRAUN, V. S. 1948. Commercial caves of Texas. Bull. Natl. Speleol. Soc., 10:33–45.

DALQUEST, W. W. 1962. The Good Creek formation, Pleistocene of Texas, and its fauna. J. Paleo., 36:568–582.

———. 1965. New Pleistocene formation and local fauna from Hardeman County, Texas. J. Paleo., 39:63–79.

———. 1977. Mammals of the Holloman local fauna, Pleistocene of Oklahoma. Southwestern Nat., 22:255–268.

———. 1978. Early Blancan mammals of the Beck Ranch local fauna of Texas. J. Mamm., 59:269–298.

DALQUEST, W. W., E. ROTH, and F. JUDD. 1969. The mammal fauna of Schulze Cave, Edwards County, Texas. Bull. Florida State Mus., 13:205–276.

DAVIS, L. C. 1969. The biostratigraphy of Peccary Cave, Newton County, Arkansas. Proc. Arkansas Acad. Sci., 23:192–196.

DAVIS, M. B. 1976. Pleistocene biogeography of temperate deciduous forests. Geoscience and Man, 13:13–26.

DAVIS, W. B. 1978. The mammals of Texas. Texas Parks and Wildlife Dept., Austin, Bull. 41, 294 pp.

DELCOURT, P. A., and H. R. DELCOURT. 1977. The Tunica Hills, Louisiana-Mississippi: late glacial locality for spruce and deciduous forest species. Quaternary Res., 7:218–237.

DICE, L. R. 1943. The biotic provinces of North America. Univ. Michigan Press, Ann Arbor, 78 pp.

DOBZHANSKY, T., F. J. AYALA, G. L. STEBBINS, and J. W. VALENTINE. 1977. Evolution. W. H. Freeman, San Francisco, xiv + 572 pp.

DORF, E. 1959. Climatic changes of the past and present. Contrib. Mus. Paleo., Univ. Michigan, 13:181–210.

EINSOHN, S. D. 1971. The stratigraphy and fauna of a Pleistocene outcrop in Doniphan County, northeastern Kansas. Unpubl. M.S. thesis, Univ. Kansas, Lawrence, 83 pp.

ELLIOT, D. G. 1899. Descriptions of apparently new species and subspecies of mammals from the Indian Territory. Field Columbian Mus., Zool. Ser., 1:285–288.

ELLIS, L. S., V. E. DIERSING, and D. F. HOFFMEISTER. 1978. Taxonomic status of short-tailed shrews (Blarina) in Illinois. J. Mamm., 59:305–311.

ESHELMAN, R. E. 1971. The paleoecology of Willard Cave, Delaware County, Iowa. Unpubl. M.S. thesis, Univ. Iowa, Iowa City, 72 pp.

———. 1972. Faunal analysis of the Schmitt Site. Proc. Iowa Acad. Sci., 79:59–61.

———. 1975. Geology and paleontology of the early Pleistocene (late Blancan) White Rock fauna from north-central Kansas. Mus. Paleo., Univ. Michigan Papers on Paleo. (Claude W. Hibbard Mem. Vol. 4), 13:iv + 60 pp.

ESHELMAN, R. E., and C. W. HIBBARD. 1981. Nash local fauna (Pleistocene: Aftonian) of Meade County, Kansas. Contrib. Mus. Paleo., Univ. Michigan, 25:317–326.

ETHERIDGE, R. 1961. Late Cenozoic glass lizards (Ophisaurus) from the southern Great Plains. Herpetologica, 17:179–186.

EVANS, G. L. 1961. The Friesenhahn Cave. Bull. Texas Mem. Mus., 2:1–22.

EVERNDEN, J. F., D. E. SAVAGE, G. H. CURTIS, and G. T. JAMES. 1964. Potassium-argon dates and the Cenozoic mammalian chronology of North America. Amer. J. Sci., 262:145–198.

FAY, L. P. 1980. Mammals of the Garrett Farm and Pleasant Ridge local biotas (Holocene), Mills County, Iowa. AMQUA 6th Bien. Mtg., Abstracts with Program, pp. 74–75.

FORD, A. L., and O. W. VAN AUKEN. 1982. The distribution of woody species in the Guadalupe River floodplain forest in the Edwards Plateau of Texas. Southwestern Nat., 27: 383–392.

FRANKEL, L. 1963. The biota of a pre-Illinoian pond in eastern Nebraska. J. Paleo., 37:249–253.

FRAZIER, M. K. 1977. New records of Neofiber leonardi (Rodentia: Cricetidae) and the paleoecology of the genus. J. Mamm., 58:368–373.

FRENCH, T. W. 1981. Notes on the distribution and taxonomy of short-tailed shrews (genus Blarina) in the southeast. Brimleyana, 6:101–110.

FULLINGTON, R. W. 1978. Supplementary data on the Rex Rodgers site: Mollusca. Pp. 108–113, in Archeology at Mackenzie Reservoir (J. T. Hughes and P. S. Willey, eds.), Texas Historical Commission, Archeological Survey Report 24, Austin, xiv + 296 pp.

GALBREATH, E. C. 1955. An avifauna from the Pleistocene of central Kansas. Wilson Bull., 67:62–63.

GAZIN, C. L. 1933. A new shrew from the Upper Pliocene of Idaho. J. Mamm., 14:142–144.

GENOWAYS, H. H., and J. R. CHOATE. 1972. A multivariate analysis of systematic relationships among populations of the short-tailed shrew (genus Blarina) in Nebraska. Syst. Zool., 21:106–116.

GENOWAYS, H. H., J. C. PATTON III, and J. R. CHOATE. 1977. Karyotypes of shrews of the genera Cryptotis and Blarina (Mammalia: Soricidae). Experientia, 33:1294–1295.

GEORGE, S. B., J. R. CHOATE, and H. H. GENOWAYS. 1981. Distribution and taxonomic status of Blarina hylophaga Elliot (Insectivora: Soricidae). Ann. Carnegie Mus., 50:493–513.

GEORGE, S. B., H. H. GENOWAYS, J. R. CHOATE, and R. J. BAKER. 1982. Karyotypic relationships within the short-tailed shrews, genus Blarina. J. Mamm., 63:639–645.

GETZ, L. L. 1961. Factors influencing the local distribution of shrews. Amer. Midland Nat., 65:67–88.

GETZ, L. L., and C. W. HIBBARD. 1965. A molluscan faunule from the Seymour Formation, of Baylor and Knox counties, Texas. Papers Michigan Acad. Sci., Arts, and Letters, 50: 275–297.

GIDLEY, J. W., and C. L. GAZIN. 1933. New Mammalia in the Pleistocene fauna from Cumberland Cave. J. Mamm., 14: 343–357.

———. 1938. The Pleistocene vertebrate fauna from Cumberland Cave, Maryland. Bull. U.S. Natl. Mus., 171:1–99.

GOULD, S. J., and N. ELDREDGE. 1977. Punctuated equilibria: the tempo and mode of evolution reconsidered. Paleobiology, 3:115–151.

GRAHAM, R. W. 1972. Biostratigraphy and paleoecological significance of the Conard Fissure local fauna with emphasis on the genus Blarina. Unpubl. M.S. thesis, Univ. Iowa, Iowa City, 90 pp.

———. 1976. Late Wisconsin mammalian faunas and environmental gradients of the eastern United States. Paleobiology, 2:343–350.

———. 1979. Paleoclimates and late Pleistocene faunal provinces in North America. Pp. 49–69, in Pre-Llano cultures of the Americas: paradoxes and possibilities (R. L. Humphrey and D. Stanford, eds.), The Anthropological Society of Washington, 150 pp.

GRAHAM, R. W., and H. A. SEMKEN. 1976. Paleoecological significance of the short-tailed shrew (Blarina), with a systematic discussion of *Blarina ozarkensis.* J. Mamm., 57: 433–449.

GRANT, V. 1971. Plant speciation. Columbia Univ. Press, New York, x + 435 pp.

GUILDAY, J. E. 1957. Individual and geographic variation in *Blarina brevicauda* from Pennsylvania. Ann. Carnegie Mus., 35:41–68.

———. 1962a. Notes on Pleistocene vertebrates from Wythe County, Virginia. Ann. Carnegie Mus., 36:77–86.

———. 1962b. The Pleistocene local fauna of the Natural Chimneys, Augusta County, Virginia. Ann. Carnegie Mus., 36:87–122.

———. 1971. The Pleistocene history of the Appalachian mammal fauna. Pp. 233–262, *in* The distributional history of the biota of the southern Appalachians, Part III: Vertebrates (P. C. Holt, ed.), Res. Div. Monogr. 4, VPI and SU, Blacksburg.

———. 1977. Sabertooth cat, *Smilodon floridanus* (Leidy), and associated fauna from a Tennessee cave (40 Dv 40), the First American Bank site. J. Tennessee Acad. Sci., 52:84–94.

GUILDAY, J. E., and H. W. HAMILTON. 1973. The late Pleistocene small mammals of Eagle Cave, Pendleton County, West Virginia. Ann. Carnegie Mus., 44:45–58.

———. 1978. Ecological significance of displaced boreal mammals in West Virginia caves. J. Mamm., 59:176–181.

GUILDAY, J. E., H. W. HAMILTON, E. ANDERSON, and P. W. PARMALEE. 1978. The Baker Bluff Cave deposit, Tennessee, and the late Pleistocene faunal gradient. Bull. Carnegie Mus. Nat. Hist., 11:1–67.

GUILDAY, J. E., H. W. HAMILTON, and A. D. MCCRADY. 1966. The bone breccia of Bootlegger Sink, York County, PA. Ann. Carnegie Mus., 38:145–163.

———. 1969. The Pleistocene vertebrate fauna of Robinson Cave, Overton County, Tennessee. Palaeovertebrata, 2:25–75.

———. 1971. The Welsh Cave peccaries (*Platygonus*) and associated fauna, Kentucky Pleistocene. Ann. Carnegie Mus., 43:249–320.

GUILDAY, J. E., H. W. HAMILTON, and P. W. PARMALEE. 1975. Caribou (*Rangifer tarandus* L.) from the Pleistocene of Tennessee. J. Tennessee Acad. Sci., 50:108–111.

GUILDAY, J. E., P. S. MARTIN, and A. D. MCCRADY. 1964. New Paris No. 4: a Pleistocene cave deposit in Bedford County, Pennsylvania. Bull. Natl. Speleol. Soc., 26:121–194.

GUILDAY, J. E., and P. W. PARMALEE. 1972. Quaternary periglacial records of voles of the genus *Phenacomys* Merriam (Cricetidae: Rodentia). Quaternary Res., 2:170–175.

GUILDAY, J. E., P. W. PARMALEE, and H. W. HAMILTON. 1977. The Clark's Cave bone deposit and the late Pleistocene paleoecology of the central Appalachian mountains of Virginia. Bull. Carnegie Mus. Nat. Hist., 2:1–87.

GUT, H. J. 1959. A Pleistocene vampire bat from Florida. J. Mamm., 40:534–538.

GUT, H. J., and C. E. RAY. 1963. The Pleistocene vertebrate fauna of Reddick, Florida. Quart. J. Florida Acad. Sci., 26: 315–328.

HALL, E. R. 1981. The mammals of North America. John Wiley and Sons, New York, 2nd ed., 1:xviii + 1–600 + 90 and 2: vi + 601–1181 + 90.

HALL, E. R., and K. R. KELSON. 1959. The mammals of North America. Ronald Press, New York, 1:xxx + 1–546 + 79.

HALL, S. A. 1982. Late Holocene paleoecology of the Southern Plains. Quaternary Res., 17:391–407.

HALLBERG, G. R., H. A. SEMKEN, and L. C. DAVIS. 1974. Qua-

ternary records of *Microtus xanthognathus* (Leach), the yellow-cheeked vole, from northwestern Arkansas and southwestern Iowa. J. Mamm., 55:640–645.

HAMON, J. H. 1964. Osteology and paleontology of the passerine birds of the Reddick, Florida, Pleistocene. Bull. Florida Geol. Surv., 44:vii + 210.

HANDLEY, C. O., JR. 1979. Mammals of the Dismal Swamp: a historical account. Pp. 297–357, *in* The Great Dismal Swamp (P. W. Kirk, Jr., ed.), Univ. Press Virginia, Charlottesville, xii + 427 pp.

HAWKSLEY, O., J. F. REYNOLDS, and R. L. FOLEY. 1973. Pleistocene vertebrate fauna of Bat Cave, Pulaski County, Missouri. Bull. Natl. Speleol. Soc., 35:61–87.

HAY, O. P. 1920. Descriptions of some Pleistocene vertebrates found in the United States. Proc. U.S. Natl. Mus., 58:83–146.

HELWIG, J. T., and K. A. COUNCIL (eds.). 1979. SAS user's guide. SAS Institute, Raleigh, North Carolina, 494 pp.

HIBBARD, C. W. 1943. The Rezabek fauna, a new Pleistocene fauna from Lincoln County, Kansas. Univ. Kansas Sci. Bull., 29:235–247.

———. 1944. Stratigraphy and vertebrate paleontology of Pleistocene deposits of southwestern Kansas. Bull. Geol. Soc. Amer., 55:707–754.

———. 1950. Mammals of the Rexroad formation from Fox Canyon, Kansas. Contrib. Mus. Paleo., Univ. Michigan, 8: 113–192.

———. 1952. Vertebrate fossils from late Cenozoic deposits of central Kansas. Univ. Kansas Paleo. Contrib. Vertebrata, 2: 1–14.

———. 1953a. The insectivores of the Rexroad fauna, upper Pliocene of Kansas. J. Paleo., 27:21–32.

———. 1953b. The Saw Rock Canyon fauna and its stratigraphic significance. Papers Michigan Acad. Sci., Arts, and Letters, 38:387–411.

———. 1955. The Jinglebob interglacial (Sangamon?) fauna from Kansas and its climatic significance. Contrib. Mus. Paleo., Univ. Michigan, 12:179–228.

———. 1956. Vertebrate fossils from the Meade formation of southwestern Kansas. Papers Michigan Acad. Sci., Arts, and Letters, 41:145–203.

———. 1957. Notes on late Cenozoic shrews. Trans. Kansas Acad. Sci., 60:327–336.

———. 1958. Summary of North American Pleistocene mammalian local faunas. Papers Michigan Acad. Sci., Arts, and Letters, 43:3–32.

———. 1960. An interpretation of Pliocene and Pleistocene climates in North America. Ann. Rep. Michigan Acad. Sci., Arts, and Letters, 62:5–30.

———. 1963. Paleontology. A late Illinoian fauna from Kansas and its climatic significance. Papers Michigan Acad. Sci., Arts, and Letters, 48:187–221.

———. 1967. New rodents from the late Cenozoic of Kansas. Papers Michigan Acad. Sci., Arts, and Letters, 52:115–131.

———. 1970. Pleistocene mammalian local faunas from the Great Plains and Central Lowland provinces of the United States. Pp. 395–433, *in* Pleistocene and Recent environments of the central Great Plains (W. Dort, Jr., and J. K. Jones, Jr., eds.), Univ. Press Kansas, Lawrence, 433 pp.

HIBBARD, C. W., and P. R. BJORK. 1971. The insectivores of the Hagerman local fauna, upper Pliocene of Idaho. Contrib. Mus. Paleo., Univ. Michigan, 23:171–180.

HIBBARD, C. W., and W. W. DALQUEST. 1966. Fossils from the

Seymour formation of Knox and Baylor counties, Texas, and their bearing on the late Kansan climate of that region. Contrib. Mus. Paleo., Univ. Michigan, 21:1–66.

———. 1973. *Proneofiber,* a new genus of vole (Cricetidae: Rodentia) from the Pleistocene Seymour Formation of Texas, and its evolutionary and stratigraphic significance. Quaternary Res., 3:269–274.

HIBBARD, C. W., and D. W. TAYLOR. 1960. Two late Pleistocene faunas from southwestern Kansas. Contrib. Mus. Paleo., Univ. Michigan, 16:1–223.

HIBBARD, C. W., D. E. RAY, D. E. SAVAGE, D. W. TAYLOR, and J. E. GUILDAY. 1965. Quaternary mammals of North America. Pp. 509–525, *in* The Quaternary of the United States (H. E. Wright, Jr., and D. G. Frey, eds.), Princeton Univ. Press, Princeton, 922 pp.

HIBBARD, C. W., R. J. ZAKRZEWSKI, R. E. ESHELMAN, G. EDMOND, C. D. GRIGGS, and C. GRIGGS. 1978. Mammals from the Kanopolis local fauna, Pleistocene (Yarmouth) of Ellsworth County, Kansas. Contrib. Mus. Paleo., Univ. Michigan, 25: 11–44.

HOFFMANN, R. S., and J. K. JONES, JR. 1970. Influence of late-glacial and post-glacial events on the distribution of Recent mammals on the northern Great Plains. Pp. 355–394, *in* Pleistocene and Recent environments of the central Great Plains (W. Dort, Jr., and J. K. Jones, Jr., eds.), Univ. Press Kansas, Lawrence, 433 pp.

HOLMAN, J. A. 1959a. Amphibians and reptiles from the Pleistocene (Illinoian) of Williston, Florida. Copeia, 1959:96–102.

———. 1959b. Birds and mammals from the Pleistocene of Williston, Florida. Bull. Florida State Mus., 5:1–24.

———. 1967. A Pleistocene herpetofauna from Ladds, Georgia. Bull. Georgia Acad. Sci., 25:154–165.

———. 1972. Herpetofauna of the Kanopolis local fauna (Pleistocene: Yarmouth) of Kansas. Michigan Academician, 5:87–98.

HOOD, C. H., and O. HAWKSLEY. 1975. A Pleistocene fauna from Zoo Cave, Taney County, Missouri. Missouri Speleol., 15:1–42.

HOYER, B. E. 1980. The geology of the Cherokee Sewer Site. Pp. 21–66, *in* The Cherokee excavations. Holocene ecology and human adaptations in northwestern Iowa (D. C. Anderson and H. A. Semken, Jr., eds.), Academic Press, New York, xvi + 277 pp.

JAMMOT, D. 1972. Relationships between the new species *Sorex scottensis* and the fossil shrews *Sorex cinereus* Kerr. Mammalia, 36:449–458.

JENKINS, J. T., JR. 1972. The Pleistocene geology and paleontology of the Fremont County Quarry, Fremont County, Iowa. Unpubl. M.S. thesis, Univ. Iowa, Iowa City, 54 pp.

JONES, J. K., JR., and B. P. GLASS. 1960. The short-tailed shrew, Blarina brevicauda, in Oklahoma. Southwestern Nat., 5:136–142.

JONES, J. K., JR., and R. B. LOOMIS. 1954. Records of the short-tailed shrew and least shrew from Colorado. J. Mamm., 35: 110.

JONES, J. K., JR., D. C. CARTER, H. H. GENOWAYS, R. S. HOFFMANN, and D. W. RICE. 1982. Revised checklist of North American mammals north of Mexico, 1982. Occas. Papers Mus., Texas Tech Univ., 80:1–22.

KAPP, R. O. 1970. Pollen analysis of pre-Wisconsin sediments from the Great Plains. Pp. 143–155, *in* Pleistocene and Recent environments of the central Great Plains (W. Dort, Jr.,

and J. K. Jones, Jr., eds.), Univ. Press Kansas, Lawrence, 433 pp.

KENNERLY, T. E., JR. 1956. Comparisons between fossil and Recent species of the genus *Perognathus.* Texas J. Sci., 8: 74–86.

KIRKLAND, G. L., JR. 1978. The short-tailed shrew, Blarina brevicauda (Say), in the central mountains of West Virginia. Proc. Pennsylvania Acad. Sci., 52:126–130.

KLIPPEL, W. E., and P. W. PARMALEE. 1982. Diachronic variation in insectivores from Cheek Bend Cave and environmental change in the Midsouth. Paleobiology, 8:447–458.

KRAMER, T. L. 1972. The paleoecology of the postglacial Mud Creek biota, Cedar and Scott counties, Iowa. Unpubl. M.S. thesis, Univ. Iowa, Iowa City, 69 pp.

KRIVDA, W. 1957. New Manitoba record for the short-tailed shrew. Canadian Field-Nat., 71:83.

KÜCHLER, A. W. 1974. A new vegetation map of Kansas. Ecology, 55:586–604.

KURTÉN, B. 1965. The Pleistocene Felidae of Florida. Bull. Florida State Mus., 9:215–273.

KURTÉN, B., and E. ANDERSON. 1980. Pleistocene mammals of North America. Columbia Univ. Press, New York, 442 pp.

LaROCQUE, A. 1967. Pleistocene Mollusca of the Ladds deposit, Bartow County, Georgia. Bull. Georgia Acad. Sci., 25:167–187.

LASALLE, P., and J. E. GUILDAY. 1980. Caverne de Saint-Elzear-de-Bonaventure. Rapport preliminaire sur les fouilles de 1977 et 1978. Ministere Ener. et Res., Quebec, 31 pp.

LEE, M. R., and E. G. ZIMMERMAN. 1969. Robertsonian polymorphism in the cotton rat, *Sigmodon fulviventer.* J. Mamm., 50:333–339.

LEONARD, A. B. 1950. A Yarmouthian molluscan fauna in the midcontinent region of the United States. Univ. Kansas, Paleo. Contrib. Mollusca, 3:1–48.

LIGON, J. D. 1965. A Pleistocene avifauna from Haile, Florida. Bull. Florida State Mus., 10:127–158.

LINDSAY, E. H., N. M. JOHNSON, and N. D. OPDYKE. 1975. Preliminary correlation of North American land mammal ages and geomagnetic chronology. Mus. Paleo., Univ. Michigan Papers Paleo., 12:111–119.

LIPPS, L., and C. E. RAY. 1967. The Pleistocene fossiliferous deposit at Ladds, Bartow County, Georgia. Bull. Georgia Acad. Sci., 25:113–119.

LUNDELIUS, E. L., JR. 1960. *Mylohyus nasutus,* long-nosed peccary from the Texas Pleistocene. Bull. Texas Mem. Mus., 1: 1–40.

———. 1967. Late-Pleistocene and Holocene faunal history of central Texas. Pp. 287–319, *in* Pleistocene extinctions: the search for a cause (P. S. Martin and H. E. Wright, Jr., eds.), Yale Univ. Press, New Haven, 453 pp.

MARTIN, R. A. 1968. Late Pleistocene distribution of *Microtus pennsylvanicus.* J. Mamm., 49:265–271.

———. 1969. Taxonomy of the giant Pleistocene beaver *Castoroides* from Florida. J. Paleo., 43:1033–1041.

———. 1973a. Description of a new genus of weasel from the Pleistocene of South Dakota. J. Mamm., 54:924–929.

———. 1973b. The Java local fauna, Pleistocene of South Dakota: a preliminary report. Bull. New Jersey Acad. Sci., 18: 48–56.

———. 1974. Fossil mammals of the Coleman IIA fauna, Sumter County. Pp. 35–99, *in* Pleistocene mammals of Florida (S. D. Webb, ed.), Univ. Press Florida, Gainesville, x + 270 pp.

———. 1975. *Allophaiomys* Kormos from the Pleistocene of North America. Mus. Paleo., Univ. Michigan Papers Paleo. (Claude W. Hibbard Mem. Vol. 3), 12:97–100.

———. 1977. Late Pleistocene *Eumops* from Florida. Bull. New Jersey Acad. Sci., 22:18–19.

MARTIN, R. A., and S. D. WEBB. 1974. Late Pleistocene mammals from the Devil's Den fauna, Levy County. Pp. 114–145, *in* Pleistocene mammals of Florida (S. D. Webb, ed.), Univ. Press Florida, Gainesville, x + 270 pp.

MATTHEWS, W. H., III. 1963. The geologic story of Longhorn Cavern. Bureau of Economic Geol., Univ. Texas, Austin, Guidebook, 4:1–50.

MAYR, E. 1969. Principles of systematic zoology. McGraw-Hill, New York, xi + 428 pp.

MCMULLEN, T. L. 1975. Shrews from the late Pleistocene of central Kansas, with the description of a new species of *Sorex*. J. Mamm., 56:316–320.

MEADE, G. E. 1961. The saber-toothed cat, *Dinobastis serus.* Bull. Texas Mem. Mus., 2:23–60.

MECHAM, J. S. 1958 (printed 1959). Some Pleistocene amphibians and reptiles from Friesenhahn Cave, Texas. Southwestern Nat., 3:17–27.

MEHRINGER, P. J., JR., C. E. SCHWEGER, W. R. WOOD, and R. B. MCMILLAN. 1968. Late-Pleistocene boreal forest in the western Ozark highlands? Ecology, 49:567–568.

MEHRINGER, P. J., JR., J. E. KING, and E. H. LINDSAY. 1970. A record of Wisconsin-age vegetation and fauna from the Ozarks of western Missouri. Pp. 173–183, *in* Pleistocene and Recent environments of the central Great Plains (W. Dort, Jr., and J. K. Jones, Jr., eds.), Univ. Press Kansas, Lawrence, 433 pp.

MENGEL, R. M. 1952. White pelican from the Pleistocene of Oklahoma. Auk, 69:81–82.

MERRIAM, C. H. 1895. Revision of the shrews of the American genera *Blarina* and *Notiosorex*. N. Amer. Fauna, 10:1–34.

MEYLAN, A. 1967. Formules chromosomiques et polymorphisme Robertsonien chez Blarina brevicauda (Say) (Mammalia: Insectivora). Canadian J. Zool., 45:1119–1127.

MILLER, B. B. 1966. Five Illinoian molluscan faunas from the southern Great Plains. Malacologia, 4:173–260.

———. 1975. A sequence of radiocarbon-dated Wisconsinan nonmarine molluscan faunas from southwestern Kansas-northwestern Oklahoma. Mus. Paleo., Univ. Michigan Papers Paleo. (Claude W. Hibbard Mem. Vol. 3), 12:9–18.

MILLER, R. A. 1974. The geologic history of Tennessee. Tennessee Div. Geol. Bull., 74:1–36.

MILSTEAD, W. W. 1956. Fossil turtles of Friesenhahn Cave, Texas, with the description of a new species of *Testudo.* Copeia, 1956:162–171.

MONCRIEF, N. D., J. R. CHOATE, and H. H. GENOWAYS. 1982. Morphometric and geographic relationships of short-tailed shrews (genus *Blarina*) in Kansas, Iowa, and Missouri. Ann. Carnegie Mus., 51:157–180.

NATIONAL OCEANIC AND ATMOSPHERIC ADMINISTRATION (NOAA). 1974. Climates of the states. Water Information Center, Inc., Port Washington, New York, 2 vols., 982 pp.

NEFF, N. A. 1975. Fishes of the Kanopolis local fauna (Pleistocene) of Ellsworth County, Kansas. Mus. Paleo., Univ. Michigan Papers on Paleo. (Claude W. Hibbard Mem. Vol. 3), 12:39–48.

NEFF, N. A., and L. F. MARCUS. 1980. A survey of multivariate methods for systematics. New York, privately published, 243 pp.

NELSON, R. S., and H. A. SEMKEN. 1970. Paleoecological and stratigraphic significance of the muskrat in Pleistocene deposits. Bull. Geol. Soc., 81:3733–3738.

NERO, R. W. 1960. Short-tailed shrew north of the North Saskatchewan River. Blue Jay, 18:41–42.

NICHOLAS, G. 1954. Pleistocene ecology of Cumberland Bone Cave. American Caver, 16:29–39.

OBER, L. D. 1978. The Monkey Jungle, a late Pleistocene fossil site in southern Florida. Plaster Jacket, 28:1–13.

OESCH, R. D. 1967. A preliminary investigation of a Pleistocene vertebrate fauna from Crankshaft Pit, Jefferson County, Missouri. Bull. Natl. Speleol. Soc., 29:163–185.

OLSEN, S. J. 1958. The bog lemming from the Pleistocene of Florida. J. Mamm., 39:537–540.

PARMALEE, P. W. 1967. A Recent cave bone deposit in southwestern Illinois. Bull. Natl. Speleol. Soc., 29:119–147.

PARMALEE, P. W., and R. D. OESCH. 1972. Pleistocene and Recent faunas from the Brynjulfson caves, Missouri. Illinois State Mus. Rep. Invest., 25:1–52.

PARMALEE, P. W., R. D. OESCH, and J. E. GUILDAY. 1969. Pleistocene and Recent vertebrate faunas from Crankshaft Cave, Missouri. Illinois State Mus. Rep. Invest., 14:1–37.

PATTON, T. H. 1963a. Fossil remains of southern bog lemming in Pleistocene deposits of Texas. J. Mamm., 44:275–277.

———. 1963b. Fossil vertebrates from Miller's Cave, Llano County, Texas. Bull. Texas Mem. Mus., 7:1–41.

PAULSON, G. R. 1961. The mammals of the Cudahy fauna. Papers Michigan Acad. Sci., Arts, and Letters, 46:127–153.

PETERSON, O. A. 1926. The fossils of the Frankstown Cave, Blair County, Pennsylvania. Ann. Carnegie Mus., 16:249–315.

PETTUS, D. 1956. Fossil rabbits (Lagomorpha) of the Friesenhahn Cave deposit, Texas. Southwestern Nat., 1:109–115.

PRUITT, W. O., JR. 1953. An analysis of some physical factors affecting the local distribution of the shorttail shrew (*Blarina brevicauda*) in the northern part of the Lower Peninsula of Michigan. Misc. Publ. Mus. Zool., Univ. Michigan, 79:1–39.

———. 1959. Microclimates and local distribution of small mammals on the George Reserve, Michigan. Misc. Publ. Mus. Zool., Univ. Michigan, 109:1–27.

PYLE, K. B. 1980. The Cherokee large mammal fauna. Pp. 171–196, *in* The Cherokee excavations: Holocene ecology and human adaptations in northwestern Iowa (D. C. Anderson and H. A. Semken, Jr., eds.), Academic Press, New York, xvi + 277 pp.

QUINN, J. H. 1972. Extinct mammals in Arkansas and related C^{14} dates circa 3000 years ago. 24th Inter. Geol. Cong., 12:89–96.

RAY, C. E. 1965. A new chipmunk, *Tamias aristus,* from the Pleistocene of Georgia. J. Paleo., 39:1016–1022.

———. 1967. Pleistocene mammals from Ladds, Bartow County, Georgia. Bull. Georgia Acad. Sci., 25:120–150.

RAY, C. E., S. J. OLSEN, and H. J. GUT. 1963. Three mammals new to the Pleistocene fauna of Florida, and a reconsideration of five earlier records. J. Mamm., 44:373–395.

REESE, J. L. 1972. The Schmitt Cave. Proc. Iowa Acad. Sci., 79:56–58.

REPENNING, C. A. 1967. Subfamilies and genera of the Soricidae. U.S.G.S. Prof. Paper, 565:iv + 74 pp.

RHODES, R. S., II. 1982. Mammalian paleoecology of the Farmdalian Craigmile and the Woodfordian Waubonsie local faunas, southwestern Iowa. Unpubl. Ph.D. dissert., Univ. Iowa, Iowa City, vi + 132 pp.

RICHMOND, N. D. 1963. Evidence against the existence of crocodiles in Virginia and Maryland during the Pleistocene. Proc. Biol. Soc. Washington, 76:65–67.

RINKER, G. C., and C. W. HIBBARD. 1952. A new beaver, and associated vertebrates, from the Pleistocene of Oklahoma. J. Mamm., 33:98–101.

ROBERTSON, J. S., JR. 1974. Fossil *Bison* of Florida. Pp. 214–246, *in* Pleistocene mammals of Florida (S. D. Webb, ed.), Univ. Press Florida, Gainesville, x + 270 pp.

ROSS, H. H. 1970. The ecological history of the Great Plains: evidence from grassland insects. Pp. 225–240, *in* Pleistocene and Recent environments of the central Great Plains (W. Dort, Jr., and J. K. Jones, Jr., eds.), Univ. Press Kansas, Lawrence, 433 pp.

ROTH, E. L. 1972. Late Pleistocene mammals from Klein Cave, Kerr County, Texas. Texas J. Sci., 24:75–84.

SAUNDERS, J. J. 1977. Late Pleistocene vertebrates of the western Ozark highland, Missouri. Illinois State Mus. Rep. Invest., 33:1–118.

SCHMIDLY, D. J., and W. A. BROWN. 1979. Systematics of short-tailed shrews (genus *Blarina*) in Texas. Southwestern Nat., 24:39–48.

SCHULTZ, C. B., and L. D. MARTIN. 1970. Quaternary mammalian sequence in the central Great Plains. Pp. 341–353, *in* Pleistocene and Recent environments of the central Great Plains (W. Dort, Jr., and J. K. Jones, Jr., eds.), Univ. Press Kansas, Lawrence, 433 pp.

SCHULTZ, C. B., and L. G. TANNER. 1957. Medial Pleistocene fossil vertebrate localities in Nebraska. Bull. Univ. Nebraska State Mus., 4:59–81.

SCHULTZ, C. B., L. D. MARTIN, L. G. TANNER, and R. G. CORNER. 1978. Provincial land mammal ages for the North American Quaternary. Trans. Nebraska Acad. Sci., 5:59–64.

SCHULTZ, G. E. 1967. Four superimposed late-Pleistocene faunas from southwest Kansas. Pp. 321–336, *in* Pleistocene extinctions: the search for a cause (P. S. Martin and H. E. Wright, Jr., eds.), Yale Univ. Press, New Haven, 453 pp.

———. 1969. Geology and paleontology of a late Pleistocene basin in southwest Kansas. Geol. Soc. Amer., Spec. Paper 105:viii + 1–85.

———. 1978. Supplementary data on the Rex Rodgers site: Micromammals. P. 114, *in* Archeology at Mackenzie Reservoir (J. T. Hughes and P. S. Willey, eds.), Texas Historical Commission, Archeological Survey Report 24, Austin, xiv + 296 pp.

SEALANDER, J. A. 1979. A guide to Arkansas mammals. River Road Press, Conway, Arkansas, x + 313 pp.

SELF, H. 1978. Environment and man in Kansas. Regents Press Kansas, Lawrence, 288 pp.

SELLARDS, E. H. 1919. The geology and mineral resources of Bexar County. Univ. Texas Bull., 1932:1–202.

———. 1937. The Vero finds in the light of present knowledge. Pp. 193–210, *in* Early man (G. G. MacCurdy, ed.), J. P. Lippincott Co., Philadelphia, 362 pp.

SEMKEN, H. A., JR. 1961. Fossil vertebrates from Longhorn Cavern, Burnet County, Texas. Texas J. Sci., 13:290–310.

———. 1966. Stratigraphy and paleontology of the McPherson Equus beds (Sandahl local fauna) McPherson County, Kansas. Contrib. Mus. Paleo., Univ. Michigan, 20:121–178.

———. 1980. Holocene climatic reconstructions derived from the three micromammal bearing cultural horizons of the Cherokee Sewer Site, northwestern Iowa. Pp. 67–99, *in* The Cherokee excavations: Holocene ecology and human adaptations in northwestern Iowa (D. C. Anderson and H. A. Semken, Jr., eds.), Academic Press, New York, xvi + 277 pp.

SHELDON, C. 1936. The mammals of Lake Kedgemakooge and vicinity, Nova Scotia. J. Mamm., 17:207–215.

SHUTLER, R., JR., D. C. ANDERSON, L. S. TATUM, and H. A. SEMKEN, JR. 1980. Excavation techniques and synopsis of results derived from the Cherokee project. Pp. 1–20, *in* The Cherokee excavations: Holocene ecology and human adaptations in northwestern Iowa (D. C. Anderson and H. A. Semken, Jr., eds.), Academic Press, New York, xvi + 277 pp.

SIMPSON, G. G. 1928. Pleistocene mammals from a cave in Citrus County, Florida. Amer. Mus. Novitates, 328:1–16.

———. 1929a. Pleistocene mammalian fauna of the Seminole Field, Pinellas County, Florida. Bull. Amer. Mus. Nat. Hist., 56:561–599.

———. 1929b. The extinct land mammals of Florida. Florida State Geol. Survey Annual Report, 20:229–279.

SKINNER, M. F., C. W. HIBBARD, E. D. GUTENTAG, G. R. SMITH, J. G. LUNDBERG, J. A. HOLMAN, J. A. FEDUCCIA, and P. V. RICH. 1972. Early Pleistocene preglacial and glacial rocks and faunas of north-central Nebraska. Bull. Amer. Mus. Nat. Hist., 148:1–148.

SLAUGHTER, B. H. 1967. Animal ranges as a clue to late-Pleistocene extinctions. Pp. 155–167, *in* Pleistocene extinctions: the search for a cause (P. S. Martin and H. E. Wright, Jr., eds.), Yale Univ. Press, New Haven, 453 pp.

SLAUGHTER, B. H., and B. R. HOOVER. 1963. Sulphur River formation and the Pleistocene mammals of the Ben Franklin local fauna. J. Grad. Res. Center, 31:132–148.

SMITH, C. L. 1954. Pleistocene fishes of the Berends fauna of Beaver County, Oklahoma. Copeia, 1954:282–289.

———. 1958. Additional Pleistocene fishes from Kansas and Oklahoma. Copeia, 1958:176–180.

SMITH, G. R. 1963. A late Illinoian fish fauna from southwestern Kansas and its climatic significance. Copeia, 1963:278–285.

SMITH, R. L. 1959. Abundance of small mammals in conifer plantations. J. Mamm., 40:253–254.

STARRETT, A. 1956. Pleistocene mammals of the Berends fauna of Oklahoma. J. Paleo., 30:1187–1192.

STEPHENS, J. J. 1960. Stratigraphy and paleontology of a late Pleistocene basin, Harper County, Oklahoma. Bull. Geol. Soc. Amer., 71:1675–1702.

TAMSITT, J. R. 1957. *Peromyscus* from the late Pleistocene of Texas. Texas J. Sci., 9:355–363.

TATE, C. M., J. F. PAGELS, and C. O. HANDLEY, JR. 1980. Distribution and systematic relationship of two kinds of short-tailed shrews (Soricidae: *Blarina*) in south-central Virginia. Proc. Biol. Soc. Washington, 93:50–60.

TAYLOR, D. W. 1954. A new Pleistocene fauna and new species of fossil snails from the high plains. Occas. Papers Mus. Zool., Univ. Michigan, 557:1–16.

———. 1966. Summary of North American Blancan nonmarine molluscs. Malacologia, 4:1–172.

TAYLOR, D. W., and C. W. HIBBARD. 1955. A new Pleistocene fauna from Harper County, Oklahoma. Oklahoma Geol. Survey, 37:1–23.

VAN DER MEULEN, A. J. 1978. *Microtus* and *Pitymys* (Arvicolidae) from Cumberland Cave, Maryland, with a comparison of some New and Old World species. Ann. Carnegie Mus., 47:101–145.

VOORHIES, M. R. 1974. Pleistocene vertebrates with boreal affinities in the Georgia Piedmont. Quaternary Res., 4:85–93.

WATTS, W. A. 1970. The full-glacial vegetation of northwestern Georgia. Ecology, 51:17–33.

———. 1980. The late Quaternary vegetation history of the southeastern United States. Ann. Rev. Ecol. Syst., 11:387–409.

WEBB, S. D. 1974. Chronology of Florida Pleistocene mammals. Pp. 5–31, *in* Pleistocene mammals of Florida (S. D. Webb, ed.), Univ. Press Florida, Gainesville, x + 270 pp.

WEBB, T., III. 1981. The past 11,000 years of vegetational change in eastern North America. Bioscience, 31:501–506.

WEEMS, R. E. 1972. Vertebrate remains from a (?) post-Wisconsin fissure deposit near Pearisburg, Virginia. Virginia J. Sci., 23:137 (Abst.).

WEEMS, R. E., and B. B. HIGGINS. 1977. Post-Wisconsinan vertebrate remains from a fissure deposit near Ripplemead, Virginia. Bull. Natl. Speleol. Soc., 39:106–108.

WEIGEL, R. D. 1962. Fossil vertebrates of Vero, Florida. Florida Geol. Surv., Spec. Publ., 10:1–59.

WELLS, P. V. 1970. Vegetational history of the Great Plains: a post-glacial record of coniferous woodland in southeastern Wyoming. Pp. 185–202, *in* Pleistocene and Recent environments of the central Great Plains (W. Dort, Jr., and J. K. Jones, Jr., eds.), Univ. Press Kansas, Lawrence, 433 pp.

WENDLAND, W. M. 1980. Holocene climatic reconstructions on the prairie peninsula. Pp. 139–148, *in* The Cherokee excavations. Holocene ecology and human adaptations in northwestern Iowa (D. C. Anderson and H. A. Semken, Jr., eds.), Academic Press, New York, xvi + 277 pp.

WETMORE, A. 1927. A record of the ruffed grouse from the Pleistocene of Maryland. Auk, 44:561.

———. 1967. Pleistocene Aves from Ladds, Georgia. Bull. Georgia Acad. Sci., 25:151–153.

WILEY, E. O. 1978. The evolutionary species concept reconsidered. Syst. Zool., 27:17–26.

WRIGHT, H. E., JR. 1968. The roles of pine and spruce in the forest history of Minnesota and adjacent areas. Ecology, 49:937–955.

———. 1970. Vegetational history of the Central Plains. Pp. 157–172, *in* Pleistocene and Recent environments of the central Great Plains (W. Dort, Jr., and J. K. Jones, Jr., eds.), Univ. Press Kansas, Lawrence, 433 pp.

———. 1971. Late Quaternary vegetational history of North America. Pp. 425–464, *in* The late Cenozoic glacial ages (K. K. Turekian, ed.), Yale Univ. Press, New Haven, 606 pp.

———. 1976. Pleistocene ecology—some current problems. Geoscience and Man, 13:1–12.

———. 1981. Vegetation east of the Rocky Mountains 18,000 years ago. Quaternary Res., 15:113–125.

YATES, T. L., and D. J. SCHMIDLY. 1977. Systematics of Scalopus aquaticus (Linnaeus) in Texas and adjacent states. Occas. Papers Mus., Texas Tech Univ., 45:1–36.

ZAKRZEWSKI, R. J. 1967. The primitive vole, *Ogmodontomys,* from the late Cenozoic of Kansas and Nebraska. Papers Michigan Acad. Sci., Arts, and Letters, 52:133–150.

———. 1969. The rodents from the Hagerman local fauna, upper Pliocene of Idaho. Contrib. Mus. Paleo., Univ. Michigan, 23:1–36.

———. 1975a. Pleistocene stratigraphy and paleontology in western Kansas: the state of the art, 1974. Mus. Paleo., Univ. Michigan Papers Paleo. (Claude W. Hibbard Mem. Vol. 3), 12:121–128.

———. 1975b. The late Pleistocene arvicoline rodent *Atopomys.* Ann. Carnegie Mus., 45:255–261.

Address (Jones and Choate):   Museum of the High Plains, Fort Hays State University, Hays, Kansas 67601.
Present address (Jones):   Department of Zoology, University of Florida, Gainesville, Florida 32611.
Address (Genoways):   Section of Mammals, Carnegie Museum of Natural History, 4400 Forbes Avenue, Pittsburgh, Pennsylvania 15213.

APPENDIX 1

Pliocene, Pleistocene, and Holocene sites from which Blarina *have been reported. Carbon-14 and geomagnetic dates, when provided, are in years before present. Samples examined in this study are numbered as in Fig. 2. Computer acronyms are indicated in parentheses.*

Site	County	Age	Taxon	Reference
ARKANSAS				
Rancholabrean				
46 Peccary Cave (PECCAR)	Newton	Wisconsinan (16,700 ± 250)	*B. brevicauda*	Graham, 1976; Graham and Semken, 1976
Irvingtonian				
45 Conard Fissure (CONARD)	Newton	Kansan	*B. brevicauda*	Graham, 1972
FLORIDA				
Rancholabrean				
Devil's Den	Levy	Holocene (possibly 7,000 to 8,000)	"*B. brevicauda*"	Martin and Webb, 1974
Melbourne	Brevard	Late Wisconsinan	"*B. brevicauda*"	Webb, 1974
Monkey Jungle Hammock	Dade	Late Pleistocene	"*B. cf. brevicauda*"	Martin, 1977; Ober, 1978, and in litt.

APPENDIX 1
Continued.

Site	County	Age	Taxon	Reference
79 Vero 2 and 3 (INDIAN)	Indian River	Late Wisconsinan	*B. carolinensis*	Webb, 1974
Kendrick IA	Marion	Wisconsinan	"*B. brevicauda*"	Webb, 1974
75 Waccasassa River IIB and III	Levy	Wisconsinan	*B. carolinensis*	Webb, 1974
72 Arredondo IB, IIA (ARREDO)	Alachua	Late Sangamonian	*B. carolinensis*	Webb, 1974
78 Bradenton 51st St.	Manatee	Sangomonian	*B. carolinensis*	Webb, 1974
71 Haile XIB, XIIIA (HAILEX)	Alachua	Sangamonian	*B. carolinensis*	Webb, 1974; Martin and Webb, 1974
73 Reddick IA (REDDIC)	Marion	Sangamonian	*B. carolinensis*	Webb, 1974
Waccasassa River VIA	Levy	Sangamonian	"*B. brevicauda*"	Webb, 1974
74 Williston IIIA	Levy	Sangamonian	*B. carolinensis*	Martin, 1974; Webb, 1974
Irvingtonian				
77 Coleman IIA	Sumter	Late Kansan to early Illinoian	*B. carolinensis*	Martin, 1974; Webb, 1974
76 Inglis IA	Citrus	Nebraskan—Aftonian	*B. carolinensis*	Webb, 1974

GEORGIA

Rancholabrean

70 Ladds Quarry (LADDSQ)	Bartow	Late Wisconsinan	*B. brevicauda* and *B. carolinensis*	Ray, 1967; Kurtén and Anderson, 1980

ILLINOIS

Rancholabrean

48 Meyer Cave (MEYERC)	Monroe	Early Holocene	*B. brevicauda* and *B. carolinensis*	Parmalee, 1967

IOWA

Rancholabrean

37 Thurman	Fremont	Holocene (980 ± 150)	*B. brevicauda* and *B. b. brevicauda*	Jenkins, 1972
35 Pleasant Ridge	Mills	Holocene (1,450)	*B. b. brevicauda*	Fay, 1980
30 Willard Cave (WILLAR)	Delaware	Holocene (1,255 ± 55; 1,605 ± 65; 3,500 ± 60)	*B. brevicauda*	Eshelman, 1971 and in litt.
31 Schmitt Cave (SCHMIT)	Dubuque	Holocene (ca. 1,300–1,900)	*B. brevicauda*	Eshelman, 1972
34 Garrett Farm	Mills	Holocene (3,400)	*B. brevicauda*	Fay, 1980
32 Mud Creek	Cedar and Scott	Holocene (6,220 ± 110)	*B. brevicauda*	Kramer, 1972
29 Cherokee Sewer Site	Cherokee	Holocene (ca. 6,350–8,400)	*B. b. brevicauda*	Hoyer, 1980; Semken, 1980
36 Waubonsie	Mills	14,830 + 1,060, −1,220	*B. brevicauda*	Rhodes, 1982
33 Craigmile	Mills	23,240 ± 535	*B. b.` brevicauda*	Rhodes, 1982

KANSAS

Rancholabrean

13 Robert	Meade	Late Wisconsinan (11,100 ± 390)	*B. b. brevicauda*	Schultz, 1969
11 Jinglebob	Meade	Early Wisconsinan	*B. carolinensis*	Kapp, 1970; Zakrzewski, 1975*a*
8 Williams	Rice	Illinoian	*B. brevicauda*	Zakrzewski, 1975*a*
12 Mt. Scott	Meade	Late Illinoian	*B. carolinensis*	Hibbard, 1963
Irvingtonian				
6 Rezabek	Lincoln	Yarmouthian or Illinoian	*B. b. brevicauda*	Hibbard and Dalquest, 1973; Zakrzewski, 1975*a*

APPENDIX 1
Continued.

Site	County	Age	Taxon	Reference
7 Kanopolis	Ellsworth	Yarmouthian	*B. brevicauda* (and possibly *B. b. brevicauda*)	Hibbard et al., 1978
9 Kentuck	McPherson	Kansan	*Blarina* sp.	Semken, 1966; Zakrzewski, 1975*a*
10 Cudahy	Meade	Late Kansan (600,000)	*B. brevicauda*	Paulson, 1961; Lindsay et al., 1975
5 Wathena	Doniphan	Aftonian or Kansan	*B. carolinensis*	Einsohn, 1971; Van der Meulen, 1978
Blancan				
14 Dixon	Kingman	Nebraskan	*B. brevicauda*	Hibbard, 1956, 1970; Skinner et al., 1972
4 White Rock	Republic	Pre-Nebraskan	*B. brevicauda*	Eshelman, 1975
KENTUCKY				
Rancholabrean				
50 Welsh Cave	Woodford	Wisconsinan (12,950 ± 550)	*B. b. brevicauda*	Guilday, Hamilton, and McCrady, 1971
MARYLAND				
Irvingtonian				
59 Cumberland Cave (CUMBER)	Allegany	Possibly late Kansan or early Illinoian	*B. brevicauda* and *B. carolinensis*	Van der Meulen, 1978; Guilday, in litt.
MISSISSIPPI				
Rancholabrean				
49 Catalpa Creek	Lowndes	Pleistocene to Holocene	*B. carolinensis*	Frazier, in litt.
MISSOURI				
Rancholabrean				
39 Brynjulfson Cave No. 2 (BRYNJU)	Boone	Holocene (2,460 ± 230)	*B. brevicauda*	Parmalee and Oesch, 1972
38 Brynjulfson Cave No. 1	Boone	Late Wisconsinan (9,440 ± 760)	*B. brevicauda*	Parmalee and Oesch, 1972
42 Crankshaft Cave (CRANKS)	Jefferson	Late Wisconsinan—Holocene	*B. b. brevicauda, B. brevicauda, B. hylophaga*	Parmalee, Oesch, and Guilday, 1969
44 Zoo Cave (ZOOCAV)	Taney	Wisconsinan—Holocene	*B. brevicauda*	Hood and Hawksley, 1975
43 Bat Cave	Pulaski	Late Wisconsinan	*B. brevicauda*	Hawksley, Reynolds, and Foley, 1973
40 Boney Spring	Benton	Late Wisconsinan (13,700 ± 600; 16,580 ± 220)	*B. brevicauda*	Mehringer, King, and Lindsay, 1970; Saunders, 1977
41 Trolinger Spring	Hickory	Mid-Wisconsinan (25,650 ± 700; 32,200 ± 1,900, −1,600)	*B. brevicauda*	Mehringer, King, and Lindsay, 1970; Saunders, 1977
NEBRASKA				
Irvingtonian				
2 Bartek Brothers	Saunders	Yarmouthian	*B. b. brevicauda*, possibly *B. brevicauda*	Frankel, 1963
3 Angus	Nuckolls	Late Yarmouthian or Early Illinoian	*B. brevicauda*	Schultz and Martin, 1970; Kurtén and Anderson, 1980

APPENDIX 1
Continued.

Site	County	Age	Taxon	Reference
OKLAHOMA				
Rancholabrean				
16 Doby Springs	Harper	Illinoian	*B. b. brevicauda*	Stephens, 1960
Irvingtonian				
15 Berends	Beaver	Early Illinoian	*B. b. brevicauda* and possibly *B. brevicauda*	Hibbard, 1970; Starrett, 1956
PENNSYLVANIA				
Rancholabrean				
Meadowcroft Rockshelter	Washington	Holocene	"*B. brevicauda*"	Guilday, in litt.
57 Bootlegger Sink	York	Late Wisconsinan—Holocene (11,550 ± 100)	*B. b. brevicauda*	Guilday, Hamilton, and McCrady, 1966
Frankstown	Blair	Wisconsinan	"*B. brevicauda*"	Peterson, 1926; Kurtén and Anderson, 1980
55 New Paris No. 4 (NEWPAR)	Bedford	Wisconsinan (9,540 to 11,300 ± 1,000)	*B. b. brevicauda, B. brevicauda*	Guilday, Martin, and McCrady, 1964
Irvingtonian				
56 Hanover Quarry Fissure	Adams	Yarmouthian	*B. carolinensis*	Guilday, in litt., 1982
58 Port Kennedy Cave	Montgomery	Aftonian, Kansan, or Yarmouthian	*B. brevicauda*	Hibbard, 1958; Guilday, 1971; Kurtén and Anderson, 1980
SOUTH DAKOTA				
Irvingtonian				
1 Java	Walworth	Kansan	*B. carolinensis*	Martin, 1973*b*
TENNESSEE				
Rancholabrean				
First American Bank Site	Davidson	Late Pleistocene—Holocene (9,410 to 10,034)	"*B. brevicauda*"	Guilday, 1977; Zakrzewski, personal communication
54 Riverside Cave (RIVERS)	Sullivan	Late Wisconsinan and Holocene	*B. brevicauda*	Guilday, personal communication
Cheek Bend Cave	Maury	Late Wisconsinan and Holocene	"*B. brevicauda*"	Klippel and Parmalee, 1982
52 Baker Bluff Cave (BAKERB)	Sullivan	Late Wisconsinan (10,560 ± 220 to 19,100 ± 850)	*B. b. brevicauda* and *B. brevicauda* (?*kirtlandi*)	Guilday et al., 1978
53 Guy Wilson Cave (GUYWIL)	Sullivan	Late Wisconsinan (19,700 ± 600)	*B. brevicauda*	Guilday, 1971
51 Robinson Cave (ROBINS)	Overton	Wisconsinan	*B. b. brevicauda*	Guilday, Hamilton, and McCrady, 1969
TEXAS				
Rancholabrean				
28 Barton Springs Road	Travis	Holocene (1,015 ± 105)	*B. carolinensis*	Lundelius, 1967
23 Hall's Cave	Kerr	Holocene	*B. carolinensis*	Lundelius, in litt., 1983
21 Longhorn Cavern	Burnet	Holocene	*B. carolinensis*	Lundelius, 1967
20 Miller's Cave	Llano	Holocene (7,200 ± 300)	*B. carolinensis* and *B. hylophaga*	Patton, 1963*b*
24 Klein Cave (KLEINC)	Kerr	Holocene (7,683 ± 643)	*B. carolinensis* and *B. hylophaga*	Roth, 1972

APPENDIX 1
Continued.

Site	County	Age	Taxon	Reference
22 Felton Cave	Sutton	Holocene (7,770 ± 130)	*B. carolinensis*	Lundelius, 1967; Graham, 1976
Rex Rodgers Site	Briscoe	Mid to late Wisconsinan	"*Blarina* sp."	Fullington, 1978; Schultz, 1978
25 Schulze Cave (SCHULZ)	Edwards	Late Wisconsinan—early Holocene (11,000–8,000)	*B. hylophaga*	Dalquest et al., 1969
27 Cave Without a Name (CAVEWI)	Kendall	Wisconsinan (10,900 ± 190)	*B. carolinensis*	Lundelius, 1967
Ben Franklin	Delta	Late Wisconsin (9,550 ± 375; 11,135 ± 450)	"*Blarina* sp."	Slaughter and Hoover, 1963
17 Howard Ranch	Hardeman	Wisconsinan	*B. brevicauda*	Dalquest, 1965
26 Friesenhahn Cave (FRIESE)	Bexar	Wisconsinan	*B. carolinensis*	Evans, 1961; Lundelius, 1960
18 Easley Ranch	Foard	Sangamonian	*B.* cf. *brevicauda*	Dalquest, 1962
Irvingtonian				
19 Vera	Baylor and Knox	Late Kansan (ca. 600,000)	*B. brevicauda*	Hibbard and Dalquest, 1966; Dalquest, 1977
Unnamed locality	Swisher	Kansan	"*Blarina* sp."	G. E. Schultz, in litt., 1983
VIRGINIA				
Rancholabrean				
69 Meadowview Cave	Washington	Late Wisconsinan or early Holocene	*B. brevicauda*	Guilday and Parmalee, 1972
67 Ripplemead Quarry	Giles	Late Wisconsinan or early Holocene	*B. brevicauda*	Guilday, in litt.
64 Natural Chimneys (CHIMNE)	Augusta	Early post-Wisconsinan	*B. b. brevicauda*	Guilday, 1962*b*
65 Back Creek Cave No. 2 (BACKCR)	Bath	Wisconsinan	*B. b. brevicauda* and *B. brevicauda*	Guilday, Parmalee, and Hamilton, 1977
66 Clark's Cave (CLARKS)	Bath	Late Wisconsinan	*B. b. brevicauda*	Guilday, Parmalee, and Hamilton, 1977
63 Strait Canyon (STRAIT)	Highland	Early Wisconsinan (29,870, +1,800–1,400)	*B. brevicauda*	Guilday, in litt.; Zakrzewski, personal communication
Unknown				
Early's Cave	Wythe	Mid-Pleistocene	"*B.* cf. *brevicauda*"	Guilday, 1962*a*
68 Jasper Saltpeter Cave (JASPER)	Lee	Unknown	*B. brevicauda*	Guilday, in litt.
WEST VIRGINIA				
Rancholabrean				
61 Hoffman School Cave	Pendleton	Late Wisconsinan—early Holocene	*B. b. brevicauda*	Guilday and Hamilton, 1978
Mandy Walters Cave	Pendleton	Late Wisconsinan—early Holocene	"*B. brevicauda*"	Guilday and Hamilton, 1978
60 Eagle Cave	Pendleton	Wisconsinan	*B. brevicauda*	Guilday and Hamilton, 1973
Irvingtonian				
62 Trout Cave (TROUTC)	Pendleton	Irvingtonian (possibly late Kansan)—Wisconsinan	*B. brevicauda* and *B. b. brevicauda*	Zakrzewski, 1975*b*; Guilday, in litt.

APPENDIX 1
Continued.

Site	County	Age	Taxon	Reference
WISCONSIN				
Rancholabrean				
47 Moscow Fissure (MOSCOW)	Iowa	17,050 ± 1,500	*B. b. brevicauda*	Semken, personal communication, 1982
ONTARIO				
Rancholabrean				
Dickson Limestone Quarry	Halton	Minimum age = 215 years	"*B. brevicauda*"	Churcher and Fenton, 1968
QUEBEC				
Rancholabrean				
80 Caverne de Sainte-Elzear-de-Bonaventure (STELZE)	Bonaventure	Holocene (5,110 ± 150)	*B. brevicauda*	LaSalle and Guilday, 1980; Zakrzewski, personal communication

ARMADILLOS IN NORTH AMERICAN LATE PLEISTOCENE CONTEXTS

Walter E. Klippel and Paul W. Parmalee

ABSTRACT

Remains of climatically incompatible mammal species frequently occur together in North American late Pleistocene contexts. An apparent incongruence of Neotropical *Dasypus* remains with those of temperate and boreal mammal species is evidenced from the faunal assemblages from Cheek Bend Cave, Tennessee, and other contemporaneous sites. A size comparison of band scutes from both late Pleistocene (*Dasypus bellus*) and modern (*Dasypus novemcinctus*) armadillos indicate that many specimens from the Midsouth are intermediate in size between the large extinct form and the smaller, modern, nine-banded armadillo.

Elements of *Dasypus* are recorded from 50 late Pleistocene deposits in 12 states and all locations except one in southwestern Iowa are located south of the 40th parallel. It is suggested that the distributional limits of the extinct species or form of armadillo was, like the nine-banded armadillo, limited by aridity in the west and cold extremes in the north. Utilization of natural denning sites such as limestone caves during periods of extreme cold may have been one means which enabled *D. bellus* to survive in northern latitudes.

INTRODUCTION

Animals represented in late Pleistocene faunal assemblages whose present-day counterparts have markedly different habitat requirements (for example, Neotropical versus boreal mammals) have posed serious problems when attempting credible interpretations of paleoenvironments in portions of North America. Such associations are frequently attributed to posthumous mixing of faunal-bearing deposits. However, Ray (1967:123) noted that "although it is difficult to imagine such seemingly ecologically disparate animals as *Dasypus bellus* and *Martes pennanti* as members of the same fauna, the possibility should not be excluded out of hand. A growing body of evidence indicates the impropriety of rigidly interpreting Pleistocene communities in terms of present ones . . ." Guilday et al. (1978:61) observed and reported Wisconsinan ". . . contemporaneity of both temperate and boreal species . . ."

as far south as Tennessee and suggested that *D. bellus* remains in Appalachian late Pleistocene sites indicated milder winter extremes. Over the past two decades a number of other authors (for example, Dalquest et al., 1969; Graham and Semken, 1976; Hibbard, 1960; Holman, 1980, 1981; Lundelius, 1974; Martin and Gilbert, 1978; Martin and Neuner, 1978; Morgan, 1972; Slaughter, 1961, 1967, 1975) have argued that extralimital southern and northern species that are climatically incompatible today could represent late Pleistocene unit faunas rather than heterochronic assemblages resulting from posthumous mixing.

Dasypus remains, recovered with those of temperate and boreal mammals from a stratified cave deposit in Middle Tennessee, have prompted the following consideration of armadillos from late Pleistocene contexts in North America.

CHEEK BEND CAVE MAMMALS AND LATE PLEISTOCENE ENVIRONMENTS IN THE MIDSOUTH

During the summer of 1978 several rockshelters situated along the Duck River were tested for archaeological deposits. Excavations in one of these, Cheek Bend Cave located ca 13 km ESE of Columbia, Maury County, Tennessee, indicated that stratified deposits extended to a depth of more than 4.5 m. Three 1 by 2 m test units were excavated along the east wall near the mouth of the cave. Two major, distinct fill episodes were delineated. These major divisions were further subdivided into eight strata

based on physical differences in the deposits (Klippel and Parmalee, 1982b).

The lower strata (I–III) included faunal species whose present day ranges are far from the cave site. The five strata closest to the surface (IV–VIII) contained mostly remains of animals extant in the area today. Strata V through VIII contained evidence of prehistoric human occupation. Stone, bone, and ceramic artifacts indicate that the cave was occupied sporadically for more than 7,000 yr. Radiocarbon dates on charcoal from Strata V, VI, and VII indicate that these units were aggrading at 7,505 ± 440 (GX 7855), 4,655 ± 75 (UGa 2775), and 2,630 ± 255 (UGa 2859), respectively.

The remains of only four mammalian species whose present distributions are generally restricted south of the 40th parallel were recovered from the cave deposits and all of those occurred in Holocene strata. *Dasypus*, whose present distribution is also restricted south of the 40th parallel today, was represented in late Pleistocene strata along with remains of two mammals whose present-day distributions are entirely restricted north of the 40th parallel (that is, *Sorex arcticus* and *Microtus xanthognathus*). Numerous other boreal species represented in the late Pleistocene strata only occur south of the 40th parallel at high elevations in the Smoky and Rocky mountains (for example, *Sorex palustris, Martes americana, Mustela rixosa, Glaucomys sabrinus, Clethrionomys gapperi, Phenacomys intermedius*, and *Napaeozapus insignis*). The remains of several temperate mammal species (for example, *Synaptomys cooperi, Tamias striatus, Geomys bursarius*) from late Pleistocene strata also occur with the boreal yellow-cheeked vole (*M. xanthognathus*). These species (temperate versus boreal) are also allopatric under present-day climatic conditions and conform to the observation by Guilday et al. (1978) that boreal and temperate species were sympatric during the late Pleistocene as far south as Tennessee. However, our primary interest here is in the presence of armadillos with clearly boreal mammal species whose remains occur in late Pleistocene context at Creek Bend Cave (Table 1).

Boreal mammal species represented at Cheek Bend Cave are congruent, in many respects, with the evidence that jack pine-spruce-fir forests dominated Middle Tennessee during the full glacial (approximately 22,000 to 16,500 Y.B.P.) and were probably most similar to forests of southern Manitoba today (Delcourt, 1979:291). Boreal-like conditions were followed by Late Wisconsinan mixed coniferous-

deciduous forests which lasted from 16,500 to 12,500 years ago. Delcourt (1979:270) has suggested that at 16,500 B.P. vegetation of Middle Tennessee was similar to modern conditions in northeast Minnesota, and by 12,500 conditions were similar to those found in northeast Wisconsin today. During the early Holocene (ca. 12,500–8,000 Y.B.P.) cool temperate mixed mesic forest prevailed throughout this portion of the Midsouth; vegetation was comparable in many respects to that of the Allegheny Plateau region of Ohio and West Virginia at 12,000 and 10,000 years ago, respectively.

While boreal mammals from late Pleistocene strata at Cheek Bend Cave generally fit expectations based on Delcourt's (1979) reconstruction of Wisconsinan environments for Middle Tennessee (Klippel and Parmalee, 1982a; Parmalee and Klippel, 1981), armadillos seem out of place in this context unless *D. bellus* either had strikingly different tolerance limits than the modern nine-banded armadillo (*Dasypus novemcinctus*) or periods of equable climates allowed the Neotropical *Dasypus* to range into areas inhabited by boreal mammals.

"It is unclear whether *D. novemcinctus* evolved directly from *D. bellus* or whether it ecologically replaced *D. bellus*" (McNab, 1980:624). Slaughter (1961:311) has suggested that *D. bellus* and *D. novemcinctus* are osteologically identical except for the greater size of *D. bellus*. Auffenberg (1957) emphasized this point in his description of a nearly complete *D. bellus* from Mefford Cave in Florida. Kurten and Anderson (1980:131) contend that "even though definite records of *Dasypus novemcinctus* do not appear until Holocene times, there is no doubt that it is the ecological equivalent of *Dasypus bellus,* the larger Pleistocene armadillo." Others have also implied or argued for the ecological equivalents of *D. bellus* and *D. novemcinctus* (for example, Lundelius, 1967:297; Slaughter, 1961:313; Ray, 1967: 123; Guilday et al., 1978:32; Hood and Hawksley, 1975:22).

THE NINE-BANDED ARMADILLO IN NORTH AMERICA

Historic records indicate that *D. novemcinctus* did not exist in North America beyond the southernmost portion of Texas (Fig. 1) during the 19th century (Baird, 1857; Cope, 1880). Numerous subsequent considerations of its range expansion (for example, Bailey, 1905; Cleveland, 1970; Fitch et al.,

1952; Kalmbach, 1943: Taber, 1939; Strecker, 1926) attest to the observation by Buchanan and Talmage (1954:142) that "one of the most interesting phenomena concerning the armadillo . . . is the speed with which it is extending its North American range."

In a recent summary of the zoogeography of *D.*

Table 1.—*Mammal remains (numbers of elements) recovered from stratified deposits at Cheek Bend Cave (40MU261). Only specimens identified to genus and species are included: specimens from equivocal contexts (e.g., Stratum III/IV) are excluded (from Klippel and Parmalee 1982 b; Tables 19, 21–27; late Pleistocene strata = I, II, III; Holocene strata = IV, V, VI, VII, and VIII). Superscript* [a] = five elements in an early Holocene stratum possibly intrusive;* [b] = one element in a late Pleistocene stratum probably intrusive;* * = mammals with present day ranges south of the 40th parallel.*

Species	Late Pleistocene strata	Late Pleistocene and Holocene strata	Holocene strata
Sorex palustris, water shrew	5		
Sorex arcticus, arctic shrew	27		
Microsorex hoyi, pygmy shrew	12		
Condylura cristata, star-nosed mole	567		
Spermophilus tridecemlineatus, 13-lined ground squirrel	139		
Glaucomys sabrinus, northern flying squirrel	53		
Clethrionomys gapperi, red-backed vole	423		
Phenacomys intermedius, heather vole	9		
Microtus xanthognathus, yellow-cheeked vole	1,419		
Geomys bursarius, plains pocket gopher	1,474[a]		
Napaeozapus insignis, woodland jumping mouse	6		
Ursus americanus, black bear	1		
Martes americana, pine martin	1		
Mustela rixosa, least weasel	58		
Mustela vison, mink	14		
Dasypus bellus, "beautiful" armadillo	69		
Megalonyx jeffersonii, Jefferson's ground sloth	1		
Blarina brevicauda, short-tailed shrew		4,777	
Sorex longirostris/cinereus, southeastern/masked shrew		58	
Sorex fumeus, smoky shrew		16	
Scalopus aquaticus, eastern mole		503	
Myotis spp., myotis bats		683	
Eptesicus fuscus, big brown bat		3,352	
Nycticeius humeralis, evening bat		17	
Tamias striatus, eastern chipmunk		950	
Tamiasciurus hudsonicus, red squirrel		42	
Microtus pennsylvanicus, meadow vole		1,125	
Microtus spp., vole		16,352	
Ondatra zibethica, muskrat		25	
Synaptomys cooperi, southern bog lemming		483	
Peromyscus spp., deer/white-footed mouse		722	
Zapus hudsonius, meadow jumping mouse		310	
Erethizon dorsatum, porcupine		5	
Sylvilagus cf. *floridanus*, rabbit		301	
Mustela frenata, long-tailed weasel		59	
Odocoileus virginianus, white-tailed deer		78	
Cryptotis parva, least shrew			4,719[b]
Pipistrellus subflavus, eastern pipistrel			32
Lasiurus borealis, red bat			2
Plecotus sp., big-eared bat*			41
Marmota monax, woodchuck			10
Sciurus carolinensis, gray squirrel			6
Sciurus niger, fox squirrel			6
Sciurus spp., squirrel			140
Glaucomys volans, southern flying squirrel			1,050
Reithrodontomys cf. *humulis*, harvest mouse*			3
Oryzomys palustris, marsh rice rat*			12
Sigmodon hispidus, hispid cotton rat*			18
Neotoma floridana, eastern woodrat			186
Castor canadensis, beaver			5
Sylvilagus aquaticus, swamp rabbit*			1
Urocyon cinereoargenteus, gray fox			1
Canis sp., canid			1
Procyon lotor, raccoon			4
Mylohyus nasutus, long-nosed peccary			3

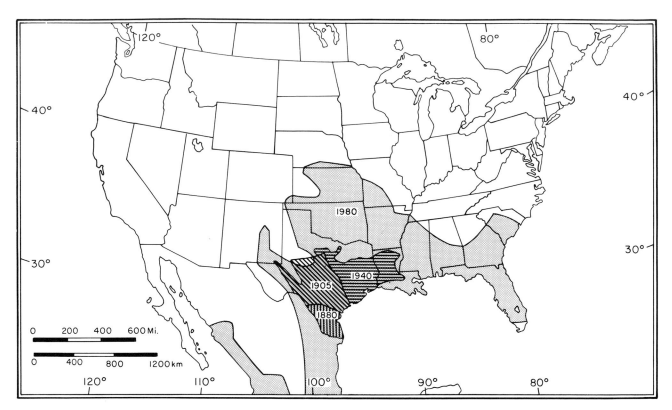

Fig. 1.—Distributions of *Dasypus novemcinctus* showing the rapid Historic Period spread of the nine-banded armadillo in North America. Distributions for 1880–1940 taken from Kalmbach (1943); ca. 1980 distribution from Hall (1981).

novemcinctus in the United States, Humphrey (1974: 457) describes the armadillo as having a strong pioneering capability. Invasion from Rio Grande City, Texas, toward the northeast occurred at roughly 10 km per year between 1880 and 1972. Between 1905 and 1914 expansion to the northeast occurred at an even greater speed of over 25 km per year (Humphrey, 1974:fig. 4). Pioneering by the nine-banded armadillo is most successful in river valleys (riparian environments are preferred habitats for the armadillo, Fitch et al., 1952) which serve as dispersal conducts where invasion can be especially rapid.

Barriers to the armadillo's continued expansion in North America appear to be aridity and cold temperatures. Arid conditions limit the armadillo's primary food source of soil and surface macroinvertebrates; these are obtained in mesic areas by probing under rotted logs and in soil rich in humus and leaf litter. In xeric areas they are restricted to valley floors where invertebrates are generally more abundant in riparian habitats. Humphrey (1974) has suggested a minimum annual precipitation of 380 mm for armadillo survival.

Heat, per se, apparently is not a deterrent to survival in that armadillos adjust their above ground foraging activities according to the seasons and time of day (Fitch et al., 1952). "High temperatures seem effective only in causing their activities to be confined largely to the cooler hours of darkness; low temperatures have the opposite effect, bring them abroad during the hottest part of the day in search of food" (Taber, 1945:212, 213). Clark (1951) suggests that armadillo burrows and natural dens (crevices in limestone bluffs) serve as limited sources of insects during drought periods. Cold temperatures also affect insect availability in that " . . . foraging at air temperatures below freezing would be of little value especially if the soil were frozen or covered with snow" (McNab, 1980:621).

Additionally, *D. novemcinctus* has a high thermal conductance which "results from the replacement of a fur coat by protective armor" (McNab, 1980: 616). It can only withstand short periods of near freezing temperature (that is, adults of ca. 5 kg, 9 to 10 days at near 0°C). Cold avoidance for short periods of time by this non-hibernating animal is usually accomplished by burrowing into the ground or seeking shelter in caves and crevices in limestone bluffs.

McNab (1980) has demonstrated that armadillos with greater body mass have a longer survival time during cold weather and suggested the 20 to 30 kg *D. bellus* that occupied portions of North America during the Pleistocene might be expected to have had an increased "survival time" of 12 days at near 0°C. An even longer period of survival would be anticipated at 10°C. However, burrowing as a means of protection from cold by such large animals would be nearly impossible; natural shelters (caves) might have served as the only refuge under such conditions. Taber's (1945) observation of *D. novemcinctus* from the Edwards Plateau in Texas indicates that where caves and crevices are available and abundant in limestone cliffs few burrows are excavated, thus indicating a preference for such features even when the alternative of burrowing is possible for the small nine-banded armadillo.

PLEISTOCENE *DASYPUS* IN NORTH AMERICA

Dasypus remains in faunal assemblages are especially apparent because of the large number of diagnostic elements that are potentially recoverable. *D. novemcinctus*, for example, has over 3,300 scutes per animal (Davis, 1969), including between 500 and 600 band scutes (UT #4489). These are in additon to elements such as teeth, mandibles, and limb bones that are usually employed in the identification of most mammals.

Remains of the "highly visible" *Dasypus* have been recovered from Pleistocene sites throughout much of southern North America from Florida and the Carolinas to New Mexico (Fig. 2, Table 2). With one known exception in southwestern Iowa (Rhodes, 1982), all Pleistocene *Dasypus* remains have been found south of the 40th parallel. However, a relatively large number have been recovered from outside the present range of the nine-banded armadillo (Fig. 1) as delineated by Hall (1981).

Deposits in North America from which *Dasypus* remains have been recovered are predominantly Rancholabrean in age (Slaughter, 1961; Kurten and Anderson, 1980) or from equivocal contexts. However, *Dasypus* from Florida have been assigned to both the Blancan (Haile XVA and Santa Fe R.IB) and Irvingtonian (Inglis IA, Santa Fe R.IIA, and Coleman IIA) Land Mammal Ages (Webb, 1974).

Some evidence suggests a general increase in size of North American *Dasypus* from Blancan through the Rancholabrean (Kurten and Anderson, 1980). *D. bellus* moveable dermal plates from Irvingtonian contexts are smaller than many of those from Rancholabrean deposits in Florida (Martin, 1974a:43). However, Guilday and McCrady (1966) reported specimens from a site in West Virginia and one from Tennessee that appear intermediate in size between Rancholabrean *D. bellus* from Florida and the nine-banded armadillo that inhabits Florida today. They noted that "perhaps the relative harshness of the environment is reflected in the smaller size of the Tennessee and West Virginia specimens as compared to those from Missouri and Florida Pleistocene sites" (Guilday and McCrady, 1966:183).

Since the mid-1960s, a number of other *Dasypus* remains have been recovered from the Midsouth. We have chosen to assess the Guilday and McCrady (1966) observation concerning size on a number of these specimens by measuring band scutes. We have somewhat modified a procedure described by Martin (1974a:42) and have recorded maximum thickness at the slight dorsal ridge where band scutes overlap; maximum scute width was taken at this point of maximum thickness. We excluded lateral scutes (the first three from each side of each band row) in that they introduce significant size variation when included with interior bands. A number of these lateral scutes in a small sample could unnecessarily make thickness measurements for an assemblage appear erroneously small. We have also excluded scutes (most from the ninth band) that are not straight or only slightly convex at the distal end. These modifications to Martin's (1974a) method tend to result in reduced sample sizes. However, they also exclude considerable measurement variation in animals we compared (that is, lateral versus interior band scutes of a nine-banded armadillo, UT #4489, as well as those of the large sample of *D. bellus* from Crankshaft Cave). Since few samples from any given site are large enough to treat with more than basic statistics (that is, ca. <10), we have excluded obvious sources of variation at the expense of having only slightly reduced samples.

Results of scute measurements of both modern and Pleistocene *Dasypus* are presented in Table 3 and Fig. 3. Two Rancholabrean sites (Crankshaft Cave in Missouri and Melbourne in Florida) pro-

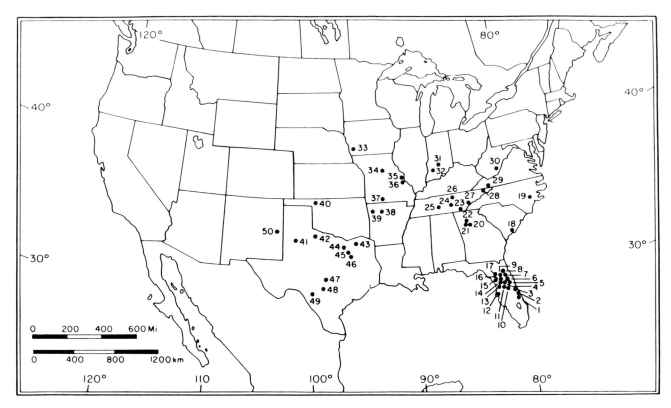

Fig. 2.—Locations of Pleistocene sites that contain remains of *Dasypus bellus*. Locality numbers correspond to numbers and information provided in Table 2.

duced relatively large samples of measurable band scutes; one standard deviation (n − 1) for each sample is illustrated in Fig. 3. Random samples of 51 band scutes from a modern nine-banded armadillo of unknown weight (UT #4489) and another of 56

band scutes from a 4.3 kg armadillo from Mississippi (UT #6479) are also illustrated (Hall, 1981, suggests that modern armadillos in Texas may reach 7.7 kg but usually range around 5 kg).

There is no overlap in size (thickness or width)

Table 2.—*Pleistocene sites in North American that contain* Dasypus *remains. Numbers at the left of this table correspond to locations in Fig. 2.*

Location (see Fig. 2)	State/site	County	Citation
	Florida		
1	Vero Beach	Indian River	Weigel, 1962
2	Melbourne	Brevard	C. E. Ray: unpublished
3	Sebastion Canal	Brevard	Webb, 1974
4	Mefford Cave	Marion	Auffenberg, 1957
5	Reddick	Marion	Gut and Ray, 1963
6	Haile XVA	Alachua	Webb, 1974
7	Kendrick IA	Marion	Webb, 1974
8	Haile XIVA	Alachua	Martin, 1974*b*
9	Arredondo IB	Alachua	Webb, 1974
10	Coleman IIA	Sumter	Webb, 1974
11	Inglis IA	Citrus	Webb, 1974
12	Seminole Field	Pinellas	Webb, 1974
13	Saber-Tooth Cave	Citrus	Simpson, 1928

Table 2.—Continued.

Location (see Fig. 2)	State/site	County	Citation
14	Wilhlacoochee River VIIA	Citrus	Webb, 1974
15	Waccasassa River IIB & III	Levy	Webb, 1974
16	Williston IIIA	Levy	Webb, 1974
17	Sante Fe River IB	Gilchrist	Webb, 1974
	South Carolina		
18	Edisto Island	Charleston	A. Sanders: unpublished
	North Carolina		
19	Harriett's Pit	Jones	C. E. Ray: unpublished
	Georgia		
20	Ladds Quarry	Bartow	Ray, 1967
21	Kingston Saltpeter Cave	Bartow	Sneed, 1983
22	Yarborough Cave	Bartow	Sneed, 1983
	Tennessee		
23	Lookout Mountain Cave	Hamilton	R. Wilson: unpublished
24	Cumberland Caverns	Warren	Smyre, 1978
25	Cheek Bend Cave	Maury	Klippel and Parmalee, 1982b
26	Robinson Cave	Overton	Guilday and McCrady, 1966
27	Finger Quarry	Blount	Robison, 1981
28	Baker Bluff Cave	Sullivan	Guilday et al., 1978
	Virginia		
29	Meadowview Cave	Washington	A. Guilday: unpublished
	West Virginia		
30	Organ-Hedricks Cave	Greenbrier	Guilday and McCrady, 1966
	Indiana		
31	Anderson Pit Cave	Monroe	Richards, 1980
32	Prairie Creek	Daviess	Tomak, 1982
	Iowa		
33	Craigmile	Mills	Rhodes, 1982
	Missouri		
34	Brynjulfson Caves	Boone	Parmalee and Oesch, 1972
35	Cherokee Cave	St. Louis	Simpson, 1949
36	Crankshaft Cave	Jefferson	Parmalee et al., 1969
37	Zoo Cave	Taney	Hood and Hawksley, 1975
	Arkansas		
38	Peccary Cave	Newton	Davis, 1969
39	10 Mile Rock Site	Washington	Sealander, 1979
	Oklahoma		
40	Bar M	Harper	Slaughter, 1967
	Texas		
41	Slaton Quarry	Lubbock	Dalquest, 1967
42	Easley Ranch	Ford	Slaughter, 1967
43	Ben Franklin	Fannin/Delta	Slaughter and Hoover, 1963
44	Clear Creek	Denton	Slaughter and Ritchie, 1963
45	Hill-Shuler	Dallas/Denton	Slaughter et al., 1962
46	Moore Pit	Dallas	Slaughter, 1966
47	Miller's Cave	Llano	Patton, 1963
48	Cave Without a Name	Kendall	Lundelius, 1967
49	Kincaid Shelter	Uvalde	Lundelius, 1967
	New Mexico		
50	Brown Sand Wedge	Roosevelt	Slaughter, 1964

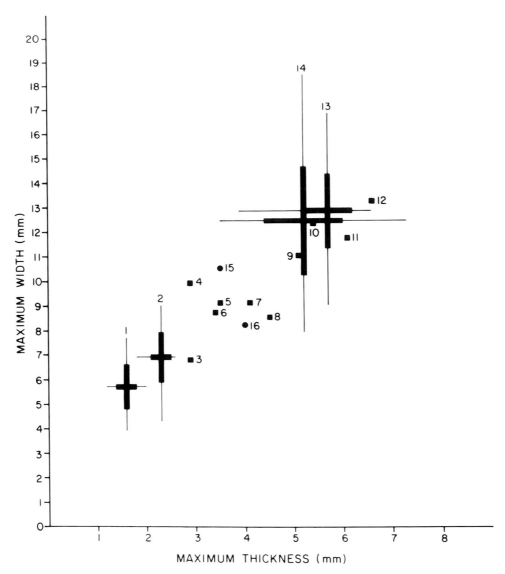

Fig. 3.—Mean band scute size of *D. novemcinctus* and *D. bellus*: 1) modern *D. novemcinctus*, UT #4489, Florida; 2) modern *D. novemcinctus*, UT #6479, Mississippi; 3) Rancholabrean *D. bellus*, Robinson Cave, Tennessee; 4) Rancholabrean *D. bellus*, Cheek Bend Cave, Tennessee; 5) Rancholabrean *D. bellus*, Kingston Salt Peter Cave, Georgia; 6) Rancholabrean *D. bellus*, Organ-Hedrick Cave, West Virginia; 7) Rancholabrean *D. bellus*, Lookout Mt., Tennessee; 8) Ranchlobrean *D. bellus*, Baker Bluff Cave, Tennessee; 9) Rancholabrean *D. bellus*, Meadow View Cave, Virginia; 10) Rancholabrean *D. bellus*, Anderson Pit Cave, Indiana; 11) Rancholabrean *D. bellus*, Reddick, Florida; 12) Rancholabrean *D. bellus*, Ladds Quarry, Georgia; 13) Rancholabrean *D. bellus*, Melbourne, Florida; 14) Rancholabrean *D. bellus*, Crankshaft Cave, Missouri; 15) Blancan *D. bellus*, Haile XVA, Florida; 16) Irvingtonian *D. bellus*, Coleman IIA, Florida. One standard deviation (n − 1) and observed ranges are illustrated for samples with over 50 scutes (that is, 1, 2, 13, and 14).

between *D. bellus* (Crankshaft Cave and Melbourne) and the nine-banded armadillo of unknown weight from Florida. Neither do scutes from within the first standard deviation of the modern 4.3 kg specimen from Mississippi and *D. bellus* from Crankshaft Cave have overlapping ranges. However, their total ob-

served ranges do overlap in maximum width. When means (or single values) of several smaller samples from the Midsouth are plotted (including the specimens from Robinson Cave and Organ-Hedricks Cave reported by Guilday and McCrady, 1966), they fall intermediate between *D. novemcinctus* and

Table 3.—Dasypus *band scute measurements (mm) from modern and Pleistocene contexts in eastern North America. Superscript* [a] = *maximum dimensions for any or all band scutes;* [b] = *maximum dimensions of inner band scutes at slight dorsal ridge where scutes overlap;* [c] = *n − 1 standard deviation;* [d] = *observed range.*

Species/locality	Maximum width				Maximum thickness			
	N	Mean	SD[c]	OR[d]	N	Mean	SD[c]	OR[d]
Recent								
D. novemcinctus (FL)								
Martin (1974:43)	235[a]	5.4	—	3.5–8.1	235[a]	1.6	—	0.9–2.4
D. novemcinctus (FL)								
UT Zooarchaeology (#4489)	51[b]	5.7	0.87	3.9–7.7	51[b]	1.6	0.22	1.2–2.0
D. novemcinctus (MS)								
UT Zooarchaeology (#6479)	56[b]	6.9	0.98	4.3–9.0	56[b]	2.3	0.21	1.8–2.6
Rancholabrean								
D. bellus (FL)								
Martin (1974:43)	163[a]	11.4	—	6.6–16.5	163[a]	3.9	—	1.9–6.1
D. bellus (FL)								
Reddick IA	4[b]	11.7	1.97	8.8–13.0	4[b]	6.2	0.63	5.6–7.1
D. bellus (FL)								
Melbourne	55[b]	12.8	1.53	9.0–16.8	55[b]	5.7	0.54	3.9–6.6
D. bellus (GA)								
Ladds Quarry	2[b]	13.2	—	12.5–13.9	2[b]	6.6	—	6.6
D. bellus (GA)								
Kingston Salt Peter Cave	6[b]	9.1	2.66	6.5–12.5	6[b]	3.5	0.48	2.8–3.9
D. bellus (IN)								
Anderson Pit	3[b]	12.4	0.93	11.3–13.0	4[b]	5.4	0.66	4.5–6.0
D. bellus (MO)								
Crankshaft Cave	221[b]	12.4	2.27	7.9–18.4	221[b]	5.2	0.86	3.5–7.3
D. bellus (TN)								
Cheek Bend Cave	10[b]	9.9	1.33	7.6–12.0	2[b]	2.9	—	2.5–3.2
D. bellus (TN)								
Robinson Cave	5[b]	6.8	0.61	5.9–7.5	5[b]	3.0	0.54	2.0–3.3
D. bellus (TN)								
Lookout Mt.	1[b]	—	—	9.1	1[b]	—	—	4.1
D. bellus (TN)								
Baker Bluff Cave	1[b]	—	—	8.5	1[b]	—	—	4.5
D. bellus (VA)								
Meadow View Cave	1[b]	—	—	11.0	1[b]	—	—	5.1
D. bellus (WV)								
Organ-Hendricks	1[b]	—	—	8.7	1[b]	—	—	3.4
Irvingtonian								
D. bellus (FL)								
Coleman IIA; Martin (1974a:43)	21[a]	8.2	—	6.2–10.3	21[a]	4.0	—	2.7–6.8
Blancan								
D. bellus (FL)								
Haile XVA; Martin (1974a:43)	10[a]	10.5	—	8.2–12.4	10[a]	3.5	—	2.9–4.2

"typical" large *D. bellus* which is considered to have been at least twice as large as the nine-banded armadillo. It is also interesting to note that the means of measured scutes (measured with slightly more variance included in samples) of specimens from the earlier Blancan (Haile XVA) and Irvingtonian (Coleman IIA sites), reportedly smaller than Rancholebrean *D. bellus* (Martin, 1974a), fall in this intermediate area as well.

There is considerable variation in the size of adult nine-banded armadillos (Russell, 1953; Wetzel and Mondolfi, 1979). According to Hall (1981) armadillos in Texas attain weights nearly twice that of the 4.3 kg specimen from Mississippi reported here

(Fig. 3:2, UT #6479). One cannot help but surmise that scutes from 7 to 8 kg armadillos might be as large, if not larger, than those falling between *D. novemcinctus* (from Florida and Mississippi) and the large Pleistocene *Dasypus* (for example, from Crankshaft Cave and Melbourne) represented in Fig. 3. In fact, if we were to assume that twice the weight equated with the twice the scute size, large *D. novemcinctus* mean scute size would clearly fall within the range of the first standard deviation of these

from Crankshaft Cave *D. bellus*. Our intent is not to attempt to demonstrate that *D. bellus* from Crankshaft Cave weighed only slightly over 8 kg, but rather to point up the considerable variation in nine-banded armadillo size. Judging from evidence presented here (Table 3), Rancholabrean *Dasypus* must have also had a considerable range in size which, in some instances, approached the upper size limits of *D. novemcinctus*.

SUMMARY

As is the case at Cheek Bend Cave in Middle Tennessee, *Dasypus* remains are frequently found in late Pleistocene contexts with those of other mammals that are presently restricted to boreal habitats. On the surface, this association seems unlikely, but when habits, tolerance limits, and the ecology of the nine-banded armadillo (modern analogue) are taken into account along with evidence for more equable climates during the late Pleistocene, the association becomes considerably more credible.

Dasypus has been in North America (at least as far north as central Florida) since Blancan times. Armadillos are notorious at pioneering and can spread rapidly under suitable environmental conditions. If modern records for *D. novemcinctus* can be used as an indication of the potential rate of dispersal (that is, ca. 25 km/yr) for the larger Pleistocene form, this mammal could have spread from the Florida Peninsula to Middle Tennessee in considerably less than a half century. From the Gulf Coast (for example, Pensacola, Florida, or Mobile, Alabama) it could have reached Middle Tennessee in less than three decades. Equable climatic conditions characterized by cool summers and winters without prolonged periods below freezing are not entirely unthinkable over the many thousands of years represented by the Rancholabrean. Such periodic conditions would undoubtedly have been conducive to large populations of insects and other invertebrates upon which *D. bellus* probably fed.

The stratigraphic position of faunal remains in Cheek Bend Cave suggests that the armadillo inhabited Middle Tennessee at least during the late Wisconsinan. Palynological evidence suggests that mixed coniferous-deciduous forests were replacing the jack pine-spruce-fir forests of the full Wisconsinan in Middle Tennessee during this period.

It is probable that during the late glacial (Wisconsinan) armadillos were able to rapidly extend their range northward during even relatively short intervals with above freezing temperatures. The preference of this mammal for natural den sites (as suggested by numerous occurrences of its remains) could account in part for its prevalence in cave contexts. It is also probable that armadillos no larger than the individual represented in Cheek Bend Cave, as well as those from other sites in the Midsouth (Fig. 3), could have burrowed for protection for short periods of cold temperatures as do modern nine-banded armadillos today. These smaller animals could represent a late Wisconsinan adaptation in the Midsouth that was not conducive to the survival of larger individuals of *D. bellus* if the latter were not able to find sufficient protection in caves or provide their own protection by burrowing. If this is the case then larger armadillos whose remains have been recovered from more northern latitudes may represent *Dasypus* expansions during more moderate interglacial periods (for example, Sangamonian) of the Late Pleistocene.

ACKNOWLEDGMENTS

To the following individuals we express our appreciation for the loan of *Dasypus bellus* specimens and for information relative to this material housed or curated under their care: Russell W. Graham, Alice M. Guilday, Oscar Hawksley, Larry D. Martin, Clayton E. Ray, Ronald L. Richards, Albert E. Sanders, Joel M. Sneed, and Ronald C. Wilson. We are especially grateful to Terry Faulkner for the preparation of the figures. Excavations at Cheek Bend Cave (40MU261) and identification of the faunal materials were undertaken under the auspices of the Tennessee Valley Authority (contract No. TVA TV-49244A and TVA TV-53013A).

LITERATURE CITED

AUFFENBERG, W. 1957. A note on an unusually complete specimen of *Dasypus bellus* (Simpson) from Florida. Quar. J. Florida Acad. Sci., 20:233–237.

BAILEY, V. 1905. Biological survey of Texas. North Amer. Fauna, 25:1–222.

BAIRD, S. F. 1857. Mammals of North America. Pacific R.R. Rept., 8:1–757.

BUCHANAN, G. D., and R. V. TALMAGE. 1954. The geographical distribution of the armadillo in the United States. Texas Acad. J. Sci., 6:142–150.

CLARK, W. K. 1951. Ecological life history of the armadillo in the eastern Edwards Plateau region. Amer. Midland Nat., 46:337–358.

CLEVELAND, A. G. 1970. The current geographic distribution of the armadillo in the United States. Texas Acad. Sci., 22:90–93.

COPE, E. D. 1880. Zoological position of Texas. Bull. U.S. Nat. Mus., 17:1–51.

DALQUEST, W. W. 1967. Mammals of the Pleistocene Slaton Local Fauna of Texas. Southwestern Nat., 12:1–30.

DALQUEST, W. W., E. ROTH, and F. JUDD. 1969. The mammal fauna of Schulze Cave, Edwards County, Texas. Bull. Florida State Mus., Biol. Ser., 13:205–276.

DAVIS, L. C. 1969. The biostratigraphy of Peccary Cave, Newton County, Arkansas. Proc. Arkansas Acad. Sci. 23:192.

DELCOURT, H. R. 1979. Late Quaternary vegetation history of the eastern Highland Rim and adjacent Cumberland Plateau of Tennessee. Ecol. Monogr., 49:255–280.

FITCH, H. S., P. GOODRUM, and C. NEWMAN. 1952. The armadillo in the southeastern United States. J. Mamm., 33:21–37.

GRAHAM, R. W., and H. A. SEMKEN. 1976. Paleoecological significance of the short-tailed shrew (*Blarina*), with a systematic discussion of *Blarina ozarkensis*. J. Mamm., 57:433–449.

GUILDAY, J. E., H. W. HAMILTON, E. ANDERSON, and P. W. PARMALEE. 1978. The Baker Bluff Cave deposit, Tennessee, and the late Pleistocene faunal gradient. Bull. Carnegie Mus. Nat. Hist., 11:1–67.

GUILDAY, J. E., and A. D. MCCRADY. 1966. Armadillo remains from Tennessee and West Virginia caves. Bull. Nat. Speleo. Soc., 28:183–184.

GUT, H. J., and C. E. RAY. 1963. The Pleistocene vertebrate fauna of Reddick, Florida. Quar. J. Florida Acad. Sci., 26:315–328.

HALL, E. R. 1981. The mammals of North America. John Wiley and Sons, New York. 2 vols., 1,181 pp.

HIBBARD, C. W. 1960. Pliocene and Pleistocene climates in North America. Annual Rept. Michigan Acad. Sci., Arts, and Letters, 62:5–30.

HOLMAN, J. A. 1980. Paleoclimatic implications of Pleistocene herpetofaunas of eastern and central North America. Trans. Nebraska Acad. Sci., 8:131–140.

———. 1981. A review of North American Pleistocene snakes. Michigan State Univ. Paleontol. Ser., 1:263–306.

HOOD, C. H., and O. HAWKSLEY. 1975. A Pleistocene fauna from Zoo Cave, Taney County, Missouri. Missouri Speleo., 15:1–42.

HUMPHREY, S. 1974. Zoogeography of the nine-banded armadillo (*Dasypus novemcinctus*) in the United States. BioScience, 24:457–462.

KALMBACH, E. R. 1943. The armadillo: its relation to agriculture and game. Game, Fish, and Oyster Commission, Austin, Texas, 61 pp.

KLIPPEL, W. E., and P. W. PARMALEE. 1982a. Diachronic variation in insectivores from Cheek Bend Cave and environmental change in the Midsouth. Paleobiology, 8:447–458.

———. 1982b. The paleontology of Cheek Bend Cave: Phase II Report. Report submitted to the Tennessee Valley Authority, Norris, Tennessee, xiii + 249 pp.

KURTEN, B., and E. ANDERSON. 1980. Pleistocene mammals of North America. Columbia Univ. Press, 442 pp.

LUNDELIUS, E. L., JR. 1967. Late Pleistocene and Holocene faunal history of central Texas. Pp. 287–319, *in* Pleistocene extinctions: the search for a cause (P. S. Martin and H. E. Wright, eds.), Yale Univ. Press, New Haven, x + 453 pp.

———. 1974. The last fifteen thousand years of faunal change in North America. Pp. 141–160, *in* History and prehistory of the Lubbock Lake Site, The Mus. J., 15:1–160.

MARTIN, L. D., and B. M. GILBERT. 1978. Excavations at Natural Trap Cave. Trans. Nebraska Acad. Sci., 6:107–116.

MARTIN, L. D., and A. M. NEUNER. 1978. The end of the Pleistocene in North America. Trans. Nebraska Acad. Sci., 6:117–126.

MARTIN, R. A. 1974a. Fossil mammals from the Coleman IIA fauna, Sumter County. Pp. 35–99, *in* Pleistocene mammals of Florida (S. D. Webb, ed.), Univ. Florida Press, Gainesville, x + 270 pp.

———. 1974b. Fossil vertebrates from the Haile XIVA Fauna, Clachua County. Pp. 100–113, *in* Pleistocene mammals of Florida (S. D. Webb, ed.), Univ. Florida Press, Gainesville, X + 270 pp.

McNAB, B. K. 1980. Energetics and the limits to a temperate distribution in armadillos. J. Mamm., 61:606–627.

MORGAN, J. M. 1972. An analysis of periglacial climatic indicators of late glacial time in North America. Unpublished Ph.D. dissert., Univ. Wisconsin, Madison, 160 pp.

PARMALEE, P. W., and W. E. KLIPPEL. 1981. A late Pleistocene record of the heather vole (*Phenacomys intermedius*), in the Nashville Basin, Tennessee. J. Tennessee Acad. Sci., 56:127–129.

PARMALEE, P. W., and R. D. OESCH. 1972. Pleistocene and recent faunas from the Brynjulson Caves, Missouri. Illinois State. Mus. Repts. Investigations, 25:1–52.

PARMALEE, P. W., R. D. OESCH, and J. E. GUILDAY. 1969. Pleistocene and Recent vertebrate faunas from Crankshaft Cave, Missouri. Illinois State Mus. Repts. Investigations, 14:1–37.

PATTON, T. H. 1963. Fossil vertebrates from Miller's Cave, Llano County, Texas. Bull. Texas Mem. Mus., 7:41.

RAY, C. E. 1967. Pleistocene mammals from Ladds, Bartow County, Georgia. Bull. Georgia Acad. Sci., 25:120–150.

RHODES, R. S. 1982. Mammalian paleoecology of the Farmdalian Craigmile and the Woodfordian Waubonsie local faunas, southwestern Iowa. Unpublished Ph.D. dissert., Univ. Iowa, Iowa City, 132 pp.

RICHARDS, R. L. 1980. Rice rat (*Oryzomys* cf. *palustris*) remains from southern Indiana caves. Proc. Indiana Acad. Sci., 89:425–431.

ROBISON, N. D. 1981. A description of the Pleistocene faunal remains recovered from Finger Quarry, Blount County, Tennessee. J. Tennessee Acad. Sci., 56:68–71.

RUSSELL, R. J. 1953. Description of a new armadillo (*Dasypus novemcinctus*) from Mexico with remarks on geographic variation of the species. Proc. Biol. Soc. Washington, 66:21–26.

SEALANDER, J. A. 1979. A guide to Arkansas mammals. River Road Press, Conway, Arkansas, pp. 313.

SIMPSON, G. G. 1928. Pleistocene mammals from a cave in Citrus County, Florida. Amer. Mus. Novitates, 328:1–16.

———. 1949. A fossil deposit in a cave in St. Louis. Amer. Mus. Novitates, 1408:1–46.

SLAUGHTER, B. H. 1961. The significance of *Dasypus bellus* (Simpson) in Pleistocene local faunas. Texas J. Sci. 13:311–315.

———. 1964. An ecological interpretation of the Brown Sand Wedge local fauna, Blackwater Draw, New Mexico, and a hypothesis concerning late Pleistocene extinction. Paleoecol. Llano Estracado, 2:1–37.

———. 1966. The Moore Pit local fauna, Pleistocene of Texas. J. Paleont., 40:70–91.

———. 1967. Animal ranges as a clue to late Pleistocene extinctions. Pp. 155–167, *in* Pleistocene extinctions: the search for a cause (P. S. Martin and H. E. Wright, eds.), Yale Univ. Press, New Haven, 453 pp.

———. 1975. Ecological interpretation of the Brown Sand Wedge local fauna. Pp. 179–192, *in* Late Pleistocene environments of the southern High Plains, Publ. Burgwin Res. Center, Rancho de Taos, New Mexico, 9.

SLAUGHTER, B. H., and B. R. HOOVER. 1963. Sulphur River formation and the Pleistocene mammals of the Ben Franklin local fauna. Southern Methodist Univ., Grad. Res. Center J., 31:132–148.

SLAUGHTER, B. H., W. W. CROOK, JR., R. K. HARRIS, D. C. ALLEN, and M. SEIFERT. 1962. The Hill-Shuler local faunas of the Upper Trinity River, Dallas and Denton Counties, Texas. Rept. Investigations Univ. Texas Bur. Econ. Geol., 48:1–75.

SLAUGHTER, B. H., and R. RITCHIE. 1963. Pleistocene mammals of the Clear Creek local fauna, Denton County, Texas. Southern Methodist Univ., Grad. Res. Center J., 31:117–131.

SMYRE, J. 1978. Recent bone finds at Cumberland Caverns. Speleonews (NSS Nashville and Chattanooga Grottoes), 22(1):3–7.

SNEED, J. M. 1983. Another Pleistocene site in Georgia. Monthly Breakdown (NSS Grotto No. 285), 3:9–12.

STRECKER, J. K. 1926. Extension of the nine-banded armadillo. J. Mamm., 7:206–210.

TABER, F. W. 1939. Extension of the range of the armadillo. J. Mamm., 20:489–493.

———. 1945. Contribution on the life history and ecology of the nine-banded armadillo., J. Mamm., 26:211–226.

TOMAK, C. H. 1982. *Dasypus bellus* and other extinct mammals from the Prairie Creek Site. J. Mamm., 63:158–160.

WEBB, S. D. 1974. Chronology of Florida Pleistocene mammals. Pp. 5–31, *in* Pleistocene mammals of Florida (S. D. Webb, ed.), Univ. Florida Press, Gainesville, x + 270 pp.

WEIGEL, R. D. 1962. Fossil vertebrates of Vero, Florida. Florida Geol. Surv., Spec. Publ., 10:1–59.

WETZEL, R. M., and MONDOLFI. 1979. The subgenera and species of long-nosed armadillos, genus *Dasypus* L. Pp. 43–63, *in* Vertebrate ecology in the northern Neotropics (J. F. Eisenberg, ed.), Smithsonian Institution Press, Washington, D.C.

Address (Klippel): Department of Anthropology, The University of Tennessee, Knoxville, Tennessee 37916.

Address (Parmalee): Frank H. McClung Museum, The University of Tennessee, Knoxville, Tennessee 37916.

A PLEISTOCENE OCCURRENCE OF *GEOMYS* (RODENTIA: GEOMYIDAE) IN WEST VIRGINIA

FREDERICK GRADY

ABSTRACT

Remains of the pocket gopher, *Geomys* sp., have been recovered from the lower levels of an excavation in New Trout Cave, Pendleton County, West Virginia. This is the first record of *Geomys* from West Virginia and is some 480 km east of the present range of *Geomys bursarius* and 500 km north of the present range of *Geomys pinetis*. The New Trout Cave *Geomys* elements occur 60 or more cm below a level that has been dated by C^{14} at 29,400 $\pm$ 1,700 B.P. The associated fauna includes *Ochotona* sp. but lacks several extinct small mammal taxa characteristics of the late Irvingtonian faunas of Trout and Cumberland Caves.

INTRODUCTION

In 1979 a rich Pleistocene bone deposit was discovered in New Trout Cave, Pendleton County, West Virginia. The cave is located at 30°36'10"N Latitude, 79°22'08"W Longitude and is at 548.6 m altitude (Grady and Garton, 1981). The deposit, consisting of bone-rich matrix, was collected in 30 cm intervals to a depth of 220 cm and wet-screened through 5 mm and 1.5 mm mesh. A fauna including fish, amphibians, reptiles, birds, and at least 60 species of mammals was recovered. The specimens have been deposited in the collections of the Department of Paleobiology of the National Museum of Natural History.

DISCUSSION

The upper 90 cm of the New Trout Cave deposit contain a fairly typical late Pleistocene fauna including northern species such as *Dicrostonyx hudsonius, Microtus xanthognathus,* and *Sorex arcticus* (Grady and Garton, 1981). Also present are two species now found in the Midwest, *Spermophilus tridecemlineatus* and *Taxidea taxus.*

Among the more interesting remains in the three levels from 120–21 cm in the New Trout Deposit, were 11 specimens of the pocket gopher, *Geomys* sp., numbered USNM 336244–336254. The specimens include six parts of upper incisors, one partial lower incisor, a distal part of a tibia, and an ungual phalanx. Identification to species level was not possible due to the fragmentary nature of the material. The New Trout Cave specimens probably are either the plains pocket gopher, *Geomys bursarius,* or the southeastern pocket gopher, *Geomys pinetis.* These two species are differentiated osteologically by the shape of the nasal bones, which are highly variable (Guilday et al., 1971) and not represented in the New Trout collection. The upper incisor parts ranged in width from 2.2 to 2.8 mm, within the ranges of incisor widths of both *G. bursarius* and *G. pinetis.* G. S. Morgan (personal communication) suggested that the New Trout tibia was closer to *Geomys bursarius* in size though Parmalee and Klippel (1981) demonstrated considerable overlap in the postcranial elements of the two species.

There have been a number of fossil records of *Geomys* in the eastern United States outside of the current ranges of *Geomys bursarius* and *Geomys pinetis* (Fig. 1) summarized by Parmalee and Klippel (1981). The localities in Illinois, Indiana, Kentucky, and Tennessee are all, except possibly Harrodsburg, Indiana, late Pleistocene or Holocene in age (Guilday et al., 1971; Guilday, 1977; Parmalee and Klippel, 1981). The New Trout Cave locality is some 480 km east of the current range of *Geomys bursarius* and 500 km north of *Geomys pinetis* (Fig. 1) (Hall, 1981). The *Geomys* specimens from New Trout Cave are the first known from West Virginia. They occur in levels 60 or more centimeters below a C^{14} date of 29,400 $\pm$ 1,700 BP. The associated fauna in the levels of New Trout Cave containing *Geomys* lacks several of the northern species present in the upper levels and includes *Ochotona* sp., previously associated with late Irvingtonian faunas of the Central Appalachians (Guilday, 1979). A number of extinct small mammal taxa present in the late

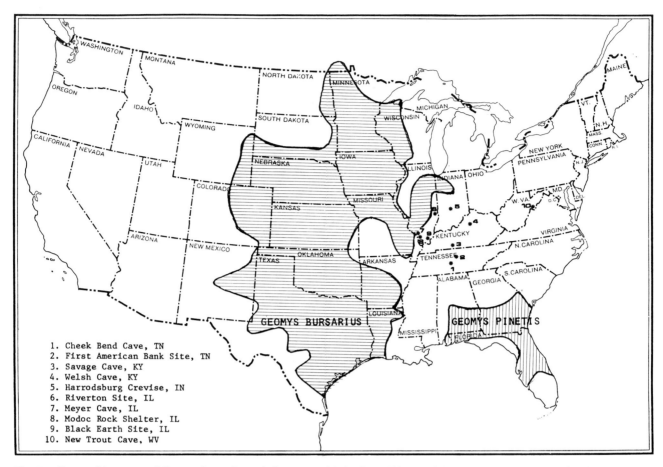

1. Cheek Bend Cave, TN
2. First American Bank Site, TN
3. Savage Cave, KY
4. Welsh Cave, KY
5. Harrodsburg Crevise, IN
6. Riverton Site, IL
7. Meyer Cave, IL
8. Modoc Rock Shelter, IL
9. Black Earth Site, IL
10. New Trout Cave, WV

Fig. 1.—Geographic ranges of *Geomys bursarius* and *G. pinetis* with fossil localities outside the present range (modified from Parmalee and Klippel, 1981).

Irvingtonian faunas of Trout and Cumberland Caves are not present in New Trout Cave. It seems likely that the lower levels of New Trout Cave are early to middle Rancholabrean in age.

The climatic implications of *Geomys* in West Virginia are unclear because the material is not specifically identifiable. Historically the vegetation in West Virginia consisted of mixed coniferous and deciduous forest (McKeever, manuscript). *Geomys bursarius* inhabits prairie and semi-prairie from Canada to Texas, whereas *G. pinetis* inhabits open pine woods with sandy soil in the Southeast (Guilday et al., 1971). Based on specimens from Kentucky, Guilday et al. (1971) suggested there was a continuous distribution of *Geomys* from the Midwest to Florida. It seems likely that *Geomys* expanded its range into West Virginia during a dry, probably glacial interval prior to the latest Wisconsin Glaciation. The absence of *Geomys* in the rodent-rich upper levels of the New Trout Cave deposit and in other late Rancholabrean faunas of the Central Appalachians indicates that *Geomys* did not survive to this interval in West Virginia. A thomomine pocket gopher, *Thomomys potomacensis,* is present in the Irvingtonian fauna of nearby Trout Cave and became extinct prior to Rancholabrean time (Kurten and Anderson, 1980).

Much work remains to be done on the New Trout Cave Fauna. Despite removal of some four and a half tons of matrix, much more remains for future investigators. New localities in nearby Trout and Hamilton Caves give promise of further additions to the Pleistocene faunal record of West Virginia.

ACKNOWLEDGMENTS

John Guilday supported the New Trout Cave project in many ways from 1979 to 1982. Ray Garton was co-director of the New Trout Cave project and assisted in all phases of the collection and processing of the materials. The National Speleological Society, current owner of New Trout Cave has provided continuing support for paleontological research in the cave. Members of two cave clubs, D. C. Grotto and Monongahela Grotto, provided the volunteers that made possible the collection of some four and a half tons of matrix. Allen McCrady and Harold Hamilton provided assistance in the processing of matrix from New Trout Cave at the New Paris field laboratory. Gary Morgan assisted in the identification of the tibia. Robert Emry and Clayton Ray critically read this paper.

LITERATURE CITED

GRADY, F., and E. R. GARTON. 1981. The collared lemming *Dicrostonyx hudsonius* (Pallas) from a Pleistocene cave deposit in West Virginia. Proc. Internat. Congress of Speleol., 8:279–281.

GUILDAY, J. E. 1977. Sabertooth cat, *Smilodon floridanus* (Leidy) and associated fauna from a Tennessee Cave (40 Dv 40), the First American Bank Site. J. Tennessee Acad. Sci., 52:84–94.

———. 1979. Eastern North American Pleistocene *Ochotona* (Lagomorpha: Mammalia). Ann. Carnegie Mus., 48:435–444.

GUILDAY, J. E., H. W. HAMILTON, and A. D. MCCRADY. 1971. The Welsh Cave peccaries (*Platygonus compressus*) and associated fauna, Kentucky Pleistocene. Ann. Carnegie Mus., 43:249–320.

HALL, E. R. 1981. The mammals of North America. John Wiley and Sons, New York, 2 vols., 1,180 pp.

KURTEN, B., and E. ANDERSON. 1980. Pleistocene mammals of North America. Columbia Univ. Press, New York, 442 pp.

MCKEEVER, S. (MS). Ecology and distribution of the mammals of West Virginia. Unpublished PhD thesis, Univ. Microfilms, Ann Arbor, 335 pp.

PARMALEE, P. W., and W. E. KLIPPEL. 1981. A late Pleistocene population of the pocket gopher, *Geomys* cf. *bursarius*, in the Nashville Basin, Tennessee. J. Mamm., 62:831–835.

Address: Department of Paleobiology, National Museum of Natural History, Smithsonian Institution, Washington, D.C. 20560.

NEOTOMA IN THE LATE PLEISTOCENE OF NEW MEXICO AND CHIHUAHUA

ARTHUR H. HARRIS

ABSTRACT

Selected characters apt to be present in fossil specimens were studied in modern woodrats (*Neotoma*) in an effort to determine discriminatory features. Emphasis was on the lower first molar. Most modern specimens of the eight species studied (*N. albigula, N. cinerea, N. floridana, N. goldmani, N. lepida, N. mexicana, N. micropus,* and *N. stephensi*) can be identified correctly to species by the use of standard statistical methods and discriminant analysis. Particularly important in preliminary separation into major groups is the absence or near-absence (≤2 mm in height) of the lateral dentine tract of the first lower molar in one group (*N. albigula, N. floridana,* some *N. lepida,* and *N. micropus*) versus presence (≥2 mm in height) in a second group (*N. cinerea, N. goldmani,* most *N. lepida, N. mexicana,* and *N. stephensi*). Within the group possessing the developed tract, *N. goldmani* and *N. lepida* tend to have the tract lower in height than do the other members of the group.

Application of the discriminatory data to over 500 fossil specimens from 24 late Pleistocene and early Holocene sites located in New Mexico and southern Chihuahua reveals profound differences between interstadial, stadial, and early Holocene woodrat faunas. Interstadial faunas (ca. 25,000 to 33,000 B.P.) were characterized by presence of two undescribed species (apparently related to *N. cinerea* and *N. goldmani*) along with *N. albigula* and *N. micropus*. Most stadial sites were dominated by *N. cinerea*, with other species represented being *N. albigula, N. ?goldmani, N. floridana,* and *N. micropus*. Of these, *N. floridana* and possibly *N. micropus* appeared only toward the end of stadial times. Jimenez Cave, in southern Chihuahua, is not certainly stadial in age; *N. lepida* was common. Early Holocene sites had woodrat faunas similar to those of today except for the additional presence of *N. mexicana*.

INTRODUCTION

Virtually every area of the western United States and northern Mexico is inhabited by one or more species of woodrats, genus *Neotoma*. Their remains often are abundant in late Pleistocene cave faunas and their contribution to our knowledge of Pleistocene ecology by means of their preserved middens is well known. Their own skeletal remains have contributed relatively little to our knowledge of the Pleistocene, however, because identification to species often is difficult and, to the skeptical, suspect. Reasonably sure identifications would add sig-

nificantly to our knowledge of Pleistocene biogeography and, to a lesser degree, ecology. The aims of this study were to produce discriminating criteria usable with commonly preserved fossil elements for those species apt to be found in late Pleistocene of the Southwest and northern Mexico, to apply these criteria to late Pleistocene/early Holocene specimens available to me, and to interpret the finding in terms of systematics, biogeography, and paleoecology.

MATERIALS AND METHODS

Modern comparative material was assembled for species currently living in the southwestern region and for species judged to have possibly occurred in the area during the late Pleistocene. These species are *Neotoma albigula* Hartley (white-throated woodrat), *N. cinerea* (Ord) (bushy-tailed woodrat), *N. floridana* (Ord) (eastern woodrat), *N. goldmani* Merriam (Goldman's woodrat), *N. lepida* Thomas (desert woodrat), *N. mexicana* Baird (Mexican woodrat), *N. micropus* Baird (southern plains woodrat), and *N. stephensi* Goldman (Stephens' woodrat). Limitations of time and scarcity of specimens in collections (*N. goldmani*) have resulted in several samples of less than ideal size for statistical treatment.

Most effort has been directed toward the dentary with its teeth, particularly m1. This in part reflects commonness of recovery in fossil faunas and in part the potential for identification.

Measurements on both modern and fossil material were taken with an ocular micrometer to the nearest 0.1 mm except for the

greatest width of loph 2 of m1 and depth of incisor, which were taken to 0.01 mm with dial calipers. Measurements consistently taken on lower jaw elements were: 1) length of alveolar cheek-tooth row (LG-ALV); 2) mid-length of m1 (LG-M1); 3) greatest width of loph 2 of m1 (WD-M1); 4) height of antero-lateral dentine tract of m1 (TRACT); 5) distance from base of lingual fold 1 to base of fold 2 of m1 (F1–F2); 6) distance from base of lingual fold 2 to anterior face of m1 (ANT-F2); 7) development of the antero-internal reentrant fold of loph 1, m1 (FOLD); and 8) a ratio comparing depth of the m1 anterointernal reentrant fold to the width of m1 (RATIO). These measurements are indicated in Fig. 1.

LG-M1 was taken from the lingual side at a level estimated to lie between one-third and one-half the height of an unworn tooth and perpendicular to the vertical axis of the tooth (not parallel to the wear surface). Occlusal length proved to be too variable with wear and angle of wear to be as useful a measurement.

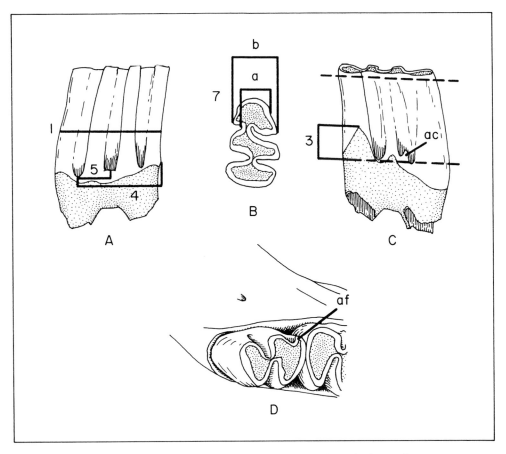

Fig. 1.—Qualitative characters and methods of taking measurements. A) Lingual view of left m1 of *Neotoma cinerea,* showing method of taking LG-M1, ANT-F2, and F1–F2. B) Occlusal view of left m1 of *Neotoma albigula,* showing measurements for calculation of RATIO (a/b). C) Labial view of left m1 of *N. cinerea,* showing method of measuring dentine tract (TRACT) and presence of accessory cusp (ac). D) Occlusal view of left m3 of *Neotoma goldmani,* showing accessory fold (af).

WD-M1 was taken with caliper blades approximately vertical. The bulge present in some teeth at the base of F2 was avoided. A series of measurements on the same specimen occasionally will vary by several hundreths millimeter; heavily worn teeth tend to give an underestimate. Very young teeth (roots broadly open) also often give smaller measurements than comparable adult teeth of the same species even when the enamel appears complete.

The anterolateral dentine tract (Fig. 1) was, to the best of my knowledge, first noted by Lundelius (1979) in *N. cinerea* and *N. mexicana.* Differing in presence and development within the genus, it is an extension of the enamel-less area ventral to the crown onto the lateral surface of loph 1 (lesser developed tracts may occur on the other lophs and on those of m2 and m3). Height was measured from a line parallel to the wear surface extended anteriorly from the base of fold 1 (Fig. 1). In most cases, a sharp demarcation between the normal enamel surface and the dentine was present and used for the upper limit. In some specimens, the transition from dentine to normal enamel is attenuate, without sharp limits; the uppermost appearance of dentine was used in these cases. With wear, the uppermost portions of the tract may be worn away and a conservative estimate of tract height

was made. Tract measurements of the high-dentine forms thus are biased slightly to the low side.

Measurements involving lingual folds were taken from the most ventral point of the fold, which may not be the median point. For ANT-F2, the measurement was made to the anterior wall of the tooth above any abrupt basal constriction.

Development of the anterointernal reentrant fold was coded as shown in Fig. 2. Except for the 0.1 coding, depth may vary greatly (though fairly typical examples are shown)—closeness of the fold base to the tooth base is the criterion used. This is somewhat subjective and wear may obliterate categories 0.2 and 0.3.

RATIO was used in an attempt to quantify the depth of the anterointernal fold in relation to the width of loph 1 (Fig. 1). Corrections for wear were made. Wear categories are 1) very light (depressions on occlusal surface not yet obliterated or enamel walls not fully worn to a flat surface); 2) light (base of labial folds not erupted to level of lateral alveolar wall, or judged to have not reached that stage in the case of isolated teeth; tooth usually notably tapered toward the top); 3) moderate (fold bases erupted but wear not sufficiently close to base as to distort enamel pat-

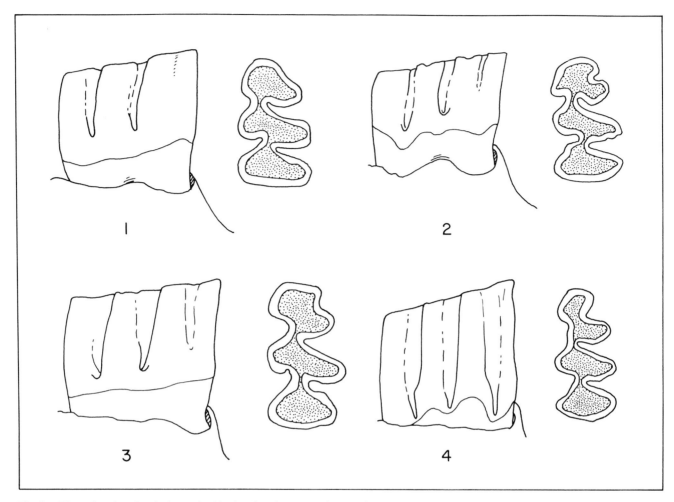

Fig. 2.—Lingual and occlusal views of m1's showing character of antero-internal grooves used for coding. 1) *Neotoma albigula,* coding 0.1. 2) *Neotoma micropus,* coding 0.2. 3) *Neotoma goldmani,* coding 0.3. 4) *Neotoma mexicana,* coding 0.4.

terns); 4) heavy (enamel patterns obviously distorted in proportions); and 5) very heavy (portions of enamel wall other than at the dentine tract missing, often fold bases isolated as islands of enamel surrounded by dentine). Very heavy wear specimens were not used in the analysis. Correctives for wear are given in Table 1 and were obtained by calculating the factors necessary to correct to the average moderate-wear category in the modern samples.

In general, more subjective weight was given to LG-M1, WD-M1, TRACT, and FOLD than to the other characters.

Additional data taken include samples of lower incisor depth (greatest distance from dorsal to ventral surface perpendicular to the tooth axis) and several qualitative characters. These included the presence of accessory cusps at the fold bases in m1 and m2 (Fig. 1), presence of an accessory fold in m3 (Fig. 1), the development of the capsule at the base of the incisor (Fig. 3), the character of the mandibular foramen (Fig. 3), and the relative depths of the external and internal reentrant folds of m3.

The quantitative data were subjected to standard statistical

Table 1.—*Correction for wear for measurement RATIO. The raw measurement is multiplied by the wear corrective.*

Taxon	Wear category			
	Very light	Light	Moderate	Heavy
Neotoma cinerea, N. mexicana, N. goldmani	1.89	1.31	1.00	0.96
Neotoma lepida	1.00	1.31	1.00	0.96
Neotoma albigula, N. floridana, N. micropus	—	1.22	1.00	0.96

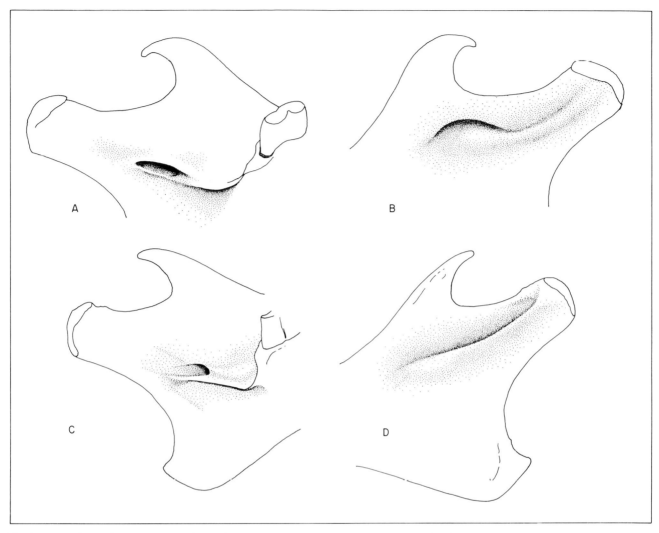

Fig. 3.—Mandibular foramen types (left) and incisor capsule types (right). A) *Neotoma albigula,* showing ventrally oriented mandibular foramen. B) *Neotoma albigula,* showing enlarged incisor capsule. C) *Neotoma goldmani,* showing more laterally directed mandibular foramen. D) *Neotoma goldmani,* showing reduced incisor capsule.

treatment, but also multivariate techniques were utilized, including discriminant analysis (BMDO7M, SPSS) and clustering techniques (NT-SYS).

Although the sample statistics and scattergrams can be used for identification, a somewhat more objective method is the use of discriminant analysis. This procedure is most powerful when used between pairs of taxa, in that the weighted characters that best separate three or more taxa are not necessarily the best for separation of any single pair of taxa. In practice, unknown specimens that might belong to any of several taxa are processed with samples of the several known taxa, the results being used to segregate the unknowns to fewer probable choices. The data given in the text and in Table 2 and Figs. 5 to 7 generally should be sufficient to make a preliminary reduction to two or three probable taxa. The data given in Tables 3 and 4 may then be used to

assign discriminant scores to individual specimens and make identifications based on these scores.

Several cautions should be observed: 1) if the specimen belongs to a taxon not represented in the discriminant analysis, a spurious identification will result; 2) some specimens of one taxon overlap with members of another taxon to the point that they are indistinguishable on the basis of the characters used, again resulting in an incorrect or equivocal identification; 3) *Neotoma* has a nasty habit of occasionally having an otherwise rather stable character state shift to the mode seen in a different taxon in rare individuals; if this is a heavily weighted character, assignment may be incorrect. For this reason, measurements should be scanned for reasonableness. As an example, several unknown specimens were assigned to *N. micropus,* but WD-M1 was considerably less than minus two standard deviations from the *N. micropus* mean.

Table 2.—*Basic statistics for samples of modern* Neotoma *specimens.*

Taxon	Mean	SD	Observed range	N
Neotoma albigula				
LG-M1	3.09	0.136	2.7–3.3	36
WD-M1	1.815	0.073	1.63–1.97	72
TRACT	0.01	0.028	0.0–0.1	72
ANT-F2	2.29	0.126	2.0–2.6	36
F1-F2	0.85	0.091	0.7–1.0	36
FOLD	0.18	0.059	0.1–0.3	36
RATIO	0.671	0.062	0.55–0.83	36
Neotoma cinerea				
LG-M1	3.52	0.122	3.3–3.7	16
WD-M1	1.954	0.102	1.79–2.14	16
TRACT	1.14	0.242	0.8–1.6	16
ANT-F2	2.68	0.181	2.4–3.0	16
F1-F2	1.03	0.101	0.9–1.2	16
FOLD	0.32	0.098	0.1–0.4	16
RATIO	0.481	0.144	0.26–0.71	16
Neotoma floridana				
LG-M1	3.46	0.130	3.3–3.8	22
WD-M1	2.070	0.084	1.93–2.26	22
TRACT	0.02	0.053	0.0–0.2	22
ANT-F2	2.57	0.109	2.4–2.8	22
F1-F2	0.98	0.102	0.9–1.2	22
FOLD	0.27	0.089	0.1–0.4	22
RATIO	0.479	0.076	0.32–0.59	22
Neotoma goldmani				
LG-M1	2.78	0.128	2.6–3.0	8
WD-M1	1.639	0.071	1.54–1.79	8
TRACT	0.35	0.120	0.2–0.5	8
ANT-F2	1.93	0.158	1.7–2.1	8
F1-F2	0.80	0.053	0.7–0.9	8
FOLD	0.20	0.093	0.1–0.3	8
RATIO	0.645	0.083	0.56–0.75	8
Neotoma lepida				
LG-M1	3.04	0.137	2.8–3.3	28
WD-M1	1.708	0.081	1.54–1.83	28
TRACT	0.28	0.188	0.0–0.7	28
ANT-F2	2.30	0.129	2.1–2.5	28
F1-F2	0.84	0.057	0.7–0.9	28
FOLD	0.25	0.079	0.1–0.3	28
RATIO	0.569	0.087	0.42–0.70	28
Neotoma mexicana				
LG-M1	3.17	0.145	2.9–3.5	42
WD-M1	1.723	0.063	1.61–1.85	42
TRACT	1.32	0.345	0.7–2.2	42
ANT-F2	2.39	0.144	2.1–2.7	42
F1-F2	0.87	0.077	0.7–1.1	42
FOLD	0.39	0.034	0.2–0.4	42
RATIO	0.546	0.081	0.39–0.77	42
Neotoma micropus				
LG-M1	3.17	0.106	2.9–3.4	26
WD-M1	1.998	0.090	1.84–2.21	26
TRACT	0.004	0.020	0.0–0.1	26
ANT-F2	2.29	0.141	2.0–2.6	26
F1-F2	0.94	0.110	0.6–1.1	26

Table 2.—*Continued.*

Taxon	Mean	SD	Observed range	N
FOLD	0.069	0.014	0.1–0.3	26
RATIO	0.609	0.072	0.47–0.77	26
Neotoma stephensi				
LG-M1	2.79	0.178	2.5–3.1	12
WD-M1	1.713	0.087	1.62–1.93	12
TRACT	1.17	0.328	0.6–1.7	12
ANT-F2	2.14	0.144	1.9–2.4	12
F1-F2	0.84	0.051	0.8–0.9	12
FOLD	0.29	0.051	0.2–0.4	12
RATIO	0.564	0.110	0.43–0.77	12

Some idea of the trustworthiness of discriminant analysis for identifying members of two taxa can be gained by using discriminant criteria to assign identifications to individual members of the modern samples. Results are given in Tables 3 and 4. The assignment error of non-sample specimens is expected to be greater, particularly when the discriminant analysis has available as "knowns" only samples of small size.

Final identifications of the fossil material were by an amalgamation of all multivariate and statistical techniques as tempered by characteristics of the specimens and samples themselves (wear, preservation, identifications of other *Neotoma* in the site, etc.).

Table 3.—*Pair-wise unstandardized canonical discriminant function coefficients for computing discriminant scores of non-dentine tract* Neotoma *m1's apt to be confused. To determine score for an unidentified individual, each measurement is multiplied by the respective coefficient and these are summed and added to the constant. The specimen is assigned to species according to its position along the discriminant function (greater or less than the division point). Some measurements are not of value for some species and are omitted. Taxa are identified by initial of species name.*

| Variable | Coefficients for species pairs | | | |
	A/F	A/L	A/M	F/M
LG-M1	—	—	—	5.483
WD-M1	8.892	−7.236	−11.757	—
TRACT	—	6.146	—	9.495
ANT-F2	—	2.134	6.830	7.058
F1-F2	—	—	−4.515	−8.473
FOLD	—	—	5.189	—
RATIO	−9.851	−5.748	11.771	−2.622
Constant	−11.044	10.674	1.941	−25.741
Division point	−0.514	0.182	−0.267	0.156
Species above point	*N.f.*	*N.l.*	*N.a.*	*N.f.*
Modern sample discriminated	100%	91%	100%	96%

Table 4.—*Pair-wise discriminant function coefficients for dentine tract* Neotoma. *See caption of Table 3 for explanation.*

Variable	Coefficients for species pairs					
	C/M	C/S	G/L	G/S	L/S	M/S
LG-M1	−2.270	6.431	—	—	−3.120	5.685
WD-M1	13.109	3.642	—	7.779	3.408	−4.880
TRACT	—	—	—	3.908	3.913	—
ANT-F2	5.825	—	7.378	—	—	—
F1-F2	—	—	—	—	—	−4.038
FOLD	−16.030	−4.952	—	—	—	15.803
RATIO	—	—	—	—	—	—
Constant	−24.390	−25.844	−16.333	−16.375	1.296	−11.519
Division point	1.148	−0.387	−0.761	−0.377	0.853	−0.998
Species above point	*N.c.*	*N.c.*	*N.l.*	*N.s.*	*N.s.*	*N.m.*
Modern sample discriminated	100%	100%	89%	100%	100%	94%

More than 500 specimens of *Neotoma* from the late Pleistocene and early Holocene of Chihuahua and New Mexico were available for study. These specimens are preserved in the Resource Collections, Laboratory for Environmental Biology, University of Texas at El Paso (UTEP). Sites and site data are given in Table 5; site localities are shown in Fig. 4.

RESULTS AND DISCUSSION

The basic statistics of the modern species samples are given in Table 2 and in part displayed in Figs. 5 to 7.

Specimens of *Neotoma* can be divided into two major groups on the basis of the development of the lateral dentine tract on m1. Most specimens of *N. albigula*, *N. floridana*, and *N. micropus* basically lack the tract, though a few individuals have the tract developed to a height of ca. 0.1 mm (0.2 mm in *N. floridana*). The remaining species have a tract height of ≥0.2 mm except for some *N. lepida* (ca. 36% of the modern sample).

Qualitative characters of the dentary differing between the two groups include generally greater development of the incisor capsule and a somewhat more ventrally oriented mandibular foramen in the non-dentine tract group (Fig. 3). Neither character is entirely clearcut in all cases, and *N. lepida* is more similar to the non-dentine tract species.

Within the group possessing a dentine tract, there is a dichotomy between those that possess a relatively low tract (*N. goldmani* and *N. lepida*) and those with a high tract (*N. cinerea*, *N. mexicana*, and *N. stephensi*) (Fig. 7).

Table 6 shows the identifications of the *Neotoma* found in each site treated.

Fig. 4.—Sketch of New Mexico and Chihuahua, showing late Pleistocene sites. 1) Isleta Caves; 2) Howell's Ridge Cave; 3) Baldy Peak Cave; 4) Khulo Site; 5) Conkling's Cavern and Shelter Cave; 6) Anthony Cave; 7) Dry Cave; 8) Dark Canyon Cave; 9) Jimenez Cave.

Table 5.—*Sites from which fossil* Neotoma *were examined. Those marked with an asterisk are in the Dry Cave system.*

UTEP no.	Name	Age (BP)	Reference
1, 17	*Lost Valley	29,290 ± 1,060	Harris, 1977
4	*Bison Chamber	<14,470, >10,730	Harris, 1977
5	*Sabertooth Camel Maze	25,160 ± 1,730	Harris, 1977
6	*Harris' Pocket	14,470 ± 250	Harris, 1977
21	Khulo Site	<10,000	Harris, 1977
22	*Animal Fair	15,030 ± 210	Harris, 1977
23	*Stalag 17	11,880 ± 250	Harris, 1977
24	*Entrance Chamber	<11,880	Harris, 1980
25	*Camel Room	Est. >12,000	Harris, 1977
26, 27	*Rm Vanishing Floor	33,590 ± 1,500	Harris, 1977
28	*Rick's Cenote	Est. 11,000	Unpublished
29	Anthony Cave	stadial	Harris, 1977
30	Shelter Cave	stadial, Holocene	Harris, 1977
32	Howell's Ridge Cave	stadial, Holocene	Harris, 1977; Van Devender and Wiseman, 1977
41	Isleta Cave No. 1	stadial, Holocene	Harris and Findley, 1964
46	Isleta Cave No. 2	stadial, Holocene	Harris and Findley, 1964
54	*TT II	10,730 ± 150	Harris, 1977
75	Dark Canyon Cave	stadial	Harris, 1977
90	Conkling Cavern	stadial	Harris, 1977
91	Jimenez Cave	?stadial and ?Holocene	Unpublished
94	Baldy Peak Cave	stadial and ?Holocene	Unpublished
122	*Pit N & W Animal Fair	?early stadial	Harris, 1977

Table 6.—*Taxa identified from each site.*

UTEP Loc.	Neotoma albigula	Neotoma cinerea	Neotoma floridana	Neotoma goldmani	Neotoma lepida	Neotoma mexicana	Neotoma micropus	A	B
Interstadial									
1	X						X	X	X
5	X								
17	X							X	X
26								X	X
27	X								X
Stadial									
4							?		
6	X	X					cf.		
22	X	X		?					
23		X	X						
25		X							
28		cf.							
29	X	X		?					
30		X							
32	X	X							
41	X	X					?		
46	X	X					?		
54		cf.	cf.				X		
75	X	X							
90	X	X							
91	X	?	cf.		X		X		
94		X				X			
122		X							
Holocene									
21	X					X		?	
24	X					X	X		

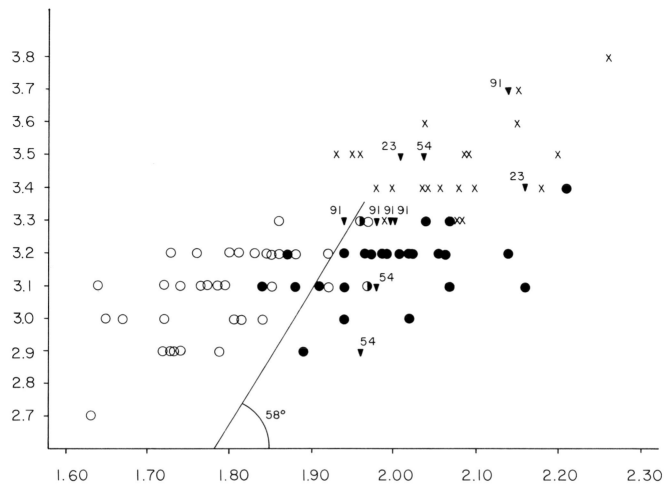

Fig. 5.—Scattergram of LG-M1 (ordinate) against WD-M1 (abscissa). Circles, modern *Neotoma albigula*; dots, modern *Neotoma micropus*; X, modern *Neotoma floridana*; triangles, selected fossil specimens from sites shown by numbers. Line shows best separation between *N. albigula* and *N. micropus*.

SPECIES ACCOUNTS

Neotoma albigula

The white-throated woodrat now occurs at or near all sites. It apparently was as ubiquitous in the past, occurring in interstadial, stadial, and early Holocene sites.

This rat is easily separable from other *Neotoma* except for *N. micropus* and some individuals of *N. lepida* on the basis of m1. Dalquest et al. (1969) and Lundelius (1979) have discussed separation criteria between *N. albigula* and *N. micropus*. Dalquest et al. found no overlap in width of loph 2 of m1, *N. albigula* having a width of <1.94 mm and *N. micropus* a width of >1.94 mm. Lundelius, in a sample of 32 *N. albigula* and 30 *N. micropus,* found 9.4%

of the former with breadths >1.94 mm and 16.7% of the latter with widths <1.94 mm. In the present study, a sample of 72 *N. albigula* showed three (4.2%) exceeding 1.94 mm, and five of 26 (19.2%) *N. micropus* with measurements of <1.94 mm. A bivariate scattergram of LG-M1 and WD-M1 (Fig. 5) does a somewhat better job of separation, but some misidentifications on the basis of these measurements are inevitable.

Confusion between *N. albigula* and *N. lepida* results from the inability to separate those *N. lepida* with dentine tracts <0.2 mm from *N. albigula*.

There is some variability within the species (Fig. 8), with some tendency for smaller size in the stadial sites and with the length of m1 from Jimenez Cave (UTEP 91) and the Khulo Site (UTEP 21) being

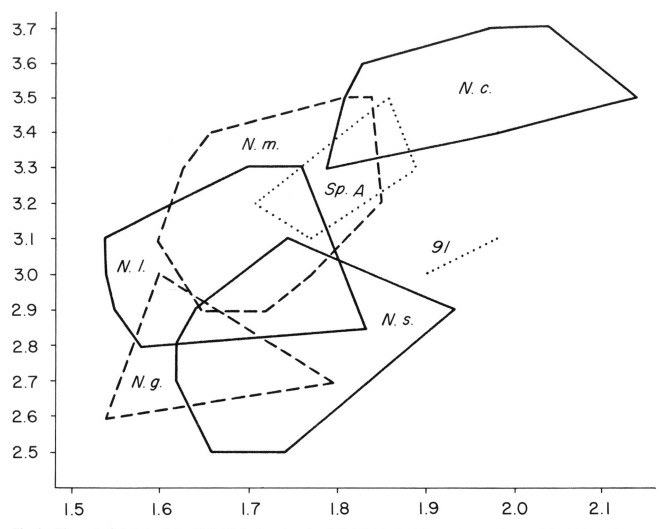

Fig. 6.—Dispersal of plotted points of LG-M1 (ordinate) against WD-M1 (abscissa) for modern specimens of five high dentine tract species, fossil specimens of extinct Species A, and two fossil specimens from UTEP 91 (species identified by initials). Polygraphs were constructed by connecting marginal points.

somewhat large. Jimenez Cave lies within the current range of *N. albigula durangae,* which Anderson (1972) notes as being somewhat intermediate between *N. a. albigula* and *N. micropus*; this may be showing up here. No such explanation is available for the Khulo Site sample, nor do width measurements indicate much chance of *N. micropus* biasing the sample by wrongful inclusion.

Today, *N. albigula* occurs in a variety of habitats from desert to pinyon-juniper woodlands and ponderosa pine forests, though seldom found in pure grassland. In various parts of its range it may be associated with all of the species considered here, though only marginally with some.

Neotoma cinerea

The bushy-tailed woodrat occurs in every stadial fauna for which there is at least a moderate-sized sample, and often makes up the majority of the recovered specimens.

There is remarkably little variation from site to site and from Pleistocene populations to modern populations, no significant differences being found among the three characters looked at closely (LG-M1, WD-M1, TRACT). The two specimens (queried) from Jimenez Cave, however, are exceptionally short, though of normal width and dentine tract height (Fig. 6); these suggest intraspecific variation at this far southern extension of the geographic range,

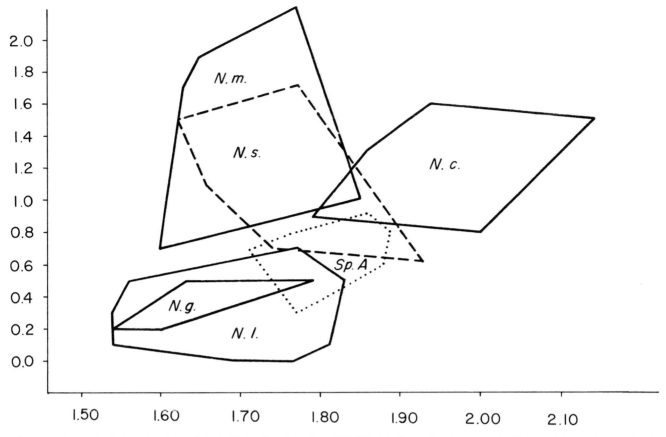

Fig. 7.—Dispersal of plotted points of TRACT (ordinate) against WD-M1 (abscissa). Specimens as in Fig. 6, without the two fossil specimens from UTEP 91.

but may represent an unknown taxon. Table 7 gives the basic statistics for three measurements from the three largest samples.

Lundelius (1979) identified *N. cinerea* from Pratt Cave (located in the Texas portion of the Guadalupe Mountains just south of the New Mexico border; age is considered Holocene by Lundelius, but some material may be Pleistocene) and from Dark Canyon Cave (Texas Memorial Museum specimens). He suggested that these specimens represented populations with larger m1's than those of modern *N. cinerea* from New Mexico; that they were of a size more like modern specimens from Wyoming. However, there is no significant difference between the present small sample of Recent *Neotoma* from New Mexico (n = 9) and the UTEP sample from Dark Canyon Cave (n = 9 adults). The Dark Canyon sample of Lundelius averaged 0.41 mm larger than his small sample (n = 4) of modern New Mexican specimens, whereas the difference in the present study

is 0.1 mm (the Lundelius measurement, however, was occlusal length of m1 rather than mid-length).

Neotoma cinerea apparently was absent from the

Table 7.—*LG-M1, WD-M1, and TRACT statistics for* Neotoma cinerea *from UTEP 22, 29, and 75.*

Locality	Mean	SD	Observed range	N
UTEP 22				
LG-M1	3.54	0.214	3.2–4.1	29
WD-M1	1.962	0.115	1.78–2.21	29
TRACT	1.10	0.219	0.7–1.6	29
UTEP 29				
LG-M1	3.49	0.151	3.2–3.7	11
WD-M1	1.986	0.071	1.88–2.12	11
TRACT	1.16	0.401	0.4–1.9	11
UTEP 75				
LG-M1	3.56	0.167	3.3–3.8	9
WD-M1	1.970	0.105	1.85–2.14	9
TRACT	1.04	0.255	0.6–1.4	9

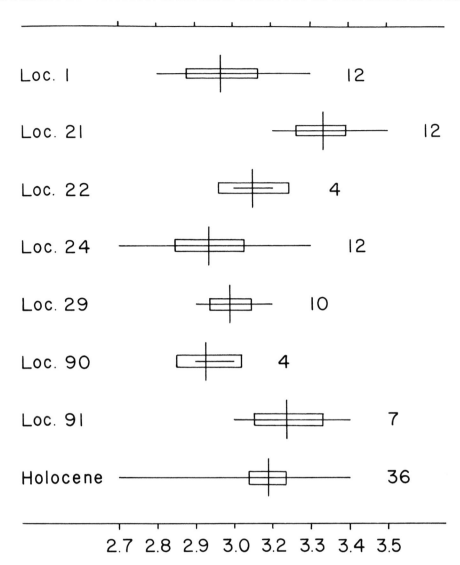

Fig. 8.—m1 lengths of fossil and modern *Neotoma albigula*. Vertical lines, sample means; horizontal lines, observed range; box, 95% confidence interval of the mean; numbers to right of figures, sample size.

interstadial deposits of Dry Cave (but see comments below, in the "Species A, undescribed" account) and from deposits that are surely Holocene, but as noted above, was virtually ubiquitous during stadial times with little morphological variation. Likely there was an essentially continuous population throughout the region. Holocene occurrence is possible at several sites, including the Isleta Caves, Howell's Ridge Cave (UTEP 32, possibly representing *N. mexicana*), and Pratt Cave, but its absence and replacement by *N. mexicana* at the presumably uncontaminated Entrance Chamber deposit (UTEP 24) at Dry Cave and the Khulo Site (UTEP 21) suggest otherwise.

Some earlier workers have interpreted presence of *N. cinerea* at low elevations south of their present range as indicative of the forests where they are found now at their point of nearest occurrence in the Sangre de Cristo Mountains of northern New Mexico (for example, Murray, 1957). This, however, ignores their occurrence under far different conditions and at lower elevations in northwestern New Mexico (extending into pinyon-juniper woodland and occasionally lower) and their widespread habitation of sagebrush and the like farther north (Cary, 1917). Likely *N. cinerea* inhabited all vegetational zones from timberline down to, and including, steppe-woodland (terminology after Harris, in press), an open woodland with well developed growth of grasses, herbs, and shrubs. Pure grassland, however, probably was not inhabited. Within these vegetational types, they likely were limited as today (Armstrong, 1972) to areas of cliffs, jumbled rocks, and caves.

Neotoma floridana

The most convincing evidence of the eastern woodrat in the southwestern Pleistocene is the central portion of a skull (UTEP 23–79) from the Stalag 17 site in Dry Cave. Qualitative features, including the presence of a well-forked palatine spine, fit *N. floridana* rather than *N. micropus*. Dalquest et al. (1969) suggest that the two species are separable on the basis of breadth of the maxillary molar rows (*N. micropus*, 8.0 to 8.7 mm; *N. floridana*, 8.7 to 9.2 mm). The method of taking molar row breadth is not described. In the present study, measurement was done by applying caliper blades to the outer sides of the toothrows at the alveolar line (measurements to the most lateral tooth surfaces are much larger). In 26 *N. micropus*, breadth measurements ranged from 7.67 to 8.64 mm; in 26 *N. floridana*, 8.06 to 9.48 mm, with 10 of the specimens <8.7 mm. Molar row breadth in 23–79 is 8.85 mm, well beyond any *N. micropus* measured by myself, Dalquest et al. (1969), or Lundelius (1979). Two dentaries with m1's from the same site are identified as *N. floridana* by discriminant analysis (see also Fig. 5). Both m1's have the deep anterointernal fold that occurs with fair frequency in *N. floridana* but is rare in *N. micropus*.

Two specimens from the early Holocene Khulo Site (UTEP 21) could be *N. floridana*, but are assigned here to *N. ?micropus*. If *N. floridana*, they would be at the small end of their size range. A similar situation occurs in the Holocene Entrance Chamber deposits (UTEP 24) of Dry Cave, but here *N. micropus* also seems present and the possible *N. floridana* again would have to be at the lower limits of its size distribution. Other equivocal specimens come from UTEP 46; they are identified here as *N. ?micropus*.

A much better candidate for *N. floridana* is specimen 28-1, an m1 from Rick's Cenote (UTEP 28) within Dry Cave. Identified by discriminant analysis as this species, its length (3.5 mm) is beyond any in the modern sample of *N. micropus* (longest, 3.4 mm). This site should be very late Pleistocene, as is another site in Dry Cave, UTEP 54 (approxi-

Table 8.—*Basic statistics for combined sample of* Neotoma ?goldmani *from UTEP 22 and 29.*

Measure-ment	Mean	SE	Observed range	N
LG-M1	2.86	0.052	2.8–2.9	8
WD-M1	1.698	0.055	1.60–1.77	8
TRACT	0.23	0.116	0.0–0.4	8
ANT-F2	2.04	0.074	1.9–2.1	8
F1-F2	0.79	0.064	0.7–0.9	8
FOLD	0.18	0.071	0.1–0.3	8
RATIO	0.673	0.059	0.57–0.76	8

mately 10,730 radiocarbon years, but with extinct fauna). Two of the three available specimens seem to be *N. micropus*, but the third clearly falls into the *N. floridana* area (Fig. 5). Four m1 specimens from Jimenez Cave (UTEP 91) in Mexico are identified as *N. floridana* by discriminant analysis, and another as *N. micropus*. Four of the specimens, including the latter, fall into the intermediate length-width area, but the fifth specimen is unequivocally within the *N. floridana* area (Fig. 5).

Thus, in summary, *N. floridana* seems defendably identified from near the Pleistocene-Holocene boundary in UTEP 23, and the evidence is rather strongly suggestive of presence in two other Dry Cave sites of latest Pleistocene age. Evidence also suggests presence at the undated Jimenez Cave, in southern Chihuahua, Mexico. Occurrence at two other sites (UTEP 21 and 46), both sites with early Holocene materials, is possible but judged unlikely.

Neotoma ?goldmani

Five m1's from the Animal Fair site (UTEP 22) of Dry Cave and three from Anthony Cave (UTEP 29) are closer in most measurements (Table 8) to *N. goldmani* than to *N. lepida*. Width and dentine height are slightly more similar to *N. lepida* than to *N. goldmani*, nearly reaching a significant difference from *N. goldmani* on dentine height in the combined sample (n = 8). In all other traits, however, *N. lepida* is more dissimilar (significantly so for M1-LG, $P \leq 0.001$; ANT-F2, $P \leq 0.001$; F1–F2, $P = 0.033$; GROOVE, $P = 0.028$; and RATIO, $P = 0.003$). In addition, one of two m3's associated with the m1's has an accessory fold and the other a suggestion of such. Six of eight modern *N. goldmani* possess this fold and it was probably developed on a seventh (the foldlet becomes unclear and then disappears with moderate to heavy wear). In 30 *N. lepida* from east of California, only one definitely has such a fold. Of 15 specimens of *N. lepida* from California and Baja California, however, six show folds (this population, however, is much larger in size than interior *N. lepida*).

Today, *N. goldmani* occurs in eastern Mexico almost north to the Big Bend of Texas. A continuous distribution northward through the mountain ranges of Trans-Pecos Texas and into southeastern New Mexico would not be startling, and this rare rat may well occur today north of its recognized range. It probably is an inhabitant today of rocky areas in arid mountains.

Neotoma lepida

The m1 of this animal seems to differ from that of *N. albigula* primarily in the possession in about two-thirds of the individuals of a dentine tract ≥0.2 mm in height; length averages significantly less, but with much overlap. Six of 72 m1's of *N. albigula* had a dentine tract height of 0.1 mm; none had a higher tract. Thus possession of a tract ≥0.2 mm rules out *N. albigula* beyond reasonable doubt. Basic statistics for UTEP 91 are given in Table 9. A comparison of dentine tract heights between the UTEP 91 sample and that of modern *N. lepida* shows a significantly lower tract in modern *N. lepida* ($P = 0.017$); this probably is biased, however, because fossil *N. lepida* m1's with a dentine tract that is <0.2 mm cannot be separated from those of *N. albigula* except when the teeth are notably smaller, and thus most such low-height tract teeth would not appear in the sample. The UTEP 91 population also differs significantly from modern *N. lepida* in slightly shorter length of m1 ($P = 0.023$) and the measurement ANT-F2 ($P \leq 0.001$), and in larger RATIO measurement ($P = 0.001$). These differences are interpreted as geographic/chronologic variation, Jimenez Cave being far from any modern population of *N. lepida*. A modern sample identified as *N. lepida* from California and Baja California shows far more difference in m1 length, for example (mean = 3.25, n = 13).

Neotoma mexicana

Demonstrable *N. mexicana* are surprisingly rare in the sites considered, and those that do occur do so in generally unexpected circumstances. Unless a few individuals are mixed in with the large number of *N. cinerea* identified from stadial sites (and this is a possibility), *N. mexicana* is absent from all stadial sites with the exception of Baldy Peak Cave (UTEP 94). This site had little fill, and post-Pleistocene remains would be inseparable from older fos-

Table 9.—*Basic statistics for* Neotoma lepida *from UTEP 91.*

Measure-ment	Mean	SD	Observed range	N
LG-M1	2.95	0.112	2.8–3.2	23
WD-M1	1.733	0.061	1.61–1.82	23
TRACT	0.39	0.122	0.1–0.5	23
ANT-F2	2.07	0.118	1.9–2.3	23
F1-F2	0.83	0.076	0.7–0.9	23
FOLD	0.25	0.095	0.1–0.4	23
RATIO	0.665	0.117	0.48–0.90	23

sils. This is the only site where both *N. cinerea* and *N. mexicana* are solidly identified.

Other sites with *N. mexicana* are Holocene. The largest sample is from UTEP 21, the Khulo Site (n = 7), and it seems comparable to modern New Mexican *N. mexicana*. From the Entrance Chamber of Dry Cave (UTEP 24), however, two of the three specimens are notably small for *N. mexicana*, but resemble no other known taxon.

Neotoma micropus

This large woodrat occurs from interstadial time to the present, though not found in large numbers. It may have been absent during full stadial conditions. Difficulties in discrimination from *N. albigula* and *N. floridana* have been described in those accounts.

Species A, undescribed

An interstadial population of relatively large woodrats with moderate development of the dentine tract is represented by 12 m1's as well as by additional material. The specimens clearly are separable from all other interstadial populations (Figs. 6 and 7). They do show a relationship in morphological characters to *N. mexicana* and *N. cinerea*. They differ significantly from these species as a population (Table 10), though some individual specimens of *N. cinerea* and *N. mexicana* cannot be discriminated. Although intermediate in several characters between those two species, the height of the dentine tract averages conspicuously lower than in either. A qualitative character, the presence of small accessory cusps at the base of fold 2 of m2 on three specimens, suggests closer relationship to *N. cinerea* than to *N. mexicana*.

Species B, undescribed

A woodrat from the interstadial deposits of Dry Cave (UTEP 1, 17, 26, 27) currently is being described.

Table 10.—*Comparison of UTEP 1 "cinerea-type" specimens (=Species A) (n = 10) with modern* N. cinerea *(n = 16) and* N. mexicana *(n = 42).* A = mean UTEP 1 (1 SD). B = mean N. cinerea *(1 SD).* C = t *between UTEP 1 and* N. cinerea. D = mean N. mexicana *(1 SD).* E = t *between UTEP 1 and* N. mexicana. *, *significant at* P ≤ 0.05; **, P ≤ 0.01; ***, P ≤ 0.001.

Character	A	B	C	D	E
LG-M1	3.29 (0.110)	3.52 (0.122)	4.81***	3.17 (0.145)	2.37*
WD-M1	1.814 (0.056)	1.954 (0.102)	3.95***	1.723 (0.063)	4.23***
TRACT	0.65 (0.158)	1.14 (0.242)	5.71***	1.32 (0.345)	5.97***
ANT-F2	2.50 (0.149)	2.68 (0.181)	2.56**	2.39 (0.144)	2.24*
F1-F2	0.92 (0.9)	1.03 (0.101)	2.82**	0.87 (0.077)	1.65
FOLD	0.40 (0.0)	0.32 (0.098)	2.60*	0.39 (0.034)	0.66
RATIO	0.487 (0.082)	0.481 (0.144)	0.11	0.546 (0.081)	2.07*

This taxon is significantly different from *N. lepida* ($P \leq 0.001$) in all but GROOVE and RATIO, and from *N. goldmani* in TRACT ($P \leq 0.001$), ANT-F2 ($P = 0.042$), and RATIO ($P = 0.003$). The most meaningful difference from *N. goldmani* probably is in the dentine tract height, which averages almost 0.2 mm higher in Species B (0.54 mm versus 0.35 mm). Three of four m3's have accessory folds preserved.

Species B presumably is most closely related to *N. goldmani* among all extant *Neotoma*. It is more than twice as common in the interstadial deposits than the next most common species, *N. albigula*.

GENERAL DISCUSSION

Interstadial, stadial, and early Holocene *Neotoma* faunas all differ radically from the modern condition, emphasizing the uniqueness of modern climate and biology.

The recognized interstadial faunas from the region are all from Dry Cave, in southeastern New Mexico. These faunas indicate conditions with greater effective moisture than today, though probably less than during stadial times (or differently distributed), an absence of cold winter temperatures, and rather warm summer temperatures (Harris, 1977; Van Devender et al., 1976; Harris and Crews, 1983).

The interstadial woodrat faunas include the mod-

ern forms *N. albigula* and *N. micropus*. These two species occur today in grassland and desert situations, usually with brush, cacti, or rocky areas available. They show local differences in habitat preference, but occur in close physical proximity. *N. micropus* becomes rare to the west now, apparently not reaching west as far as the New Mexico-Arizona border.

The other two species in the interstadial faunas appear to represent undescribed, extinct species. One, the Species A of this study, seems to be allied with *N. mexicana* or *N. cinerea*, more probably the latter. A reasonable scenario would be isolation in the southeastern New Mexican moderate to high elevations following an earlier stadial expansion of *N. cinerea*. This would allow a period of differentiation during isolation, perhaps even from early Wisconsin stadial conditions. Late Pleistocene re-expansion of *N. cinerea* could then cause extinction by competition or genetic swamping.

The other species (Species B) seems to have its affinity with the *N. goldmani-lepida* group. The rarity of *N. goldmani* in collections and the large geographic variation seen in modern nominal *N. lepida* make assessment of exact relationships difficult. Species B might well be ancestral to *N. goldmani* or both *N. goldmani* and *N. lepida*, or of course be a distinct taxon becoming extinct without leaving issue.

By mid-stadial times, the woodrat fauna had changed considerably. An *N. goldmani*-like form occurs at Dry Cave, a possible descendant of Species B (the occurrence of the *goldmani*-type woodrat at Anthony Cave, north of El Paso on the Texas-New Mexico border, is undated, but apparently full-stadial, at least in part). *N. albigula* remains are widespread, but *N. cinerea* of modern character now is the most commonly represented species in the cave faunas. As the complete faunas make clear (Harris, in press), some of the sites considered here were below the coniferous forest zone, lying in sagebrush-grasslands or steppe-woodlands. Notably cooler summers, cold winters (but probably not to the extremes of today), and more effective moisture allowed invasion of a vast area by the bushy-tailed woodrat, even possibly to southern Chihuahua.

Changing conditions near the end of the Pleistocene allowed invasion by *N. floridana* into the Dry Cave area, probably in the context of better grassland habitat. *Neotoma cinerea* still is present, however, as are other species of mammals now extirpated from the lower elevations of southern New Mexico (Harris, 1977). Jimenez Cave, in southern

Chihuahua, may show presence of *N. floridana,* but the age is not definitely known to be late stadial.

Neotoma micropus may have been present during full stadial times (UTEP 6, cf.; UTEP 4, ?), but is not reasonably surely present until latest Pleistocene (UTEP 54). It also occurred at Jimenez Cave. As with *N. floridana,* it may denote increasing summer temperatures and possibly increased emphasis on summer precipitation.

At the close of the Pleistocene, a major change occurs—*N. cinerea* abruptly disappears, replaced by *N. mexicana* (this assumes presence of *N. cinerea* at Pratt Cave is Pleistocene in age). With the exception of *N. mexicana,* the *Neotoma* fauna is modern. Midden evidence indicates the early Holocene maintained woodland in the lowlands of the Southwest until at least 8,000 B.P. (Van Devender and Spaulding, 1979). It seems evident that *N. mexicana* could replace *N. cinerea* under those conditions, either by out-competing it or by moving in as *N. cinerea* succumbed to other factors. *N. mexicana* appears equipped to survive under such conditions, hanging on even today in a few jumbled-rock areas far below its more common elevational and vegetational range (Findley et al., 1975).

Of the original list of species considered, only *N. stephensi,* now occurring from western New Mexico across central and northern Arizona, has not been identified. With the present study as a base, it can be hoped that examination of *Neotoma* from other sites in the western United States and Mexico can clarify the geographic and chronologic distribution of late Pleistocene *Neotoma.*

ACKNOWLEDGMENTS

I wish to thank T. Yates and W. Barber of the Museum of Southwestern Biology, The University of New Mexico; D. Hoffmeister of the Illinois Natural History Museum, The University of Illinois at Urbana-Champaign; and R. Fisher of the Fish and Wildlife Service Museum Section, National Museum of Natural History, for the loan of modern comparative material. Other comparative material was collected under permits from the New Mexico Department of Game and Fish. Many of the fossil specimens reported on here were collected under a grant from the National Geographic Society with the permission of the Bureau of Land Management. H. Messing allowed study of *Neotoma* from the Jimenez Cave fauna, currently under study by him.

LITERATURE CITED

ANDERSON, S. 1972. Mammals of Chihuahua, taxonomy and distribution. Bull. Amer. Mus. Nat. Hist., 148:149–410.

ARMSTRONG, D. M. 1972. Distribution of mammals in Colorado. Mus. Nat. Hist., Univ. Kansas Monogr., 3:1–415.

CARY, M. 1917. Life zone investigations in Wyoming. N. Amer. Fauna, 42:1–95.

DALQUEST, W. W., E. ROTH, and F. JUDD. 1969. The mammal fauna of Schulze Cave, Edwards County, Texas. Bull. Florida State Mus., Biol. Sci., 13:205–276.

FINDLEY, J. S., A. H. HARRIS, D. E. WILSON, and C. JONES. 1975. Mammals of New Mexico. Univ. New Mexico Press, Albuquerque, 360 pp.

HARRIS, A. H. 1977. Wisconsin age environments in the northern Chihuahuan Desert: Evidence from the higher vertebrates. Pp. 23–52, *in* Transactions of the symposium on the biological resources of the Chihuahuan Desert Region, United States and Mexico (R. H. Wauer and D. H. Riskind, eds.), Natl. Park Serv. Trans. Proc. Ser., 3:1–658.

———. 1980. The paleoecology of Dry Cave, New Mexico. Natl. Geogr. Soc., Res. Rep., 12:331–338.

———. In press. Late Pleistocene paleoecology of the West. Univ. Texas Press, Austin.

HARRIS, A. H., and C. R. CREWS. 1983. Conkling's roadrunner—a subspecies of the California roadrunner? Southwestern Nat., 28:407–412.

HARRIS, A. H., and J. S. FINDLEY. 1964. Pleistocene-Recent fauna of the Isleta Caves, Bernalillo County, New Mexico. Amer. J. Sci., 262:114–120.

LUNDELIUS, E. L., JR. 1979. Post-Pleistocene mammals from Pratt Cave and their environmental significance. Pp. 239–258, *in* Biological investigations in the Guadalupe Mountains National Park, Texas (H. H. Genoways and R. J. Baker, eds.), Natl. Park Serv. Proc. Trans. Ser., 4:1–442.

MURRAY, K. F. 1957. Pleistocene climate and the faunal record of Burnet Cave, New Mexico. Ecology, 38:129–132.

VAN DEVENDER, T. R., and W. G. SPAULDING. 1979. Development of vegetation and climate in the southwestern United States. Science, 204:701–710.

VAN DEVENDER, T. R., and F. M. WISEMAN. 1977. A preliminary chronology of bioenvironmental changes during the paleoindian period in the monsoonal Southwest. Pp. 13–27, *in* Paleoindian lifeways (E. Johnson, ed.), West Texas Mus. Assoc., The Mus. J., 17:1–197.

VAN DEVENDER, T. R., K. B. MOODIE, and A. H. HARRIS. 1976. The desert tortoise (*Gopherus agassizi*) in the Pleistocene of the northern Chihuahuan Desert. Herpetologica, 32:298–304.

Address: Laboratory for Environmental Biology, University of Texas at El Paso, El Paso, Texas 79968.

THE EVOLUTION OF COTTON RAT BODY MASS

Robert A. Martin

ABSTRACT

Body mass of fossil *Sigmodon* is estimated by an equation generated from a relationship between length of the first lower molar and body mass in extant cricetines. Replacement sequences of fossil cotton rats in four limited geographical regions of the continental United States manifest the Cope-Depéret Rule. Explanatory hypotheses are briefly examined.

INTRODUCTION

Body mass (or weight) of mammals is the key to estimating other meaningful biological parameters in extinct species. Mass is highly correlated in living species with metabolic rate (Kleiber, 1961; Martin, 1980), home range (McNab, 1963; Harestad and Bunnell, 1979), population density (Martin, 1981; Damuth, 1981), and other variables. As I showed in an earlier paper (Martin, 1980), certain morphometric variables are tightly correlated with body mass in mammals and therefore can be mathematically manipulated to estimate any other variables correlated with mass. This simple procedure will allow paleobiologists to test new hypotheses dealing with mammalian extinction, morphological evolution, and the evolution of faunal energetics.

The purpose of this essay is to provide some baseline information for cotton rats, genus *Sigmodon,* and to speculate on the evolution of body size in this genus. This paper is part of a larger project which will examine the overall evolution of cotton rat energetics. In an earlier treatment (Martin, 1980) I had used two morphometric variables, the greatest width across the occipital condyles and the greatest width of the femoral head, to estimate body mass. As skulls are rare paleontologic finds, and femora not always available or readily identifiable, another measurement, length of the first lower molar (M_1) was used in this study. First lower molars are common in deposits containing *Sigmodon,* and are also taxonomically useful (Martin, 1979). As will be demonstrated, the length of this tooth is highly correlated with body mass in *Sigmodon* and other cricetine rodents.

MATERIALS AND METHODS

Length of the M_1 (first lower molar) was measured with an American Optical (AO) filar micrometer eyepiece coupled to a Spencer binocular dissecting microscope. The eyepiece was calibrated with an AO two millimeter micrometer slide. Measurements were taken from 33 individuals representing six cricetine species. Mass (W) data were taken from skins associated with each mandible. Species, number of individuals, range of M_1 length, and range of observed W are as follows: *Reithrodontomys humulis* (1; 1.27 mm; 7.1 g), *Peromyscus floridanus* (2; 1.87–2.02 mm; 40.3–43.6 g), *P. gossypinus* (3; 1.58–1.72 mm; 17.4–19.5 g), *P. polionotus* (6; 1.30–1.39 mm; 9.1–18.3 g), *P. leucopus* (10; 1.38–1.70 mm; 11.6–24.0 g), *Sigmodon hispidus* from the Florida peninsula (11; 2.44–2.72 mm; 79.8–120.6 g). A least squares straight line was fitted to the logarithm of W plotted against the log of M_1 length (Fig. 1). Body mass data were then calculated for extinct and extant cotton rat species based on the equation that was generated. With the exception of the *S. hispidus* sample noted above and a sample of *Sigmodon medius* from the Blanco fauna, these calculations are made from measurements published by Martin (1979).

RESULTS AND DISCUSSION

The relationship between M_1 length and body mass in cricetines is illustrated by the least squares line in Fig. 1. The equation for this line, converted to its exponential form, is

$$W = 4.05L^{3.33} \qquad (1)$$

where L = length of the first lower molar in millimeters and mass is in grams. The two variables are highly correlated (r = 0.98) and the 95% confidence limits of β are calculated as 3.33 ± 0.35. Ninety-five percent confidence limits of antilog $\hat{y}$ are 9.0 ± 1.1 (at minimum x), 30.0 ± 1.1 (at $\bar{x}$), and 111.4 ± 1.1 (at maximum x). The confidence interval contracts from 12.5 percent of $\hat{y}$ at minimum x to one

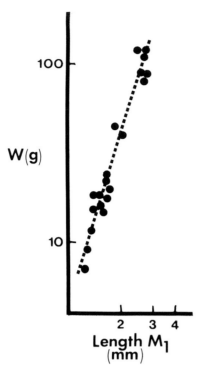

Fig. 1.—Least squares regression line running through scatter diagram of body mass plotted against length of the first lower molar in extant cricetines.

Table 1.—*Body mass of extinct and extant species of* Sigmodon. *Mass data estimated by equation (1) in text.*

Taxon	N	Mean length of first lower molar (mm)	Mean body mass (g)	Observed range of body mass (g)
Extinct				
Sigmodon medius[1]	90	2.06	45.0	28.7–63.9
Sigmodon minor Curtis Ranch, Borchers)	47	1.90	34.4	24.7–55.1
Sigmodon curtisi[2]	19	2.38	72.8	52.6–84.5
Sigmodon libitinus (Haile XVIA)	26	2.00	51.0	28.7–77.9
Sigmodon hudspethensis (Red Light, Hudspeth)	4	1.83	68.7	55.1–95.1
Sigmodon bakeri (Coleman IIA, Williston IIIA)	23	1.84	69.0	43.5–98.8
Extant				
Sigmodon hispidus (Florida Recent)	11	2.63	101.4	78.8–113.4
Sigmodon hispidus (Reddick IA)	18	2.49	84.5	59.4–113.4
Sigmodon hispidus[3]	8	2.35	69.7	54.3–84.5
Sigmodon ochrognathus	5	2.26	61.2	47.9–84.5
Sigmodon fulviventer	8	2.27	62.1	55.9–73.7
Sigmodon alleni	5	2.41	75.8	62.1–93.9
Sigmodon mascotensis	3	2.22	57.7	54.3–60.3
Sigmodon arizonae	4	2.45	80.1	76.8–83.4
Sigmodon leucotis	9	2.54	90.3	79.0–117.6
Sigmodon peruanus	9	2.68	107.9	65.8–162.4

[1] Mean of means from the following localities: Benson, Tusker, Rexroad Loc. 3, Sanders, Sand Draw, Haile XVA, Wendell Fox Pasture, Blanco.
[2] Curtis Ranch, Kentuck, Inglis IA combined.
[3] *S. h. berlandieri*; a small extant subspecies from Texas and Mexico.

percent when W > 100 g. Predicted values of body mass for *Sigmodon,* in the upper levels of mass, average less than 10% (usually closer to 1%) different from values observed on specimen tags. The average mass of *S. hispidus* predicted by equation (1) from the first lower molars measured in this study was 101.4 g. The observed mean value was 101.6 g.

Body mass data for extinct and extant *Sigmodon* are presented in Table 1 and Fig. 2. Within all of the *Sigmodon* replacement chronologies (*not* phylogenies) illustrated in Fig. 2, there is a general trend towards increased body size. Reversals are seen in the *S. medius* to *Sigmodon minor* lineage in Kansas and Arizona, and from the Inglis IA *Sigmodon curtisi* to Haile XVIA *Sigmodon libitinus* in Florida, but these smaller animals are subsequently replaced by larger species.

Based upon mass data generated by equation (1), living *Sigmodon* species average around 78 g (range of averages; 58–108 g). This level of average body size was attained during the early Irvingtonian land mammal age (approximately 1.8 million years ago) in the southwest and Great Plains regions by *Sigmodon hudspethensis* and *S. curtisi.* Nevertheless,

for those areas from which we have a replacement chronology, the trend towards increased size continued; albeit with the reversals noted above. I have previously suggested (Martin, 1979) that the smaller average size of *S. minor* relative to *S. medius* is the result of ecological separation (character displacement) away from the larger, more progressive *S. curtisi* with which it was contemporaneous.

Why is it that the fossil history of *Sigmodon,* like so many other mammalian phylogenies, demonstrates evolution towards increased size? This tendency is often referred to as Cope's Rule, although Stanley (1973) indicates that it was first explicitly stated by Depéret (1909). Two hypotheses have been presented. The first, which we may label as the *fundamental advantage* hypothesis, suggests that, all other things being equal, it is more beneficial to be large than small. Some of these advantages are listed

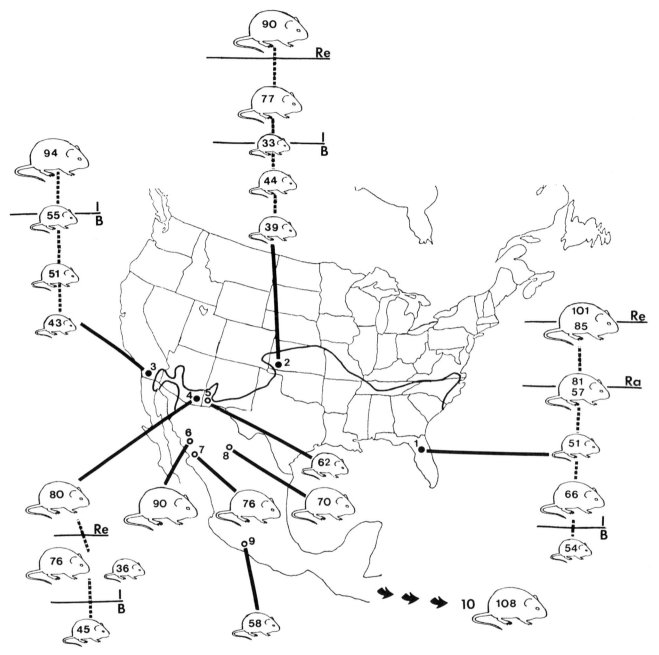

Fig. 2.—Replacement chronologies of cotton rats, comparing the evolution of body mass in *Sigmodon* with the distribution of body mass seen in modern species. The horizontal lines within each chronology separate land mammal ages. B = Blancan, I = Irvingtonian, Ra = Rancholabrean, Re = Recent. Solid circles represent fossil deposits; open circles are areas from which samples of extant species only were captured. The solid black line on the United States map depicts the present limits of cotton rat distribution. Numbers within illustrations are average mass values for the samples. At each locality, replacement sequences are listed from oldest to youngest sample. 1 = northcentral Florida: *S. medius* (Haile XVA)—*S. curtisi* (Inglis IA)—*S. libitinus* (Haile XVIA)—*S. bakeri* (2 samples; Coleman IIA and Williston IIIA)—*S. hispidus* (2 samples; Reddick IA and Recent). 2 = Meade Basin: *S. medius* (Rexroad Loc. 3)—*S. medius* (Sanders)—*S. minor* (Borchers)—cf. *S. curtisi* (Kentuck)—*S. hispidus* (Recent). 3 = Vallecito-Fish Creek Beds: *S. medius* (Layer Cake)— *S. medius* (Arroyo Seco)—*S. medius* (transition zone between Arroyo Seco and Vallecito Creek faunas)—cf. *S. curtisi* (Vallecito Creek). 4 = San Pedro Valley: *S. medius* (Benson and Tusker combined)—*S. minor* (36 g; Curtis Ranch) and *S. curtisi* (76 g; Curtis Ranch)— *S. arizonae* (Recent). 5 = *S. fulviventer.* 6 = *S. leucotis.* 7 = *S. alleni.* 8 = *S. hispidus berlandieri.* 9 = *S. mascotensis.* 10 = *S. peruanus,* a large South American species. Not included on this illustration is *S. (Sigmomys) alstoni,* an average-sized South American *Sigmodon* with grooved upper incisors.

by Stanley (1973), and include ". . . improved ability to capture prey or ward off predators, greater reproductive success, increased intelligence (with increased brain size), better stamina, expanded range of acceptable food, decreased annual mortality, extended individual longevity, and increased heat retention per unit volume." The second hypothesis, proposed by Stanley (1973), can be for convenience referred to as the *structural limitation* hypothesis. It is a probabilistic model which can be summarized as follows: (1) large species of a given body plan are more specialized than are small ones, (2) large scale adaptive breakthroughs occur at small body size, and (3) the tendency of taxa in a lineage to evolve larger size represents movement towards an optimum size for the adaptive zone being filled, not evolution away from small body size.

According to this hypothesis, size increase is not innately favored as a particular adaptive strategy, but simply represents the normal inertial sequence during diversification. By this principle, the diminutive *Sigmodon minor* and *S. libitinus* are no more special than the larger species which appeared before and after them (unless, as small species, they were ancestral to larger, more advanced ones!). Van Valen (1975, 1978) implies that the structural limitation hypothesis should pertain only to the start of a trend towards large size, not to its maintenance. That is, why does a trend towards large size continue in a phylogeny after the initial stages of diversification? If body size is not being regulated, should not there be an equal tendency towards dwarfing or gigantism in a given lineage following an initial explosion of taxa? I am not certain that this follows. It would seem to depend on how close a species is to an optimum size for its new adaptive zone. Stanley (1973) does not deny that selection may continue to favor large size or that large size is not adaptive (for one or more of the reasons listed as fundamental advantages); merely that dwarfing is not nearly as likely as size increase because of the structural constraints associated with largeness.

The late Pliocene to Recent evolution of cotton rats, spanning a time period of about four million years, is clear in one regard—each local replacement chronology moves inexorably towards animals of increased average body mass (Fig. 3). Further, dwarfing appears to occur only just prior to and during invasion and replacement by larger species occupying similar adaptive zones. I do not know if this is a general trend for other mammalian groups, but it should be examined. The results of this study

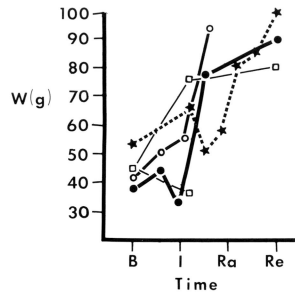

Fig. 3.—Change of *Sigmodon* body mass through late Pliocene and Pleistocene time. W = mass, B = Blancan, I = Irvingtonian, Ra = Rancholabrean, Re = Recent. Solid circles = Meade Basin, Open circles = Vallecito-Fish Creek Beds, solid stars = north-central Florida, open squares = San Pedro Valley. See Fig. 2 and text for further detail.

generally support Stanley's (1973) structural limitation hypothesis. A single species, *Sigmodon medius,* exists unchanged for almost two million years in pastoral habitats on the central Great Plains and southern United States from coast to coast. Towards the end of Blancan time it undergoes a relatively rapid dwarfing throughout most of its range. By the end of the Blancan it is sufficiently diminished in size to be recognized as other species, *S. minor.* At least in Arizona, it is seen briefly cohabiting with the advanced *Sigmodon curtisi.* From late Blancan through Irvingtonian time two or three *Sigmodon* species are recognizable during a given temporal interval. Four species—*S. hispidus, S. fulviventer, S. ochrognathus,* and *S. arizonae*—are recognized by neomammalogists as extant in these areas today. It would be very difficult to separate dentitions of these species if interred in the same accumulation, and I suspect that only two or three would be accepted as valid paleospecies. One could argue that, having made a behavioral-anatomical breakthrough in changing from a granivorous, browsing mode of foraging to grazing, a trend of evolution towards larger size then began, continuing into present time. This increase in body size was complemented by an increase in dental hypsodonty and the nature by which the tooth crown is supported (see Martin, 1979, for details). It is likely that an optimum size for *Sigmodon* has yet to be reached. Some of the dental specializations seen as variants in *S. hispidus* (such as lamination and root capture) suggest to me that this group is moving structurally in a direction reminiscent of the large Central and South American hystricomorphs. While Stanley's structural limitation hypothesis may seem to be a tautology (that is, today's specialization may be tomorrow's generalization), it can be supported by available data.

LITERATURE CITED

DAMUTH, J. 1981. Population density and body size in mammals. Nature, 290:699–700.

DEPÉRET, C. 1909. The transformations of the animal world. D. Appleton and Co., New York, 360 pp.

HARESTAD, A. S., and F. L. BUNNELL. 1979. Home range and body weight—a reevaluation. Ecology, 60:389–402.

KLEIBER, M. 1961. The fire of life. Wiley, New York, 454 pp.

MARTIN, R. A. 1979. Fossil history of the rodent genus *Sigmodon*. Evol. Monogr. 2:1–36.

———. 1980. Body mass and basal metabolism of extinct mammals. Comp. Biochem. Physiol., 66A:307–314.

——— 1981. On extinct hominid population densities. J. Human Evol., 10:427–428.

McNAB, B. K. 1963. Bioenergetics and the determination of home range size. Amer. Nat., 97:133–140.

STANLEY, S. M. 1973. An explanation for Cope's Rule, Evolution, 27:1–26.

VAN VALEN, L. 1975. Group selection, sex, and fossils. Evolution, 29:87–94.

———. 1978. [Review of] Patterns of evolution, illustrated by the fossil record, edited by A. Hallam. Paleobiology, 4:210–216.

Address: Department of Biological and Allied Health Sciences, Fairleigh Dickinson University, Madison, New Jersey 07940.

BIOSTRATIGRAPHY AND BIOGEOGRAPHY OF QUATERNARY MICROTINE RODENTS FROM NORTHERN YUKON TERRITORY, EASTERN BERINGIA

RICHARD E. MORLAN

ABSTRACT

Recent work in northern Yukon Territory has produced samples of microtine rodent fossils from well documented stratigraphic contexts spanning much of late Pleistocene and Holocene time. *Clethrionomys, Microtus, Lemmus,* and forms of the Dicrostonychini are present throughout the sequence which is also punctuated by two appearances of *Phenacomys.* Evolutionary changes can be recognized within the *Dicrostonyx* lineage, the fossils of which are analyzed in terms of morphotypes defined by Agadjanian and von Koenigswald (1977). Vole fossils bearing similarities to *Microtus paroperarius* are labelled "*Microtus* sp.

X" and are separated from both the former and the morphologically similar tundra vole, *M. oeconomus,* by means of measurements proposed by van der Meulen (1973, 1978).

The Yukon sequence is arranged in terms of intervals defined by Hopkins (1982), and possible correlations with Alaskan and Siberian sediments and faunas are discussed. The Beringian microtine record is seen to be woefully incomplete, poorly dated, and comprised of samples that are often very small. Nonetheless, some tentative proposals are put forth concerning biostratigraphy and biogeography, and gaps in the record are identified.

INTRODUCTION

The pivotal role of Beringia in the formation of Holarctic biota is well known (see Hopkins et al., 1982, for a recent overview). Not only has this region been an important pathway between the Palearctic and Nearctic, but also it may have been a center of evolution for many taxa. Of practical importance to paleontologists is the fact that much of Beringia, from the Kolyma Lowland of Siberia to the Mackenzie valley of northwestern Canada (Fig. 1), was not glaciated and therefore offers opportunities to study the evolution of faunal and floral communities over long periods. The non-glacial and periglacial environments of Beringia served as refugia for both plants and animals during glacials and were sources for the dispersal of life forms following deglaciation (Hughes et al., 1981:330).

For several reasons, however, primary vertebrate paleontological data have accumulated slowly and have seldom been adequate for defining long sequences of faunas in datable stratigraphic contexts. The permafrost that confers excellent preservational environments in much of Beringia also inhibits access to Pleistocene sediments. Many areas are remote, accessible only by boat or helicopter, and are logistically difficult or very expensive to study. In central Alaska and Yukon, most vertebrate remains have been recovered from perennially frozen colluvial silts ("mucks") exposed in placer mines where good stratigraphic controls are difficult to maintain and where concentrations of microfaunal remains are often low (for example, Guthrie, 1968:Fig. 5).

Only recently have fossil remains been recovered from other kinds of commercial excavations such as roadcuts (for example, Weber et al., 1981).

In northern interior areas of Siberia, Alaska, and Yukon, many streams have exposed fossiliferous sediments along their banks, and these have begun to produce small samples of vertebrate remains in primary stratigraphic contexts. One of the most productive streams is Old Crow River, a northern Yukon tributary of Porcupine River, where large concentrations of vertebrate fossils on the modern sand and gravel bars reflect the presence of abundant fossils yet to be eroded or excavated from the river banks (Harington, 1977, 1978:62–63; Morlan, 1980a, 1980b; Jopling et al., 1981:table 2). Work in this area has only begun to reveal the potential for such studies. The samples are too small, too poorly dated, and too few in number to support a thorough paleoecological analysis, but they are adequate to define taxonomic, biostratigraphic and biogeographic problems and to suggest a few solutions to some of the problems.

In addition to research in the lowlands, excavations at an upland site, the Bluefish Caves, have produced large samples of late Pleistocene and Holocene vertebrate remains (Cinq-Mars, 1979, 1982; Morlan and Cinq-Mars, 1982; Morlan, 1983a). These samples will be mentioned briefly in order to represent the critical millennia at the close of the Pleistocene, and their paleoecological implications will be explored elsewhere.

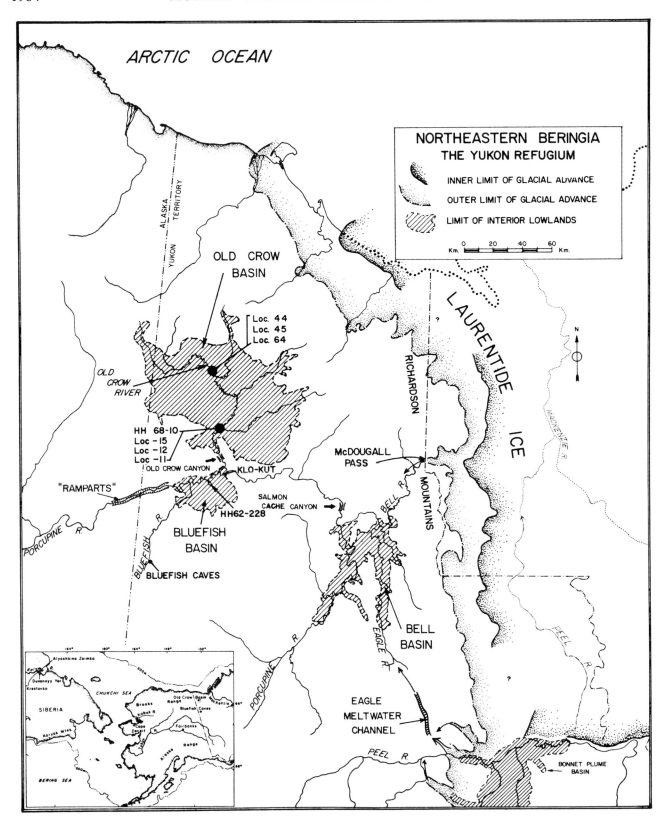

Fig. 1.—Map of northern Yukon Territory showing localities mentioned in the text. See inset for other Beringian localities.

Table 1.—*Summary of taphonomic variables and references to associated fossils for microtine rodent samples from the northern Yukon Territory. Letters at left margin refer to Fig. 2.*

	Samples	Enclosing sediment	Redeposition	Recovery techniques	References to associated fossils
Recent					
K.	Klo-kut	Overbank alluvium	Nil	Trowelling	Morlan, 1973
J.	Bluefish Caves, Holocene	Organic-rich rubble	Nil	Trowelling, dry and wet screening	Cinq-Mars, 1979, 1982; Morlan and Cinq-Mars, 1982; Ritchie et al., 1982
Duvanny Yar Interval					
I.	Bluefish Caves, Pleistocene	Loess	Nil	Trowelling, dry and wet screening	Cinq-Mars, 1979, 1982; Morlan and Cinq-Mars, 1982; Ritchie et al., 1982
Boutellier Interval					
H.	HH62-228, "Unit 2b"	Lacustrine	Unknown	Unknown	Delorme, 1968; Harington, 1977, 1978; Lichti-Federovich, 1974; Matthews, 1975; McAllister and Harington, 1969
G.	HH68-10, Unit 2b	Loess(?)	Nil	Bulk removal, handpicked	(None)
F.	Loc. 12, Disconformity A	Overbank alluvium	Significant	Trowelling, wet screen (1.6 mm, 50 liters)	Jopling et al., 1981 (Fauna 7); Morlan, 1981; this report
E.	Loc. 15, Disconformity A	Overbank alluvium	Significant	Trowelling, wet screen (1.6 mm, 560 liters)	Bobrowsky, 1982; Cumbaa et al., 1981; Janssens, 1981; Lichti-Federovich, 1973; Morlan, 1980a, 1980b; Morlan and Matthews, 1983
pre-Happy Interval					
D.	Loc. 15, Unit 2a	Point bar(?) alluvium	Significant	Trowelling	Cumbaa et al., 1981; Morlan, 1980a, 1980b; this report
C.	Loc. 11, Unit 2a	Point bar alluvium	Significant	Trowelling, wet screen (1.6 mm, 300 liters)	Jopling et al., 1981 (Fauna 6); this report
B.	Loc. 12, Unit 2a	Point bar(?) alluvium	Significant	Trowelling, wet screening	Jopling et al., 1981 (Fauna 2)
A.	Locs. 44, 45, 64, Unit 2a	Channel-bottom alluvium	Significant	Trowelling, hosing, wet screening	Harington, 1977, 1978; Lichti-Federovich, 1973; Matthews, 1975

This paper will focus primarily on the Old Crow Basin of northern Yukon Territory and will present new data on microtine rodents from several stratigraphic and taphonomic contexts. Many of the northern Yukon localities have produced wide varieties of fossils in addition to the microtine rodents (see Table 1 for references). The multi-disciplinary research needed to integrate the northern Yukon fossil record has already produced a substantial volume of published material (summaries in Hughes et al., 1981:331; Jopling et al., 1981), and additional reports are in preparation.

METHODS

Recovery techniques are listed in Table 1 with mesh size of screen and volume of sample shown in parentheses when known.

At the Bluefish Caves, Cinq-Mars (1979:5) recovered microtine rodents by trowelling and dry screening all matrix on a 3 mm mesh (ca. 4 mm diagonal sieve openings) and by wet screening (0.18 mm sieve openings) ca. 5 kg of residue that passed through the coarse screen from each excavation level; many other specimens will be recovered in the laboratory from bulk samples that have not yet been analyzed.

Identifications were made by means of direct comparisons with reference specimens of extant taxa and by means of literature searches for extinct taxa. Drawings of many molars were made at approximately 15× with a drawing tube on a Wild-Leitz M5 microscope. The drawing tube facilitated reconstruction of some fragmentary teeth.

Locality designations on Old Crow River were assigned by Harington (1977) except for HH68-10, studied by O. L. Hughes who also designated HH62-228 on Porcupine River. Specimens from Locs. 44, 45, 64 and HH62-228 are kept in the Paleobiology Division, National Museum of Natural Sciences, Ottawa, and are catalogued with five-digit numbers following the prefix "NMC-." Specimens reported by Jopling et al. (1981) are currently housed in the Department of Anthropology, University of Toronto, and have not been examined for the purposes of this report. All others are kept in the Archaeological Survey of Canada, National Museum of Man, Ottawa, under catalogue designations based on the Borden system. The catalogue numbers appear in parentheses in the following list: Loc. 11 (MkVl-9); Loc. 12 (MkVl-10); Loc. 15 (MlVl-2); HH68-10 (MlVl-8); Bluefish Caves (MgVo-1, MgVo-2); Klo-kut (MjVl-1).

GROSS STRATIGRAPHY AND CHRONOLOGY

Microtine rodent samples have been recovered from seven localities in Old Crow Basin, two localities in Bluefish Basin, and two caves in the uplands immediately south of Bluefish Basin (Fig. 1). None of these areas was glaciated during the Pleistocene, but all were indirectly affected by glaciers that reached the eastern flanks of Richardson Mountains. During late Wisconsinan time, meltwater from Laurentide ice that occupied Bonnet Plume Basin and McDougall Pass created large lakes in Old Crow, Bluefish, and Bell basins (Hughes et al., 1981). As the lakes drained, silt from the newly exposed lake bottom in Bluefish Basin supplied loess that accumulated in the Bluefish Caves to the south (Cinq-Mars, 1979, 1982). The drainage of the lakes took place through a canyon newly cut by Porcupine River through bedrock near the Alaska/Yukon border (Hughes, 1972; Thorson and Dixon, 1983), and the lower base level created by downcutting enabled Porcupine River and its tributary, Old Crow River, to dissect thick sequences of Quaternary sediment in Bluefish and Old Crow basins, respectively.

Exposures along Old Crow River can be characterized stratigraphically in terms of four gross units (see Hughes, 1972; Morlan and Matthews, 1978; Morlan, 1979, 1980a, 1983b): Unit 1, a lacustrine silty clay at the base, largely concealed by the river; Unit 2, approximately 18–20 m of sand, silt, and clay representing an interlacustrine sequence consisting primarily of alluvium, subdivided into Units 2a and 2b at some sections; Unit 3, approximately 5 m of glaciolacustrine clay; and Unit 4, silt and peat, usually about 1–3 m thick, that forms the modern surface (Fig. 2). Some of these units can be related to a series of intervals defined by Hopkins (1982) as a framework for integrating Beringian stratigraphy, chronology, and paleogeography. For example, the glaciolacustrine clay of Unit 3 represents the Duvanny Yar Interval and can be correlated with various regional late Wisconsinan glacial advances in Beringia as well as with oxygen isotope stage 2 (Hopkins, 1982:fig. 2; Shackleton and Opdyke, 1973). A radiocarbon date of 25,170 ± 630 B.P. (NMC-1232) on a proboscidean tusk only 50 cm below the base of the clay in Bluefish Basin provides a maximum age for glacial lake formation in both basins. A minimum age for drainage of the lakes is provided by radiocarbon dates of approximately 12,000 B.P. on *Bison crassicornis* bones recovered from a channel incised into the top of the lake clay in Old Crow Basin (Harington, 1977:827–834, table 5, 1978:55; Morlan, 1980a:fig. 9.3).

As the glacial lakes drained, the newly exposed lake bottom in Bluefish Basin supplied loess that buried bones and artifacts in the Bluefish Caves, but dates as early as 18,000 years ago at the latter site (Cinq-Mars, personal communication, 1982) suggest that the level of the glacial meltwater lake was lowered considerably or episodically during the last millennia of its existence.

Where Unit 2 can be subdivided, the boundary between units 2a and 2b is marked by an erosional contact informally known as Disconformity A. Radiocarbon dates on Unit 2b range from >51,000 B.P. (GSC-2559-2) on an autochthonous peat near the base of the unit in Old Crow Basin to 25,170 B.P. (on the proboscidean tusk mentioned above). The erosional contact was created after an episode of climatic warming that thawed ice wedges and replaced them with sediment-filled pseudomorphs that are truncated by Disconformity A. The disconformity and Unit 2b together represent middle Wis-

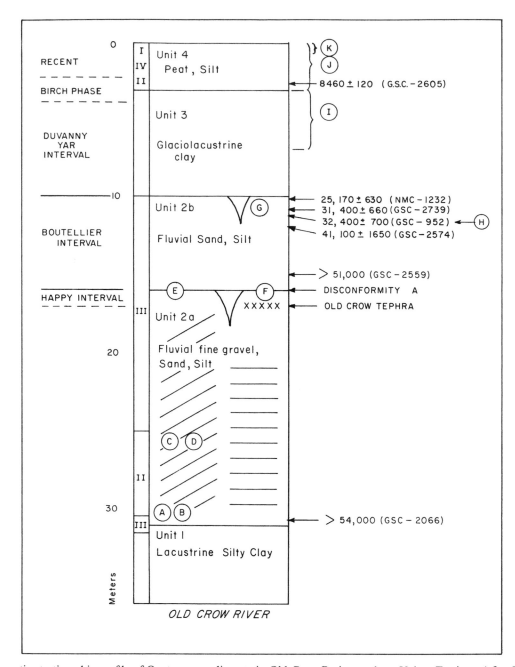

Fig. 2.—Schematic stratigraphic profile of Quaternary sediments in Old Crow Basin, northern Yukon Territory (after Hughes, 1972; Morlan and Matthews, 1978; Morlan, 1983b). Circled letters (A–G) mark approximate positions of microtine rodent samples from Old Crow Basin. Letters outside the column (H–K) refer to samples from Bluefish Basin and Bluefish Caves that are correlated with Old Crow Basin on the basis of radiocarbon dates. Roman numerals at left edge of column refer to pollen assemblage types defined by Lichti-Federovich (1973). Lines in Unit 2a represent bedding planes, and "xxx" marks the position of Old Crow tephra. V-shaped marks at the upper contacts of Units 2a and 2b represent ice wedge pseudomorphs. Named intervals are after Hopkins (1982).

consinan time, the Boutellier Interval of Hopkins (1982), approximately equivalent to oxygen isotope stage 3.

Presumably the cold Happy Interval, to which

many regional Beringian glacial advances are assigned (Hopkins 1982:fig. 2), is represented in Old Crow Basin by the growth of ice wedges mentioned above. The Happy Interval is the oldest named interval in Hopkins' framework and is approximately equivalent to oxygen isotope stage 4. The lower boundary of the interval is not yet defined, but it includes Old Crow tephra (Hopkins, 1982:6). At five sections in Old Crow Basin, Old Crow tephra has been found just below Disconformity A, and fission track analysis indicates that this tephra is ≤120,000 years old (Naeser et al., 1982; Westgate et al., 1983). Old Crow tephra has also been found in southern Yukon and at several Alaskan localities (Westgate, 1982), and Schweger and Matthews (n.d.) have collated paleoenvironmental and chronometric evidence from several of these sites to narrow the possible time of deposition of the tephra to between 87,000 and 105,000 years B.P. Although this range does not identify the "instant" of deposition, it suggests that Old Crow tephra was deposited sometime during oxygen isotope stage 5 (Shackleton and Opdyke, 1973:table 3).

Ten or more meters of Unit 2a alluvium underlie Old Crow tephra at each of the sections where the tephra has been observed, but variable amounts of time may be represented by such sediment thickness. Some sections expose the dipping beds of point bars that could have accumulated rapidly, but in other exposures all the bedding is finely laminated and flat-lying and could represent a long time period. Some of the fossils, such as spotted skunk, from the lower part of Unit 2a suggest an age within the Sangamon Interglaciation (Harington, 1978:62–63) equivalent to some part of isotope stage 5e and therefore older by definition than the Happy Interval. Even greater age is implied by four uranium series dates (160,000 years ago or more) on bones from approximately the middle of Unit 2a (Morlan, 1983b:table 4), and some of the microtine rodent remains to be described below are suggestive of a longer time scale than has usually been contemplated for the sequence exposed along Old Crow River.

Obviously, no precise age can be given for the Unit 1 clay at the base of the Old Crow sections. A thick lacustrine unit in Bluefish Basin is reversely magnetized and "probably lies within the Matuyama Epoch or an earlier epoch" (Pearce et al., 1982:926). Hughes (1972) correlated this unit with Unit 1 in Old Crow Basin, believing both to be glaciolacustrine in origin. Subsequent discovery of high percentages of pine pollen in the Bluefish Basin unit (Lichti-Federovich, 1974) cast doubt on that mode of origin, but no additional data are available either to deny or to confirm the correlation with Old Crow Basin Unit 1. Either a paleomagnetic study of the Unit 1 clay or a date on the recently discovered Little Timber tephra (just upsection from the clay) could provide a critical limit on the chronological framework for Old Crow Basin (Schweger and Matthews, n.d.).

Detailed analysis of stratigraphy in the northern Yukon basins is now in preparation, but the foregoing outline is adequate for the purposes of this paper. The gross stratigraphic context of each microtine rodent sample from Old Crow Basin is shown by a letter (A–G) on the composite sketch in Fig. 2, and samples from Bluefish Basin and Bluefish Caves (H–K) are positioned outside the stratigraphic column on the basis of radiocarbon dating. Taphonomic variables and references to associated fossils are summarized in Table 1. It is noteworthy that there are no samples to represent the Happy Interval, because the fossils associated with Disconformity A are actually contained in the alluvium that covers the erosional contact and therefore belong to the Boutellier Interval. Because the glaciolacustrine clay of Unit 3 is devoid of fossils, only upland sites, such as Bluefish Caves, can provide a record for the Duvanny Yar Interval. The Klo-kut archaeological site is included here for two reasons: (1) as a lowland counterpart to the Holocene sample from Bluefish Caves; and (2) to correct misidentifications published 10 years ago (for example, *Dicrostonyx* did not occur there).

SYSTEMATICS AND MORPHOLOGY

Of the arvicoline ("microtine") rodents known from northern Yukon, the voles and lemmings are described here, and muskrat (*Ondatra zibethicus*) specimens will be presented elsewhere. Systematic study of the northern Yukon fossils is essentially complete for most taxa but is in a preliminary stage for some. Those requiring further study are fossils of *Microtus* and *Dicrostonyx* that resemble extinct species described from other areas. For these, the analysis has been based entirely on published literature rather than on comparisons with original fossil materials. Hence this is an interim report in which current knowledge is summarized, and a more detailed study is planned.

Table 2.—*Summary of microtine rodent remains from the northern Yukon Territory.*

Sample	Clethrionomys cf. rutilus	Phenacomys cf. intermedius	Microtus sp. X	M. xanthognathus	M. miurus	M. miurus or pennsylvanicus	M. pennsylvanicus	M. oeconomus	Microtus spp.	Lemmus sibiricus	cf. Dicrostonyx	D. cf. simplicior	D. gulielmitorquatus	D. torquatus	Total NISP
Recent															
K. Klo-kut				18	1*		5	10	4	1					39
J. Bluefish Caves, Holocene	164			478					325	119					1,086
Duvanny Yar Interval															
I. Bluefish Caves, Pleistocene	160			322	223			12	513	346				329	1,905
Boutellier Interval															
H. HH62-228, "Unit 2b"[1]										+					
G. HH68-10, Unit 2b														41	41
F. Loc. 12, Disconformity A[2]		+		+	+		+			+			?		
F. Loc. 12, Disconformity A					3		2		3	15			4		27
E. Loc. 15, Disconformity A	7	1		6	7	6	5		20	102			33		187
pre-Happy Interval															
(Unit 2a, middle)															
C. Loc. 11, Unit 2a	16		22	39	5*	7	11*		85	194		42			421
C. Loc. 11, Unit 2a[2]	+	+		+*	+			?		+	+				
D. Loc. 15, Unit 2a										4	2				6
(Unit 2a, lower)															
B. Loc. 12, Unit 2a[2]				+	+			?		+	+				
A. Loc. 44, Unit 2a	1		6	2	4*	1	2*			29	17				62
A. Loc. 45, Unit 2a							1*			1	1				3
A. Loc. 64, Unit 2a			8	8						9	8				33

[1] Harington (1977:121–126); [2] Jopling et al. (1981:Table 2); ? Identification questioned in this report; + Present; * Tentative species identification (cf.); letters at left margin refer to Fig. 2. NISP, Number of Identified Specimens.

The distribution of northern Yukon microtine rodent taxa is summarized in Table 2. It is believed that *Clethrionomys* and *Phenacomys* are monotypic in the northern Yukon, but no dental characters have been found to distinguish the eastern Beringian species from others that live in the temperate zone (Rausch and Rausch, 1975a; Guilday and Parmalee, 1972). *Clethrionomys,* the red-backed vole, is present throughout the sequence but is rare in Pleistocene deposits in Old Crow Basin. Its absence from the Klo-kut record is probably a sampling error in view of its modern abundance at the site (Savage, 1977). *Phenacomys,* the heather vole, is extremely rare in the fossil assemblages and does not live in the northern Yukon today (Youngman, 1975:map 26).

Lemmus sibiricus, the brown lemming, is conspecific with the Siberian *L. obensis* but cannot be separated on the basis of tooth characters from the more westerly *L. lemmus* (Rausch and Rausch, 1975b). It is abundant throughout the northern Yukon fossil record (Table 2).

Of the five species of *Microtus* that now live in the Yukon and on the Alaskan mainland, only *M. longicaudus,* the long-tailed vole, has not been recognized among the fossils reported here. The taiga vole, *M. xanthognathus,* is distinctive for its large size as well as its occlusal patterns on lower first and upper third molars, and it is prominent in most of the northern Yukon fossil assemblages.

Other species of voles are often difficult to distinguish from one another. In referring the *Microtus* fossils to species, I have made extensive use of van der Meulen's (1973, 1978) measurements and ratios and have enlarged his techniques to encompass the highly complex teeth of *M. miurus* and *M. pennsylvanicus,* the singing and meadow voles, respectively. Despite such careful study, I have been able to refer only some of the fossils to species (see Guilday et al., 1964:166, fig. 22, for a similar problem) and have left others under the label "*M. miurus* or *pennsylvanicus.*" Future studies of Beringian microtine fossils should focus on methods of distinguishing these species from one another as well as from their Palearctic counterparts (*M. gregalis, M. agrestis*). The diagnostic upper second molar of *M.*

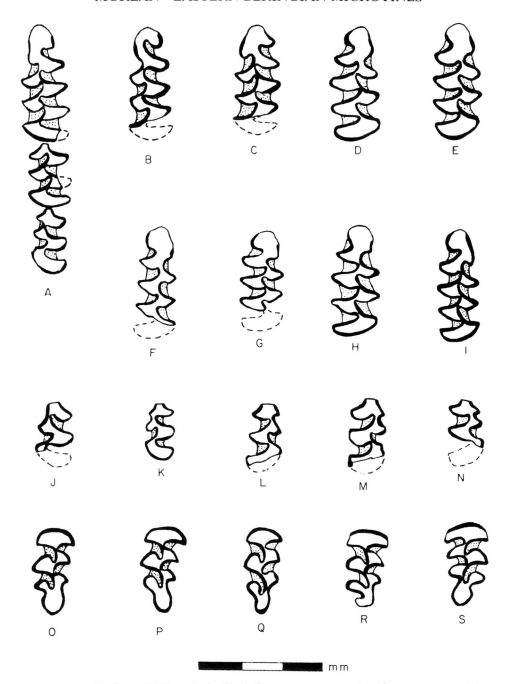

Fig. 3.—Molars of "*Microtus* sp. X" from Old Crow Basin. Black lines represent enamel, white areas represent dentine, and stippled areas represent crown cementum. Dashed lines are reconstructions based on complete teeth. A, Loc. 64 (NMC-25111); B–S, Loc. 11 (MkVl-9:44). A, Lm1–Lm3; B–C, Lm1; D–I, Rm1; J–K, Lm3; L–N, Rm3; O–R, LM3; S, RM3.

pennsylvanicus is present in samples from Unit 2a (Loc. 11), Disconformity A (Locs. 12 and 15), and Klo-kut. It is not present in the large Bluefish Caves sample where all lower first molars with 5–6 triangles that lack taiga vole characteristics have been referred to *M. miurus* (Morlan, 1983a). *M. miurus* disappears abruptly in the latest Pleistocene levels at Bluefish Caves and is now confined to mountainous areas of the Yukon (Youngman, 1975:map 31).

Table 3.—*Morphotypes of maxillary molars of varying lemmings from Old Crow Basin (after Agadjanian and von Koenigswald, 1977, except for "<I" as explained in text). "?" means not recorded. See Fig. 4.*

Sample	Upper M1					Upper M2				Upper M3		
	<I	I	II	II/III	III	I	II	II/III	III	I	II	III
Modern, n. Yukon			2	2	26	1			29	?	?	?
BFC, Pleistocene			3	3	29	2	2		39	3	5	8
HH68-10, Unit 2b					10		2		5		2	6
Loc. 15, Disc. A			2	1	1		3		1		4	2
Loc. 11, Unit 2a		6				7					1	1
Loc. 44, Unit 2a	2	4			1	1						1
Loc. 64, Unit 2a	2											

The tundra vole, *M. oeconomus,* is unique among living Beringian voles in having only four closed triangles on the lower first molar. Good examples of this species have been found in the Pleistocene levels of the Bluefish Caves and at Klo-kut, but no specimens are known for the Boutellier Interval. Both the middle and lower portions of Unit 2a have yielded four-triangled lower first molars that are listed under "*Microtus* sp. X" in Table 2. These teeth frequently exhibit a fourth buccal salient angle that is extremely rare in *M. oeconomus* (Fig. 3A–I), and they have lower A/L ratios and higher B/W ratios (see van der Meulen, 1978) than the modern tundra vole. One mandible from Loc. 64 contains all its teeth (Fig. 3A) and shows that the distinctive first molar is associated with an unusual third molar on which the apex of the first buccal re-entrant angle bisects the second lingual salient angle (rather than meeting the apex of the first lingual re-entrant angle as in most species of this genus); six isolated lower third molars of this type are present in the sample from Loc. 11 (Fig. 3J–N). Also from Loc. 11, there are six isolated upper third molars that have more primitive occlusal patterns than any living species (Fig. 3O–S).

The Beringian taxon most similar to "*Microtus* sp. X" is the ancient *M. deceitensis* from the Cape Deceit Formation in western Alaska (Guthrie and Matthews, 1971), but the Cape Deceit vole is much larger than the specimens from Old Crow Unit 2a. The most similar taxa outside Beringia are *M. paroperarius* (Hibbard, 1944:fig. 11; Guthrie, 1965:fig. 2; van der Meulen, 1978:fig. 11A–G) and *Microtus* sp. C as described by van der Meulen (1973:87–88) from Layer 10 at the southern Hungarian site of Villány-8. Van der Meulen (1978:124) argues that *M. paroperarius* is an immigrant from Eurasia, directly descending from *Microtus* sp. C, and I suggest that the "*Microtus* sp. X" fossils could be Beringian

representatives of that immigration. I have labelled the fossils "*Microtus* sp. X" without intending to imply that they represent a formal taxon; small sample sizes and considerations of significant redeposition history (Table 1) preclude the formal definition of a species at this time, and future study must entail direct comparisons with other fossils in addition to the literature review mentioned here. Even in this preliminary stage of the study, however, it is reasonable to question the report of *M. oeconomus* fossils from Unit 2a (Jopling et al., 1981:table 2, Faunas 2 and 6) and to suggest that they might represent "*Microtus* sp. X." If so, the modern tundra vole would appear to have arrived in eastern Beringia in late Wisconsinan time.

Study of varying lemming fossils from Old Crow Basin has benefited from the concepts advanced by Agadjanian (1972, 1976; Agadjanian and von Koenigswald, 1977) and Zazhigin (1976), but the resulting sequence of taxa should be regarded as tentative (Table 2). The morphotypes defined for the maxillary dentition of *Dicrostonyx* (Agadjanian and von Koenigswald, 1977) are adequate to characterize the Old Crow sequence from the middle of Unit 2a onward through time, but some of the fossils from the lower part of Unit 2a are even more "primitive" than Morphotype I and have been labelled "<I" in Table 3 (see Fig. 4H–K). The latter match published drawings of *Predicrostonyx* maxillary teeth (Guthrie and Matthews, 1971:fig. 12; Zazhigin, 1976: fig. 1), but there are too few specimens in the samples to lend much confidence to an identification.

If larger samples were to support the patterns seen in Table 3, we could conclude that lower Unit 2a (at Locs. 44 and 64) contains fossils of *Predicrostonyx* and/or a primitive species of *Dicrostonyx* (Fig. 4H–M), but Loc. 44 has also yielded a single upper first molar (Fig. 4N) that represents a more advanced form. The Loc. 11 sample, from the middle

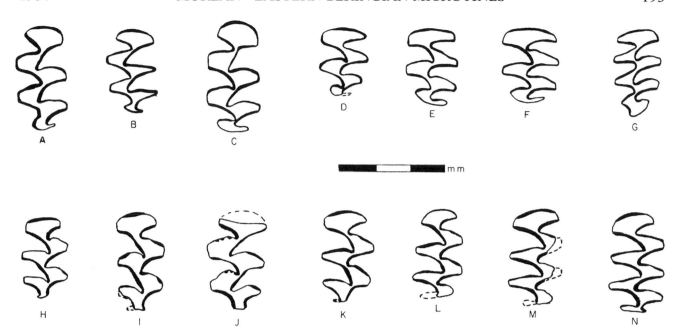

Fig. 4.—Maxillary molars of Dicrostonychini from Old Crow Basin, Unit 2a. Black lines represent enamel and white areas represent dentine. Dashed lines are reconstructions based on features preserved below the plane of the occlusal surfaces. A–G, Loc. 11 (MkVl-9:54); H–I, Loc. 64 (NMC-25101, -25135); J–N, Loc. 44 (NMC-25640, -25637, -25638, -25641, -25639). A–C, LM1, Morphotype I; D–F, LM2, Morphotype I; G, RM3, Morphotype II; H–I, RM1, Morphotype <I; J, LM1, Morphotype <I; K, RM1, Morphotype <I; L–M, RM1, Morphotype I; N, RM1, Morphotype III.

of Unit 2a, appears to represent *D*. cf. *simplicior* (or else an early manifestation of *D. hudsonius*; Fig. 4A–G). All teeth from younger samples are more advanced than Morphotype I. Those from Disconformity A exhibit a range of variation seen in *D. gulielmi* (Agadjanian and von Koenigswald, 1977), and I have borrowed a semantic device from Russian colleagues (Sher et al., 1979:55) in referring them to *D. gulielmi-torquatus*. Specimens from HH68-10 (upper Unit 2b) and the Bluefish Caves do not differ significantly from the modern northern Yukon population and are referred to *D. torquatus*.

All these identifications are based on maxillary molar morphology. The mandibular dentition in the samples from Unit 2a exhibits more variation than I had expected on the basis of published statements and illustrations. It would appear that the presence or absence of anterior spurs on the lower second molar permits discrimination between *Predicrostonyx* and *Dicrostonyx* (compare Guthrie and Matthews, 1971:figs. 5–6 and Zazhigin, 1976:Fig. 1), but the specimens from Loc. 11 exhibit the entire range of variation. Similar variations are seen in the samples from Locs. 44 and 64 (Morlan, 1983c), but all samples younger than Unit 2a contain lower molars

like those of *D. torquatus* (that is, both buccal and lingual spurs on the lower second molar and on most lower third molars).

Jopling et al. (1981:21) concluded that "the earliest *Dicrostonyx* in Old Crow Basin were of the *Misothermus* line which was replaced by *D. torquatus* in early Wisconsinan time." This conclusion may be premature in that all the available samples are small and were derived from fluvial sediments that could contain mixtures of different species representing various time periods. Furthermore, we must account for the very primitive tooth patterns that resemble *Predicrostonyx,* bearing in mind that this genus supposedly evolved into *Dicrostonyx* by at least early Irvingtonian time (Repenning, 1980:43). There are several possible explanations for the mixture of morphotypes seen at Loc. 44 (Fig. 4J–N): (1) the "primitive" teeth belong there, and the single "advanced" specimen (Fig. 4N) is intrusive because the sample was derived from sediments below the seasonal high water mark of Old Crow River; (2) the advanced specimen belongs there, and the more primitive teeth were redeposited from older sediments; or (3) the sequence of morphotypes and named taxa that has been inferred from Eurasian evidence has limited utility in eastern Beringia. I favor the first alternative, because we have documented the problem of intrusive specimens below the high water mark elsewhere in the basin; all maxillary teeth in the Loc. 11 sample from the middle of Unit 2a represent Morphotype I; and the weight of evidence from Old Crow Basin supports the Eurasian-based sequence for this lineage.

The presence of *D. torquatus* in early Wisconsinan time cannot yet be demonstrated, but this species is definitely present near the top of Unit 2b in sediments that immediately underlie the glaciolacustrine clay of the late Wisconsinan (=Duvanny Yar Interval). Varying lemmings disappear abruptly at the end of the Pleistocene at the Bluefish Caves, and they are now confined to the Arctic Coastal Plain and to mountainous areas of northern and central Yukon (Youngman, 1975:map 35).

BIOGEOGRAPHY

Youngman (1975:table 1) summarized the refugial origins of microtine rodents now living in Yukon Territory. The Beringian refugium was home to *Clethrionomys rutilus, Microtus miurus, M. oeconomus, Lemmus sibiricus trimucronatus,* and *Dicrostonyx torquatus.* A Rocky Mountain refugium held populations of *Lemmus sibiricus helvolus,* and all others are seen as postglacial immigrants from south of Wisconsinan ice limits—*Phenacomys intermedius, Microtus pennsylvanicus, M. longicaudus, M. xanthognathus,* and *Synaptomys borealis.* Of those he elected to discuss, Macpherson (1965) agreed with all these assessments. Guthrie (1968:239) saw the distribution of *M. xanthognathus* as indicating "post-glacial expansions from Alaska rather than to it," and Rausch and Rausch (1975b:27) considered *Lemmus sibiricus helvolus* to be "the result of post-glacial extension from the north" (see Burns, 1980).

The northern Yukon fossil record sheds new light on the histories of some taxa but remains mute on others. *Microtus longicaudus* has not been recognized, and no specimens of *Synaptomys,* the bog lemming, have been found in eastern Beringia (given that the specimen from Tofty, Alaska, Repenning et al., 1964, may have been incorrectly identified and has since been lost; Harington, 1978:76). Of extant taxa represented by northern Yukon fossils, all except *Microtus pennsylvanicus* and *Phenacomys intermedius* were present at the Bluefish Caves during the Duvanny Yar Interval which embraces the late Wisconsinan glaciation (Table 2). The two exceptions are interesting, because their remains have been found in sediments representing earlier intervals, and they may have been extirpated from eastern Beringia during the last glaciation. Unfortunately we do not yet have samples representing the relatively cold Happy Interval in the Yukon, but present evidence suggests that the meadow and heather voles were present there during the warm Boutellier Interval and during pre-Happy Interval episodes that may represent relatively warm periods of isotope stage 5.

The modern range of *M. xanthognathus* forms a large trapezoid with apices near Churchill, Manitoba, the Mackenzie Delta, west-central Alaska, and west-central Alberta (Banfield, 1974:map 92). Eleven Rancholabrean faunal assemblages in the eastern and central United States include *M. xanthognathus* (Hallberg et al., 1974), and it is the dominant form in some of them (for example, Guilday et al., 1964:164, 1977:64). Because some of these occurrences have been known for 20 years or more, whereas the late Wisconsinan record from the Bluefish Caves was only recently discovered (Cinq-Mars, 1979), the taiga vole has usually been interpreted as a postglacial southern immigrant that followed retreating glaciers northward (for example, Youngman, 1975:99; Guilday et al., 1977:65–67). The Bluefish Caves samples now show that the taiga vole was also resident in Beringia during late Wisconsinan glaciation, and it seems remarkable that two large demes of the modern population could have remained isolated from one another during the late Wisconsinan without differentiation to at least the subspecific level. The northernmost of the U.S. fossil occurrences is 1,800 km south of Churchill, and none of the U.S. fossils has been dated to less than 11,000 years ago (cf. Hallberg et al., 1974:640, fig. 1; Graham, 1979: fig. 6). Unless more northerly and younger midcontinental fossils are found, it seems likely that the southern deme of *M. xanthognathus* became extinct and that the modern population was derived from the northern deme that, on the basis of the Bluefish Caves evidence, persisted in the northern Yukon throughout the late Wisconsinan and Holocene.

Varying lemming fossils have also been found south of the Wisconsinan ice terminus in both eastern and western North America. The eastern form, as represented at New Paris No. 4, Pennsylvania (Guilday et al., 1964:160), has been identified as *Dicrostonyx hudsonius,* and western fossils have been referred to *D. torquatus* (Guilday, 1968; Martin et al., 1979; Burns, 1980). Guilday (1968) noted that the late Pleistocene occurrence of separate western and eastern forms indicated either that the periglacial tundra belt was discontinuous or that a continuous tundra belt was occupied by varying lemmings that exhibited clinal variations in tooth form. The recent discovery of midwestern *Dicrostonyx* fossils (for example, Voorhies, 1981; Rhoades, 1982) may hold promise for choosing between these alternatives. The northern Yukon fossils show that pre-Wisconsinan *Dicrostonyx* populations contained all the variations needed to produce a cline, but they shed no light on the speciation of the two modern varying lemmings. Given the complete glacial cover of Ungava Peninsula and Hudson Bay (Prest, 1969), it seems logically impossible to derive the modern *D. hudsonius* population from any other source than northward postglacial migration, but, as Burns (1980: 1510) has noted, the southern population of *D. torquatus* did not necessarily move northward following deglaciation. It could have become extinct, as suggested here for the taiga vole, with the modern northern population descending from Beringian residents that are known to have lived in the Yukon, and in Alaska (Guthrie, 1968; Guthrie and Matthews, 1971), during the Duvanny Yar Interval (Table 2).

BIOSTRATIGRAPHY

Repenning (1980, n.d.) has shown that microtine rodents can provide chronometric evidence because of their rapid evolutionary changes punctuated by episodes of widespread dispersal. The data presented here from Old Crow Basin (Table 2) indicate some promise in this regard but do not yet permit unequivocal conclusions. For example, the very primitive varying lemming molars in the lower part of Unit 2a (Table 3) suggest that Quaternary exposures in Old Crow Basin may span the period of time during which *Dicrostonyx* evolved from *Predicrostonyx*. This evolutionary change is documented in the Olyor Suite of the Kolyma Lowland in Siberia, but the dating of the Olyor Suite is open to more than one interpretation. A paleomagnetic reversal at the Unit IIIa/IIIb contact within the Olyor Suite at the Krestovka section is interpreted by Sher et al. (1979) and Giterman et al. (1982:52) to represent the Matuyama-Bruhnes boundary of approximately 700,000 years ago, and the earliest specimens of *Dicrostonyx* appear just below that contact. Repenning (n.d.) calibrates the paleomagnetic record of the Kolyma Lowland on a much longer time scale, placing the reversely magnetized Olyor IIIa prior to the Reunion Subchron, more than 2 m.y. ago, and the normal Olyor IIIb in the Olduvai Subchron (1.7–1.8 m.y. ago). As mentioned above, a paleomagnetic reversal of late Matuyama or older age has been identified in Bluefish Basin (Pearce et al., 1982), but it is not known whether the Unit 1 clay in Old Crow Basin should be correlated with the reversed lacustrine unit in Bluefish Basin. Hence the implications of the varying lemmings from Old Crow Basin cannot be understood until three problems are solved: (1) confirmation of the presence or absence of *Predicrostonyx* in the lower part of Unit 2a; (2) correlation of Unit 1 with Bluefish Basin and in turn with the Kolyma Lowland sequence; and (3) consensus on calibration of the paleomagnetic record from Kolyma Lowland.

Subsequent changes within the *Dicrostonyx* lineage in Old Crow Basin parallel those recorded in Siberia and Europe (Zazhigin, 1976; Agadjanian and von Koenigswald, 1977). The stages also match meager evidence from Alaska. The "*henseli* zone" in Unit 1 of the Deering Formation at Cape Deceit can now be understood to have yielded a *D. gulielmi-torquatus* type of lemming that represents the Happy Interval. Superjacent Unit 2 in the Deering Formation represents the Duvanny Yar Interval and contains *D. torquatus* (Guthrie and Matthews, 1971; Matthews, 1974; Hopkins, 1982; Giterman et al., 1982). The report of *D. torquatus* from an Alaskan deposit representing the pre-Illinoian Kotzebuan transgression cannot be taken at face value, because the single specimen is a mandible containing m1–m3 that was not seen in situ (Péwé and Hopkins, 1967:268–269).

The possible chronometric significance of "*Microtus* sp. X" in Old Crow Unit 2a is also uncertain.

Even if we assume that the fossils placed in this category represent a valid taxon related to the immigration of *Microtus paroperarius,* the picture remains unclear. Van der Meulen (1978:143) dates the arrival of *M. paroperarius* at "just prior to 0.7 million years" in the mid-continent of North America, but apparently it reached the Wellsch Valley of Saskatchewan at about the time of the Olduvai Event (1.7–1.8 m.y. ago; Repenning, n.d.; Stalker et al., 1982).

The primitive *Dicrostonyx* and "*Microtus* sp. X" fossils might be taken to imply that Unit 2a contains an Irvingtonian fauna, but the associated fossils of the meadow vole comprise a classic marker of the beginning of the Rancholabrean when "*M. pennsylvanicus* appears as a flood in almost every appropriate fossil locality east of the Rocky Mountains . . ." (Repenning, 1980:42). Obviously, it should not be assumed that the land mammal ages of mid-continental North America have the same temporal limits and species composition as those that might be defined in eastern Beringia.

The data reported here represent the first step toward defining land mammal ages for eastern Beringia, and even this preliminary framework may aid in solving an interesting problem. The fauna from Loc. 12 reported by Jopling et al. (1981; letter "B" in Fig. 2 and Tables 1–2 of this report) was obtained from dipping beds for which two markedly different interpretations have been offered. To Jopling and his colleagues, a "pingo origin is the most rational explanation for this particular occurrence" (Jopling et al., 1981:16–17). According to this interpretation, some of the vertebrate fossils were deposited in alluvium while others occurred in colluvial deposits, and the resulting assemblage is little younger than the lake clay of Unit 1 (said to be Illinoian; Jopling et al., 1981:18, 29–30). An alternate interpretation of the dipping beds is that they represent the foreset beds of point bars laid down by a meandering precursor of Old Crow River (O. L. Hughes, personal communication, 1978, 1981). The entire suite of beds might be attributed to a single fluvial cycle with vertical relief of approximately 15 m between channel bottom and floodplain (similar in scale to the modern river). This interpretation would derive the vertebrate assemblage from a stream channel much younger in age than the Unit 1 clay and possibly similar in age to Disconformity A. Interest in the age of the assemblage is heightened by the recovery of altered bones that are interpreted as ancient artifacts (Jopling et al., 1981:26–29). The microtine rodent remains could aid in determining the approximate age of the assemblage. If "*Microtus* sp. X" and primitive species of *Dicrostonyx* are associated, a pre-Happy Interval age would be suggested. If all the *Microtus* specimens resemble modern taxa and the varying lemmings exhibit maxillary morphotypes II and III, a Happy Interval or early Boutellier Interval age would be indicated.

CONCLUSION

This summary has shown that substantial progress has been made in the recovery and interpretation of microtine rodent fossils in eastern Beringia, but it has shown even more clearly that research should focus more specifically and more deliberately on the problems and prospects posed by small mammals. The need for larger samples and better chronometric controls restricts not only our views of biostratigraphy but also the use of the record for paleoecological reconstruction. This should now become a primary goal, because the relatively restricted and well defined habitat requirements, dietary preferences, and intra- and interspecific interactions of microtine rodents (for example, Batzli and Jung, 1980) will provide a powerful tool in paleoenvironmental analysis, especially if their fossil record can be integrated with that of other animals and plant remains.

ACKNOWLEDGMENTS

Laboratory work on the microtine rodent samples from Locs. 11, 12, and 15 was begun in 1978. I sent a number of troublesome specimens and a draft manuscript on the samples from Disconformity A to Dr. John Guilday who kindly replied at length and encouraged me to continue the study. An early draft of this paper was nearly ready for mailing to Dr. Guilday when I learned of his death, and I am grateful to Drs. Dawson and Genoways for the opportunity to publish the report in this memorial volume.

Catherine Craig-Bullen undertook the laborious task of picking microtine teeth and other fossils from samples that were collected during field work of the Yukon Refugium Project, sponsored by the National Museum of Man, the Geological Survey of Canada,

SHER, A. V., T. N. KAPLINA, R. E. GITERMAN, A. V. LOZHKIN, A. A. ARKHANGELOV, S. V. KISELYOV, YU. V. KOUZNETSOV, E. I. VIRINA, and V. S. ZAZHIGIN. 1979. Late Cenozoic of the Kolyma Lowland. XIV Pacific Science Congress, Guide to Tour XI, 116 pp.

STALKER, A. MacS., C. S. CHURCHER, and R. S. HILL. 1982. Ice age deposits and animals from the southwestern part of the Great Plains of Canada. Geol. Surv. Canada Miscel. Report 31 (wall chart).

THORSON, R. M., and E. J. DIXON, JR. 1983. Alluvial history of the Porcupine River, Alaska: role of glacial-lake overflow from northwest Canada. Bull. Geol. Soc. Amer., 94:576–589.

VAN DER MEULEN, A. J. 1973. Middle Pleistocene smaller mammals from the Monte Peglia, (Orvieto, Italy) with special reference to the phylogeny of Microtus (Arvicolidae, Rodentia). Quaternaria, 17:1–144.

———. 1978. Microtus and Pitymys (Arvicolidae) from Cumberland Cave, Maryland, with a comparison of some New and Old World species. Ann. Carnegie Mus., 47:101–145.

VOORHIES, M. R. 1981. A fossil record of the porcupine (Erethizon dorsatum) from the Great Plains. J. Mamm., 62:835–837.

WEBER, F. R., T. D. HAMILTON, D. M. HOPKINS, C. A. REPENNING, and H. HAAS. 1981. Canyon Creek: a late Pleistocene vertebrate locality in interior Alaska. Quaternary Res., 16:167–180.

WESTGATE, J. A. 1982. Discovery of a large-magnitude, late Pleistocene volcanic eruption in Alaska. Science, 218:789–790.

WESTGATE, J. A., T. D. HAMILTON, and M. P. GORTON. 1983. Old Crow tephra: a new late Pleistocene stratigraphic marker across north-central Alaska and western Yukon Territory. Quaternary Res., 19:38–54.

YOUNGMAN, P. M. 1975. Mammals of the Yukon Territory. Publ. Zool., Nat. Mus. Nat. Sci., 10:1–192.

ZAZHIGAN, V. S. 1976. Rannie etapy evoliutsii kopytnykh lemmingov (Dicrostonychini, Microtinae, Rodentia)—kharakternykh predstavitelei sybarkticheckoi fauny Beringii (Early evolutionary stages of collared lemmings (Dicrostonychini, Microtinae, Rodentia)—as characteristic representatives of Beringian subarctic fauna). Pp. 280–288, in Beringiia v Kainozoe (V. L. Kontrimavichus, ed.), Khabarovsk, Nauka.

Address: Archaeological Survey of Canada, National Museum of Man, Ottawa, Ontario K1A 0M8.

NEW ARVICOLINES (MAMMALIA: RODENTIA) FROM THE BLANCAN OF KANSAS AND NEBRASKA

Richard J. Zakrzewski

ABSTRACT

Study of arvicolines from five local faunas of Blancan age in Kansas (Dixon, White Rock) and Nebraska (Sand Draw, Mullen, Big Springs) permits the redefinition of the genus *Pliophenacomys,* the naming of two new genera, *Guildayomys* and *Hibbardomys,* and six new species—*P. dixonensis, G. hibbardi, H. marthae, H. skinneri, H. fayae,* and *H. voorhiesi.* The genera are distinguished on the basis of the occlusal pattern of m1 and M3 and the species on the relative development of dentine tracts on the posterior loop and anteroconid complex of m1.

INTRODUCTION

While studying the small mammals from the Big Springs local fauna (l.f.) from the Blancan of north-eastern Nebraska (Voorhies and Zakrzewski, manuscript) a problem arose in attempting to describe the specimens that had been tentatively assigned to *Pliophenacomys osborni.* Based on the range of variation established for the m1 by previous studies (Martin, 1972; Eshelman, 1975) one highly-variable hypsodont seven-triangled arvicoline was present, but two distinct morphotypes of the m1 could be observed. One in which the sixth alternating triangle was normally developed and the dentine tract found on the side of the triangle interrupted the occlusal pattern when a millimeter or more of wear had occurred; the second had a sixth triangle that was generally reduced and rounded in appearance and the dentine tract found on the side of the triangle interrupted the occlusal pattern very early in life.

Eshelman (1975) working with a larger sample (N = 69) than Martin (N = 5) observed that the specimens from the White Rock l.f. (Blancan of north-central Kansas) resembled those from the Mullen l.f. (Blancan of central Nebraska) but may have been aware of the differences I stated when he suggested that the specimens might belong in a taxon other than *Pliophenacomys* as the m1s from the White Rock looked more like specimens that Guthrie and Matthews (1971) had described as *Pliomys* from the Pleistocene of Alaska. Eshelman questionably assigned his specimens to *Pliophenacomys osborni* along with specimens of an arvicoline from the Dixon l.f. (Blancan of south-central Kansas) that Hibbard had cataloged as new but never described.

Continued examination of the Big Springs material along with material from the local faunas mentioned above showed that two morphotypes could be discerned among each of the other molars, although it was not obvious at first to which m1 morphotype the other molars belonged. With additional study it became apparent that not one but three different taxa were represented among the Big Springs, Mullen, and White Rock specimens considered to be *Pliophenacomys osborni.* One of these is *P. osborni.* The other two are new genera. A redefinition of *P. osborni* and a description of the new genera and their contained species form the basis for this report.

METHODS

The specimens consist mainly of isolated teeth. The kind and amount are listed under the appropriate taxon. Standard measurements (Zakrzewski, 1967, Fig. 1) were made using a Gaertner Measuring Microscope. The terminology used in the description of the m1 (Fig. 1A) follows van der Meulen (1973); terminology for the M3 (Fig. 1B) follows Zakrzewski (1967). Abbreviations used with this terminology are given below. These abbreviations are used throughout the text.

ABBREVIATIONS

ACC = anteroconid complex
AC = anterior cap
AL = anterior loop
BRA = buccal reentrant angle

LRA = lingual reentrant angle
PL = posterior loop
T = triangle

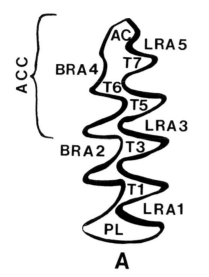

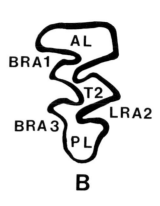

Fig. 1.—Occlusal patterns of arvicoline teeth (A, Lm1; B, RM3) listing terminology used in text. See list of abbreviations in text for explanation.

SYSTEMATIC PALEONTOLOGY

Guildayomys, new genus

Diagnosis.—Arvicoline with evergrowing teeth, lacking cement in the reentrant angles. Dentine tracts extending the entire height of the crown are found on both sides of the PL and the anterior face of the lower teeth, and on both sides of the AL and the posterior face of the upper teeth. Dentine tracts may also be present on T6 and 7 of the m1 and T1 of the M1.

Type species.—*Guildayomys hibbardi*, new species.

Etymology.—Named for the late John E. Guilday, Carnegie Museum of Natural History.

Guildayomys hibbardi, new species
(Fig. 2)

Pliophenacomys osborni Martin, 1972 (in part), Bull. Univ. Nebraska St. Mus., 9:174–176.
Pliophenacomys? osborni Martin; Eshelman, 1975 (in part), Univ. Michigan Papers Paleo., 13:41–43.

Holotype.—UNSM 51839, partial left dentary with I, m1–m2.

Horizon and type locality.—Long Pine formation, Big Springs l.f., Antelope Co., Nebraska.

Paratypes.—Big Springs l.f., UNSM 52700, 90800, partial right dentary with m1–m2; 52794, partial right dentary with m1; 90801, partial right dentary with I; 52695, 52701, 52705, 52712–52714, 52721, 52726–52727, 52729–52730, 52733, 52735–52739, 52741, 52744–52745, 52748, 52751, 52754, 52761–52762, 52764–52766, 52769, 52774, 52776–52778, 52781, 52783–52784, 52786, 90802–90807, 43 m1s; 90808–90815, 8 m2s; 90816–90817, 2 m3s; 90818–90823, 6 M1s; 90824–90827, 4 M2s; 52893–52897, 52899, 52901, 52903, 90828–90831, 12 M3s. Mullen l.f., UNSM 39569, m1. White Rock l.f., UM-K1-66: UM 60593, m1.

Diagnosis.—Same as for genus.

Etymology.—Named for the late Claude W. Hibbard.

Description of holotype.—The holotype (Fig. 2A) represents an individual in the adult stage of development. The m1 is composed of a PL, seven alternating triangles, and a simple AC3. A thin strip of dentine connects the alternating triangles to each other and in turn to the PL and AC. Triangles 6 and 7 are nearly confluent and open broadly into the AC. Dentine tracts are present on both sides of the PL, T6 and 7, and the anterior face of the AC. They are high enough so that the enamel on the occlusal surface is interrupted at each of these points. In addition to being interrupted, the enamel on the occlusal surface is differentiated into thick and thin segments. Except for where it is interrupted the enamel is thicker on the AC and the anterior faces of the triangles and PL. The enamel remains thick around the apex of the alternating triangles onto the posterior face to a point not quite halfway to the midline of the tooth where it begins to thin until it reaches the apex of the reentrant angle. The enamel on the posterior face of the PL is thin. The reentrant

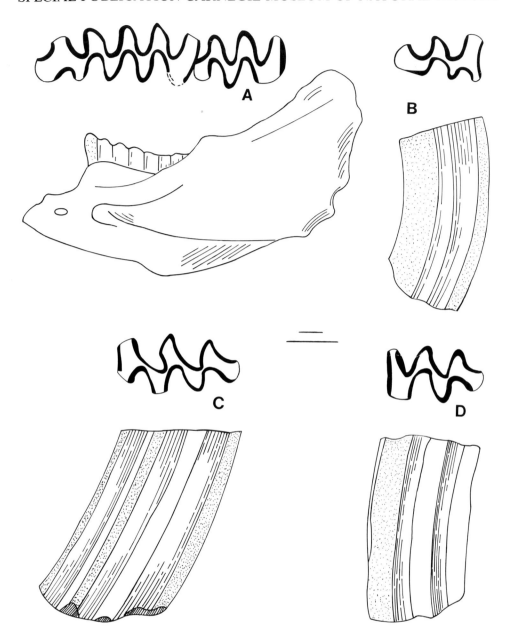

Fig. 2.—*Guildayomys hibbardi*. A) UNSM 51839, partial left dentary with m1–m2, type, occlusal view of teeth, buccal view of dentary. B) UNSM 52897, LM3, occlusal and buccal view. C) UNSM 90818, LM1, occlusal and lingual view. D) UNSM 90825, LM2, occlusal and lingual view. Short bar equals 1 mm on dentary; long bar equals 1 mm on other specimens.

angles are broad and rounded at their apices. They do not constrict near the midline and the direction defined by a line which bisects the apex is variable.

The m2 is composed of a PL and four alternating triangles. The fourth triangle is oval shaped. Dentine tracts are present on both sides of the PL and the anterior face of T4. Characteristics of the enamel, triangle closure, and reentrant angles are as in m1.

The occlusal length of the m1–m2 is 4.52 mm.

The length of m1 is 3.01 mm, width is 1.21 mm. For m2 the same measurements are 1.68 and 1.12 mm. Crown height could not be measured.

The dentary is nearly complete. The masseteric ridges and crests are well developed. The mental foramen is lateral. There is no capsular process for the reception of the incisor. There is no temporal fossa between the m3 and ascending ramus and the base of the alveoli for m3 is not greatly exaggerated.

Description of paratypes.—The remaining dentaries are similar to the holotype. The characters mentioned above and size make the dentaries of *Guildayomys* almost identical to those of *Phenacomys intermedius* Merriam. Dentaries of *Pliophenacomys* are also similar except that the mental foramen is more ventrally located. This difference can be seen easily when examining the dentary in a dorsal view. Dentaries of *Pitymys* and *Microtus* have a well developed temporal fossa and the base of the alveoli of m3 is greatly enlarged to accomodate the tooth.

The m1s are similar in occlusal pattern and most characteristics to that of the holotype. The occlusal pattern in terms of number, closure, and relative size of alternating triangles is *Microtus*-like. The shape of the reentrant angles are similar to *Pliophenacomys*. Dentine tracts are present on both sides of the PL and on the anterior face of the AC in all the teeth. The enamel on the occlusal surface is also interrupted at all these sites with the exception of one tooth (UNSM 52774) from an immature individual, where only the buccal tract on the PL is interrupted. The majority of m1s (41/48) also have dentine tracts on T6 and 7 and in most cases (36/41) the tracts cause an interruption of the enamel on the occlusal surface. Five specimens (including the m1 from White Rock) have tracts on both triangles that do not extend the entire length of the crown; three have tracts only on T6; two only on T7; and two (including the Mullen specimen) lacks tracts on both T6 and 7.

The paratypic m2s are similar to that of the holotype. The m3 is composed of a posterior loop and three alternating triangles. Except for size and one less triangle the m3 is like the m2 in the position and length of dentine tracts, differentiation of enamel, shape of reentrants, and closure and size of triangles.

The M1 is composed of an AL, three closed alternating triangles, and a fourth triangle that expands confluently into a PL (Fig. 2C). The buccal triangles are slightly larger than the lingual triangles. The enamel of the occlusal surface tends to be thicker on the posterior faces of the triangles and thinner on the anterior faces. The enamel of the anterior face of the PL is thick. This enamel is interrupted on the posterior face of the PL, the apex of T1, and both sides of the AL as dentine tracts are present along the sides of the crown, throughout its entire length, at these points. The apices of the BRAs are directed posteriorly, whereas those of the LRAs tend to be variable.

The M2 is composed of an AL, two closed alternating triangles, and a third triangle that expands confluently into a PL (Fig. 2D). Smaller size, one less triangle, and the development of dentine tracts only on the loops distinguishes the M2 of *Guildayomys* from the M1.

The M3 (Fig. 2B) is composed of an AL that opens broadly into a small buccal triangle. This small triangle is generally closed off from a large lingual triangle, which in turn is generally closed off from a PL. This pattern most closely resembles that in *Pliophenacomys* and *Pliolemmus*. Like the other molars the enamel of the occlusal surface is differentiated into thick and thin segments. The enamel is thinner on the anterior face of the AL and T2 and the posterior face of LRA2. The enamel pattern of the occlusal surface is interrupted by dentine tracts on both the buccal and lingual side of the AL and on the posterior face of the PL. In three of the seven specimens the tract on the PL is wide enough to cover the buccal side of the loop as well. The direction that the apices of the reentrant angles take with respect to the midline of the tooth is variable, but they tend to be perpendicular to it for the most part. In two specimens, a small BRA3 is present. Measurements of various parameters of the teeth of *Guildayomys* are given in Table 1.

Table 1.—*Measurements of teeth in* Guildayomys *from the Big Springs.*

Tooth	Character	N	Mean	SD	Observed range
m1	GL	23	2.99	0.19	2.65–3.35
	GW	23	1.99	0.07	1.00–1.35
	GH	19	3.71	0.50	2.45–4.99
m2	GL	7	1.74	0.04	1.68–1.79
	GW	7	1.10	0.05	1.03–1.20
	GH	4	3.16	0.27	2.86–3.42
m3	GL	2	1.36	0.05	1.32–1.39
	GW	2	0.80	0.09	0.73–0.86
	GH	2	2.21	0.28	2.01–2.41
M1	GL	4	2.50	0.11	2.38–2.61
	GW	4	1.33	0.11	1.24–1.48
	GH	4	3.83	0.09	3.70–3.90
M2	GL	4	1.93	0.04	1.89–1.96
	GW	4	1.19	0.06	1.12–1.27
	GH	4	3.53	0.43	2.92–3.83
M3	GL	7	1.74	0.08	1.63–1.81
	GW	7	1.02	0.05	0.96–1.08
	GH	7	2.88	0.37	2.42–3.50
	LBRA1	7	0.28	0.04	0.23–0.33
	LBRA3	7	0.04	0.07	0.00–0.16

See Table 3 for abbreviations not listed in text.

Discussion.—The occlusal pattern of the teeth in *Guildayomys* most closely resemble those of *Pliophenacomys*. *Guildayomys* can be easily distinguished from *Pliophenacomys* and all other pre-Irvingtonian voles except for *Pliolemmus*, by the fact that the molars are evergrowing. *Pliolemmus*, the only other Blancan vole with evergrowing molars, can be distinguished from *Guildayomys* by the fact that in many specimens of *Pliolemmus* all of the alternating triangles and loops of the teeth possess dentine tracts. Because of this feature the tips of the triangles and loops in *Pliolemmus* tend to be blunt. In addition, the M3 of *Pliolemmus* has an additional triangle and it and *Pliophenacomys* have a shorter BRA1 on the M3. The PL of the upper molars tend to be more elongate in *Guildayomys* than in either *Pliophenacomys* or *Pliolemmus*.

Guildayomys differs from all the other voles with evergrowing teeth except for the extant steppe lemming *Lagurus*, and the extinct *Eolagurus* and *Prolagurus*, by the lack of cement in the reentrant angles. These latter three taxa known from Asia, have a M3 with an additional triangle, a deeper BRA1, and more elongate PL. The sage brush vole, extant in western North America and generally assigned to *Lagurus*, has a similar M3 but has cement in the reentrants. It also has a dentine tract on T7 of the m1.

The similarities of occlusal pattern between *Guildayomys* and *Pliophenacomys* suggests that perhaps the latter may stand in an ancestral position to the former. *Guildayomys* relationship to *Pliolemmus*, advanced voles with cement in their reentrants, and *Lagurus* is not as obvious. Except for the two isolated m1s from the Mullen and White Rock, the taxon is known only from the material in the Big Springs l.f. Additional material from North America and a detailed study of the Asian material is necessary before any definitive statement on relationships can be made.

Hibbardomys, new genus
(Figs. 3–5)

Diagnosis.—*Hibbardomys* is a primitive vole with rooted teeth, lacking cement in the reentrant angles. The lingual reentrants of the lower molars are narrower than in *Pliophenacomys*. The reentrants tend to constrict and turn anteriorly as they approach the midline of the tooth. The enamel on the occlusal surface of the lower molars is differentiated into thick and thin segments. The anterior faces of the alternating triangles, the ACC, and both faces of the PL exhibit thick enamel, whereas the enamel is thin on the posterior faces of the alternating triangles.

The m1 is composed of a PL, three to seven alternating triangles, and an AC. The complexity of the AC varies. It is more complex in specimens with fewer triangles and becomes simpler as triangles are added. A well developed sixth triangle is seldom, if ever, present. In primitive species what would be counted as the sixth triangle is rectangular in shape and appears to be an outgrowth of the AC; a condition generally referred to as trilobate when seen in other Blancan voles (Hibbard and Zakrzewski, 1967). In advanced species T6 tends to be blunt or rounded. In a few specimens T6 may even appear to be lacking, but no matter what the configuration a well developed dentine tract is always present where T6 would form. The buccal side of the AC generally lacks a dentine tract but a very thin tract may occur on the anterior face of the AC.

The M3 is composed of an AL, three alternating triangles, and a PL. Though smaller than T2, T1 and 3 are better developed than in *Pliophenacomys*. The BRA1 is well developed (≥ 0.35 mm). The posterior lingual reentrants of M1 (LRA3) and M2 (LRA2) tend to be shallower than in *Pliophenaco-*

mys. Reentrant pits (Zakrzewski, 1969, fig. 7c) are variously developed on both sides of the last triangle on the M1 and M2. The height of the dentine tract on the lingual side of the AL on the M1 is significantly smaller than the tract on the buccal side.

Type species.—*Hibbardomys voorhiesi,* new species.

Etymology.—Named for the late Claude W. Hibbard.

Recognized species and occurrences.—*H. marthae*, new species, White Rock and Big Springs; *H. skinneri*, new species, Sand Draw; *H. fayae*, new species, Dixon; *H. voorhiesi*, new species, White Rock, Mullen, and Big Springs.

Hibbardomys marthae, new species
(Fig. 3A–B)

Ogmodontomys sp.; Eshelman, 1975, Papers Paleo., Univ. Michigan, 13:38–39.
Pliophenacomys cf. *P. primaevus* Hibbard; Eshelman, 1975, Papers Paleo. Univ. Michigan, 13:43–44.

Holotype.—UNSM 51279, partial right dentary with i, and m1.

Horizon and locality.—Long Pine formation, Big Springs l.f., Antelope Co., Nebraska.

Paratypes.—Big Springs, UNSM 52770, 52905, 52906, 52908, 52910, 52912, 90832–90836, 11 mls; 90837, M1; 90838–90844, 7 M2s; 52898, 52900, 2 M3s. White Rock—UM-K1-66: UM 61657, m1; 81750, M3; UM-K3-69: 61858, 3 mls; UM-K4-72: 61743, M1, 81751, 2 M2s; UM-K5-72: 61744, 5 M1s, 81752, 4 M2s.

Diagnosis.—*Hibbardomys marthae* is distinguished from other members of the genus by its shorter dentine tracts and the complexity of the ACC on the m1.

Etymology.—Named for my wife, Martha.

Description of holotype.—The m1 (Fig. 3A) consists of a PL, three alternating triangles and a complicated ACC. The ACC is composed of a fourth triangle that is modified by a *Mimomys*-ridge and prism fold and a fifth triangle that is confluent with a very small third lobe at the position of the sixth triangle. These in turn open into a small AC.

The enamel of the occlusal surface is interrupted at the third lobe = T6 by a well developed dentine tract on the buccal side of the ACC. Dentine tracts are slightly developed on the buccal side of the tooth on PL and T2. The actual height of the tracts and crown height of the tooth could not be measured because the base of the tooth is covered by the dentary. The occlusal length of the tooth is 2.90 mm, occlusal width 1.41 mm.

The m1 is two rooted and lacks cement in the reentrant angles. The majority of the reentrant angles are narrow and their apices are oriented anteriorly. However, BRA3 is round, shallow and its "apex" is directed posteriorly. The apex of LRA4 is perpendicular to the midline of the tooth.

With the exception of the PL the enamel of the posterior faces of the remaining triangles is thinner than that of the anterior faces. The enamel of the anterobuccal face of the AC appears to be thinner as well.

The dentary is broken posterior to the alveoli for the m3. Muscle attachments for the masseter complex on the buccal side of the jaw are well developed. There is no evidence of a temporal fossa between the tooth row and ascending ramus.

Description of paratypes.—The other m1s are generally similar in occlusal pattern and dentine tract development. The major differences between the teeth are in the relative complexity of the ACC. The complexity is determined by the presence or absence of a prism fold, enamel ridge, or a trilobed pattern. All m1s have at least one of these characters developed, some have two. A summary of the variation seen in the ACC is listed in Table 2. Twelve of the 16 m1s have the enamel interrupted at the position of T6. The tooth that shows enamel interruption at the earliest age is one with a crown height of 2.74 mm. The shortest tract where the crown is uninterrupted is at 1.70 mm. Measurements of other parameters of the m1 are listed in Table 3.

The M1 and M2 are assigned to this taxon because of the development of reentrant pits (Fig. 3B). The M3s are assigned because of occlusal pattern.

Hibbardomys skinneri, new species
(Fig. 3F–H)

Pliophenacomys primaevus Hibbard; Hibbard, 1972 (in part), Bull. Amer. Mus. Nat. Hist., 148:102–105.

Table 2.—*Variation of ACC of m1 in* Hibbardomys marthae.

Locality	Character				
	Prism fold	Enamel ridge	Third lobe	PF 3L	ER 3L
Big Springs	1	1	4	5	1
White Rock	0	0	0	4	0

Holotype.—UM 57192, Rm1.

Horizon and type locality.—Keim formation, UM-Nb3-67 Sand Draw l.f., Brown Co., Nebraska.

Paratype.—Magill: UM 57017, m1; 81753, 2 m2s; 81754, m3; 81755, 3 M1s; 81756, 2 M2s; 81757, M3; Owl Pellet: 81758, m3: 81759, M2.

Diagnosis.—*Hibbardomys skinneri* is distinguished from *H. marthae* by its slightly larger size, higher dentine tracts, and less complicated ACC on m1. It differs from other members of the genus by its lower dentine tracts and slightly complex ACC on m1.

Etymology.—Named for Morris F. Skinner, American Museum of Natural History.

Description of holotype.—The holotype represents an individual in an early adult stage of development. The tooth (Fig. 3F–H) is composed of a PL, three alternating triangles, and a slightly complex ACC. The ACC is composed of a fourth and fifth alternating triangle; T5 is confluent with what could be considered a sixth triangle but the latter is rectangular in shape. These two triangles open slightly into a simple AC. The occlusal length of the tooth is 2.87 mm, the width is 1.32 mm. Crown

Table 3.—*Measurements of m1 in* Hibbardomys marthae.

Locality	N	Mean	SD	Observed range	N	Mean	SD	Observed range
		GL				GW		
Big Springs	12	2.74	0.14	2.46–3.00	12	1.36	0.07	1.23–1.47
White Rock	2	2.61	0.06	2.56–2.65	2	1.31	0.01	1.30–1.31
		GH				DTH T6		
Big Springs	11	1.99	0.55	1.15–2.74	3	2.18	0.52	1.70–2.74*
White Rock	2	2.07	0.85	1.47–2.67	1	2.16		
		HBDT				HLDT		
Big Springs	11	0.68	0.07	0.60–0.82	11	0.37	0.04	0.32–0.45
White Rock	2	0.69	0.08	0.63–0.74	2	0.32	0.01	0.31–0.33

N = number of specimens; SD = standard deviation; GL = greatest length; GW = greatest width; GH = greatest height; DTH T6 = dentine tract height of triangle 6; HBDT = height of buccal tract on PL; HLDT = height of lingual tract on PL; * = occlusal pattern interrupted, therefore, may not be representative of maximum height; if on lower value, may not be minimum height.

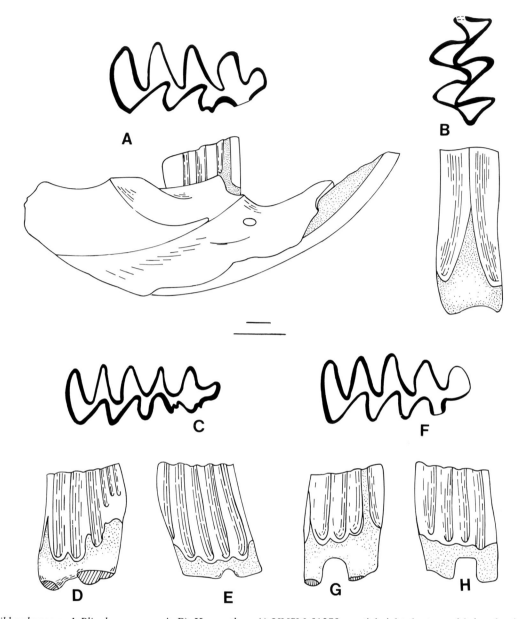

Fig. 3.—*Hibbardomys* and *Pliophenacomys*. A–B) *H. marthae*; A) UNSM 51279, partial right dentary with i and m1, type, occlusal view of tooth, buccal view of dentary; B) UNSM 90837, LM1, occlusal and posterior views. C–E) *P. primaevus,* UM 57183, Rm1, C) occlusal view, D) buccal view, E) lingual view. F–H) *H. skinneri,* UM 57192, Rm1, type, F) occlusal view, G) buccal view, H) lingual view. Short bar equals 1 mm on dentary and side views; long bar equals 1 mm on occlusal and posterior views.

height is 2.82 mm. Dentine tract height on the PL is 0.83 mm on the buccal side and 0.30 mm on the lingual. The height of the dentine tract on T6 is sufficient to interrupt the enamel of the occlusal surface at this point.

Description of paratypes.—The other m1 (UM 57017) differs from the type in that T6 though not as well developed is highly crenulated. In addition a shallow LRA5 is developed on the ACC

that delimits a small seventh triangle. Measurements of this tooth are listed in Table 4. The other teeth are assigned to *Hibbardomys* because they have the generic characters. Their measurements are listed in Tables 5–9.

Hibbardomys fayae, new species
(Fig. 4A–F)

Pliophenacomys meadensis Hibbard; Hibbard, 1956 (in part), Papers, Michigan Acad. Sci., Arts & Letters, 41:164–167.

Table 4.—*Measurements of ml in* Hibbardomys *and* Pliophenacomys.

Taxon	N	Mean	SD	Observed range	N	Mean	SD	Observed range
			GL				GW	
H. skinneri	2	2.93	0.08	2.87–2.99	2	1.40	0.11	1.32–1.47
H. fayae	8	2.92	0.25	2.45–3.16	7	1.26	0.08	1.11–1.37
H. voorhiesi[1]	15	2.89	0.15	2.58–3.03	15	1.27	0.07	1.15–1.37
H. voorhiesi[2]	12	2.96	0.19	2.66–3.55	12	1.30	0.07	1.19–1.38
H. voorhiesi[3]	1	3.00			1	1.36		
P. primaevus	1	2.86			1	1.22		
P. dixonensis	5	2.91	0.12	2.78–3.03	5	1.22	0.08	1.10–1.30
P. osborni[1]	10	2.93	0.12	2.75–3.13	10	1.19	0.05	1.11–1.26
P. osborni[2]	7	3.12	0.18	3.00–3.40	9	1.22	0.09	1.11–1.32
P. osborni[3]	2	3.08	0.35	2.83–3.32	2	1.25	0.23	1.09–1.41
			GH				HLDT	
H. skinneri	2	2.72	0.13	2.62–2.81	2	0.30	0.01	0.29–0.30
H. fayae	8	2.52	0.95	1.24–4.11	8	0.63	0.15	0.37–0.88
H. voorhiesi[1]	15	3.58	0.79	1.60–4.43	I			
H. voorhiesi[2]	10	2.86	0.86	1.34–4.02	I			
H. voorhiesi[3]	1	3.00			I			
P. primaevus	1	3.63			1	0.74		
P. dixonensis	5	2.67	0.32	2.39–3.20	5	1.96	0.30	1.68–2.39*
P. osborni[1]	10	3.50	0.57	2.44–4.30	I			
P. osborni[2]	9	2.67	0.84	1.34–3.54	I			
P. osborni[3]	2	3.35	0.08	3.29–3.41	I			

[1] Big Springs.
[2] White Rock.
[3] Mullen.
I specimens have interrupted enamel, except for immature, see text for measurements. See Table 3 for other abbreviations.

Gen. et *sp.* indet.; Hibbard and Zakrzewski, 1967, Contr. Mus. Paleo. Univ. Michigan, 21:262.
Pliophenacomys? *osborni* Martin; Eshelman 1975, Univ. Michigan Papers Paleo., 13:41–43.

Holotype.—UM 54948, Lm1.

Horizon and type locality.—"Belleville" formation (Hibbard, 1972), Dixon l.f., Kingman Co.; Kansas.

Paratypes.—UM 54945, 54947, 55445, 12 m1s; 54950, 55446, 55449, 11 m2s; 54954, 55448, 55453, 4 m3s; 54951, 54953, 54954, 55455, 14 M1s; 55450, 55451, 55456, 10 M2s; 54952, 55458, 5 M3s.

Diagnosis.—Hibbardomys *fayae* has dentine tracts and anatomical roots that are intermediate in development between *H. skinneri* and *H. voorhiesi.*

Etymology.—Named for Faye Ganfield Hibbard, wife of the late Claude W. Hibbard.

Description of the holotype.—The holotype represents an individual that is in an early adult stage of development. The tooth (Fig. 4A–C) consists of a PL, three closed, alternating triangles, and a complex ACC. The ACC consists of five additional triangles and a small AC. Triangles 4 and 5 form a pair in continuity with T1–3. Triangle 6 has a rect-

angular rather than a triangluar shape. The anterior edge of the enamel on the occlusal surface of T6 is interrupted by a dentine tract along its side. Triangles 7 and 8 are confluent and open broadly into the AC. Apices of the reentrant angles are directed anteriorly except for the most anterior three, which are directed posteriorly. BRA4 is shallower than it is in *H. voorhiesi.*

The tooth has an occlusal length of 2.98 mm, width of 1.22 mm and a crown height of 3.26 mm. In addition to the dentine tract on T6, tracts are present on both sides of the PL. The buccal tract is 2.80 mm, the lingual 0.88 mm.

Description of paratypes.—The other m1s are similar to the holotype, except that their ACCs are not as complex. Six of the seven paratypic m1s, wherein T6 could be observed, have the enamel of the occlusal surface interrupted at that position. The other specimen represents an immature individual with a crown height of 4.11 mm. The height of the tract on T6 in this individual is 3.59 mm. Five of the m1s also have the enamel interrupted on the buccal side of the PL. It seems that the enamel will always be interrupted on the buccal side when wear on the crown reduces its height to approximately 2.35 mm. The highest crowned tooth in which the enamel is interrupted measures 2.37 mm, and the shortest tract that can be measured on a tooth that does not possess interrupted enamel is also 2.37 mm. The highest buccal tract that can be measured is on the immature individual men-

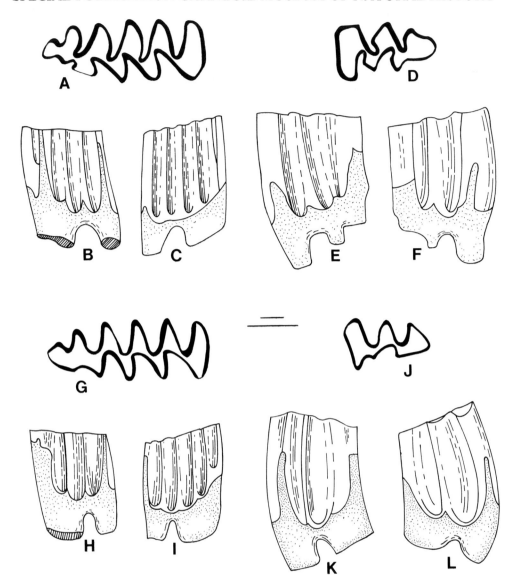

Fig. 4.—*Hibbardomys* and *Pliophenacomys*. A–F) *H. fayae*; A–C) UM 54948, Lm1, type; A) occlusal view, B) buccal view, C) lingual view. D–F) UM 55458, RM3; D) occlusal view, E) buccal view, F) lingual view. G–L) *P. dixonensis*; G–I) UM 55444, Lm1, type, G) occlusal view, H) buccal view, I) lingual view. J–L) UM 55452, RM3, J) occlusal view, K) buccal view, L) lingual view. Short bar equals 1 mm on side views of m1s; long bar equals 1 mm on occlusal view of all teeth and side views of M3s.

tioned above, 2.85 mm. The enamel on the lingual side of the PL will not be interrupted until crown height wears down to 0.88 mm, which is the height of this tract on the type. Measurements on the m1s are listed in Table 4. The remaining teeth were assigned on the basis of generic characters. Six of the 11 m2s have the enamel interrupted on the buccal side of the PL, whereas only two teeth have the enamel interrupted on the lingual side. Measurements of the m2 are listed in Table 5, those of m3 in Table 6.

The M1s of *Hibbardomys* can be easily distinguished from those of *Pliophenacomys* in the Dixon l.f. by possessing a shal-

lower LRA3, shorter dentine tracts (although there is no appreciable difference between the crown heights of the two taxa no *Hibbardomys* specimens exhibits interrupted enamel on the occlusal surface), and anatomical roots are better developed (all 14 *Hibbardomys* M1s have three roots, whereas two of six *Pliophenacomys* M1s have two roots). Measurements of M1 are listed in Table 7.

The M2s of *Hibbardomys* are distinguished from *Pliophenacomys* by a shallower LRA2 and less developed reentrant pits. Unlike the M1s the dentine tracts on the AL of the M2s in *H. fayae* are high enough so that two specimens have an interrupted

Table 5.—*Measurements of m2 of* Hibbardomys *and* Pliophenacomys.

Taxon	N	Mean	SD	Observed range	N	Mean	SD	Observed range
			GL				GW	
H. skinneri	1	1.82			2	1.22	0.10	1.15–1.29
H. fayae	11	1.69	0.12	1.55–1.91	11	1.20	0.09	0.98–1.29
H. voorhiesi[1]	5	1.72	0.08	1.61–1.81	5	1.20	0.08	1.11–1.32
H. voorhiesi[2]	15	1.68	0.08	1.60–1.79	16	1.17	0.08	1.05–1.32
H. primaevus	3	1.74	0.03	1.71–1.77	3	1.13	0.06	1.07–1.17
P. dixonensis	5	1.72	0.05	1.67–1.79	5	1.09	0.05	1.02–1.15
P. osborni[1]	2	1.76	0.19	1.62–1.89	2	1.12	0.11	1.04–1.20
P. osborni[2]	5	1.66	0.09	1.56–1.75	5	1.06	0.09	0.92–1.14
			GH					
H. skinneri	2	2.39	0.51	2.03–2.75				
H. fayae	11	2.18	0.80	1.08–3.12				
H. voorhiesi[1]	5	2.76	0.56	2.35–3.68				
H. voorhiesi[2]	15	2.41	0.64	1.65–3.59				
P. primaevus	3	2.54	0.43	2.07–2.92				
P. dixonensis	5	2.09	0.57	1.13–2.57				
P. osborni[1]	2	3.01	0.31	2.79–3.23				
P. osborni[2]	5	2.55	0.28	2.30–2.89				
			HBDT				HLDT	
H. skinneri	2	1.00	0.21	1.15–1.29	2	0.52	0.28	0.32–0.71
H. fayae	5	2.23	0.22	1.90*–2.44	10	0.93	0.28	0.59–1.34*
H. voorhiesi[1]	2	2.96	1.02	2.24–3.68*	2	2.64	1.48	1.59–3.68*
H. voorhiesi[2]	3	3.14	0.40	2.78*–3.57	6	2.40	0.80	1.54–3.35
P. primaevus	3	1.47	0.21	1.28–1.70	2	0.99	0.15	0.88–1.09
P. dixonensis	2	2.35	0.32	2.12*–2.57*	4	1.44	0.17	1.23*–1.65
P. osborni[1]	2	2.00	1.90	0.66–3.34*	2	1.88	2.07	0.41–3.34*
P. osborni[2]	2	2.25	0.35	2.00–2.49	3	2.23	0.35	1.89–2.58

See Table 3 for abbreviations.

enamel pattern on the buccal side of the AL and two specimens have both sides interrupted. Measurements of M2 are listed in Table 8.

The M3s of *Hibbardomys fayae* are distinguished by their occlusal pattern (Fig. 4D–F). Two of the M3s have interrupted enamel on both sides of the AL. One of these specimens has a crown height of 2.04 mm. This is the highest tract present on the lingual side. Measurements of the M3 are listed in Table 9.

Hibbardomys voorhiesi, new species
(Fig. 5)

Pliophenacomys osborni Martin, 1972 (in part), Bull. Nebraska St. Mus., 9:174–176.
Pliophenacomys? osborni Martin; Eshelman, 1975 (in part), Univ. Michigan Mus. Paleo., Papers Paleo., 13:41–43.

Holotype.—UM 55523, partial left dentary with m1–m2.

Horizon and type locality.—Belleville fm., White Rock l.f., Loc. UM-K1-66, Republic Co., Kansas.

Paratypes.—White Rock l.f.—UM-K1-66: 81760, 3 m1s; 81761, 4 m2s; 81762, m3; 60590, 6 M1s; 81763, 2 M2s; 81764, M3. UM-K3-69: 61736, 24 m1s; 81765, 20 m2s; 81766, 4 m3s; 81767, 29 M1s; 81768, 17 M2s; 81769, 8 M3s. UM-K4-72: 61737, 2

m1s; 81770, 2 m2s; 81771, 2 m3s; 81772, 3 M1s; 81773, 2 M2s. UM-K4-72 + 12′ SE: 81774, 2 m1s; 81775, m2; 81776, 2 M1s. UM-K4-72 + 200′ NW: 81777, m1; 61739, 2 M1s. UM-K5-72: 61740, 11 m1s; 81788, 10 m2s; 81779, 2 m3s; 81780, 10 M1s; 81781, 3 M2s; 81782, 7 M3s. UM-K6-72: 81783, 2 m1s; 81784, 2 m2s; 81785, 2 M3s. UM-K7-72: 61742, M1. Mullen l.f.— UNSM Coll Loc. Cr. 10, Pit 3: 39570–39571, 2 m1s. Big Springs l.f.: 52697–52699, 52703–52704, 52706–52711, 52715–52718, 52722–52725, 52728, 52732, 52742, 52750, 52753, 52755–52759, 52768, 52771, 52773, 52775, 52779–52780, 52782, 52785, 52788–52789, 52791–52793, 52795, 90845, 44 m1s; 90846–90850, 5 m2s; 90851, m3; 90852–90902, 51 M1s; 90903–90910, 8 M2s; 52874–52875, 52877–52880, 52883–52886, 52888–52889, 52891–52892, 90911, 15 M3s.

Diagnosis.—*Hibbardomys voorhiesi* is distinguished from the other members of the genus by its greater hyposodonty and higher dentine tracts.

Etymology.—Named for Michael R. Voorhies, University of Nebraska-Lincoln.

Description of holotype.—The holotype (Fig. 5A) represents an individual in an adult stage of development. The back of the jaw posterior to the alveoli

Table 6.—*Measurements of m3 in* Hibbardomys *and* Pliophenacomys.

Taxon	N	Mean	SD	Observed range	N	Mean	SD	Observed range
		GL				GW		
H. skinneri	2	1.66	0.18	1.53–1.79	2	1.11	0.04	1.08–1.13
H. fayae	4	1.63	0.20	1.34–1.79	4	1.00	0.10	0.85–1.08
H. voorhiesi[1]	1	1.56			1	0.92		
H. voorhiesi[2]	4	1.55	0.14	1.37–1.71	4	0.96	0.08	0.87–1.03
P. primaevus	2	1.56	0.02	1.54–1.57	2	0.97	0.07	0.92–1.02
P. dixonensis	1	1.33			1	0.86		
P. osborni[1]	1	1.46			1	0.90		
P. osborni[2]	5	1.46	0.10	1.36–1.61	5	0.93	0.05	0.85–0.98
		GH						
H. skinneri	2	1.90	0.20	1.34–1.79				
H. fayae	4	1.68	0.36	1.15–1.92				
H. voorhiesi[1]	1	3.40						
H. voorhiesi[2]	4	2.04	0.88	1.41–3.21				
P. primaevus	2	1.82	0.33	1.58–2.05				
P. dixonensis	1	2.40						
P. osborni[1]	1	1.87						
P. osborni[2]	5	2.26	0.10	1.57–2.24				
		HBDT				HLDT		
H. skinneri	2	0.65	0.11	0.57–0.73	2	0.44	0.16	0.32–0.55
H. fayae	3	1.25	0.48	0.77–1.73	2	0.97	0.18	0.84–1.10
H. voorhiesi[1]	1	2.42			1	2.42		
H. voorhiesi[2]	2	1.98	0.33	1.75–2.21*	3	1.59	0.28	1.27–1.80
P. primaevus	2	0.64	0.31	0.42–0.86	2	0.61	0.06	0.57–0.65
P. dixonensis	1	1.90			1	0.88		
P. osborni[1]	1	1.12			1	0.83		
P. osborni[2]	5	1.80	0.26	1.57–2.24*	4	1.48	0.23	1.27–1.88

See Table 3 for abbreviations.

for m3 is missing. The m1 consists of a PL, seven alternating triangles, and a simple AC. The PL is closed off from T1. A thin strip of dentine successively joins each of the first six alternating triangles. Triangle 6 opens broadly into T7. Seven, in turn, opens into the AC. Dentine tracts are present on both sides of the PL, T6, and the anterobuccal edge of the AC. The tracts are high enough so that the enamel of the occlusal surface is interrupted at these points. The tract on T6 is so developed that the triangle has a rounded appearance.

The apices of the reentrant angles are directed anteriorly except for LRA 5, which is perpendicular.

The m2 consists of a PL, three alternating triangles, and a small oval-shaped anterior loop or fourth triangle. Details of enamel closure and thickness, reentrant direction, and dentine tract development are similar to that of the m1. The major difference is that the entire anterior face of the fourth triangle lacks enamel.

The occlusal length of m1–m2 is 4.68 mm. The length of m1 is 3.10 mm and m2, 1.58 mm. The width of the teeth is 1.30 and 1.20 mm, respectively, crown and dentine tract height could not be measured.

The dentary is too fragmentary to provide for an adequate description. The diastemal region is short and the crests for muscle attachment are poorly developed. But whether these are primary or secondary conditions is not determinable. The dental foramen is in a more ventral position than it is in *Pliophenacomys*. It appears that a very slight masseteric fossa may have been present between the m3 and the ascending ramus, but a dentary in which this area is better preserved is needed before a more positive statement can be made.

Description of paratypes.—Although variation is present, the remaining m1s are well characterized in the description of the holotype. The pattern of the ACC with its rounded T6 on which the enamel is interrupted because of the dentine tract height appears to be so consistent that m1s of this taxon can be separated from all others so far known with relative ease.

Only one (UNSM 52755) of the 27 m1s that could be measured does not have an interrupted occlusal surface at the site of T6.

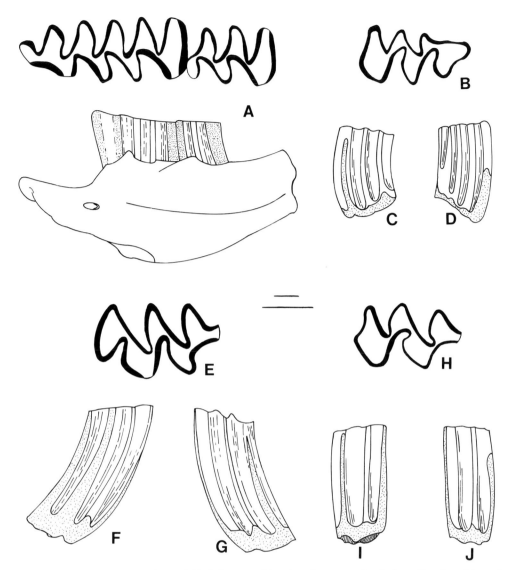

Fig. 5.—*Hibbardomys voorhiesi.* A) UM 55523, partial left dentary with m1–m2, type, occlusal view of teeth and buccal view of dentary. B–D) UNSM 90911, LM3; B) occlusal view, C) lingual view, D) buccal view. E–G) UNSM 90852, LM1, E) occlusal view, F) lingual view, G) buccal view. H–J) UNSM 90903, LM2, H) occlusal view, I) lingual view, J) buccal view. Short bar equals 1 mm on side views, long bar equals 1 mm on occlusal views.

This specimen represents an immature individual that has a crown height of 4.32 mm. The tract height of T6 is 4.08 mm. This measurement may represent a minimum height for this tract in the taxon as all m1s with a crown height of less than 4.00 mm have interrupted enamel, and three m1s that have a crown height equal to or greater than 4.32 mm have the enamel interrupted at the site of T6. These three teeth represent immature individuals as well and are from the Big Springs l.f. Three of the four m1s mentioned above and an immature individual from the White Rock l.f. lack interrupted enamel on the PL as well. The crown height and the tract height of the buccal tract and lingual tract, respectively are 4.06, 3.71, 3.21; 4.32, 3.90, 3.78; 4.32, I, 4.25;

4.43, 3.85, 3.70 mm. The immature specimen from the Big Springs that exhibits interrupted enamel on the PL has a crown height of 4.37. These measurements may represent the potential range in tract height at these sites. Other measurements of the m1s are listed in Table 4.

Isolated m2s and m3s were considered to pertain to this taxon if they met the generic criteria of constricted and anteriorly directed reentrant angles, and thick enamel on the posterior face of the PL. Four of 20 m2s do not exhibit interrupted enamel on the buccal side of the PL and seven of 20 do not exhibit this condition on the lingual side. Measurements of tract height and other parameters of the m2s are listed in Table 5. Ranges of tract

Table 7.—*Measurements of M1 in* Hibbardomys *and* Pliophenacomys.

Taxon	N	Mean	SD	Observed range	N	Mean	SD	Observed range
		GL				GW		
H. skinneri	2	2.49	0.18	2.36–2.62	2	1.57	0.12	1.48–1.65
H. fayae	14	2.40	0.13	2.17–2.64	14	1.50	0.09	1.32–1.66
H. voorhiesi[1]	10	2.40	0.09	2.27–2.52	10	1.48	0.08	1.36–1.61
H. voorhiesi[2]	22	2.33	0.13	2.07–2.55	22	1.44	0.11	1.14–1.58
P. dixonensis	6	2.39	0.05	2.33–2.46	6	1.40	0.07	1.31–1.50
P. osborni[1]	1	2.42			1	1.43		
P. osborni[2]	9	2.37	0.08	2.25–2.50	9	1.41	0.05	1.30–1.48
		GH				LLRA3		
H. skinneri	2	3.25	0.07	3.20–3.30	2	0.25	0.02	0.23–0.26
H. fayae	14	3.07	0.78	2.11–4.73	14	0.21	0.03	0.17–0.27
H. voorhiesi[1]	10	4.71	0.55	3.38–5.22	10	0.24	0.03	0.20–0.29
H. voorhiesi[2]	22	4.00	0.56	2.50–4.85	22	0.23	0.03	0.18–0.29
P. dixonensis	6	2.79	0.82	1.65–3.90	5	0.35	0.03	0.32–0.39
P. osborni[1]	1	4.37			1	0.39		
P. osborni[2]	9	3.73	0.98	1.68–4.94	8	0.35	0.03	0.27–0.37
		HBDT				HLDT		
H. skinneri	2	0.65	0.13	0.55–0.74	2	0.29	0.12	0.20–0.37
H. fayae	14	1.24	0.33	0.63–1.67	14	0.54	0.14	0.37–0.84
H. voorhiesi[1]	5	4.57	0.59	3.73–5.19*	9	3.34	0.89	1.44–4.26*
H. voorhiesi[2]	15	3.38	0.88	1.42–4.59*	22	1.21	0.80	0.32–2.81
P. dixonensis	4	2.60	0.31	2.37*–3.06	6	1.77	0.25	1.33–2.03*
P. osborni[1]	1	3.79			1	2.52		
P. osborni[2]	4	3.74	0.25	3.45–4.04	7	2.61	0.94	1.23–3.66

See Table 3 for abbreviations.

height on the PL of m3 cannot be established with certainty because of the small sample (N = 5). Two teeth do not exhibit interrupted enamel on the buccal side of the PL; whereas only one has the enamel interrupted on the lingual side of the PL. This tooth has a crown height of 1.41 mm. The potential range of tract heights and other measurements of m3 are listed in Table 6.

The M1 is composed of an AL and four alternating triangles (Fig. 5E–G). Triangle 1 tends to be rounded rather than triangular. The AL and triangles are either closed off from each other or a very thin strip of dentine may connect successive triangles. The enamel of the anterior faces of the triangles is thinner than on the posterior faces of the AL. Dentine tracts are present on both sides of the AL, T1, and the posterior end of T4. The tracts are shorter on the lingual side of the AL so that the enamel pattern is interrupted later in the wear of the tooth than on the buccal side. None of the M1s from White Rock exhibit an interrupted enamel pattern on the lingual side of the AL and only two of ten from the Big Springs show interruption at this site. Potential range of tract heights as well as other measurements are given in Table 7. Apices of the reentrant angles tend to be directed posteriorly except for LRA3, which, if present, is very shallow and directed anteriorly. Anatomical roots do not develop until relatively late in the life of an individual. The majority of teeth that have developed anatomical roots have two, a few have three. The third root, if present, lies under T1 and is greatly reduced. More generally it fuses with the anterior root, which underlies the AL, to form a single large root.

The M2 is similar in occlusal pattern to the M1 except that it has one less triangle and the differences between the dentine tract height on the sides of the AL do not appear to be as great (Fig.

5H–J). All seven M2s from the Big Springs exhibit an interrupted enamel pattern on the buccal side of the AL. The maximum crown height from this sample is 4.91 mm. Three of the seven teeth do not exhibit an interrupted pattern on the lingual side. But the heights of these tracts fall within the range exhibited by those teeth that are interrupted. Nine of the 13 M2s exhibit interrupted enamel on the buccal side of the AL and eight of 13 are interrupted on the lingual side. The range of tract heights are listed in Table 8 along with other measurements of the M2s. The tooth has two roots.

The M3 consists of an AL, three alternating triangles and a PL. Infrequently, a shallow LRA 3 is present on the PL so that a very small fourth triangle is formed (Fig. 5B–D). Dentine tracts are present on the sides of the AL and the posterior face of the PL. The enamel of the triangles is thinner on the anterior faces and thicker on the posterior, whereas the enamel of the loops is of equivalent thickness on both faces. Measurements of M3 are found in Table 9. The tooth has two roots.

Discussion.—That the specimens assigned to *Hibbardomys* might represent a distinct taxon has been known for some time (Hibbard and Zakrzewski, 1967). As mentioned above, Eshelman (1975) questionably placed the specimens from the White Rock into *Pliophenacomys* and stated that the White Rock and Mullen specimens appeared to be more similar to *Pliomys deeringensis* Guthrie and Matthews from the Cape Deceit l.f. (Irvingtonian of Alaska) than to

Table 8.—*Measurements of M2 in* Hibbardomys *and* Pliophenacomys.

Taxon	N	Mean	SD	Observed range	N	Mean	SD	Observed range
		GL				GW		
H. skinneri	3	2.11	0.18	1.91–2.22	3	1.40	0.14	1.24–1.49
H. fayae	10	1.98	0.05	1.91–2.08	10	1.31	0.06	1.22–1.45
H. voorhiesi[1]	7	2.00	0.07	1.88–2.07	7	1.34	0.06	1.28–1.42
H. voorhiesi[2]	13	1.99	0.06	1.86–2.07	13	1.33	0.08	1.23–1.47
P. primaevus	4	2.11	0.10	2.00–2.23	4	1.37	0.12	1.22–1.49
P. dixonensis	1	1.95			1	1.16		
P. osborni[1]	1	2.05			1	1.28		
P. osborni[2]	7	1.98	0.06	1.88–2.07	7	1.27	0.05	1.21–1.34
		GH				LLRA2		
H. skinneri	3	2.74	0.33	2.36–2.93	3	0.22	0.05	0.17–0.25
H. fayae	10	3.17	0.66	1.73–3.77	9	0.21	0.03	0.17–0.27
H. voorhiesi[1]	7	3.94	0.54	3.34–4.91	7	0.24	0.03	0.22–0.28
H. voorhiesi[2]	13	3.02	0.82	1.44–4.08	13	0.23	0.03	0.20–0.28
P. primaevus	4	2.82	0.96	1.52–3.81	4	0.30	0.02	0.28–0.33
P. dixonensis	1	3.26			1	0.30		
P. osborni[1]	1	4.54			1	0.39		
P. osborni[2]	7	2.66	0.45	2.16–3.48	7	0.34	0.03	0.29–0.38
		HBDT				HLDT		
H. skinneri	3	0.77	0.06	0.72–0.82	3	0.80	0.12	0.69–0.93
H. fayae	6	2.70	0.45	1.86*–3.10	7	2.28	0.23	1.97*–2.65
H. voorhiesi[1]	7	I			4	3.75	0.16	3.52–3.90*
H. voorhiesi[2]	5	2.77	1.16	0.76–3.77*	6	3.04	0.40	2.60–3.77
P. primaevus	4	0.83	0.22	0.58–1.08	4	0.87	0.12	0.74–1.02
P. dixonensis	1	2.40			1	2.13		
P. osborni[1]	1	I			1	I		
P. osborni[2]	1	2.87			2	2.49	0.44	2.18–2.80*

See Table 3 for abbreviations.

Table 9.—*Measurements of M3 in* Hibbardomys *and* Pliophenacomys.

Taxon	N	Mean	SD	Observed range	N	Mean	SD	Observed range
		GL				GW		
H. skinneri	1	1.63			1	1.05		
H. fayae	5	1.94	0.23	1.65–2.23	5	1.17	0.08	1.08–1.30
H. voorhiesi[1]	10	1.89	0.20	1.53–2.10	10	1.07	0.09	0.93–1.21
H. voorhiesi[2]	9	1.99	0.10	1.82–2.10	9	1.18	0.06	1.08–1.22
P. primaevus	4	1.90	0.08	1.79–1.99	4	1.08	0.14	0.96–1.27
P. dixonensis	4	1.73	0.11	1.62–1.87	4	1.08	0.07	1.01–1.09
P. osborni[2]	7	1.76	0.10	1.63–1.87	7	1.05	0.10	0.90–1.15
		GH				LLRA1		
H. skinneri	1	2.74			1	0.25		
H. fayae	5	2.32	0.62	1.49–2.94	5	0.48	0.11	0.34–0.63
H. voorhiesi[1]	10	2.85	0.67	1.72–3.95	10	0.47	0.05	0.40–0.55
H. voorhiesi[2]	9	2.23	0.62	1.26–2.98	9	0.51	0.04	0.45–0.58
P. primaevus	4	1.99	0.59	1.12–2.39	4	0.27	0.04	0.22–0.30
P. dixonensis	4	2.16	0.02	2.14–2.19	4	0.26	0.09	0.18–0.35
P. osborni[2]	7	2.33	0.50	1.35–2.75	7	0.26	0.04	0.20–0.31
		HBDT				HLDT		
H. skinneri	1	0.72			1	0.76		
H. fayae	3	1.58	0.55	1.13–2.19	4	1.45	0.40	1.20–2.04*
H. voorhiesi[1]	6	2.20	0.77	0.80–3.06	6	2.31	0.64	1.48–3.07
H. voorhiesi[2]	3	2.11	0.49	1.56–2.48*	2	2.57	0.55	2.18–2.96*
P. primaevus	4	0.82	0.13	0.68–1.00	4	0.91	0.12	0.82–1.09
P. dixonensis	4	1.61	0.15	1.45–1.74	4	1.55	0.26	1.35–1.93
P. osborni[2]	5	2.31	0.15	2.11–2.49	3	2.37	0.23	2.16–2.62

See Table 3 for abbreviations.

the *Pliophenacomys* from the Fox Canyon l.f. based on the height of the dentine tracts and thickness of enamel. I was of a similar opinion until I had the opportunity to examine the specimens in detail.

As can be seen from their diagnoses *Hibbardomys* is distinct from *Pliophenacomys* on the basis of the occlusal pattern of the teeth, especially m1 and M3. *Hibbardomys* is distinct from *Pliomys* for the same reason. *Pliophenacomys* and *Pliomys* are similar in the occlusal pattern of the m1 and M3. *Pliophenacomys* and *Pliomys* may be congeneric but I did not have a large enough sample of *Pliomys* on which to base a conclusion.

Based on the small sample of *Pliomys deeringensis* I examined it appears that the species belongs to neither *Hibbardomys* nor *Pliomys* but, again, a larger sample is needed before a judgement can be made.

Characteristics of *Hibbardomys* such as the occlusal pattern of M3, reentrant pits on the posterior triangles of M1 and M2, prism folds on the m1 of the primitive species, and constricted, anteriorly-directed apices on the reentrants of the lower molars suggest that it may have evolved from some species of *Cosomys*. Only in *Cosomys* are all of the above characters found, but no *Cosomys* has a well developed dentine tract at the position of T6. *Ogmodontomys* and *Ophiomys* are other possibilities, but I know of no *Ophiomys* with reentrant pits on the upper teeth and although primitive species have a complex ACC, prism folds and pits seem to be lost fairly early in phylogeny of the genus. A similar statement can be made regarding *Ogmodontomys* and in addition it retains a three rooted M3 and the lingual reentrants of the m1 in the most advanced species tend to be broad and perpendicular to the midline of the tooth. At present *Hibbardomys* is known from only four sites in the northern and eastern parts of the central Plains. Though not the most common of the arvicolines in the faunas where it is found it appears to be replacing *Pliophenacomys,* which it resembles in gross dental morphology (m1 with five to seven triangles). *Hibbardomys* becomes more abundant with respect to *Pliophenacomys* in the younger White Rock and Big Springs local faunas.

Pliophenacomys Hibbard, 1938
(Fig. 3–4, 6)

Phenacomys (*Pliophenacomys*) Hibbard, 1938; Trans. Kansas Acad. Sci., 40:248–249.
Pliophenacomys Hibbard, 1950; Contr. Mus. Paleo. Univ. Michigan, 8:150–157.

Emended diagnosis.—*Pliophenacomys* is a primitive vole with rooted teeth, lacking cement in the reentrant angles. The lingual reentrants of the lower molars tend to be broader than in *Hibbardomys* with their apical angles directed nearly perpendicular to the midline of the tooth. The apices of the reentrants do not constrict as they approach the midline.

The enamel of the occlusal surface tends to be differentiated into thick and thin segments. The anterior faces of the alternating triangles and posterior loop in the lower molars exhibit thick enamel, whereas the posterior faces of the same elements exhibit thin enamel.

The m1 is composed of a PL, five to seven alternating triangles, and an AC. Dentine tracts on the ACC are confined generally to the buccal side (AC and T6).

The M3 is composed of an AL, two alternating triangles, and a PL. Triangle 1 is poorly developed, confluent with the AL, and generally closed off from T2. The BRA1 is very shallow (≤ 0.35 mm). The M3 has two roots. The posterior lingual reentrants of M1 (LRA3) and M2 (LRA2) tend to be deeper (≥ 0.27 mm) than in *Ogmodontomys, Ophiomys,* or *Hibbardomys.*

Recognized species and occurrence.—*P. finneyi* Hibbard and Zakrzewski, 1972; Fox Canyon; *P. primaevus* Hibbard, 1938; Rexroad Loc. 2 and Sand Draw; *P. dixonensis,* new species; Dixon; *P. osborni* Martin 1972, Mullen, Big Springs, and White Rock.

Pliophenacomys primaevus Hibbard, 1938
(Fig. 3C–E)

Phenacomys (*Pliophenacomys*) *primaevus* Hibbard, 1938; Trans. Kansas Acad. Sci., 40:248–249.
P. primaevus Hibbard, Hibbard 1972 (in part); Bull. Amer. Mus. Nat. Hist., 148:102–105.

Holotype.—KU 3905, right dentary with m1–m2.
Type locality.—Rexroad formation, Rexroad Loc. 2, Meade County, Kansas.

Referred material.—Rexroad Loc. 2, KU 5976 m1, m3. Sand Draw l.f., Magill: UM 57183, m1; UM 81814, m2; UM 81816, 4 M2s; UM 81817, 2 M3s; Owl Pellet: UM 81813, 2 partial m1s; UM 59815, 2 m2s; UM 81815, 2 m3s; UM 81818, 2 M3s; UM 57181, M1–M3.

Emended diagnosis.—*Pliophenacomys primaevus* is slightly more hypsodont and has higher dentine tracts than *P. finneyi.* It has shorter dentine tracts than the remaining taxa in the genus.
Description.—Remains of *Pliophenacomys primaevus* (Fig. 3C–E) possess the characteristics listed

in the generic diagnosis, therefore, additional detailed description is not warranted. Measurements of various parameters of the teeth are listed in Tables 4–9. These measurements suggest that the four species of *Pliophenacomys* are similar in size and separated on the development of the dentine tract on the PL of m1. Although the dentine tract on the buccal side of the posterior loop is much higher in *P. primaevus* than it is in *P. finneyi*, in no specimen is the tract high enough to interrupt the enamel on the occlusal surface until substantial wear has occurred. The dentine tract on the lingual side (Fig. 3E) of the PL is much shorter than that on the buccal (Fig. 3D), a characteristic that can be used to separate *P. primaevus* from *P. osborni*.

Pliophenacomys dixonensis, new species
(Fig. 4G–L)

Holotype.—UM 55444, Lm1.

Paratypes.—UM 54946, 54949, 5 m1s; 55447, 5 m2s; 54946, m3; 81787, 6 M1s; 81788, M2; 55452, 55457, 4 M3s.

Horizon and type locality.—"Belleville" formation (Hibbard, 1972), Dixon l.f., Kingman County, Kansas.

Diagnosis.—*Pliophenacomys dixonensis* is distinguished by the relative height of its dentine tracts, which are higher than those of *P. primaevus* but shorter than those of *P. osborni*.

Etymology.—This species is named for its occurrence in the Dixon l.f.

Description of holotype.—The holotype (Fig. 4G–I) possesses the characters of the genus. It represents an individual in the adult stage of development. The tooth measures 2.99 mm in length, 1.20 mm in width, and 2.69 mm in crown height. The maximum height of the buccal dentine tract on the PL cannot be determined, as the crown has been reduced enough so that the enamel pattern of the occlusal surface is interrupted at this point. The height of this tract was at least 2.69 mm, whereas the tract on the lingual side of the PL is 2.13 mm high. The tract on the anterior face of the AC is also high enough to interrupt the enamel of the occlusal surface at this point.

Description of paratypes.—The remaining m1s are similar to the holotype. Only one of the m1s has a dentine tract on the buccal side of the PL that does not interrupt the enamel of the occlusal surface. This specimen is in an immature stage of development (no anatomical roots have developed). It has a crown height of 3.20 mm and the height of the buccal tract is 2.80 mm.

The remaining m1s have a crown height below 2.70 mm. Measurements of m1 are listed in Table 4.

Only one of the m2s has a dentine tract on the buccal side of the PL that does not interrupt the enamel of the occlusal surface. The height of the tract is 2.12 mm, whereas the crown height is 2.44 mm. A height of 2.12 mm is not the maximum height as m2s with crown heights of 2.25 and 2.57 mm have interrupted enamel patterns on the buccal side. Interrupted enamel is not established on the lingual side of the PL until substantial wear takes place. A m2 with the crown height of 1.13 mm exhibits an interrupted enamel pattern on the lingual side of the PL. The highest tract of 1.65 mm does not exhibit an interrupted pattern. *Pliophenacomys dixonensis* is similar to *P. primaevus* in that the buccal tract is better developed than the lingual tract on the PL in the lower molars.

In addition to having the LRA3 deeper, M1s of *P. dixonensis* can be distinguished from those of *Hibbardomys* by the height of the dentine tracts on the sides of the AL. Buccal tracts are probably greater than 2.00 mm and lingual tracts greater than 1.30 mm. Only one of six M1s had three distinct roots. Three of the six had the small median root fused to the large anterior one, and the remaining two had only two roots.

The only M2 that is known has two roots like *P. primaevus*. Other characters of the M2 and those of the M3 (Fig. 4J–L) are typical of the genus; except for the tract height which is typical for the species. Measurements for the teeth can be found in Tables 5–9.

Pliophenacomys osborni Martin
(Fig. 6)

Pliophenacomys osborni Martin, 1972 (in part), Bull. Nebraska St. Mus., 9:174–176.
Pliophenacomys? osborni Martin; Eshelman, 1975 (in part), Papers Paleo. Univ. Michigan, 13:41–43.

Holotype.—UNSM 39216, partial left dentary with m1–m3.

Type locality.—Unnamed formation, UNSM Coll. Loc. Ho-103, Hooker County, Nebraska.

Referred material.—UNSM Coll. Loc. Cr. 10, Pit 3, 39568, m1; Big Springs l.f., UNSM 52707, 52731, 52734, 52740, 52747, 52749, 52752, 52763, 52772, 52787, 90912, 90913, 12 m1s; 90914, 90915, 2 m2s; 90916, m3; 90917, 90918, 2 M1s; 90919, 90920, 2 M2s; White Rock l.f., UM-K1-66: UM 81789, 3 m1s; 81790, 2 m2s; 81791, M2; 81792, M3. UM-K3-69: 81793, 11 m1s; 81794, 8 m2s; 81795, 4 m3s; 81796, 11 M1s; 81797, 5 M2s; 81798, 2 M3s. UM-K4-72: 81799, 2 m1s; 81800, m3; 81801, 5 M1s; 80802, 3 M2s; 81803, 2 M3s. UM-K4-72 + 12′ SE: 61738, m1; 81804, m2; 81805, M1; 81806, 2 M2s. UM-K5-72: 81807, 2 m2s; 81808, 4 m3s; 81809, 9 M1s; 81810, M2; 81786, 81811, 3 M3s. UM-K6-72: 61741, 2 m1s; 81812, M1.

Emended diagnosis.—*Pliophenacomys osborni* is distinguished from the other species in the genus by having the highest dentine tracts on the sides of the PL in the lower molars and by the fact that the lingual tract is generally as high as the buccal tract.

Description.—With the exception of the characters mentioned in the emended diagnosis *P. osborni*

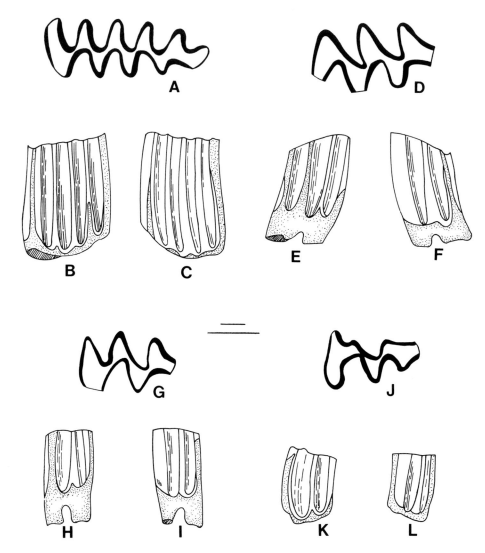

Fig. 6.—*Pliophenacomys osborni.* A–C) UNSM 52772, Rm1; A) occlusal view, B) buccal view, C) lingual view. D–F) UM 81812, LM1, D) occlusal view, E) lingual view, F) buccal view. G–I) UM 81810, LM2, G) occlusal view, H) lingual view, I) buccal view. J–L) UM 81786, LM3, J) occlusal view, K) lingual view, L) buccal view. Short bar equals 1 mm on side views, long bar equals 1 mm on occlusal views.

does not differ significantly from *P. primaevus* or *P. dixonensis.* Although specimens of *P. osborni* appear to be more hypsodont than those of *P. primaevus* or *P. dixonensis,* measurements of crown height (Tables 4–9) do not readily support this view.

Only two of the m1s (Fig. 6A–C) of *P. osborni* do not have an interrupted enamel pattern on the PL. These teeth (UNSM 52752 and 52731) are both from the Big Springs l.f. and in a young adult stage of development. The crown height, buccal tract height, and lingual tract height of these specimens are 3.48, 3.38, 2.60 and 3.26, 3.09, 3.13 respec-

tively. Ten specimens with crown heights greater than these have interrupted patterns.

The two m2s from the Big Springs l.f. are problematical. UNSM 90915 has a buccal tract height of 0.66 mm and a lingual tract height of 0.41 mm, whereas UNSM 90914 with a crown height of 3.34 mm has an interrupted occlusal pattern. These teeth were assigned to *Pliophenacomys* on the shape of the reentrant angles and the presence of thin enamel on the posterior face of the PL. Though variable the five m2s from the White Rock appear to represent a more coherent sample. The two specimens from the Big Springs either represent the extremes in population variation or pertain to different taxa.

The m3 from Big Springs is substantially different in tract height. It could represent a different taxon or the Big Springs shows a clinal relationship in tract height with the White Rock.

No M1s (Fig. 6D–F) of *Pliophenacomys osborni* were found that exhibit three distinct roots. Of those specimens that exhibited anatomical roots, the majority had a slight groove or depression on the median lingual side of the anterior root suggesting some fusion had taken place. The remaining teeth were at a stage of development in which no anatomical roots had formed. Crown height and wear on the occlusal surface suggests that anatomical roots develop later in life in *P. osborni* than they do in the other species of the genus. Except for a higher dentine tract height and a slight increase in hypsodonty with the correlative retardation of the development of anatomical roots the M2s (Fig. 6G–I) and M3s (Fig. 6J–K) of *P. osborni* do not differ from those of other species of *Pliophenacomys* described in this paper.

Discussion.—After its occurrence in the Fox Canyon l.f. where *Pliophenacomys* is represented by at least two and perhaps as many as three thousand individuals it becomes a rare element in subsequent Blancan faunas. The reason for this rarity is not intuitively obvious, because when it first appears *Pliophenacomys* is relatively advanced in terms of dentition when compared to coeval arvicolines. The advancement continues as shown by the increase in hypsodonty, increase in dentine tract height, decrease in number of roots on M1 and 2, and the formation of roots later in life. Perhaps the rarity is a function of sampling, with one exception, Loc. 2A, all non-*finneyi Pliophenacomys* are found well to the north and east of the well-sampled Meade County area. In addition to geography perhaps *Pliophenacomys* was an upland dweller and, therefore, not well represented in fossiliferous deposits that are primarily lowland in origin.

Also of interest is the decrease in number of *Pliophenacomys* with respect to *Hibbardomys* through time. When the two genera occur together for the first time in the Sand Draw l.f., though both taxa are rare, they are represented by an approximately equal number of individuals. However, in the later faunas *Hibbardomys* outnumbers *Pliophenacomys* approximately 2 to 1. This relationship could be a function of sampling, reflect habitat preference, or replacement of one by the other. Additional material at other localities will be necessary to determine which of these alternatives is correct.

ACKNOWLEDGMENTS

I thank the following individuals for their aid in this study; for the loan of specimens in their care Drs. Michael R. Voorhies, Vertebrate Paleontology Section, University of Nebraska State Museum, Lincoln (UNSM), Philip D. Gingerich, Museum of Paleontology, University of Michigan, Ann Arbor (UM), Charles A. Repenning, United States Geological Survey, Menlo Park, California, Jerry R. Choate, Museum of the High Plains, Fort Hays State University; for critical reading of the manuscript Drs. Choate, Repenning, and Voorhies; for typing of the manuscript and preparation of the line drawings Gwenne Cash.

LITERATURE CITED

Eshelman, R. E. 1975. Geology and paleontology of the early Pleistocene (late Blancan) White Rock fauna from the north-central Kansas. Univ. Michigan Papers Paleo., 13 (C. W. Hibbard Mem. Vol. 4):1–60.

Guthrie, R. D., and J. V. Matthews. 1971. The Cape Deceit fauna—early Pleistocene mammalian assemblage from the Alaskan Arctic. Quaternary Res., 1:474–510.

Hibbard, C. W. 1972. Class Mammalia. *In* Early Pleistocene pre-glacial and glacial rocks and faunas of north-central Nebraska (M. F. Skinner and C. W. Hibbard et al.) Bull. Amer. Mus. Nat. Hist., 148:1–148.

Hibbard, C. W., and R. J. Zakrzewski. 1967. Phyletic trends in the late Cenozoic microtine *Ophiomys gen. nov.,* from Idaho. Contr. Mus. Paleo., Univ. Michigan, 21:255–271.

Martin, L. D. 1972. The microtine rodents of the Mullen Assemblage from the Pleistocene of north central Nebraska. Bull. Univ. Nebraska St. Mus., 9:173–182.

Meulen, A. J. van der. 1973. Middle Pleistocene smaller mammals from the Monte Peglia, (Orvieto, Italy) with special reference to the phylogeny of *Microtus* (Arvicolidae, Rodentia). Quaternaria, 17:1–144.

Zakrzewski, R. J. 1967. The primitive vole, *Ogmodontomys,* from the late Cenozoic of Kansas and Nebraska. Papers Michigan Acad. Sci., Arts, & Letters, 52:133–150.

———. 1969. The rodents from the Hagerman local fauna, upper Pliocene of Idaho. Contr. Mus. Paleo., Univ. Michigan, 23:1–36.

Address: Sternberg Memorial Museum and Department of Earth Sciences, Fort Hays State University, Hays, Kansas 67601.

GEOGRAPHIC DIFFERENTIATION IN THE RANCHOLABREAN DIRE WOLF (*CANIS DIRUS* LEIDY) IN NORTH AMERICA

BJÖRN KURTÉN

ABSTRACT

The late Rancholabrean dire wolf of California and Mexico differs from the nominate subspecies in details of dental proportions and especially in the limb bones which average more than 10% shorter, and is referred to *C. dirus guildayi,* new subspecies. Within the nominate subspecies, which inhabited North America east of the Rocky Mountains, microevolution on the deme level may be observed in the course of the Rancholabrean.

INTRODUCTION

The discovery of the extinct dire wolf, *Canis dirus* Leidy, dates back to 1854, but it was only with the excavations at Rancho La Brea that adequate material of this species became known (Merriam, 1912). The species has now been found at about 100 sites in North America (Nowak, 1979), ranging from southern Canada to Mexico. It is also known from South America (Churcher, 1959), but has not been recorded from Alaska or northern Canada.

With such a range in space, and also an appreciable stratigraphic range, some degree of local and/ or temporal differentiation may well be expected. In fact a number of species of the dire wolf group have been proposed from time to time. However, study of such infraspecific differentiation has been hampered by lack of adequate statistical data. Obviously, the great collection from Rancho La Brea would be an ideal sample for population studies. To date, only a number of limb elements have been studied in this manner (Nigra and Lance, 1947; Stock and Lance, 1948). Comparison with limb bones from other sites indicates that geographic differentiation

existed (Hawksley et al., 1963; Galbreath, 1964; Lundelius, 1972; Kurtén and Anderson, 1980). However, limb bones are not particularly common except at the major fossil-producing sites. Still, when limb bones from a large number of sites were pooled, a very marked difference between the dire wolves of Rancho La Brea on one hand, and those from North America east of the Rocky Mountains on the other, became apparent.

A metric study of dire wolf dentitions revealed that concomitant differences between western and eastern-central dire wolves existed, as well as geographic or temporal differentiation on the deme level. Unfortunately, lack of time prevented a study of the main Rancho La Brea collections in Los Angeles, but a respectable sample was obtained in other collections, the largest being that of the Field Museum of Natural History, Chicago. A second important dire wolf mass occurrence is at San Josecito Cave, Mexico; this material was studied in the Natural History Museum of Los Angeles County.

METHODS AND MATERIALS

Material from a total of about 40 sites was studied in the course of 1970–1972, and additional material seen in 1981. The work was partially assisted by National Science Foundation Grant No. GB 31297 to Bryan Patterson, Harvard University. Support was also given by the American Educational Foundation and by the Societas Scientiarum Fennica.

Many persons have made material and/or information available, among them E. Anderson, W. Auffenberg, C. S. Churcher, W. W. Dalquest, T. Downs, D. Fortsch, J. E. Guilday, O. Hawksley, C. W. Hibbard, R. Hoffman, E. L. Lindsay, E. L. Lundelius, R. M. Nowak, P. Parmalee, B. Patterson, C. E. Ray, K. Richey, H. A. Richmond, D. E. Savage, C. B. Schultz, H. Semken, G. G. Simpson, B. Slaughter, J. Soiset, R. Tedford, W. B. Turnbull, S. D. Webb, D. Whistler, and J. A. White.

Institutions and collections are identified by the following acronyms:

AMNH	American Museum of Natural History
ANSP	Academy of Natural Sciences at Philadelphia
BM(NH)	British Museum (Natural History)
CIT	California Institute of Technology, collection now in LACM
CM	Carnegie Museum of Natural History
CMS	Central Missouri State University
FM	Field Museum of Natural History
FSM	Florida State Museum, University of Florida
ISUM	Idaho State University Museum
KUM	University of Kansas Museum

218

LACM	Los Angeles County Museum of Natural History
MCZ	Museum of Comparative Zoology, Harvard University
SMU	Southern Methodist University Museum of Paleontology
TMM	Texas Memorial Museum
UCMP	University of California Museum of Paleontology
UMMP	University of Michigan Museum of Paleontology

| UNSM | University of Nebraska State Museum |
| USNM | National Museum of Natural History |

All measurements are in millimeters. Most are self-explanatory. Palatal length was measured along midline from prosthion to internal narial opening. Width of upper molars is maximum width from anteroexternal base of crown to most distant lingual point; M^1 length was measured parallel to buccal face of crown. Limb bone lengths are maximum lengths.

TAXONOMY

Genus *Canis* Linaeus, 1758
Subgenus *Aenocyon* Merriam, 1918

Type species.—*Canis dirus* Leidy, 1858.

Diagnosis (mainly adapted from Merriam, 1912).—Large *Canis* with very big head; jaws elongate; carnassial teeth larger and more massive than in other *Canis*; palate, frontal region, and zygomatic arches relatively broad; sagittal crest high; inion overhanging; nasal bones extending relatively far back; nasal processes of frontals relatively short; postpalatine foramina opposite to posterior ends of P^4; optic foramen and foramen lacerus anterior in common pit; M^1 with reduced hypocone; P^4 with somewhat reduced protocone.

Canis dirus Leidy, 1858

Synonyms under subspecies.

Diagnosis.—Sole known species of the subgenus *Aenocyon*.

Stratigraphic range.—Rancholabrean Land-Mammal Age.

Geographic range.—North America from southern Canada to the south; western South America.

Canis dirus dirus Leidy, 1858

Canis primaevus Leidy, 1854 (not *C. primaevus* Hodgson, 1833).
Canis dirus Leidy, 1858.
Canis indianensis Leidy, 1869.
Canis mississippiensis Allen, 1876.
Canis ayersi Sellards, 1916.

Type.—ANSP 11614, left maxilla.
Type locality.—Evansville, Indiana; age late Wisconsin (9,400 ± 250 B.P., Hester, 1967).
Diagnosis.—Limb bones longer, especially distally, than in *C. d. guildayi*; P^2 relatively short.
Stratigraphic range.—Rancholabrean.
Geographic range.—Range of species east of North American continental divide; possibly Idaho, Utah.

Material referred to C. d. dirus[1].—Arredondo, Alachua Co., Florida. FSM 2571 tibia, 2977 P^4. Age Sangamon.

Aucilla River, Jefferson Co., Florida. Maxilla, teeth. Age Wisconsin.

Bat Cave, Pulaski Co., Missouri. CMS, remains of at least four individuals as listed in Hawksley et al. (1963, 1973). Age late Wisconsin, probably about 16,000–10,000 B.P.

Blue Mounds, Dane Co., Wisconsin. MCZ 10988 humerus, tibia. Type material of *C. mississippiensis* Allen (1876). Age presumably Rancholabrean.

Brynjulfson Caves, Boone Co., Missouri. CMS, radius, radius fragment, MC 2, MC 3–5, MC 4, 2 MT 4, MT 5. Age late Wisconsin to Early Holocene, 9,440 ± 760 B.P.

Carroll Cave, Camden Co., Missouri. Partial skeleton, data from Hawksley et al. (1963, 1973). Age Wisconsin.

Cherokee Cave, St. Louis Co., Missouri. Metapodials, data from Simpson (1949). Age late Wisconsin.

Cragin Quarry, Meade Co., Kansas. KUM 4613 mandible, M^1. Age Sangamon.

Eichelberger Cave, Marion Co., Florida. FSM 1622 mandible, 1624 M_1, 2177 M^1. Age probably Sangamon.

*Flamingo Waterway, Florida. FM 23671 M_1. Age probably Wisconsin.

Frankstown Cave, Blair Co., Pennsylvania. CM, remains of 4 individuals as listed by Peterson (1926). Age Wisconsin.

Friesenhahn Cave, Bexar Co., Texas. TMM 933, 2 mandibles, 2 C^1, P^3, M^1, 2 C_1, 3 humeri, 3 radii, 3 femora, tibia, 2 MC 3, MT 4, fragments. Age Wisconsin.

Herculaneum, Jefferson Co., Missouri. FM WC 1736, P^4, M^2. Age Wisconsin.

Hermit's Cave, Eddy Co., New Mexico. UNSM 19212 maxilla. Age Wisconsin, 12,900 ± 350 and 11,850 ± 350 B.P.

Hornsby Sink, Alachua Co., Florida. FSM 3986 P^4, 3987 mandible, 3988 maxilla. Age Wisconsin.

---

[1] This and subsequent lists record material seen or used on the basis of published material; in the latter case, the source is indicated. For most sites, further references may be found in Nowak (1979) and Kurtén and Anderson (1980). Records not in Nowak (1979) or substantially augmenting his data are denoted with an asterisk.

Ichetucknee River, Columbia Co., Florida. FSM 8005 mandible, 8006 maxilla, 8208 mandible, 8209 maxilla, 8210 MC 2, 8211 MT 4, 12899 mandible, 17717 mandible. Age Wisconsin.

Ingleside, San Patricio Co., Texas. TMM, remains of at least two individuals as listed by Lundelius (1972). Age early Wisconsin.

Laubach Cave, Williamson Co., Texas. SMU 61172 femur, 61173 humerus. Age may be Wisconsin interstadial, 45,000–25,000 B.P.

*Merritt Island, Brevard Co., Florida (Kurtén, 1965). FSM 8217 P^4. Age Wisconsin.

Moore Pit, Dallas Co., Texas. SMU 60979 MT 4. Age Wisconsin, 50,000–25,000 B.P.

Orr Cave, Beaverhead Co., Montana. CM 12088 mandible. Age Wisconsin.

Powder Mill Creek Cave, Shannon Co., Missouri. Skeleton, data from Galbreath (1964). Age late Wisconsin (13,170 ± 600 B.P).

Reddick, Marion Co., Florida. FSM 2903 humerus, radius, femur, MC 2–5, MT 2–5, 2923 skull with mandible, 3081 skull, 6595 M_1, unnumbered palate, mandible, teeth. Age Sangamon.

Renick Cave, Greenbrier Co., West Virginia. CM 24327 mandible. Age Wisconsin.

Sabertooth (Lecanto) Cave, Citrus Co., Florida. AMNH 23404 MC 5, MT 3, MT 5. Age Sangamon or Wisconsin.

Vero, Indian River Co., Florida. MCZ 17945 M_1. Type skull of *C. ayersi* Sellards (not seen) also from this site. Age late Wisconsin.

Welsh Cave, Woodford Co., Kentucky. CM 12625 casts of skull, mandible, various bones (2 individuals). Age late Wisconsin, 12,950 ± 550 B.P.

Zoo Cave, Taney Co., Missouri. Remains of one individual, data from Hood and Hawksley (1975). Age late Wisconsin, 13,000–9,000 B.P.

In addition, numerous sites have yielded material probably referable to *C. dirus dirus* (lists in Nowak, 1979:112–115, Arkansas, Florida, Illinois, Indiana, Kansas, Louisiana, Missouri, Nebraska, New Mexico, Oklahoma, Tennessee, Texas, Virginia, Wisconsin).

Canis dirus guildayi, new subspecies

Type.—CIT 10834, skull, figured Merriam (1912, Figs. 1–4), and associated mandible (Merriam, 1912, Figs. 1, 5–6).

Type locality.—Rancho La Brea, Los Angeles Co., California; age late Wisconsin.

Diagnosis.—Limb bones shorter, especially distally, than in nominate subspecies; P^2 relatively long.

Stratigraphic range.—Sangamon?—Wisconsin.

Geographic range.—California, Mexico.

Material referred to C. d. guildayi.—Coal Quarry, Eldorado Co., California (Univ. California Loc. No. V-4805). UCMP mandible. Age Wisconsin.

*El Tajo Quarry, Mexico (Univ. California Loc. No. V-2501). UCMP 26650, skull and mandible. Age Rancholabrean.

McKittrick, Kern Co., California. UCMP, 2 maxillae, 2 mandibles, teeth as listed by Schultz (1938). Age Wisconsin.

*Potrecito near Cienega, Mexico. AMNH 67302 mandible. Age Pleistocene.

Rancho La Brea, Los Angeles Co., California. FM 5 skulls, 2 palates, 24 maxillae, 12 mandibles, teeth, long bones, metapodials; UMMP skull, 6 maxillae, 3 mandibles, teeth; USNM 6 mandibles; AMNH, BM(NH), CM, MCZ, UCMP, 10 skulls, 9 mandibles, teeth, limb bones. Age mostly late Wisconsin.

Samwel Cave, Shasta Co., California. UCMP 9566 mandible, teeth. Age Wisconsin.

San Josecito Cave, Nuevo León, Mexico. LACM 192, 5 skulls, 4 palates, 11 maxillae, 46 mandibles, teeth. Age Wisconsin.

Material referred to Canis dirus, *subspecies unknown.*—*American Falls, Power Co., Idaho. ISUM 4800/000052 skull, 47001/17004, 48001/1563, 49001/2623 mandibles. While confirming the tentative identification by Gazin (1935, as *Canis* cf. *dirus*), the allocation to subspecies of this and other Idaho material remains uncertain. Age Sangamon or early Wisconsin.

*Dam, Power Co., Idaho. ISUM 26798 mandible with P_{2-4}, 27776 C^1. Robust proportions indicate *C. dirus*. Age Wisconsin, 26,500 ± 3,500 B.P.

*Fossil Lake, Lake Co., Oregon. UCMP 26910 mandible. It verifies the tentative recognition of *C.* cf. *dirus* based by Elftman (1931) on a carnassial fragment. The dentition is extensively damaged and the specimen is not subspecifically determinable. Geographically, it is closer to *C. dirus guildayi* than to the nominate subspecies. Age early or middle Wisconsin.

Rainbow Beach, Power Co., Idaho. ISUM 2623 mandible. Age Wisconsin, between 31,300 ± 2,300 and 21,500 ± 700 B.P.

DISCUSSION

LIMB PROPORTIONS

Galbreath (1964), Hawksley et al. (1963, 1973), and Hood and Hawksley (1975) have commented upon the large size of dire wolf limb bones, especially metapodials, from Missouri (see also Nowak, 1979:114). Such large bones were found in Bat, Brynjulfson, Carroll, Cherokee and Powder Mill Creek caves. Similar characters are found in other samples from east of the Rocky Mountains, that is, from Ichetucknee River and Reddick (Florida), Blue Mounds (Iowa), Welsh Cave (Kentucky), Frankstown Cave (Pennsylvania), Friesenhahn Cave, Ingleside, Laubach Cave, Moore Pit (Texas), as noted by Kurtén and Anderson (1980:171). Occasional smaller specimens, that is, from Arredondo and Lecanto Cave (Florida), Cragin Quarry (Kansas), Zoo

Table 1.—*Limb bone lengths in* Canis d. dirus *compared with means for* C. d. guildayi *from Rancho La Brea (RLB).*

Bone	N	Observed range	Mean	SD	RLB[1]
Humerus	11	224–254	243.7 ± 3.0	9.9	217.0 ± 0.4
Radius	9	222–259	243.0 ± 4.0	12.0	209.4 ± 0.3
Femur	6	256–278	266.3 ± 3.2	7.7	241.8 ± 0.5
Tibia	9	230–267	250.0 ± 4.9	14.7	231.6 ± 0.3
MC 2	15	80–98	91.1 ± 1.3	4.9	77.2 ± 0.09
MC 3	14	94–111	102.4 ± 1.5	5.2	88.1 ± 0.10
MC 4	13	95–110	103.6 ± 1.3	4.6	97.1 ± 0.10
MC 5	15	76–100	89.5 ± 1.6	6.2	73.8 ± 0.09
MT 2	9	88–112	99.8 ± 2.2	6.7	83.3 ± 0.10
MT 3	9	96–116	106.3 ± 2.8	8.4	94.0 ± 0.11
MT 4	15	102–127	112.9 ± 1.7	6.7	96.2 ± 0.11
MT 5	10	88–111	100.6 ± 2.4	7.5	88.1 ± 0.10

[1] Data from Stock and Lance (1948) and from Nigra and Lance (1947, for left metapodials; standard error emended).

Table 2.—*Lengths of front limb (humerus + radius + MC 3) and hind limb (femur + tibia + MT 3) in samples of* Canis lupus *and* Canis dirus.

Sample	Front	Hind	Front in percent of hind
C. lupus, MCZ 267	498	540	92.2
C. lupus, sample CL	567	617	91.9
C. dirus guildayi, RLB	515.4	567.4	90.8
C. d. dirus	589.1	622.6	94.6

Cave (Missouri), and Slaton (Texas) may be regarded as distal variants of this population.

In Table 1 all of this material has been pooled and compared with mean values for Rancho La Brea as published by Nigra and Lance (1947) and Stock and Lance (1948). As might be expected in a sample so scattered geographically and temporally (ranging from the Sangamon to the late Wisconsin), the *C. d. dirus* material shows a somewhat greater variability than that from the Californian site. Whereas the coefficients of variation for the very large Rancho La Brea sample (in which the number of specimens ranges from 313 femora to more than 2,500 MC 5) average 4.05, that for *C. d. dirus* averages 5.6. The latter figure, although only slightly higher, may suggest moderate (deme-level) local and/or temporal differentiation.

The limb bones of *C. d. dirus* average significantly longer than those of *C. d. guildayi* throughout. The joint overlap, which may be determined by the method outlined by Mayr et al. (1953), is minimal. For radius length, for instance, it is only about 4%, whereas the corresponding figure for length of MC 4 is only about 1%. In effect, then, 99% of a mixed sample of MC 4 can be allocated to the correct taxon on this basis, which greatly exceeds the subspecific differentiation limit (90%) suggested by Mayr et al.

Relative proportions of limb bones also differ markedly in the two subspecies. Data on the three main segments of the front and hind limb are summarized in Fig. 1, which is a ratio diagram (Simpson, 1941) in which Rancho La Brea means serve as the standard of comparison. Means for other samples

are expressed as log and percentage differences from the standard. Apart from the greater length of all the limb segments, the nominate subspecies is characterized by relatively longer radii and metapodials (more than 15% longer than in *C. d. guildayi*), whereas the excess of the length in the humeri, femora, and tibiae averages only 10%.

Also, it is evident that the front limb as a whole is longer, in relation to the hind limb, in *C. d. dirus*. The sum humerus + radius + MC 3 lengths in *C. d. dirus* is 14.3% greater than in *C. d. guildayi*; for femur + tibia + MT 3, the difference is only 9.7%. A further difference is seen in the fact that the radius of *C. d. dirus* is longer relative to the humerus,

Table 3.—*Cranial and mandibular dimensions in samples of* Canis d. guildayi *and* C. d. dirus.

Sample	N	Observed range	Mean	SD
Condylobasal length				
Rancho La Brea	14	252–282	262.0 ± 2.0	7.9
San Josecito Cave and El Tajo	3	258–270	264.7 ± 2.9	—
C. d. dirus	6	265–306	284.3 ± 5.7	15.3
Palatal length				
Rancho La Brea	15	136–155	143.9 ± 1.2	4.8
San Josecito Cave and El Tajo	4	134–152	144.0 ± 3.2	—
C. d. dirus	5	147–168	154.6 ± 3.4	7.6
Width over P⁴–M¹, inclusive				
Rancho La Brea	14	88–102	93.8 ± 0.9	3.5
Rancho La Brea[1]	62	88–104	96.2 ± 0.5	3.9
San Josecito Cave and El Tajo	5	95–103	99.2 ± 1.6	3.5
C. d. dirus	4	98–101	100.0 ± 0.6	—
Mandibular length				
Rancho La Brea	13	197–224	209.1 ± 2.0	7.3
San Josecito Cave and El Tajo	9	206–218	211.0 ± 1.2	3.8
C. d. dirus	6	205–245	221.2 ± 5.6	15.0

[1] Data from Nowak (1979:149).

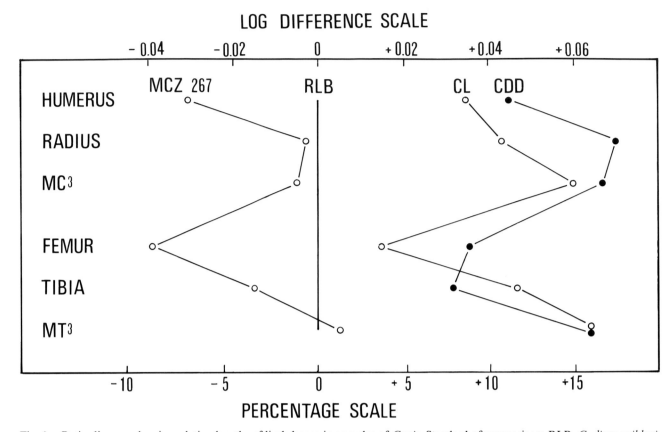

Fig. 1.—Ratio diagram showing relative lengths of limb bones in samples of *Canis.* Standard of comparison: RLB, *C. dirus guildayi,* Rancho La Brea (means). Other samples are: CDD, *C. d. dirus* (means); CL, *C. lupus,* Recent (means); MCZ 267, *C. lupus,* Recent (small single specimen).

whereas the tibia is not elongated relative to the femur.

I have included data on Recent *Canis lupus* in the ratio diagram. MCZ 267 is a relatively small specimen, whereas CL denotes mean values for a sample of eight Recent wolves published by Stock and Lance (1948). The pattern in both cases deviates from *C. d. guildayi* of Rancho La Brea in the same manner as that of *C. d. dirus,* except that the tibia is more elongated in *C. lupus.* In absolute size, the Stock and Lance sample (CL) is close to *C. d. dirus.*

The relationships between the lengths of the front limb (humerus + radius + MC 3) and hind limb (femur + tibia + MT 3) are set down in Table 2. In *Canis lupus* and in *C. d. guildayi,* the front limb averages 91–92% of the length of the hind. In *C. d. dirus* it is markedly longer, almost 95%. Such an elongation of the front limb is commonly seen in large mammals of the open plains, which typically have a sloping back.

SKULL AND MANDIBLE

Measurements of skulls and mandibles are summarized in Table 3. Samples are Rancho La Brea (RLB), Mexico (SJC + ET), and *Canis d. dirus* (CDD). Length measurements for Californian and Mexican samples (*C. d. guildayi*) coincide closely, whereas those for the nominate subspecies average significantly longer. Muzzle width (across P^4–M^1) is relatively great in the Mexican sample and the difference from my RLB sample is significant. However, the larger sample (62 specimens) measured by Nowak (1979) gave a somewhat higher mean for Rancho La Brea. Age differences may affect the relative width of the skull.

DENTITION

Table 4 shows measurements of the teeth of *Canis dirus.* The material was split into four samples— Rancho La Brea (RLB), San Josecito Cave (SJC),

Table 4.—*Tooth dimensions in samples of* Canis d. guildayi *and* C. d. dirus.

Measure-ment	Sample	N	Observed range	Mean	SD
Length C[1]	Rancho La Brea	11	13.7–16.8	14.86 ± 0.29	0.97
	San Josecito Cave	10	14.0–18.0	15.65 ± 0.36	1.13
	C. d. dirus (Sangamon–Early Wisconsin)	3	16.5–18.3	17.17 ± 0.57	—
	C. d. dirus (Late Wisconsin)	5	14.5–16.0	15.02 ± 0.28	0.63
Length P[2]	Rancho La Brea	12	14.3–17.1	15.73 ± 0.21	0.74
	San Josecito Cave	7	15.8–17.2	16.40 ± 0.23	0.60
	C. d. dirus (Sangamon–Early Wisconsin)	5	15.0–16.6	15.98 ± 0.29	0.64
	C. d. dirus (Late Wisconsin)	7	14.1–15.5	14.71 ± 0.19	0.51
Width P[2]	Rancho La Brea	12	6.5–8.1	7.40 ± 0.15	0.52
	San Josecito Cave	7	7.3–7.9	7.53 ± 0.09	0.25
	C. d. dirus (Sangamon–Early Wisconsin)	5	6.7–8.3	7.62 ± 0.27	0.61
	C. d. dirus (Late Wisconsin)	6	6.2–7.8	6.87 ± 0.25	0.62
Length P[3]	Rancho La Brea	15	16.4–19.7	17.95 ± 0.25	0.98
	San Josecito Cave	10	16.4–21.1	18.52 ± 0.39	1.22
	C. d. dirus (Sangamon–Early Wisconsin)	6	17.9–21.2	19.10 ± 0.47	1.16
	C. d. dirus (Late Wisconsin)	11	16.6–18.9	17.85 ± 0.28	0.92
Width P[3]	Rancho La Brea	15	6.9–9.7	7.95 ± 0.12	0.45
	San Josecito Cave	10	7.4–9.0	8.36 ± 0.15	0.46
	C. d. dirus (Sangamon–Early Wisconsin)	6	8.0–8.9	8.48 ± 0.16	0.38
	C. d. dirus (Late Wisconsin)	11	7.0–9.0	7.91 ± 0.17	0.58
Length P[4]	Rancho La Brea	46	28.5–35.4	31.32 ± 0.23	1.54
	San Josecito Cave	17	29.4–34.1	31.88 ± 0.31	1.28
	C. d. dirus (Sangamon–Early Wisconsin)	8	30.7–34.8	31.99 ± 0.45	1.28
	C. d. dirus (Late Wisconsin)	13	29.6–32.5	31.52 ± 0.27	0.96
[1]Width BP[4]	Rancho La Brea	48	10.9–14.5	12.25 ± 0.12	0.82
	San Josecito Cave	17	11.5–14.2	12.65 ± 0.18	0.76
	C. d. dirus (Sangamon–Early Wisconsin)	8	12.1–14.2	13.14 ± 0.22	0.63
	C. d. dirus (Late Wisconsin)	13	11.8–13.6	12.73 ± 0.16	0.58
Width M[1]	Rancho La Brea	56	22.8–28.3	25.40 ± 0.18	1.32
	San Josecito Cave	22	22.7–26.7	25.11 ± 0.19	0.90
	C. d. dirus (Sangamon–Early Wisconsin)	10	24.5–28.5	26.08 ± 0.32	1.02
	C. d. dirus (Late Wisconsin)	10	24.4–27.9	26.25 ± 0.37	1.17
Length M[1]	Rancho La Brea	56	17.3–22.0	19.48 ± 0.16	1.18
	San Josecito Cave	23	17.6–20.9	19.38 ± 0.19	0.91
	C. d. dirus (Sangamon–Early Wisconsin)	9	18.5–21.3	19.99 ± 0.29	0.87
	C. d. dirus (Late Wisconsin)	10	18.7–20.5	19.99 ± 0.17	0.54
Width M[2]	Rancho La Brea	33	13.2–17.0	14.98 ± 0.17	0.98
	San Josecito Cave	15	13.6–16.2	14.85 ± 0.17	0.67
	C. d. dirus (Sangamon–Early Wisconsin)	9	14.3–17.8	15.94 ± 0.38	1.15
	C. d. dirus (Late Wisconsin)	8	14.4–16.7	15.60 ± 0.28	0.79
Width C[1]	Rancho La Brea	19	10.1–13.7	11.51 ± 0.18	0.79
	San Josecito Cave	16	10.1–12.4	11.51 ± 0.17	0.68
	C. d. dirus (Sangamon–Early Wisconsin)	3	12.0–12.3	12.17 ± 0.09	—
	C. d. dirus (Late Wisconsin)	7	10.5–11.9	11.00 ± 0.25	0.66
Length P[2]	Rancho La Brea	20	14.4–17.9	15.72 ± 0.20	0.89
	San Josecito Cave	30	13.2–16.4	15.22 ± 0.13	0.69
	C. d. dirus (Sangamon–Early Wisconsin)	4	15.2–16.1	15.70 ± 0.23	0.47
	C. d. dirus (Late Wisconsin)	9	12.2–15.8	14.78 ± 0.40	1.19
Width P[2]	Rancho La Brea	20	6.3–8.2	7.43 ± 0.09	0.39
	San Josecito Cave	30	6.5–8.1	7.21 ± 0.07	0.39
	C. d. dirus (Sangamon–Early Wisconsin)	4	6.5–7.6	7.15 ± 0.25	0.51
	C. d. dirus (Late Wisconsin)	10	5.2–7.5	6.70 ± 0.21	0.67

Table 4.—*Continued*

Measurement	Sample	N	Observed range	Mean	SD
Length P₃	Rancho La Brea	20	14.1–18.7	16.27 ± 0.22	1.00
	San Josecito Cave	32	15.0–17.3	16.04 ± 0.11	0.62
	C. d. dirus (Sangamon–Early Wisconsin)	6	15.8–17.5	16.70 ± 0.31	0.75
	C. d. dirus (Late Wisconsin)	13	13.5–17.4	16.06 ± 0.30	1.08
Width P₃	Rancho La Brea	20	7.3–9.2	8.15 ± 0.12	0.52
	San Josecito Cave	34	7.0–8.9	7.97 ± 0.08	0.44
	C. d. dirus (Sangamon–Early Wisconsin)	6	7.1–8.9	8.15 ± 0.27	0.67
	C. d. dirus (Late Wisconsin)	14	6.7–9.0	7.69 ± 0.23	0.86
Length P₄	Rancho La Brea	26	17.6–21.8	19.95 ± 0.19	0.97
	San Josecito Cave	32	17.5–20.7	19.60 ± 0.13	0.75
	C. d. dirus (Sangamon–Early Wisconsin)	7	18.5–22.0	20.27 ± 0.46	1.23
	C. d. dirus (Late Wisconsin)	16	17.8–21.2	19.36 ± 0.32	1.28
Width P₄	Rancho La Brea	25	9.4–11.1	10.28 ± 0.10	0.52
	San Josecito Cave	33	9.3–11.9	10.20 ± 0.10	0.60
	C. d. dirus (Sangamon–Early Wisconsin)	8	9.0–10.8	10.15 ± 0.23	0.64
	C. d. dirus (Late Wisconsin)	16	8.3–10.8	9.61 ± 0.18	0.71
Length M₁	Rancho La Brea	33	32.9–40.2	35.54 ± 0.27	1.57
	San Josecito Cave	44	31.5–39.3	35.50 ± 0.23	1.52
	C. d. dirus (Sangamon–Early Wisconsin)	10	32.5–39.1	36.09 ± 0.61	1.93
	C. d. dirus (Late Wisconsin)	14	32.1–37.4	35.16 ± 0.52	1.93
Width M₁	Rancho La Brea	29	12.3–14.9	13.31 ± 0.13	0.71
	San Josecito Cave	40	12.4–15.0	13.62 ± 0.10	0.61
	C. d. dirus (Sangamon–Early Wisconsin)	11	12.8–15.5	14.10 ± 0.30	0.98
	C. d. dirus (Late Wisconsin)	22	11.8–16.3	13.40 ± 0.24	1.13
[1] Length TM₁	Rancho La Brea	27	22.3–27.0	24.36 ± 0.20	1.02
	San Josecito Cave	33	22.0–27.9	24.64 ± 0.19	1.09
	C. d. dirus (Sangamon–Early Wisconsin)	6	23.5–26.3	24.53 ± 0.44	1.08
	C. d. dirus (Late Wisconsin)	15	21.0–26.4	24.01 ± 0.44	1.72
Length M₂	Rancho La Brea	25	11.7–15.5	13.33 ± 0.16	0.79
	San Josecito Cave	33	11.2–14.2	12.76 ± 0.14	0.80
	C. d. dirus (Sangamon–Early Wisconsin)	7	11.3–15.8	13.74 ± 0.55	1.46
	C. d. dirus (Late Wisconsin)	13	12.4–14.8	13.49 ± 0.22	0.81
Width M₂	Rancho La Brea	23	9.2–10.5	9.86 ± 0.10	0.47
	San Josecito Cave	34	8.9–10.4	9.60 ± 0.06	0.34
	C. d. dirus (Sangamon–Early Wisconsin)	5	9.7–11.7	10.48 ± 0.34	0.75
	C. d. dirus (Late Wisconsin)	12	8.9–11.5	10.05 ± 0.21	0.73

[1] B, blade; T, trigonid.

Canis d. dirus of probably Sangamonian and early Wisconsinan age (CDD(1)), and *Canis d. dirus* of late Wisconsinan age (CDD(2)). In the ratio diagram (Fig. 2) the principal dimensions of the teeth are compared.

To test for possible heterogeneity within samples, the coefficients of variation for the dental variates was averaged, with results as given in Table 5. In an analogous study of four samples of *Canis lepophagus, C. arnensis,* and *C. latrans,* the mean CV (coefficient of variation) varied between 5.6–5.9 (Kurtén, 1974). In the present case, only the sample CDD(2) gave a slightly higher average; this late Wis-

consinan sample is widely scattered geographically. (Individual CV values for the two CDD samples are variable but this appears to be due to small-sample effects.)

Table 5.—*Averages and observed ranges for coefficient of variation in 22 dental variates of* Canis dirus.

Taxon	Sample	Observed range	Mean CV
C. d. guildayi	Rancho La Brea	4.2–6.5	5.6
C. d. guildayi	San Josecito Cave	3.3–7.2	4.9
C. d. dirus	Sangamon–Early Wisconsin	1.2–10.6	5.5
C. d. dirus	Late Wisconsin	3.0–11.2	6.4

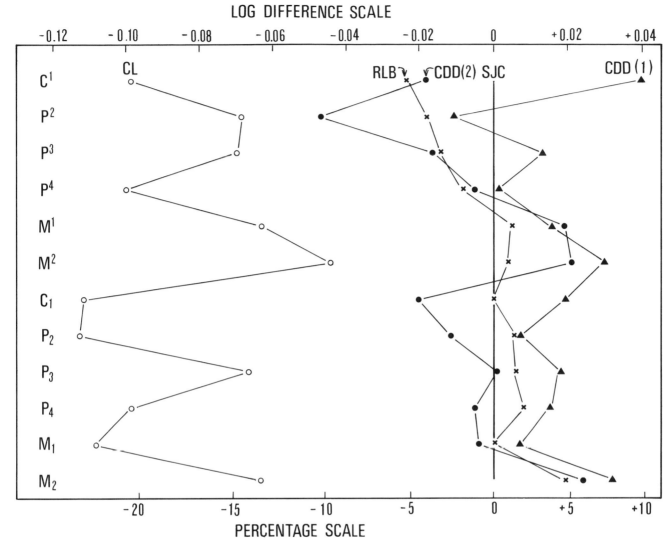

Fig. 2.—Ratio diagram showing relative size of teeth (widths of M^{1-2}, C_1, lengths of other teeth) in samples of *Canis* (means). Standard of comparison: SJC, *C. dirus guildayi*, San Josecito Cave. Other samples are: RLB, *C. d. guildayi*, Rancho La Brea; CDD(1) and CDD(2), *C. d. dirus*, early and late samples respectively (see text); CL, *C. lupus*, late Wisconsinan to early Holocene.

The Californian and Mexican samples differ only moderately from each other and none of the differences reaches the significance level of $P = 0.01$ (see Table 6). However, there are probably slight differences on the deme level. The two samples of the nominate subspecies show more consistent differences from *C. d. guildayi*, for instance in the relatively small size of P^2 and the large size of M^1 and M^2. Most measurements for late Wisconsinan *C. d. dirus* (CDD(2)) are smaller than those for the earlier (CDD(1)) sample which suggests a slight phyletic dwarfing within the nominate subspecies. Other de-

tails may be gleaned from the ratio diagram and from Table 6 in which *P*-values for significant or probably significant differences have been collected.

It is of interest to compare the means for *C. dirus* with those for *C. lupus*. For this purpose, a late Wisconsinan/early Holocene sample of *C. lupus* (mainly from Moonshiner Cave, Idaho) was studied and the means recorded in the ratio diagram (CL). The pattern for *C. lupus* differs in several respects from those for the various dire wolf samples, notably in the small size of the canines and carnassials, and the large size of the postcarnassial teeth.

Table 6.—*Selected comparisons between dental dimensions in samples of* Canis dirus. *For data and samples see Table 4. Body of table gives values for* P *(two-sided).*

Rancho La Brea/ San Josecito Cave		San Josecito Cave/ C. d. dirus (Sangamon– Early Wisconsin)	
LP^2	<0.05	WM^1	<0.02
LM_2	<0.02	WM^2	<0.02
		WC_1	<0.01
Rancho La Brea/C. d. dirus (Sangamon–Early Wisconsin)		San Josecito Cave/ C. d. dirus (Late Wisconsin)	
LC^1	<0.01	LP^2	<0.01
LP^3	<0.05	LM_2	<0.01
WBP^4	<0.01		
WC_1	<0.01		
Rancho La Brea/C. d. dirus (Late Wisconsin)		C. d. dirus (Sangamon– Early Wisconsin)/ C. d. dirus (Late Wisconsin)	
LP^2	<0.01	LC^1	<0.05
WP_2	<0.01	LP^2	<0.01
WP_4	<0.01	WC_1	<0.01

CONCLUSIONS

The taxa *Canis d. dirus* and *C. d. guildayi* differ in body size, as reflected in the size of the head and limbs, in limb proportions, and to some extent in dental proportions. The adaptive significance of the differences in limb proportions is evident; presumably the nominate subspecies was somewhat fleeter of foot than *C. d. guildayi* and this may reflect certain differences in the mode of life.

Although the two taxa differ in many respects, the similarity in many characters, especially dental, is such that, in my opinion, they should be regarded as subspecies of a single species.

The history of the two subspecies is incompletely known. Large *C. d. dirus* with long metapodials date back to the Sangamon, whereas the first appearance of the shorter-limbed *C. d. guildayi* may be slightly more recent. This, and the resemblance between *C. d. dirus* and *C. lupus* in limb proportions, might suggest that the condition in *C. d. guildayi* is the derived one. As long as the origin of the dire wolf is unknown such considerations must remain speculative.

LITERATURE CITED

ALLEN, J. A. 1876. Description of some remains of an extinct species of wolf and an exinct species of deer from the Lead Region of the Upper Mississippi. Amer. J. Sci., 40:47–51.

CHURCHER, C. S. 1959. Fossil *Canis* from the Tar Pits of La Brea, Peru. Science, 130:564–565.

ELFTMAN, H. O. 1931. Pleistocene mammals of Fossil Lake, Oregon. Amer. Mus. Nov., 481:1–21.

GALBREATH, E. C. 1964. A dire wolf skeleton and Powder Mill Creek Cave, Missouri. Trans. Illinois State Acad. Sci., 57:224–242.

GAZIN, C. L. 1935. Annotated list of Pleistocene Mammalia from American Falls, Idaho. J. Washington Acad. Sci., 25:297–307.

HAWKSLEY, O., J. F. REYNOLDS, and R. L. FOLEY. 1973. Pleistocene vertebrate fauna of Bat Cave, Pulaski County, Missouri. Bull. Nat. Spel. Soc., 35:61–87.

HAWKSLEY, O., J. F. REYNOLDS, and J. McGOWAN. 1963. The dire wolf in Missouri. Missouri Speleol., 5:63–72.

HESTER, J. J. 1967. The agency of man in animal extinctions. Pp. 169–192, *in* Pleistocene extinctions: the search for a cause (P. S. Martin and H. E. Wright, eds.), Yale Univ. Press, New Haven and London, x + 453 pp.

HOOD, C. H., and O. HAWKSLEY. 1975. The Pleistocene fauna from Zoo Cave, Taney County, Missouri. Missouri Speleol., 15:1–42.

KURTÉN, B. 1965. The Pleistocene Felidae of Florida. Bull. Florida State Mus., 9:215–273.

———. 1974. A history of coyote-like dogs (Canidae, Mammalia). Acta Zool. Fennica, 140:1–38.

KURTÉN, B., and E. ANDERSON. 1980. Pleistocene mammals of North America. Columbia Univ. Press, New York, xvii + 443 pp.

LEIDY, J. 1854. Notice of some fossil bones discovered by Mr. Francis A. Lincke, in the banks of the Ohio River, Indiana. Proc. Acad. Nat. Sci. Philadelphia, 7:199–201.

———. 1858. Notice of remains of extinct Vertebrata, from the Valley of the Niobrara River, collected during the exploring expedition of 1857, in Nebraska, under the command of Lieut. G. K. Warren, U.S. Top. Eng., by Dr. F. V. Hayden. Proc. Acad. Nat. Sci. Philadelphia, 1858:20–29.

———. 1869. The extinct mammalian fauna of Dakota and Nebraska, including an account of some allied forms from other localities, together with a synopsis of the mammalian remains of North America. J. Acad. Nat. Sci. Philadelphia, ser. 2, 7:1–472.

LUNDELIUS, E. L., JR. 1972. Fossil vertebrates from the late Pleistocene Ingleside fauna, San Patricio County, Texas. Rep. Invest. Bur. Econ. Geol., Austin, 77:1–74.

MAYR, E., E. G. LINSLEY, and R. L. USINGER. 1953. Methods and principles of systematic zoology. McGraw-Hill New York, 328 pp.

MERRIAM, J. C. 1912. The fauna of Rancho La Brea. Part II. Canidae. Mem. Univ. California, 1:215–272.

NIGRA, J. O., and J. F. LANCE. 1947. A statistical study of the metapodials of the dire wolf group from the Pleistocene of Rancho La Brea. Bull. So. California Acad. Sci., 46:26–34.

NOWAK, R. M. 1979. North American Quaternary *Canis*. Monogr. Mus. Nat. Hist., Univ. Kansas, 6:1–154.

PETERSON, O. A. 1926. The fossils of Frankstown Cave, Blair County, Pennsylvania. Ann. Carnegie Mus., 16:249–315.

SCHULTZ, J. R. 1938. A late Quaternary mammal fauna from the tar seeps of McKittrick, California. Publ. Carnegie Inst. Washington, 487:111–215.

SELLARDS, E. H. 1916. Human remains and associated fossils from the Pleistocene of Florida. Rep. Florida Geol. Surv., 8:121–160.

SIMPSON, G. G. 1941. Large Pleistocene felines of North America. Amer. Mus. Nov., 1136:1–27.

———. 1949. A fossil deposit in a cave in St. Louis. Amer. Mus. Nov., 1408:1–46.

STOCK, C., and J. F. LANCE. 1948. The relative length of limb elements in *Canis dirus*. Bull. So. California Acad. Sci., 47: 79–84.

Address: Department of Geology, University of Helsinki, 00170 Helsinki 17, Finland.

USE AND ABUSE OF DOGS

ELIZABETH S. WING

ABSTRACT

Dog remains are abundant in most archeological sites in Mexico. Their remains have been recovered from human burials and from contexts which are interpreted as food refuse. Early Spanish chronicles describe the use of dogs for food and ritual. Whether these dogs were hairless, as is frequently stated, is the question. Genetically hairless dogs can be identified among osteological collections by the extreme abnormalities in the dentition which accompanies the hairless condition. No such abnormalities have been reported in faunal samples. Artificially produced hairless dogs may account for some of the early reports of hairless dogs but these can not be detected archeologically. Evidence of another example of dogs which were altered is the sample of dog burials from West Mexico. A majority of these dogs had intentionally broken canines and incisors.

INTRODUCTION

Extraordinary interdependence has developed between man and dog during the 12,000 years of this close association. Throughout the millenia people have modified dogs through selection and when the results of selection did not satisfy, dogs were tailored to fit a need or ideal. They can with some justification be called man-made (Clutton-Brock, 1977) though the relationship between man and dog is a mutual one.

Dog remains occur commonly in North American archeological sites and are a hallmark of Mexican sites. Dogs are also frequently mentioned in the early Spanish chronicles describing Aztec and Maya ways of life (Maudslay trans. Diaz del Castillo, 1956; Duran, 1967; Anderson and Dibble trans. Sahagún, 1950; Tozzer, 1941). Duran (1967) describes the marketplace at Acolman in 1539, where over 400 dogs were for sale at higher prices than other meat. The dog meat was consumed for special feasts. Diaz del Castillo (Maudslay trans., 1956) also mentions that small dogs were fattened for consumption in the home and for sale at the market at Tlatelolco. Landa (Tozzer, 1941) describes the place held by dogs in Maya sacrificial ceremonies and after their ceremonial offering they were eaten. Dogs also played a part in burial rites of born Aztecs and Maya (Tozzer, 1941: 143). Sahagún (Anderson and Dibble, 1950) has illustrated a dog called a chichi with long yellow hair, erect ears, and curly tail which is just the kind of dog used to carry a dead one "across the place of the nine rivers in thc land of the dead." Sahagún applies the name xoloitzcuintli to the hairless dogs which he claims are made permanently hairless by rubbing turpentine on their skin while they are puppies. In addition, he notes that some say dogs are born without hair.

RESULTS AND DISCUSSION

Remains from archeological sites must be relied upon to verify these observations. One type of archeological information which is often described in the context of prehistoric dog use are the famous Colima dog statues. These have been found in great numbers associated with human burials in the present state of Colima on the West Coast of Mexico. The Colima dogs are uniformly fat with proportions similar to that of a puppy and with what appears to be a full complement of teeth though they are too uniform in size and shape to be accurate representations of dog dentition. This context has suggested the interpretation that these statues represent the dogs described by Sahagún as the guides for the dead (Wright, 1960). They have also been interpreted as hairless dogs intentionally fattened for a food animal (Coe, 1962; Wright, 1960).

The other type of archeological material that can provide information about dog use is the remains of dogs themselves. Three of the questions which may be answered by these remains are: whether dogs are associated with human burials, in which case they may represent the guide dog to the land of the dead; whether dogs were used for food; and whether either of these roles were played by hairless dogs. Dog remains have been found in clear association with human burials (Wing, manuscript). This association is particularly frequent in the sites of the

Table 1.—*Recorded crosses of hairless dogs.*

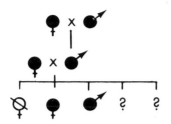

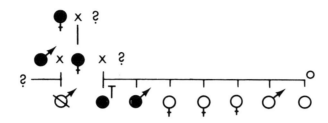

♂♀ - male or female normal hair, normal dentition

● - male or female hairless with abnormal dentition

⚥♂ - male or female normal hair, abnormal dentition

T - animal born dead

? - condition of hair and dentition unknown

O - gender unknown, normal hair and dentition

o - litter too young to determine dental abnormalities

**Data collected in Peru by Coppelia Hays,
Zoology Department, University of Florida**

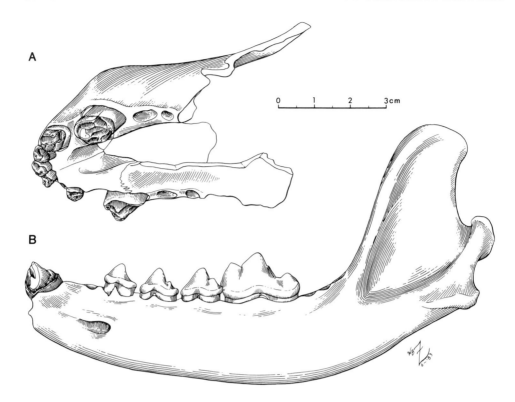

Fig. 1.—Muzzle of dog (78-100) associated with human burial at the site designated N12/1, Marismas Nacionales. A) Oblique view of palate showing broken incisors and canines; B) Lateral view of left dentary showing broken canine. Drawings by Wendy Zomlefer.

Marismas Nacionales in the states of Sinaloa and Nayarit on the West Coast of Mexico just north of Colima. Dog remains are also very frequently associated with other food remains and share with them butchering marks and burned skeletal elements characteristic of food remains (Flannery, 1976; Hamblin, 1980; Pohl, 1976; Pohl and Feldman, 1982; Wing, 1978). Dogs constitute as much as 25% of the animals used for food (Wing, 1981). Whether any of these dogs was hairless is open to question and warrants some discussion of the characteristics of the congenital hairless condition.

Hairless dogs are known from many places including Mexico, the Greater Antilles, Peru, Paraguay as well as in the Old World (Allen, 1920; Braund, 1975; Weiss, 1976; Wright, 1960). The condition of hairlessness, when not induced by rubbing the skin with turpentine, is a genetic disorder passed to following generations by an autosomal dominant pattern with incomplete penetrance and variable expressivity (Graber, 1978; Hutt, 1979). Both the degree of hairlessness and the degree of abnormality and congenital absence of the teeth are affected by

this condition. The results of experimental matings of hairless dogs indicates that the individuals which are homozygous for this trait are usually born dead with extreme abnormalities, and the heterozygous individuals are hairless (Hutt, 1979; Letard, 1930). The ratio of normal to hairless viable to hairless dead from experimental crosses is 9:12:4 which fits the expected 1:2:1 from the segregation of dominant genes from two heterozygous parents (Letard, 1930, reported by Hutt, 1979). The crosses presented on Table 1 are incomplete but indicate the penetrance of dental abnormalities even in animals with normal coats. The extent of congenital tooth absence is profound (Table 2). The American Kennel Club in writing standards for the breed recognize the absence of the premolars associated with the hairlessness and do not penalize an animal for the absence of the incisors as well.

If these hairless dogs represent the same genetic condition as the xoloitzcuintli then the remains of hairless dogs can be recognized with confidence by their abnormal dentition even though the degree of abnormality varies. As far as I know the degree of

Table 2.—*Documented absence of teeth in hairless dogs or normally haired dogs from hairless ancestors.*

Name	Gender	Age in years	Teeth in upper and lower jaw										
			I_1^1	I_2^2	I_3^3	C	P_1^1	P_2^2	P_3^3	P_4^4	M_1^1	m_2^2	m_3^0
Flor	F	2½–3	+		+	+	+	+	+			+	
				+			+	+	+			+	+
Chispi	F	1½						+	+	+			
					+				+	+			
Chaski	F	4			+	+	+	+	+	+		+	
					+		+	+	+	+		+	+
Nanau	F	7			+	+	a	a	+	+		+	
					+		+	+	+	+			+
Mitsuko	M	7					+		+	+		+	
			+				+	+	+	+			+
Nanpu	M	4	+	+		+	+	+	+	+		+	
					+	+	+	+	+	+	+		
No name	M	11	+	+	+	+	+	+	+	+		+	
			+	+	+		+	+	+	+		+	+
Lalau	M	8	+	+	+	+	+	+	+	+	+	+	
			+	+	+	+	+	+	+	+	+	+	+
Juanacho	M	4–5			+	+	+		+				
					+			+		+			
Tali normal hair	M	¾					+						
							+						
Sumac normal hair	F	4									+		
							+	+					+

blank = tooth present.
+ = tooth absent.
a = tooth present but abnormal.
Data collected in Peru by Coppelia Hays, Zoology Department, University of Florida.

tooth absence permitted in the breed by the American Kennel Club standards or seen by Coppelia Hays in the Peruvian examples has never been reported in dog remains from Mexico. The Colima dog statues clearly have a full complement of teeth so their interpretation as hairless dogs is open to question. The large sample of dog burials from the post classic sites of the Marismas Nacionales north of Colima also have normal though modified dentition and would by this criterion have normal hair. None of the many dog remains described as associated with food remains are reported to have abnormal dentition of the extent seen in known hairless dogs. This lack of archeological evidence for the presence of hairless dogs does not mean that hairless dogs did not exist. It does, however, suggest that genetically hairless dogs were not as abundant as normal dogs.

Artificially hairless dogs, those whose hair was intentionally removed with turpentine, can not be detected archeologically. Thus, no way exists of determining how wide spread this practice reported by Sahagún (Anderson and Dibble, 1950) was.

Though probably not related to artificial simulation of the hairless dog, many of the dog burials from the five sites in the Marismas Nacionales have broken incisors and canines (Fig. 1 and Table 3). These broken teeth are darkened by the blood pigments typically seen in dead teeth that are retained in the mouths of live animals. Two of the broken canines show slight wear subsequent to the damage to the crown. No evidence of fractured bone is apparent. Fifty-three of the total sample of 96 canines from the dog burials are broken. The most frequently damaged incisor is the third which is the largest and the closest to the canine. In addition to the dog burials, four raccoons were included with the human burials and only one of these raccoons escaped the tooth breaking treatment. These data suggest that the front teeth particularly the canines were intentionally broken during the animal's life prior to their burial with a human. Furthermore, this was a fre-

Table 3.—*Broken and unbroken teeth from dogs and raccoons (from Chalpa only) associated with human burial sites in the Marismas Nacionales, West Mexico.*

	Broken		Unbroken canines
Sites	Incisors	Canines	
N12/1	5	3	3
Tecualillo	2	6	7
Cristo Rey	9	11	6
Rincon	3	4	18
Chalpa			
Canis	6	24	6
Procyon	0	5	3
Total	25	53	43

quent practice and extended to raccoons as well as dogs. Raccoons apparently had a role similar to dogs in the minds of the people inhabiting the estuary. What motivated these people to mutilate the animals, which may have been intended as guides to the after life, can only be a matter of speculation.

Though it may not be possible to reconstruct motivations, the results of prehistoric action may be revealed. Clearly, dogs were very important in the lives and perhaps after lives of the prehistoric people of Mexico. There can be no doubt that dogs were used for food and for ceremonial purposes related to burial rites. The role of hairless dogs, either genetically or artificially hairless, remains a tantalizing question.

ACKNOWLEDGMENTS

This paper is gratefully dedicated to John Guilday for his pioneering work in Zooarcheology and setting goals and standards for the growing numbers of zooarcheologists to strive towards.

Thanks are also due Coppelia Hays for collecting data on the dental condition of hairless dogs and to Wendy Zomlefer for her drawing of the brutalized muzzle. I am also most grateful to Sylvia Scudder for help in searching the literature.

LITERATURE CITED

ALLEN, G. M. 1920. Dogs of the American aboriginies. Bull. Mus. Comp. Zool., 63:431–517;478–481.

BRAUND, K. 1975. The uncommon dog breeds. Arco Publ. Co. Inc., New York, pp. 194–229.

CLUTTON-BROCK, J. 1977. Man-made dogs. Science, 197:1340–1342.

COE, M. D. 1962. Mexico. Ediciones Lara, Mexico City, Mexico, 245 pp.

DIAZ DEL CASTILLO, B. 1517–1521. The Discovery and Conquest of New Mexico. Translated by A. P. Maudslay London 1908–16 (American ed., Farrar, Straus and Cudahy, New York, 1956).

DURAN, F. D. 1967. Historia de Las Indias de Nueva Espana y Islas de Tierra Firma. Vol. II, pp. 218–219. Introduction by J. F. Ramirez. Editora Nacional, Mexico.

FLANNERY, K. V. 1976. The early Mesoamerican village. Academic Press, New York, 377 pp.

GRABER, L. W. 1978. Congenital absence of teeth: a review with emphasis on inheritance patterns. Jada, 96:266–275.

HAMBLIN, N. L. 1980. Animal Utilization by the Cozumel Maya: Interpretation through faunal analysis. Unpublished Ph.D. dissert. Univ. Arizona, Tucson, 349 pp.

HUTT, F. B. 1979. Genetics for dog breeders. W. H. Freeman, San Francisco, 245 pp.

LETARD, E. 1930. Le Mendélisme expérimental Expériences sur hérédité mendelienne du caractère "peau nue" dans l'espèce chien. Rev. Vét. et J. Méd. Vét., Toulouse, 82:553–570.

POHL, M. E. D. 1976. Ethnozoology of the Maya: an analysis of fauna from five sites in the Peten, Guatemala. Unpub-lished Ph.D. dissert., Harvard Univ., Cambridge, Massachusetts, 317 pp.

POHL, M., AND L. H. FELDMAN. 1982. The traditional role of women and animals in lowland Maya economy in Maya subsistence. Pp. 295–311, in Studies in memory of Dennis E. Puleston (K. V. Flannery, ed.), Academic Press, New York, 368 pp.

SAHAGÚN, F. B. DE. 1950. General history of the things of New Spain translated from the Nahuatl by Arthur Jo Anderson and Charles E. Dibble. Sante Fe, Nos. 1–5; 7–9; 12.

TOZZER, A. M. 1941. Landa's relacion de Las Cosas de Yucatan. Papers Peabody Mus. Amer. Arch. Ethnol., Harvard Univ., 18:1–394.

WEISS, P. 1976. El Perro Peruano Sin Pelo. Museo Nac. Antropologia y Arqueologia ser.: Paleobiologia No 1 reproducido de la revista 1970 Acta Herediana. Univ. Peruana Cayetano, Heredia, 3:33–54.

WING, E. S. 1978. Use of dogs for food: an adaptation to the coastal environment in prehistoric coastal adaptations edited by B. Stark and B. Voorhies. Academic Press, New York, pp. 29–41.

———. 1981. A comparison of Olmec and Maya food ways in the Olmec and their neighbors edited by Elizabeth P. Benson. Dumbarton Oaks Research Library and Collections, Harvard Univ., Washington D.C., pp. 20–28.

———. ms. Vertebrate remains from the Archeological sites in the Marismas Nacionales.

WRIGHT, N. P. 1960. El enigma del xoloitzcuntli. Inst. Nac. Antropologia e Historia, Mexico, 102 pp.

Address: Florida State Museum, University of Florida, Gainesville, Florida 32611.

TIME OF EXTINCTION AND NATURE OF ADAPTATION OF THE NOBLE MARTEN, *MARTES NOBILIS*

Donald K. Grayson

ABSTRACT

The noble marten, *Martes nobilis,* is known from at least 12 sites in the western United States and the Yukon. This extinct marten is generally thought to have become extinct at the end of the Pleistocene, and to have been adapted to cool, if not boreal, conditions. Two sites, however, have yielded specimens that date to between 3,000 and 3,500 B.P., one of which is reported here for the first time. It is argued that the noble marten may not have become extinct until late in the Holocene. In addition, the vertebrate taxa with which *Martes nobilis* has now been found in the western United States suggest that it was adapted to a broad variety of environmental settings, and that it could not have suffered from competition with the American marten (*Martes americana*) throughout its range.

INTRODUCTION

In 1926, E. R. Hall described a new subspecies of marten, *Martes caurina* (=*americana*) *nobilis* on the basis of eight maxillae and mandibles from the Pleistocene deposits of Samwel and Potter Creek caves, northern California (Hall, 1926). Although Hall (1936) later concluded that the material he had described did not differ sufficiently from the local *M. americana sierrae* to merit subspecific recognition, Anderson (1970) was able to demonstrate consistent differences between the skeletons of modern *M. americana* and a sizeable series of marten cranial and postcranial remains from four western United States cave faunas. Accordingly, she assigned the paleontological material to *Martes nobilis,* the noble marten (see taxonomic review in Anderson, 1970). Since Anderson's revision, *M. nobilis* has been identified from a number of additional paleontological and archaeological faunas. This large, extinct marten is now known from at least 12 sites in the western United States (Colorado, Wyoming, Idaho, Nevada, and California) and the Yukon (Webster, 1978; Kurtén and Anderson, 1980, Grayson, 1982a).

The noble marten is generally thought to have become extinct at the end of the Pleistocene, and to have been adapted to cool, if not boreal, conditions (for example, Anderson, 1970; Hager, 1972; Ziemens and Walker, 1974; Miller, 1979; Kurtén and Anderson, 1980). Given this apparent adaptation and apparent date of extinction, terminal Pleistocene climatic change has often been assigned a role in explaining the extinction of the noble marten (for example, Anderson, 1970; Kurtén and Anderson, 1980). The recent discovery of *Martes nobilis* in the prehistoric fauna of Hidden Cave, Nevada, coupled with a previously published but generally overlooked record from Dry Creek Rockshelter, Idaho (Webster, 1978), suggests that current hypotheses concerning the nature of adaptation and timing of extinction of *M. nobilis* may be incorrect.

HIDDEN CAVE AND THE NOBLE MARTEN

Hidden Cave (Nv-Ch-16; elevation 1,251 m) is located on the northern face of Eetza Mountain in the southern Carson Desert of western Nevada, approximately 27 km southeast of the town of Fallon (see Figs. 1–2). The site has been the scene of three professional archaeological excavations, the first conducted by the Nevada State Park Commission in 1940, the second by the University of California, Berkeley in 1951, and the third by the American Museum of Natural History (AMNH) in 1979 and 1980 under the direction of David H. Thomas

(Thomas, 1982). Although the Berkeley excavations provided a great deal of information on the archaeology and stratigraphy of the Hidden Cave deposits (Morrison, 1964; Roust and Clewlow, 1968), the AMNH excavations were much more extensive, and provided, among other things, a stratified sequence of vertebrate faunal remains that span roughly the past 21,000 years (see Fig. 3). To date, 6,603 bones and teeth belonging to 45 taxa (genera and species) of mammals have been identified from the 14 defined strata within the site.

Fig. 1.—The general location of Hidden Cave in western Nevada.

Today, the area surrounding Hidden Cave is arid. Vegetation on the slopes adjacent to the site is dominated by little greasewood (*Sarcobatus baileyi*) and shadscale (*Atriplex confertifolia*), while the Carson Desert lowlands to the immediate north of Eetza Mountain are dominated by black greasewood (*Sarcobatus vermiculatus*). Although now quite arid,

however, the desertification of the Eetza Mountain area did not occur long ago. All strata within Hidden Cave with sizeable faunal samples, including deposits above a layer of Mono tephra dated to ca. 1,500 years before the present (B.P.), contain yellow-bellied marmots (*Marmota flaviventris*) and bushy-tailed wood rats (*Neotoma cinerea*), species

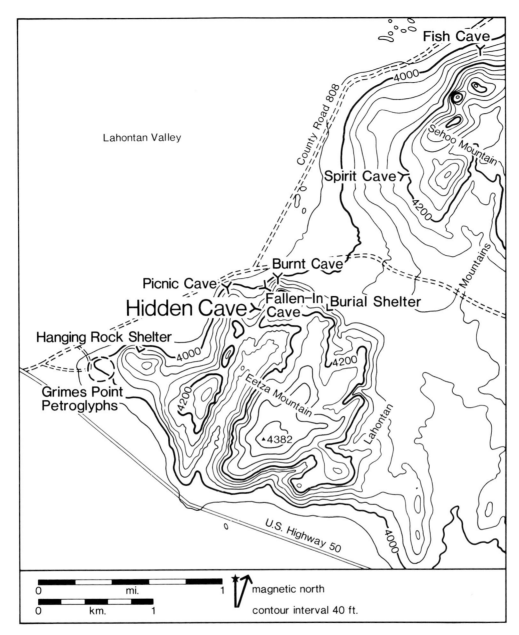

Fig. 2.—The location of Hidden Cave and selected other archaeological sites in the Eetza Mountain area.

that require relatively moist environments and that are no longer found on or near Eetza Mountain. In addition, cat-tail (*Typha latifolia* and *T. angusti-folia*) pollen is abundant in the site, accounting for approximately 20% of the identified pollen above Mono tephra (Wigand and Mehringer, 1982). The precise timing of the aridification of the Eetza Mountain area is not known, but is currently under detailed investigation.

Among the identified mammalian bones and teeth from Hidden Cave is a single specimen of *Martes nobilis,* a left M^1 (AMNH HC-184; Fig. 4). This tooth is much larger than the corresponding tooth of *M. americana*; compared to *M. pennanti*, the inner lobe of the tooth is expanded, producing a deeper constriction between inner and outer lobes. These are characters in which the tooth agrees with *M. nobilis* (Anderson, 1970). Plotting the width of

Fig. 3.—An internal view of Hidden Cave, late in the 1979 excavations. Photograph by Albert A. Alcorn, Fallon, Nevada.

M^1 against the length of the inner lobe of the tooth on the scattergram provided by Anderson (1970) shows the Hidden Cave specimen to lie well above the distribution of *M. americana,* at the lower end of the distribution for *M. pennanti,* and at the upper end of that for *M. nobilis.* The measurements of the tooth (width M^1 = 10.0 mm, length M^1 inner = 6.3 mm, length M^1 mid = 4.4 mm, length M^1 outer = 4.4 mm; see Anderson, 1970 for measurement definitions) fall within known ranges for *M. nobilis* (Anderson, 1970; Hager, 1972). Identification of the specimen as *M. nobilis* has been confirmed by Anderson (personal communication).

This single tooth was recovered from Hidden Cave stratum II. Although the stratum II fauna is small (279 specimens identified to date), the composition of this fauna is similar to that of other Hidden Cave strata in that it contains mesic species, including *Marmota flaviventris* and *Neotoma cinerea,* that do not exist on Eetza Mountain today (Table 1). The precise dating of stratum II is somewhat problematic, though the general age of this stratum is not.

Two radiocarbon dates are available for this stratum—810 ± 80 B.P. (WSU-2457) and 3,850 ± 110 B.P. (WSU-2458). The younger of these determinations is clearly too young, as it underlies Mono

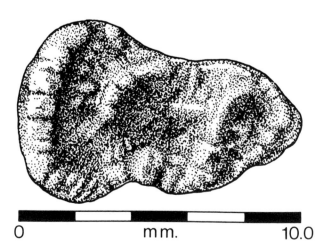

Fig. 4.—The Hidden Cave *Martes nobilis* M^1.

Table 1.—*Numbers of identified specimens (NISP) per taxon—Hidden Cave stratum II mammals.*

Taxon	NISP
Sylvilagus sp.	72
Sylvilagus nuttallii	1
Lepus sp.	119
Marmota flaviventris	9
Spermophilus sp.	1
Spermophilus cf. *townsendii*	3
Spermophilus townsendii	8
Thomomys sp.	2
Thomomys cf. *bottae*	3
Thomomys bottae	1
Perognathus sp.	1
Perognathus longimembris	1
Dipodomys sp.	11
Dipodomys microps	3
Peromyscus maniculatus	1
Neotoma sp.	11
Neotoma cf. *lepida*	7
Neotoma lepida	1
Neotoma cf. *cinerea*	2
Neotoma cinerea	6
Microtus sp.	9
Canis latrans	2
Canis lupus	1
Martes nobilis	1
Mustela vison	1
Taxidea taxus	1
Spilogale gracilis	1
Total	279

tephra dated to ca. 1,500 B.P. The older date, in turn, may be too old, because it overlies younger dates. Eight radiocarbon determinations from the deeper stratum IV fall between ca. 2,000 and 5,400 B.P., with four of these dates lying between ca. 3,500 and 3,800 B.P. Basing his interpretation on the internal fit of radiocarbon determinations, on geological considerations, and on the archaeological evidence, Jonathan O. Davis (personal communication; see also Davis, 1982) estimates that stratum II was deposited between about 3,700 and 3,600 B.P. Stratum II seems to have formed the surface of the site between ca. 3,600 and 1,500 B.P., during which time very little deposition occurred (Davis, personal communication). Stratum I dates to between ca. 1,500 B.P. and historic times.

Unfortunately, some of the Hidden Cave sediments have been disturbed by rodent activity, and by the activities of prehistoric peoples. As a result, the stratigraphic integrity of the deposits from which the *M. nobilis* tooth came cannot be given unquestioning faith. Indeed, part of the unit from which the tooth came (Lambda 10, level 3; see Fig. 5 for the location of the AMNH Hidden Cave excavations units) was disturbed by rodents. Because the tooth was discovered during the screening of deposits from that unit, it is not possible to know whether the specimen was originally located in the disturbed area. All that can be concluded from the Hidden Cave *M. nobilis* tooth is that either the noble marten survived much longer than has been thought, or that this tooth has been displaced upwards from Pleistocene strata. Both options are open.

THE NOBLE MARTEN: TIME OF EXTINCTION AND ADAPTATION

Although the stratigraphic placement of the Hidden Cave *Martes nobilis* specimen cannot be regarded as secure, it is also true that this does not represent the first late Holocene record for the noble marten. Kurtén and Anderson (1972), for instance, identified 45 specimens of *M. nobilis* from Jaguar Cave, southern Idaho, of which two came from a stratum that provided a single radiocarbon date of ca. 4,000 B.P. (Sadek-Kooros, 1972). These specimens, however, may also have reached this position through disturbance, whereas Sadek-Kooros (1972) believes the 4,000 B.P. date may be too young. More important, then, is the fact that S. J. Miller has identified a single tooth of *M. nobilis* from Dry Creek Rockshelter in the Boise River Valley of southwestern Idaho (in Webster, 1978), an identification that has also been verified by Anderson (personal communication). Bone from the stratum that provided the *M. nobilis* tooth was dated to 3,270 ± 110 B.P. (WSU-1574), whereas a date of 3,530 ± 85 B.P. (WSU-1486) is available for the stratum beneath, and of 2,090 ± 80 B.P. (WSU-1503) for the next dated stratum above. These dates are fully in line with artifactual stylistic markers from the same strata. Significantly, the stratum that provided WSU-1486 sits directly on the bedrock floor of the cave, so that the *M. nobilis* specimen in Dry Creek Rockshelter could not have been derived from deeper Pleistocene strata within the site (see the stratigraphic profile in Webster, 1978). The Dry Creek Rockshelter *M. nobilis*, then, seems to provide firm support for the survival of this animal into the late

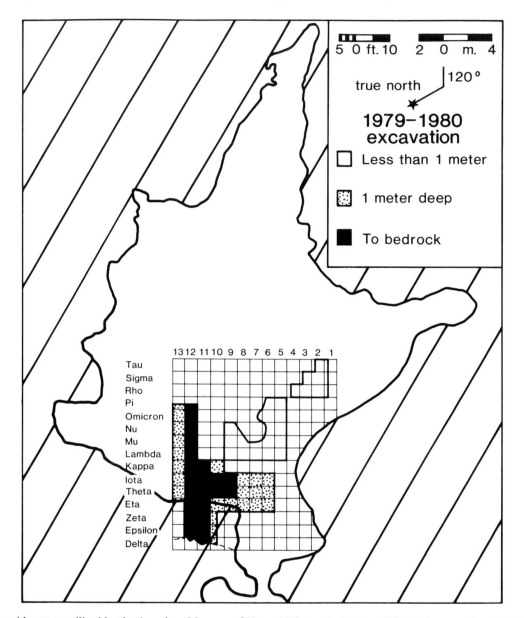

Fig. 5.—The grid system utilized by the American Museum of Natural History during the 1979–1980 excavations. The *Martes nobilis* specimen came from unit Lambda 10.

Holocene. Given that this is the case, the Hidden Cave *M. nobilis* may, in fact, indicate the late survival of the noble marten in western Nevada as well.

The likelihood that *M. nobilis* survived into late Holocene times in the arid west suggests that subfossil specimens of *Martes* from this area should be examined extremely closely. For instance, Spiess (1974) reported two specimens of *Martes americana* from Bronco Charlie Cave, Ruby Range, eastern

Nevada. One of these specimens was a lower right canine for which Spiess did not provide provenience. The second specimen, however, was a lower right mandible retaining M_1. Although precise information on the provenience of this specimen was not provided by Spiess, it did come from deposits that also contained archaeological materials. The oldest projectile points in Bronco Charlie Cave are Elko points (Casjens, 1974), a style that dates to between ca. 3,500 and 1,200 B.P. in the central and western Great Basin (O'Connell, 1967; Heizer and Hester, 1973; Thomas, 1981), although earlier dates are available to the east (Aikens, 1970). Thus, assuming no disturbance, the Bronco Charlie *Martes* mandible came from sediments that would seem to be late Holocene in age. Although *Martes americana* is no longer found on the Ruby Range (the closest known modern population are some 325 km to the east: Durrant, 1952), the extinction of boreal mammals in the Great Basin during Holocene times is now extremely well-documented (Grayson, 1977, 1981, 1982*b*; Thompson and Mead, 1982), and the Bronco Charlie *Martes* may represent another such instance. However, it seems equally possible that the specimen was misidentified and actually pertains to *Martes nobilis*. It should be reexamined to see whether this is the case.

I have noted that *Martes nobilis* is generally thought to have been adapted to cool conditions. This adaptation has been inferred from the fact that noble marten remains have been found associated with the remains of boreal taxa at a number of sites. At both Little Box Elder Cave, Wyoming, and Jaguar Cave, Idaho, for instance, late Pleistocene sediments incorporated specimens of both *M. nobilis* and the collared lemming, *Dicrostonyx* cf. *torquatus* (Anderson, 1968, 1970; Kurtén and Anderson, 1972). In contrast, however, and as Anderson (1970) has discussed, no boreal taxa were associated with *M. nobilis* at Samwel and Potter Creek caves in Shasta County, California; with the possible exception of the extinct large mammals, the species identified from those caves do not require an environment much different from that which characterizes the area today (Sinclair, 1903; Furlong, 1906).

Whether or not the Hidden Cave *M. nobilis* came from an undisturbed stratigraphic setting, it was not associated with a boreal fauna—there are no boreal species of mammals or birds in the Hidden Cave fauna. In fact, excluding the remains of extinct horse (*Equus* sp.), the set of mammalian and avian species represented in the Hidden Cave fauna identified to date (and the identifications are nearly complete) can be duplicated by extant faunas in many well-watered parts of the Great Basin today. Similarly, Dry Creek Rockshelter lacked boreal taxa. When arrayed alongside such sites as Samwel, Potter Creek, and Little Box Elder caves, Hidden Cave and Dry Creek Rockshelter suggest that the noble marten was adapted to a broad variety of environmental settings, and certainly cannot be taken as indicative of boreal conditions. In addition, it would seem to have occupied a wider variety of environments than *M. americana*. If so, the suggestion that competition with the American marten was in part responsible for the extinction of the noble marten throughout its range (Anderson, 1970; Kurtén and Anderson, 1972, 1980) loses much of its force.

In short, accumulating information on the noble marten suggests that it was adapted to a broad variety of environmental settings, that it would not have suffered from competition with the American marten throughout its range, and that it did not become extinct until late in the Holocene, perhaps after 3,000 B.P. If these hypotheses are correct, it must be asked why the noble marten survived the major environmental changes associated with the end of the Pleistocene at about 10,000 B.P. and those associated with the establishment of approximately modern environmental conditions in the arid west between about 8,000 and 7,000 B.P. (Mehringer, 1977; Van Devender and Spaulding, 1979), only to become extinct during late Holocene times.

ACKNOWLEDGMENTS

I thank Elaine Anderson for verifying my identification of the Hidden Cave *Martes nobilis* specimen, and for so freely sharing her thoughts on the adaptation and extinction of the noble marten. I thank E. Anderson, J. O. Davis, C. Maser, P. J. Mehringer, Jr., and D. H. Thomas for providing critical comments on this manuscript, and Brian Hatoff for making the reexcavation of Hidden Cave possible. Figures 1, 2, 4, and 5 were drawn and provided by Dennis O'Brien of the American Museum of Natural History; his help is once again gratefully acknowledged. Casts of the Hidden Cave *Martes nobilis* specimen have been deposited in the collections of the Burke Memorial Museum, University of Washington, the Frank H. McClung Museum, University of Tennessee, and the Department of Geosciences, University of Arizona. The specimen itself is housed in the collections of the Department of Anthropology, American Museum of Natural History. The research reported herein was supported by the American Museum of Natural History and the Bureau of Land Management (Carson City, Nevada), with additional support from the Richard Lounsbury and Speidel Foundations. The final report on the results of the Hidden Cave excavations will be issued as an Anthropological Paper of the American Museum of Natural History.

LITERATURE CITED

AIKENS, C. M. 1970. Hogup Cave. Univ. Utah Anthro. Papers., 93:1–286.

ANDERSON, E. 1968. Fauna of the Little Box Elder Cave, Converse County, Wyoming. The Carnivora. Univ. Colorado Studies, Ser. Earth Sci., 6:1–59.

———. 1970. Quaternary evolution of the genus *Martes* (Carnivora, Mustelidae). Acta Zool. Fennicae, 130:1–132.

CASJENS, L. 1974. The prehistoric human ecology of the southern Ruby Valley, Nevada. Unpublished Ph.D. dissert., Harvard Univ., Cambridge, Massachusetts 593 pp.

DAVIS, J. O. 1982. Sediments and geological setting of Hidden Cave. Pp. 183–242, *in* The archaeology of Hidden Cave (D. H. Thomas, ed.), Report submitted to the Bureau of Land Management, Carson City, Nevada, 886 pp.

DURRANT, S. D. 1952. The mammals of Utah: taxonomy and distribution. Univ. Kansas Publ., Mus. Nat. Hist., 6:1–549.

FURLONG, E. L. 1906. The exploration of Samwel Cave. Amer. J. Sci., 172:235–237.

GRAYSON, D. K. 1977. On the Holocene history of some northern Great Basin lagomorphs. J. Mamm., 58:507–513.

———. 1981. A mid-Holocene record for the heather vole, *Phenacomys* cf. *intermedius,* in the central Great Basin and its biogeographic significance. J. Mamm., 62:115–121.

———. 1982a. The paleontology of Hidden Cave: birds and mammals. Pp. 286–363, *in* The archaeology of Hidden Cave (D. H. Thomas, ed.), Report submitted to the Bureau of Land Management, Carson City, Nevada.

———. 1982b. Toward a history of Great Basin mammals during the past 15,000 years. Pp. 82–101, *in* Man and environment in the Great Basin (D. B. Madsen and J. F. O'Connell, eds.), Soc. Amer. Archaeol. Papers, 2:1–242.

HAGER, M. W. 1972. A late Wisconsin-Recent vertebrate fauna from the Chimney Rock Animal Trap, Laramie County, Wyoming. Univ. Wyoming Contrib. Geol., 2:63–71.

HALL, E. R. 1926. A new marten from the Pleistocene cave deposits of California. J. Mamm., 7:127–130.

———. 1936. Mustelid mammals from the Pleistocene of North America with systematic notes on some recent members of the genera *Mustela, Taxidea,* and *Mephitis.* Carnegie Inst. Washington Publ., 473:41–119.

HEIZER, R. F., and T. R. HESTER. 1973. Review and discussion of Great Basin projectile points: forms and chronology. Archaeological Research Facility, Dept. Anthro., Univ. California, Berkeley, ii + 39 pp.

KURTÉN, B., and E. ANDERSON. 1972. The sediments and fauna of Jaguar Cave. II. The fauna. Tebiwa, 15(1):21–45.

———. 1980. Pleistocene mammals of North America. Columbia Univ. Press, New York, 442 pp.

MEHRINGER, P. J., JR. 1977. Great Basin late Quaternary environments. Pp. 113–167, *in* Models and Great Basin prehistory (D. D. Fowler, ed.), Desert Res. Inst. Publ. Soc. Sci., 12:1–213.

MILLER, S. J. 1979. The archaeological fauna of four sites in Smith Creek Canyon. Pp. 272–329, *in* The archaeology of Smith Creek Canyon, eastern Nevada (D. R. Tuohy and D. L. Rendall, eds.), Nevada State Mus. Anthro. Papers, 17:1–394.

MORRISON, R. B. 1964. Lake Lahontan: geology of southern Carson Desert, Nevada. U.S. Geol. Surv. Prof. Papers, 401:1–156.

O'CONNELL, J. 1967. Elko eared/Elko corner-notched points as time markers in the Great Basin. Univ. California Archaeol. Surv. Report, 70:129–140.

ROUST, N. L., and C. W. CLEWLOW, JR. 1968. Projectile points from Hidden Cave (Nv-Ch-16), Churchill County, Nevada. Univ. California Archaeol. Surv. Report, 71:103–116.

SADEK-KOOROS, H. 1972. The sediments and fauna of Jaguar Cave. I. The sediments. Tebiwa, 15(1):1–20.

SINCLAIR, W. J. 1903. A preliminary account of the exploration of Potter Creek Cave, Shasta County, California. Science, 17:708–712.

SPIESS, A. 1974. Faunal remains from Bronco Charlie Cave (26EK801), Elko County, Nevada. Pp. 452–486, *in* The prehistoric human ecology of the southern Ruby Valley, Nevada, by L. Casjens. Unpublished Ph.D. dissert., Harvard Univ., Cambridge, 593 pp.

THOMAS, D. H. 1981. How to classify the projectile points from Monitor Valley, Nevada. J. California and Great Basin Anthro., 3:7–43.

———. 1982. Previous research at Hidden Cave. Pp. 80–109 *in* The archaeology of Hidden Cave (D. H. Thomas, ed.), Report submitted to the Bureau of Land Management, Carson City, Nevada, 886 pp.

THOMPSON, R. S., and J. I. MEAD. 1982. Late Quaternary environments and biogeography in the Great Basin. Quaternary Res., 17:39–55.

VAN DEVENDER, T., and W. G. SPAULDING. 1979. Development of vegetation and climate in the southwestern United States. Science, 204:701–710.

WEBSTER, G. S. 1978. Dry Creek Rockshelter: cultural chronology in the western Snake River region of Idaho ca. 4,150 B.P.–1,300 B.P. Tebiwa, Misc. Papers Idaho State Univ. Mus. Nat. Hist., 15:1–35.

WIGAND, P. E., and P. J. MEHRINGER, JR. 1982. Pollen and seed analysis for Hidden Cave. Pp. 252–285, *in* The archaeology of Hidden Cave (D. H. Thomas, ed.), Report submitted to the Bureau of Land Management, Carson City, Nevada, 886 pp.

ZIEMENS, G., and D. N. WALKER. 1974. Bell Cave, Wyoming: preliminary archaeological and paleontological investigations. Pp. 88–90, *in* Applied archaeology and geology: the Holocene history of Wyoming (M. Wilson, ed.). Geol. Surv. Wyoming Rept. Inv., 10:1–127.

Address: Department of Anthropology and Burke Memorial Museum, University of Washington, Seattle, Washington 98195.

LAVA BLISTERS AS CARNIVORE TRAPS

JOHN A. WHITE, H. GREGORY MCDONALD, ELAINE ANDERSON, AND
JAMES M. SOISET

ABSTRACT

Analysis of the bones of 58 kinds of vertebrate animals found in two Idaho lava blister caves indicates that in one cave, Moonshiner, 30% (42% of biomass) of the animal remains are of carnivorous mammals, such as wolves, coyotes, foxes, badgers, wolverines, martens, and weasels. Carnivores make up 42% (40% of biomass) in the other cave, Middle Butte. In contrast, the percentage of carnivores in the Recent regional fauna or in a "normal" fossil cave fauna, normally ranges from 2 to 5%. This strongly suggests that both caves functioned as selective carnivore traps. Evidently the bone deposit in Moonshiner Cave began accumulating at an earlier time than the deposit in Middle Butte Cave.

INTRODUCTION

One of the major problems in interpreting the paleontology of a fossil fauna is the recognition of bias. An accurate and feasible interpretation of the paleoecology of a fossil fauna requires that all processes, biological, climatological, or geological, which alter the representation of animals in a fossil deposit from that which occurred when those animals were living, should be identified. An ideal fauna for paleoecological analysis is one in which the preservation of individuals is proportional to their abundance in the living community so that the relative abundance of different species is preserved. Unfortunately, this is generally not the case and studies have identified various types of bias in the fossil record (Voorhies, 1969). Even though a random unbiased sample is ideal, studies of biased samples are also informative and may highlight interesting events that took place in the past and shed light on the paleoethology of a species. Large accumulations of bones of single species are presumably nonrandom. Such deposits include—the *Menoceras* and *Stenomylus* quarries of Miocene age at Agate Springs, Nebraska; the Pliocene Hagerman horse quarry in Idaho; the accumulation of bones of cave bear, *Ursus spelaeus,* in Pleistocene caves in Europe; and the asphalt deposits at Rancho La Brea, California. The over-representation of *Menoceras* bones at Agate Springs is attributed to stream transport of the bones (Peterson, 1923), whereas that of *Stenomylus* is interpreted as a catastrophic event which overwhelmed a herd while on its bedding ground (Brown, 1929). The wealth of cave bear bones is attributed to the gradual accumulation in dens over many generations in bear defended caves (Kurtén, 1976; Soergel, 1940), whereas the over-representation of large carnivores such as the sabertooth, *Smilodon fatalis,* and the dire wolf, *Canis dirus,* in tar pits, is interpreted as selective trapping as a result of scavenging food habits by individuals as they attempted to feed on entrapped herbivores (Stock, 1972). We propose that the large accumulation of carnivore bones in the lava caves of Idaho, discussed in this paper, is due to carnivores entering the caves to feed on entrapped herbivores and becoming entrapped themselves.

METHODS

The procedures used in excavating and collecting the bones from the two lava blister caves were as follows.

MOONSHINER CAVE

A test trench was excavated across the cave floor near the confluence of the northeast and southeast tunnels (Fig. 1), revealing the fine-grained silt to be deposited in pockets in the floor of the cave. The deepest of the latter pockets was less than 0.5 m deep. Thus the sediments ranged in thickness from 0 to 0.5 m. Almost no bones were found to be in articulation.

All medium to large-sized bones were collected from the surface of the floor of the cave, and almost none of the large bones were completely buried in the sediments. An estimated 10% of the bone-bearing sediments was removed at random and screened to produce approximately 1.5 metric tons of concentrate.

MIDDLE BUTTE CAVE

Two test trenches were excavated through the sediments down to the basalt floor. One trench reached a depth of 3.7 m and the other 2.6 m. The deeper trench was located 2 m west of the cave opening and the shallower one, 2 m north of the rock pile (Fig. 2). A shallow trench directly below the cave entrance revealed the rock fall, which is related to the formation of the cave entrance, to be confined to the upper meter of the sediments. This

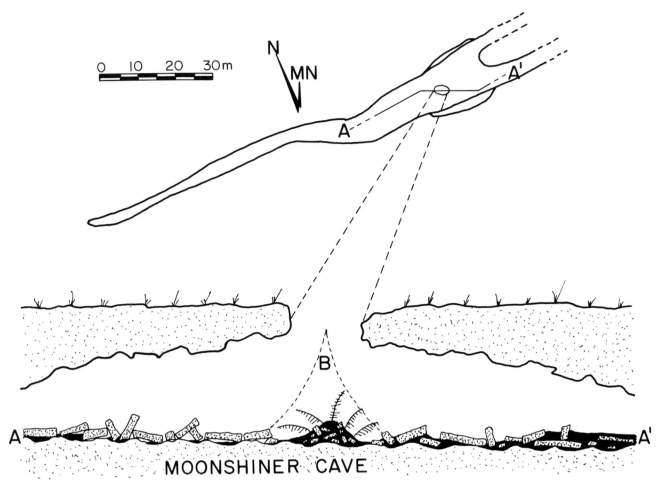

Fig. 1.—Map of Moonshiner Cave with a restored cross section (not to scale).

suggests that the sediments entered via another opening and that most of the sedimentation had occurred prior to the formation of the present opening.

Most of the bones (99%) were encountered in the upper 0.3 m of the sediments in the deeper trenches mentioned above. The sediments are in the clay to silt-size range, are very cohesive, and can be removed in well-defined blocks. A fluviatile origin is indicated by the fine laminations present throughout the entire deposit and by the presence of cross-bedding characteristic of stream deposits. The surface bones were intermixed with dung from *Neotoma* and plant material presumably brought into the cave by packrats.

While excavating the deeper trenches two layers of ash were found. The upper ash was at a depth of 25 cm and the lower ash was at a depth of 120 cm. The upper ash is Mazama and the lower Glacier Peak (Owen Davis, personal communication, 1982). Mazama ash is dated at 6,600 years B.P. and Glacier Peak 12,000 years B.P. This gives a rate of deposition of 95 cm in 5,400 years or one cm every 57 years. Using this average rate of sedimentation the 25 cm above the Mazama ash represents approximately 1,425 years. Because the majority of the bone was collected above the top of the consolidated sediments a maximum age for most of the bones would be approximately 5,175 years B.P.

All medium to large-sized bones were collected from the surface of the cave floor, and almost none of the large bones were completely buried in the sediments. Almost no bones were in articulation. An estimated 10% of the bone-bearing sediments were removed at random and screened to produce approximately one metric ton of concentrate.

GEOLOGIC SETTING

The lava fields of the Snake River Plain of Idaho contain numerous lava tubes (tunnels) and lava blisters (tumuli of Daly, 1914:133) as well as deep fissures such as the Great Rift near Aberdeen, Idaho. Tumuli are formed by the expansion of entrapped gases within the molten rock, and pressure ridges

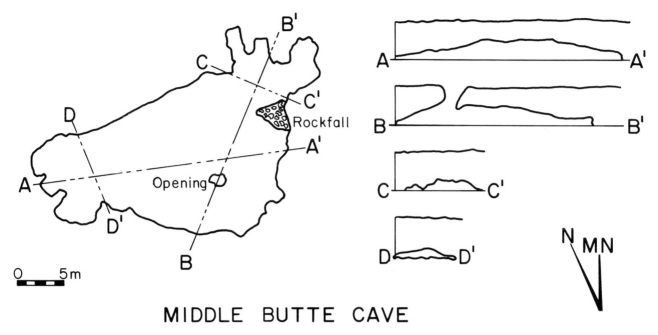

MIDDLE BUTTE CAVE

Fig. 2.—Map of Middle Butte Cave with cross sections.

and cracks appear on the ground surface over the cavities. These surface features can be traced to determine the underground extent of a tumulus. Lava tubes or tunnels are formed by the cooling of the surface of the molten lava prior to the deeper portions of the flow. The underlying still molten lava then flows out from beneath the hardened crust leaving a cylindrical cavity. Cavities in the lava or basalt, whether formed by gas or by flowing lava, range from a few centimeters to tens of meters in diameter and/or length. The collapse of a portion of the roof of these structures provides the entrance through which debris can accumulate in the caves. Debris accumulation can be attributed to climatic events such as eolian or fluviatile sedimentation, to raptorial birds, carnivore denning, human habitation, or some other source. The type and amount of debris accumulation ranges from many meters to none at all and depends on the number, size, and position of openings, external and internal drainage patterns, time of exposure to the outside, and the nature of the prevailing winds.

Although lava blisters may have but a single opening formed by collapse of part of the roof, the cracks formed on the surface of the ground above the roof enables water to trickle into the cave below.

THE CAVES

MOONSHINER CAVE

This cave, which was inhabited by a moonshiner during Prohibition, is a lava blister formed in undifferentiated basalts of Pleistocene age (Nace et al., 1975). It is located 48 km west of Idaho Falls, Bingham County, Idaho, and 0.8 km east of East Butte, an extinct volcano.

The cave entrance is approximately 1.5 m wide and 2 m long. A pile of breakdown is present 2.5 m below the entrance; it was formed by the collapse of the portion of the roof which formed the entrance.

A careful search of the surface along pressure ridges and of the inside of the cave revealed only one entrance, although the numerous cracks in the pressure ridges would permit water to trickle into the cave. The cave is Y-shaped (Fig. 1) with the tail projecting 90 m to the west, one arm projecting to the northeast and the other to the southeast. All portions of the floor lie essentially in the same plane with a slight sloping to the east. The highest portion of the cave ceiling is at the entrance. From this point, the ceiling slopes domelike in all directions. Ten meters west of the entrance the ceiling is 1 m high and it maintains this height to the end of the tunnel. The ceiling of the northeast arm decreases to 0.1 m only 15 m from the entrance. The same reduction in ceiling height is reached in the southeast arm 20 m from the entrance. Immediately under the cave entrance there is, from spring to late summer, a luxuriant growth of a fern, *Pteridium aquilinium* (Fig. 1). In winter and early spring, a snow cone forms which has been observed to reach to within 0.5 m of the cave entrance (Fig. 1B). The largest concentration of bone is situated near the entrance of the southeast arm.

The small amount of sediments in the cave suggests an eolian origin for the following reasons: 1) the cave opening evidently has been present for thousands of years; 2) water entered the cave through the cave opening and seeped through the cracks in the pressure ridges; 3) the pressure ridges are almost denuded of sediments and water tends to drain away to the sides. Thus it can be inferred that few sediments were washed into the cave. Wakefield Dort (personal communication) suggested there may have been some washing of sediments down slope into the northeast and southeast arms (each of which was traced over 100 m on the surface by following the pressure ridges), thus removing the sediments from the main chamber.

MIDDLE BUTTE CAVE

Middle Butte Cave is a lava blister which is connected to a lava tube. Evidently the lava blister was formed before the underlying lava tube was formed making the lava blister an outpocketing of the lava tube. The surface features on the ground above the lava blister and the connected lava tube can be traced 70 m northeast of the entrance to the cave to a large collapsed feature. A portion of the lava tube that has not collapsed extends south and west of the collapsed feature. This preserved portion has filled to within 1.5 m of the ceiling with fine laminated sediments. This seems to be the entrance by which water entered to deposit sediments in Middle Butte Cave. Sedimentation had essentially closed off this opening farther back in the lava tube prior to the formation of the entrance into the cave.

Middle Butte Cave is located 8 km east of Atomic City, Bingham County, Idaho, and 1.6 km south of Middle Butte, another extinct volcano, from which it gets its name. The distance between Middle Butte Cave and Moonshiner Cave is 8 km.

This cave is roughly triangular in outline when viewed from above (Fig. 2). The ceiling is domed, the highest point being approximately 2.5 m west of the entrance. The cave entrance is 2 m above the floor. Unlike the vertical entrance of Moonshiner Cave, the opening of Middle Butte Cave is sloped at about a 45 degree angle. The passage from the surface into the cave is about 2.5 m long. The sides of the cave are smooth except in areas where rock fall from the ceiling has occurred.

The temperature inside both caves remains consistently around 8° C.

THE FAUNAS

The bones from both caves are essentially unaltered. Some of the specimens have been encrusted with a carbonate deposit. This mineral encrustation is present also on the ceilings, marking places where water seeped through cracks in the basalt. A few mummified remains of animals that had recently fallen into the caves were found, but the majority of the bones were disarticulated and scattered on the cave floor. Although many bones had been gnawed by rodents, none of the limb bones exhibited unhealed fractures. This would suggest that the fall into the cave was not sufficient to cause breakage of limb bones and that not much scavenging occurred. A few pathological specimens are present in the samples but most are old and healed injuries, not related to entrapment.

The outstanding feature common to the faunas of the two caves is the abnormally high number of carnivores (Figs. 3 and 4), a condition that is comparable to that described for the fauna from the Rancho La Brea tar pits (Marcus, 1960; Stock, 1972).

The majority of the bones from both caves are of pygmy rabbits, cottontails, jack rabbits, sagebrush voles, red foxes, coyotes, badgers, and long-tailed weasels, all of which occur in the sagebrush-grass community found in the region today. *Martes americana* is is present in Middle Butte Cave, whereas *M. nobilis* is in the Moonshiner Cave fauna. The absence of *M. nobilis* from Middle Butte Cave suggests that entrapment in this cave occurred later than in Moonshiner Cave. Grizzly bear, wolf, kit fox, marten, black-footed ferret, wolverine, wapiti, and bison, although present in the cave faunas, are not found in the area today. The current absence of

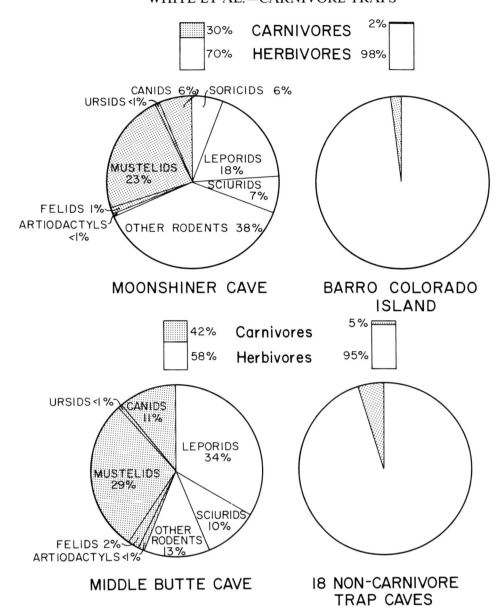

Fig. 3.—Pie diagrams, based on minimum numbers of individuals, comparing the ratios of carnivores to herbivores in the two selective trap caves, a Recent fauna, and 18 combined non-selective trap caves (See Table 2).

bison, wapiti, and wolf is probably related to the agency of man, whereas the disappearance of marten, black-footed ferret, and kit fox may be related to environmental changes. A detailed study of all faunal elements is in progress and will be published at a future date. A faunal list is given in Table 1.

CAVES AS CARNIVORE TRAPS

Observing the caves as they are today, one can hypothesize how the caves may have functioned as selective carnivore traps. The preponderance of carnivores in the faunas of both caves suggests that there must have been a bias of some type occurring during the accumulation of the bones. The use of caves as dens comes to mind. The difficulty to exit after entry into the caves precludes this possibility.

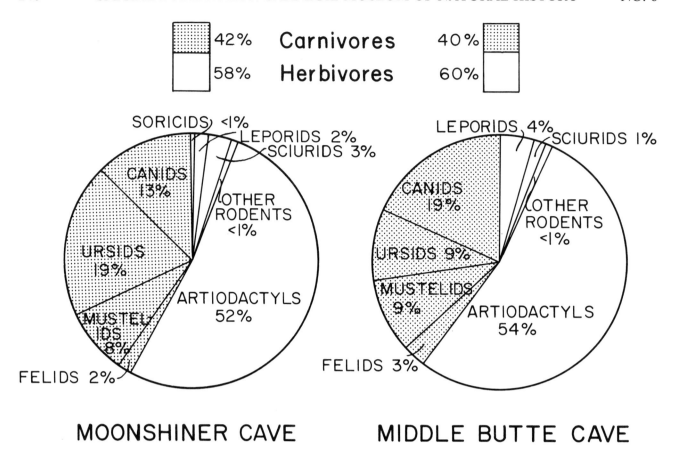

Fig. 4.—Pie diagrams, based on biomass, comparing the ratios of carnivores to herbivores in the two selective trap caves (See Table 1).

If the accumulation of animal bones had taken place in a cave with an entrance 7 m or more above the floor and creating a deadfall situation, or a cave with an entrance that provided easy ingress and egress, the resulting fauna probably would have a normal carnivore-herbivore distribution.

In Moonshiner Cave, from late spring to early autumn, the luxuriant growth of ferns immediately under the entrance gives the illusion that the distance to the floor is 1.5 m, an easy jump for carnivores such as coyotes and bobcats. An herbivore caught in the cave would have had little chance to escape, and a hungry carnivore, enticed by the sight, smell, or noise of an entrapped animal, could have jumped into the cave after it. Although an easy jump into the cave, the 2.5 m height of the entrance would require running in a straight line to obtain the necessary impetus to jump out. The rubble resulting from rock fall from the ceiling would have made running in a straight line difficult, thus trapping most

carnivores. Weakened or dead from starvation, a carnivore could become additional bait for the trap. Carnivores are opportunistic, and as Ricklefs (1973: 531) puts it, "One would expect extreme selectivity by predators that spend much time searching for and pursuing prey; the predators should choose the easiest catch, the individual with a relatively small chance of escape . . . ."

In winter, at Moonshiner Cave, a soft and precipitous snow cone develops which reaches to within a half meter of the ground level (Fig. 1B). This cone is formed by falling and drifting snow and is easily broken down by even a small object falling on it. Small herbivores or carnivores could easily jump onto the cone which would collapse under them and thus trap them in the cave. This kind of entrapment could occur by voluntary action of both herbivores and carnivores or as the result of pursuit.

In contrast, the entrance into Middle Butte Cave slants through approximately 2 m of basalt. The

Table 1.—*Faunal list with minimum number of individuals for Moonshiner Cave and Middle Butte Cave, Bingham County, Idaho. "P" indicates a group to be represented in the fauna, for which an accurate identification and census was not made. "Wt" indicates the average weight for each taxon, which was used to calculate the biomass.*

Taxon	Moonshiner Cave	Middle Butte Cave	Wt
Class Reptilia—reptiles			
Order Lacertilia—lizards	P	P	—
Order Ophidia—snakes			
Family Colubridae—racers, king snakes, and others	P	P	—
Family Crotalidae—rattlesnakes	P	P	—
Class Aves—birds			
Order Falconiformes—raptors			
Family Cathartidae—New World vultures			
Cathartes aura—Turkey Vulture	1	—	—
Family Accipitridae—eagles and hawks			
Aquila chrysaetos—Golden Eagle	1	—	—
Order Galliformes—grouse, ptarmigan, and others			
Family Tetraonidae—pheasants, quail, and others			
Centrocercus urophasianus—Sage Grouse	4	6	—
Order Strigiformes—owls			
Family Strigidae—owls			
Asio flammeus—Short-eared Owl	—	1	—
Order Piciformes—woodpeckers			
Family Picidae—woodpeckers			
Colaptes auratus—Common Flicker	1	—	—
Order Passeriformes—perching birds			
Family Corvidae—crows and jays	—	1	—
Family Sturnidae—starlings			
Sturnus vulgaris—Starling	1	—	—
Class Mammalia—mammals			
Order Insectivora—moles, shrews, and others			
Family Soricidae—shrews			
Sorex cf. *S. cinereus*	163	—	7 g
Order Chrioptera—bats			
Family Vespertilionidae—vespertilionid bats			
unidentified bats	21	1	—
Order Lagomorpha—rabbits, hares, pikas			
Family Ochotonidae—pikas			
Ochotona cf. *O. princeps*—pika	1	—	142 g
Family Leporidae—rabbits and hares			
Brachylagus idahoensis—pygmy rabbit	230	139	440 g
Sylvilagus nuttallii—Nuttall's cottontail	211	82	755 g
Lepus americanus—snowshoe hare	9	17	1.4 kg
Lepus cf. *L. californicus*—black-tailed jack rabbit	2	3	2.3 kg
Lepus cf. *L. townsendii*—white-tailed jack rabbit	15	12	3.4 kg
Order Rodentia—rodents			
Family Sciuridae—squirrels			
Eutamias minimus—least chipmunk	27	—	35 g
Spermophilus townsendii—Townsend's ground squirrel	27	—	120 g
Spermophilus richardsoni—Richardson's ground squirrel	4	—	410 g
Spermophilus lateralis—golden-mantled ground squirrel	6	—	226 g
Spermophilus sp.—ground squirrel	57	69	200 g
Marmota flaviventris—yellow-bellied marmot	56	4	6 kg
Family Geomyidae—pocket gophers			
Thomomys cf. *T. talpoides*—northern pocket gopher	123	17	150 g
Family Heteromyidae—pocket mice, and others			
Perognathus cf. *P. parvus*—silky pocket mouse	23	—	24 g
Family Castoridae—beavers			
Castor canadensis—beaver	1	—	22 kg

Table 1.—*Continued*

Taxon	Moonshiner Cave	Middle Butte Cave	Wt
Family Cricetidae—New World rats and mice			
Peromyscus maniculatus—deer mouse	561	3	35 g
Neotoma cinerea—bushy-tailed woodrat	48	75	395 g
Microtus montanus—montane vole	1	—	60 g
Microtus longicaudus—long-tailed vole	1	—	60 g
Lagurus cf. *L. curtatus*—sage vole	1	—	30 g
Clethrionomys cf. *C. gapperi*—red-backed vole	1	—	30 g
Phenacomys cf. *P. intermedius*—heather vole	1	—	35 g
Unidentified voles	229	5	35 g
Family Erethizontidae—porcupines			
Erethizon dorsatum—porcupine	1	1	5 kg
Order Artiodactyla—even-toed hoofed mammals			
Family Cervidae—deer, wapiti, and others			
Cervus cf. *C. elephas*—wapiti	2	2	275 kg
Family Antilocapridae—pronghorns			
Antilocapra americana—pronghorn	4	1	50 kg
Family Bovidae—bison, cattle, sheep			
Bos taurus—cattle	P	P	—
Bison bison—American bison	P	P	—
Bos or *Bison*	7	2	900 kg
Ovis aries—domestic sheep	1	1	75 kg
Order Carnivora—carnivores			
Family Canidae—dogs			
Canis lupus—timber wolf	11	4	51 kg
Canis latrans—coyote	32	23	18 kg
Vulpes vulpes—red fox	120	46	4.8 kg
Vulpes macrotis—kit fox	2	11	2 kg
Family Ursidae—bears			
Ursus arctos—grizzly bear	13	2	205 kg
Family Mustelidae—mustelids			
Mustela frenata—long-tailed weasel	293	153	300 g
Mustela erminea—ermine	35	13	190 g
Mustela sp. (*M. frenata* or *M. erminea*)	109	25	245 g
Mustela nivalis (=*M. rixosa*)—least weasel	8	—	40 g
Mustela nigripes—black-footed ferret	2	—	700 g
Mustela vison—mink	1	—	1 kg
Martes nobilis—extinct pine marten	25	—	1.5 kg
Martes cf. *M. americana*—pine marten	—	2	850 g
Gulo gulo—wolverine	25	5	21 kg
Taxidea taxus—badger	55	15	8.2 kg
Spilogale gracilis—spotted skunk	24	6	600 g
Mephitis mephitis—striped skunk	1	—	1.7 kg
Family Felidae—cats			
Lynx canadensis—Canada lynx	12	—	10.1 kg
Lynx rufus—bobcat	12	—	11.2 kg
Lynx sp. (*L. canadensis* or *L. rufus*)	—	14	10.2 kg

sides of this opening are smooth and almost without crevices for feet or claws to obtain footholds. The opening, which is 1 m² and passes through 2.5 m of basalt, permits but a few minutes of direct sunlight to enter the cave each day and provides for an abrupt change from light to darkness. Dim light near the entrance may have affected the depth perception of a carnivore making ready to jump into the cave, or, in the case of pursuit, there would be insufficient time to adjust to the difference of light intensity. Any attempt to leap out of the cave would be made considerably more difficult without "footholds" on the sides of the entrance, and by the presence of rubble on the floor.

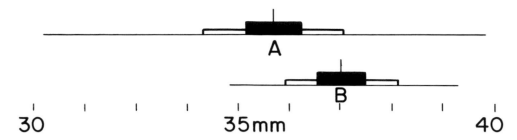

Fig. 5.—Modified Dice-Leraas diagram comparing the basilar length of skull (of Hensel) of male, extant *Mustela erminea invicta* from Idaho (A), with *M. erminea* from Moonshiner Cave (B). Data for "A" from E. R. Hall (personal communication).

AGE OF THE FAUNAS IN MOONSHINER AND MIDDLE BUTTE CAVES

The presence of *Martes nobilis,* the only extinct species found in Moonshiner Cave, suggests that it may have been acting as a trap longer than Middle Butte Cave. Jaguar Cave in Lemhi County, Idaho, has *M. nobilis* as a member of the fauna with a date of 10,370 ± 530 years B.P. (Kurtén and Anderson, 1972). Webster (1978) reported *M. nobilis* from a level dated between 3,300 and 2,550 years B.P. in Dry Creek Rock Shelter, Ada County, Idaho. There are no extinct taxa from Middle Butte Cave. As discussed above, the fauna from Middle Butte Cave cannot be any older than 5,175 years B.P. as estimated from the rate of sedimentation and is probably much younger. Despite the evidence that *M. nobilis* may have survived until relatively recent time in the state, we think the fauna from Moonshiner Cave is older than that from Middle Butte Cave and that it started functioning as a trap at an earlier date.

The presence of pine marten, wolverine, ermine, least weasel, and Canada lynx indicates a cooler climate. Specimens of *Mustela erminea* from Moonshiner Cave are significantly larger than specimens of this species occurring north of the area today, *M. e. invicta,* and are about the size of *M. e. richardsoni* from near Fort Franklin, Mackenzie, some 2,400 km to the north (Fig. 5) (Hall, 1951). This situation is akin to that in the Conard Fissure in Arkansas. Regarding the latter, Hall (1951:167) writes, "It may be significant that the cranial characters of the female ermine from there are most nearly approximated among Recent weasels by those which live along the southern edge of the frozen tundra."

The bone accumulation in both caves must have taken place over a long period of time, because it seems unlikely that 30 wolverines and 446 weasels could exist in the same vicinity at the same time. The caves are separated by 8 km and Moonshiner Cave was open at the time Middle Butte Cave began functioning as a trap.

COMPARISONS WITH OTHER CAVES

Comparisons with other cave deposits provide information pertinent to our carnivore trap hypothesis. The fauna from Meyer Cave, Illinois (Parmalee, 1967), includes a large number of carnivores; however, the percentage (5.9%) is not so disproportionate as in our two Idaho cave faunas. The Meyer Cave bone deposit is at the bottom of a vertical fissure with a drop of 4.2 m. The bone deposit terminated at a depth of 6.4 m below the top of the fissure. The opening into the fissure is connected to the outside of the bluff by a 4.5 m horizontal crawlway. Easy access to the crawlway is made possible by a talus slope. Remains of mountain lion kittens, raccoon pups, numerous bobcat kittens, and fox pups in Meyer Cave suggest that it was used as a den, and that the accumulation of carnivore bones in the bottom of the fissure was due to occasional accidental falls. The relatively small number of adult carnivores from the bottom of the fissure suggests that the opening to the pitfall was generally avoided.

There is a striking similarity in the composition of the fauna and distance from the cave opening to the floor, between our two Idaho caves and the Chimney Rock Animal Trap Cave, Larimer County,

Colorado (Hager, 1972). In the latter cave the distance from the cave entrance to the floor is approximately 2.5 m and the fauna has a ". . . preponderance of carnivores . . ." (Hager, 1972:68).

The bone deposit at New Paris No. 4, Pennsylvania (Guilday et al., 1964), accumulated in a dome pit formed in a narrow fissure connected to the surface and adjacent to a sterile sinkhole, 9 m in diameter. The large size of the sinkhole makes it an obvious obstacle to avoid, and the rubble formed by the collapse of the roof formed a slope too gentle to function as an efficient trap. The small adjacent dome pit with its restricted opening at the surface functioned as an efficient pitfall trap which resulted in the collection of a random sample of the fauna in the area; it included only five species of carnivores, all mustelids. The vertical distance of 9 m from the top of the fissure to the bottom seems too large to induce carnivores to enter, despite the wealth of carcasses accumulated on the bottom. The depth and restricted dimensions of the fissure may have prevented carnivores from knowing of the trapped animals.

A final comparison can be made with an open sink hole accumulation, Nichol's Hammock, Florida (Hirschfeld, 1968). The sinkhole is 3.7 m to 4.6 m deep and 3 m wide. Unlike the large sinkhole at New Paris, the walls are vertical with a slight inward curvature at the top. This configuration insures that any animal trapped in the pit would not be able to get out. The mammalian fauna from the sinkhole was composed of a minimum of 44 individuals of 13 species, 28% (12 individuals) of which were carnivores. Perhaps the trapped herbivores in the sinkhole served to attract carnivores to the edge, and in an attempt to find a way down they might have accidentally fallen into the sinkhole. Unlike Moonshiner and Middle Butte Caves, the drop into the plainly visible sinkhole would be far too great to assume that any would jump in voluntarily.

Rather than extending these comparisons to include all known Pleistocene cave faunas, a synopsis is presented in Table 2. Caves for which MNI's have been published are listed along with the percentage of carnivores in each one. The high representation of carnivores in Moonshiner and Middle Butte Caves can easily be seen. At the bottom of the table, the percentage of carnivores for all faunas is only 4.7 in contrast to 32.8 for the two Idaho caves. The percentage of carnivores in the fauna generally ranges from zero to 11%. Three of these faunas have higher than expected numbers of carnivores, although not so high as in the Idaho caves. Guilday and Adam (1967) attribute the high number of carnivores at Jaguar Cave to its use as a denning site. The high percentage of carnivores in the other two faunas, Nichol's Hammock and the First American Bank Site, are a result of small samples, 44 and 31 MNI, respectively. In smaller samples, the representation of rarer individuals (in this case the carnivores) will be exaggerated (Grayson, 1978; Payne, 1972). The samples from Moonshiner and Middle Butte Caves are large enough to overcome this problem. It must be admitted that continued counting of small rodents in the Idaho cave samples would mathematically decrease the percentage of medium and large-sized carnivores; but, at the same time, it would not greatly affect the percent biomass they represent (see Fig. 5). An increased number of mice in the faunal count would not alter the absolute number of carnivores from the caves. To explain the large number of carnivores in the Idaho caves, we offer the hypothesis of selective entrapment based on the peculiar configuration of the two caves and the behavior of the carnivores.

FACTORS RELATING TO CAVES AS CARNIVORE TRAPS

AGE DISTRIBUTION

The age distribution of medium to large-sized carnivores, especially in Moonshiner Cave, suggests that juveniles and very old adults were more often entrapped than prime adults. This may be related to the smaller size of younger animals and the overall weakened condition of the older adults. Thus, prime adult animals may have been able to escape.

DIVERSITY INDEX

The calculation of the diversity index (H') and equilibrity (E) for each of the faunas in Table 2 also reflects the degree of bias toward carnivores in the fauna. Species diversity (H') of each fauna was computed with the formula:

$$H' = \sum_{i=1}^{s} p_i \ln p_i$$

Table 2.—*The numbers of mammalian carnivores and herbivores recorded from several North American caves. Bats are excluded. See text for explanations of diversity index, maximum diveristy, and equilibrity.*

Locality	Number of species	MNI	MNI carnivores	Percent carnivores	Diversity index (H')	Maximum diversity (H$_{max}$)	Equili-brity (E)	Diversity index carnivores	Percent diveristy carnivores
Arizona:									
Papago Springs cave (Skinner, 1942)	27	217	14	6.5	2.297	3.296	0.697	0.306	13.3
Florida:									
Nichol's Hammock (Hirschfeld, 1968)	14	44	10	22.7	2.227	2.639	0.844	0.646	29.0
Idaho:									
Jaguar Cave (Guilday and Adam, 1967; Kurtén and Anderson, 1972)	39	549	88	16.0	2.679	3.664	0.371	0.679	25.3
Middle Butte Cave (this paper)	29	756	319	42.2	2.458	3.367	0.730	1.031	42.0
Moonshiner Cave (this paper)	43	2,591	780	30.1	2.666	3.761	0.709	0.893	33.5
Illinois:									
Meyer Cave (Parmalee, 1967)	40	8,870	523	5.9	1.763	3.689	0.478	0.224	12.7
Kentucky:									
Welsh Cave (Guilday et al., 1971)	23	127	6	4.7	2.149	3.135	0.685	0.207	9.6
Missouri:									
Brynjulfson Cave 1 (Parmalee and Oesch, 1972)	45	247	27	10.9	3.130	3.807	0.822	0.511	16.3
Brynjulfson Cave 2 (Parmalee and Oesch, 1972)	37	781	63	8.1	2.424	3.611	0.671	0.351	14.5
Crankshaft Cave (Parmalee et al., 1969)	55	1,926	47	2.4	2.454	4.007	0.612	0.148	6.0
Pennsylvania:									
Bootlegger Sink (Guilday et al., 1966)	32	63	9	12.0	3.145	3.466	0.908	0.464	14.8
Hosterman's Pit (Guilday, 1967)	8	12	0	0.0	1.814	2.079	0.873	0	0
New Paris No. 4 (Guilday et al., 1964)	36	1,363	5	0.3	2.469	3.584	0.689	0.024	1.0
Sheep Rock Shelter (Guilday and Parmalee, 1965)	32	287	29	10.3	2.429	3.466	0.701	0.349	14.4
Tennessee:									
First American Bank Site (Guilday, 1977)	20	31	7	22.6	2.891	2.996	0.965	0.669	23.1
Robinson Cave (Guilday et al., 1969)	41	725	21	2.9	2.653	3.714	0.714	0.160	6.0
Virginia:									
Clark's Cave (Guilday et al., 1977)	45	2,754	18	0.7	2.537	3.807	0.666	0.044	5.1
Natural Chimneys (Guilday, 1962)	50	523	14	1.9	3.130	3.912	0.800	0.159	5.1
West Virginia:									
Eagle Cave (Guilday and Hamilton, 1973)	26	53	2	3.3	2.928	3.258	0.899	0.150	5.1
Wyoming:									
Little Box Elder Cave (Anderson, 1968)	48	2,917	120	4.1	2.327	3.871	0.601	0.204	8.7
Subtotal: (excluding Moonshiner and Middle Butte Caves)	—	21,489	1,003	4.7*	—	—	—	—	11.9*
Subtotal (Moonshiner and Middle Butte Caves)	—	3,347	1,099	32.8*	—	—	—	—	37.8*
Grand total:		24,836	2,102						

* = means.

where p_i represents the proportion of the total number contributed by the ith species, s represents the number of species in the fauna (Pielou, 1966), and

$$E = H/H_{max}$$

where H_{max} is the natural logarithm of the number of observed species. The carnivores in the faunas from Moonshiner and Middle Butte Caves contribute 33.5 and 42.0, respectively, to the faunal diversity. None of the carnivores in the other cave faunas come close to these numbers with respect to their contribution toward faunal diversity. Nichol's Hammock, Jaguar Cave, and the First American Bank Site do stand out in that the carnivores contribute greatly to the total faunal diversity but they still do not approach those of the carnivores in the Idaho caves. Because of energy flow and their trophic position, carnivores do not form a major portion of a modern community in relation to other animals, neither in number of individuals nor in number of species (Odum, 1968, 1971).

Few complete studies of modern vertebrate faunas are available. The study of nonvolant mammals on Barro Colorado Island by Eisenberg and Thorington (1973) lists the carnivores as comprising 2.1% of the individuals and 2.7% of the biomass. They also cite a study of Suriname in which the carnivores contribute 2.06% of the individuals and 1.16% of the biomass. Many of the caves listed in Table 2, such as Meyer, Welsh, Crankshaft, Robinson, Eagle, and Little Box Elder caves, compare with these percentages for modern faunas.

That the carnivores in Moonshiner and Middle Butte Caves contribute such a large percentage toward total numbers of individuals in the faunas and to overall faunal diversity, attests to the uniqueness and effectiveness of the caves as traps.

CLOSED CAVE SYSTEM

In addition to the trapping of animals, the configuration of the Idaho caves also prevents postmortem removal of bones. In stream deposits there may be a selective winnowing of bones (Voorhies, 1969) and on the surface there may be a scattering of the bones by scavengers resulting in the loss of skeletal elements (Hill, 1980). In cave systems such as we have described for the selective traps, theoretically, there is no loss of bone. Although carcasses may be torn apart and scattered by the entrapped carnivores, they still remain in the cave. Specimens can only be lost by breakage due to chewing or gnawing, being eaten and completely digested, being sub-

jected to leaching and decomposition resulting from cave dampness, or removal by humans prior to the time we made the collections.

POPULATION DENSITY AND BODY SIZE

A major consideration is the population density of a particular species. The absolute number of individuals within a given geographical area is related to body size. It is a standard axiom in ecology that the larger the body size of an animal, the fewer the numbers of individuals within a given area. For example, the large number of long-tailed weasels, *Mustela frenata* (293 in Moonshiner Cave and 178 in Middle Butte Cave), versus the number of grizzly bears, *Ursus arctos* (13 and 2, respectively) fits this axiom.

TERRITORIALITY

Territoriality is also a factor. The wolverine, *Gulo gulo,* for example, is a solitary animal; a single male may range over an area of 3,000 km² which it will share with two or three females. Covering a large territory requires a great deal of time and the chances are slim that a wolverine would discover an herbivore shortly after it had fallen into the cave. There would be a time lag before a new individual would move into the newly vacated territory. Even after the invasion of the new territory, it is still possible that the new animal would not come into the vicinity of the cave. Because wolverines are solitary, antisocial animals, the presence of 25 individuals in Moonshiner Cave would argue not only for the effectiveness of the trap but also for the considerable length of time over which it functioned.

RARE SPECIES

The smallest carnivore in the cave faunas and in North America today is the least weasel, *Mustela rixosa* (*M. nivalis* of some authors). It is also one of the rarest mammals, both in fossil and modern faunas. This pattern is duplicated in Moonshiner Cave where only eight individuals were recovered. The presence of eight individuals of a rare carnivore preserved in the cave strengthens the idea that the cave was acting as a baited trap over a long period of time.

BODY SIZE AND HEIGHT OF CEILING

One of the main criteria for the selectivity of the Idaho caves as traps is the floor to ceiling height. This criterion applies especially to large carnivores because the cave opening to floor distance is small in proportion to body size, but it does not account for the abundance of *Mustela frenata.* The average size of the individuals of this species is less than one tenth of the distance from the cave floor to the entrance. Therefore it is doubtful that weasels would jump down into the cave from late spring to early autumn, although some may have fallen in while trying to find a way down into the cave.

The long-tailed weasel is known for its voracious feeding behavior. It has a well-developed sense of smell and follows the trails of its prey by scent, loping along until it overtakes its victim, then kills it by biting the base of the skull. It may kill more than it can eat but often caches the excess for future meals, especially in late summer and autumn. Studies on the food habits of *M. frenata* (Hall, 1951; Quick, 1951) show that mice, especially *Microtus* and *Peromyscus,* are favored prey. When mice are scarce, long-tailed weasels take pocket gophers, chipmunks, ground squirrels, rabbits, shrews, and other animals. Thus, almost any mammal, ranging in size from a rabbit down to a shrew, could serve to bait the caves as traps for weasels. Weasels eat about one third of their weight every 24 hrs. Population density ranges from a low of one per 1.3 km² (Quick, 1951) to as many as one per 0.25 km² (Burt and Grossenheider, 1976). They have a small home range and a short cruising radius.

As noted above, from late autumn to early spring, snow drifts through the entrance of Moonshiner Cave and forms a large cone that almost reaches up to ground level. This snow cone reduces the floor to ceiling height to less than a meter and it seems probable that smaller-sized carnivores could make the relatively easy jump onto the snow cone. Since the cone is formed by drifting snow, it probably would break down to some extent by the weight of the jumping animal, making it difficult to jump out again. Entrapment of large carnivores in Moonshiner Cave, therefore, probably occurred in summer and winter, whereas smaller carnivores were probably only trapped in winter.

The age distribution of the numerous individuals of *Mustela frenata* in the two caves support our contention of its winter entrapment. Employing the criteria used by Hall (1951: 25), the crania of *M. frenata* from the caves were segregated into age groups, as follows: *juveniles*—individuals with milk teeth, birth to 3 months; *young*—individuals with open sutures between the maxilla and premaxilla and between premaxilla and nasals, 3 to 7.5 months; *subadults*—individuals with sutures between max-

illa and premaxilla visible but indistinct, 7.5 to 10 months; *adults*—individuals with bones of rostrum coalesced and no traces of sutures visible to the naked eye, over 10 months.

Of 293 individuals from Moonshiner Cave, 66.6% were adults, 26.6% were subadults, and 6.8% fell into the young age group. No juveniles were present. The sample of 178 crania from Middle Butte Cave produced 71.3% adults, 23.4% subadults, 5.3% young, and no juveniles.

Generally, weasels are born in April (Wright, 1948). It would be highly improbable for a juvenile to become entrapped, because by the time the young weasels moved away from the den, the snow cone would have melted away from underneath the cave entrance (Fig. 6). By the time they had matured enough to hunt for themselves, it would be late autumn or early winter, at which time they would have achieved the young or early subadult age class. This coincides with the first snow fall in the area of the caves between late October and early December and the first appearance of a snow cone under the entrance (Fig. 1). Although one might assume that more of the younger, inexperienced animals would be trapped, this is not the case. Most of the individuals are in the adult age class.

At Middle Butte Cave, the slope of the entrance precludes the formation of a snow cone and a growth of ferns, and it reduces the amount of light in the cave, thus insuring an abrupt change in light intensity. In daytime, weasels or any other predator in pursuit of prey entering the cave would have insufficient time to adjust to the abrupt change in light intensity and, consequently, would fall into the cave. At night, a pursuing predator may not have been able to stop in time on the smooth, inclined surface of the entrance. The relatively large number of leporid bones in this cave (Fig. 3), coupled with the knowledge that rabbits often go into any available shelter to escape a predator, suggests that many carnivores were trapped while in pursuit of rabbits. In this cave, we cannot account for the absence of juvenile weasels.

CARNIVORE HABITS

The biased representation of carnivores in the Idaho selective-trap caves is related essentially to two factors: 1) the configuration of the caves, which has already been discussed and can be considered a constant; 2) the behavior and habits of carnivores leading to entrapment in the caves. The latter factor

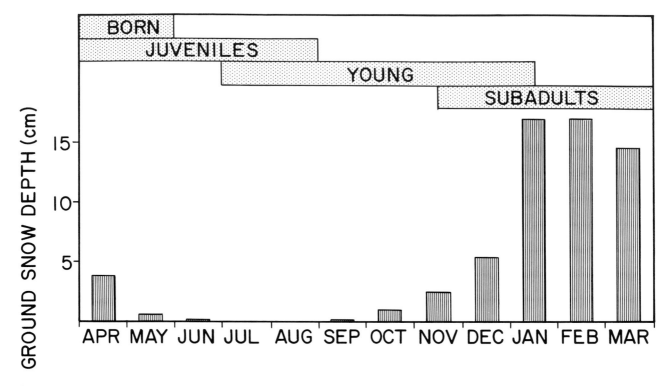

Fig. 6.—Diagram with the age-distribution of extant weasels by month, and the average depth per month of snow on the ground at the Idaho National Engineering Laboratory (data from O. Doyle Markham, personal communication).

is extremely variable and depends on the behavior and habits of the species in question.

A brief description of the habits of some carnivores (Burt and Grossenheider, 1976; Walker, 1975) will round out this discussion. The spotted skunk, *Spilogale gracilis,* is carnivorous, but its food varies seasonally. In summer, it feeds mainly on vegetable matter and insects, whereas in winter it feeds on rodents and other small mammals. They are nocturnal and have a population of more than one per ha and a home range of 64 ha. The small size of these skunks suggests that they probably were trapped in Moonshiner in winter.

Although the pine marten, *Martes nobilis,* from Moonshiner Cave is extinct, its habits probably were similar to those of the living *M. americana* which is present in Middle Butte Cave. This small, active, mainly nocturnal carnivore feeds on rodents and rabbits but takes carrion in the winter. A population density of fewer than one per km² is probably high, and the presence of 25 MNI in Moonshiner Cave contributes to our interpretation of selective entrapment.

The badger, *Taxidea taxus,* is fossorial, preys on rodents and rabbits, and uses carrion as an important part of its diet. As are wolverines, badgers are solitary animals but have a smaller home range of about 850 ha (Long, 1973). The relatively larger number of badgers in Moonshiner Cave (55 badgers and 25 wolverines) correlates with the smaller home range, thus increasing the chances for an individual badger to find the cave, certainly better than for wolverines.

The canids, *Vulpes vulpes, Canis latrans,* and *C. lupus,* have keen senses of smell and hearing, hunt at night, and often feed on carrion. They provide another example of smaller carnivores with small home ranges being more numerous than larger ones with larger home ranges. In Moonshiner Cave, there were 120 red foxes, 32 coyotes, and 11 wolves, and in Middle Butte Cave there were 46 red foxes, 23 coyotes, and 4 wolves.

A final example is the genus *Lynx* (*L. rufus* and *L. canadensis*), solitary and nocturnal cats that feed on rabbits, hares, rodents, and birds. A minimum number of 22 *Lynx* were identified at Moonshiner and 14 at Middle Butte Cave.

Several points are evident from the above:

1. Most carnivores hunt by scent and feed on carrion.
2. Most carnivores are nocturnal.
3. Small carnivores are generally more abundant and have smaller home ranges than larger ones.
4. Based on what is known of home range, population density, and food habits of modern carnivores, the proportion of carnivores to carnivores in the caves is the same as in a modern regional fauna.

All these factors lead us to conclude that the caves operated over a long period of time as baited traps which were easy to enter but difficult to escape.

ACKNOWLEDGMENTS

We greatly acknowledge the financial support of Mr. and Mrs. E. Hadley Stuart, Jr., in the excavation and field study especially of Middle Butte Cave. For permission to collect materials from Moonshiner Cave, we thank the officials of the Idaho National Engineering Laboratory. Tom Amberson, Penn Gildersleeve, Mark Johnson, Scott McDonald, and Jim White aided in the collection of the cave materials. Wakefield Dort and William H. Burt provided many invaluable comments on the geology and zoology of the caves, respectively. We thank O. Doyle Markham, Department of Energy, Idaho Operations Office, for providing data on snowfall during 14 years at the Idaho National Engineering Laboratory. E. Raymond Hall kindly provided measurements of extant *Mustela erminea invicta* from Idaho. John E. Guilday, Barry L. Keller, Ernest L. Lundelius, Jr., and Michael R. Voorhies critically read the manuscript and made many valuable suggestions. John Moeller made the illustrations.

LITERATURE CITED

ANDERSON, E. 1968. Fauna of the Little Box Elder Cave, Converse County, Wyoming: The Carnivora. Univ. Colorado Stud. Ser., Earth Sci., 6:1–59.

BROWN, B. 1929. A Miocene camel bed-ground. Nat. Hist., 29:658–662.

BURT, W. H., and R. P. GROSSENHEIDER. 1976. A field guide to the mammals. Houghton Mifflin Co., Boston, 3rd ed., 289 pp.

DALY, R. A. 1914. Igneous rocks and their origin. McGraw-Hill, New York, 563 pp.

EISENBERG, J. F., and R. W. THORINGTON, JR. 1973. A preliminary analysis of a Neotropical fauna. Biotropica, 5:150–161.

GRAYSON, D. K. 1978. Minimum numbers and sample size in vertebrate faunal analyses. Amer. Antiq., 43:53–65.

GUILDAY, J. E. 1962. The Pleistocene local fauna of the Natural Chimneys, Augusta County, Virgina. Ann. Carnegie Mus., 36:87–122.

———. 1967. The climatic significance of the Hosterman's Pit local fauna, Centre County, Pennsylvania. Amer. Antiq., 32:231–232.

———. 1977. Sabertooth cat, *Smilodon floridanus* (Leidy), an associated fauna from a Tennessee Cave (40 DV 40), the First American Bank Site. J. Tennessee Acad. Sci., 52:84–94.

GUILDAY, J. E., and E. K. ADAM. 1967. Small mammal remains from Jaguar Cave, Lemhi County, Idaho. Tebiwa, 10:26–36.

GUILDAY, J. E., and H. W. HAMILTON. 1973. The Late Pleistocene small mammals of Eagle Cave, Pendleton County, West Virginia. Ann. Carnegie Mus., 44:45–58.

GUILDAY, J. E., and A. D. MCCRADY. 1966. The bone breccia of Bootlegger Sink, York County, Pennsylvania. Ann. Carnegie Mus., 38:145–163.

———. 1969. The Pleistocene vertebrate fauna of Robinson Cave, Overton County, Tennessee. Palaeovertebrata, 2:25–75.

———. 1971. The Welsh Cave peccaries (Platygonus) and associated fauna, Kentucky, Pleistocene. Ann. Carnegie Mus., 43:249–320.

GUILDAY, J. E., and P. W. PARMALEE. 1965. Animal remains from the Sheep Rock Shelter (36 Hu 1), Huntingdon County, Pennsylvania. Pennsylvania Archael., 35:34–49.

GUILDAY, J. E., P. S. MARTIN, and A. D. MCCRADY. 1964. New Paris No. 4: a Late Pleistocene cave deposit in Bedford County, Pennsylvania. Bull. Natl. Speleol. Soc., 26:121–194.

GUILDAY, J. E., P. W. PARMALEE, and H. W. HAMILTON. 1977. The Clark's Cave bone deposit and the Late Pleistocene paleoecology of the central Appalachian Mountains of Virginia. Bull. Carnegie Mus. Nat. Hist., 2:1–87.

HAGER, M. W. 1972. A Late Wisconsin-Recent vertebrate fauna from the Chimney Rock Animal Trap, Larimer County, Colorado. Univ. Wyoming Contrib. Geol., 11:63–71.

HALL, E. R. 1951. American weasels. Univ. Kansas Publ., Mus. Nat. Hist., 4:1–466.

HILL, A. P. 1980. Early postmortem damage to the remains of some contemporary East African mammals. Pp. 131–152, *in* Fossils in the making, vertebrate taphonomy and paleoecology, Univ. Chicago Press, Chicago, Illinois, 338 pp.

HIRSCHFELD, S. E. 1968. Vertebrate fauna of Nichol's Hammock, a natural trap. Quart. J. Florida Acad. Sci., 31:177–189.

KURTÉN, B. 1976. The cave bear story. Columbia Univ. Press, New York, 163 pp.

KURTÉN, B., and E. ANDERSON. 1972. The sediments and fauna of Jaguar Cave: II the fauna. Tebiwa, 15:21–45.

LONG, C. A. 1973. *Taxidea taxus*. Mammalian Species, 26:1–4.

MARCUS, L. 1960. A census of the abundant large Pleistocene mammals from Rancho La Brea. Los Angeles County Mus., Contrib. Sci., 38:1–11.

NACE, R. L., P. T. VOEGELI, J. R. JONES, and M. DEUTSCH. 1975. Generalized geologic framework of the National Reactor Testing Station, Idaho. U.S.G.S. Prof. Paper, 725B:119–147.

ODUM, E. P. 1968. Energy flow in ecosystems: a historical review. Amer. Zool., 8:11–18.

———. 1971. Fundamentals of ecology. Saunders, Philadelphia, 3rd ed., 574 pp.

PARMALEE, P. W. 1967. A Recent cave bone deposit in southwestern Illinois. Bull. Natl. Speleol. Soc., 29:119–147.

PARMALEE, P. W., and R. O. OESCH. 1972. Pleistocene and Recent faunas from the Brynjulfson Caves, Missouri. Illinois St. Mus. Rep. Invest., 25:1–52.

PARMALEE, P. W., R. O. OESCH, and J. E. GUILDAY. 1969. Pleistocene and Recent vertebrate faunas from Crankshaft Cave, Missouri. Illinois St. Mus. Rep. Invest., 14:1–37.

PAYNE, S. 1972. On the interpretation of bone samples from archaeological sites. Pp. 65–82, in Papers in economic prehistory (E. S. Higgs, ed.), Cambridge Univ. Press, Cambridge, 219 pp.

PETERSON, O. A. 1923. A fossilbearing slab of sandstone from the Agate Springs Quarries of western Nebraska exhibited in the Carnegie Museum. Ann. Carnegie Mus., 15:91–93.

PIELOU, E. C. 1966. The measurement of diversity in different types of biological collections. J. Theor. Biol., 13:131–144.

QUICK, H. F. 1951. Notes on the ecology of weasels in Gunnison County, Colorado. J. Mamm., 32:281–290.

RICKLEFS, R. E. 1973. Ecology. Chiron Press, Newton, Massachusetts, 861 pp.

SKINNER, M. F. 1942. The fauna of Papago Springs Cave, Arizona. Bull. Amer. Mus. Nat. Hist., 80:143–220.

SOERGEL, W. 1940. Die Massenvorkommen des Hohlenbaren. G. Fisher. Jena, 112 pp.

STOCK, C. 1972. Rancho La Brea, a record of Pleistocene life in California. Los Angeles Co. Mus. Sci., Ser. 20 (Paleo. no. 11, 7th printing): 1–81.

VOORHIES, M. R. 1969. Taphonomy and population dynamics of an Early Pliocene vertebrate fauna, Knox County, Nebraska. Univ. Wyoming Contrib. Geol., Spec. Paper, 1:1–69.

WALKER, E. P. 1975. Mammals of the World. Johns Hopkins Press, Baltimore, 3rd ed., 1500 pp.

WEBSTER, G. S. 1978. Dry Creek Rockshelter: Cultural Chronology in the Western Snake River Region of Idaho. Tebiwa, 15:1–35.

WRIGHT, P. L. 1948. Breeding habits of captive long-tailed weasels (Mustela frenata). Amer. Midland Nat., 39:338–344.

Address (White):   Idaho Museum of Natural History, Idaho State University, Pocatello, Idaho 83209.

Address (McDonald):   Vertebrate Paleontology, Royal Ontario Museum, 100 Queen's Park, Toronto, Ontario M5C 2C6, Canada.

Address (Anderson):   730 Magnolia Street, Denver, Colorado 80220.

Address (Soiset):   1284 South Fisher Avenue, Blackfoot, Idaho 83221.

REVIEW OF THE SMALL CARNIVORES OF NORTH AMERICA DURING THE LAST 3.5 MILLION YEARS

ELAINE ANDERSON

ABSTRACT

During the last 3.5 million years 30 genera and about 65 species of small carnivores (mustelids, small canids, procyonids, and small felids) have inhabited North America. The number of genera has remained constant (18–19) during each land mammal age, but the number of species has fluctuated between 25 and 36. At the present time 18 genera and 30 mainland species are recognized. Extinction patterns between small and large carnivores are compared.

INTRODUCTION

Smilodon, Arctodus, Canis dirus—to most people those animals are the Pleistocene carnivores. The smaller species are usually neglected or overlooked, yet the weasels, martens, badgers, otters, skunks, foxes, coyotes, raccoons, lynxes, and small cats have played and continue to play an important role in the ecosystem. In this paper I will discuss briefly the small carnivores, those ranging in size from the least weasel, *Mustela rixosa* (called *nivalis* by some workers), the smallest known carnivore weighing only 38 to 63 g to the sea otter, *Enhydra lutris* that may reach a weight of 37 kg. The time span covered is approximately 3.5 million years, the Blancan, Irvingtonian and Rancholabrean land mammal ages and the Holocenc period in North America.

A population explosion of microtine rodents occurred in the early Blancan, and this event can be correlated with a radiation of small carnivores. They filled all available niches and many were adapted to feed on the hordes of rodents. Helping control rodent populations continues to be a major role for the small carnivores.

John Guilday often kidded me about my carnivores eating his mice. He frequently sent me isolated teeth and fragmentary bones of carnivores from his Appalachian faunas to identify. In the Baker Bluff fauna (Guilday et al., 1978) I was delighted to find fragmentary remains of four pine martens, a fisher, 10 weasels, a badger, and four skunks; he patiently tabulated hundreds of microtines. At Strait Canyon Fissure, Virginia (manuscript), I recorded 61 mustelids and 30 procyonids; he counted 656 cricetids and 44 sciurids. This brings out a significant point, namely carnivores are rare in most Blancan and Pleistocene faunas. Only at carnivore trap sites (Rancho La Brea, Moonshiner Cave) where they are attracted to the site by the scent and struggles of entrapped animals, are the remains numerous, and these localities represent deposition over a long period of time. At such sites the bones of young and old animals predominate. At raptor sites (Clark's Cave) the relative scarcity of carnivore remains reflects the collecting bias of the birds of prey (Guilday et al., 1977). Due to the alertness and agility of most small carnivores, they seldom fall in sinkholes (New Paris No. 4). Caves used as dens may contain more complete material of animals that died there or were brought in as prey (Little Box Elder Cave). At some sites, carnivore remains are almost non-existent, for example, at Hansen Bluff, a new Irvingtonian locality in southern Colorado, only a canid phalanx has been found.

Tables 1–3 list Blancan, Irvingtonian, and Rancholabrean species of small carnivores and give their geologic and geographic ranges, number of reported faunas, size, habitat, and food. Additional data such as new faunal records and taxonomic changes are given in the brief species accounts that follow. For more detailed information see Kurtén and Anderson (1980).

SPECIES ACCOUNTS

Family Mustelidae

*Martes diluviana

The Irvingtonian fisher, the earliest known American species of *Martes,* has been found at Port Kennedy, Cumberland, and Conard Fissure.

Martes pennanti

The extant fisher had a wider range in the late Rancholabrean than it does today and was found throughout the Appalachians as well as in the South and Midwest. It has recently been identified at Strait Canyon Fissure, Virginia.

Table 1.—*Small carnivores of the Blancan Land Mammal Age in North America.* † = *extinct genus.* * = *extinct species. Superscript[1]* = *first appearance of species.*

Species	Geologic range	Geographic range Blancan	No. Bl faunas	Size	Habitat	Food
Mustelidae						
*Mustela rexroadensis	M Bl[1]	W U.S.	4	med. weasel	savanna	small rodents
Mustela frenata	L Bl[1]-R	Kansas	1	85–340 g	savanna	small rodents
*Mustela meltoni	M Bl[1]	SW Kansas	1	mink size	swampy areas	microtines
†Ferinestrix vorax	Bl[1]	Idaho	1	>Gulo	savanna	carrion
†Trigonictis macrodon	Bl[1]	U.S.	12	ca. 2–5.5 kg	various	small animals
†Trigonictis cookii	Bl[1]	W U.S.	5	2/3 size of T. macrodon	various	small rodents
†Sminthosinus bowleri	M Bl[1]	Idaho, Nebraska	2	<T. cookii	marshy areas	small microtines
†Canimartes cumminsi	L Bl[1]	N Texas	1	mod. large	open plains	small animals
Taxidea taxus	Bl[1]	W U.S.	7	5.8–11.2 kg	prairies	rodents, rabbits
†Satherium piscinarium	Bl[1]	U.S.	5	>Lutra	aquatic	aquatic animals
†Buisnictis breviramus	E[1]–M Bl	W U.S.	5	ca. 85–340 g	swampy areas	omnivorous
†Buisnictis burrowsi	M Bl[1]	Nebraska	1	ca. 85–340 g	savanna	omnivorous
*Spilogale rexroadi	Bl	W U.S.	3	size of ♀ S. pygmaea	savanna	omnivorous
*Spilogale(?) microdens	E Bl[1]	Texas	1	454–999 g	savanna	mice, insects
Spilogale putorius	L Bl[1]-R	Kansas	1	363–567 g	savanna	mice, insects
Mephitis mephitis	M Bl[1]-R	Nebraska	1	2.7–6.3 kg	brushy areas	omnivorous
†Brachyopsigale dubius	E Bl[1]	Kansas, Texas	2	Mephitis-size	near water	omnivorous
Canidae						
*Canis lepophagus	Bl[1]-E Irv	U.S.	15	ca. 9–22 kg	various	small-med. animals
*Urocyon progressus	Bl[1]	Kansas, Texas	2	ca. 6 kg	brushy areas	rodents, rabbits
Procyonidae						
*Nasua pronarica	E Bl[1]	Texas	1	>N. narica	open forests	omnivorous
*Procyon rexroadensis	M Bl[1]	Kansas, Texas	2	>P. lotor	swampy areas	omnivorous
*Bassariscus casei	E[1]–M Bl	W U.S.	3	900–1,300 g	savanna	small animals
†Parailurus angelicus	M Bl	Washington, Europe	1	>Ailurus	gallery forest	omnivorous
Felidae						
*Felis lacustris	Bl[1]-E Irv	W U.S.	8	large Lynx size	near water	small-med. animals
*Felis rexroadensis	E Bl-E Irv	Kansas	1	lynx size	savanna	rodents, rabbits
Lynx rufus	L Bl[1]-R	Texas	1	6.7–15.7 kg	wooded areas	rabbits, rodents

*Martes nobilis

Remains of the noble marten have recently been identified at Little Canyon Creek Cave, Washakie County, Wyoming ^{14}C dated 10,170 ± 250 B.P. (Walker, D. P., personal communication) and at two Holocene archeological sites, Hidden Cave, Churchill County, Nevada, dated 3,700 years B.P. and Dry Cave, Ada County, Idaho ^{14}C dated 3,270 ± 110 B.P. (Grayson, this volume). At Little Canyon Creek Cave, it was associated with boreal and plains species, but there are no boreal species at the other two sites. A more equable climate probably explains these occurrences.

Martes americana

Pine marten remains are rare in Pleistocene faunas. A late immigrant from Eurasia, it first appears in North America in the late Rancholabrean. It has recently been identified at Strait Canyon Fissure.

*Mustela rexroadensis

This medium sized Blancan weasel has been found at Hagerman, Rexroad, Fox Canyon, and White Rock. It may have been ancestral to *Mustela frenata*.

Mustela frenata

First recognized in the late Blancan Borchers fau-

Table 2.—Small carnivores of the Irvingtonian Land Mammal Age in North America. † = extinct genus. * = extinct species. Superscript[1] = first appearance of species.

Species	Geologic range	Geographic range Irvingtonian	No. Irv. faunas	Size	Habitat	Food
Mustelidae						
*Martes diluviana	Irv[1]	E U.S.	3	ca. 2.0–3.5 kg	wooded areas	omnivorous
Mustela frenata	L Bl-R	E U.S.	2	85–340 g	various	small rodents
Mustela erminea	L Irv[1]-R	Kansas, Arkansas	2	28–170 g	wooded areas	mice
Mustela vison	L Irv[1]-R	Kansas, Maryland, Arkansas	3	567–1,362 g	near water	aquatic animals
†Tisisthenes parvus	M Irv[1]	South Dakota	1	weasel size	grassy areas	small rodents
*Gulo schlosseri	E Irv[1]	E U.S., Europe	2	ca. 8% smaller than *G. gulo*	forested areas	carrion, rodents, rabbits
Taxidea taxus	Bl-R	E U.S.	2	5.8–11.2 kg	prairies	rodents
Lutra canadensis	Irv[1]-R	E U.S.	2	4.5–11.2 kg	inland waterways	aquatic animals
*Enhydra macrodon	Irv[1]-RLB	California	2	>*E. lutris*	marine	marine invertebrates
Spilogale putorius	L Bl-R	Arizona, E U.S.	6	363–567 g	open woods	mice, insects
†Brachyprotoma obtusata	Irv[1]-RLB	E U.S.	3	small *Spilogale*	forests	hard-shelled insects
†Osmotherium spelaeum	E Irv[1]	Pennsylvania	1	ca. 3–6 kg	wooded areas	omnivorous
Mephitis mephitis	M Bl-R	Florida, Arkansas	3	2.7–6.3 kg	semi-open areas	omnivorous
Conepatus leuconotus	Irv[1]-R	Florida	1	0.9–2.7 kg	wooded areas	insects
Canidae						
*Canis lepophagus	Bl-E Irv	W U.S.	3	ca. 9–22 kg	prairies, woods	small-med. animals
Canis latrans	M Irv[1]-R	U.S.	5	9–22 kg	prairies, woods	rodents, rabbits
Canis rufus	E Irv[1]-R	Pennsylvania, Florida	2	ca. 18–31 kg	savanna	small animals
Urocyon cinereoargenteus	Irv[1]-R	E U.S.	3	3.2–5.8 kg	wooded areas	omnivorous
Vulpes velox	L Irv[1]-R	Nebraska, Oklahoma	2	1.8–2.9 kg	open country	small animals
Procyonidae						
Procyon lotor	L Irv[1]-R	Florida	1	5.4–15.8 kg	wooded areas	omnivorous
Felidae						
*Felis lacustris	Bl-E Irv	Arizona	1	Lynx-size	near water	rodents, rabbits
*Felis rexroadensis	E Bl-E Irv	Nebraska	1	Lynx-size	savanna	rodents, rabbits
Felis yagouroundi	?Irv-R	Pennsylvania	1	6.7–8 kg	brushy areas	small animals
Lynx issiodorensis	E Irv[1]	Nebraska	1	>*L. canadensis*	savanna	med. animals
Lynx rufus	L Bl-R	U.S.	6	6.7–15.7 kg	forests, brush	rodents, rabbits

Table 3.—*Small carnivores of the Rancholabrean Land Mammal Age in North America.* † = *extinct genus.* * = *extinct species. Superscript[1]=first appearance of species. Superscript[2] = survives outside of North American today.*

Species	Geologic range	Geographic range Rancholabrean	No. RLB faunas	Size	Habitat	Food
Mustelidae						
Martes pennanti	M RLB[1]-R	Yukon Terr., E U.S.	11	2.0–5.5 kg	mixed hardwoods	small-med. animals
*Martes nobilis	L RLB[1]-E H	W U.S.	11	ca. 0.9–3 kg	forests, brush	small animals
Martes americana	L RLB[1]-R	Yukon Terr., W U.S., Appalachians	11	681–1,248 g	coniferous forests	small animals, fruit
Mustela frenata	L Bl-R	S Canada, U.S.	35	85–340 g	various	small animals
Mustela erminea	L Irv-R	Alaska, Yukon Terr., U.S. S to Texas	14	28–170 g	tundra, forests	mice
Mustela rixosa	L RLB[1]-R	U.S.	9	38–63 g	meadows, brush	small mice
Mustela vison	L Irv-R	Alaska, U.S.	20	567–1,362 g	near water	aquatic animals
Mustela eversmanni[2]	L RLB[1]	* ssp. Alaska	1	>*M. nigripes*	steppes	rodents
Mustela nigripes	RLB[1]-R	Canada, W U.S.	17	0.681–1.6 kg	prairies	rodents, *Cynomys*
Gulo gulo	RLB[1]-R	Alaska, Canada, W U.S.	10	16–27 kg	tundra, taiga	carrion, mammals
Taxidea taxus	Bl-R	W U.S., Alaska, prairie corridor E U.S.	39	5.8–11.2 kg	prairies	rodents
Lutra canadensis	Irv-R	E U.S.	15	4.5–11.2 kg	inland waterways	aquatic animals
*Enhydra macrodonta	Irv-RLB	California	1	>*E. lutris*	marine	marine invertebrates
Enhydra lutris	M RLB[1]-R	N Pacific coast	3	13.5–37 kg	marine kelpbeds	marine invertebrates
Spilogale putorius	L Bl-R	U.S., S Canada	46	363–567 g	prairies, brush	mice, insects, eggs
†*Brachyprotoma obtusata*	Irv-RLB	E Cent. U.S.	3	small *Spilogale*	forests	hard-shelled insects
Mephitis mephitis	M Bl-R	U.S.	48	2.7–6.3 kg	semiopen areas	omnivorous
Conepatus leuconotus	Irv-R	U.S., Mexico	10	0.9–2.7 kg	semiwooded areas	insects
Canidae						
Canis latrans	M Irv-R	S Canada, U.S.	60+	9–22 kg	prairies, woods	rabbits, rodents
Canis rufus	E Irv-R	S U.S., Mexico	8	18–31.5 kg	brushy areas	small animals, crabs
*Canis cedazoensis	M RLB[1]	Mexico	1	<*C. latrans*	prairies	small animals, carrion
Canis familiaris	L RLB[1]-R	Canada, U.S.	14	2 size classes	various	small animals, carrion
Cuon alpinus[2]	L RLB[1]	Beringia, Mexico	3	10–20 kg	forest, steppes	med.-large animals
Urocyon cinereoargenteus	Irv-R	U.S.	30	3.2–5.8 kg	brushy areas	omnivorous
Vulpes velox	L Irv-R	Great Plains	7	1.8–2.9 kg	open plains	small animals
Vulpes macrotis	L RLB[1]-R	California, SW U.S.	11	1.4–2.7 kg	desert scrub	rodents, rabbits
Vulpes vulpes	M RLB[1]-R	North America	28	4.5–6.7 kg	mixed forests	rodents, rabbits
Alopex lagopus	L RLB[1]-R	Yukon Terr.	1	3.2–6.7 kg	tundra	carrion
Procyonidae						
Procyon lotor	L Irv-R	U.S.	58	5.4–15.8 kg	wooded areas	omnivorous
*Bassariscus sonoitensis	L RLB[1]	Arizona, Mexico	2	>*B. astutus*	rocky areas	small animals
Bassariscus astutus	L RLB[1]-R	SW U.S., California	22	900–1,130 g	rocky areas	small animals
Felidae						
Felis pardalis	M RLB[1]-R	Florida	1	9–18 kg	forests	mammals
*Felis amnicola	RLB[1]	Florida	7	ca. 6.5–8 kg	brushy areas	small animals
Felis yagouroundi	?Irv-R	Texas, Mexico	2	6.7–8.1 kg	thorn thickets	small animals
Lynx canadensis	M RLB[1]-R	Alaska, S Canada, Idaho, Utah	5	6.7–13.5 kg	boreal forests	small animals
Lynx rufus	L Bl-R	U.S.	57	6.7–15.7 kg	forests, brush	rodents, rabbits

na, the long-tailed weasel has the longest stratigraphic record and the greatest geographic range of any American weasel. Found in almost all terrestrial habitats except deserts, *Mustela frenata* is often the most abundant small carnivore in a fauna (see White et al., this volume). New records include Strait Canyon Fissure, Upper Sloth Cave, Texas, and Muskox Cave, New Mexico.

Mustela erminea

The circumboreal ermine reached America in the late Blancan or early Irvingtonian. Geographical variation in size and sexual dimorphism are pronounced, and there is size overlap between the larger *M. frenata* and the smaller *M. rixosa*. The ermine had a wider range in the late Pleistocene and it has recently been identified at Strait Canyon Fissure and Smith Creek Cave, Nevada.

Mustela rixosa

Smallest of the Pleistocene and extant carnivores, the least weasel is probably an evolutionary offshoot of *M. nivalis* and reached America in the late Rancholabrean. It has recently been recognized in the Strait Canyon Fissure fauna.

*Mustela meltoni

This poorly known Blancan species has only been found at Rexroad.

Mustela vison

Youngman (1982) showed that *M. vison* and *M. lutreola,* the European mink, are not closely related and placed the American species in a separate Nearctic, endemic subgenus (*Vison*) that also includes the extinct sea mink, *M. macrodon.* Amphibious in habits, the mink is a good indicator of nearby permanent water. New records of the species include Smith Creek Cave, Strait Canyon Fissure, and Christensen Bog, Indiana.

*Mustela macrodon

The largest known American mink, this species has only been found along the northeast coast from the Bay of Fundy to Massachusetts where its remains have been recovered from Indian middens. It probably survived until a few hundred years ago.

Mustela nigripes

Additional late Rancholabrean records of the black-footed ferret include Little Canyon Creek Cave, Wyoming, and January Cave, Alberta; both sites are within the historic range of the species. A population of this endangered mammal has recently been discovered in northern Wyoming.

Mustela eversmanni

An extinct subspecies, *M. e. beringiae,* of the Eurasian steppe ferret has been recognized in the Fairbanks fauna. Apparently it did not extend its range beyond the unglaciated steppes of Beringia.

†Tisisthenes parvus

This small, poorly known weasel-like mustelid is known only from the Irvingtonian Java fauna, South Dakota.

*Gulo schlosseri

Schlosser's wolverine, the earliest recognized species of *Gulo,* has been found in contemporaneous-age deposits in eastern North America and central Europe. It was ancestral to *Gulo gulo.*

Gulo gulo

This rare circumboreal species first appears in late Rancholabrean faunas. Late Pleistocene wolverines were larger than early Holocene and Recent specimens.

†Ferinestrix vorax

This poorly known Blancan species is known only from two specimens found at Hagerman. Bjork (1970) placed it in the subfamily Mellivorinae.

†Trigonictis macrodon

A recent revision of *Trigonictis* (Ray et al., 1981) showed that a large galictine mustelid, now called *T. macrodon* was widespread in the Blancan. It is known from Charles County, Maryland (type locality), Hagerman, Rexroad, White Bluffs, Grand View, Broadwater-Lisco, Sand Draw, Deer Park, near Channing, Texas, Vallecito Creek, near Safford Arizona, Smith Mill River in Wayne County, North Carolina, and Santa Fe River VIIIA. *Trigonictis* probably reached North America from Eurasia in the early Blancan, perhaps somewhat earlier.

†Trigonictis cookii

Differing from *T. macrodon* by only slightly smaller size, *T. cookii* has been found at Hagerman, Grand View, Sand Draw, Broadwater, Red Corral, and Haile XVIA. In addition, *Trigonictis* sp. is recorded from Cita Canyon, Arroyo Seco, and Inglis IA. To have two closely related galictine mustelids occurring together in at least four faunas has puzzled all who have studied them. Until more material becomes available to clarify their relationship, they are regarded as distinct species occupying slightly different niches.

†*Sminthosinus bowleri*

This small galictine mustelid has only been found at Hagerman and Broadwater. It may only be sub-specifically distinct from *Trigonictis*.

†*Canimartes cumminsi*

This species is only known from one specimen found in the Blanco fauna. It may not even be a mustelid, because M^2 is apparently present.

Taxidea taxus

The fossorial badger first appears in the Blancan, and by the late Rancholabrean they were common in western faunas. Prairie corridors provided an eastern extension of their range (Baker Bluff, Welsh, Peccary, Bootlegger Sink), and they also spread northward to unglaciated eastern Beringia (Fairbanks, Gold Run Creek, Dominion Creek) where steppe conditions prevailed in the late Pleistocene. Badger remains have recently been found at Dutton, Little Canyon Creek Cave, and Conkling Cavern, New Mexico.

†*Satherium piscinarium*

The Blancan otter has recently been recognized in the Vallecito Creek fauna. It was probably ancestral to *Pteroneura brasiliensis,* the flat-tailed otter of South America.

Lutra canadensis

The aquatic river otter has recently been recognized at Warm Mineral Springs, Florida, and Strait Canyon Fissure, Virginia. River otters probably reached North America from China in the early Irvingtonian.

Enhydra macrodonta

Specimens from three sites in California and Oregon have been referred to this species. It is not known if it was ancestral to the extant species, *Enhydra lutris.*

Enhydra lutris

In the Pleistocene the range of the sea otter extended from Point Barrow, Alaska, to southern California. Although strictly protected today, they oc-cupy only about one-fifth of their former range and are regarded as a threatened species.

†*Buisnictis breviramus*

This small, short-faced skunk has been found in early and middle Blancan faunas in Kansas, Texas, and Idaho. Its relationship to other small skunks is uncertain for its P^4 is *Mustela*-like but the M^1 resembles *Mephitis* and the M_1 is four-rooted as in typical mephitines (Bjork, 1974).

†*Buisnictis burrowsi*

More advanced than *B. breviramus,* this small skunk has only been found at Sand Draw.

Spilogale rexroadi

Ancestral to later species of *Spilogale,* the primitive Rexroad skunk has been found at Rexroad, Blanco, and Beck Ranch.

Spilogale(?) *microdens*

This skunk is known only from a single well preserved mandible found in the early Blancan Beck Ranch, Texas fauna. As large as a male *S. putorius,* it has relatively small teeth. Until upper dentitions are found, its taxonomic position remains uncertain.

Spilogale putorius

Spotted skunks are common in Rancholabrean faunas throughout the country. New records include Monkey Jungle, Florida; Strait Canyon Fissure, Virginia; Riverside Cave, Tennessee; Howell's Ridge and Muskox caves, New Mexico. The eastern and western populations are now generally regarded as distinct species—*S. putorius* in the eastern half of the county does not have delayed implantation; *S. gracilis* the western form has a long (210–260 days) delayed implantation (Mead, 1968, 1981). Of course, these physiological differences are not reflected in the fossil material, and size and morphological differences between the two populations are small. Unknown is when or why delayed implantation developed in these skunks.

†*Brachyprotoma obtusata*

The short-faced skunk has been found at several sites in east-central United States. *Brachyprotoma* is the only genus of small carnivores to have become extinct in the late Rancholabrean/early Holocene; why, no one knows.

†Osmotherium spelaeum

This skunk is only known from the early Irvingtonian Port Kennedy fauna. It may not be generically distinct from *Mephitis*.

Mephitis mephitis

Striped skunks make their first appearance in the late Blancan Broadwater fauna, and their remains are common in Rancholabrean faunas. This is partly explained by their use of caves as dens and by their propensity for falling down fissures. *M. mephitis* has recently been identified in the Strait Canyon Fissure fauna.

Mephitis macoura

The hooded skunk, primarily a Mexican species that barely ranges into the United States today, has not been found in any U.S. Pleistocene fauna.

Conepatus leuconotus

In the Pleistocene the range of the hog-nosed skunk extended to Florida and Georgia; today its eastern limit is Texas. It has recently been identified at Muskox Cave, New Mexico.

†Brachyopsigale dubius

This poorly known skunklike mustelid is known only from two mandibular fragments recovered from the early Blancan Fox Canyon and Beck Ranch faunas. Characterized by an extremely short jaw and crowded teeth, it was about the size of *Mephitis*.

Family Canidae

*Canis lepophagus

This ancestral coyote was widespread in the Blancan, and it survived into the early Irvingtonian when it was replaced by *C. latrans*.

Canis latrans

Remains of coyotes have been found at more than 100 Pleistocene localities making it one of the most common species. It is especially numerous at trap sites (Rancho La Brea, Moonshiner Cave) because of its scavenging habits.

*Canis cedozoensis

Smaller than the coyote, this Mexican canid has counterparts at several sites in the U.S. Their relationships are uncertain.

Canis rufus

Nowak (1979) in his study of Quaternary *Canis* referred specimens from the following Pleistocene sites to *C. rufus*: Port Kennedy, Inglis IA, Vero, Melbourne, Crystal River Power Plant, Haile VIIA, Devil's Den, Miller's Cave, Eddy Bluff, Arkansas, and the Upper Becerra Formation in Mexico, and from archeological sites in Florida, Alabama, Arkansas, Virginia, West Virginia, Pennsylvania, and Ohio. Today the endangered red wolf is only found in parts of southeastern United States.

Canis familiaris

Domestic dog remains are rare in Rancholabrean faunas and may be difficult to distinguish from those of coyote or wolf. The specimen from Old Crow River 11A has not been dated yet but is probably as old or older than the ones from Jaguar Cave ^{14}C dated 10,370 ± 350 B.P. (Beebe, 1980).

Cuon alpinus

Remains of the Asian dhole with its distinctive trenchant dentition have been found in Rancholabrean faunas in Beringia (Fairbanks, Old Crow River 14N) and Mexico (San Josecito). Its New World extirpation is unexplained.

*Urocyon progressus

The earliest known gray fox is this Blancan species that has been found at Rexroad, Red Light, and probably Broadwater. It was ancestral to *U. cinereoargenteus*.

Urocyon cinereoargenteus

This extant species is first recognized in the Irvingtonian, and its remains are quite common in Rancholabrean faunas. New records include Dutton, Monkey Jungle, Warm Mineral Springs, and Harrodsburg Crevice, Indiana.

Vulpes velox

Swift and kit foxes have usually been united in a single species, *V. velox*. Recent studies (Egoscue, 1979; McGrew, 1979) have demonstrated that these two small foxes have distinct, non-overlapping (except possibly in eastern New Mexico) ranges and can be separated by several external and cranial characters. This geographic separation is apparent in Pleistocene faunas containing swift fox, *V. velox* (prairies east of the Rocky Mountains) or kit fox, *V. macrotis* (deserts and semideserts of western

North America). I am tentatively referring specimens from the following sites to *V. velox*: Cragin Quarry, Angus, Chimney Rock, Berends, Lubbock Lake, and Klein. Dalquest (1978) tentatively referred some small canid remains from Beck Ranch (early Blancan) to *V. velox*; if this identification is correct, it is the earliest known occurrence of the species.

Vulpes macrotis

The desert-dwelling kit fox differs from *V. velox* in having a longer tail, larger ears, and a narrower skull with a more slender rostrum (McGrew, 1979). Specimens from the following faunas are tentatively referred to *V. macrotis*: Ventana, Burnet, Gray Sand fauna of Blackwater Draw, Isleta, Dry, Smith Creek, Schulze, Moonshiner, and Middle Butte caves.

Vulpes vulpes

The Holarctic red fox reached North America from Eurasia in the Rancholabrean. It has recently been identified at Natural Trap, Smith Creek Cave, and Monkey Jungle.

Alopex lagopus

Remains of the arctic fox have only been found at Old Crow River. It apparently did not reach the Nearctic until the late Rancholabrean.

Family Procyonidae

*Nasua pronarica

A single lower fourth premolar found in the early Blancan Beck Ranch fauna is the only known specimen of *N. pronarica*. Because coatimundi remains have not been identified at other Blancan sites, Beck Ranch may represent the northern limits of this South American representative.

Nasua narica

Although the coatimundi is now found in southern Arizona, the extant species is not known from any Pleistocene locality in the United States. It is probably descended from *N. pronarica*.

*Procyon rexroadensis

The Blancan Rexroad raccoon has been found at Rexroad, Fox Canyon, Cita Canyon, and the Anza Borrego faunas.

Procyon lotor

Raccoons are common in Pleistocene faunas from the late Irvingtonian onward. Its adaptability to a wide variety of habitats and its omnivorous habits contribute to its success. It has recently been identified at Monkey Jungle, Warm Mineral Springs, Rancho La Brea, Strait Canyon Fissure, and Christensen Bog.

*Bassariscus casei

This Blancan species has been found at Rexroad, Beck Ranch, and Vallecito Creek. It was ancestral to *B. astutus*.

*Bassariscus sonoitensis

The Sonoita ringtail is known only from skull and postcranial material found in the late Rancholabrean faunas of Papago Springs and San Josecito.

Bassariscus astutus

Remains of the extant ringtail have recently been identified in the following late Rancholabrean southwestern cave faunas: Vulture, Stanton's, Pratt, Upper Sloth, and Muskox.

†Parailurus angelicus

The only record of the English panda in North America is an isolated M^1 found in the early Blancan Taunton fauna in southwestern Washington.

Family Felidae

*Felis lacustris

This lynx-sized cat was widely distributed in the Blancan and survived until the early Irvingtonian.

*Felis rexroadensis

About the same size as *F. lacustris,* this distinct Blancan species shows affinities to *Lynx*.

Felis pardalis

The ocelot has only been identified in the Reddick fauna. Hunted for its fur, it is now classified as an endangered species.

*Felis amnicola

Remains of several small cats in the margay-jaguarundi size range have been referred to this extinct species found in Rancholabrean faunas in Florida.

Felis yagouroundi

Fossils of the jaguarundi have been found at Schulze and San Josecito caves, and an early form may be represented at Port Kennedy.

Felis wiedii

The Neotropical margay has only been reported from a middle Holocene fauna in coastal Texas. Its range extends into southern Texas today.

*Lynx issiodorensis

This species, the earliest known *Lynx*, was widespread in Old World faunas ranging in age from early Pliocene through the Villafranchian; it has tentatively been identified from Mullen I, an early Irvingtonian fauna. In appearance it resembles *Felis* rather than recent lynxes with its shorter legs and relatively larger skull. Werdelin (1981) believes it was ancestral to later species.

Lynx canadensis

An immigrant from Eurasia, this boreal species is rare in Rancholabrean faunas. Its population fluctuates with that of *Lepus americanus*.

Lynx rufus

The bobcat is common in Pleistocene faunas throughout the country. It has recently been identified at Monkey Jungle, Harrodsburg Crevice, Conkling Cavern, Muskox Cave, and Strait Canyon Fissure.

DISCUSSION

The number of genera of small carnivores has remained constant during the last 3.5 million years while the number of species increased from 25 in the Irvingtonian to 36 in the Rancholabrean, mainly due to new immigrants. The most successful species have a long stratigraphic range, a wide geographic range, live in a variety of habitats, utilize a number of food sources, and are adaptable to changing conditions. These species include *Mustela frenata*, *Taxidea taxus*, *Spilogale putorius*, *Mephitis mephitis*, *Canis latrans*, *Procyon lotor*, and *Lynx rufus*. All of them are autochthonous, native American species. In general, species having a restrictive range, narrow habitat requirements, and limited food sources became extinct or are threatened with extinction today.

In contrast to these common species, 25 species are poorly known—a few fragmentary specimens found in three or fewer faunas. Whether their scarcity is due to chance collecting or actual rarity of the animal is unknown. The rarity of some of the Blancan and Irvingtonian species makes their taxonomic relationships uncertain—this is especially true for the Blancan skunks. Hopefully more material will be found so that comparisons can be made and the range of variation and sexual dimorphism ascertained. This may result in the number of genera and species being reduced or perhaps increased.

Table 4 shows differences in extinction patterns between small and large carnivores from the Blancan to the present. During that period of time, there were 30 genera and 65 species of small carnivores and 16 genera and 27 species of large carnivores

(>40 kg, a mustelid, large canids, ursids, large felids, and a hyaenid). Of these, 11 genera (36%) and 34 species (52%) of the small carnivores became extinct, whereas 10 genera (62%) and 21 species (74%) of the large carnivores disappeared. At the present time 18 genera and 33 species of small carnivores inhabit North America, whereas only four genera and six species of large carnivores (*Canis lupus*, *Ursus americanus*, *Ursus arctos*, *Ursus maritimus*, *Felis concolor*, and *Panthera onca*) survive here.

More small carnivores became extinct in the Blancan than in the following mammal ages. In contrast more large carnivores succumbed to Rancholabrean extinctions. *Brachyprotoma* was the only ge-

Table 4.—*Comparison of extinction patterns between small and large carnivores during the last 3.5 million years.*

Land mammal age and size	No. genera	No. species	No. extinct genera	No. extinct species
Blancan				
Small carnivores	19	26	8	21
Large carnivores	9	10	3	5
Irvingtonian				
Small carnivores	18	26	3	11
Large carnivores	11	16	3	7
Rancholabrean				
Small carnivores	19	36	1	6
Large carnivores	9	13	5	8
Holocene				
Small carnivores	18	30	0	2
Large carnivores	4	6	0	0

nus of small carnivore to vanish at the end of the Rancholabrean. Disappearance of prey, competition, climatic change, and man's activities in the late Rancholabrean have usually been cited as causes of carnivore extinction, but the small carnivores were less affected by these factors than the large ones, so the cause(s) of their extinction remains an enigma.

ACKNOWLEDGMENTS

My thanks go to the late John Guilday for many stimulating discussions on Pleistocene mammals and to Bjorn Kurtén for steering me to the small carnivores. Donald Grayson, Greg McDonald, Horace Quick, Sam Erlinge, Mikael Sandell, and Carolyn King provided data for this paper.

LITERATURE CITED

BEEBE, B. 1980. A domestic dog (*Canis familiaris* L.) of probable Pleistocene age from Old Crow, Yukon Territory, Canada. Canadian J. Archaeol., 4:161–168.

BJORK, P. R. 1970. The Carnivora of the Hagerman local fauna (late Pliocene) of southwestern Idaho. Trans. Amer. Phil. Soc., ns, 60(7):1–54.

———. 1974. Additional carnivores from the Rexroad Formation (Upper Pliocene) of southwestern Kansas. Trans. Kansas Acad. Sci., 76:24–38.

DALQUEST, W. W. 1978. Early Blancan mammals of the Beck Ranch local fauna of Texas. J. Mamm., 59:269–298.

EGOSCUE, H. J. 1979. *Vulpes velox*. Mammalian Species, 122:1–5.

GUILDAY, J. E., H. W. HAMILTON, E. ANDERSON, and P. W. PARMALEE. 1978. The Baker Bluff Cave deposit, Tennessee, and the late Pleistocene faunal gradient. Bull. Carnegie Mus. Nat. Hist., 11:1–67.

GUILDAY, J. E., P. W. PARMALEE, and H. W. HAMILTON. 1977. The Clark's Cave bone deposit and the late Pleistocene paleoecology of the central Appalachian Mountains of Virginia. Bull. Carnegie Mus. Nat. Hist., 2:1–88.

KURTÉN, B., and E. ANDERSON. 1980. Pleistocene mammals of North America. Columbia Univ. Press, New York, 442 pp.

McGREW, J. C. 1979. *Vulpes macrotis*. Mammalian Species, 123:1–6.

MEAD, R. A. 1968. Reproduction in western forms of the spotted skunk (genus *Spilogale*). J. Mamm., 49:373–390.

———. 1981. Delayed implantation in mustelids, with special emphasis on the spotted skunk. J. Reprod. Fert., Suppl., 29:11–24.

NOWAK, R. M. 1979. North American Quaternary *Canis*. Monogr. Mus. Nat. Hist., Univ. Kansas, 6:1–154.

RAY, C. E., E. ANDERSON, and S. D. WEBB. 1981. The Blancan carnivore *Trigonictis* (Mammalia, Mustelidae) in the eastern United States. Brimleyana, 5:1–36.

WERDELIN, L. 1981. The evolution of lynxes. Ann. Zool. Fennici, 18:37–71.

YOUNGMAN, P. M. 1982. Distribution and systematics of the European mink (*Mustela lutreola* Linnaeus, 1761). Acta Zool. Fennica, 166:1–48.

Address: 730 Magnolia Street, Denver, Colorado 80220.

THE PLEISTOCENE DUNG BLANKET OF BECHAN CAVE, UTAH

Owen K. Davis, Larry Agenbroad, Paul S. Martin, and
Jim I. Mead

ABSTRACT

Boluses of dung rich in graminoid stems dropped by a large herbivore and comparable in size and content to African elephant (*Loxodonta*) dung were discovered recently in Bechan Cave, southeastern Utah. Two boluses were radiocarbon dated at 11,670 and 12,900 yr B.P., respectively. They are embedded in a 255 m³ dung blanket along with fecal remains and hair of ground sloths, artiodactyls, and small mammals. While no diagnostic bones of mammoth were recovered, the deposit yielded long coarse hair attributed to mammoth.

This unusual discovery supports the previous report by Hansen (*in* Jennings, 1980) of mammoth dung at Cowboy Cave, Utah.

Both deposits constitute major finds for paleoecologists and are the most impressive discovery of their kind since the discovery in the 1930s of dry caves yielding ground sloth dung in Nevada, Arizona, New Mexico, and west Texas. The Bechan Cave deposit accumulated during an interval of vegetation change when blue spruce and water birch declined, oak became more abundant, and the regional vegetation was sagebrush steppe. Blue spruce and water birch currently occupy higher elevations in southern Utah. Megaherbivore occupation of Bechan Cave evidently ended hundreds of years before mammoth and ground sloth (*Nothrotheriops*) extinction in the region, ca. 11,000 yr B.P.

INTRODUCTION

In November of 1982 a National Park Service team undertaking a grazing survey in Glen Canyon National Recreation Area entered a large dry cavern in a narrow valley north of the Colorado River. The west facing cave (Fig. 1), unusually large for caverns in the Navajo Sandstone, lies above a relatively steep slope. Unlike certain other shelters in the vicinity it contained no surface deposit of cow manure or other evidence of feral cattle.

Attention of the exploring party, which included Larry Belli and Charles Berg of the National Park Service and NPS consultants Steve Carruthers and Lauren Haury of Flagstaff, was drawn to dry dung fragments exposed in the walls of shallow holes dug by pot hunters in the sandy floor of the cave. Aware of the interest in fossil ground sloth coprolites found in other caves in the region, Carruthers forwarded a sample to Tucson via Dr. Art Phillips of the Museum of Northern Arizona. Inspection revealed that the fragment was rich in graminoid culms, rather than in browse fragments as is typical of Shasta ground sloth dung. In shape and texture the sample was unlike the numerous samples of Shasta ground sloth dung in the University of Arizona Paleoenvironmental Laboratory collection.

Finely chewed ruminant dung could be eliminated from comparison. While of a coarser texture and dominated by fragments 0.3–2.0 mm in length (Spaulding and Martin, 1979), equid droppings lack large grass culms. The unknown sample resembled droppings of *Loxodonta* (African elephant). A visit to the cave was planned.

In February 1983 guided by Belli and other NPS personnel, accompanied by Utah State Archaeologist Dave Madsen and Utah Paleontologist Jim Madsen, Jim and Emilee Mead and Larry Agenbroad made a brief reconnaissance. The cave, designated "Bechan Cave" from the Navajo word for excrement, is roughly 173 ft (52.8 m) deep, 103 ft (31.4 m) wide, up to 30 ft (9.1 m) in height and well lit during the daytime through its cavernous mouth. The dung blanket proved extensive. To determine its thickness a small pit was dug close to the north wall of the cave (see Fig. 2). The contents of the pit ("pit M") included two large boluses of unbroken dung. The larger, designated M-1, measures 230 by 170 by 85 mm (Fig. 3); the smaller, M-2, is subspherical and 225 m in diameter (Fig. 4). Both closely resemble boluses of African elephant dung (Fig. 5). They lack twigs typical of dung balls of the Shasta ground sloth (Fig. 6). The M-1 and M-2 dung balls were adjoining in a small pit and their pollen contents (Table 2) are very similar. While their carbon fourteen ages (Table 1) are significantly different (<0.01), their carbon thirteen values are identical (Table 1) and we infer that they were dropped by the same animal.

These remarkable specimens may not be the first record of fossil proboscidean dung found in Utah. A similar dung blanket was found in Cowboy Cave

Fig. 1.—Mouth of Bechan Cave within Navajo Sandstone. Utah juniper, Indian rice grass, and various shrubs grow on sandy soils within the canyon.

immediately west of Canyonlands National Park (Jennings, 1980). Fecal remains from a variety of herbivores, living and extinct and including elephant dung fragments were reported by Hansen (1980) from Cowboy Cave.

Subsequent excavations and screening of test pits at Bechan Cave have yielded fragments but no additional intact boluses of the quality of M-1 and M-2. Mead and Agenbroad recovered fecal pellets comparable to elk or Harrington's mountain goat, including some pellets 26 by 19 by 18 mm in size which resemble unknown herbivore droppings ("elk-camel") reported in Hansen (1980:182). In addition they found twigs and other plant macrofossils apparently transported by *Neotoma* and a strand of

coarse hair 150 mm long, and 1.15 mm in diameter, too coarse to be attributed to living members of the Utah megafauna (that is, elk, pronghorn, mountain sheep, mule deer, black bear, wolf, and mountain lion).

In view of the rich finds made in sample pit "M" more extensive work in the cave was proposed and approved by National Park Service managers. A protective policy was adopted which we seek to follow in not identifying the site by its map coordinates. Such information can be requested from the Glen Canyon National Recreation Area, Page, Arizona. Despite these efforts unauthorized digging within the cave continues.

In March 1983, Agenbroad and Mead returned

Fig. 2.—Excavation of sample pit "M" by Agenbroad and Mead (kneeling), February 1983.

for five days with two Northern Arizona University students, Richard Ryan and Deborah Meier to be joined by P. S. Martin and O. K. Davis. The cave floor was mapped (Fig. 7), "Test Pit 1" was dug adjacent to sample pit "M," contents of the test pit were screened with window screen (2 mm mesh), and samples were removed for radiocarbon dating and paleobotanical study. At 35 cm Test Pit 1 yielded a ground sloth dung ball packed with twigs, including an acorn of *Quercus gambelii.* Davis and Martin began a plant survey outside the cave while Mead and Agenbroad undertook geological reconnaissance of the alluvial fill exposed in deep cuts within the canyon.

In May 1983 Agenbroad, Davis, and Martin accompanied by Brett and Finn Agenbroad and Mary Kay O'Rourke, returned to Bechan Cave for four days of field work. Geological and botanical reconnaissance continued. Thickness of overburden and thickness of the dung blanket itself was measured in a series of 49 auger holes positioned on a 5-m grid, yielding an isopatch map (Fig. 8). Test Pit 2 (position shown on Fig. 7) was dug to sterile sand near the back of the cave and its contents screened. The trampled edge of the dung blanket at the mouth of the cave was faced, photographed, and back-filled. Future visitors will be able to examine details of dung blanket texture and stratigraphy by removing the backfill rather than digging within the cave. An auger hole to 5 m near the cave mouth ("N" on Fig. 7) indicated dry sterile sand becoming wet at 1.5 m beneath the dung blanket.

Fig. 3.—Bolus M-1 from sample pit "M"; plant stems are of graminoids.

CAVE CONTENTS

Based on auger probes the dung blanket of Bechan Cave is estimated to contain 370 m³ of organic materials. Overburden near the cave walls constitutes 10–20 cm of loose sand thickening to 90 cm or more toward the center of the cave and exceeding that near the mouth where dislodged sandstone blocks prevented augering. In 2.1 m³ of screened test pit material (less than one percent of the deposit) and in auger probes no diagnostic bones of large mammals were found. Large bone fragments within the dung blanket near the mouth of the cave were not suitable for taxonomic purposes. Occasional jaws and other small mammal bones were collected. Packrats use and evidently have used the cave for 12,000 years at least. A packrat midden near the back of the cave yielded archeological corn cobs.

Fig. 4.—Bolus M-2 from sample pit "M"; plant stems are of graminoids.

Evidently Paiute or Navajo placed a brush shelter of juniper branches against the north wall. Corn cobs lay near rock cysts at the mouth of the cave; other contents, if any, had been plundered before our visits. No pottery and only a few stone artifacts including metates were found; layers of charcoal 10 cm thick and plant debris in sand above the dung blanket indicate Holocene occupation. It is possible that a rich Archaic record comparable to that at Cowboy Cave lies beneath the ceiling collapse near the mouth of the cave. Filled voids (krotovinas) dug by rat-size rodents were examined in the cleared exposure at the mouth of the cave. Charcoal was abundant in the krotovinas but we encountered no *in situ* charcoal and no artifacts in the dung blanket itself.

In the absence of diagnostic bones we sought to identify the ancient occupants by two means—comparison with fecal samples of living and extinct megaherbivores, and by identification of hair. Sev-

Fig. 5.—*Loxodonta* (African elephant) dung from Amboseli, Kenya.

Table 1.—*Bechan Cave radiocarbon dates on megaherbivore dung from dung blanket.*

Sample no.	Age	$\delta^{13}C$	Comment
A-3212	11,670 ± 300	−23.2%	Sample pit "M," bolus M-1 of mammoth(?) dung, coll. Mead and Agenbroad, February 1983
A-3213	12,900 ± 160	−23.2%	Sample pit "M," bolus M-2 of mammoth(?) dung, coll. Mead and Agenbroad, February 1983
A-3298	12,620 ± 220	−18.4%	Test Pit 1, 10N 2E; fragment of bolus of mammoth(?) dung from base of unit; M-4 coll. Mead, March 1983
A-3297	12,400 ± 250	−21.9%	Test Pit 1, 11N 1E, top of dung unit, loose material, coll. Mead, March 1983
A-3296	11,850 ± 160	−25.7%	Test Pit 1, 11N 1E, middle to base of dung unit, fragment of mammoth(?) dung, M-3 coll. Mead, March 1983
GX-9371	13,505 ± 580		Sample pit "M," mixed dung blanket material, coll. Agenbroad, February 1983

Fig. 6.—*Nothrotheriops* (Shasta ground sloth) dung from Rampart Cave, Arizona; plant stems are of various shrubs, not graminoids.

eral dozen individual hairs were removed during screening of the dung blanket for macrofossils, and occasional hairs or hair masses were found during facing of the dung blanket at the mouth of the cave. Preliminary study of the collection by Charles Bolen, a graduate student in Anthropology at Northern Arizona University, has revealed the presence of hair comparable in thickness to mammoth, hair of ground sloths (cf. *Glossotherium* and *Nothrotheriops*), several artiodactyls, *Equus,* and various small mammals.

Four radiocarbon dates, in addition to samples

from M-1 and M-2, were obtained: one (GX-9371) from the sample pit "M" and the other three from Test Pit 1 (see Table 1). A-3297 was collected from the top of the dung blanket stratigraphically above A-3296. A-3298, a fragment of cf. mammoth dung, was collected at the base of the dung blanket on another corner of Test Pit 1. All three dates fall within the range established by M-1 and M-2 (A-3212 and A-3213). The sample of dung blanket material collected by Agenbroad from sample pit "M" overlaps within one sigma the age of A-3213.

Thus the radiocarbon samples can be viewed as

Table 2.—*Concentration of identifiable remains in Bechan dung layer (11N, 1E, NE corner).*

Taxa of plants	Plant part identified	Depth in dung			
		0–10	10–20	20–30	30–40
Betulaceae (361)					
Betula occidentalis	bracts	9	6	12	8
	nutlets	7	2	3	3
	twigs	55	122	83	51
Cactaceae (519)					
Opuntia sp.	spines	162	148	109	96
	epidermis			1	
O. polyacantha	seed		1		
Sclerocactus sp.	seeds		2		
Caprifoliaceae (76)					
Sambucus sp.	twigs	30	7	3	24
Symphoricarpos sp.	seeds		2	3	2
	twigs		1	3	1
Chenopodiaceae (154)					
Atriplex sp.	twigs	38	55	34	7
Atriplex cf. *canescens*	fruits	5	2		
	leaf	1			
A. cf. *confertifolia*	fruits				2
Chenopodium sp.	seeds	1		2	1
Corispermum sp.	seeds	2	1		3
Cleomaceae (5)					
Cleome sp.	seeds	1		2	2
Compositae (97)					
Artemisia cf. *tridentata*	leaves	3	2	1	3
	twigs	30	25	16	10
Cirsium sp.	achene	1	1	2	2
cf. *Heleanthus*	achene	1			
Cornaceae (13)					
Cornus stolonifera	stones	4	4	2	3
Cupressaceae (4)					
Juniperus sp.	male bract			1	
Juniperus communis	seeds	3			
Cyperaceae (707)					
Carex cf. *lenticularis*	nutlets	75	16	28	14
Carex cf. *interior*	nutlets	210	52	44	38
Carex lasiocarpa	nutlets	99	33	41	54
Scirpus sp.	nutlets	3			
Fagaceae (12)					
Quercus sp.	wood		1	8	3
Gramineae (204)					
cf. *Agropyron*	florets	3	8	12	5
Oryzopsis hymenoides	caryopses	53	37	49	28
Panicum	caryopses	2			1
Spartina gracilis	spikelet	1	1	1	2
Sporobolis	spikelet	1			
Gramineae + Cyperaceae (708)	culms	127	406	109	66
Labiatae (1)					
cf. *Marrubium*	seed			1	
Leguminosae (8)					
cf. *Dalea*	seeds		4	2	2

Table 2.—*Continued.*

Taxa of plants	Plant part identified	Depth in dung			
		0–10	10–20	20–30	30–40
Liliaceae (1)					
cf. *Smilacina*	seed				1
Malvaceae (2)					
Sphaeralcea sp.	seeds			2	
Pinaceae (59)					
Picea pungens	needles	1	16	22	14
	twigs	2	1	2	1
Potamogetonaceae (1)					
Potamogeton diversifolium	nutlet	1			
Rosaceae (342)					
Amelanchier sp.	achenes	1	2	1	
Purshia sp.	leaf	1			
Rosa sp.	thorns	33	20	31	25
	twigs	45	62	53	26
	achenes	16	7	8	10
Rubus sp.	achene	1			
Saxifragaceae (3)					
Ribes sp.	seeds				3
Total (3,277)					
Total Graminoids (1,619) = 49.4%					

ranging from 11,700 to 12,900 years in age and the dung blanket appears to have been deposited, perhaps intermittently, over roughly 1,000 years. Various authors (Long and Martin, 1974; Thompson et al., 1980) have stressed the 9th millennium B.C. as the critical time for extinction of Shasta ground sloths and other animals in this region. On the basis of the six radiocarbon dates in Table 1 it would appear that Bechan Cave was not occupied by Pleistocene herbivores after 11,700 years ago, which is slightly but significantly older than the time of extinction of Shasta ground sloths in western Arizona as determined by Long and Martin (1974).

Abandonment can be explained in at least two ways, either temporary access to the cave ended with erosion of a dune deposit, left as a stabilized remnant 100 m southwest of the cave, or the vegetation adjacent to the cave may have been attractive to the megafauna for a brief interval only.

GEOLOGIC SETTING

Bechan Cave occupies a canyon cut into the Navajo Sandstone and, in places, the upper units of the Kayenta Formation. The cave itself appears to have been formed by the spalling of the cross bedded sandstone, perhaps aided by moisture infiltrating from joint controlled surface drainage, spray from an adjacent plunge pool, and mechanical forces caused by crystal growth of calcite, due to evaporating ground water.

Entrance to the cave is steep and hazardous. There is geomorphic suggestion of a ramp in the past, with access provided by a climbing dune. A remnant of this dune still exists, although much of it has been removed by the plunge pool and surface drainage. The stabilized climbing and falling dunes within the canyons testify to abundant eolian deposition in the past. Terrace remnants indicate a former valley fill of wind blown sand and silt.

A side valley below Bechan Cave contains up to 40 m of alluvium. One measured section indicates

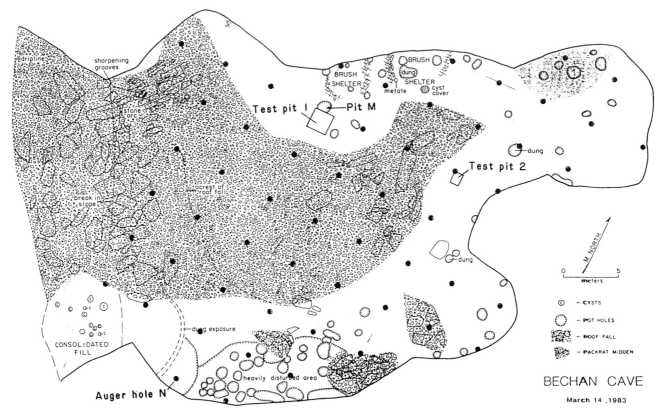

Fig. 7.—Floor plan, Bechan Cave.

black cienega soils and a water table fluctuating during the accumulation of the alluvium. The valley fill has been dissected subsequently and the main stem head cut has migrated up gradient to the terminus of the box canyons. For large quadrupeds seeking access to the valley harboring Bechan Cave there is only one entry point—across steeply sloping bare sandstone.

MODERN VEGETATION

The vegetation near Bechan Cave occupies a variety of habitat types ranging from bare walls and ridge tops of Navajo sandstone vegetated only along cracks or in pockets to rich riparian growth along a semipermanent stream. The region is also noted for its hanging "gardens" (Welsh and Taft, 1981), face and foot wall gardens and plunge basins fed by seepage that may harbor primose (*Primula specula*), maiden-hair fern (*Adiantum*), rock spirea (*Petrophytum*), cardinal flower (*Lobelia*), death camas (*Zigadenus*), orchids (*Epipactis*), monkey flowers (*Mimulus*), columbine (*Aquilegia*), and other aquatic or mesic species unusual in this arid region (mean annual ppt. at Page, Arizona = 14.6 cm). Above the

canyon bottoms waterpockets (deep plunge pools) supporting willows (*Salix exigua*) maintain a permanent supply during the summer season. The riparian habitat near Bechan Cave includes *Populus fremontii* (Fremont cottonwood), *Rhamnus betulifolia* (birch leaf buckthorn), *Carex* spp. (sedge), poison ivy (*Toxicodendron rydbergii*), and other aquatic or mesic species. *Fraxinus anomala* (single leaf ash) occupies cracks in the bedrock.

Scattered trees along the canyon bottom include *Juniperus osteosperma* (Utah juniper), *Quercus gambelii* (Gambel oak), and *Amelanchier utahensis* (Utah service berry). Shrubs include *Coleogyne ramosissima* (Black brush), *Ephedra* spp. (Mormon

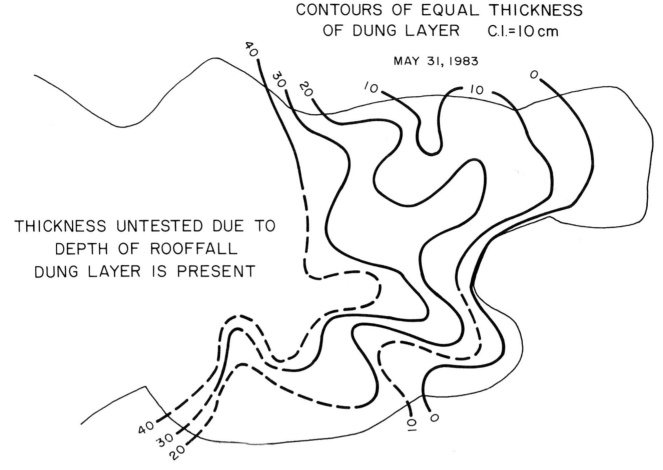

Fig. 8.—Isopatch map of the dung blanket, Bechan Cave.

tea), *Atriplex canescens* (four-winged salt bush), *Grayia spinosa* (spiny hop sage), and *Opuntia phaeacantha major* (prickly pear). Common grasses include *Oryzopsis hymenoides* (Indian rice grass), *Stipa comata* (needle grass), and *Sporobolus* sp.

(sacaton). Above the canyon and away from bare cliffs the vegetation is dominated by *Coleogyne ramosissima, Juniperus osteosperma,* and *Oryzopsis hymenoides.*

PALEOECOLOGICAL ANALYSIS

Pollen was extracted from samples collected from sample pit "M" by Mead and Agenbroad in February (Fig. 1) and from samples collected from the east wall of pit "1" excavated on 17 March 1983. The first set of samples was from within the dung layer and included samples from the two large dung boluses (M-1 and M-2). The second set included samples taken at 2 cm intervals from the sand above

and below the dung layer, which was at 23 to 39 cm.

Pollen samples were soaked in dilute HCl and *Lycopodium* tracer tablets were added to allow calculation of pollen concentrations. The samples were then screened to remove coarse debris and transferred to test tubes. After treatment with concentrated HCl and HF, the samples were acetylized

(Faegri and Iversen, 1975), treated with 10% KOH, and transferred to silicone oil. Over 500 pollen grains were counted in each sample at 400× and 1,000×. Identification of the pollen types is based on the reference collection and library of the Paleoenvironmental Laboratory at the University of Arizona. Preservation was excellent in all samples.

Bulk samples of dung for macrofossil analysis were taken from a 10 by 10 cm column in the northeast corner (11N-1E) of the test pit excavated on 17 March 1983. The dung layer was 40 cm thick at that locality and the four samples, each 10 cm thick, were approximately 1,000 cc in volume.

Macrofossil samples were soaked in water for 24 hrs, then screened through #6 (3.35 mm) and #20 (0.85 mm) ASTM screens. After the residue had dried, the identifiable remains were removed for examination. Identifications are based on the macrofossil collections at the Paleoenvironmental Laboratory and the University of Arizona Herbarium. All spruce (*Picea*) needles were either thin-sectioned or examined in cross-section to determine the size and placement of the resin canals. Twigs that lacked external identifying characters (thorns, buds, leaf-scars, distinctive bark) were examined in cross-section.

RESULTS AND INTERPRETATIONS

The dung samples are composed primarily of what appear to be crushed culms and leaves of graminoids (Gramineae and Cyperaceae), and most identifiable macrofossils represent these two families (Table 2). Apart from cactus spines, nutlets of sedges are the most abundant type of fossil in the dung layer. The concentration of graminoids is greatest in the uppermost samples, which is true for the concentrations of most other macrofossil types.

The relationship of the fossils in the column to the diet of any one organism is difficult to establish because the samples are probably from the dung of more than one animal. Four different types of dung were found in the pit. *Neotoma* (packrat) fecal pellets are present in the dung layer and this animal favors cactus pads in house construction. Abundant spines of *Opuntia* (prickly pear) in the dung layer may result primarily from the activities of *Neotoma*. However, packrats do not gather grass culms in quantity. As these are evident in the two large dung balls (Fig. 2, 3), the organism that deposited them, cf. *Mammuthus*, may be responsible for the prevalence of graminoids in the identifiable remains.

With a few notable exceptions, the macrofossil composition in samples from different depths is very similar. The lack of substantial differences among samples may result from the short period of accumulation of the dung layer, which is suggested by the narrow range of radiocarbon dates (Table 1). Lack of a clearcut stratigraphic sequence within the millennium embraced by the radiocarbon dates could reflect trampling and disturbance of the dung blanket by those mammals that deposited it.

Although specific dietary information cannot be gained from the column macrofossils, the data provide evidence of the nature of local vegetation. The fossil assemblage includes plants from several types of habitats. As is true today, the riparian community was then an important vegetation type in the narrow canyon. However, its members were different from those of today. Three species of sedges were identified based on their perigynia and nutlets (Fig. 9). The southern limits of two of these, *Carex lasiocarpa* and *C.* cf. *lenticularis,* are currently in Idaho and Montana. The ancient community contained water birch (*Betula occidentalis*) which at the latitude of the cave occurs today only at higher elevations.

Several of the shrub species found in the dung layer were probably more abundant in the riparian community than in the uplands—*Sambucus* (elderberry), *Symphoricarpos* (snowberry), *Rosa* (wild rose), *Rubus* (raspberry), and *Ribes* (currant). Given the small amounts of *Picea* pollen in the dung samples (Table 3) and spruce's tendency to reach its lowest elevational limits along stream sides, we suspect that spruce too may have favored the riparian community. We have collected blue spruce (*Picea pungens*) in stream bottoms in the adjacent Henry Mountains about 1,890 m where it associates with *Betula* and is parasitized by nymphs of the gall aphid, *Chermes cooleyi* Gill. The distinctive cone galls caused by this parasite were found in the dung blanket (Fig. 9). *Juniperus communis,* present in the uppermost fossil sample, is also presently restricted to higher elevations.

Dune vegetation represents another type of habitat reflected in the fossil assemblage. Several species characteristic of arenaceous habitats today were recovered in the ancient dung layer. Fossil members

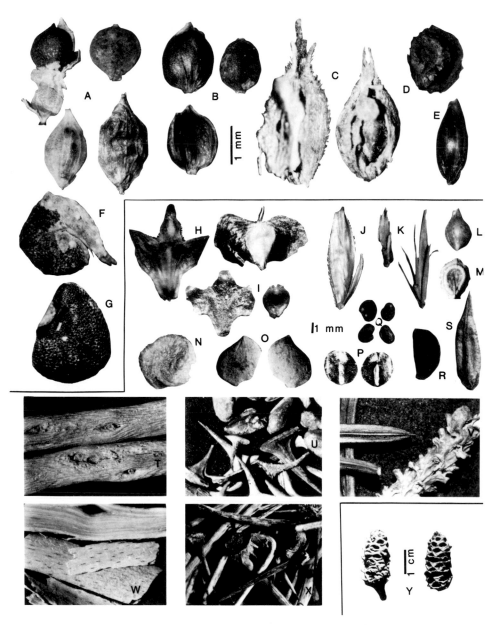

Fig. 9.—Selected plant macrofossils from Bechan Cave. (A) *Carex* cf. *lenticularis* modern achene and perigynium (left), fossil achene and perigynium (right). (B) *Carex* cf. *interior* fossil perigynium front and back (left), fossil achene (right). (C) *Carex lasiocarpa* modern perigynium (left), fossil perigynium (right). (D) *Potamogeton* cf. *diversifolium* fossil. (E) *Panicum* fossil. (F) *Cleome* fossil. (G) *Sclerocactus* fossil. (H) *Betula occidentalis* modern bract (left), achene (right). (I) *Betula occidentalis* fossil bract (left), achene (right). (J) *Spartina* fossil. (K) cf. *Agropyron* fossil. (L) *Oryzopsis* fossil. (M) *Corispermum* (fossil). (N) *Opuntia polyacantha* fossil. (O) *Cornus stolonifera* fossils. (P) cf. *Smilacina* fossil. (Q) cf. *Dalea* fossils. (R) *Amelanchier* fossil. (S) *Symphoricarpos* fossil. (T) cf. *Atriplex* fossil wood. (U) *Rosa* fossil achenes and thorns. (V) *Picea pungens* fossil needles and twig. (W) *Artemisia* modern *A. bigelovii* (top), fossil *A. tridentata* (middle two pieces), modern *A. tridentata* (bottom). (X) *Opuntia* fossil spines. (Y) Spruce cone galls modern from *Picea pungens* (left), fossil (right). Photos are shown at three different magnifications, note length bar in three different groupings (A–G, H–X, and Y) separated by solid lines.

Table 3.—*Pollen percentages for more common pollen types in Bechan Test Pit 1, 11N–1E.*

Pollen types	0–2	10–12	20–22	Dung layer top	Dung layer bottom	40–42	78–80	Dung balls M1	Dung balls M2
Deteriorated	1.4	5.4	8.3	10.4	12.9	13.7	26.5	7.8	8.5
Unknown	2	.8	.4	.4	1.2	0	.3	.2	0
Juniperus	41.6	12.1	13	11.8	8.5	4.2	.9	.4	1
Picea	.4	0	.2	.2	2.2	8.9	.6	0	0
Pinus	17.3	22.9	33.7	6.1	14.1	27.6	29.8	4	2.2
Ostrya	0	0	0	.8	.4	10.3	1.8	.2	.2
Quercus	5.2	7	6.1	6.3	1.8	0	0	0	.2
Ephedra	2.6	2.6	0			0	0		
Chenopodiaceae-Amaran.	7.2	8.2	4.5	5.7	15.4	9.3	8.1	33.2	39.4
Artemisia	11	18.5	21.5	12.4	30.9	19.3	18.7	20.8	21.4
Ambrosia	3.8	8.7	9.5	7.9	1.4	1	2.1	11.2	7.5
Cirsium	0	.4	.2	.2	.4	1	1.5	0	0
Other compositae	3.4	10.3	2	.4	1.8	1.8	6.9	6	6
Gramineae	1.2	.8	.4	34.1	7.5	2.4	1.5	12.6	9.2
Sphaeralcea	0	0	0	2.2	.2	0	0	1.2	1.5
Umbelliferae	0	0	0	.2	0	0	0	2.2	2.5
Acer negundo	0	0	0	.4	0	0	0	0	0
Betula	0	0	0	0	8.3	3	1.5	.8	.2
Cyperaceae	0	0	0	.4	2.8	1.2	0	.6	.7
Typha	0	0	0	6.7	0	0	0	11.8	11
Sporormiella	.2	0	0	0	0	0	0	16	16.2
Other fungal	1	1	.2	38.4	8.1	6	.6	5.8	6.2
Tracers added (×1,000)	56.6	56.6	56.6	90.6	90.6	56.6	56.6	90.6	90.6
Tracers recovered	20	9	7	47.5	4.5	134	621	21.6	23.7
Vol. sedim. used	15	15	15	.5	2	15	15	1.4	1
Conc. (×1,000)	94.029	211.051	273.606	484.9	20	14.175	2.019	3	1.6
Pollen sum	498	503	507	508	495	503	332	500	401

of this community include *Opuntia polyacantha* (a different species occurs in the modern vegetation), *Sclerocactus* sp., *Atriplex canescens,* and *Oryzopsis hymenoides.*

The natural upland vegetation growing on Navajo Sandstone during the time of dung accumulation is difficult to reconstruct from the macrofossil remains. Today the regional vegetation at the elevation of the cave is black brush steppe. The relatively high percentages (30.9%) of *Artemisia* pollen in the dung layer (Table 3) may indicate that sagebrush steppe was the regional vegetation type while the dung accumulated. *Artemisia* cf. *tridentata* macrofossils are present in the dung layer; the species is not now found near Bechan Cave.

Although the dung layer may have experienced some mixing by trampling, some trends in the macrofossils (Table 2) deserve mention because they may indicate vegetation change during the period of dung accumulation. The abundance of aquatic plants at the top of the dung is greater than in lower samples. The three-fold increase in sedge achenes

in the 0–10 cm sample is far greater than the increase in concentration of any other type, and this sample contains two other aquatic taxa—*Scirpus* sp. and *Potamogeton diversifolium* that do not occur in lower samples. Other taxa (*Symphoricarpos, Quercus,* and *Picea pungens*) show the opposite trend—being more abundant in the three lower samples. *Betula occidentalis* (Fig. 9) is most abundant in the two middle samples.

Preliminary analysis of pollen from the dung unit ('Top' and 'Bottom' samples in Table 3) only partially confirms this trend. The percentage of *Picea* pollen is greater in the sample from the bottom of the dung, as is the abundance of *Betula* pollen. However, *Quercus* pollen is more abundant in the top sample, and *Symphoricarpos* pollen was not recovered.

Although macrofossils and pollen from the dung layer provide limited evidence for vegetation change, preliminary analysis of pollen in cave fill above and below the dung layer (Table 3) suggests strong temporal differences. The sand surrounding the dung should be free from the effects of trampling by large mammals. Differences in *Picea* pollen percentages above and below the dung layer are in the same direction as the macrofossil concentrations and dung pollen percentages. Immediately below the dung layer *Picea* pollen reaches 8.9% and above the layer the percentages drop to 0.2% (Table 3).

Several other taxa show changes with hop hornbeam (*Ostrya*), box elder (*Acer negundo*), Gramineae, and Cyperaceae all more abundant below the layer than above. *Juniperus, Quercus,* and *Ambrosia* showed the opposite trend and are more abundant above the dung layer than below (Table 3). Although the pollen data are preliminary they are sufficient to indicate that the dung layer accumulated during an interval of major vegetation change when spruce

and birch became less abundant and oak became more abundant.

One last fossil is worthy of note: *Sporormiella* spores were common only in the two dung balls, M-1 and M-2 (Table 3). *Sporormiella* is a dung fungus whose abundance in certain lake sediments has been correlated with disturbance by grazing animals (Davis et al., 1977). Its restricted distribution in the dung layer reflects its poor dispersal, and its appearance in the surface sample probably results from the introduction of livestock by European immigrants. The abundance of *Sporormiella* spores in the two dung balls is of considerable interest. It suggests that an abundance of the spore type in ancient lake sediments may be used to indicate the presence of an abundant megafauna in the surrounding range.

DISCUSSION AND SUMMARY

Around the 10th millennium B.C. and within a few hundred years of the last records of Shasta ground sloths and Columbian mammoths in the region, several species of large extinct herbivores occupied Bechan Cave in southern Utah. They left a distinct stratigraphic unit, a dung blanket containing abundant plant material and animal droppings up to 40 cm in thickness and estimated at 255 m³ in volume. No diagnostic bones of the large mammals have been found to date; however, based on the preliminary examination of dry boluses of herbivore dung and collections of hair we believe the cave was visited by mammoth. In addition at least one and probably two kinds of ground sloth, and various unknown species of ungulates ranging in size from that of mountain goats to elk were present. Subsequent use of Bechan Cave by large mammals was largely restricted to *Homo sapiens*; charcoal from hearths commonly occurs above the dung blanket. The dung blanket at Bechan Cave is larger than a similar deposit of comparable age found beneath abundant archeological material at Cowboy Cave in eastern Utah. Neither site has yielded evidence of human occupation during the time of dung blanket deposition around 12,000 years ago.

As determined by pollen and plant macrofossil analysis the environment outside Bechan Cave roughly 12,000 years ago constituted a riparian gallery of water birch, blue spruce, elderberry, snowberry, currant, sedges, and cattail. Other species present in the sandy valley or along the cliff-rimmed

canyon beyond the cave include (in order of abundance of fossil remains), big sagebrush (*Artemisia* cf. *tridentata*), prostrate juniper (*Juniperus communis*), Utah cactus (*Sclerocactus* sp.), prickly pear (*Opuntia* sp.), salt bush (*Atriplex* sp.), thistle (*Cirsium*), oak (*Quercus* sp.), Indian rice grass (*Oryzopsis hymenoides*) and other grasses, and hop hornbeam (*Ostrya* sp.). Many of these species do not occur in the region today; presently the extra-local species can be found at higher elevations or higher latitudes such as at 1,800 m in canyons of the Henry Mountains or Aquarius Plateau. The regional vegetation around 12,000 years ago may have been primarily sagebrush rather than blackbrush (*Coleogyne*) steppe as it is now.

The canyon outside the cave continues to support large herbivores as shown by the small herds of feral cattle and horses dating from the establishment of the Glen Canyon National Recreation Area. The potential of the region to support the variety of species present 11,600 to 12,900 years ago is less certain. Did their habitat fail? All the plant taxa present in the dung layer can be found today at higher elevations in the Colorado Plateau. As in the case of the Shasta ground sloth (Long and Martin, 1974; Hansen, 1980), dietary information on the extinct fauna indicates that plant taxa ingested by the vanished large herbivores continue to grow at no great distance from the fossil site.

The most remarkable items found to date in Bechan Cave are the two dung boluses, M-1 and M-

2. They are too large to represent most of the 39 genera of megaherbivores known from the late Pleistocene (Rancholabrean) of North America. While we cannot be certain that they were not voided by a large grazing ground sloth such as *Glossotherium,* they do not resemble fragments of *Glossotherium* dung from Mylodon Cave, Chile. Within the proboscidea the American mastodont, genus *Mammut,* is known in Utah from the late Pleistocene. However, late Pleistocene mastodonts are rare in this part of North America and they are thought to have been browsers, not grazers. The dung boluses in question are packed with large graminoid stems. They most closely resemble African elephant (*Loxodonta*) dung in size and texture and we attribute them to mammoth (*Mammuthus*), a common Pleistocene fossil in western North America.

Finally, it is of interest that the dung balls harbor spores of the coprophilous fungus *Sporormiella.* Davis et al. (1977) reported an increase in *Sporormiella* spores at Wildcat Lake, Washington, coincident with the introduction and heavy grazing impact of sheep. With the extinction of two-thirds of the large mammal species in western North America 11,000 years ago a decline in *Sporormiella* may be expected. The decline in *Sporormiella* and possibly other coprophilous fungi should be sought in suitable spore-rich sediments or cave deposits. Independently of fossil animal remains, the change could reflect the late Pleistocene extinction of megafauna.

ACKNOWLEDGMENTS

This study began during John Guilday's last month; we think it would have attracted his wide ranging paleontological interest. Mr. John Lancaster, Superintendent of Glen Canyon National Park, Dr. Adrienne Anderson, Larry Belli and Vic Vierra, all of the National Park Service, provided outstanding help and valuable logistic aid. With their encouragement and that of Emil Haury of the University of Arizona, we sought and received National Geographic research support to continue our work on Bechan Cave. Early phases of the work were funded by NSF grant #BSR82-14939 to P. S. Martin and T. R. Van Devender. Cave maps and geological data were drafted by Agenbroad, pollen and plant macrofossil analyses were contributed by Davis, with grass identification by Larry Toolin; Mead analyzed fossil dung samples; Charles Bolen identified hair samples. Bonnie Fine Jacobs responded to an appeal for African elephant dung from Kenya and Cynthia Moss of the World Wildlife Fund, Nairobi, verified our opinion about the similarity between the Bechan Cave samples and *Loxodonta* dung balls.

LITERATURE CITED

DAVIS, O. K., D. A. KOLVA, and P. J. MEHRINGER, JR. 1977. Pollen analysis of Wildcat Lake, Whitman County, Washington: the last 1,000 years. Northwest Science, 51:13–30.

FAEGRI, K., and J. IVERSEN. 1975. Textbook of modern pollen analysis. Hafner Press, New York, 295 pp.

HANSEN, R. M. 1980. Late Pleistocene plant fragments in the dungs of herbivores at Cowboy Cave. Pp. 179–189, *in* Cowboy Cave (J. D. Jennings), Univ. Utah Anthro. Papers, Salt Lake City, 104:1–224.

JENNINGS, J. D. 1980. Cowboy Cave. Univ. Utah Anthro. Papers, Salt Lake City, 104:1–224.

LONG, A., and P. S. MARTIN. 1974. Death of American ground sloths. Science, 186:636–640.

SPAULDING, W. G., and P. S. MARTIN. 1979. Ground sloth dung of the Guadalupe Mountains. Pp. 259–269, *in* Biological Investigations in the Guadalupe Mountains National Park, Texas (H. H. Genoways and R. J. Baker, eds.), National Park Service Proc. and Trans. Ser., 4:xvii + 1–442.

THOMPSON, R. S., T. R. VAN DEVENDER, P. S. MARTIN, A. LONG, and T. FOPPE. 1980. Shasta ground sloth (*Nothrotheriops shastensis* Hoffstetter) at Shelter Cave, New Mexico: environment, diet, and extinction. Quaternary Res., 14:360–376.

WELSH, S. L., and C. A. TAFT. 1981. Biotic communities of hanging gardens in southeastern Utah. National Geographic Soc. Res. Rept., 1972 Projects, pp. 663–681.

Address (Davis and Martin):   Department of Geosciences, University of Arizona, Tucson, Arizona 85721.
Address (Agenbroad):   Department of Geology, Northern Arizona University, Flagstaff, Arizona 86011.
Address (Mead):   Center for the Study of Early Man, University of Maine, Orono, Maine 04473.

PLEISTOCENE TAPIRS IN THE EASTERN UNITED STATES

CLAYTON E. RAY AND ALBERT E. SANDERS

ABSTRACT

The name *Tapirus haysii* is reinstated for the larger Pleistocene tapirs, its type locality in North Carolina reestablished, topotypic and other material reported, and its distribution reviewed. A nearly complete skull of *T. veroensis* from South Carolina is described, and other material recorded, especially as it supplements previously known distribution.

INTRODUCTION

Simpson (1945) laid the foundation upon which has been built all subsequent research on North American Pleistocene tapirs. Although Leidy (1855: 200; 1859:106, for example), with his usual acumen, understood that the fossils generally fell into two size groups, one similar in size to the living *T. terrestris* (=*T. americanus*) and another, larger form, his *T. haysii,* specific assignments through the years frequently seemed subjective, if not capricious, prior to Simpson's careful analysis. The notorious morphological conservatism of tapirs has hampered specific identification of the most common specimens (partial dentitions and isolated teeth) and contributes to a continuing suspicion, already clearly expressed by Leidy (1869:391), that phantom taxa may be concealed within the ostensibly homogeneous, stereotyped dentitions. The suspicion is reinforced by the co-occurrence today in northern Colombia of three species (*T. terrestris* and *T. bairdii* sympatrically, and *T. pinchaque* nearby at higher elevations; Hershkovitz, 1954:490–491, 494; fig. 62) that are well demarcated in life and in cranial characters, but broadly overlapping in most dental characters, including size. There would seem to be no a priori reason why as many taxa could not have left their remains within a small area of eastern North America given its altitudinal range and the time and changing environments of the Pleistocene. The widespread and fairly abundant fossil material is generally inadequate for rigorous specific identification, and it must be understood that our concept of the number of species in the Pleistocene of North America and of their geographic and temporal distribution remains very insecurely founded.

So few skulls are known that each one discovered (Sellards, 1918; Simpson, 1945; Lundelius and Slaughter, 1976; Bogan et al., 1980) has facilitated significant advancement in understanding. Thus, a primary purpose of this communication is to make known an essentially complete skull of *T. veroensis* from South Carolina, rivaled in quality only by the holotype.

Until well preserved cranial material of a larger North American Pleistocene tapir is available, its relationship to *T. veroensis,* and more broadly within the Tapiridae, will remain equivocal. Meanwhile, *T. haysii* is resurrected for the large eastern tapir on the basis of identification of its type locality, availability of topotypes, and Simpson's (1945:65) recognition of the holotype as a P_4 rather than M_2.

We also discuss new or obscure specimens of both taxa. As it has not been feasible to reexamine by any means all recorded material, we have concentrated on that which extends or fills in the geographic distribution and/or aids in understanding temporal distribution.

In addition to *Tapirus haysii* and *T. veroensis,* all nominal taxa from the Pleistocene of North America are reviewed.

METHODS AND MATERIALS

The following abbreviations for institutions are used throughout:

AAS	Arkansas Archeological Survey, State University
AMNH	American Museum of Natural History, New York
ANSP	Academy of Natural Sciences of Philadelphia
BM(NH)	British Museum (Natural History), London
CAS	California Academy of Sciences, San Francisco
CM	Carnegie Museum of Natural History, Pittsburgh
ChM	Charleston Museum, Charleston, South Carolina
MCZ	Museum of Comparative Zoology, Harvard University, Cambridge, Massachusetts
NCSM	North Carolina State Museum of Natural History, Raleigh

NYSM   New York State Museum, Albany
SCMC   South Carolina Museum Commission, Columbia
UF   Florida State Museum, University of Florida, Gainesville
UF/FGS   Florida State Museum/Florida Geological Survey Collections, Gainesville
USGS   United States Geological Survey
USNM   [United States] National Museum of Natural History, Smithsonian Institution, Washington, D.C.

All measurements are in millimeters; abbreviations for measurements of cheek teeth are: L, anteroposterior diameter of crown; W, maximum transverse diameter of crown; AW, width of crown across anterior lobe; PW, width of crown across posterior lobe.

For comparative material of the living species we have relied primarily on the collections of ANSP and USNM, which together have provided 50 skulls of *T. terrestris*, 6 of *T. pinchaque*, 61 of *T. bairdii*, and 12 of *T. indicus*. Adequate representation was available for all except *T. pinchaque*, for which additional specimens of all ages are needed. Three of the most important studies of tapir systematics (Hatcher, 1896; Simpson, 1945; Radinsky, 1965) relied on a single skull of *T. pinchaque* for critical comparisons. This situation unfortunately is not unusual for large mammals, museum collections of which frequently are inadequate in quantity and quality for study of individual, sexual, ontogenetic, and geographic variation.

NOMINAL TAXA

All nominal taxa of North American Pleistocene tapirs are discussed here in chronological sequence. In much of the nineteenth century literature the fossils were not clearly distinguished from living tapirs, and in many instances were recorded under the name *Tapirus americanus,* a junior synonym then current for *T. terrestris.*

Tapirus mastodontoides was described by Harlan (1825:223–225) on the basis of an isolated tooth from Big Bone Lick, Kentucky. Because deinotheres were then considered to be giant tapirs, he regarded his new form as very small, only a little larger than living tapirs. The tooth was recognized almost immediately as an anterior milk molar of American mastodon (Cooper, 1831:163; Hays, 1834:324). In rebuttal to Cooper, Harlan (1835:265–267), having made further comparisons with Cuvier's collections in Paris, insisted that the tooth was that of a tapir, in fact the P^1, and that its size and the structure of its roots distinguished it from teeth of mastodon. After Harlan's death, when his collection ultimately came to ANSP, Hays identified the holotype and Leidy (1858:12) confirmed the conclusion of Cooper and Hays that "the specimen was a first milk molar of the Mastodon." Many years later however, Simpson (1942:162) again raised the possibility "that it is really a tapir, in which case the name is valid and antedates any other for our fossil tapirs," a supposition presumably quickly abandoned as the name is not mentioned by Simpson (1945).

Until or unless some totally unexpected evidence comes to light, Hays' recognition of the holotype must be accepted at this late date. The specimen, ANSP 13261, illustrated here for the first time (Fig. 1A, B, I, M), appears to be a left DP_2 of *Mammut americanum* (DP^2 according to Gillette and Colbert, 1976:33). The anteroposterior length of the crown is approximately 33.4 mm and the maximum height of the tooth 48 mm. It is deeply worn, and much mutilated, lacking the posterolingual corner, and most of the enamel of the anterolabial corner and posterior margin. This damage must have occurred after the original description, when Harlan (1825: 224) characterized the crown as "nearly quadrangular," which would have made it more similar in appearance to DP_2 in *Mammut* than it now is. This tooth is not common in collections, and seemingly is rather variable.

An unworn right DP_2, CM 2332F, from the Frankstown Cave, Pennsylvania, with roots broken away, is rather smaller (crown length 28.1 mm) and the crown more quadrangular in plan and the cross lophs more transversely oriented (Peterson, 1926: 275, pl. 23). An unworn presumed left DP_2, USNM 12062, from Cumberland Cave, Maryland (Gidley and Gazin, 1938:70), is similar in crown shape and pattern to that from Frankstown, but is larger (length 32.6 mm) and has essentially complete roots that match those of *T. mastodontoides* almost exactly (Fig. 1E, F, K, O). A left DP_2, MCZ 11106, described and figured (but unfortunately subsequently damaged) in the classic monograph by Warren (1855: 67; pl. 2, fig. 1; pl. 8, fig. 1) is very similar to Harlan's holotype (Fig. 1G, H, L, P). Harlan's specimen is matched best, however, by USNM 4986, an isolated tooth thought to be a right DP_2, from Kimmswick, Missouri, with crown length of 33.9 mm, and maximum height as preserved of 51.1 mm. It was described and illustrated by Hay (1912:679–680; pl. 17, figs. 6 and 7; 1914:346; pl. 47, figs. 9 and 10; text-fig. 115), and is refigured here (Fig. 1C, D, J, N).

The detailed similarity between the holotype of *Tapirus mastodontoides* and teeth of known mas-

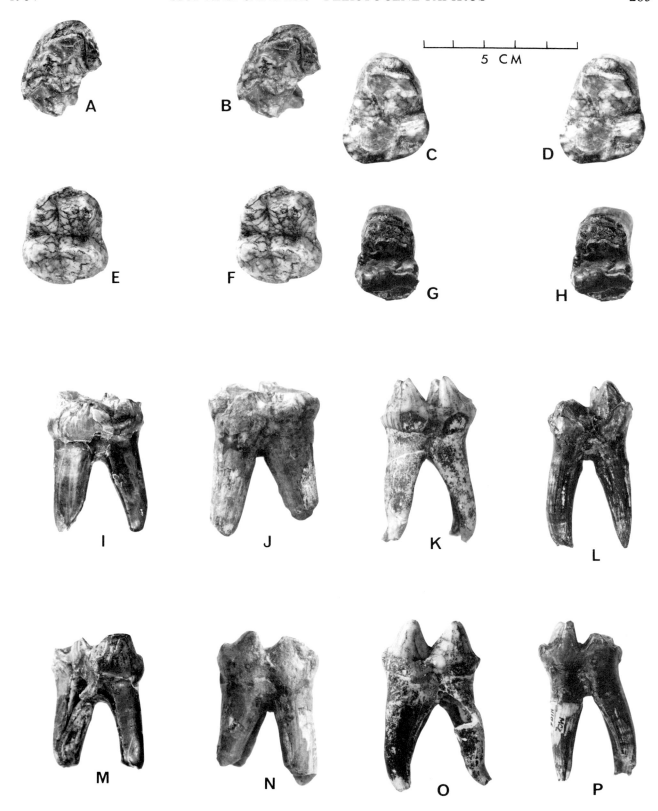

Fig. 1.—Holotype of *Tapirus mastodontoides* Harlan 1825 and some examples of left DP$_2$ of *Mammut americanum,* in occlusal (A–H; stereo pairs), labial (I–L), and lingual (M–P) aspects, ×0.8. *T. mastodontoides*: ANSP 13261, A–B, I, M; *M. americanum*: USNM 4986, C–D, J, N (all figures reversed); USNM 12062, E–F, K, O; MCZ 11106, G–H, L, P.

todonts seems sufficient to lay *T. mastodontoides* to rest forever as a junior synonym of *Mammut americanum*. It seems probable however that *T. mastodontoides* is not without significance for fossil tapirs, as it may help to explain the otherwise groundless linkage between Big Bone Lick (the source of *T. mastodontoides*) and *Tapirus haysii,* beginning with Leidy's lapsus (1859:106) and carried forward to the present. Although Leidy (1859:107, footnote) had the opportunity to examine the type of *T. mastodontoides,* and recognized it as a mastodon milk tooth, prior to publication of his paper in 1859, at the time of writing the body of that paper he said of *Tapirus haysii* that "it may be questioned if it had not already been noticed by Dr. Harlan, under the name of *Tapirus mastodontoides.* The specimen described by Dr. Harlan, on which the latter was founded, is also stated to have been a lower molar, from Big-bone-lick."

Tapirus americanus fossilis was introduced into the North American paleontological literature by Leidy (1849) with reference to: a left P_4 from the vicinity of Opelousas, Louisiana, reported and figured by Carpenter (1842); to an immature left mandibular ramus and a mature left maxilla with P^4–M^3, both from the Brazos River near San Felipe, Texas, also reported and figured by Carpenter (1846: 245, 247–249; figs. 3 and 4); and to an isolated, deeply worn tooth regarded as a left DP_3, from near Natchez, Mississippi. Leidy (1859: pl. 17, figs. 1 and 6) also refigured the specimens from Texas. This name has been generally neglected and would best remain so had it not been suggested (in unpublished notes) that it qualifies as a valid senior synonym of *T. veroensis,* a suggestion with which we do not concur.

In the first place, it is questionable whether Leidy intended that appellation as a formal trinomial, for, in both the original (1849) and subsequent (1859: 106; 1869:391) publications, he characterized the specimens as nearly or entirely indistinguishable from their counterparts in living *T. americanus* (=*T. terrestris*). Furthermore, although "*fossilis*" has been used as a formal species-group epithet through the years, it was also used at times in the nineteenth century in quite another manner, simply to label as fossils specimens otherwise indistinguishable from their modern counterparts. Leidy (1853:9–10) himself employed this system in an annotated list of North American fossil mammals, where this Pleistocene tapir and at least five other taxa listed under the names of living species were listed in the format,

Tapirus americanus (fossilis), with the word *fossilis* not only parenthetical, but printed in a different and smaller type-face than the binomial.

If it is insisted that the more rigid modern procedural standards be applied retroactively, and that Leidy therefore established a species-group taxon perhaps unintentionally, we would argue that *Tapirus fossilis* (Leidy 1849) can be ignored as a forgotten name, because we can find no instance of its use in publications of the past 50 years, including Gillette and Colbert's (1976) Catalogue of Types in ANSP. (Simpson's, 1945:65, mention does not constitute usage.)

In the event that others wish to consider the subject, we hereby designate the immature left mandibular ramus from the Brazos River, Texas, ANSP 11500, as the lectotype. Further, we note that Lund earlier (1841:264) employed a similar notation "*Tapirus foss.*") for remains from Brazilian caves, at least in a species list. However, we have found nothing to qualify as a description, and Winge (1906) did not perpetuate Lund's usage. In any case, more than one effective suppressant seems to be available against resurrection of the name for North American Pleistocene tapirs.

Tapirus haysii, introduced as a nomen nudum in 1852 and validated in 1859, both by Leidy, but set aside as indeterminate by Simpson (1945), is the subject of detailed attention elsewhere in this paper, where we propose its reinstatement.

Tapirus californicus was described by Merriam (1913) as *T. haysii californicus* on the basis of an isolated tooth crown, probably a left M_2, although possibly a P_4, from near Sonora, Tuolumne County, California. In reviewing its status and literature, Simpson (1945:67–69) gave the dimensions of the holotype as L 25.3, AW 17.8, PW 17.5, noted that the tooth is relatively narrow but not clearly separable in size or structure from corresponding teeth of *T. veroensis, T. terrestris,* or *T. bairdii,* and pointed out that it is significantly smaller than M_2 of *T. haysii* (his *T. copei*). Although the weakness of Sellards' (1918:65–66) effort to distinguish *T. veroensis* from *californicus* was confirmed by Simpson's observations, he wisely chose not to pursue the taxonomic implications and instead to recognize this smaller west coast tapir tentatively as *T. californicus,* pending discovery of more diagnostic material. This remains the most satisfactory course, although new material from the vicinity of Los Angeles, listed by genus but not described by Miller (1971:36, 52–54), should aid in resolving the problem. We have had available the right maxilla with M^1–M^3 from the Elk River fauna of Oregon, CAS 101, generally referred to *T. californicus,* most recently by Leffler (1964). Its dental dimensions are L M^1, 21.9; AW M^1, 27.5; PW M^1, 24.9; L M^2, 25.4; AW M^2, 30.8; PW M^2, 26.8; L M^3, 26.3; AW M^3, 29.5; PW M^3, 24.1. On the basis of size and structure, the specimen would be assigned routinely to *T. veroensis* had it been found in the eastern United States.

It should be noted that *T. californicus,* whatever its ultimate taxonomic fate, falls within the smaller, *T. veroensis* size group, not the larger, *T. haysii-T. copei* size group, contrary to Lundelius and Slaughter (1976:228, 239) and Kurtén and Anderson (1980: 292).

Tapirus veroensis, described by Sellards (1918) on the basis of a fine skull from Florida, was the first species of North American Pleistocene tapir to be founded on fully satisfactory, confidently diagnostic material, and it retains this distinction. The species is discussed at length elsewhere in this paper, where we record new material, most notably an additional fine skull.

Tapirus merriami was established by Frick (1921: 311–314) on the basis of the crowns of two right lower molars, each broken through the transverse valley between anterior and posterior lophids, and the fragments disarranged but embedded in matrix with an associated mandibular fragment (Frick, 1921: fig. 26). The specimen came from beds of the Bautista Creek badlands in southern California, later assigned an Irvingtonian age. The tooth fragments were reassociated to form a slightly worn, nearly complete crown of an anterior tooth thought to be M_1, and an unworn, presumably unerupted, incomplete crown of a posterior tooth thought to be M_2. No reason was given for identifying the teeth as M_1 and M_2 rather than M_2 and M_3. The proportions of the teeth are at least as compatible with the latter assignment, if not more so, and the sizes of the teeth would then be more compatible with Simpson's (1945) sample of *T. copei.* The measurements of the teeth (from Frick, 1921:313) are L M_1(M_2?), (32.5); AW M_1(M_2?), 23.7; PW M_1(M_2?), 22; L M_2(M_3?), (32.5); AW M_2(M_3?), 23.6; PW M_2(M_3?), 20.5.

The lengths of the crowns were regarded as imprecise by Frick, but, judging from his illustrations, it seems also that the widths, especially PW of the unerupted posterior tooth, might well be suspect. Simpson (1945:69) accepted Frick's placement of the teeth and thought that the species would prove to be valid because M_1 was significantly larger than any other from the North American Pleistocene. He did, however, point out that differences between anterior and posterior widths of the teeth were within the range for *T. copei,* and that the two forms were closer to one another than to any other on the basis of the meager comparisons possible, though he refrained from suggesting special relationship.

A small quantity of scattered, mostly fragmentary, disparate elements of large Pleistocene tapirs has accumulated since 1921 and mostly since 1945. This material is reviewed under Distribution of *Tapirus haysii* below. Meager as it is, it goes far toward closing the gap in size, morphology, and distribution between Port Kennedy and Bautista Creek, and suggests to us that the twain indeed shall meet within the foreseeable future.

Tapirus tennesseae was named by Hay (1920:88–90) on the basis of 10 isolated teeth, USNM 8949, probably representing at least two individuals, from a cave near Whitesburg, Tennessee. Simpson (1945: 65) considered it "a possible synonym [of *T. veroensis*], essentially indeterminate at present, and properly ignored." Simpson's conclusions are reinforced by additional material from the region, notably the skull from Claiborne Cave, Tennessee, reported by Bogan et al. (1980) and described further herein. We feel that *T. tennesseae* should not be ignored but properly regarded as a junior synonym of *T. veroensis.*

Tapirus copei was proposed by Simpson (1945: 66) for the large tapir in the Irvingtonian fauna of Port Kennedy, Pennsylvania. This material had previously been referred to *T. haysii,* which Simpson (1945) regarded as indeterminate. Elsewhere in this paper we argue for reinstatement of *T. haysii,* and the relegation of *T. copei* to its junior synonymy.

Tapirus excelsus was named by Simpson (1945: 70) on the basis of a juvenile skull with partial skeleton and another partial younger skeleton, from a sinkhole near Enon, Missouri. He recognized its general similarity to *T. veroensis,* and Lundelius and Slaughter (1976) have shown that the supposed cranial distinctions are attributable to ontogenetic change, a conclusion reinforced by our additional material. Lundelius and Slaughter (1976) left open the possibility of recognizing *excelsus* as a large, midcontinental subspecies of *T. veroensis.* We feel that subspecies are not usefully employed in the present state of knowledge of fossil tapirs. Nevertheless, as additional material accumulates, especially from the midcontinent, the possibility should be kept in mind that *T. excelsus* may yet prove to represent a distinguishable geographic entity. Remains with rather large (for *T. veroensis*), rather broad teeth continue to be found in the region (Parmalee et al., 1969:26–30; Lundelius, 1972:60–63).

SYSTEMATIC ACCOUNTS

Tapirus haysii Leidy 1859

Type description.—The specific name was introduced by Leidy (1852*a*) as a nomen nudum, and additional material was referred to it in the same year (Leidy, 1852*b*), still without description. Later, in referring yet another specimen to the species, he alluded (Leidy, 1855:200) indirectly to the large size of *T. haysii,* but not until 1859 did he provide measurements and illustrations of the holotype, in what is generally regarded as the valid type description (Leidy, 1859:106; pl. 17, figs. 7 and 8). This last publication is usually cited as 1860. It was published in Holmes' "Post-Pleiocene Fossils of South-Carolina," which appeared in parts from 1858 to 1860. Blackwelder and Ward (in press) have presented persuasive evidence that the parts including *Tapirus haysii* were published in 1859.

Type specimen.—There has never been any question about the holotype being the isolated tooth crown with remnants of roots catalogued as ANSP 11504, although Hay (1912:591; figs. 38 and 39) in his figure captions mistakenly identified his reproduction of Leidy's (1859, pl. 17, figs. 9 and 10) illustrations of the Evansville, Indiana, tooth (ANSP 11505) as representing the type, an error that he later corrected (Hay, 1927:76).

Leidy (1859:106) regarded the holotype as a second lower molar, but Simpson (1945:65) pointed out that the proportions of the crown correspond much better to those of the fourth lower premolar. We regard it as a right P_4 (Fig. 2D).

Type locality.—Simpson (1945:65) stated that "the locality is not now known and can never be established with any degree of certainty." The uncertainty arose from Leidy's (1859:106) assertion that the type specimen was "supposed to have been obtained from Big-bone-lick, Kentucky." As indicated previously in the discussion of *Tapirus mastodontoides,* the association between the two in Leidy's mind at the time of writing his most substantial account of *T. haysii* may help to explain the spurious connection of the latter with Big Bone Lick. There is in fact no evidence for occurrence of any tapir at Big Bone Lick in any of the massive collections made there (Schultz et al., 1963). Although Leidy was arguably the best vertebrate paleontologist that this country has produced, this would be by no means a unique instance of his garbling data (cf. Emry and Purdy, in press).

In the first published reference to the type (Hays, 1852), "Dr. Hays stated that the tooth of the fossil Tapir presented by him this evening [16 March 1852], was found in the bed of a canal in North Carolina. It had been in his possession for several years, and was the first fossil Tapir tooth found in North America." Also, in the "Donations to Museum" (of ANSP) for that date is listed "Molar of a fossil Tapir, from North Carolina; and mould in plaster, with several casts of the same. Presented by Dr. Isaac Hays." (The mould and casts are still preserved with the specimen.)

Fortunately, evidence does exist not only to support the derivation of the tooth from North Carolina, but also to localize its source to the north bank of the Neuse River, 16 mi below New Bern (Fig. 3, locality 1). During the first quarter of 1832 Thomas Nuttall made a southern field trip funded ($260) by Harvard University, and in his report of 21 June to the Corporation he indicated that "the principal part of my time was spent in the vicinity of Newbern in North Carolina" and that "on the banks of the river Neuse, while in the neighbourhood of Newbern, I carefully examined for fossils of wh. an immense stratum is here exposed to view by the washing of the river" (Graustein, 1967:260–261). Nuttall went on to discuss the fossils and their implications at some length and promised that "an arranged collection of these fossils will be placed in the museum of the University." Recent efforts to locate such a collection in the MCZ have failed.

Graustein (1967:261) went on to state that "He was fortunate in some of his fossil finds. At a meeting of the Academy on June 19, Dr. Hays exhibited the anterior molar of a mastodon that Nuttall got and more than a year later announced from the same source a tooth of 'A Tapir,' the first evidence found in North America of this animal." The hand-written, unpublished minutes of meetings of the ANSP preserved in their archives record the following verbal communications: for 19 June 1832, "Dr. Hays

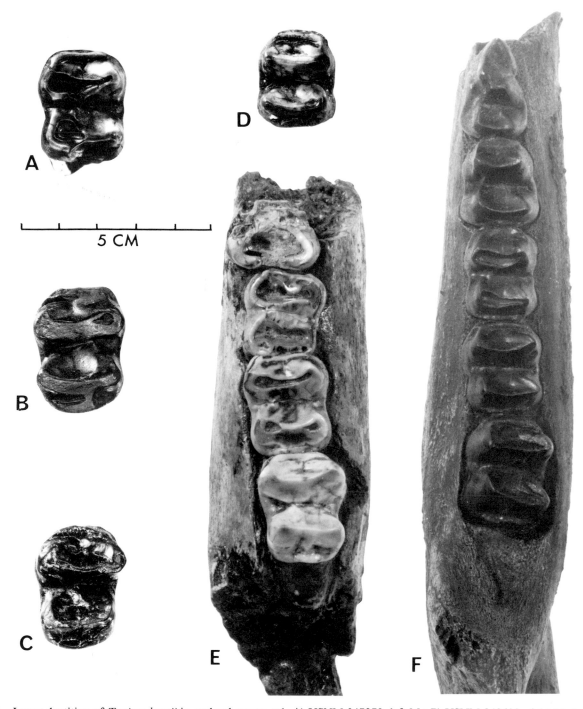

Fig. 2.—Lower dentition of *Tapirus haysii* in occlusal aspect. ×1. A) USNM 347270, left M_2; B) USNM 243692, right M_2 (cast); C) USNM 215061, right M_3; D) ANSP 11504, right P_4 (Holotype); E) AAS 69-658, left mandibular ramus with M_1–M_3 and posterior lophid of P_4; F) USNM 244439, left mandibular ramus with P_2, P_3, DP_4, M_1, and M_2 (cast). A–D) Neuse River, North Carolina; E) Hancock Ditch, Arkansas; F) Hanover Quarry, Pennsylvania.

exhibited the anterior molar of mastodon found by Mr. Nuttal (sic) near Newbern N. Ca."; for 6 August 1833, "Dr. Hays announced that he possesses the fossil tooth of a Tapir, from Newbern N. Ca. found by Mr. Nuttall, which the Dr. states to be the first instance of the detection of the remains of this animal on the North American continent." Obviously Hays (no friend to Harlan; Simpson, 1942:164–165)

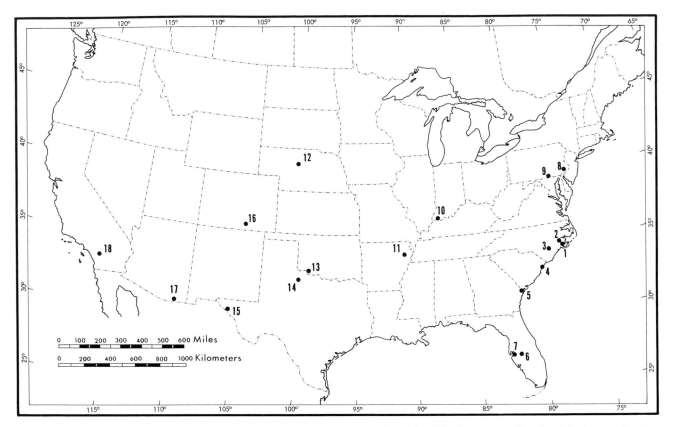

Fig. 3.—Distribution map for *Tapirus haysii*. Localities correspond to numbered localities in the text. Locality 1 is the type locality; locality 18 is the type locality for *T. merriami*.

had by that time already rejected *T. mastodontoides* Harlan 1825 as a tapir. One can only surmise the satisfaction with which Hays recognized the Nuttall specimen as tapir, especially plucked from under Harlan's nose, as Harlan had noted as early as 1835 (p. 267) that "Mr. T. Conrad possesses specimens [of *Equus*] from the *Pliocine* on the river Neuse, 16 miles below Newbern, N. C." and later wrote the following note: "April 19th, 1837. Examined a collection of fossil bones collected by Mr. Nuttall near Newbern, North Carolina," and listed the animals identified, which did not include tapir (*in* Conrad, 1838:11[13], footnote).

Unfortunately, Nuttall's field journals seem to have been destroyed after having survived for more than a century (Graustein, 1967:397), else we surely would have had much more explicit data direct from the collector, who never carried out his plan to publish extensively on the collection. However, as Graustein (1967:262) noted, Timothy Conrad and Richard Harlan reported on items from Nuttall's

collections, and he became acquainted in New Bern with Hardy B. Croom, a lawyer, plantation owner, fellow botanist, and fossil enthusiast.

Croom (1835:168–171), in a letter dated 12 September 1833, was the first to report on the richly fossiliferous deposits along the lower Neuse River, "on the estate of Mr. Benners, occupying the north bank of Neuse River, sixteen miles below Newbern." In his annotated list of specimens Croom mentioned two as having been given to Mr. Nuttall.

Timothy Conrad (1835:107–110) reported on a collection sent to him by Croom after the latter's publication of the same year, listing 67 species of mollusks, and later (Conrad, 1838:x[12]) stated, "on the left bank of the Neuse, about 15 miles below Newbern, in North Carolina, there is a vast collection of the rolled and water-worn bones . . . specimens of which are preserved in my cabinet, most of them a present from my kind friend, the distinguished Nuttall, who thoroughly explored this remarkable locality." He went on to quote Harlan's note reproduced in part above, and Harlan (1842: 143) later alluded briefly to "a collection of fossils obtained by Mr. Nuttall, in Newbern."

Mitchell (1842:128) stated that "We have also the grinder of an Elephant, found in the marl pits of the late Lucas Benners, Esq., 16 miles below Newbern, and other teeth not yet determined."

Charles Lyell (1843:39) recorded that "Mr. Nuttall discovered, on the Neuse 15 miles below Newburn, in South Carolina [corrected in reprint to Newbern, in North Carolina; Lyell, 1844:323], a large assemblage of mammalian bones. . . . Mr. Conrad presented Mr. Lyell with the tooth of a horse covered with barnacles, from this locality. Professor Owen has examined it . . . ." In the account of his second visit to the United States Lyell (1849:259) noted that "fossil horses were found by Mr. Nuttall on the banks of the Neuse, fifteen miles below Newbern, in North Carolina," and in a footnote, that "Mr. Conrad intrusted me with Mr. Nuttall's collection, and Mr. Owen has found among them the three species of Equidae . . . ." Some horse teeth from the locality found their way to the BM(NH) (Lydekker, 1886:89).

Although the Benners locality has fallen into latterday obscurity, it clearly attracted great attention in the 1830's and 1840's, and as late as 1923 (p. 358) Hay was able to say, "Doubtless the locality in North Carolina, the most important to the student of Pleistocene vertebrate palaeontology, is that reported long ago on the northern shore of Neuse River, 16 miles below Newbern."

There is no evidence that Nuttall collected fossils elsewhere in North Carolina, and the context of all contemporary reports indicates that reference to "Newbern" as a locality is merely shorthand for the Benners locality, 16 mi downstream. The one mention (Hays, 1852) of the source as the "bed of a canal" in North Carolina can be dismissed as yet another lapsus in view of the weight of evidence in favor of the Benners estate. Some Pleistocene vertebrate remains had been encountered in excavations for the Clubfoot and Harlow canal on the south side of the Neuse and slightly downriver, but very little was salvaged, the date (1817, Mitchell, 1842: 128) was too early, and there is no hint that Nuttall obtained material from there. Thus, the evidence seems sufficiently conclusive to establish the type locality for *Tapirus haysii* as the estate of Lucas Benners, river bank and adjacent marl diggings on north side of Neuse River, 16 mi below New Bern,

North Carolina, in what was then Craven Co., but since 1872 has been Pamlico Co. (Corbitt, 1950: 169).

Records of real estate transactions housed in the Craven County Courthouse, New Bern, show a 272-acre tract of land known as the "Bluff of Beards Creek" purchased by John Benners on 27 March 1767; a purchase of 100 acres on 16 December 1790 on the north side of Neuse River on both sides of the "old road that leads to Wilkinsons Point" and referring to "Smith Branch" as one landmark; an additional purchase in December 1792 of 228 acres on the north side of the Neuse "being part of a patent granted to Captain James Beard for 650 acres"; and finally, on 1 May 1804 the purchase by Lucas Benners of 640 acres at the head of Beard Creek. Richards (1950:41) also identified the Benners plantation as lying on the "east side of Bairds (sic) Creek." Richards (1936:1634) was mistaken in supposing that Mansfield's locality 15 mi below New Bern was near the Benners locality, as the two are on opposite sides of the Neuse River.

Although excavations have not been available in modern times, beginning in the early 1960's enormous numbers of vertebrate fossils, and lesser quantities of invertebrates, have been collected from the north bank and adjacent shelving bottom of the Neuse River by P. J. Harmatuk and sons, Mrs. Thelma Bennett, Raymond Douglas, Mr. and Mrs. Philip J. Kennel, and others. These collections have been made for the most part from the vicinity of the mouth of Beard Creek, at approximately 35°N, 76°52′W, on the Arapahoe 7.5-minute quadrangle, USGS, and downstream some 2.3 mi (3.7 km) to the vicinity of Smith Gut, approximately 34°59′N, 76°50′W, just upstream from Wilkinson Point, on the Cherry Point 7.5-minute quadrangle, USGS. This is undoubtedly the same segment of river in which Nuttall collected. Among the thousands of specimens recovered (some 1,400 horse teeth for example) are a few of tapir, representing animals the size of both *T. veroensis* and *T. haysii*. Thus, more than 150 years after the holotype was collected, we have available topotypes of *Tapirus haysii*.

Geologic age.—Simpson (1945:65) stated that "It is much the most reasonable assumption that the [type] specimen is from the Pleistocene, but even this is not absolutely certain, and of course it cannot be assigned to any particular part of that epoch." No reason was given for raising doubt as to its Pleistocene age, and we are aware of no evidence for pre-

Pleistocene tapirs of such large size in North America (unless all of the Blancan is regarded as Pliocene; cf. Savage and Russell, 1983:345, fig. 7–4).

Although stratigraphic distinctions were recognized as important from the beginning (Croom, 1835: 169) and considerable attention has been given to Neogene biostratigraphy along the Neuse below New Bern during the modern era, beginning with Dall (1892), generally accepted answers to critical questions on number and age of discrete stratigraphic units, faunal associations, stratigraphic relations on opposite sides of the Neuse, and stratigraphic nomenclature, are yet to be forthcoming.

The marl pits dug by Mr. Benners reached depths up to 25 feet, or 10 feet below the river's surface; the upper levels produced shark teeth, bones of marine fish, and sea shells; the lower, terrestrial vertebrates and sea shells of great variety (Croom, 1835: 169). In a collection from these pits sent by Croom at that time, Conrad (1835:107–110) identified 67 species of mollusks, including a few indicative of age no younger than the Yorktown Formation according to Druid Wilson (personal communication). Unfortunately, recent efforts to locate this collection at ANSP have not been successful. Interestingly, the Yorktown indicators were not included in a later list by Conrad (1842:191–192), but that list apparently was based on a collection made by Conrad himself, and he noted that the older of the two "formations" exposed in digging the marl was not accessible at the time of his visit as the pits had been refilled. He mentioned also that Mr. Benners had told him that all of the bones came from the younger deposit. He further noted that many of the bones are waterworn, black, and mineralized, and presumably transported down the ancient Neuse River to be deposited among marine shells. Foster (1857: 166), attributing his information to conversations with Conrad, stated that the terrestrial mammal remains came from an upper Tertiary stratum covered by a deposit of Pleistocene shells up to 15 ft thick, and that the much waterworn bones may have been redeposited from an older stratum. Stephenson (1912:268, 289) regarded both levels at the Benners locality as Pleistocene.

Unquestionable pre-Pleistocene fossils are known in modern collections from the immediate area and from the region. The large collections of vertebrate fossils made by Mr. Harmatuk and others from the Neuse River include marine vertebrates indistinguishable from those occurring in the Yorktown Formation at the Lee Creek Mine and in the Superior Stone Company Quarry at New Bern. This latter locality exhibits a basal rubble of Yorktown age including marine vertebrates similar in taxonomic composition and in preservation (mutilated, rolled, bored, encrusted, blackened). Marine mollusks of Pliocene age are known from the south side of the Neuse River across from the Benners locality, either redeposited (Du Bar and Solliday, 1963:226) or in place (Fallaw and Wheeler, 1969:52). Ward and Blackwelder (1980:25, 57) recognized evidence of their pre-Yorktown Eastover Formation, Cobham Bay member (upper Miocene), at their locality 48, based on mollusks collected by Mr. Harmatuk within the section of river of concern here.

Considering the restricted geographic area of the Neuse River estuary below New Bern and the number of geologists who have devoted attention to it, the interpretation of its geologic history seems inordinately confusing. The undoubtedly real biostratigraphic complexity is now buried beneath a manmade overburden of biostratigraphic nomenclature, which continues to be deepened and reworked. Complete review of the extensive and conflicting literature is beyond the scope of this paper and would in any case be inconclusive until the critical field work and mature analysis have been carried out.

For immediate purposes relating to the occurrence of tapirs and other terrestrial mammals at the Benners Estate, there is no evidence of pre-Pleistocene deposits in place and naturally exposed at surface.

The Pleistocene beds here were assigned to the Pamlico Formation of Stephenson (1912:286) by Richards (1936:1634; 1950:41) and would have been included in the original concept of the Croatan Formation of Dall (1892:205–206, 209). Most recently these beds have been referred to the Beard Creek member (Mixon and Pilkey, 1976:16) of the Flanner Beach Formation of Du Bar and Solliday (1963). The Beard Creek member is equivalent to the Neuse Formation of Fallaw and Wheeler (1969), both having as type locality the bluff adjacent to the mouth of Beard Creek. However, it now seems to be generally agreed that the name Neuse Formation should be suppressed in deference to the name Flanner Beach Formation (Wheeler et al., 1979:47; Blackwelder, 1981:12). Mixon and Pilkey (1976: fig. 4) showed the base of the Flanner Beach Formation (the base of their Quaternary, but not that of most authors) at zero to 10 ft below sea level along the Neuse River at the Benners locality. They regarded the Flanner Beach Formation as early Pleistocene, whereas others (for example, McCartan et al., 1982) regarded it as late Pleistocene. Literature cited in papers cited here will provide entree to the numerous publications on the geology of the area. Obviously, no conclusive statement about the source bed or its exact age can be made as yet, but it may be concluded tentatively that the terrestrial vertebrates are derived from the Beard Creek member of the Flanner Beach Formation of early or late (but not latest) Pleistocene age.

The field context is such as to inspire confidence that conclusive evidence lies fallow in the ground of the Benners plantation adjacent to the Neuse River. An excavation comparable to the marl pits of 150 years ago could yield a described and measured section, with stratigraphically controlled samples of vertebrates, macroinvertebrates, and microfossils. Such a study not only would close the gap between the meager, practically lost, promising beginnings of the early nineteenth century and the extensive (and expensive), relatively less fruitful work of the modern era, but also would aid in resolving much inconclusive contention.

Associated fauna.—The remains of quadrupeds identified among the early collections, mostly Nuttall's, included "teeth of the great mastodon . . . the hoof, horns, and vertebrae of an elk of great size and the teeth of an animal supposed to be the hyena" reported by Croom (1835:169; the presence of hyena was questioned at the time by the editor and has not been verified since, though of course quite possible). Harlan (in Conrad, 1838:xi[13]; repeated by Harlan, 1842:143) identified "hippopotamus, mastodon, elephas, elk, deer, horse, sus, seal, cetacea, tortoise, shark (several species), skate, snake, and fish." Lyell (1849:259) reported "*Equus curvidens, E. plicidens,* and a third species of the size of *E. asinus*" identified by Owen from Nuttall's collections.

No conclusive evidence has ever been presented of precolumbian hippopotamus or *Sus* anywhere in North America. Otherwise, with the exception of elk and snake, the modern collections from the Neuse River correspond well with these early indications. Preliminary study of the collections deposited in USNM by Mr. Harmatuk yields the following terrestrial mammals: *Megalonyx, Eremotherium, Castoroides, Hydrochoerus, Mammut, Cuvieronius*?, *Mammuthus, Tapirus* spp., *Equus* spp., *Nannipus, Platygonus,* a camelid, *Odocoileus,* and *Bison.* These are identified from float collections and very well may not constitute a unit fauna. Bruce MacFadden (personal communication) has identified as *Nannipus phlegon* some 38 teeth from the collection, indicative of an age no younger than Blancan.

The same segment of river bottom and bank has yielded marine vertebrates, including sharks, bony fishes, pinnipeds, and cetaceans, at least most of which are indicative of an earlier fauna, of Yorktown aspect.

On Harlan's behalf it may be mentioned that teeth of *Sus* are included in modern collections from this and other southeastern coastal localities, and some of them are blackened and very similar in appearance to teeth of known Pleistocene species. Further, the modern collections from the Neuse River include fragments of walrus tusks (thought to pertain to the Yorktown part of the assemblage). Similar specimens well might have provided the basis for inclusion of hippopotamus in Harlan's list, not an egregious error for the time.

Measurements.—Authors from Cope (for example, 1871:95) onward have remarked that *T. haysii* was distinguished from *T. terrestris* (=*T. americanus*) and later from *T. veroensis* by "size only," presumably implying inadequacy. However, size alone is sufficient if truly distinctive, and affords a convenient, testable criterion especially useful for isolated teeth of tapirs, the stereotyped character of which may be strongly suspected of masking species that would be readily separable on skulls. Thus, although individual variation in tooth size (to say nothing of the difficulty of correct placement in the dental series of many isolated teeth) and generally unknown variation with time and geography undoubtedly swamp out real distinctiveness in size by introducing spurious overlap from multiple sources, there yet seems to be two clusters in size that are sufficiently extreme to militate against assignment to a single extraordinarily variable species.

Simpson's (1945) unified samples of *T. copei* and *T. veroensis* showed almost no overlap in tooth measurements, but Lundelius and Slaughter's (1976) measurements of material referred to *T. veroensis* from various localities in Florida and Texas ostensibly resulted in expanding the observed size range of *T. veroensis* upward to engulf that of *T. copei* more or less completely, especially for the lower dentition (Tables 1 and 2). On the basis of their data and identifications the conclusion seems inevitable that only one species can be recognized in North America. However, Lundelius and Slaughter did not take that step. We have not been able to substantiate their data for the upper limits of their observed ranges. Further, it might appropriately be asked on what basis the larger specimens were assigned to *T. veroensis* rather than to *T. copei* (=*T. haysii*). In at least one instance a mandibular fragment (UF 8225) was included in their table of specimens of *T. veroensis* that had previously been identified as *T. copei* by Ray (1964). Although the distinctions in size between teeth of *T. veroensis* and *T. haysii* thus seem blurred at present, we feel that size probably is still a usable criterion and prefer to await confirmation of Lundelius and Slaughter's findings. The time is perhaps ripe for a new, unified statistical study of the teeth of North American Pleistocene tapirs, with particular attention to geographic and chronologic variation. There is for example, a large sample from Melbourne, Florida, divided among MCZ, USNM, and Amherst College, that has never been thoroughly studied, and there is a still growing sample from the Cooper River, South Carolina, that is suitable for quantitative study. A second sample comparable to that from Port Kennedy is much to be desired.

In addition to measurements given in Tables 1–3, dimensions of selected individual teeth are given in the text where those teeth are discussed.

Distribution.—As Simpson (1945:65) emphasized, citation of the name *T. haysii* in the literature cannot be relied upon as evidence that a large tapir is present at a given locality. This applies in large part to the post-1945 literature as well, in which large tapirs generally are listed as *T. copei.* We have attempted to compile a reasonably comprehensive record of distribution (Fig. 3) based on direct examination of new and old specimens and on critical evaluation of published accounts.

None of the Rancholabrean localities for *T. copei* listed by

Table 1.—*Dimensions (mm) of upper cheek teeth of* Tapirus veroensis *and* T. haysii. *Asterisks (*) indicate possibly mixed samples including deciduous predecessors or adjacent teeth. Parentheses indicate that the tooth is actually DP⁴, not P⁴.*

Dimension		ChM PV4257 left/right	Claiborne Cave, Tennessee, left (Bogan et al., 1980)	UF/FGS V277, holotype (Lundelius and Slaughter, 1976)	T. veroensis, Texas observed range (Lundelius and Slaughter, 1976)	T. veroensis, Florida observed range (Lundelius and Slaughter, 1976)	T. veroensis, Seminole field observed range (Simpson, 1945)	T. copei Port Kennedy observed range (Simpson, 1945)
P¹	L	16.2/16.2	19.2	17.5	19.2–20.4	17.5–20.8	16.9–20.4*	22.4–24.9
	W	16.1/15.6	17.8	14.9	14.5–19.6	14.9–18.6	14.5–17.4*	19.6–21.5
P²	L	17.8/17.7	20.3	18.7	19.8–20.5	18.7–21.1	18.5–20.8	21.9–24.0
	AW	20.0/20.0	20.0	20.9	21.1–23.9	19.6–23.2	19.3–22.9	25.5–26.5
	PW	21.0/21.3	24.0	22.8	22.4–25.9	22.8–25.8	21.0–25.5	27.4–27.9
P³	L	18.9/19.1	21.1	19.1	20.6–23.5	19.0–22.0	19.3–22.2*	22.7–24.5
	AW	22.7/22.6	24.0	24.5	25.8–27.6	22.0–26.3	22.8–27.6*	27.0–29.5
	PW	22.4/22.2	25.0	24.1	25.0–27.6	24.1–26.5	23.0–27.5*	26.1–29.0
P⁴	L	19.6/19.4	(21.7)	20.7	21.0–24.5	20.0–22.9	19.3–22.2*	24.1–26.4
	AW	24.3/23.9	(23.1)	26.9	29.4–30.3	24.2–28.6	22.8–27.6*	29.9–31.8
	PW	23.5/22.9	(21.7)	26.2	29.1–29.3	22.5–28.9	23.0–27.5*	28.4–30.1
M¹	L	21.1/21.2	22.8	22.4	22.6–25.6	20.2–23.8	19.6–27.1*	25.8–26.4
	AW	24.7/24.7	25.7	26.8	25.2–29.2	24.2–28.4	22.8–30.9*	28.9–31.1
	PW	23.0/23.2	23.3	24.0	24.0–27.2	22.3–26.2	21.3–27.7*	25.8–27.9
M²	L	24.0/23.6	25.6	25.2	25.7–27.4	23.5–27.0	19.6–27.1*	27.3–29.7
	AW	27.4/27.1	27.7	27.8	30.3–31.1	26.3–31.1	22.8–30.9*	31.3–34.9
	PW	23.9/23.7	24.4	26.1	26.5–29.0	23.8–28.4	21.3–27.7*	28.0–31.5
M³	L	—	—	25.1	23.7–28.5	23.5–26.1	22.4–26.0	26.8–29.2
	AW	—	—	29.0	27.6–32.9	28.0–32.1	25.5–31.4	31.0–34.1
	PW	—	—	24.6	22.3–29.1	23.1–28.1	20.4–24.8	26.5–29.0

Table 2.—*Dimensions (mm) of lower cheek teeth of* Tapirus veroensis *and* T. haysii. *Parentheses indicate uncertain measurements due to breakage. Asterisks (*) indicate possible mixed samples including deciduous predecessors. Measurements of the Hanover Quarry specimen, left ramus, were made by John E. Guilday on the original, not on the casts in CM and USNM.*

Dimension		AAS 69-658, Weona, Arkansas	Lehner Site, Arizona (Lance, 1959)	USNM 24588, Ladds, Georgia left/right	USNM 214663, Island Creek, North Carolina left/right	CM 30230, USNM 244439 Hanover Quarry, Pennsylvania	T. veroensis, Florida observed range (Lundelius and Slaughter, 1976)	T. veroensis, Seminole Field observed range (Simpson, 1945)	T. copei, Port Kennedy observed range (Simpson, 1945)
P₂	L	—	(26.7)	−/22.8	−/−	26.0	22.3–28.5	21.1–26.9*	24.7–27.0
	W	—	17.9	−/15.4	−/−	17.3	13.8–19.1	13.2–15.6*	15.5–17.6
P₃	L	—	(26.7)	−/19.9	19.6/20.2	23.1	20.0–24.5	20.6–20.9	23.0–25.1
	AW	—	(16.8)	−/15.7	15.0/15.2	18.3	15.1–17.5	16.8–17.6	16.1–18.0
	PW	—	21.4	−/16.8	16.6/16.6	20.4	16.0–21.4	18.3–18.9	17.8–20.2
DP₄	L	—	—	−/−	−/−	24.4		20.9–22.8	
	AW	—	—	−/−	−/−	19.0		15.2–16.5	
	PW	—	—	−/−	−/−	18.8		14.6–16.1	
P₄	L	—	29.7	20.4/20.3	19.6/20.2	—	20.2–25.2	22.0–24.6	24.1–24.9
	AW	—	20.4	17.7/17.9	17.9/18.1	—	15.9–20.6	17.7–20.9	18.3–21.7
	PW	24.3	(20.7)	18.0/18.4	18.4/18.3	—	15.9–22.0	19.5–22.2	19.5–22.8
M₁	L	22.6	25.4	21.6/21.9	20.2/21.0	24.2	20.0–26.9	21.7–25.0	23.4–27.0
	AW	21.4	20.8	18.0/18.6	17.9/17.8	21.5	17.0–22.0	17.6–19.3	19.8–22.9
	PW	19.0	20.6	16.3/16.9	16.3/16.3	20.2	17.5–22.8	16.1–18.9	18.7–20.8
M₂	L	26.0	(30.6)	23.6/24.1	23.4/23.5	29.1	22.5–29.0	22.6–26.6	27.4–30.8
	AW	22.6	(22.6)	18.6/18.8	18.8/18.8	22.1	18.4–23.4	18.9–19.7	20.5–24.1
	PW	21.4	(22.3)	17.4/17.7	17.4/17.5	21.9	17.8–22.8	18.0–19.7	18.3–22.8
M₃	L	29.1	33.7	25.1/25.8	25.0/−	—	25.4–32.2	25.7–28.3	30.1–31.5
	AW	23.0	24.2	18.6/19.1	19.1/19.2	—	19.0–23.1	19.5–20.5	21.7–23.8
	PW	19.9	21.1	16.9/17.0	17.0/−	—	17.0–21.7	17.4–18.6	19.2–20.5

Table 3.—*Measurements (mm) of some skulls of* Tapirus ver- *oensis. Dimensions A–H are those of Simpson (1945), I is that of Hershkovitz (1954). Values of A–E and G–I for holotype are after Simpson (1945:75). Figures in parentheses are minima, owing to damage.*

Dimension	UF/ FGS V 277, holo- type	ChM PV4257 Cooper River, South Caro- lina	SCMC SC 75.31.173 Cooper River, South Caro- lina	Clai- borne Cave, Ten- nessee
A. Prosthion-foramen magnum	379	345	—	—
B. Prosthion-choana	199	189	—	—
C. Prosthion-orbit	174	168	—	—
D. Sagittal crest-basisphenoid	113	112	108	114
E. Postauditory breadth	122	(110)	117	—
F. Oblique nasal length	(68)	(60)	—	—
G. Nasal breadth	(75)	63	—	—
H. Maxillary diastema	44	43	—	—
I. Gnathion-nuchal crest	407	367	—	—

Kurtén and Anderson (1980:293) is conclusive, and most are doubtful or incorrect. Gazin (1950:403) listed both *Tapirus* cf. *haysii* and *T. veroensis* for Melbourne, Florida, in his study of the USNM collection of that fauna, but provided no measurements, and neither we nor Lundelius and Slaughter (1976) have found evidence in it for a tapir other than *T. veroensis.* The Moore Pit, Texas, record (Slaughter, 1966:88) may prove to be valid, but as yet it is somewhat equivocal in that the M_2 (L 27.6, AW ca. 21, PW 21.7) on which it is based is within the expanded size range for *T. veroensis* as conceived by Lundelius and Slaughter (1976) as well as the observed range for *T. copei* (Table 2), and the same beds produced a metapodial smaller than its homolog in *T. terrestris.* The Mullen II, Nebraska, tapir was assigned to *Tapirus* cf. *excelsus,* not *T. copei,* by Schultz et al. (1975:12). We have been unable to find the El Paso Co., Texas, specimen, reported by Richardson (1907, 1909). Leidy (1859:106), who was well attuned to the significance of size in this context, stated that the isolated molar from Natchez, Mississippi, "corresponds in form and size with its homologue in the recent Tapir." Simpson (1945:63) reidentified as *T. veroensis* the tapir from Saber-tooth Cave, Florida, that he earlier reported as *Tapirus* cf. *haysii.* The Apollo Beach, Florida, specimen (Ray, 1964) seems properly assigned to *T. haysii,* but is not demonstrably Rancholabrean. Other, very late records, notably those for the Hancock Ditch, Arkansas, and the Lehner Mammoth Site, Arizona, are not so readily dismissed. These and other more or less reliably founded records known to us are noted here and are mapped by number in Fig. 3, proceeding generally east to west, starting with the type locality.

Benners Estate, north bank of Neuse River, Pamlico Co., 16 mi below New Bern, North Carolina (Fig. 3, locality 1). ANSP 11504, the holotype, a right P_4, L 26.3, AW 21.2, PW 21.6 (Fig. 2D). Contrary to Lundelius and Slaughter (1976:227), Simpson did not demonstrate that the tooth "was probably within the size range of *T. veroensis.*" In fact, it is a bit large for *T. copei* of Port Kennedy, but Simpson's (1945:66) expectation has been fulfilled that the observed range in size of teeth for the species as a whole would be greater than that in the homogeneous sample from Port

Kennedy. Topotypic specimens here assigned to *T. haysii* include: USNM 215062, left M^2 or M^3, L 28.9, AW 34.1, PW 30.1, Fig. 8E; USNM 215202, left M^3 (cast), L 28.6, AW 33.5, PW 30.4; USNM 347270, left M_2, L 29.8, AW 24.3, PW 23.0, Fig. 2A; USNM 215061, right M_3, L 33.2, AW 24.0, PW 21.4, Fig. 2C; USNM 243692, right M_2 (cast), L 33.3, AW 25.1, PW 23.0, Fig. 2B. Other fragments of teeth, including a posterior loph of a P_4, are of appropriate size for *T. haysii.* In addition some specimens in the assemblage are clearly referable to *T. veroensis,* but it cannot be asserted that the two occurred together in life.

Fort Barnwell, Craven Co., North Carolina (Fig. 3, locality 2). NCSM 132, right P_4, L 26.1, AW 20.5, PW 20.5. Richards (1950: 26, 42) reported "teeth" of *Tapirus haysii* along with remains of mastodon and a cetacean from a Pleistocene muck overlying the Yorktown Formation in marl pits on the property of Z. B. Broadway, one mi north of Fort Barnwell. We have been able to locate only the single tooth in NCSM and none in ANSP.

Billy B. Fussell Co., Inc., marl pit, 1 mi south and 0.5 mi west, of Rose Hill, Duplin Co., North Carolina, 34°48′00″N, 78°01′40″W (Fig. 3, locality 3). The Pleistocene vertebrate fauna here has been collected primarily by P. J. Harmatuk from overburden being cleared in preparation for marl excavation. The fauna as a whole is under study by R. E. Eshelman and C. E. Ray. Specimens assigned to *T. haysii* include: USNM 306473, left DP^1 (cast), L 21.3, W 21.3, Fig. 8A, B (original specimen); USNM 347321, left P^4, L 24.8, AW 30.9 min., PW 30.6; USNM 347321, left M^1, L 26.1, AW 29.1, PW—; USNM 347321, left M^2, L 29.4, AW 34.5, PW—; USNM 347322, left M^3, L 28.1, AW 33.1, PW 29.2; USNM 305213, left M^3, L 28.2, AW 34.2, PW 29.0. USNM 306473 is in a fragment of bone revealing the vestige of a crypt for P^1 beneath the tooth. The three teeth, P^4–M^2, assigned a single catalog number are thought to represent a single individual on the basis of matching pressure facets, compatible wear, and having been collected together on the same day in a small area of spoil. USNM 347322, though collected at the same time by Mr. Harmatuk, is from an older animal and is slightly different in coloration. All of these teeth are suitable in size for *T. haysii.* Two fragments of crowns from the same locality appear to represent smaller teeth, but their differing color and character suggest milk teeth.

Myrtle Beach area, Horry Co., South Carolina (Fig. 3, locality 4). This is a classic area for beach collecting. Material comes from the vicinity of Garden City Beach, Surfside Beach, and Myrtle Beach. Little has been done by vertebrate paleontologists regarding provenience and faunal associations in these deposits, but possibilities may exist for obtaining material in place. Formations of varied Pleistocene age occur in the area (McCartan et al., 1982).

Isolated teeth, mostly somewhat tumbled, polished, and fragmented, certainly span the size range of both *T. veroensis* and *T. haysii.* Specimens more or less certainly vouching for occurrence of *T. haysii* include: USNM 244383, right M_3, L 32.3, AW 24.0, PW 21.6; USNM 347323, right M_2, L 30.8, AW 22.7, PW 21.6.

Savannah, Chatham Co., Georgia (Fig. 3, locality 5). Material dredged from bottom of Savannah River. The specimens listed below are among the many vertebrate fossils placed in museum collections over a period of many years, at least from the 1930's to the 1960's, by the late Ivan R. Tomkins, an avid and able amateur naturalist, active mostly in the vicinity of Savannah. Specimens of *T. veroensis* size are known also from dredgings at Savannah, but the following are among those thought to represent *T. haysii:* ChM 55.103.113, right M^3?, L 29.3, AW 34.4, PW

27.5 min.; ChM 55.103.82, left M_1 or M_2, L 29.4, AW 22.3, PW 21.1.

Phosphoria Mine, IMC, 3.5 mi SW of Bartow, Polk Co., Florida; Sec. 2, T 31S, R 24E, Bradley Junction 7.5-minute quadrangle, USGS (Fig. 3, locality 6). UF 40060, right M^2, L 29.0, AW 33.3, PW 30.3. This well preserved, slightly worn tooth with roots intact was collected from spoil in a phosphate mine by Ben Torres, Jr., in 1975.

Apollo Beach, Hillsborough Co., Florida (Fig. 3, locality 7). UF 8225, fragmented mandible with left M_2 and right P_2–M_3; from dredgings of uncertain age. Reported by Ray (1964). Lundelius and Slaughter (1976:241) apparently saw only the left mandibular fragment, and regarded the tooth as M_1.

There are other specimens from Florida that fall within the size range of *T. haysii,* but the records thus far are unsatisfactory. For example, Simpson (1945:64) noted a fragment of right mandibular ramus, AMNH 23110, with part of M_2 and unerupted M_3 (L 30.5, AW 21.5, PW 19.6), in the Holmes collection, but lacking locality data other than Florida. Another specimen is NCSM 277, a moderately worn right lower tooth, probably M_2, L 30.6, AW 23.0, PW 21.9, from somewhere in Sarasota County. Undoubtedly more specimens of large Pleistocene tapirs from Florida are already in collections, or will come to light as new faunas, especially of pre-Wisconsin age, are found. The richness of the record in Florida and its southerly latitude make it the most likely source of the material needed for significant improvement of understanding not only of *T. haysii* but of fossil tapirs generally.

Port Kennedy, Montgomery Co., Pennsylvania (Fig. 3, locality 8). This locality provided the classic ANSP collection, still the only population sample of large Pleistocene tapirs from North America, and the basis for Simpson's (1945:66) *Tapirus copei.* The fauna is early Irvingtonian in age.

Hanover Quarry, York Co., Pennsylvania (Fig. 3, locality 9). In 1977 quarrymen recovered vertebrate fossils from a large fissure encountered in the Hanover Quarry. The find was reported to USNM from which it was referred to John E. Guilday at CM, who arranged to borrow selected specimens for casting, photography, and measurement. Remains of mastodon, horse, and tapir are known to have been recovered. The tapir material consists of a partial left mandibular ramus with P_2, P_3, DP_4 M_1, and M_2, and a right mandibular fragment with M_2. The original specimens were returned to the finders, who sold them to a private collector, but casts are housed in CM (nos. 30230 and 30231) and USNM (nos. 244439, represented in Fig. 2F, and 244440). Although left and right mandibular rami are given separate numbers, they almost certainly represent a single individual, judging from size, details of ossification, and stage of tooth eruption and wear.

All comparable dental dimensions (Table 2) place the specimen within the observed range for the Port Kennedy sample, and 12 of 14 measurements are beyond the observed range for Simpson's (1945) sample of *T. veroensis.* We assign the material tentatively to *T. haysii.*

Pigeon Creek, near Evansville, Vanderburg Co, Indiana (Fig. 3, locality 10). ANSP 11505, left P_4, L 22.1, AW 20.3, PW 22.9. This tooth is deeply worn, with extensive interdental pressure facets anteriorly and posteriorly that undoubtedly reduce L significantly. On this basis and on that of its great PW (cf. Table 2), it is referred tentatively to *T. haysii.* The tooth was reported by Leidy (1855:200), illustrated by him (1859: pl. 17, figs. 9 and 10) and the illustrations repeated by Hay (1912: figs. 38 and 39) who mistakenly identified the specimen as the type in his figure caption, but later (Hay, 1927:76) corrected his error.

A radiocarbon date of 9,400 ± 250 years (Rubin and Alexander, 1958:1484) has been cited repeatedly as the youngest date for tapir remains in the Quaternary of North America. This date was obtained from wood said to have been collected in 1870 from the same river alluvial stratum near Evansville from which Lincke, at least 16 years earlier, collected vertebrate fossils including the tapir tooth discussed here. This seems a tenuous association on which to base any critical conclusion.

Hancock Ditch (ditch no. 39), near Weona, Poinsett Co., Arkansas (Fig. 3, locality 11). AAS 69-658, left mandibular fragment with posterior half of P_4, and M_1–M_3 (Fig. 2E). The M_1 is deeply worn and thus secondarily reduced in length. The specimen was collected by Phyllis Morse on 30 May 1970 with mastodon and possibly sloth from a clay deposit in a braided stream system thought to be late Pleistocene in age (Dan F. Morse, personal communication). The specimen was identified as tapir by the late C. W. Hibbard and reported by Morse (1970), who also described the field context.

The measurements of this specimen (Table 2) make its specific assignment somewhat equivocal. The length of crown in all three molars is too small for *T. haysii,* but all transverse dimensions are quite large enough, and the bony ramus is very robust. We assign the specimen tentatively to *T. haysii.* This and other specimens make it very clear that much remains to be learned about ranges in size and time of North American Pleistocene tapirs.

Mallory Sand and Gravel Pit, near Thedford, Thomas Co., Nebraska (Fig. 3, locality 12). Incomplete mandibular ramus with roots of M_2 and M_3, recorded as *Tapirus* sp., "a very large species, possibly *T. copei,*" by Schultz et al. (1975:13). They indicated that early Pleistocene fossils had been collected at the site, but that the horizon of the tapir was uncertain.

Holloman gravel pit, one mi north of Frederick, Tillman Co., Oklahoma (Fig. 3, locality 13). A fragment of right mandibular ramus with DP_3 (L 25, W 18.5), DP_4 (L 26, W 21.4), P_3 (L 27), P_4 (L 26), and M_1 (L 27.5, W 23.5), was reported and illustrated by Hay and Cook (1930:17–18; pl. 4, fig. 3). Their measurements indicate an animal larger than any in the Port Kennedy sample (cf. Table 2). Dalquest (1977:262) reported that the specimen is now in the collection of Midwestern State University. The Holloman local fauna is regarded as earliest Irvingtonian, comparable in age to the Gilliland local fauna (Dalquest, 1977:256).

Gilliland local fauna, Knox Co., Texas (Fig. 3, locality 14). Two partial mandibular rami of suitable size for *T. copei* were reported by Hibbard and Dalquest (1966:39–40; fig. 5E), who remarked on the similarity in geologic age of the Gilliland and Port Kennedy faunas (both indicative of early Kansan, Irvingtonian age) and on the wide geographic distribution of *T. copei* in early middle Pleistocene time.

Campo Grande Arroyo, Hudspeth Co., Texas (Fig. 3, locality 15). A weathered, partial palate of an old individual with deeply worn, incomplete dentition was reported as *Tapirus* cf. *copei* by Strain (1966:48–50; pl. 13), seemingly on good grounds. The specimen was found in the lower part of the Camp Rice Formation, regarded as Aftonian, late Blancan, in age (Strain, 1966: 21).

Donnelly Ranch, Las Animas Co., Colorado (Fig. 3, locality 16). Four upper teeth were referred to *T. copei* on the basis of their large size by Hager (1974:13), who regarded the fauna as of pre-Nebraskan, late Blancan age.

Lehner Mammoth Site, near Hereford, Cochise Co., Arizona (Fig. 3, locality 17). Lance (1959:37) identified as *Tapirus* sp. a crushed left mandibular ramus with P$_3$–M$_3$, an isolated right P$_2$, and a fragment of upper molar, representing a very large tapir. As Lance pointed out, the Lehner specimen exceeds in some dental dimensions the upper limit observed in the Port Kennedy sample of *T. copei* (cf. Table 2). However, it can be matched or exceeded in all available dental measurements among specimens here referred to *T. haysii*. Radiocarbon dates for the Lehner Site, hovering about 11,000 years, provide a seemingly reliable late date for this large tapir. Lance suggested an adaptation to drier conditions to explain the occurrence of tapirs in the Quaternary of the Southwest.

Bautista Creek badlands, northeast of Hemet, Riverside Co., California (Fig. 3, locality 18). As already noted, this is the type locality of *T. merriami*, a very large tapir known from a single fragment of mandible with fragments of two teeth (Frick, 1921), which we regard as probably conspecific with *T. haysii*. The fauna is Irvingtonian in age (Kurtén and Anderson 1980:26).

Additional records, at least one already in other hands for study, will be needed to bridge the geographic gaps in distribution of large Pleistocene tapirs between the Pacific coast and the remainder of the conterminous United States. Of course much more and better material will be needed to resolve problems of specific relationships, and better documentation and faunal associations will be needed to resolve time-stratigraphic distribution, which is puzzling at present.

Much of the limited evidence now available suggests that *T. haysii* is a species at least especially characteristic of, if not limited to, the early and middle Pleistocene (late Blancan and Irvingtonian; Savage and Russell, 1983, include all of the Blancan in the Pliocene). Most of its best-documented occurrences to date fit this pattern and most other records are permissive. However, there are a few records (notably the Hancock Ditch, Arkansas, and Lehner Site, Arizona) that are both very late Pleistocene and more or less conclusively identified. Hibbard and Dalquest (1966: 40) and Hibbard et al. (1978:38) commented on the suggestively early (pre-Yarmouth) distribution, and the latter authors also suggested the possibility of niche partitioning, with *T. haysii* (=*T. copei*) occupying drier, more open country, and *T. veroensis*, wetter, more forested habitat. This would be compatible with Lance's (1959:39) suggested explanation of occurrence at the Lehner Site. Smaller tapirs more or less certainly assigned to *T. veroensis* are the common form in Wisconsinan faunas. If the large tapir of the late Blancan and Irvingtonian is the same species as that of the terminal Wisconsinan, where was it during most of Rancholabrean time?

Taxonomic status.—The name *Tapirus haysii* was widely and often loosely applied to North American Pleistocene tapirs until Simpson (1945:66) concluded "that this type is essentially indeterminate and that the species to which it belongs, and consequently the species properly called *T. haysii*, cannot at present be identified." He then erected the new species, *T. copei*, based on the sample from Port Kennedy previously referred to *T. haysii*, and diagnosed it as follows: "essentially the species called *T. haysii* by Cope (1899) and most later authors. Larger than any recent species or than any other North American Pleistocene species except *T. merriami*. Significantly smaller than *T. merriami*. P^1 a large, robust, relatively transverse and complex tooth with the protocone relatively far forward, a large, heavily ridged basin between this and the ectoloph. P^2 advanced in molarization, difference in anterior and posterior widths slight, protoloph fully developed." He has been followed almost unanimously in applying the name *T. copei* to remains of large tapirs found subsequently in North America.

Simpson recognized the probable fallibility of both tooth size and premolar characteristics in distinguishing species. As discussed elsewhere in this paper, his expectations have been realized. The continuing poor representation of anterior upper premolars of large tapirs has meant that size necessarily has been the practical criterion by which *T. copei* and *T. veroensis* have been distinguished in every study since 1945. Its efficacy has been widely doubted, but its application continues.

All workers on fossil and modern tapirs have been appropriately impressed by their conservatism (see Radinsky, 1965:69, 101). Their differentiation in Quaternary time is little reflected in their dentition. The modern occurrence of three species in northern Colombia should instill caution in all paleontologists attempting to distinguish species on the basis of isolated teeth and fragments of jaws. A strict constructionist could make the case that only *T. veroensis* among described North American Pleistocene tapirs is presently identifiable, and even that, only by virtue of cranial characters.

Thus, in spite of the relatively rich sample, *T. copei* could be faulted by standards similar to those on which *T. haysii* was rejected, because skull characters are not available. It would be convenient to have better type specimens for many paleotaxa, but stability is served best by conserving named taxa if at all possible. General application of standards comparable to those on which *T. haysii* was supplanted by *T. copei* would lead to rejection of many, perhaps most, nineteenth century types in vertebrate paleontology. Such summary rejection based on rigorous criteria would result in chaos. Paleotaxa are generally more or less provisional pending discovery of new material. Skulls of large tapirs from the east and skulls of any size from the far west will be required to place the specific taxonomy on a sound basis, and these will surely be forthcoming sooner or later. Meanwhile, *Tapirus haysii* seems to be usable, pro tem.

Tapirus veroensis Sellards 1918

Skull.—As Simpson (1945:56) observed, Sellards (1918) established the species securely on the basis of the very fine holotype skull, now UF/FGS V 277. Since then several skulls and partial skulls have been found, including those of the closely related, probably conspecific *T. excelsus.* Each of these has stimulated studies contributing in turn to progressively better understanding of North American Pleistocene tapirs (Simpson, 1945; Parmalee et al., 1969; Lundelius and Slaughter, 1976). Most recently, Bogan et al. (1980) have reported the major part of a skull from a cave in Tennessee. Through the kindness of its owner, Mr. Donn Claiborne, and of Paul F. Parmalee, we have had this skull available for further study in conjunction with previously unreported new specimens from South Carolina.

In recent years collecting by divers in the rivers of South Carolina in the vicinity of Charleston has yielded numerous fine specimens of fossil vertebrates, many of which have been deposited in the collections of ChM and SCMC. The finest specimen known to us, an essentially complete skull, now ChM PV 4257, was found by Susan A. Wallace on the bottom of the West Branch of the Cooper River in Berkeley Co., South Carolina, approximately 20 mi (32.2 km) due north of Charleston, during the summer of 1976 (Fig. 4). Miss Wallace, a Charleston SCUBA-diving instructor, discovered the skull while diving for fossils near the south bank of an eastward course of the river at 33°4.4′N, 79°56.2′W (USGS Kittredge 7.5-minute quadrangle), approximately 1.1 mi (1.8 km) southeast of Strawberry Landing. According to Miss Wallace, the specimen was found in nearly 10 ft of water on a terrace that gently slopes to the main channel of the river, which is bedded in the Ashley Member (upper Oligocene) of the Cooper Formation, a highly-indurated, calcareous marine formation that underlies the entire Charleston area. The terrace has resulted from subaqueous erosion of overlying sediments from the surface of the Cooper Formation between the main channel and the south bank of the river.

Miss Wallace recalls that the skull was lying directly on the surface of the "marl" (the Cooper Formation), where it had come to rest after having been washed from its matrix. The stratigraphic origin of this specimen is highly uncertain. Recent studies by the USGS indicate that in the immediate area of the discovery the Cooper Formation is unconformably overlain by early (?) Holocene sediments and that the nearest Pleistocene deposits are those of the Wando Formation (upper Pleistocene; McCartan et al., 1980:114–115), which has been mapped at a location 0.15 mi (0.2 km) upstream from the tapir skull site (Robert E. Weems, personal communication). There being abundant evidence of *Tapirus* in the Pleistocene of the eastern United States and (to our knowledge) none from the Holocene of this region, we might assume a priori that the skull was washed out of the aforementioned Wando Formation deposits and subsequently transported by the river currents downstream to its point of recovery. However, Susan Wallace, who has logged many hours of diving in this stream, has observed that the currents are swiftest near the surface of the river and are of such negligible velocity at the bottom that objects as light as bottles usually remain at their initial point of contact with the bottom. She considers it quite unlikely that an object of the size of the tapir skull would have been transported by bottom currents over the distance from the Wando deposits upstream. Her opinion is supported by consideration of the excellent condition in which the skull has been preserved, showing no signs of having been rolled about on the hard surface of the Cooper Formation at the bottom of the river. Thus, it seems almost certain that PV4257 came from a stratum of the river bank in the immediate vicinity of its discovery by Miss Wallace. That conclusion suggests an early Holocene date for this specimen based on USGS studies of the post-Cooper sediments in the Charleston area. There is, however, a distinct possibility, if not a probability, that a lens of Wando-age (Sangamonian) sediments is concealed beneath the Holocene deposits bordering the river at the tapir skull site and that it was that stratum that yielded the specimen (R. E. Weems, personal communication). Until the fact of the matter can be determined we can state only that PV4257 is of late Pleistocene or early Holocene age. Whatever its age, we are extremely grateful to Susan Wallace for making this splendid specimen permanently available for scientific study and for her crucial information about the stream in which she found it.

A well preserved braincase, also from the Cooper River, was made available to us by Rudolph Mancke from the SCMC collections, specimen number SC 75.31.173 (Fig. 5). SCMC also has several maxillae and mandibles not utilized in this study, but which

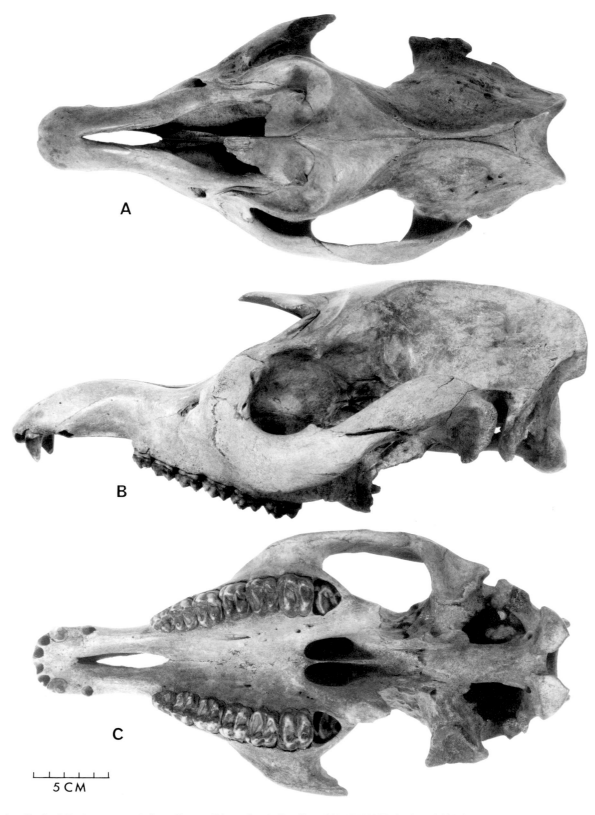

5 CM

Fig. 4.—Skull of *Tapirus veroensis* from Cooper River, South Carolina, ChM PV4257, in dorsal (A), lateral (B), and ventral (C) aspects, ×0.4.

constitute a very useful population sample for future study.

The Claiborne Cave (Fig. 6) and Cooper River specimens provide the basis for further refinement of our understanding of skull characters in *T. veroensis,* in particular with regard to ontogenetic and individual variation. There is now a fairly continuous ontogenetic series of skulls available, beginning with the holotype of *T. excelsus,* a juvenile retaining all deciduous premolars, and with M^1 in use and M^2 unerupted (Simpson, 1945:71). A fragmented, incomplete skull from Branford, Florida, UF 14056, had P^1–P^3 in use, DP^4 retained but well worn with advanced P^4 underneath, M^1 slightly worn, and unworn M^2 above the alveolar margin but not fully erupted (this specimen is dentally younger than suggested in the caption to Fig. 1B of Lundelius and Slaughter, 1976:228, and is recorded there under the catalog number for an Itchtucknee River specimen). The dentition of the Claiborne Cave, Tennessee, specimen is at almost exactly the same stage of wear and development (Fig. 8F). DP^4 is the only cheektooth in which the enamel has been breeched by wear. The incompletely formed crown of M^3 is visible through a small window in its small crypt. Next in dental ontogenetic stage are the skulls from Crankshaft Cave, Missouri (Parmalee et al., 1969), and from the Cooper River, South Carolina, reported here (Fig. 7). These two skulls have permanent P^1–M^2 fully erupted and worn, with M^3 well formed and visible through a large window in its large crypt. Finally, the holotype of *T. veroensis* is a mature skull with all adult teeth fully erupted and worn. The skull from the Livingston Dam Site, Texas, is similar in age or very slightly younger (Lundelius and Slaughter, 1976:228). This series is especially valuable for interpretation of development of the temporal crests.

In the discussion of the skull that follows, primary attention is given to the fine skull from the Cooper River, ChM PV4257, with comparison to other specimens and commentary on previous studies as appropriate. Several excellent studies of Quaternary tapir skulls have been published. Those found especially useful for our purposes are Hatcher (1896), Sellards (1918), Simpson (1945), Colbert and Hooijer (1953), Hershkovitz (1954), and Lundelius and Slaughter (1976). The work of Radinsky (for example, 1965), focused primarily on earlier tapiroids, has been useful especially for the broader context. In testimony to the perceptiveness of these investigators, we have found few characters in the tapir

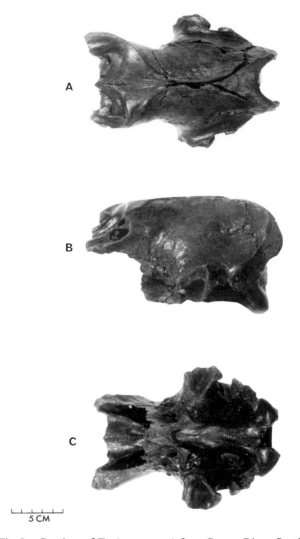

Fig. 5.—Cranium of *Tapirus veroensis* from Cooper River, South Carolina, SCMC SC 75.31.173, in dorsal (A), lateral (B), and ventral (C) aspects.

skull that had not been thoroughly described or at least touched upon by our predecesors. Thus, originality is claimed for little that follows, even though, for brevity, prior work is cited generally only where necessary, for example where our observations are contradictory or supplementary. Discussion of skull characteristics is organized according to the normal aspects of the skull in which they are most clearly revealed, in the sequence dorsal (Figs. 4A, 5A, 6A), lateral (Figs. 4B, 5B, 6B), and ventral (Figs. 4C, 5C, 6C), in each case proceeding from anterior to posterior.

Hershkovitz (1954:469) indicated that the living Neotropical tapirs, with the possible partial excep-

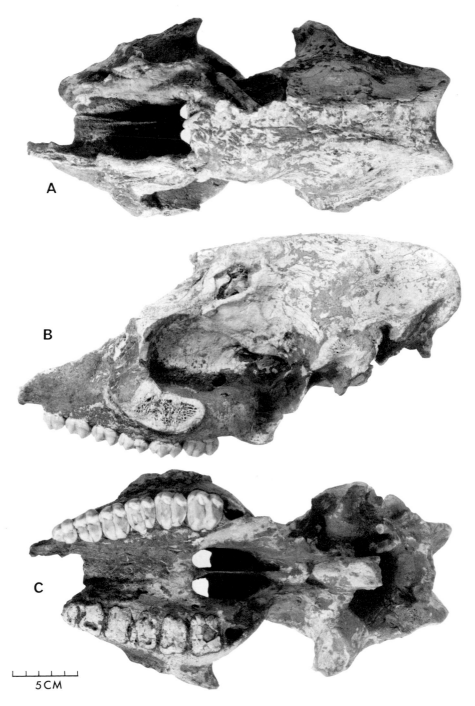

5 CM

Fig. 6.—Partial skull of *Tapirus veroensis* from Claiborne Cave, Tennessee, private collection of Donn Claiborne, in dorsal (A), lateral (B), and ventral (C) aspects.

tion of *T. bairdii,* have reached their definitive adult size in terms of skull length by the time that M² has erupted. The only two skulls of *T. veroensis* on which Hershkovitz's measurement is possible do not fit this pattern (Table 3). The Cooper River specimen, with M² in use is 367 mm long, and the holotype, with M³ in use, is 407 mm long. In handling numerous specimens of fossil (mostly fragmentary) and

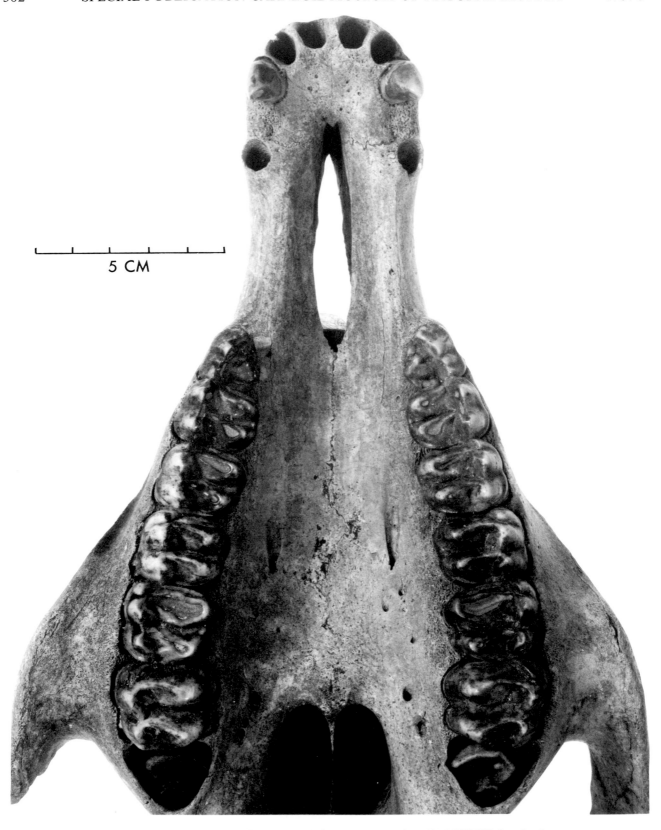

5 CM

Fig. 7.—Dentition of *Tapirus veroensis* from Cooper River, South Carolina, ChM PV4257, in palatal aspect, ×1.

modern tapirs through the years, one of us (Ray) has formed the subjective impression that specimens with all permanent teeth in use except M^3 and M_3 are disproportionately well represented in collections. This further suggests that this is a period (young adulthood) in the life cycle either of unusual vulnerability or of slower dental development (thus representing a longer opportunity for sampling).

In dorsal aspect (Fig. 4A), the posterodorsally ascending process of the premaxilla terminates posteriorly in an acute point above the anterior margin of P^1, as in the holotype. The details of shape and posterior extent of this process are variable in the living species, but are generally similar to *T. veroensis* in *T. terrestris, T. pinchaque,* and *T. indicus*. *T. bairdii* is dissimilar to all in having a shorter posterior process and in having its medial border oriented highly obliquely, resulting in a bluntly transverse posterior termination. Hatcher (1896: 172–173) used these differences as key characters in distinguishing *bairdii* (with *dowi*) generically from the other living tapirs.

The maxilla is exposed narrowly posteromedial to the posterior process of the premaxilla, less broadly than in the holotype. This is the "superior branch of the maxillary" of Hatcher (1896:173), but its characteristics do not seem to be as profoundly diagnostic as implied in his key.

There is no suggestion of a dorsal flange of the maxilla for embracement of the vertical mesethmoid cartilage as seen in extreme development in *T. bairdii,* in which this flange rises dorsomedially to roof over the entire bony muzzle posteriorly as far as the infraorbital foramen, leaving only a uniformly narrow sagittal slit occupied by the mesethmoid. This flange is fully formed even in very young animals, and in older animals may coossify in part with the mesethmoid (for example, ANSP 18873, USNM 6938, 14217, and 283704). Colbert and Hooijer (1953:84) mistakenly stated that this flange is absent in American tapirs, whereas it finds its extreme development in *T. bairdii,* in which the flange develops a vertical, planar, medial border to fit against the mesethmoid. The flange is modestly developed in *T. indicus* and in *T. augustus*. In *T. indicus* it is developed only anteriorly, medial to the posterior process of the premaxilla, as a low, vertical flange with a feathered or serrated edge. A similar flange is weakly suggested in some individuals of *T. terrestris,* in which a slight, feather-edged, upturned flange is developed on the maxilla medial to the posterior process of the premaxilla (for example,

USNM 239966 and 406847, neither superannuated individuals). Most individuals of *T. terrestris,* however, show little or none of the maxilla medial to the posterior process of the premaxilla. We have seen no suggestion of a medial flange in any individual of *T. pinchaque,* which most closely resembles *T. veroensis* in this region.

Posterior to the termination of the premaxilla the dorsomedial border of the maxilla is smoothly rounded, directed medially, and posteriorly divergent. It is more extensive in the holotype, possibly related to greater individual age. *T. indicus* is most like *T. veroensis* in this region, differing primarily in the tendency to develop a very slight ventrally directed edge, not a rolled edge. *T. bairdii* is of course very different, as described above. In *T. terrestris* and to a lesser degree in *T. pinchaque* this maxillary border is rolled ventrally and the laterally adjacent body of the maxilla is modestly inflated, to define together a semi-enclosed chamber on either side of the bony snout anterior to the infraorbital foramen.

A long, strong, posterodorsally ascending process of the maxilla borders the narial aperture along its anteroventral half, anterior to the orbit. This process is widely exposed dorsally, medial to the anterior supraorbital flange of the frontal and lateral to the anteroventrally descending process of the nasal, which borders the narial aperture along its posterodorsal half. The ascending process of the maxilla is very similar to that of *T. veroensis* in *T. terrestris, T. pinchaque,* and *T. indicus,* although in *T. terrestris* it tends to twist more strongly laterally in its dorsal portion where it borders the descending process of the nasal and forms the medial wall of the lateral groove. *T. bairdii* is very different in that the posterodorsally ascending process of the maxilla has no broad, anterodorsally exposed surface, but is instead a vertically oriented lamina lying almost entirely inside the narial aperture, of which it forms the lateral border along with the frontal. A descending process of the frontal takes the place of that of the nasal, which is usually absent in *T. bairdii* (observed on the right side only of one young individual, USNM 102522).

A detail of the ascending process of the maxilla seemingly unique to *T. veroensis* is the presence of two well marked creases extending from above the infraorbital foramen posterodorsally, anterodorsal and medial to the lacrimal bone. In *T. indicus* this area is occupied by a single, wide trough, not two grooves. In *T. bairdii* there is generally a single lateral crease developed on the lacrimal and extending posterodorsally between the frontal and the ascending process of the maxilla. There are no similar creases in *T. terrestris* and *T. pinchaque.*

Everyone who has worked with tapir skulls has devoted attention appropriately to the conspicuous groove or channel which accommodates the dorsolateral nasal diverticulum on each side in life. This groove is surely important, but highly variable and difficult to characterize. In the Cooper River specimen it is very similar to that in the holotype of *T. veroensis.* In plan, each groove resembles nothing so much as an inverted bass clef of the musical staff. The anterior outlet lies on the posteriorly ascending process of the maxilla on either side of the cavernous narial aperture. From there it passes posterolaterally onto the supraorbital flange of the frontal bone, thence posteromedially onto the dorsal table of the frontal bone, on which lies its greatest part and on which it swirls posteromedially and then anteriorly to terminate on the dorsal surface of the nasal at its posterior end. Efforts have been made to use this scroll in taxonomic distinctions, usually based on depth and position of the groove, proximity of approach to one another at the midline, and extent to which the termination lies on the frontal or nasal bone. In spite of great variation, some general features or tendencies seem usable. In *T. veroensis* the scrolls approach one another very closely at the midline (but do not meet as Simpson, 1945:74, asserted), especially in SCMC SC 75.31.173 (Fig. 5A). In this they resemble *T. bairdii* and *T. indicus,* and differ from *T. terrestris* and *T. pinchaque,* in which they are widely separated.

The depth of the groove is highly variable in modern tapirs, and somewhat so in the few specimens of *T. veroensis.* Throughout its length it is somewhat more deeply impressed in the holotype than in ChM PV4257, in which it is shallowly but distinctly demarcated except at its anterointernal termination on the nasal, where it fades away anteriorly. In SCMC SC 75.31.173 it is strongly impressed as far as preserved, especially laterally, where it forms a deep channel supraorbitally. *T. bairdii* exhibits extraordinary variation in this region. One specimen, USNM 179064, has no lateral groove at all, and only a faint depression on the frontal and nasal where the internal swirl should be. Other individuals collected at the same time and in the same area have a very deep groove. Two extremes of development of the lateral groove may be observed in two skulls taken at the same time and place in Costa Rica; one a young adult with M² erupted, USNM 14218, has

no lateral border at all, making the feature a lateral shelf (more as in *T. indicus*), not a groove; the other, an adult with M³ in place, USNM 14219, has a high lateral wall formed by the supraorbital flange of the frontal, creating a deep lateral groove. The few sexed specimens available give no suggestion of sexual dimorphism in this character, nor does age seem to be a factor. Presence or absence of the bony wall simply may not be functionally significant.

T. indicus never has a lateral, supraorbital wall to the lateral groove, which is developed as a supraorbital shelf. The groove terminates internally in *T. indicus* in a very deep, rounded cup at the end of the scroll, developed mainly on the posterior part of the nasal.

The nasals are incomplete in ChM PV4257, but the dorsoventral thinning anteriorly of the left nasal suggests that very little is missing from its anterior tip. To the extent preserved, it is very similar to the nasals of the holotypes of *T. veroensis* and *T. excelsus.* In all of these, the outline of the complete paired nasals would resemble a stylized St. Valentine's Day heart. As Hatcher (1896: pl. 3, figs. A–E) and Simpson (1945:49–50, 73) have shown, the nasals of modern tapirs are exceedingly variable, and difficult to use taxonomically. Our observations confirm this at least for *T. bairdii* and *T. terrestris.* In *T. bairdii* some individuals have only vestigial nasals; some have asymmetrical nasals ossified from multiple centers; some old individuals have long nasals that blend indistinguishably into the ossified mesethmoid (for example, USNM 6938, 14217, and 283704). In some individuals of *T. bairdii* there is no indication of a nasal at all, its place being taken apparently by an anterior projection of the frontal. This condition was observed both in older (USNM 11636 and 14485) and younger individuals (USNM 11281 and 266851) and in one apparent newborn (USNM 21088). In view of the variability in this region in *T. bairdii,* it would not be surprising if those individuals had "floating" nasal ossifications that were lost in preparation of the skulls. In *T. terrestris,* nasal length varies greatly, at least from 80 mm (USNM 270353) to 118 mm (ANSP 2982) in adults. Nevertheless, some tendencies may be usable. For example, all *T. pinchaque* skulls known to us have very long, slender nasals; all *T. indicus* have uniformly short, lyriform nasals. The few observations possible thus far suggest that *T. veroensis* consistently has shorter nasals than any other American tapir.

Hershkovitz (1954:470, 478, 489) described the

"descending sigmoid process" of the nasal in the living Neotropical tapirs. This is the process descending anteroventrally along the upper part of the lateral margin of the narial aperture, alluded to previously in discussion of the ascending process of the maxilla. It is present and extends forward as far as the anterior margin of the orbit in *T. veroensis*. It is generally similar in *T. pinchaque* and *T. terrestris,* although in at least one juvenile *T. terrestris* (ANSP 4725) it appears that the process comes from the frontal, as it does in some *T. indicus* (ANSP 2985) and normally in *T. bairdii*. In one juvenile *T. bairdii* there is a short descending nasal process on the right side only (USNM 102522).

The extent of the dorsal table of the frontal bones exposed anteromedial to the anteriorly diverging temporal lines or crests has been brought forward as a taxonomic character. All available specimens of *T. veroensis* resemble one another in this area, and are most similar to *T. pinchaque* among living tapirs. In both species the temporal lines of adults converge posteriorly approximately at the midsagittal frontoparietal boundary. From this point forward, the temporal lines diverge more obliquely than in *T. indicus* and much more so than in *T. bairdii,* in which they are subparallel. In younger individuals of *T. veroensis* in which the temporal crests have not converged to approximate a sagittal crest this dorsal table continues posteriorly on the parietals, as it does through life in *T. indicus* and *T. bairdii*. *T. terrestris* is grossly dissimilar in this region even in juvenile animals, owing to its extraordinary and precocious development of a high, thick, true sagittal crest, which extends forward onto the frontals, leaving only a very small dorsal frontal exposure anteromedial to the divergence of the temporal crests. Perhaps correlated with these differences, *T. terrestris* has its minimum postorbital width of the frontals far anteriorly, whereas *T. veroensis* has it far posteriorly, adjacent to the braincase, as is the tendency in living tapirs other than *T. terrestris*.

The living species of tapirs differ greatly from one another in characteristics of the sagittal crest or parasagittal lines or crests. These features are well known and thoroughly described and illustrated in Hatcher (1896), Simpson (1945), and Hershkovitz (1954).

In "normal" mammals, for example in many carnivores, it can be confidently assumed that temporal crests will approach one another more closely with increasing age of an individual and that a sagittal crest, if any, will form and increase in accentuation in adulthood. Modern tapirs are exceptional in that juveniles tend to look like miniature adults not only in general (Simpson, 1945:50), but specifically in this feature. *T. terrestris* develops its strikingly high, long, thick true sagittal crest early in life. In contrast, *T. bairdii* and *T. indicus* never develop a sagittal crest, but retain a low, broad dorsal table between widely separated parasagittal crests throughout life. Adult individuals of *T. pinchaque* have a low, relatively short sagittal crest, developed only on the parietals. Only recently has information been published on the condition in youth, a single individual with M^2 already erupted (Lundelius and Slaughter, 1976:fig. 1C). In a young female skull of *T. pinchaque* (ANSP 19161) with all deciduous premolars and M^1 in place, the sagittal crest is fully formed across the fused posterior half of the parietals, and on the anterior half closely approximated parasagittal crests are narrowly separated by a groove. In an adult male (ANSP 19160) with M^3 near eruption the parietals are not completely fused in the midline and have a sagittal groove between them; the sagittal crest is lower and shorter in anteroposterior extent than in the younger female.

Simpson (1945) founded *T. excelsus* on the basis of a juvenile skull which differed most profoundly from the holotype of *T. veroensis* in its widely separated, subparallel temporal crests separated by a broad intertemporal table. The species was named on the basis of rigorous application of sound procedure, using analogy with the most closely related taxa, in this case congeneric. Of the three adequately known living relatives, one showed that, if a sagittal crest was to develop at all, it did so very early, and two showed that retention through adulthood of widely separated temporal crests could be expected. Only with subsequent accumulation and analysis of specimens of various ages did it become clear that this region of the skull in *T. veroensis* (and probably in *T. pinchaque*) developed as in a "normal" mammal, and not as in the well known species of *Tapirus*. Lundelius and Slaughter (1976) thus made a convincing case for regarding the holotype of *T. excelsus* as a juvenile of *T. veroensis,* and our material corroborates their interpretation. In the two specimens from South Carolina the segments of the temporal crests on the parietal bones approach one another along the sagittal plane and virtually touch adjacent to the open suture between the parietals, as in the skull from Texas (Lundelius and Slaughter, 1976: 237; fig. 3B). In the holotype of *T. veroensis* the parietals are coossified sagitally and the crest is essentially sagittal. In successively younger individuals described and illustrated by Lundelius and Slaughter (1976) the crests are successively farther apart, as would be expected. The skull from Claiborne Cave, Tennessee, fits this regression in that its temporal crest condition and its dental age (Figs. 6A and 8F) are similar to those of the specimen from Florida illustrated by Lundelius and Slaughter (1976:fig. 1B). Thus the known skulls of *T. veroensis* show complete concordance in dental development and temporal crest convergence. Some degree of individual variation undoubtedly will be evinced in future finds.

Interparietal bones are notorious in mammals for their variable incidence, size, number, and shape. Their occurrence is difficult or impossible to determine in mature skulls. Among modern tapir skulls studied in ANSP and USNM, none of the six skulls of *T. pinchaque* show a demonstrable interparietal, although ANSP 19160, an adult male with M^3 near eruption, seems likely to have shown one at an earlier age. Lundelius and Slaughter (1976:fig. 1C) found a free, triangular interparietal in a young *T. pinchaque*. Of 46 skulls of *T. terrestris* with the appropriate part of the skull intact, one (USNM 220915) revealed the presence of an interparietal. This specimen has DP^{1-4} in place, and M^1 at the point of eruption. The small, rounded interparietal is manifested externally as a small irregular oval (17.1 mm long, 11 mm wide) exposed on top of the already well-formed, high, narrow sagittal crest near its posterior end. A second, very young specimen, with only DP^{1-3} in use (USNM 143861) shows a pair of incomplete sutures, bilaterally situated, extending posteromedially from each of the parietal-supraoccipital sutures. These strongly imply the existence of a discrete interparietal bone at a very early age. Of 60 skulls of *T. bairdii* with the supraoccipital-parietal region preserved, 10 show well-defined interparietals. Of 10 *T. indicus* skulls available, one (ANSP 12424), with P^{1-3} in place but unworn, DP^4 and M^1 worn, and M^2 almost fully erupted, has a completely separate, triangular interparietal. A second skull (USNM 267510) has an open transverse suture along the anterior edge of the supraoccipital and an open oblique suture on the left side, with the parietal, and continuous anteriorly with the sagittal suture between parietals, but not matched by a corresponding suture on the right side. This may indicate an independent interparietal at an earlier age, but the sagittal sutures on the skull roof tend to wander in *T. indicus*.

Thus, among living tapirs, interparietals were observed in order of decreasing frequency in *T. bairdii* (10 of 60), *T. indicus* (1 of 10, plus one possible), *T. terrestris* (1 of 46, plus one probable), and *T. pinchaque* (none of 6, one possible). These figures do not reflect true incidence of the bone, as detection is generally dependent on immaturity and open sutures.

Such is not the case in *T. veroensis*, in which every known skull except the holotype shows a discrete interparietal, and the holotype very likely had one in youth as well (Lundelius and Slaughter, 1976: 237). These include the holotype of *T. excelsus* (with two interparietals) described by Simpson (1945:74), the Texas skull reported by Lundelius and Slaughter (1976:237; fig. 3B), the specimen from Crankshaft Cave, Missouri (Lundelius and Slaughter, 1976:237; fig. 1D), the specimen from Branford, Florida (UF 14056), the Claiborne Cave, Tennessee, skull (Fig. 6A), and both specimens from South Carolina (Figs. 4A and 5A). The bone is irregularly rhomboidal in outline in some specimens (see Lundelius and Slaughter, 1976, figs. 1A and 3B), but isosceles-triangular in most (see Lundelius and Slaughter, 1976: fig. 1D, and our Figs. 4A, 5A, 6A). In the Claiborne Cave specimen, the base of the triangle is 33.5 mm long and each leg approximately 40 mm; in ChM PV4257, 15.5 and 30 mm, respectively; in SCMC SC 75.31.173, 17 and 25 mm; and in UF 14056, 28 and 34 mm. The high incidence of the interparietal and retention of independence relatively late in ontogeny would seem to be of some taxonomic value for *T. veroensis*.

The lambdoidal crests are strongly developed and posteriorly projecting in both specimens from the Cooper River, more laterally projecting in the Claiborne Cave specimen and in the holotype. Their profile in dorsal aspect is U-shaped or W-shaped, varying from deep and narrow (Fig. 5A) to shallow and wide (Fig. 6A). We have been unable to define characters of taxonomic value in the lambdoidal crests (see Simpson, 1945:74).

In lateral aspect (Fig. 4B) the third incisor is slightly opisthodont in relation to the plane of the molariform toothrow (see Hershkovitz, 1954:470, 478, 489). As in *T. pinchaque* and most *T. terrestris* the maxilla is not visible in lateral aspect above the posterior ascending process of the premaxilla, whereas it is visible as an elevated, serrated flange in all *T. indicus* and weakly in some *T. terrestris*. The blunt termination of the premaxillary process in *T. bairdii* makes this relationship irrelevant in that species.

The lateral border of the posterior premaxillary

process descends anteroventrally to a point clearly anterior to the alveolar margin of the canine, as in all living tapirs except *T. indicus* in which it essentially intersects the lateral margin of the alveolar border (see Hatcher, 1896:173; pl. 5, fig. 3), as it does also in *T. augustus* (see Colbert and Hooijer, 1953: pl. 18, fig. 1).

The lacrimal is broadly exposed as a flattened quadrangle on the face anterior to the orbit. The dorsal border does not participate in the lateral groove for the nasal diverticulum but is overlain by the anterior extremity of the supraorbital flange of the frontal. The anterior margin is applied to a thin crest of the maxilla that interrupts in lateral aspect the profile of the otherwise smooth margin of the narial vacuity. This crest continues ventrally as a line behind the infraorbital foramen, thence curving posteroventrally to connect with a crest formed along the ventral margin of the jugal. The lacrimal has a broad, flattened preorbital process lying across the middle of the orbital border of the facial part of the lacrimal. Above this process, and partially concealed by it in lateral aspect is a large lacrimal foramen; below it, and clearly visible in lateral aspect is a second, smaller lacrimal foramen.

The lacrimal appears to be distinctive in *T. veroensis* and in each living species of tapir, although *T. terrestris* and *T. pinchaque* are most similar to one another, and very different from all others. In *T. terrestris* the facial exposure of the lacrimal is high and narrow and rugose. Most individuals have a slender, strongly developed, pointed or knobby lacrimal process on the orbital border, and a second, similar process, subequal in size, arising near the anterior edge of the lacrimal. Typically there are two rather small lacrimal foramina. The larger, more dorsal one generally lies just inside the orbit, dorsomedial to the preorbital lacrimal process, where it may or may not be visible laterally. The smaller, more ventral foramen lies ventral to the process and is visible in lateral aspect. ‹

The lacrimal of *T. pinchaque* is similar to that of *T. terrestris* in shape and extent of facial exposure and in development of a spicular preorbital process. There is a tendency to develop a second, anterior process, and rugose sculpturing, but neither as strongly marked as in *T. terrestris*. The lacrimal foramina are similar in size and placement to those of *T. terrestris,* but strikingly variable in the small series available. Most individuals exhibit three foramina, of which two lie inside the orbital margin.

In *T. bairdii* the lacrimal is extensively exposed on the face, but the exposure is high and narrow, and narrows conspicuously dorsally. There is a single broad lacrimal process, and generally two foramina—a smaller, more ventral one, usually visible in lateral aspect on the orbital border below the process; an extremely large dorsomedial one situated inside the orbit and usually concealed from lateral view behind the broad lacrimal process.

In *T. indicus* the lacrimal is only narrowly exposed facially, and much of that exposure is occupied by two rather large, shallowly interconnecting foramina that are visible in lateral aspect.

In *T. veroensis* the dorsal profile of the cranium, formed by the upper margins of the nasals, frontals, parietals, and supraoccipital, is low and smoothly curved. The nasals are set down slightly below the level of the frontals, but their profiles merge without an abrupt step. Behind the nasals the profile is very low and gently convex throughout. In the holotype the entire dorsal profile is more nearly rectilinear. The dorsal profile of *T. pinchaque* is most like that of *T. veroensis,* although flatter and with longer nasals. In both *T. veroensis* and *T. pinchaque* the dorsal profile is roughly parallel to the basicranial profile (Figs. 4B, 5B, 6B).

In *T. indicus* and *T. bairdii* the nasals generally are set down abruptly below the frontal profile, and particularly in *T. indicus* the dorsal table of the frontal is inflated, giving the skull a "forehead," and making that the highest point of the dorsal profile above the basicranial line.

In *T. terrestris* the nasals are usually set well below the level of the maximum elevation of the dorsal profile, which is generally very high and strongly convex or ascending posteriorly, in either case highly elevated above the basicranium.

In *T. veroensis* the occlusal line of the cheek tooth row ascends slightly anteriorly in relation to the basicranium, which is taken as the horizontal line. This is the typical condition in all species examined other than *T. terrestris,* in which the ascent tends to be steeper, resulting in an upward flexure of the facial upon the cranial part of the skull. This effect is accentuated by the high, convex dorsal profile of the cranium.

The postglenoid and mastoid processes of *T. veroensis* converge, but do not touch, below the ventrally open, rounded external auditory meatus, describing a keyhole outline in lateral aspect. In *T. bairdii* the meatus is rounded and generally constricted ventrally in a narrow neck by complementary pointed processes from the postglenoid and mastoid processes. In *T. indicus* the postglenoid process curves posteriorly but the anterior edge of the mastoid process is fairly straight and vertical, leaving a rather wide ventral outlet from the auditory meatus. *T. augustus* seems to be the most distinctive in this region, having the two processes closely approaching or touching one another (Colbert and Hooijer, 1953:87). *T. terrestris* is highly variable in this region. In many skulls the meatus is widely open ventrally, whereas in others (for example, USNM 281390) the mastoid process sends a sharply pointed process forward to constrict the opening severely.

We have found very little that is distinctive in the ventral aspect of the skulls (Figs. 4C, 5C, 6C) other than characters of the dentition (Figs. 7 and 8F). The third incisors appear to be only slightly larger than the first two or the canine, judging from the alveoli of these latter teeth. The alveolus for the canine however, is more nearly circular in cross-section, whereas those of the incisors are elongated labiolingually. Sellards (1918:60) felt that the palate of *T. veroensis* was more concave in transverse section than that of the living species, but his comparisons were with single specimens, and, as Simpson (1945:57) noted, this is a highly variable character. The choanae terminate on a line with the posterior half of M^2, and a little farther posteriorly in the holotype. The choanal margin on the pterygoids is slightly constricted posteriorly in the holotype (Sellards, 1918:60), but not in the Cooper River skull.

The characteristics of P^1 and P^2 have been discussed in detail by Simpson (1945) and others. The essential features are those of degree of molarization, in which *T. veroensis* in general resembles *T. terrestris*. Hershkovitz (1954: fig. 60) illustrated typical P1's in *T. terrestris* and *T. pinchaque,* but noted that P^1 in some individuals of *T. pinchaque* is similar to that of *T. terrestris*. One individual of *T. terrestris* observed in this study, an adult female (ANSP 12476), has the most highly molariform P^1 of any specimen of tapir known to us. The tooth is subquadrangular in outline and has a large, high hypocone (?) almost as large as the protocone.

Measurements.—Problems in distinguishing *T. veroensis* and *T. haysii* on the basis of size of molariform teeth have been discussed under this heading for *T. haysii*. Dimensions for samples of each and for selected specimens are given in Tables 1–3. Dimensions of other specimens, mostly isolated teeth, are given in the text where the specimen in question is discussed. Additional statistical analyses of unified, well-documented samples of dentitions, already available for *T. veroensis* and much needed for *T. haysii,* undoubtedly will aid in clarification of temporal, geographic, and taxonomic variation.

Distribution.—It has not been feasible to reexamine by any means all previously reported specimens or to seek out all new or overlooked material. Nevertheless, the broad pattern of distribution of the smaller tapirs, of *T. veroensis* size, is reasonably well known from the surveys by Hay (1923:203–210), Simpson (1945:53–55), Schultz et al. (1975:17), and Bogan et al. (1980: 11, for Tennessee). Many records are based on isolated teeth that are specifically identifiable not by morphology but by default, that is, on the assumption that there is but one species of smaller tapir in the Pleistocene of North America. Until a comprehensive, critical review is undertaken, we have not felt that a new map is warranted. Accordingly, we have concentrated on new, overlooked, or obscure records known to us that either extend the periphery of, or fill blanks in, the previously mapped distribution. These records are discussed below in geographic order from west to east, and north to south.

Carthage, Hancock Co., Illinois. ANSP 11506, a right M^3, L 25.9, AW 29.3, PW 25.9. The specimen consists of a moderately worn, well preserved crown, lacking roots (Fig. 8C), attributed to "Dr. G. M. Hall," At the existing stage of wear, a posterior pressure facet would be expected if the tooth were M^2. It compares closely to M^3 in the holotype of *T. veroensis* in size, configuration, and wear. In the Proceedings of ANSP for 1867 under "Donations to the Museum" is listed "Hall, Geo. W., M.D. Fossil Tooth of a Tapir, from Illinois." Leidy (1869:391) included Illinois among the states in which *Tapirus americanus* was said to occur, presumably in allusion to this tooth. The specimen seems otherwise not to have been noticed in the literature on North American Pleistocene tapirs, nor to have been listed among the fossil mammals of Illinois (Bader and Techter, 1959).

Hollidaysburg, Blair Co., Pennsylvania. Well preserved juvenile partial dentition identified as *Tapirus veroensis*. The tapir is part of a rich late Wisconsinan fauna preserved in a fissure filling in the same hillside as the classic Frankstown Cave fauna (Guilday, personal communication). The fauna is being collected and studied by Dr. Shirley Fonda, and has been recorded in print as yet only briefly (Fonda, 1982).

Gebhards Cave, near Schoharie, Schoharie Co., New York. NYSM V-40, the major part of the crown of a left upper molar, either M^2 or M^3, little or not at all worn in life but rolled and polished postmortem, and lacking the basal lingual margin of the enamel, L 24.0, AW ca. 26.6, PW ca. 24.5 (Fig. 8D). The tooth and a supposed bone fragment were found in the bed of a small stream in the cave by Dr. Richard Veenfliet, Jr., who brought them to the New York State Museum. They were submitted to W. K. Gregory and to C. L. Gazin, who identified the tooth as tapir. The second object is not now to be found in the collections, and probably was discarded, as there is a note with the tooth, dated 31 October 1956, indicating that "the so called vertebra is inorganic and is a piece of chert." The tooth was donated to NYSM in 1962 by Mrs. Milton Burgess, Dr. Veenfliet's daughter. The specimen was recorded by Moodie (1933:115) in an addendum to a popular handbook, but has not previously found its way into the literature on fossil tapirs.

Five Fathom Bank, 25 mi SSE of Cape May, New Jersey. ANSP 15270, a fragmentary right mandibular ramus with alveoli of P_4–M_3, with vestiges of their roots, in part freshly broken (Fig. 9B). The size of the jaw is suggestive of *T. veroensis* but is inconclusive. No meaningful measurements are possible as the alveolar margins are broken away. The specimen was dredged from the sea bottom in the summer of 1956 by Preston Hawk of Rio Grande, New Jersey, and recorded by Richards (1957), who also alluded to it less explicitly later (Richards, 1959).

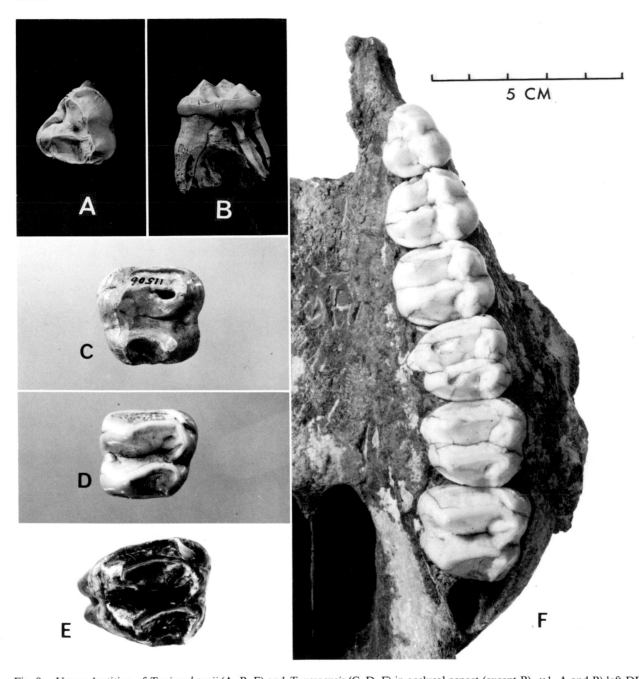

Fig. 8.—Upper dentition of *Tapirus haysii* (A, B, E) and *T. veroensis* (C, D, F) in occlusal aspect (except B), ×1. A and B) left DP¹ in occlusal and lingual aspects, original specimen whitened for photography, USNM 306473 (cast), Fussell's Pit, North Carolina; C) ANSP 11506, right M³, Carthage, Illinois; D) NYSM V-40, left M² or M³, Gebhards Cave, New York; E) USNM 215062, left M³, Neuse River, North Carolina; F) private collection of Donn Claiborne, left P¹–P³, DP⁴, M¹, M², Claiborne Cave, Tennessee.

Neither Simpson (1945: fig. 5) nor Schultz et al. (1975: fig. 12) mapped any localities for Pleistocene tapirs in West Virginia or North Carolina, and but a single locality in Virginia, presumably from Cope's Wythe Co. collection, restudied by Guilday (1962). Tapir material is uncommon in the otherwise rich Pleistocene cave and fissure faunas of West Virginia and Virginia, and consists of isolated and mostly fragmentary cheek teeth. From West

Virginia, Garton (1979:113) has recorded tapir from Buckeye Creek Cave, Greenbrier Co., and from the Bowden Cave system, Randolph Co. There is also an isolated tooth in the USNM from Wormhole Cave, Pendleton Co. From Virginia, in addition to the Wythe Co. occurrence, Clark (1938) recorded tapir from a fissure in a quarry at Limeton, 6 mi south of Front Royal, Warren Co. The material was identified at USNM by C. W. Gilmore but

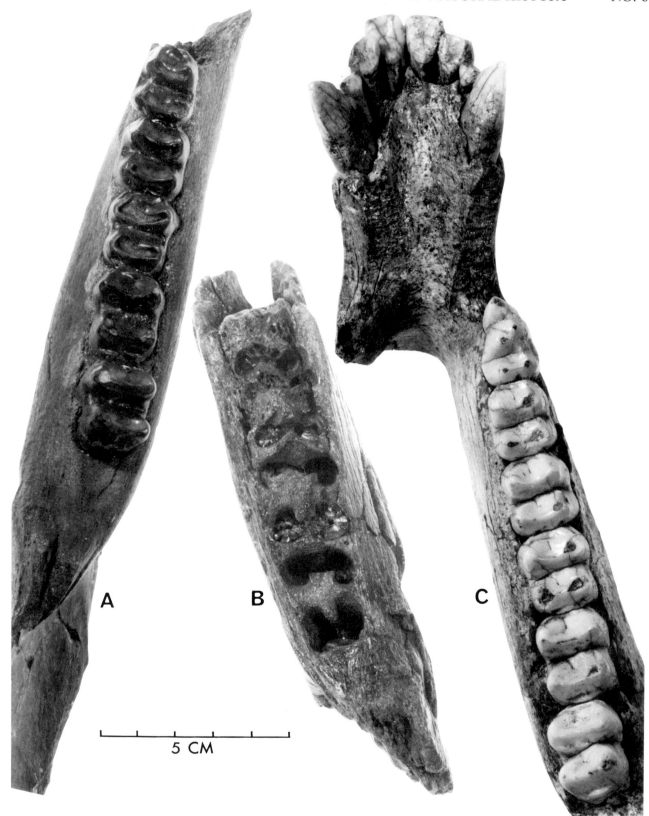

Fig. 9.—Partial mandibles of *Tapirus veroensis* in occlusal aspect, ×1. A) USNM 214663, left P₃–M₃, Island Creek local fauna, North Carolina; B) ANSP 15270, partial alveoli and roots of P₄–M₃, continental shelf off New Jersey; C) USNM 24588, left and right I₁–I₃ and canines, right P₂–M₃, Ladds, Georgia.

was not deposited in the collections. Isolated teeth, mostly fragments, of *T. veroensis* size are known from at least three fissure fillings—4.5 mi south of Harper, West Virginia, in Highland Co. (in the Strait Canyon fauna, under study by John Guilday at the time of his death); 1 mi south of Edinburg, Shenandoah Co., collected in an abandoned stone quarry by Dr. and Mrs. L. S. Bartell and Dr. Ray Gieseman, 1965; 4 mi east of Pearisburg, Giles Co., collected in an abandoned quarry in the south bank of New River by Robert Weems and Brenda Higgins, 1972.

From North Carolina, in addition to the material of *Tapirus haysii* from the Neuse River, 16 mi below New Bern, there are also isolated teeth and two incomplete maxillae referable to *T. veroensis*. P. J. Harmatuk has also collected a fragment of a tapir tooth from spoil along Mosley Creek, a tributary of the Neuse River that forms a part of the Lenoir-Craven Co. line 8.5 mi east northeast of Kinston; and several isolated teeth and fragments from disturbed overburden at the N. C. Lime Company Quarry, 4 mi west-northwest of Comfort, Jones Co. The most important material of *T. veroensis* from North Carolina is that collected by Mr. Harmatuk from windrows of Pleistocene overburden at the Ideal Cement Company Quarry, New Hanover Co., immediately south of the junction of Island Creek with the Northeast Cape Fear River (34°22′30″N, 77°50′00″W). The Pleistocene assemblage here has been designated the Island Creek fauna, and is under study by R. E. Eshelman and C. E. Ray. The principal specimens are left and right mandibular rami, USNM 214663, not making contact across the symphysis but almost certainly representing a single, rather small individual. Measurements of the teeth are given in Table 2, and the left ramus is illustrated in Fig. 9A. A fragment of left ramus, USNM 239816, has a rather large M_3, L 29.4, AW 22.2, PW 18.8. The collection also includes a partial right maxilla with P^1–P^3 (USNM 244789), an isolated right M^2? (USNM 347326), a mandibular symphysis with incomplete canine (USNM 239817), and a left astragalus (USNM 244749), all of which represent rather small individuals, certainly not of *T. haysii* size.

In South Carolina and Georgia remains of tapirs of *T. veroensis* size, mostly isolated teeth, are known from numerous coastal and near-coastal occurrences, from beaches, river bottoms, and dredgings. In addition to numerous teeth from the Myrtle Beach, South Carolina, area, there is a left mandibular ramus with P_4–M_3 (USNM 244462) from Litchfield Beach, approximately 20 mi southwest of Myrtle Beach, Georgetown Co. This specimen was collected and donated by Roy Herrick in 1978. As noted elsewhere, there are numerous specimens, mostly partial dentitions, in the ChM and SCMC collections from the Cooper River where the skull described herein was found. Several specimens were listed by Roth and Laerm (1980:18) from the Edisto Island assemblage, some 21 mi southwest of Charleston. Isolated teeth have been found in the dredgings from the Savannah River at Savannah, Georgia, and from dredgings for interstate highway construction near Brunswick. *Tapirus* cf. *veroensis* was recorded among the mammals from Ladds, Bartow Co., Georgia, on the

basis of a few fragments (Ray, 1967:142), but after that report had been published, David Bailey, then a student at Shorter College, found the major part of a mandible with all teeth on both sides except left P_2 and P_3 (USNM 24588). Measurements of the cheek teeth are given in Table 2, indicating that this is a rather small individual. The right mandibular ramus and symphyseal region are shown in Fig. 9C.

Nearly 40 years ago Simpson (1945:63) was able to state that localities for tapirs were essentially continuous over much of Florida, and that is all the more accurate today. Webb (1974:18) tabulated numerous localities for Florida, and Lundelius and Slaughter (1976) studied specimens from several localities. It should perhaps be put on record however, that several of the Floridian localities in the latter paper are in error. Most importantly, neither of the two upper dentitions listed as topotypes in their table 4 is from Vero; UF/FGS V4389 is from Rock Springs, Orange Co., and UF/FGS V5447 is from Haile, Alachua Co.

Taxonomic status.—As Simpson (1945) indicated, Sellards (1918) firmly established the specific distinctiveness of the skull of *Tapirus veroensis*. Our study confirms this conclusion. Thus, whatever eventuates regarding priorities, synonymies, and nomenclature, there is unquestionably in the Quaternary of North America a wellmarked, widespread, extinct species of tapir similar in size to the living forms. It seems that tapirs, within the confines of a rather limited and conservative repertoire of characters, have recombined these characters in various ways in speciation, so that *T. veroensis* and the four living species present a congery of variations on the traditional tapirid themes. Supraspecific taxonomy is considered under Discussion and Conclusions.

With regard to number of species, we suspect that only a single species of late Pleistocene tapir of the size of *T. veroensis* occurred in North America. With regard to specific nomenclature, *T. fossilis* may be suppressed as a forgotten name, and the characteristics of *T. californicus* are so little known that its status can scarcely be considered until additional material is collected and studied.

With regard to subspecific nomenclature (possibly utilizing the name *T. excelsus,* if its employment eventually proves useful, it will be only after variation in time and space is much better documented than it now is.

DISCUSSION AND CONCLUSIONS

Relationships among taxa of tapirs in the Quaternary cannot be established confidently on the basis of material that does not include skulls. Thus the only North American taxon that is adequately defined as yet is *Tapirus veroensis,* founded on a fine

skull, supplemented by several referred skulls of various individual ages from various localities. They demonstrate that *T. veroensis* is characterized by its own combination of variations within the overall conservative morphologic range of *Tapirus*; that it was a basic tapirine, without any extreme characteristics such as the precociously high sagittal crest of *T. terrestris* or the extensively ossified mesethmoid cartilage embraced by high maxillary flanges of *T. bairdii*. The available specimens indicate that *T. veroensis* was very widespread in the conterminous United States in late Pleistocene time, at least from Texas and Missouri eastward through Tennessee to South Carolina and Florida; that it may have extended from coast to coast and north to within the glaciated northern states, and possibly back to earlier Pleistocene time, if more fragmentary specimens are correctly referred.

All other nominal North American taxa and their distribution in time must be regarded as provisional pending recovery of well preserved, well documented skulls. Meanwhile these insecurely based taxa need not be rejected, but should be retained pending inevitable further discoveries. At present our highly tentative conclusion is that there are two species of *Tapirus* in the Pleistocene of North America—a smaller late (and probably earlier) Pleistocene form including *T. californicus, T. veroensis, T. tennesseae,* and *T. excelsus*; and a larger, early to middle (and possibly latest) Pleistocene form including *T. haysii, T. merriami,* and *T. copei*.

Supraspecific taxonomy of Quaternary tapirs cannot be meaningfully considered without reference to all adequately founded species, living and extinct. Such consideration strongly recommends grouping of all species within the single genus *Tapirus,* without subgeneric divisions. Almost any desired grouping (with the possible exception of that uniting all New World forms, see Simpson, 1945:40) could be achieved by selection of the appropriate character(s). Conversely, isolating each species in its own monotypic genus or subgenus as favored by some (see Hershkovitz, 1954:466; Colbert and Hooijer, 1953:90) defeats the grouping function of generic-group nomenclature. In order to attain nomenclatural harmony within that scheme, *T. veroensis* would require its own genus or subgenus, which we regard as unwarranted.

If one accepts that the primary function of the genus is grouping, not dividing, and considers all adequately known Quaternary tapirs, then ironically *Megatapirus augustus* and *Tapirus indicus,* of all known species, provide the least justifiable case for separation at the generic-group level. They are the one pair of species among Quaternary tapirs that are indubitably closer to one another than either is to any other. Thus, they afford the one unarguable opportunity for the generic group to fulfill its intended function.

In the absence of any overriding argument in favor of generic group subdivisions, there is also a practical value in retaining all within *Tapirus,* as Simpson (1945:41) indicated. That arrangement permits assignment of inadequately known species such as *T. haysii* to a genus without implying more than is known, and without making *Tapirus* or *Tapirus (Tapirus)* a receptacle for miscellany, as would be inevitable under a divisive arrangement.

ACKNOWLEDGMENTS

As always, vertebrate paleontology begins in the field, with recovery of significant specimens. This project, like so many in our experience in the southeastern United States, owes its initiation to collectors who are not professional paleontologists. Foremost in this case must be mentioned Susan Wallace who found and donated to the Charleston Museum the best Pleistocene tapir skull yet recovered in North America, rivaled only by the holotype of *Tapirus veroensis*. Peter J. Harmatuk's tireless pursuit of fossils along the lower Neuse River has renewed interest in this nearly forgotten classic collecting ground and provided the catalyst for reconsideration of *Tapirus haysii*. Other collectors who have provided key material and information include Mary Agnew, Donnie Bailey, L. S. Bartell, Thelma Bennett, Raymond Douglas, Fred Grady, Joe Hartsell, Roy Herrick, Mr. and Mrs. Philip Kennel, Frances Malinow, M. Sturgis, and Brandy Vasilew.

The following proprietors of specimens or collections have made specimens available for study: Larry Agenbroad, Donn Claiborne, Paul Connor, Howard Converse, Mary Dawson, Arthur Harris, Ernest Lundelius, Bruce MacFadden, Rudolph Mancke, Dan Morse, Paul Parmalee, Charles Schaff, Charles Smart, Vince and Judy Schneider, Thomas Uzzell, Gay Vostreys, and David Webb. They have also provided useful information, as have Shirley Fonda, Lewis Lipps, Gary Morgan, Carol Spawn, and Robert Weems.

The photographs were made by Victor Krantz, and the figures prepared by Lawrence Isham.

Finally we wish to express our appreciation for the support of this and other research provided through the years by the late John Guilday. His help with this small project is exemplary of the sustained professional performance that was seemingly instinctive for him, but exceptional in the world. At irregular intervals he passed along information and casts of new specimens of fossil tapirs as they came his way, always unsolicited and with no expectation of quid pro quo. His absence leaves a void in our profession, but his memory provides a lasting model to be emulated.

LITERATURE CITED

BADER, R. S., and D. TECHTER. 1959. A list and bibliography of the fossil mammals of Illinois. Nat. Hist. Miscellanea, 172:1–8.

BLACKWELDER, B. W. 1981. Stratigraphy of upper Pliocene and lower Pleistocene marine and estuarine deposits of northeastern North Carolina and southeastern Virginia. U.S. Geol. Surv. Bull., 1502-B:1–16.

BLACKWELDER, B. W., and L. W. WARD. 1983. Late Pliocene and early Pleistocene Mollusca from the James City and Chowan River formations at the Lee Creek Mine. Smithsonian Contrib. Paleobiol., in press.

BOGAN, A. E., P. W. PARMALEE, and R. R. POLHEMUS. 1980. A review of fossil tapir records from Tennessee with descriptions of specimens from two new localities. J. Tennessee Acad. Sci., 55:10–14.

CARPENTER, W. M. 1842. Notice of an interesting fossil. Amer. J. Sci. Arts, 42:390–391.

——. 1846. Remarks on some fossil bones recently brought to New Orleans from Tennessee and from Texas. Amer. J. Sci. Arts, ser. 2, 1:244–250.

CLARK, A. H. 1938. Some Pleistocene mammals from Warren County, Virginia. Science, 88:82. (Reprinted in part, 1939, Raven, 10:6–7.)

COLBERT, E. H., and D. A. HOOIJER. 1953. Pleistocene mammals from the limestone fissures of Szechwan, China. Bull. Amer. Mus. Nat. Hist., 102:1–134.

CONRAD, T. A. 1835. Observations on the Tertiary strata of the Atlantic coast. Amer. J. Sci. Arts, 28:104–111.

——. 1838. Fossils of the medial Tertiary of the United States, No. 1. Judah Dobson, Philadelphia, 52 pp. (Reprinted with introduction and annotations by W. H. Dall, 1893, Wagner Free Inst. Sci., Philadelphia.)

——. 1842. Observations on a portion of the Atlantic Tertiary region, with a description of new species of organic remains. Bull. Proc. Nat. Inst. Promotion Sci., 2:171–194.

COOPER, W. 1831. Notices of Big-bone Lick. Monthly Amer. J. Geol. Nat. Sci., 1:158–174, 205–217.

COPE, E. D. 1871. Preliminary report on the Vertebrata discovered in the Port Kennedy bone cave. Proc. Amer. Philos. Soc., 12:73–102.

CORBITT, D. L. 1950. The formation of the North Carolina counties, 1663–1943. State Dept. Archives and History, Raleigh, 323 pp.

CROOM, H. B. 1835. Some account of the organic remains found in the marl pits of Lucas Benners, Esq. in Craven County, N. C. Amer. J. Sci. Arts, 27:168–171.

DALL, W. H. 1892. Contributions to the Tertiary fauna of Florida, with especial reference to the Miocene silex-beds of Tampa and the Pliocene beds of the Caloosahatchie River. Part II. Streptodont and other gastropods, concluded. Trans. Wagner Free Inst. Sci., 3:201–473.

DALQUEST, W. W. 1977. Mammals of the Holloman local fauna, Pleistocene of Oklahoma. Southwestern Nat., 22:255–268.

DU BAR, J. R., and J. R. SOLLIDAY. 1963. Stratigraphy of the Neogene deposits, lower Neuse estuary, North Carolina. Southeastern Geol., 4:213–233.

EMRY, R. J., and R. W. PURDY. 1983. The holotype and would-be holotypes of *Hyracodon nebraskensis* (Leidy 1850). Acad. Nat. Sci. Philadelphia Notulae Naturae, in press.

FALLAW, W., and W. H. WHEELER. 1969. Marine fossiliferous Pleistocene deposits in southeastern North Carolina. Southeastern Geol., 10:35–54.

FONDA, S. 1982. Late Pleistocene vertebrates from a filled fissure-cave in central Pennsylvania. Geol. Soc. Amer., NE. SE. Sec., Abstracts with Programs, 14:18.

FOSTER, J. W. 1857. On the geological position of the deposits in which occur the remains of the fossil elephant of North America. Proc. Amer. Assoc. Adv. Sci., 10:148–169.

FRICK, C. 1921. Extinct vertebrate faunas of the badlands of Bautista Creek and San Timoteo Canon, southern California. Univ. California Publ., Bull. Dept. Geol., 12:277–424.

GARTON, E. R. 1979. Late Pleistocene and Recent mammal remains from two caves at Bowden, West Virginia. Proc. West Virginia Acad. Sci. for 1977, 49:110–116.

GAZIN, C. L. 1950. Annotated list of fossil Mammalia associated with human remains at Melbourne, Fla. J. Washington Acad. Sci., 40:397–404.

GIDLEY, J. W., and C. L. GAZIN. 1938. The Pleistocene vertebrate fauna from Cumberland Cave, Maryland. Bull. U.S. Nat. Mus., 171:1–99.

GILLETTE, D. D., and E. H. COLBERT. 1976. Catalogue of type specimens of fossil vertebrates, Academy of Natural Sciences, Philadelphia. Part II: terrestrial mammals. Proc. Acad. Nat. Sci. Philadelphia, 128:25–38.

GRAUSTEIN, J. E. 1967. Thomas Nuttall, naturalist, explorations in America, 1808–1841. Harvard Univ. Press, Cambridge, Massachusetts, 481 pp.

GUILDAY, J. E. 1962. Notes on Pleistocene vertebrates from Wythe County, Virginia. Ann. Carnegie Mus., 36:77–86.

HAGER, M. W. 1974. Late Pliocene and Pleistocene history of the Donnelly ranch vertebrate site, southeastern Colorado. Univ. Wyoming Contrib. Geol., Spec. Papers, 2(1975):1–62.

HARLAN, R. 1825. Fauna Americana: being a description of the mammiferous animals inhabiting North America. Anthony Finley, Philadelphia, i–x, 11–318 pp.

——. 1835. Critical notices of various organic remains hitherto discovered in North America. Pp. 253–313, *in* Medical and physical researches: or original memoirs in medicine, surgery, physiology, geology, zoology, and comparative anatomy (R. Harlan), Lydia R. Bailey, Philadelphia, 653 pp. (Reprinted from 1834, Trans. Geol. Soc. Pennsylvania, 1:46–112, non vid.).

——. 1842. Notice of two new fossil mammals from Brunswick Canal, Georgia; with observations on some of the fossil quadrupeds of the United States. Amer. J. Sci. Arts, 43:141–144.

HATCHER, J. B. 1896. Recent and fossil tapirs. Amer. J. Sci., ser. 4, 1:161–180.

HAY, O. P. 1912. The Pleistocene Period and its Vertebrata. Pp. 539–784, *in* E. Barrett, Indiana Dept. Geol. Nat. Res. Ann. Rept. 1911, 36:1–873.

——. 1914. The Pleistocene mammals of Iowa. Pp. 1–662, *in* G. F. Kay, Iowa Geol. Surv. Ann. Rept. 1912, 23:i–xlviii, 1–662.

——. 1920. Descriptions of some Pleistocene vertebrates found in the United States. Proc. U.S. Nat. Mus., 58:83–146.

——. 1923. The Pleistocene of North America and its vertebrated animals from the states east of the Mississippi River and from the Canadian provinces east of longitude 95°. Carnegie Inst. Washington Publ., 322:1–499.

————. 1927. The Pleistocene of the western region of North America and its vertebrated animals. Carnegie Inst. Washington Publ., 322B:1–346.

HAY, O. P., and H. J. COOK. 1930. Fossil vertebrates collected near, or in association with, human artifacts at localities near Colorado, Texas; Frederick, Oklahoma; and Folsom, New Mexico. Proc. Colorado Mus. Nat. Hist., 9(2):4–40.

HAYS, I. 1834. Descriptions of the inferior maxillary bones of mastodons, in the cabinet of the American Philosophical Society, with remarks on the genus Tetracaulodon, &c. Trans. Amer. Philos. Soc., 4:317–339. (Separate dated 1833.)

————. 1852. [Remarks on a tooth of the fossil Tapir.] Proc. Acad. Nat. Sci. Philadelphia, 6:53.

HERSHKOVITZ, P. 1954. Mammals of northern Colombia, preliminary report no. 7: tapirs (genus *Tapirus*), with a systematic review of American species. Proc. U.S. Nat. Mus., 103:465–496.

HIBBARD, C. W., and W. W. DALQUEST. 1966. Fossils from the Seymour Formation of Knox and Baylor counties, Texas, and their bearing on the late Kansan climate of that region. Univ. Michigan, Contrib. Mus. Paleontol., 21:1–66.

HIBBARD, C. W., R. J. ZAKRZEWSKI, R. E. ESHELMAN, G. EDMUND, C. D. GRIGGS, and C. GRIGGS. 1978. Mammals from the Kanopolis local fauna, Pleistocene (Yarmouth) of Ellsworth County, Kansas. Univ. Michigan, Contrib. Mus. Paleontol., 25:11–44.

KURTÉN, B., and E. ANDERSON. 1980. Pleistocene mammals of North America. Columbia University Press, New York, 442 pp.

LANCE, J. F. 1959. Faunal remains from the Lehner mammoth site. Pp. 35–39, *in* The Lehner mammoth site, southeastern Arizona (E. W. Haury, E. B. Sayles, and W. W. Wasley, eds.), Amer. Antiq., 25:2–42.

LEFFLER, S. R. 1964. Fossil mammals from the Elk River Formation, Cape Blanco, Oregon. J. Mamm., 45:53–61.

LEIDY, J. 1849. *Tapirus Americanus fossilis*. Proc. Acad. Nat. Sci. Philadelphia, 4:180–182.

————. 1852a. [Reference to a fossil tooth of a Tapir.] Proc. Acad. Nat. Sci. Philadelphia, 6:106.

————. 1852b. [Remarks on *Tapirus Haysii*.] Proc. Acad. Nat. Sci. Philadelphia, 6:148.

————. 1853. The ancient fauna of Nebraska: or, a description of remains of extinct Mammalia and Chelonia, from the mauvaises terres of Nebraska. Smithsonian Contrib. Knowledge, 6(7):1–126.

————. 1855. Notice of some fossil bones discovered by Mr. Francis A. Lincke, in the banks of the Ohio River, Indiana. Proc. Acad. Nat. Sci. Philadelphia, 7:199–201.

————. 1858. [Remarks on a cast of a mastodon tooth, &c.] Proc. Acad. Nat. Sci. Philadelphia, 10:12.

————. 1859. Description of vertebrate fossils. Pp. 99–122, *in* 1858–1860, Post-Pleiocene fossils of South-Carolina (F. S. Holmes, ed.), Russell and Jones, Charleston, 122 pp.

————. 1869. The extinct mammalian fauna of Dakota and Nebraska, including an account of some allied forms from other localities, together with a synopsis of the mammalian remains of North America. J. Acad. Nat. Sci. Philadelphia, ser. 2, 7:i–vii + 8–472.

LUND, P. W. 1841. Blik paa Brasiliens dyreverden för sidste jordomvaeltning. Tredie afhandling: fortsaettelse af pattedyrene. Kongelige Danske Videnskabernes Selskabs, Naturvidenskabelige og Mathematiske Afhandlinger, 8:217–272.

LUNDELIUS, E. L., JR. 1972. Fossil vertebrates from the late Pleistocene Ingleside fauna, San Patricio County, Texas. Univ. Texas Austin Bur. Econ. Geol. Rept. Investigations, 77:1–74.

LUNDELIUS, E. L., JR., and B. H. SLAUGHTER. 1976. Notes on American Pleistocene tapirs. Pp. 226–243, *in* Essays on palaeontology in honour of Loris Shano Russell (C. S. Churcher, ed.), Royal Ontario Mus., Life Sci. Misc. Publ., Athlon, 286 pp.

LYDEKKER, R. 1886. Catalogue of the fossil Mammalia in the British Museum, (Natural History) Cromwell Road, S. W. Part III. Containing the Order Ungulata, suborders Perissodactyla, Toxodontia, Condylarthra, and Amblypoda. British Museum (Natural History), London, 186 pp.

LYELL, C. 1843. On the geological position of the *Mastodon giganteum* and associated fossil remains at Bigbone Lick, Kentucky, and other localities in the United States and Canada. Proc. Geol. Soc. London 4:36–39. (Reprinted with minor changes, 1844, Amer. J. Sci. Arts, 46:320–323.)

————. 1849. A second visit to the United States of North America. vol. 1. Harper and Brothers, Publishers, New York, i–xii, 13–273 pp.

McCARTAN, L., J. P. OWENS, B. W. BLACKWELDER, B. J. SZABO, D. F. BELKNAP, N. KRIAUSAKUL, R. M. MITTERER, and J. F. WEHMILLER. 1982. Comparison of amino acid racemization geochronometry with lithostratigraphy, biostratigraphy, Uranium-series coral dating, and magnetostratigraphy in the Atlantic coastal plain of the southeastern United States. Quaternary Res., 18:337–359.

McCARTAN, L., R. E. WEEMS, and E. M. LEMON, JR. 1980. The Wando Formation (upper Pleistocene) in the Charleston, South Carolina, area. *In* Changes in stratigraphic nomenclature by the U.S. Geological Survey, 1979 (N. F. Sohl and W. B. Wright, eds.), U.S. Geol. Surv. Bull., 1502-A:110–116.

MERRIAM, J. C. 1913. Tapir remains from late Cenozoic beds of the Pacific coast region. Univ. California Publ., Bull. Dept. Geol., 7:169–175.

MILLER, W. E. 1971. Pleistocene vertebrates of the Los Angeles basin and vicinity (exclusive of Rancho La Brea). Bull. Los Angeles County Mus. Nat. Hist., Sci., 10:1–124.

MITCHELL, E. 1842. Elements of geology, with an outline of the geology of North Carolina: for the use of the students of the University. Publisher and place not indicated, 141 pp.

MIXON, R. B., and O. H. PILKEY. 1976. Reconnaissance geology of the submerged and emerged coastal plain province, Cape Lookout area, North Carolina. U.S. Geol. Surv. Prof. Paper, 859:1–45.

MOODIE, R. L. 1933. A popular guide to the nature and the environment of the fossil vertebrates of New York. New York State Mus. Handbook, 12:1–122.

MORSE, D. F. 1970. Preliminary notes on a recent mastodon and tapir find in northeastern Arkansas. Arkansas Archeol., Bull. Arkansas Archeol. Soc., 11:45–49.

PARMALEE, P. W., R. D. OESCH, and J. E. GUILDAY. 1969. Pleistocene and Recent vertebrate faunas from Crankshaft Cave, Missouri. Illinois State Mus. Rept. Investigations, 14:1–37.

PETERSON, O. A. 1926. The fossils of the Frankstown Cave, Blair County, Pennsylvania. Ann. Carnegie Mus., 16:249–314.

RADINSKY, L. B. 1965. Evolution of the tapiroid skeleton from *Heptodon* to *Tapirus*. Bull. Mus. Comp. Zool., 134:69–106.

RAY, C. E. 1964. *Tapirus copei* in the Pleistocene of Florida. Quart. J. Florida Acad. Sci., 27:59–66.

———. 1967. Pleistocene mammals from Ladds, Bartow County, Georgia. Bull. Georgia Acad. Sci., 25:120–150.

RICHARDS, H. G. 1936. Fauna of the Pleistocene Pamlico Formation of the southern Atlantic coastal plain. Bull. Geol. Soc. Amer., 47:1611–1656.

———. 1950. Geology of the coastal plain of North Carolina. Trans. Amer. Philos. Soc., new ser., 40:1–83.

———. 1957. Fossil mammals from the New Jersey coast. Ann. Bull. Cape May Geogr. Soc., 11:8–10.

———. 1959. Pleistocene mammals dredged off the coast of New Jersey. Bull. Geol. Soc. Amer., 70:1769 (abstract).

RICHARDSON, G. B. 1907. [*Elephas columbi, Equus complicatus*, and *Tapirus haysii* from El Paso, Texas.] Science, n.s., 25:32.

———. 1909. Description of the El Paso District. U.S. Geol. Surv., Geol. Atlas of the U.S., El Paso Folio, Texas, 166:1–11.

ROTH, J. A., and J. LAERM. 1980. A late Pleistocene vertebrate assemblage from Edisto Island, South Carolina. Brimleyana, J. North Carolina State Mus. Nat. Hist., 3:1–29.

RUBIN, M., and C. ALEXANDER. 1958. U.S. Geological Survey radiocarbon dates IV. Science, 127:1476–1487.

SAVAGE, D. E., and D. E. RUSSELL. 1983. Mammalian paleofaunas of the World. Addison-Wesley Publishing Co., Reading, Massachusetts, 432 pp.

SCHULTZ, C. B., L. D. MARTIN, and R. G. CORNER. 1975. Middle and late Cenozoic tapirs from Nebraska. Bull. Univ. Nebraska State Mus., 10:1–21.

SCHULTZ, C. B., L. G. TANNER, F. C. WHITMORE, JR., L. L. RAY, and E. C. CRAWFORD. 1963. Paleontologic Investigation at Big Bone Lick State Park, Kentucky: a preliminary report. Science, 142:1167–1169.

SELLARDS, E. H. 1918. The skull of a Pleistocene tapir including description of a new species and a note on the associated fauna and flora. Florida Geol. Surv. Ann. Rept., 10:57–70.

SIMPSON, G. G. 1942. The beginnings of vertebrate paleontology in North America. Proc. Amer. Philos. Soc., 86:130–188.

———. 1945. Notes on Pleistocene and Recent tapirs. Bull. Amer. Mus. Nat. Hist., 86:33–82.

SLAUGHTER, B. H. 1966. The Moore Pit local fauna; Pleistocene of Texas. J. Paleontol., 40:78–91.

STEPHENSON, L. W. 1912. Quaternary formations. Pp. 266–290, *in* The coastal plain of North Carolina (W. B. Clark, B. L. Miller, L. W. Stephenson, B. L. Johnson, and H. N. Parker, eds.), North Carolina Geol. Econ. Surv., 3:1–552.

STRAIN, W. S. 1966. Blancan mammalian fauna and Pleistocene formations, Hudspeth County, Texas. Bull. Texas Mem. Mus., 10:1–55.

WARD, L. W., and B. W. BLACKWELDER. 1980. Stratigraphic revision of upper Miocene and lower Pliocene beds of the Chesapeake Group, Middle Atlantic coastal plain. U.S. Geol. Surv. Bull., 1482-D:1–61.

WARREN, J. C. 1855. Description of a skeleton of the Mastodon giganteus of North America. John Wilson and Son, Boston, second ed., 260 pp.

WEBB, S. D. 1974. Chronology of Florida Pleistocene mammals. Pp. 5–31, *in* Pleistocene mammals of Florida (S. D. Webb, ed.), Univ. Presses of Florida, Gainesville, 270 pp.

WHEELER, W. H., R. B. DANIELS, and E. E. GAMBLE. 1979. Some stratigraphic problems of the Pleistocene strata in the area from Neuse River estuary to Hofmann Forest, North Carolina. Pp. 41–50, *in* Structural and stratigraphic framework for the coastal plain of North Carolina (G. R. Baum, W. B. Harris, and V. A. Zullo, eds.), Carolina Geol. Soc. and Atlantic Coastal Plain Geol. Assoc. Field Trip Guidebook, October 19–21, 1979, Wrightsville Beach, North Carolina, Geol. Surv. Sec., Dept. Nat. Res. and Community Dev., Raleigh, 111 pp.

WINGE, H. 1906. Jordfundne og nulevende Hovdyr (*Ungulata*) fra Lagoa Santa, Minas Geraes, Brasilien. Med Udsigt over Hovdyrenes indbyrdes Slaegtskab. E Museo Lundii, 3(1):1–239.

Address (Ray):   Department of Paleobiology, National Museum of Natural History, Smithsonian Institution, Washington, D.C. 20560.

Address (Sanders):   Department of Natural Sciences, Charleston Museum, Charleston, South Carolina 29403.

SANGAMONA: THE FURTIVE DEER

Charles S. Churcher

DEDICATION

John E. Guilday was not enamoured of *Sangamona fugitiva* and wrote "I hope someone someday comes up with both ends of this beast so we can really pin it down. . . . Apparently its habits were such that it didn't often get into cave deposits—and even when it did it apparently checked its lower plate before doing so" (personal communication, 14 February 1977) and "Perhaps someday someone will come up with a good specimen of *Sangamona* so that we can see what the beast looks like in its entirety. Someone should make direct comparisons someday" (personal communication, 2 March 1978). His attitude of mind is further revealed in his reference to it as "*Sangamona furtiva*" (Guilday et al., 1969:64) and gives the title to this paper. I therefore dedicate this consideration of the status of *Sangamona fugitiva* to my colleague and good friend John E. Guilday, from whom I learnt much about Pleistocene vertebrates and with whom I carried on a most cordial correspondence.

ABSTRACT

The cervid *Sangamona fugitiva* described by Hay (1921) from deposits in Tennessee, Maryland, and Illinois is compared with modern representatives of *Cervus canadensis* (wapiti), *Odocoileus virginianus* (white-tailed deer), *O. hemionus* (mule deer) and others. The type of *S. fugitiva*, a worn M², is considered indeterminate. Hay's other "typical" specimens from Maryland and Illinois resemble elements of *Odocoileus* or *Cervus*. Additional fossil specimens placed within *Sangamona sensu lato* from New Mexico, Tennessee, Missouri, Nebraska, Iowa, and Pennsylvania are also assessed. In most instances these specimens are associated with remains of *Cervus, Odocoileus, Rangifer* (caribou), and *Alces* or *Cervalces* (moose) and may be assigned to one of these taxa. The taxon *Sangamona fugitiva* Hay 1920 is therefore invalid, except for the type, which is undiagnostic and indeterminate to species and genus.

INTRODUCTION

Hay (1921:91, pl. 3, figs. 14–15) founded *Sangamona fugitiva* on a left upper second molar (USNM 8954) collected from the Pleistocene of a cave at Whitesburg, Hamblen County, Tennessee, in 1885 by Ira Sayles. In the same paper Hay attributed to his new taxon specimens from Alton, Illinois, collected by Hon. Wm. McAdams before 1883 and comprising two right mandibular fragments with partial dentitions (USNM 9002–9003) and a fragmentary and badly worn left upper molar row (USNM 9096) which cannot be profitably compared with the type specimen. He also included within *S. fugitiva* cervid remains (USNM 9192–9193) from Cavetown, Maryland, collected by Dr. Charles Peabody and Warren K. Moorehead. No upper cheekteeth occur among these chiefly post-cranial elements, but a well worn lower first or second molar and a lower first incisor (USNM 9192) are included.

Subsequent to the original description, other cervid specimens have been assigned to *Sangamona fugitiva*, few of which are directly comparable with Hay's (1921) type or "paratypic" materials described in the same paper. The descriptions of material given below include the type, paratypes, and referred specimens that have been placed within *S. fugitiva* by one or more authors. Associations with other cervid taxa found in the same deposits are noted.

ABBREVIATIONS AND MEASUREMENTS

The following abbreviations are used to indicate the institutions or collections in which fossil specimens or recent comparative skeletons are conserved:

ANSP   Academy of Natural Sciences, Philadelphia, Pennsylvania.

CM   Carnegie Museum of Natural History, Pittsburgh, Pennsylvania.

FA   Field Accession Number, Department of Mammalogy, Royal Ontario Museum, Toronto, Ontario.

MCZ   Museum of Comparative Zoology, Harvard University, Cambridge, Massachusetts.

ROM   Royal Ontario Museum (Department of Mammalogy), Toronto, Ontario.

RW   Red Willow County paleontological site, University of Nebraska State Museum, Lincoln, Nebraska.

UNSM   University of Nebraska State Museum (Division of Vertebrate Paleontology), Lincoln, Nebraska.

USNM   United States National Museum, now The National Museum of the United States (Smithsonian Institution), Washington, D.C.

All measurements are given in millimetres unless otherwise indicated; ranges of variation are given as maxima, minima, means, standard deviations and sample sizes (Max., Min., mean, SD, and N); "e" indicates an estimated measurement, "—" a measurement that cannot be obtained due to lack of reference point(s), and blanks indicate absence of information.

MATERIAL CURRENTLY ASSIGNED TO *SANGAMONA*

SPECIMENS ORIGINALLY ASSIGNED TO *SANGAMONA FUGITIVA* BY HAY (1921)

1. TYPE. Left M^2, complete but maturely worn (USNM 8954). From a cave at Whitesburg, Hamblen Co., Tennessee. Hay, 1921:91, pl. 3, figs. 14–15. Collected in 1885 by Ira Sayles. Associated with *Odocoileus virginianus* and *Cervus canadensis*.

2. Right I_1 (USNM 9192), left M_1 or M_2, distal end of left radius, right scaphoid, right innominate acetabulum, left malleolus, proximal half of astragalus, right calcaneum, two (probably metatarsal) sesamoids, and first, second, and third (probably pedal) phalanges (USNM 9193). From fissure deposits at Cavetown, Washington Co., Maryland. Hay, 1921: 102–104. Collected by Dr. Charles Peabody and Warren K. Moorehead. Associated with *Odocoileus virginianus*.

3. Two right mandibular fragments with maturely worn cheekteeth; one with M_1–M_2, remnants of P_4, and traces of mesial roots of M_3 (USNM 9002), and one with M_1–M_3 (USNM 9003); and a very well worn left upper dentition P^3–M^3 (USNM 9096). From loess nodules from Alton, Madison Co., Illinois. Hay, 1921:111, pl. 5, figs. 5–6. Collected before 1883 by Hon. Wm. McAdams. Associated with *Cervalces roosevelti* and *Rangifer muscatensis* (=*R. tarandus*).

SPECIMENS DISCOVERED LATER AND ASSIGNED TO *SANGAMONA FUGITIVA* OR *SANGAMONA* SP.

4. Metapodial (ANSP 13931), humerus (ANSP 13930), right maxillae with P^{3-4} (ANSP 13592) and P^3–M^1 (ANSP 13591), left maxilla with P^{2-4} (ANSP 13578), right and left upper molars (ANSP 14065, 13582, respectively), right adult and immature dentaries (ANSP 14060, 13579, respectively). From Burnet Cave, Eddy Co., New Mexico. Schultz and Howard, 1935:287, pl. 14, fig. 9 (ANSP 13931); referred to *Sangamona*? sp. Associated with *Odo-*

coileus virginianus, O. hemionus, Rangifer? fricki n. sp.

5. Two right P^4's, left P^2, left p_4 and right P_3 (CM 8372–8375, 8377). From Armadillo Pit, Robinson Cave, Overton Co., Tennessee. Guilday et al., 1969: 64, figs. 12b, d, e, f; referred to *Sangamona furtiva*. Associated with *Odocoileus* cf. *virginianus*.

6. Left metatarsal, left humerus, paired tibiae, and distal half of left tibia assigned to "cf. *Sangamona*" and two left dentaries with P_3–M_3 and P_4–M_3 assigned to "cf. *Sangamona* or *Odocoileus halli*." From Brynjulfson Cave No. 1, Boone Co., Missouri. Parmalee and Oesch, 1972:40–43. Associated with *Odocoileus virginianus* and cf. *Alces* or *Cervalces*.

7. Left antler fragment, pedicel and fragment of frontal (UNSM 46326), distal ends of two right humeri (UNSM 48534–48535), left metacarpal (USNM 48526), and distal end of right humerus (UNSM 46304). From gravel pits RW 101 and 102, south bank of the Republican River, Red Willow Co., Nebraska. Corner, 1977:85. Associated with *Odocoileus* sp. and *Rangifer tarandus*.

SPECIMENS REASSIGNED TO *SANGAMONA FUGITIVA* FROM OTHER TAXA

8. Right metatarsal lacking distal end, left humerus lacking proximal epiphysis, and left radius lacking distal end. From Dubuque, Dubuque Co., Iowa. Allen, 1876:49–50, assigned to *Cervus whitneyi*; Kurtén, 1975:508, referred to *Sangamona*. Associated with *Cervus canadensis*.

9. Female skull fragments with reconstructed upper toothrows (CM 11044), limb and foot bones (CM 11043). From Frankstown Cave, Blair Co., Pennsylvania. Peterson, 1926:257, assigned to *Odocoileus hemionus*; Guilday et al., 1969:64, reassigned to *Sangamona*. Associated with *Odocoileus virginianus* and *Cervalces americanus*.

10. Single P^4 (CM 29501) and left P_3 (CM 30060). From Baker Bluff Cave, Sullivan Co., Tennessee. Guilday et al., 1978:50, assigned to cf. *Sangamona*

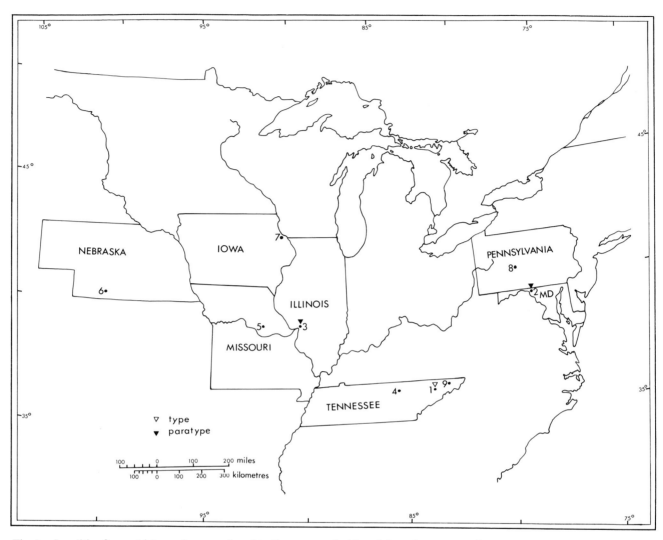

Fig. 1.—Localities from which specimens assigned to *Sangamona fugitiva* s.l. have been reported.

1. Whitesburg, Hamblen Co., Tennessee. Type specimen; Hay, 1921.
2. Cavetown, Washington Co., Maryland. Paratypic specimens; Hay, 1921.
3. Alton, Madison Co., Illinois. Paratypic specimens; Hay, 1921.
4. Armadillo Pit, Robinson Cave, Overton Co., Tennessee. Guilday et al., 1969.
5. Brynjulfson Cave, Boone Co., Missouri. Parmalee and Oesch, 1972.
6. Republican River, pits RW 101 & 102, Red Willow Co., Nebraska. Corner, 1977.
7. Dubuque, Dubuque Co., Iowa. Kurtén, 1975.
8. Frankstown Cave, Blair Co., Pennsylvania. Peterson, 1926.
9. Baker Bluff Cave, Sullivan Co., Tennessee. Guilday et al., 1978.

Burnet Cave, Eddy Co., New Mexico, and Peccary Cave, Newton Co., Arkansas, are not shown as the former represents *Navahoceros* and the latter is unconfirmed.

fugitiva. Associated with cf. *Cervus elaphus, Odocoileus virginianus* and *Rangifer tarandus.*

Kurtén and Anderson (1980:313) listed Peccary Cave, Newton Co., Arkansas, as containing remains of *Sangamona fugitiva.* However, Semken (1969) and Davis (1969) did not mention *Sangamona* among the mammals discussed. No other mention of this record has been seen.

CONSIDERATION OF HAY'S TYPE AND ASSOCIATED SPECIMENS OF *SANGAMONA FUGITIVA* (LOCALITIES 1–3)

The taxon, *Sangamona fugitiva,* was founded by Hay (1921) on an isolated second upper molar from a cave at Whitesburg, Tennessee, for which the only comparable fossil material known at present is that from Frankstown Cave, Pennsylvania (Peterson, 1926). The upper molar row (USNM 9096) from Alton, Illinois, is too worn and incomplete for any of the qualitative characters observable in the type to be compared and size comparisons are suspect because of the small amount of enamel wall remaining. Wells (1959) has expressed reservations on the use of single equid teeth as reliable taxonomic specimens and suggested that entire molar rows or extensive uniform collections of teeth only be considered as sound foundations for taxonomic identifications. Even then, because of variations within species and individuals because of sex, age, and wear, he considered that such identifications should be used with caution. I believe that such caveata apply equally to artiodactyl dentitions. Thus, *Sangamona fugitiva* as founded on an isolated and overly worn tooth as the type specimen is a *nomen vanum* in the sense of Simpson (1945:27).

Hay (1921:91) described the characters of the type (USNM 8954) of *Sangamona fugitiva* thus: "This genus differs much from our other deer in the nearly complete absence of the strong ribs which occupy the outer faces of the lobes of the upper molars" and diagnosed the species by "Styles, or ribs on paracone and metacone absent or obsolete. Size intermediate between Virginian deer and wapiti." Examination of the type specimen shows no rib on the metacone and a very weak one on the paracone, and the paracone style to be rounded and that on the metacone bluntly pointed. No distal hypostyle is present. The protocone is strong, with rugose enamel on the cingulum, and the hypocone is large. The tooth appears proportionately more brachydont than is usual in *Odocoileus virginianus.* There is a small interlophar endostyle or pillar within the protocone-hypocone valley as occurs in some *O. virginianus.* The prefossette opens lingually past the endostyle, the protocone joins the parastyle mesially, and the hypocone joins the distobuccal corner of the metacone. No plis occur within the fossettes, although the distal end of the protoloph is slightly bifurcated. The specimens from Cavetown, Maryland, are

mostly postcranial elements but include a right first incisor and a first or second lower molar. Both of these specimens are as unsuitable for specific identification as the upper molar from Whitesburg and do not provide a reliable basis on which to found a new taxon. Hay (1921:103) considered that "This [molar] tooth and the bones are entirely too large to have belonged to any known species of *Odocoileus* and too small for any known species of *Cervus.* The incisor (Cat. No. [USNM] 9192) is considerably larger than the corresponding one of the Virginian deer" and "There is a rather strong tubercle ectostylid or pillar at the mouth of the principal valley" on the lower molar. These teeth and postcranial elements are not illustrated, but Hay's measurements show the animal to be generally intermediate in size between *O. virginianus* and *C. canadensis,* but often closer to the former.

The specimens from Alton, Illinois, provide at least two associated partial right mandibular tooth rows by which the taxon may be characterized. Hay (1921:111, pl. 5, figs. 5–6) stated of the M_3's "At the outer mouth of the median valley of these teeth there is a conspicuous accessory pillar [ectostylid]. The crowns of the lower molars are higher than in *Odocoileus.* The inner faces of the lobes are flatter than in *Odocoileus,* and they are, relative to the length, much broader. They agree in size so well with the upper tooth . . . found at Whitesburg, Tennessee, and with the lower tooth found at Cavetown, Maryland, that they are referred to that species. In size, they agree well with the lower molar found at Cavetown, Maryland, and are referred to *S. fugitiva.*" He provided few measurements of the specimens and made no mention of the well worn upper molar row (USNM 9096) also from Alton, Illinois.

The problem that arises from Hay's (1921) descriptions is how one can, with any degree of certainty, compare and conclude that an isolated upper molar collected from a cave in Tennessee in 1885, two partial right mandibular molar rows collected from loess in Illinois before 1883, and an incisor and lower molar collected in 1908 or later from a fissure in Maryland belong within a single new genus and species. The type specimen is isolated and cannot be compared with the other specimens, and is not compared with the worn molar row from Illi-

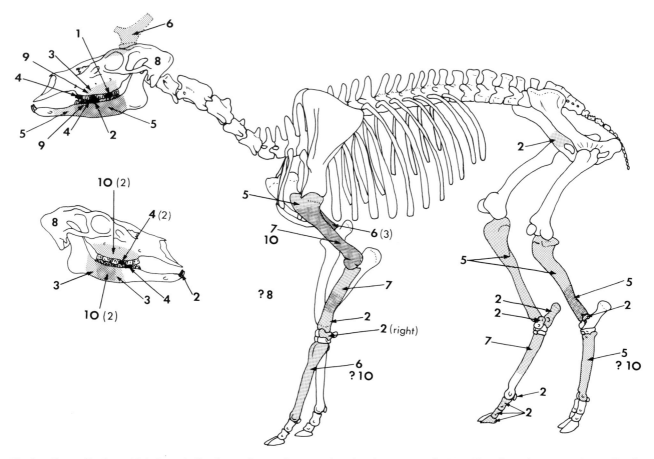

Fig. 2.—Generalized cervid skeleton indicating major specimens assigned to *Sangamona fugitiva*. Not all specimens are shown. Shading indicates elements assigned to *S. fugitiva*—light shading indicates a single specimen or element, dark shading indicates two or more specimens of an element, and black indicates isolated teeth. Figures in parentheses indicate multiple specimens.

1. Type specimen from Whitesburg, Hamblen Co., Tennessee. (Hay, 1921).
2. Paratypic specimens from Cavetown, Washington Co., Maryland. (Hay, 1921).
3. Paratypic specimens from Alton, Madison Co., Illinois. (Hay, 1921).
4. Specimens from Armadillo Pit, Robinson Cave, Overton Co., Tennessee. (Guilday et al., 1969).
5. Specimens from Brynjulfson Cave, Boone Co., Missouri. (Parmalee and Oesch, 1972).
6. Specimens from pits RW 101 and 102, south bank of Republican River, Red Willow Co., Nebraska. (Corner, 1977).
7. Reassigned specimens from Dubuque, Dubuque Co., Iowa. (Kurtén, 1975).
8. Reassigned partial skeleton from Frankstown Cave, Blair Co., Pennsylvania. (Peterson, 1926).
9. Specimens from Baker Bluff, Sullivan Co., Tennessee. (Guilday et al., 1978).
10. Specimens from Burnet Cave, Eddy Co., New Mexico. (Schultz and Howard, 1935).

nois. The isolated incisor and postcranial elements from Maryland are also unmatched. The isolated first or second molar and the damaged mandibular rows are comparable and could represent a morphology sufficiently distinct to provide a basis for the establishment of a new taxon. However, Hay contented himself by noting that the lower teeth from Illinois agree so well in size with the upper tooth and isolated lower teeth that they are referred to the same taxon. Hay (1921:91) diagnosed the genus *Sangamona* thus:

"*Diagnosis*—Upper molars of medium height, broad, outer face of anterior lobe with feebly developed style; outer face of the hinder lobe deeply concave and devoid of style. Lower molars relatively broad; inner faces of front and hinder lobes flat and with feebly developed style."

He diagnosed the species *S. fugitiva* thus:

"*Diagnosis*—Styles, or ribs, on paracone and metacone absent or obsolete. Size intermediate between the Virginian deer and the wapiti."

Effectively, these diagnoses say the same thing, that is, the upper molar teeth have small or reduced styles and reduced or absent ribs, and the lower molar teeth have flat lingual surfaces with feeble styles or ribs. These conditions are not unknown in *Odocoileus* but are rare. They are common in *Rangifer*, but its tooth pattern is different from that of the fossils. However, these conditions are more often seen in small specimens of *Cervus canadensis* than in *Odocoileus*.

In all three deposits (Whitesburg, Tennessee; Cavetown, Maryland; Alton, Illinois) *Sangamona* is found variously associated with *Cervus canadensis* (Tennessee), *Odocoileus virginianus* (Tennessee and Maryland), and *Cervalces roosevelti* and *Rangifer tarandus* (Illinois), and thus the suspicion arises that the specimens constitute elements of a well known cervid that have been misidentified. Misidentification may have come about because Hay had few specimens with which to compare his fossils and because he may not have been aware of the dental variants possible in these cervids.

In descriptions of fossil *Sangamona* Hay gives comparative measurements of single specimens of /*O. virginianus* and/or *C. canadensis*. A reinvestigation and remeasurement of original specimens and extended comparative samples generating ranges of measurements and mean values, provide insights into the probable identities of the specimens attributed to *Sangamona fugitiva* by Hay and others.

Hay's (1921) original measurements are uninfor-

mative except to show that *Sangamona*'s type and associated specimens were marginally smaller in mesiodistal and buccolingual diameters of the cheekteeth than in *Cervus canadensis*. Comparisons with more numerous specimens of *Odocoileus* show the type upper molar to be similar in length but broader, even though *Sangamona* lacks the well developed styles or ribs on the buccal faces of the ectoloph (Table 1), and is considerably smaller than comparable molars of *Cervus*.

The Alton *Sangamona* rami afford suites of measurements for M_2 and M_3. Compared with those for *Odocoileus*, the *Sangamona* teeth are always 2 to 3 mm larger in all dimensions. Compared with those for *Cervus*, they fall just within or below the minima for modern specimens.

The Cavetown postcranial elements are all more massive than those of *Odocoileus* (Table 2), although in some dimensions only marginally so, and generally smaller than those of *Cervus*, but again in some only marginally less.

These comparisons suggest that the type molar is not only indeterminable qualitatively but quantitatively, and could represent a large individual of *Odocoileus* or a small one of *Cervus*, with both of which it is associated in the cave. The Alton rami bear teeth that are too large for *Odocoileus* but suitable for *Cervus*, from which they do not differ strongly qualitatively except in the strength of the lingual styles and ribs. The Cavetown postcranial elements appear to represent an individual intermediate in size between *Odocoileus* and *Cervus*, as for the type.

COMMENTS ON OTHER SPECIMENS ASSIGNED TO *SANGAMONA FUGITIVA* (LOCALITIES 4–7)

4. Burnet Cave, Eddy Co., New Mexico.

Schultz and Howard (1935) listed nine specimens identified as "*Sangamona*? sp. Extinct Cervid" from Burnet Cave, New Mexico. These comprise two right maxillae with P^{3-4} and P^3–M^1 (ANSP 13592, 13591, respectively), a left maxilla with P^{2-4} (ANSP 13578), left and right upper molars (ANSP 13582, 14065, respectively), right adult and immature dentaries (ANSP 14060, 13579, respectively), humerus (ANSP 13930) and partial metapodial (ANSP 13931), deriving from at least three individuals. Schultz and Howard (1935:287) remarked only "Size interme-

diate between *Odocoileus* and *Cervus*. Probably equal to *Sangamona*" and refrained from giving measurements or comparisons with living *Odocoileus* or *Cervus* species, with Hay's (1921) specimens of *Sangamona*, or with their newly described and associated *Rangifer*? *fricki*.

However, Kurtén (1975) considered the various cervids from the Pleistocene of the southwestern United States and Mexico, described as *Rangifer*? *fricki* (Schultz and Howard, 1935), *Sangamona*? sp. (Schultz and Howard, 1935), *Cervus lascrucensis* (Frick, 1937), and *Odocoileus halli* (Alvarez, 1969),

Table 1.—*Measurements of the type specimens of* Sangamona fugitiva *from Hay (1921) and by Churcher, and of comparable teeth of* Recent Cervus canadensis *and* Odocoileus virginianus. *(L) or (R) indicate left or right side specimens, "e" indicates an estimated measurement, "+" a minimal measurement, SD = standard deviation, and "N" = number in samples.*

Specimens and measurements	Sangamona fugitiva		Cervus canadensis		Odocoileus virginianus				
						Churcher			
	Hay	Churcher	Hay	Churcher	Hay	Maximum— Minimum	Mean	SD	N

First lower incisor

Specimens	Cavetown, Maryland (USNM 9192)			FA 342-4 ♀					
Crown height	8	13		18.1		8.45—6.9	7.7	.56	5
Crown width (diameter)	7.5	9 Max. 6.7 Min.		14.1		8.9—6.8	7.6	1.18	3
Total length		32		50?					

First or second lower molar

Specimen	Cavetown, Maryland (USNM 9193 C [L])								
Mesiodistal length	20	19.4	24		12.5	(vide infra)			
Buccolingual width	14.8	15.0	15		9				
Crown height		8.0							
Ectostylid	present	present							

First lower molar

Specimens	Alton, Illinois USNM series				FA series 342-						
	9002 (R)	9003 (R)	9002 (R)	9003 (R)	1 ♂	2 ♀	4 ♀	Maximum— Minimum	Mean	SD	N
Mesiodistal length	15e	—	15.8e	—	25.3	22.1	20	15.0—12.5	14.2	.82	8
Distal buccolingual width	—	—	—	12.2	14.5	17.9	16.3	11.6—9.8	10.4	.77	8
Crown height			9+	10.0							
Ectostylid			present	absent	absent	present	present	present 3/8			

Second lower molar

	9002 (R)	9003 (R)	9002 (R)	9003 (R)	Hay	1 ♂	2 ♀	4 ♀	Maximum— Minimum	Mean	SD	N
Mesiodistal length	18	20	19.4	23	24	27.7	28.7	26.0	16.9—15.6	16.2	.60	8
Mesial buccolingual width	13	14.8	12.6	13.5		15.3	18.4	16.8	11.6—9.8	10.9	.69	8
Distal buccolingual width			14.0	16.0	15	14.8	20.7	17.7				
Crown height			11.5	14.5								
Ectostylid			present	present		absent	absent	absent	present 1/8			

Third lower molar

	9002 (R)	9003 (R)	9002 (R)	9003 (R)	1 ♂	2 ♀	4 ♀	Maximum— Minimum	Mean	SD	N
Total mesiodistal length	22e	27	22e	28.5	36.1	38.5	35.3	21.1—19.3	20.3	.57	8
Mesiodistal length minus hypoconulid	18	20	19.2	21.5	28.7	27.4	26.0	16.8—14.3	15.6	.81	8
Mesial buccolingual width	13		13.8	16.0	16.2	19.2	17.6	11.0—9.5	10.5	.46	8
Distal buccolingual width			12.1	13.4	15.8	19.7	16.9				
Crown height		15	11.5	16.4							
Ectostylid		present	absent	present	absent	absent	vestigial				

Table 1.—*Continued.*

| Specimens and measurements | *Sangamona fugitiva* | | *Cervus canadensis* | | *Odocoileus virginianus* | | | | |
	Hay	Churcher	Hay	Churcher	Hay	Churcher Maximum— Minimum	Mean	SD	N
				Second upper molar					
Specimens	Whitesburg, Tennessee (Type USNM 8954 [L])		FA 342-2 ♀						
Maximum buccal mesio-distal length	20	19.7	21	28.0		19.8—16.2	17.4	1.19	11
Mesiodistal length over buccal cingulum	16	18.1		25.0		15.5—12.5	14.5	1.08	11
Mesial buccolingual diameter over styles	22	21.6	26	28		18.4—14.9	16.5	1.45	11
Distal buccolingual diameter over styles	20.5	20.4		25.3		19.0—15.3	16.8	1.58	11
Mesial buccolingual diameter over ribs		19.3		27.8		18.4—14.2	15.8	1.32	11
Distal buccolingual diameter over ribs		19.6		24.1		18.5—15.6	16.5	1.09	11
Crown height buccally		13.5				13.2—5.7	11.1	2.0	11
Crown height lingually		9.1				12.5—4.9	10.7	2.0	11
Interlophar endostyle						present 3/11			

and placed them in a new taxon *Navahoceros fricki.* Kurtén (1975:537) characterized this monotypic genus as "Medium to large sized cervids with very thick-set limb-bones and short metapodials; males with simple forked antlers; females antlerless." Kurtén and Anderson (1980:312) modify this to "large deer, in the size range between mule deer and wapiti, with stocky limbs, very short and heavy metapodials, simply built, three-tined antlers, and molar teeth with weakly developed ribs." The observed range of variation in the San Josecito Cave, Mexico, sample encompasses that seen in the other specimens and thus constitutes individual variation.

Kurtén (1975) and Kurtén and Anderson (1980) described both *Sangamona* and *Navahoceros* as intermediate in size between *O. hemionus* and *C. canadensis* but that *Sangamona* has stilt-like metapodials and long, slender limbs, whereas *Navahoceros* has shortened metapodials and limb bones reminiscent of chamois and ibex. They also concluded that the two species' ranges did not overlap, with *Navahoceros* being adapted to an alpine habitat and *Sangamona* to the eastern regions, possibly as a grazer on open ground.

5. Robinson Cave, Overton Co., Tennessee. Guilday et al. (1969) listed three individuals of

Sangamona furtiva from the Armadillo Pit. These are represented by two right P⁴'s (CM 8372–8373), a left P² (CM 8374), a left milk p₄ (CM 8375) and a right P₃ (CM 8377). Two measurements are given for each of these specimens (Table 3) and Guilday et al. (1969:64, figs. 12b, d, e, f) gave line diagrams of one P⁴, and the P², p₄, and P₃. While Hay (1921) described no premolars in the original material, and thus none of these teeth is comparable to any of the paratypic specimens, he gave as a major characteristic of the cheek teeth of *Sangamona* the absence of well developed styles and ribs on the ectolophs of upper molars and on the lingual faces of lower molars. In the specimens reported by Guilday et al., 1969:64) obvious styles and ribs are present, especially on the permanent upper premolars. The milk fourth premolar (CM 8375) has stylids and ribs, and twin ectostylids, typical of a cervid. Guilday and Dr. Clayton E. Ray of The National Museum of the United States (USNM) assigned these teeth to *Sangamona,* although Guilday had previously identified them as *Rangifer.* Guilday suspected initially that these teeth derived from *Rangifer* and that *Sangamona* was a synonym for that genus. Measurements of the Robinson Cave *Sangamona* teeth (Table 3) show them to be closer in size to those of

Table 2.—*Measurements of the postcranial elements from Cavetown, Maryland, assigned to* Sangamona fugitiva *by Hay (1921) compared with those of* Cervus canadensis *and* Odocoileus virginianus.

| Measurements | Sangamona fugitiva Cavetown, Maryland USNM 9193 | | C. canadensis | | | O. virginianus | |
| | | | | Churcher | | | Churcher |
	Hay	Churcher	Hay	ROM 25241 ♂	FA 342-1 ♂	Hay	(ROM 14907 ♀)
Radius, distal end							
Maximum transverse diameter		48.6		58.5	63.2		35.1
Width above articulation less ulna	48	43.8	65	57.8	55.3	40	29.7
Transverse diameter of articulation		45		57.0	53.0		34.1
Thickness above articulation	31	32.4	45	44.5	43.1	27	26.2
Scaphoid							
Proximodistal length	26	27	?19	28.3	31.2		18.6
Anteroposterior width	32	32	26	35	38.0		24.2
Transverse diameter		17	?19	18.8	22.4		13.1
Acetabulum							
Anteroposterior diameter		45.8		56	59.8		33.6
Length	45		57			37	
Malleolar							
Anteroposterior horizontal diameter	26	25.4		33.1	31.9		19.3
Transverse diameter		17.4		16.3	17.3		9.0
Proximodistal diameter		27.8		26.1	24.0		16.2
Calcaneum (juvenile)							
Preserved maximum length	±105	102+	138	132.5	138.5	103	93.6
Height at articulation for fibula	44	43.5	56	49.2	50.7	32	30.5
Height over sustentaculum		37.1		43.0	40.5		25.5
Thickness at lateral (sustentacular) process	33	32.5	40	41.0	35.5	30	23.4
Length of cuboid facet		30.5		33.1	36.3		20.6
Width of cuboid facet		12.2		15.0	14.8		8.8
Astragalus							
Tibial (articular) width	30	30.7		37.4	36.5		26.1
Dorsoventral depth		26.1		34.6	34.6		21.1
Phalanx I							
Total length	66	66.5	68	62.8	65.1	53	53.2
Proximal height	28	28.5	35	30	31.4	21	21.3
Proximal width	23	22.5	27	23.7	25.0	17	17.4
Distal height	18	18.2	21	19.6	20.0	14	12.5
Distal width	19	18.6	26	24.4	24.3	15	12.8
Least shaft height		16.6		17.6	17.8		11.6
Least shaft width		17.0		20.3	19.1		12.3
Phalanx II							
Total length	42	42.1	48	45.0	47.3	41	38.5
Proximal height	26	25.4	35	29.8	30.5	22	20.9
Proximal width	19	17.9	25	24.3	22.4	17	14.7
Distal height	24	24.5	29	27.0	28.5	18	17.0
Distal width	16	15.8	26	19.0	20.0	12	11.0
Least shaft height		19.3		20.5	21.3		14.1
Least shaft width		14.0		18.5	15.6		10.7
Phalanx III							
Length	45e	38.4+		50.3			36.6
Proximal height	30	31.3		32.3			20.2
Proximal width	18	18.1		10.0			12.3

Table 3.—*Measurements of cheek teeth of* Sangamona fugitiva *(CM 8372–8375, 8377) from Robinson Cave, Tennessee (after Guilday et al., 1969:65) compared with those of Recent* Cervus canadensis, Odocoileus virginianus, *and* Rangifer tarandus.

Tooth and measurements	*Sangamona furtiva* Tennessee		*Cervus canadensis* British Columbia		*Odocoileus virginianus*		*Rangifer tarandus* Yukon
					Ontario and Alberta	Ontario	
P²	CM 8374		FA 342-2		Churcher Collection	ROM 14907	Churcher Collection
Mesiodistal length	17.5		20.5		11.8, 12.1, 15.4, 16.0		
Buccolingual width	13.7		17.3		10.7, 10.7, 14.0, 14.1		
P⁴	CM 8372	CM 8373					
Mesiodistal length	14.5	13.2	17.7		11.3, 10.7, 12.9, 12.8	10.2, 9.2	
Buccolingual width	18.5	18.4	22.5		12.3, 12.1, 17.8, 17.2	10.5, 13.1	
P₃	CM 8377		FA 342-3	FA 342-4			
Mesiodistal length	13.1		18.8	18.3	11.1, 11.2, 11.8, 11.8	10.4	14.3
Buccolingual width	9.1		13.0	12.3	7.7, 7.3, 7.2, 7.3	6.8	9.5
P₄	CM 8375						
Mesiodistal length	22.5				16.2, 15.8		22.5, 21.0, 22.0
Buccolingual width	11.1				8.2, 8.1		7.6, 7+, 7.5

Cervus canadensis than to *Odocoileus virginianus* and are similar to those of *Rangifer tarandus,* except in the buccolingual diameter of p₄.

The fauna from Robinson Cave includes *Odocoileus* cf. *virginianus,* for which Guilday et al. (1969) identified a right P₂ and P₃, and it is likely that the *Sangamona* teeth represent this deer because of similar size and patterns of the cusps and lophs.

6. Brynjulfson Cave No. 1, Boone Co., Missouri.

Parmalee and Oesch (1972) reported cf. *Sangamona* sp. from Brynjulfson Cave No. 1, represented by a left humerus, right and left tibiae, distal end of left tibia, and a left metatarsal. Two left rami with P₄–M₃ (Jaw A) and P₃–M₃ (Jaw B) (Parmalee and Oesch, 1972:42–43, fig. 14, table 8; erroneously labeled P₃ and P₂–P₃ in table 8) may belong to *Sangamona,* although Dr. Clayton E. Ray is cited as stating that they agree well with the two topotypic rami of *Odocoileus halli.* No teeth of *Odocoileus* are reported although remains of five individuals (85 specimens) of *O. virginianus* and a left P¹, a cervical vertebra, a partial left calcaneum, a right acetabulum, and the distal end of a right cubitus of *Alces americana* or *Cervalces* are recorded.

None of the postcranial materials is comparable to the paratypic specimens from Cavetown described by Hay (1921; Table 4). The doubtfully assigned rami with molar series possess strongly ribbed convex lingual surfaces and appear to differ from the Alton rami in the characters of the molars (compare Hay, 1921, pl. 5, figs. 5–6; Parmalee and Oesch, 1972:42, fig. 14). The pattern of the Brynjulfson P₃ differs from that from Robinson Cave but, because these patterns vary considerably in recent and extinct deer, this difference may not be significant. The

Brynjulfson *Sangamona* lower dentitions are all smaller than those of *C. canadensis* (Table 5 and compare with Table 1) and larger than those of *O. virginianus.* Thus it appears that the Brynjulfson dentitions are more properly assigned to *O. halli,* as suggested by Dr. Ray, or to *O. virginianus* which is known from the cave but not represented by dental material.

The postcranial materials are not comparable to Hay's (1921) originally described materials but may be compared with recent cervid elements. The humerus is smaller than that of *C. canadensis* and comparable to that of *O. virginianus* (Table 4). The tibiae and metatarsal are intermediate in size between those of *C. canadensis* and *O. virginianus.* The three postcranial elements may therefore be parsimoniously assigned to *Odocoileus* sp., which supports the possible identity of the two rami.

7. Gravel Pits on the South Bank of the Republican River, Red Willow Co., Nebraska.

Corner (1977) reported a left antler pedicel (USNM 46326), two distal ends of right humeri (UNSM 48534–48535) from Pit RW 101 and a left metacarpal (UNSM 48526) and the distal end of a right humerus (UNSM 46304) from Pit RW 102 from the south bank of the Republican River.

Corner (1977:85) stated "The metacarpal is referred to *Sangamona* mainly on the basis of size, being intermediate between mule deer and wapiti (Kurtén, 1975:508). *Sangamona* differs from *Navahoceros* in having longer, more slender limbs. The metacarpal is structurally similar to that of the wapiti (*Cervus canadensis*), except the posterior vascular groove is not as deep. The partial humeri were . . . found to be similar in size and structure" to a hu-

Table 4.—*Measurements of the postcranial elements of "Sangamona" from Brynjulfson Cave, Missouri, after Parmalee and Oesch (1972: 41, table 7), of* Cervus whitneyi, Odocoileus virginianus, *and* O. hemionus *after Allen (1876:50), with corrected or additional measurements of the same specimens and Recent* Cervus canadensis, Odocoileus virginianus, *and* O. hemionus.

Element and dimension	C. whitneyi Dubuque Iowa	"Sangamona" Brynjulfson Cave Missouri	C. canadensis ROM 25241 ♂ British Columbia	C. canadensis FA 342-2 ♀ British Columbia	O. virginianus MCZ 1733 Maine	O. hemionus MCZ 1781 Wyoming
Humerus						
Total length	—	236.0	308	310	220	227
Length from caput to medial condylar surface	—		282	282	200	203
Transverse width of distal condylar surfaces	48	49	62	68.8	38	42
Anteroposterior width over distal medial condyle	51		64.3	65.0	42	42
Maximum anteroposterior width in centre of shaft (A)	—	29.0	40.5	36.0	24.6	—
Lateral transverse width in centre of shaft (B)	—	23.0	33.5	29.5	19.8	—
Calculated circumference in centre of shaft (A + B)π/2	85	81.7	116.4	218.4	73	76
Radius				Allen, 1876		
Total length	—				230	242
Transverse width of proximal end	—				37	39
Transverse width of distal end	41		57.1	55.3	36	38
Least transverse diameter of shaft	29				24	25
Least circumference of shaft	80				65	68
Metatarsal						
Total length	—	314.0	326	336	254	273
Transverse width of proximal end	33				28	29
Anteroposterior width of proximal end	36	40.0	47.3	47.1	27.9	32
Least transverse width of shaft	22				18	21
Transverse width in centre of shaft		19.5	28.8	26.2	15.4	
Least circumference of shaft	67				58	66
Width of distal end	—	37.0	50.7	49.5	32.3	35
Tibia						
Total length		376, 378, —	410	420	311.2	
Width of proximal end		69.0, 72.0, —	86.2	91.2	56.2	
Transverse width in centre of shaft		29,0, 29.0, 26.0	36.6	21.5	32.4	
Maximum width of distal end		48.0, 48.0, 44.0	37.0	58.3	37.0	

Table 5.—*Measurements of the dental series from the* "Sangamona" *dentaries from Brynjulfson Cave, Missouri (after Parmalee and Oesch, 1972:43, table 8) compared with those of Recent* Cervus canadensis *and* Odocoileus virginianus.

Dimension	"Sangamona" Brynjulfson Cave, Missouri P₃	P₄	M₁	M₂	M₃	Cervus canadensis FA 342-2, 342-4, British Columbia P₂	P₃	P₄	M₁ M₂ M₃
Jaw A									
Mesiodistal length		15.0	16.0	18.5	23.5	12.1	19.3	20.5	Measurements for lower molars of
Buccolingual width	—	11.5	13.5	14.0	13.5	9.7	12.3	14.7	*Cervus canadensis* and *Odocoileus virginianus* are given in Table 1.
Alveolar length of tooth row	—	101.0				116.7			
Jaw B									
Mesiodistal length	16.5	17.5	18.0	22.0	26.5	15.5	19.5	20.3	
Buccolingual width	13.0	14.0	12.0	13.0	12.0	10.8	13.0	15.6	
Alveolar length of toothrow		115 estimated				124.5			

merus reported from Burnet Cave by Schultz and Howard (1935:287) and considered as *Navahoceros fricki* by Kurtén (1975:507). Corner (1977:85) referred the partial humeri from Pit RW 101 to "*Sangamona* and not *Navahoceros* because there is no evidence to suggest that both forms are present at Red Willow, and the complete metacarpal can be referred to *Sangamona* with some confidence."

Corner (personal communication, 9 January 1978) recognized that some Recent wapiti metacarpals approach that of the Red Willow specimen in size but are narrower at the articulations, and also that the humeri may "be referrable to the long-legged bighorn sheep (*Ovis catclawensis*)." Measurements of the metacarpal (UNSM 48526) and of Recent *C. canadensis* and *O. virginianus* metacarpals (Table 6) show that it more nearly agrees with those of the former in all dimensions. The metacarpal probably represents a small *Cervus canadensis* and the humeri possibly *Ovis catclawensis*, which is also present at RW 101.

"The antler fragment is much larger than one would expect for the small *Odocoileus*" (Corner, 1977:85) and "is intermediate in size between *O. hemionus* and *Cervus canadensis*" and "does compare nicely to the *Navahoceros* from Slaughter Canyon Cave, New Mexico, which is mounted and on display" in Lincoln, Nebraska. However, the antler

Table 6.—*Measurements of the left metacarpal of "Sangamona" from the Republican River, Red Willow Co., Nebraska, compared with those of Recent* Cervus canadensis *and* Odocoileus virginianus.

Dimension	"Sanga-mona" UNSM 48526 Nebraska	*C. canadensis* British Columbia		*O. vir-ginianus* ROM 14907 Ontario
		ROM 25241	FA 342-1	
Articular length	270	285	291	215.9
Transverse width of proximal end	45.3	47	50.3	28.9
Anteroposterior width of proximal end	30.9	33.3	35.2	20.9
Midshaft transverse width	27.0	28.5	27.3	17.3
Minimum anteroposterior shaft diameter	19.9	20.8	23.5	18.1
Transverse width of distal end	45.2	49.3	51.0	30.7
Anteroposterior width of distal end	—	33.3	34.8	20.8

pedicel (UNSM 46326) is not too small to have derived from a small individual of *Cervus canadensis* and likely represents the same taxon as the metacarpal.

The Red Willow County records of *Sangamona* are thus suspect, and likely *partim* represent *Cervus canadensis* and *partim Ovis* sp., possibly *Ovis catclawensis* (=*O. canadensis catclawensis*).

COMMENTS ON SPECIMENS LATER REFERRED TO *SANGAMONA* FROM OTHER CERVID TAXA (LOCALITIES 8–10)

8. Dubuque, Dubuque Co., Iowa; *Cervus whitneyi.*

Allen (1876) described *Cervus whitneyi* from Dubuque, Iowa, from a fauna originally reported by Wyman (1862) as comprising *Bos* (=*Bison*), *Cervus*, *Mastodon* (=*Mammut*), *Megalonyx*, *Dicotyles* (=*Platygonus*), and *Canis* and based on the cervid specimens that Wyman identified as "Red Deer (*Cervus virginianus*)." The cervid remains comprise a left humerus lacking the proximal epiphysis, and a left radius lacking the distal end, both imperfect or damaged, and a right metatarsal lacking its distal end (referred to as left and right by Allen, 1876:48, 50, respectively), all from a young animal. Wyman (1862:412) described the humerus as "closely resembling that of the red deer, and of intermediate size between this and the humerus of a Caribou." Allen (1876:48) considered that these specimens

"evidently belonged to a species different from any hitherto described, either extinct or living," and placed them in a new species "*Cervus Whitneyi*, in honor of their discoverer, Professor J. D. Whitney."

Allen (1876:50) considered that the three cervid elements indicated "a species of about the same proportions as *Cervus Virginianus* [=*Odocoileus virginianus*], but much larger, considerably exceeding in size *Cervus macrotis* [=*Odocoileus hemionus*]." He gave comparative measurements of these elements and those of "*C. macrotis*" and "*C. Virginianus*" which show the fossil form to be larger than the recent species by between 10% and 17% (Table 4).

The humerus is typically cervid in form, the radius is similar but the ulna is fused for nearly the whole radial contact, as in *O. virginianus* but not in *O. hemionus*. The metatarsal is similar to that of *O.*

virginianus, but the posterior groove continues more distad and differs from that of *O. hemionus* in being more laterally compressed distally, slenderer, and rounder in section, as in *C. canadensis.*

Allen's (1876) trivial name *"whitneyi"* for these Iowan specimens predates Hay's (1921) description of the genus *Sangamona* and its single species, *S. fugitiva.* Thus the correct form of the name is *Sangamona whitneyi* (Hay). However, Allen's name has seldom if ever been applied, except to indicate these specimens, and Hay's binomial has achieved a wide currency. Kurtén and Anderson (1980) prefer to retain *Sangamona fugitiva* for Hay's type materials and other specimens assigned to it, and probably for *C. whitneyi* also. The name *Cervus whitneyi* is thus a *nomen oblitum.*

Kurtén (1975:508) tangentially referred the Iowan specimens to *Sangamona,* but without discussion. Kurtén and Anderson (1980:313) stated that *Sangamona fugitiva* "is probably also represented by remains from Dubuque, Iowa, which were described as *Cervus whitneyi* by Allen (1876)." Kurtén (personal communication) considered that the Dubuque *C. whitneyi* elements typical of *Sangamona.* However, no specimens of humerus or metatarsal are recorded by Hay (1921) from the type locality at Whitesburg, Tennessee, or from Alton, Illinois, or Cavetown, Maryland, and only the distal end of a left radius from the latter, so comparisons are not possible. No comparable elements are recorded from Robinson Cave, Tennessee (Guilday et al., 1969), but a left humerus and a left metatarsal are recorded from Brynjulfson Cave, Missouri (Parmalee and Oesch, 1972). Comparison of measurements of these elements with those of modern *O. virginianus* and *C. canadensis* show that the elements from the three fossil sites resemble one another in size (Table 4). The humerus from Brynjulfson Cave is slightly longer than in modern *Odocoileus* but similar in section to that from Dubuque. The radial ends from Cavetown (Table 2) and Dubuque (Table 4) are intermediate in size between those of *Odocoileus* and *Cervus canadensis.* The metatarsals from Brynjulfson Cave and Dubuque (Table 4) are smaller than that from Red Willow Co. (Table 6) originally assigned to *Sangamona* by Corner (1977) but now considered to derive from *Cervus canadensis.*

9. Frankstown Cave, Blair Co., Pennsylvania; *Odocoileus hemionus.*

Peterson (1926) reported *Odocoileus hemionus* from Frankstown Cave, Pennsylvania, based on fragments of a skull (CM 11044) with complete dentition and a number of limb and foot bones (CM 11043). He (1926:257) stated "The bones are those of an animal of considerably larger size than *Odocoileus virginianus* and most closely resemble the recent species, *O. hemionus.* The inner basal cusps of the [upper] molars . . . are in the present specimen, small and rather irregularly developed, more nearly like the recent mule deer." The appendicular elements are appropriately sized to derive from a single individual of this species. These remains were associated with elements of *O. virginianus* and *Cervalces americanus.*

The development of the inner basal cusps (endostyles) between the lophs of the upper molars of deer (and most artiodactyls) is highly varied and unreliable for taxonomic use in Cervidae. Thus size alone differentiates the Frankstown specimens from those of the associated *O. virginianus* and suggests *O. hemionus.*

Guilday et al. (1969:64) remarked that, "In the process of comparing the specimens of Cervidae from Robinson Cave, Tennessee, a partial skeleton from Frankstown Cave, Pa. identified as mule deer, *Odocoileus hemionus* (Peterson, 1926), was found to be that of *Sangamona.*" Guilday et al.'s (1969) statement was made when only Hay's (1921) original and Schultz and Howard's (1935) Burnet Cave reports and specimens were available. Thus comparisons were possible only with the worn type upper M^2 from Whitesburg, Tennessee, the almost featureless and undescribed P^3–M^3 from Alton, Illinois, the partial pes and other postcranial fragments from Cavetown, Maryland, and the specimens from Burnet Cave, New Mexico, now assigned to *Navahoceros fricki.* The comparative materials thus lacked either valid points for qualitative distinction or were distinctly different.

Guilday (personal communication, in précis) gave the following details about the Frankstown Cave cervid cranium: the skull is extremely fragmentary, the preserved parts are the dorsal surfaces of the frontals and parietals, portions of the occipitals with intact condyles, two petrous temporals, and complete upper tooth rows. The animal, although an adult doe, is not small; it is markedly larger than a whitetail. The molars are large with reduced buccal ribbing, and the premolars are not only actually but relatively larger than in *O. virginianus.* The skull is so fragmentary that only two dimensions may be measured: maximum width across the condyles, 61.0, and transverse diameter of foramen magnum, 26.8 mm. He also remarked that the upper dentition was so strikingly dissimilar from that of *Odocoileus* that he did not consider them closely related.

Examination of the Frankstown cranial specimens allows little comparison as to size and morphology, except to substantiate that the animal was female and, while large, not outside the size range for *Odocoileus* and towards the lower range of size for *C. canadensis*.

10. Baker Bluff Cave, 13 km SE of Kingsport, Sullivan Co., Tennessee.

Guilday et al. (1978) reported two isolated premolars (P^4, CM 29501; left P$_3$, CM 30060) assigned to "cf. *Sangamona fugitiva* Hay—'fugitive' deer." They stated that the upper premolar was larger than that of *Odocoileus* and about the size of that of large *Rangifer* from the deposit. Both teeth have the weak buccal ribbing that Hay (1921) considered diagnostic of *Sangamona*. The P$_3$ is "identical in cusp formation with *Cervus* or *Odocoileus* and was referred to *Sangamona* because of its intermediate size" (Guilday et al., 1978:50).

Measurements of P^4 (CM 29501) are mesiodistal length, 14.0, buccolingual width, 15.8 mm, and of P$_3$ (CM 30060) are buccolingual width, 9.6 mm. These measurements agree tolerably with those for the Robinson Cave, Tennessee, specimens, but are, of course, not comparable to the type molar from Whitesburg, Tennessee (USNM 8954), or the lower molars of the paratypes from Alton, Illinois (USNM 9002, 9003).

SUMMARY DISCUSSION

Each fossil occurrence attributed to *Sangamona* has generated divergent views, but in few of them have discussions or examinations of tooth size and general morphology, or of the size of the whole animal or its long bones, been made on a comparative basis, or if they have been, then the bases for the conclusions have not been stated. In part this is probably because of the sparse and idiosyncratic nature of Hay's (1921) incomparable "type specimens" and to the incomplete skeletal examples later added or referred to *Sangamona*. When size alone appears to be the ruling criterion for assignment to a genus, it may well be unreliable, especially if the taxa compared are not represented by ranges of size, nor are all likely taxa considered.

Sangamona has thus "evolved" from an original indeterminate tooth, supported by geographically separated dental and postcranial samples, through the additions of further dental and postcranial elements and two mandibular rami. When these specimens are considered parsimoniously for comparability, it is evident that in only two sites (Brynjulfson Cave No. 1, Missouri, and Frankstown Cave, Pennsylvania) are there possibly associated cranial and postcranial elements, and in both instances these are associated with verifiable *Odocoileus virginianus* remains. No antler rack has been reported and the pedicel (Red Willow Co., Nebraska) is likely *Cervus canadensis*.

Martin and Guilday (1967:52) summed up the taxonomic situation at the time when only Hay's (1921) type materials and Schultz and Howard's (1935) erroneous report were available. They stated "An extinct deer about the size of the modern caribou, *Sangamona* was widely distributed in N.A. Remains have been found in both western and eastern U.S.A., but there is no evidence of contemporaneity with man. The type specimen is a single tooth. While additional fragmentary remains appear to uphold the validity of this late Pleistocene deer, *a definite description has yet to appear*" (my italics).

Thus *Sangamona* is known as a deer that lacks antlers (possibly only females known?), tends to be associated with other cervids in cave deposits, is intermediate in size between *O. hemionus* or *O. virginianus* and *C. canadensis*, possesses molars with weak styles, and is built in other aspects like *Odocoileus* or *Cervus*. As stated before, size is an unreliable indicator of taxon when qualitative criteria are absent, and thus the postcranial elements alone provide little basis for allocation to a taxon other than those already recovered in the deposit. Stylar variation in North American cervid teeth is wide, and anomalously heavy or nearly absent conditions occur in modern populations of *Odocoileus virginianus*, *Cervus canadensis*, and *Rangifer tarandus* examined in the Royal Ontario Museum, Toronto, the National Museum of Natural Science, Ottawa, and the British Museum (Natural History), London.

The recovery of a cranium with portions of the antler beams to the bez tines from the early postglacial (latest Pleistocene, 11,315 ± 325 B.P.) of To-

Table 7.—*Summary of occurrences of* Sangamona fugitiva, *identifications, associated cervid taxa, and revised identifications.* + = *present in deposit,* +cf = *probable presence,* +sp = *genus present, species uncertain, and* X = *wrongly identified.*

Site and original identification	Odocoileus virgini- anus	Odo- coileus hemionus	Cervus canadensis	Rangifer tarandus	Alces alces or Cervalces sp.	Revised identification
Whitesburg, Tennessee *Sangamona fugitiva*–type	+		+			Indeterminate: ?large *Odocoileus* or small *Cervus*
Alton, Illinois *Sangamona fugitiva*				+	+ *Cervalces roosevelti*	*Cervus canadensis*
Cavetown, Maryland *Sangamona fugitiva*	+					large *Odocoileus* or small *Cervus*
Burnet Cave, New Mexico *Sangamona?* sp.	+	+		X *Rangifer? fricki*		*Navahoceros fricki*— Kurtén, 1975
Robinson Cave, Tennessee *Sangamona furtiva*	+cf					*Odocoileus virginianus*
Brynjulfson Cave, Missouri cf. *Sangamona* or *Odocoileus halli*	+				+cf *Alces* or *Cervalces*	*Odocoileus ?halli*
Republican River, Nebraska *Sangamona* sp.	+sp			+		*Cervus canadensis* and *Ovis catclawensis*
Dubuque, Iowa *Cervus whitneyi*			+sp			*Cervus canadensis*
Frankstown Cave, Pennsylvania *Odocoileus hemionus*	+				+	large *Odocoileus* or small *Cervus*
Baker Bluff Cave, Tennessee cf. *Sangamona fugitiva*	+		+	+		large *Odocoileus* or small *Cervus*

ronto, named *Torontoceros hypogaeus* (Churcher and Peterson, 1982), was greeted by some as providing the missing antlers of *Sangamona*. However, the Toronto cranium has no points of comparison with Hay's (1921) original specimens and the Frankstown Cave cranium is too fragmentary and could not be reliably compared to the Toronto cranium. Because the Toronto cranium has geological significance in addition to its taxonomic information, publication of the record was desirable. To publish the cranium as *Sangamona* would only have exacerbated the situation already existing by suggesting information on the shape of the antlers, but based on a specimen that had no reason except convenience to be assigned to *Sangamona*.

What is *Sangamona fugitiva*? It appears to be derived almost equally from *Odocoileus* (probably *O. virginianus*) and *Cervus canadensis* remains (Table 7). It was inadequately founded by Hay (1921) and has been extended by attribution of unusual-sized bones or cheek teeth with reduced styles from seven other sites. Thus I respond to John Guilday's stated wishes that someone someday would make direct comparisons and pin the beast down, and conclude that *Sangamona* is mythological. *Sangamona fugitiva*—the fugitive or furtive deer—is a construct of diverse and disparate skeletal elements, all of which derive from other and usually better known taxa and, except in the minds of men, never existed.

ACKNOWLEDGMENTS

I have first to thank Dr. R. L. Peterson of the Department of Mammalogy, Royal Ontario Museum, Toronto for involving me in the problem of the validity of *Sangamona fugitiva* by asking me to identify the cranium with antlers we subsequently described as *Torontoceros hypogaeus* (Churcher and Peterson, 1982), as this was possibly *Sangamona*. Without his unknowing impetus, this work would never have been completed. Dr. Björn Kurtén of the University of Helsinki, Finland, Dr. C. Richard

Harington of the National Museum of Natural Sciences, Ottawa, Dr. R. George Corner of the University of Nebraska State Museum, Lincoln, Nebraska, and the late Mr. John E. Guilday all made freely available their experience and knowledge of fossil deer and *Sangamona*. However, the conclusions expressed here are solely my responsibility. Dr. Juliet Clutton-Brock of the Department of Mammals, and Dr. Alan W. Gentry of the Department of Palaeontology, British Museum (Natural History), London, England, and Drs. Harington and Peterson assisted me by allowing me to use the collections in their care. I have also been helped by the friendly advice of Dr. Thomas S. Parsons and Dr. James A. Burns of this department, and by Mrs. Judy Hann-Chernos, my technical assistant, who drew the diagrams and typed an earlier draft of this paper, and by Ms. Margjka Mychajlowycz, who helped proof the final version. The work was financed by National Science and Engineering Research Council Grant A 1760.

LITERATURE CITED

ALLEN, J. A. 1876. Description of some remains of an extinct species of wolf and an extinct species of deer from the Lead Region of the Upper Mississippi. Amer. J. Sci. Arts., Ser. 3, 11(Whole No. 111):47–50.

ALVAREZ, T. 1969. Restos fósiles de mamíferos de Tlapacoya, Estado de México (Pleistoceno-Reciente). Misc. Publ. Mus. Nat. Hist., Univ. Kansas, 51:93–112.

CHURCHER, C. S., and R. L. PETERSON. 1982. Chronologic and environmental implications of a new genus of fossil deer from Late Wisconsin deposits at Toronto, Canada. Quaternary Res., 18:184–195.

CORNER, R. G. 1977. A Late Pleistocene-Holocene vertebrate fauna from Red Willow County, Nebraska. Trans. Nebraska Acad. Sci., 4:77–93.

DAVIS, L. C. 1969. The biostratigraphy of Peccary Cave, Newton County, Arkansas. Proc. Arkansas Acad. Sci., 23:192–196.

FRICK, C. 1937. Horned ruminants of North America. Bull. Amer. Mus. Nat. Hist., 69:1–699.

GUILDAY, J. E., H. W. HAMILTON, and A. D. McCRADY. 1969. The Pleistocene vertebrate fauna of Robinson Cave, Overton County, Tennessee. Palaeovertebrata, 2:25–75.

GUILDAY, J. E., H. W. HAMILTON, E. ANDERSON, and P. W. PARMALEE. 1978. The Baker Bluff Cave deposit, Tennessee, and the Late Pleistocene Faunal Gradient. Bull. Carnegie Mus. Nat. Hist., 11:1–76.

HAY, O. P. 1921. Descriptions of some Pleistocene vertebrates found in the United States. Proc. U.S. Natl. Mus., 58 (2328): 83–146.

KURTÉN, B. 1975. A new Pleistocene genus of American mountain deer. J. Mamm., 56:507–508.

KURTÉN, B., and E. ANDERSON. 1980. Pleistocene mammals of North America. Columbia Press, New York, xvii + 443 pp.

MARTIN, P. S., and J. E. GUILDAY. 1967. A bestiary for Pleistocene biologists. Pp. 1–62, *in* Pleistocene extinctions: the search for a cause (P. S. Martin and H. E. Wright, Jr., eds.), Proc. VII Congress INQUA, Yale Univ. Press, New Haven, x + 453 pp.

PARMALEE, P. W., and R. D. OESCH. 1972. Pleistocene and Recent faunas from the Brynjulfson Cave, Missouri. Illinois State Mus., Rept. Invest., 25:1–52.

PETERSON, O. A. 1926. The fossils of Frankstown Cave, Blair County, Pennsylvania. Ann. Carnegie Mus., 16:249–315.

SEMKEN, H. A. 1969. Paleoecological implications of micromammals from Peccary Cave, Newton County, Arkansas. Geol. Soc. Amer., South Central Sect., Abstr. Prog., 2:27.

SIMPSON, G. G. 1945. Principles of classification and a classification of mammals. Bull. Amer. Mus. Nat. Hist., 85:1–350.

SCHULTZ, C. B., and E. B. HOWARD. 1935. The fauna of Burnet Cave, Guadalupe Mountains, New Mexico. Proc. Acad. Nat. Sci. Philadelphia, 87:273–298.

WELLS, L. H. 1959. The nomenclature of South African fossil equids. S. African J. Sci., 55:64–66.

WYMAN, J. 1862. Chapter VII, Observations upon the remains of extinct and existing species of Mammalia found in the crevices of the lead-bearing rocks, and in the superficial accumulations within the Lead Region of Wisconsin, Iowa and Illinois. Pp. 421–423, *in* Report on the Geological Survey of the State of Wisconsin (J. Hall and J. D. Whitney), 1:1–437.

Address: Department of Zoology, University of Toronto, Toronto, Ontario, Canada M5S 1A1 and Department of Vertebrate Palaeontology, Royal Ontario Museum, Toronto, Ontario, Canada M5S 2C6.

GEOLOGICAL SETTING AND PRELIMINARY FAUNAL REPORT FOR THE ST-ELZÉAR CAVE, QUEBEC, CANADA

Pierre LaSalle

ABSTRACT

The St-Elzéar cave is located in limestone terranes of Middle Silurian age. Because of its vertical entrance, the cave was a trap for large and small animals, mostly small mammals, the majority of which are still living in the area today. The cave was presumably opened to the surface only in post-glacial time. However, until excavations are terminated, no definite age can be assumed. It is virtually certain that the cave area was overridden by glacier ice at some time in the past because of the presence of erratics in the area surrounding the cave, north and south of it. The erratics appear to have been superimposed on the rubble formed in situ possibly in pre-glacial time. Uranium-thorium dates suggest that the precipitation of calcite inside the cave was already under way in pre-glacial time, probably during the Sangamon or some older inter-glacial.

INTRODUCTION

The St-Elzéar cave (65°21'30" long. W; 48°14'20" lat. N), located near St-Elzéar-de-Bonaventure, Québec, Canada (Fig. 1), was discovered (or rediscovered?) in December 1976 by local residents during a snowshoe trek. The author started working on the talus of the cave entrance in the autumn of 1977, with the hope of discovering old sediments, possibly older than the last glaciation. As many readers are possibly not aware, the glacial history of the Gaspé Peninsula has been the subject of a long debate which is still going on vigorously, and evidence for multiple glaciations is still lacking in most of the area. However, it is generally agreed that the Gaspé Peninsula has been overridden by continental glaciers several times in the past (Mailhiot, 1919; Coleman, 1922; Alcock, 1935, 1944; McGerrigle, 1952; Grant, 1977; Lebuis and David, 1977). The dispute is more with the nature of the glaciation and its extent during Wisconsinan time (Grant, 1977) than with the fact of the glaciation itself (Prest, 1970; Mayewski et al., 1981).

A preliminary bone collection was made from the talus at the foot of the cave entrance in the autumn of 1977, and was sent for study and identification to the late John E. Guilday. The author thought the bones would be mostly from bats. To the great surprise of both the author and John E. Guilday, the bone collection included remains of the collared lemming (*Dicrostonyx hudsonius* Pallas). Further studies by Guilday have revealed the presence of the remains of more than 30 species in the talus deposit of the St-Elzéar cave; most of these species are still living in the area today. A short report has been published on the fauna (LaSalle and Guilday, 1980) and this matter will be dealt with briefly as the studies are still in progress. Rather, the purpose of the present paper is to describe the bedrock geology in the area surrounding the cave entrance, and report on the glacial geology as it is known today.

BEDROCK GEOLOGY

For a review of the literature relevant to the bedrock geology of the area, the interested reader is referred to Badgley (1956) and Bourque and Lachambre (1980). St-Elzéar cave is located (Fig. 2) in the La Vieille Formation of Middle Silurian age. This formation is generally composed of three members of nodular and algal limestone. The contact with adjacent formations is gradual and determined arbitrarily (50% nodular limestone or calcareous mudstones; see Bourque and Lachambre, 1980:57).

The cave is located on the southern flank of a synclinorium, and some minor faulting is associated with the placement of the cave. This situation has presumably favored groundwater circulation and enhanced the transport of carbonate in solution. Presently, the water table is below the cave floor, and its level is linked in some way with the water level in the stream flowing in the ravine below the cave entrance.

As shown in Fig. 3, the cave, because of its vertical

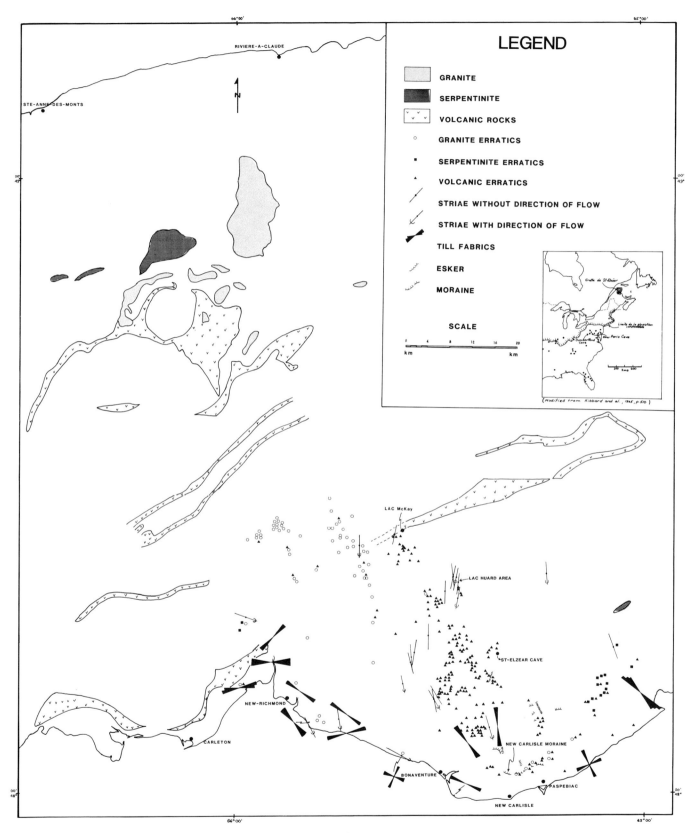

Fig. 1.—Glacial features in the area surrounding St-Elzéar Cave, Gaspé Peninsula, Quebec.

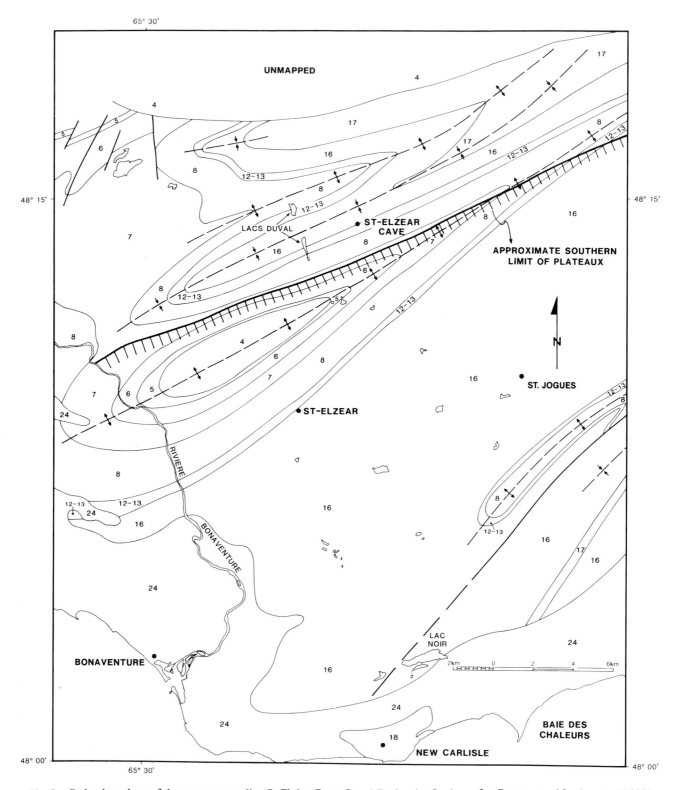

Fig. 2.—Bedrock geology of the area surrounding St-Elzéar Cave, Gaspé Peninsula, Quebec, after Bourque and Lachambre (1980).

CHRONOSTRATIGRAPHY		LITHOSTRATIGRAPHY		LITHOLOGY	
		GROUP	FORMATION		
CARBONIFEROUS		NO NAME	BONAVENTURE	24	Red conglomerates, shales and sandstones
UNCONFORMITY					
DEVONIAN	GEDINNIAN		INDIAN POINT	18	Mostly siltstones
—?—	—?—		WEST POINT	17	Reef Complex
	PRIDOLIAN				
	LUDLOVIAN		GASCONS	16	Mostly siltstones
	WENLOCKIAN		LA VIEILLE	13	Nodular and algal limestones
SILURIAN	C	CHALEURS	ANSE A PIERRE-LOISELLE	12	Mostly mudstones with limy nodules
			ANSE CASCON	8	Mostly quartz sandstones
	LLANDOVERIAN		WEIR	7	Feldspathic sandstones and conglomerates
	B		CLEMVILLE	6	Mudstones to sandstones
	A		MATAPEDIA	5	Mostly limestones and limy shales
ORDOVICIAN	ASHGILLIAN	HONORAT		4	Mostly mudstones
DICONFORMITY?					

← Key to Fig. 2.

entrance, is a trap for small and large animals. The talus itself has accumulated by layers under the influence of gravity, and some of the material (besides the bones) has been washed or blown in from the outside. The talus is thus composed of various plant and animal remains, together with blocks fallen from the roof of the chamber set in a matrix of clayey brownish material. Moonmilk is also present in small masses or layers surrounding stones or other geological objects. A crude stratification can be observed in the talus with a slope of approximately 35°.

GLACIAL GEOLOGY

The glacial geology of the Gaspé Peninsula has been the subject of discussions and controversy for almost one hundred years. The first report dealing specifically with the subject area of this paper was published by Chalmers (1887; 1906). Subsequently, several geologists (mostly bedrock geologists; see Bourque and Lachambre, 1980) have commented and added some new data concerning the glacial geology of the area. Mailhiot (1919) thought that Mont Albert had been glaciated, on the basis of its geomorphology and the presence of a "few erratic blocks of granite and hornblende schist . . . ." (Mail-

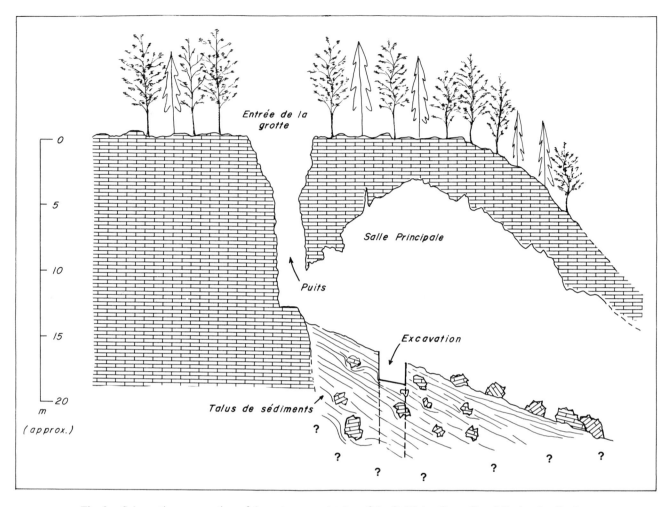

Fig. 3.—Schematic cross-section of the entrance and talus of the St-Elzéar Cave, Gaspé Peninsula, Quebec.

hiot, 1919:146). Coleman (1922) studied the glacial geology of the Gaspé Peninsula and concluded that it had been covered by local glaciers. Alcock (1935) reported the presence of granite erratics on the north shore of the Baie-des-Chaleurs, and also observed glacial striations in the same area. McGerrigle (1952) summarized the data available at the time of publication of his report (1952) and seems to have put to rest one of the most persistent hypotheses put forth by botanists, especially by Fernald (1925). The latter thought that because of the presence of arctic endemics, that is, tundra type plants in their contemporaneous floras, the higher peaks of the McGerrigle Mountains in the northern part of the Gaspé Peninsula (and other elevated regions of east-

ern North America) had been nunataks during glaciations. As McGerrigle (1952:38) concluded, quoting several authors, the nunatak theory is not needed to explain the geographic distribution of the plant endemics in the contemporaneous flora of the northern highlands of the Gaspé Peninsula. Scoggan (1950) reviewed the question of the nunataks in Gaspé from the point of view of a botanist and stated that "the presence of endemics in the arctic-alpine flora of eastern North America offers no positive evidence in the support of the nunatak theory" (Scoggan, 1950: 11). Flint et al. (1942), after a visit to the Table Top Mountains, concluded that Mont Albert had been buried by glacier ice at some time in the past. They also believed that the angular mantle of debris on the Shickshocks resulted from weathering of the local bedrock following the glaciation that had erased all traces of continental or other glaciations (Flint et al., 1942).

The idea that the ice sheet over-rode the Gaspé Peninsula at some time in the past became firmly established with the discovery of long distance erratics on the plateau and on the southern flank of the Gaspé Peninsula (McGerrigle, 1952; Prest, 1970). Among those far travelled erratics were the following lithologies: granite gneiss, specular hematite and anorthosite (Alcock, 1935; McGerrigle, 1952; Badgley, 1956; Skidmore, 1965); all of the erratics presumably coming from the north shore of the St-Lawrence River, that is, from the Canadian Shield. More recently, Lebuis and David (1977) have studied the western part of the Gaspé Peninsula and used erratics with some success to decipher the glacial history of that area. Their history (Lebuis and David, 1977) shows glacial events related to the continental ice sheet and to local ice caps that existed prior to the continental glaciation and also during its waning phase. Hughes (1981) proposed a model of continental glaciation that seems to explain well many of the glacial or other geological features of the Gaspé Peninsula, which David and Lebuis (1984, in press) seem to have demonstrated at least for the western part of the Gaspé Peninsula.

Hughes' model (1981) calls for areas of glacier ice with a frozen base and consequently no erosion and areas with a melting base with deposition and erosion. The latter statement is certainly a simplification and does not explain everything, that is, that glacier ice once it reached the higher plateau did not seem to carry much debris beside the long-distance erratics. This phenomenon is explained by the presence of a frozen ice base picking up very little local debris and depositing almost none except long distance erratics (Hughes, 1981; David and Lebuis, 1984, in press). The erratics are picked up in the higher part of the ice from terranes that stand in relief in the topography (Hughes, 1981; David and Lebuis, 1984, in press).

Indeed, there are very few accumulations of glacial sediments on the southern plateau of Gaspé and in general one finds fluvio-glacial deposits in the large river valleys where they form the higher terraces. Those sediments can be ascribed to paraglacial sedimentation (Church and Ryder, 1972) and were presumably emplaced in late-glacial time. Eskers and other fluvio-glacial deposits are found at the southern edge of the plateau and in the lowlands, at least in the area studied by the writer (see Fig. 1).

All geologists studying the bedrock geology have recognized at various occasions and locations the presence of long and short distance erratics in the interior of the Gaspé Peninsula. During the summer of 1983, the glacial geology of the area immediately surrounding the St-Elzéar cave, covering approximately four NTS Sheets (New Carlisle, 22 A/3; New Richmond, 22 A/4; Lac McKay, 22 A/5; Rivière Reboul, 22 A/6), was studied.

In that region, especially on the plateau, deposits are generally thin. They consist of an accumulation in situ of local disintegrated bedrock to which has been added erratics of various dimensions (Fig. 4). The in situ accumulation can also be referred to as a rubble or a saprolite. In other places, on the plateau and on limestone terranes, one finds only an accumulation (generally 1 m thick) of fine (silt and clay size) material, very sticky when wet, with erratics semmingly resting on it. Hence, the long distance erratics (volcanics or granites) seem to have been superimposed on the material (saprolite or rubble) already in place. In other places the rubble contains striated angular pieces of bedrock. There are also areas of bare bedrock, that is, with a remarkable absence of erratics and rubble at the surface of bedrock. This is true of the limestone area northeast of St-Jogues.

The situations described above may have been the result of: A) in situ disintegration of previously striated bedrock; this would imply that the rubble was formed following deglaciation; B) in situ disintegration of elacial material (till or other glacial diamicton) since the last glaciation; this is very unlikely as true glacial deposits are rather rare on the plateau; C) the rubble having been formed in pre-glacial time under periglacial conditions preceding the last glacial readvance. The rubble would then have been buried under some sort of glacial diamicton (till or other) as has been observed in one instance on the plateau on the southern flank of the Gaspé Peninsula. The rubble (or saprolite) would include debris formed by rock disintegration (in situ) and soil forming processes in pre-glacial time. This would be in agreement with the concept of a glacier with a frozen ice base, thus preventing erosion of the previously formed saprolite or rubble (David and Lebuis, 1984, in press).

In places, one can also observe striated bedrock undisturbed together with local rubble and long dis-

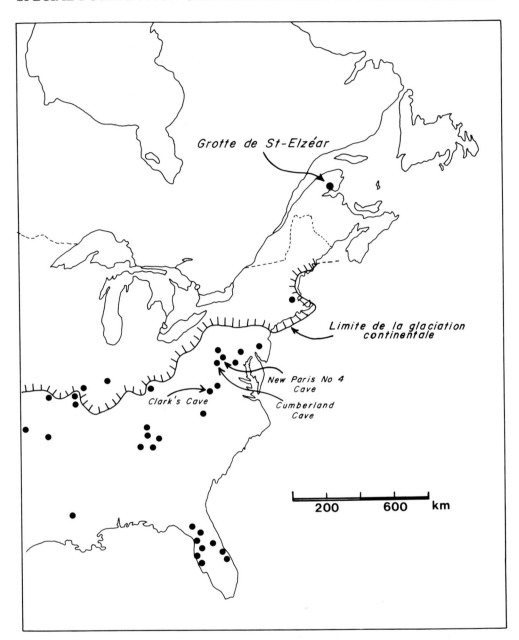

Fig. 4.—Main sites of accumulation of Pleistocene mammal assemblages for eastern North America (after Hibbard et al., 1965).

tance erratics. This is particularly true of an area north of the cave entrance (Lac Huard valley) where abundant volcanic erratics can be observed with local (or short distance transport) rubble lying on a striated bedrock surface. In the area around the cave itself, no erratics had been found by the writer until this summer (1983).

Rubble[1] and colluvium were observed in the area of the cave entrance in the fall of 1982 by the author

[1] According to the AGI glossary, p. 620, the word "rubble" refers to "A loose mass, layer, or accumulation of rough, irregular, or angular rock fragments broken from larger masses usually by physical (natural or artificial) forces, coarser than sand (diameter greater than 2 mm), and commonly but not necessarily poorly sorted; the unconsolidated equivalent of breccia."

Fig. 5.—Volcanic erratics and rubble on the plateau of southern Gaspé Peninsula, Quebec.

in company with Allen D. McCrady. In 1983, erratics (volcanics from the Lac McKay area) were found north and south of the cave entrance (Fig. 5), and there is no doubt that glacier ice overrode the area at sometime in the past. Grant (1977) has suggested that ice during the Wisconsinan maximum (some 18,000 years ago) did not entirely cover the north shore of Chaleurs Bay. This can be true only if the erratics (volcanics[2] and granites) have been scattered by glacier ice during some earlier glaciations other than the Wisconsinan. As have many authors who have studied the area (Badgley, 1956; Skidmore, 1965), volcanics have been found by the author scattered all over the area studied south of the cave entrance right up to the New Carlisle moraine. After a careful search (along favorable exposures made by new lumber roads), no volcanics

of the Lac McKay type have been observed north of Lac McKay. Hence, in the area studied there is no evidence recorded to date, that would suggest a northward ice flow, or glacial transport of erratics.

A careful search for volcanic erratics south of the cave entrance and south of St-Elzéar village has shown that the concentration of volcanic erratics decreases gradually from the outcrop area near Lac McKay to the New Carlisle moraine. In the area between St-Elzéar and the coast (above the marine limit) the surface material (excluding ice-contact drift) can hardly be called a till, except in a few isolated sections; it is generally a diamicton that is loose, that is, not very compact, and composed of local bedrock and long and short distance erratics. Generally, south of the cave entrance and south of the plateau limit (above the marine limit) volcanic erratics form less than 1% of the material in the cobble size range.

---

[2] A porphyritic facies of volcanics which is found outcropping in the Lac McKay area.

DISCUSSION

Several radiocarbon dates have been obtained on bone material within the talus and at the surface of the talus. They all give ages of about 5,000 B.P. or younger. Hence, it can be suggested tentatively that

the opening of the cave entrance is post-glacial and that glacier ice covered the area until about 12 to 13,000 B.P. (QU-83, 13,580 ± 350 B.P.; QU-84, 13,450 ± 470 B.P.; Lebuis and David, 1977). However, this is very tentative, and no precise dating of the material in the talus has been obtained because samples of charcoal are too small. Besides charcoal, bones of small mammals could be used. Apart from the reservations associated with bone radiocarbon dating, this course of action has not been followed because it would sacrifice too much bone material before the whole collection has been fully studied. Uranium-thorium dates have been performed on speleothems by Mel Cascoyne formerly of Mc Master University, Hamilton, Canada (Roberge and Gascoyne, 1978); those dates suggest that solution-precipitation processes were active inside the cave at some time between 100,000 and 225,000 B.P., possibly during the Sangamon or some other older interglacials (Fulton et al., 1984, in press; Henning et al., 1983).

The fact that a chamber with solution-precipitation features presumably dating back to 225,000 years B.P. has not been eroded suggests that the glacier was not very effective as an erosional agent in that area (unless some of those chambers were at some depth below the surface). It seems to support the concept of a frozen ice base at least in some part of the Gaspé Peninsula (David and Lebuis, 1984, in press) during the last glaciation. It is also possible that the opening of the cave entrance is due in part to glacial erosion.

St-Elzéar cave with its vertical entrance about 13 m deep constitutes an ideal trap for small and large animals. The diameter of the funnel-shaped upper part of the entrance is about 4 m. A large animal such as a moose sliding down a steep snowy slope had no chance of recovering. Hence, as Guilday (1971) remarked, to compare cave faunas, not only the type of entrance is important (its diameter and its angle to the vertical) but also its latitudinal and longitudinal location. Another important factor is the mode of accumulation, that is, by free falling of the animals or by accumulation of feces of raptors feeding on particular species (Guilday, 1971). Comparing the faunas in the deposit of New Paris No. 4 and Clark's Cave Guilday stated: "A casual size analysis of entrapped animals from the two faunas would give no indication of their differing modes of deposition. The proportion of small-to-large mammals in a raptor deposit is due to the selection bias of the birds of prey, and is independent of the time involved in forming such a deposit. The proportion of small-to-large mammals in a tumble-in trap, like a sinkhole, will increase with time. The slower the rate of infill, the greater the proportion of small-to-large mammals, because of their overwhelming majority in the surrounding fauna. This majority increases their probability of entrapment." (Guilday, 1971:71–72).

Hence the differences between the two cave faunas (New Paris No. 4 and Clark's) are found at the level of the number of individuals for each species because of raptor bias and in the presence or absence of some species because of the size of the entrance or latitudinal location. In the case of the St-Elzéar cave because of its large entrance, the large mammals are present in the faunal deposit, but also in the lower levels, one finds besides other small mammals, *Dicrostonyx hudsonius,* which has been found only in the deposit of New Paris No. 4 and St-Elzéar in eastern North America.

Dicrostonyx is found today only west of Hudson's Bay in New Quebec, beyond the limit of the boreal forest. *Microtus xanthognathus* is present at St-Elzéar, Clark's, New Paris No. 4, and at many other sites reported by Guilday (1971:64–67). The present geographical distribution of these small mammals compared with their distribution in late-Wisconsinan time is not only linked with late-glacial (late-Wisconsinan events; Guilday, 1971:64–67) but also with the particular ecology of each species (Guilday, 1971:64–67). It is interesting to note that *Microtus xanthognathus* has a present western distribution that recalls that of Fernald's endemics (Fernald, 1925; Scoggan, 1950) of the high mountains of northern Gaspé. Presumably, south of the Wisconsinan ice limits, there was a mixture of species with different ecological requirements, mixtures that are not found today (Guilday et al., 1977). Those species were dispersed into new habitats by the changing environments in late-glacial time, finally finding presumably more permanent niches in post-glacial time (Guilday, 1971).

CONCLUSIONS

The main problem in the area around the cave and in the area studied by the writer during the summer of 1983 is the differentiation of the glacial material from the local rubble. As the opinion has been expressed before (Flint et al., 1942) that the formation of the rubble is probably post-glacial this opinion is supported by the presence of pieces of striated local bedrock in the rubble in some places.

The following hypotheses can thus be formulated concerning the origin of the rubble:

1) the rubble has been formed before the glaciation and not been eroded; in this case, glacial elements have been superimposed on it; the presence of a saprolite under a glacial diamicton as observed in one instance would tend to support the hypothesis;

2) the rubble has formed since the glaciation from local bedrock; in certain places, angular pieces of local bedrock in the rubble show striations and this would suggest that bedrock was overridden by glacier ice at some time in the past and then broken in situ by weathering processes since the last glaciation;

3) a third hypothesis (undoubtedly preposterous) is that the rubble or saprolite has survived several glaciations with each time new elements being superimposed by glacial action with the final product being a mixture of glacial elements and rubble; some or all the erratics would have progressed towards the south as the result of successive glaciations; if the model of Hughes (1981) and David and Lebuis (1984, in press) is applicable to the last glaciation, it should apply to the other older ones as well, because it calls for a frozen ice base with practically no erosion in most areas on the plateau, it is hardly possible for most of the erratics to have been incorporated into the glacier more than once.

The occasional occurrence of erratics had been reported before by others (Badgley, 1956, Alcock, 1935; Skidmore, 1965). However, the systematic examination carried out during the past summer removes any doubt as to the presence of glacier ice sometime in the past on the plateaus in the New Carlisle, New Richmond, Lac McKay, and Rivière Reboul areas.

Concerning the faunal remains that have been recovered in the talus, the author hesitates to comment except to remark that the distribution of *Microtus xanthognathus* may not have been too different from that of the plant endemics of Fernald (1925) in late Wisconsinan time. Its present apparent distribution results either from a lack of information on the real geographical distribution of the species, or its distribution has become restricted in post-glacial time because of unfavorable ecological conditions. This would suggest that the elements of the fauna recovered in deeper levels in the talus were probably together south of the glacial limit during glaciation as the Quaternary periglacial records of *Phenacomys* would suggest (Guilday and Parmalee, 1972); they then migrated back north following deglaciation, and their present distribution is the result of more favorable conditions for each particular species in selected areas. Presumably, this is true of *Dicrostonyx hudsonius,* which was also present in the faunal remains of New Paris No. 4 (Guilday et al., 1964) together with *Microtus xanthognathus,* but whose (*Dicrostronyx*) southern limit today coincides with the northern limit of the boreal forest, west of Hudson Bay. This would also be true of *Phenacomys* as it is suggested by its periglacial records and its present distribution in North America (Guilday and Parmalee, 1972).

The cave entrance may have been opened to the surface by glacial action, but more probably became open to the surface in post-glacial time; however, solution and precipitation inside the cave may have occurred during the Sangamon or some older nonglacial period (Henning et al., 1983; Fulton et al., 1984, in press).

Henning et al. (1983) have shown that the formation of travertine seems to be concentrated during certain periods, mostly between 90,000 and 130,000 BP and since 15,000 years ago. The period between 90,000 and 130,000 corresponds to zone 4 and 5 of Shackleton and Opdyke (1973) and to parts of the Sangamon interglacial (Fulton et al., 1984).

ACKNOWLEDGMENTS

Allen D. McCrady, Norman Wuerthele, and Robert Cardillo of the Carnegie Museum of Natural History in Pittsburgh, PA, have participated in the excavation and their contribution is gratefully acknowledged. The author is also very thankful to the residents of St-Elzéar-de-Bonaventure too numerous to be named, who were very helpful during the excavations and have accompanied the author several times to the cave.

The author is also grateful to Michel Bouchard, Université de Montréal, who has commented on some aspects of this paper, and to P. P. David, Université de Montréal, who spent a few days in the field with the author during the summer of 1983. Published with permission of the deputy-Minister, Quebec Department of Energy and Resources.

LITERATURE CITED

ALCOCK, F. J. 1935. Géologie de la région de la Baie des Chaleurs. Commission Géologique du Canada, Mém., 183:1–165.

———. 1944. Further information on glaciation in Gaspé. Trans. Roy. Soc. Canada, 38:15–21.

BADGLEY, P. C. 1956. La région de New Carlisle. Ministère des Mines, Québec, Geol. Rept., 70:1–40.

BOURQUE, P. A. and G. LACHAMBRE. 1980. Stratigraphie du Silurien et du Devonien basal du sud de la Gaspésie. Ministère de l'Énergie et des Ressources, Québec, Spec. paper, 30:1–123.

COLEMAN, A. P. 1922. Physiography and glacial geology of Gaspé Peninsula, Québec. Bull. Geol. Surv. Canada, 34:1–52.

CHALMERS, R. 1887. Surface geology, Northern New Brunswick and Southeastern Quebec. Geological and Natural History Survey of Canada, Part M. Annual Report 1886, Dawson Brothers, Montreal, 39 pp.

———. 1906. Surface geology of eastern Quebec. Geol. Surv. Canada; Ann. Report for the year 1904 (part A):250–263.

CHURCH, M., and J. M. RYDER. 1972. Paraglacial sedimentation: A consideration of fluvial processes conditioned by glaciation. Geological Soc. America Bull., 83:3059–3072.

DAVID, P. P., and J. LEBUIS. 1984. The last glacial maximum and deglaciation of the western half of Gaspé Peninsula and adjacent areas, Québec, Canada; in press.

FERNALD, M. L. 1925. Persistence of plants in glaciated areas of boreal America. Memoirs of the Academy of Arts and Sciences, XV(III):241–342.

FLINT, R. F., M. DEMOREST, and A. L. WASHBURN. 1942. Glaciation of Shick Shock Mountains, Gaspé Peninsula. Geol. Soc. America Bull., 53:1211–1230.

FULTON, R. J., P. F. KARROW, P. LASALLE, and D. R. GRANT. 1984. Summary of Quaternary stratigraphy and history, eastern Canada. In press.

GRANT, D. R. 1977. Glacial style and ice limits, The Quaternary Stratigraphic record, and changes of land and ocean level in the Atlantic provinces, Canada. Géog. phys. et Quaternaire, 31:247–260.

GUILDAY, J. E. 1971. The Pleistocene history of the Appalachian mammal fauna. In The distributional history of the biota of the southern Appalachians, Part III (P. C. Holt, ed.) Research Monogr. Virginia Polytechnic Inst. State Univ., Blacksburg, Virginia, 4.

GUILDAY, J. E., P. S. MARTIN, and A. D. McCRADY. 1964. New Paris No. 4. A Pleistocene cave deposit in Bedford County, Pennsylvania. Bull. Nat. Speleol. Soc., 26:121–194.

GUILDAY, J. E., P. W. PARMALEE, and H. W. HAMILTON. 1977. The Clark's cave bone deposit and the late Pleistocene paleoecology of the Central Appalachian Mountains of Virginia. Bull. Carnegie Mus. Nat. Hist., 2:1–87.

GUILDAY, J. E., and P. W. PARMALEE. 1972. Quaternary periglacial records of voles of the genus *Phenacomys* Merriam (Cricetidae: Rodentia): Quaternary Research, 2:170–175.

HENNING, G. J., R. GRÜN, and K. BRUNNACKER. 1983. Speleothems, travertines and paleoclimates. Quaternary Res. 20:1–29.

HIBBARD, C. W., D. E. RAY, D. E. SAVAGE, D. W. TAYLOR, and J. E. GUILDAY. 1965. Quaternary mammals of North America. Pp. 509–526, *in* Quaternary of the United States (H. E. Wright and D. G. Frey, eds.), Princeton Univ. Press, 922 pp.

HUGHES, T. J. 1981. Numerical reconstruction of paleo-ice sheets. Pp. 222–261, *in* The last great ice sheets (G. H. Denton and T. J. Hughes, eds.), John Wiley and Son, New York.

LASALLE, P., and J. E. GUILDAY. 1980. Caverne de St-Elzéar-de-Bonaventure. Ministère de l'Énergie et des Ressources, Québec, DPV-750, 31 pp.

LEBUIS, J., and P. P. DAVID. 1977. La stratigraphie et les événements du Quaternaire de la partie occidentale de la Gaspésie, Québec. Géog. Phys. Quaternaire, 21:275–296.

MAILHIOT, A. 1919. Geology of Mount Albert, County of Gaspé, P.Q. Report on Mining Operations in the Province of Quebec for 1918, pp. 146–151.

MAYEWSKI, P. A., G. H. DENTON, and T. J. HUGHES. 1981. Late Wisconsin ice sheets of North America. Pp. 67–178, *in* The last great ice sheets (G. H. Denton and T. J. Hughes, eds.) John Wiley and Son, New York.

McGERRIGLE, H. W. 1952. Pleistocene glaciation of Gaspé Peninsula. Trans. Royal Soc. Canada, 46:37–51.

PREST, V. K. 1970. Quaternary geology of Canada. Pp. 676–764, *in* Geology and economic minerals of Canada. Economic Geology Report No. 1, Fifth ed.

ROBERGE, J., and M. GASCOYNE. 1978. Premiers résultats de datations dans la grotte de St-Elzéar, Gaspésie, Québec. Géog. Phys. Quaternaire 32:287.

SCOGGAN, H. J. 1950. The flora of Bic and the Gaspé Peninsula, Québec. Bull. Nat. Mus. Canada, 115:1–309.

SHACKLETON, N. J., and N. D. OPDYKE. 1973. Oxygen isotope and paleomagnetic stratigraphy of equatorial Pacific core V28–238: Oxygen isotope temperatures and ice volumes of a 10^5 year and 10^6 year scale. Quat. Res., 3:39–55.

SKIDMORE, W. B. 1965. Région d'Honorat-Reboul, Comté de Bonaventure. Ministère des Richesses Naturelles, Québec, Rapport Géologique, 107:1–36.

Address: Ministere de l'Energie et des Ressources, 1620 Boulevard de l'Entente Quebec City, Quebec G1S 4N6, Canada.

APPENDIX I

*Summary of Faunal Remains from
Caverne de St-Elzéar-de-Bonaventure,
1977–1978, by the late John E. Guilday*

This report covers vertebrate and invertebrate remains excavated from Caverne de St-Elzéar-de-Bonaventure by Dr. Pierre LaSalle, Geological Exploration Service, Department of Energy and Resources, Province of Quebec, assisted by A. D. McCrady, Section of Vertebrate Fossils, Carnegie Museum of Natural History, Pittsburgh, Pennsylvania, during the field seasons of 1977 and 1978. All items were from the talus cone at the base of the entrance shaft. They are currently housed at Carnegie Museum of Natural History. Insect remains were identified by Dr. George E. Ball, The University of Alberta; reptiles by Dr. Thomas Van Devender, University of Arizona, Tucson; birds by Dr. Paul W. Parmalee, University of Tennessee, Knoxville; seeds by Dr. C. R. Gunn, Plant Taxonomy Laboratory, U.S.D.A., Beltville, Maryland.

A minimum of 532 invertebrates and 4,679 vertebrates were recovered: one beetle, 531 land snails, 181 amphibians (three species), two reptiles (one species), four birds (four species), and 4,503 mammals (34 species). The vertebrate remains (97.9%) consisted primarily of skeletal elements of small mammals (smaller than a hare)—74.3% of these were from 14 species of small rodents and 24.5% from seven species of small insectivores. See faunal list, Table 1.

The minimum numbers of individual animals per level were derived from bone or tooth counts. These are only close approximations; a left lower jaw from one level and a right lower jaw from an adjacent level may conceivably have belonged to the same individual, but have been recorded as two animals when tallied stratigraphically, and some skeletons were reduced by talus abuse to unidentifiable fragments.

Because the stratigraphic levels of the 1977 excavation differ somewhat from those of the 1978 excavation due to talus slope, and because stratigraphic levels were arbitrary rather than natural, the minimum numbers of individuals from both years were lumped into four stratigraphic zones that appeared to have some biological justification. This insured that the minimum number of individuals from each zone would be large enough for the relative fluctuations between zones to have some statistical validity, and minimized the chances of counting the same animal more than once from scattered remains (the skeleton of a single black bear cub, for instance, occurred throughout a vertical depth of at least 77 cm).

The four stratigraphic zones were characterized by the presence or absence of certain distinctive genera of small mammals and were organized from the 1977 and 1978 stratigraphic levels as follows:

Zone 1: surface of talus and general cave floor.
Zone 2: 1–49 cm (1977) plus 1–60 cm (1978), talus only. *Phenacomys* and *Dicrostonyx* absent.
Zone 3: 49–119 cm (1977) plus 60–80 cm (1978). *Phenacomys* present, *Dicrostonyx* absent.
Zone 4: 119–168 cm (1977) plus 80–120 cm (1978). Both *Phenacomys* and *Dicrostonyx* present.

All amphibians, reptiles, and birds from the deposit are present in the area today. One species of mammal, the yellow-cheeked vole, *Microtus xanthognathus,* is presently a native of Alaska and Keewatin and has not been recorded east of Hudson's Bay. It was, however, a common rodent throughout the Appalachian Mountain region from Tennessee north at least to Pennsylvania during the late Wisconsinan, in boreal faunas that predated the glacial recession. Its presence at St-Elzéar (one individual so far) suggests that this now western taiga species survived to a relatively late date in the Gaspé. Five other species of mammals have not been reported from the Recent mammal fauna of the Gaspé. The Ungava collared lemming, *Dicrostonyx hudsonius,* is now confined to the tundra of northern Ungava. The Arctic hare, *Lepus arcticus,* another tundra species, reaches its southern limit in Newfoundland. Both occur rarely in Zone 4, suggesting that lower levels may hold further evidence of ecological changes. The heather vole, *Phenacomys intermedius,* a boreal forest rodent confined to Zones 3 and 4, is not found south of the Gulf of St. Lawrence today. The other two species hitherto unreported from the Gaspé, the Arctic shrew, *Sorex arcticus,* and the least weasel, *Mustela nivalis,* have been found in adjacent areas and their former presence was not unexpected. The Gaspé shrew, *Sorex gaspensis,* and (to the best of my knowledge) the Arctic hare have not hitherto been reported in the fossil state.

Table 1.—*Faunal list for Caverne de St-Elzéar-de-Bonaventure, Québec, 1977–1978 field collections.*

Scientific name	Common name	MMC	Zone 1	Zone 2	Zone 3	Zone 4	Total MNI
Vertebrates							
Class Mammalia, order Insectivora, family Talpidae							
Condylura cristata (Linnaeus)	Star-nosed mole	7	—	1	—	—	1
Family Soricidae							
Sorex arcticus Kerr	Arctic shrew	—	—	1	3	6	10
Sorex cinereus Kerr	Masked shrew	117	44	150	311	315	820
Sorex fumeus Miller	Smoky shrew	29	3	9	17	14	43
Sorex gaspensis Anthony	Gaspe shrew	16	—	1	4	8	13
Sorex palustris Richardson	Water shrew	6	—	—	1	4	5
Microsorex hoyi (Baird)	Pigmy shrew	23	5	23	52	46	126
Blarina brevicauda (Say)	Short-tailed shrew	54	8	13	41	24	86
Order Chiroptera, family Vespertilionidae							
Myotis lucifugus (LeConte)	Little brown bat	14	1	2	5	1	9
Order Lagomorpha, family Leporidae							
Lepus americanus Erxleben	Snowshoe hare	24	7	3	3	9	22
Lepus arcticus Ross	Arctic hare	—	—	—	—	1	1
Order Rodentia, family Sciuridae							
Marmota monax (Linnaeus	Woodchuck	23	11	4	3	4	22
Tamias striatus (Linnaeus)	Chipmunk	48	—	—	2	2	4
Tamiasciurus hudsonicus (Erxleben)	Red squirrel	70	1	—	1	3	5
Glaucomys sabrinus (Shaw)	Northern flying squirrel	7	1	1	3	2	7
Family Cricetidae							
Peromyscus maniculatus (Wagner)	Deer mouse	319	1	4	10	18	33
Family Arvicolidae							
Clethrionomys gapperi (Vigors)	Red-backed vole	423	57	235	366	470	1,128
Synaptomys cooperi Baird	Southern bog lemming	3	7	40	19	17	83
Synaptomys borealis (Richardson)	Northern bog lemming	1	36	125	239	105	505
Phenacomys intermedius Merriam	Heather vole	—	—	—	22	34	56
Dicrostonyx hudsonius (Pallas)	Ungava collared lemming	—	—	—	—	7	7
Microtus pennsylvanicus (Ord)	Meadow vole	158	10	48	37	36	131
Microtus chrotorrhinus (Miller)	Rock vole	22	116	279	432	383	1,210
Microtus xanthognathus (Leach)	Yellow-cheeked vole	—	—	—	—	1	1
Family Zapodidae							
Napaeozapus insignis (Miller)	Woodland jumping mouse	159	14	39	37	30	120
Zapus hudsonius (Zimmerman)	Meadow jumping mouse	25	1	1	1	—	3
Family Castoridae							
Castor canadensis Kuhl	Beaver	3	—	—	—	1	1
Family Erethizontidae							
Erethizon dorsatum (Linnaeus)	Porcupine	11	11	2	2	3	18
Order Carnivora, family Ursidae							
Ursus americanus Pallas	Black bear	6	5	1	1	2	9
Family Mustelidae							
Mustela nivalis Linnaeus	Least weasel	—	2	1	1	4	3
Mustela erminea Linnaeus	Ermine	20	—	—	—	—	—
Gulo gulo (Linnaeus)	Wolverine	—	2	1	—	1	4
Order Artiodactyla, family Cervidae							
Alces alces (Linnaeus)	Moose	1	7	1	1	2	11
Rangifer tarandus (Linnaeus)	Caribou	10	1	—	—	—	1

Table 1.—*Continued.*

Scientific name	Common name	MMC	Zone 1	Zone 2	Zone 3	Zone 4	Total MNI
			\multicolumn{4}{c	}{Stratigraphic levels}	
Mammal total (cave)			351	985	1,614	1,553	4,503
Class Aves, order Passeriformes, family Sittidae							
Sitta canadensis Linnaeus	Red-breasted nuthatch						1
Family Corvidae							
Cyanocitta cristata (Linnaeus)	Blue jay						1
Family Fringillidae							
Loxia cf. *leucoptera* Gmelin	White-winged crossbill ?						1
Species, indeterminate							
Class Amphibia (data incomplete), order Caudata, family Ambystomidae							
Ambystoma laterale Hallowell	Blue-spotted salamander						2
Order Salientia, family Bufonidae							
Bufo americanus Holbrook	American toad		14	104	20	37	175
Family Ranidae							
Rana palustris LeConte	Wood frog						4
Invertebrates							
Class Gastropoda (incomplete data)							
Iriodopsis sp.	Land snails						
Anguispira sp.	Land snails						
Succinca sp.	Land snails						
Other species of land snails to be identified							
Class Insecta, order Coleoptera, family Carabidae							
Plerostichus stypicus	Land beetle						

MMC = Modern mammal collection (i.e., specimens of living species in museum mammal collections).

The cave fauna excavated to date suggests boreal forest conditions at least as severe as those of today, with an indication of nearby tundra conditions during lower level times. There are some striking differences between the relative composition of the Recent mammalian fauna and that of the cave deposit, as well as internal changes within the deposit itself that point to more boreal conditions. For instance, a survey of Recent mammals collected in the Gaspé, in North American museum and university collections, shows that the deer mouse, *Peromyscus maniculatus,* is one of the Peninsula's commonest small mammals—20.4%. However, in the four zones of the deposit its percentage occurrence varies from 0.3% to 1.16%. This woodland rodent is rare in northern boreal forests but its high percentage today may also be due to ecological changes brought on by timbering practices. The percentage of the red-backed vole, *Clethrionomys gapperi,* a characteristic boreal forest vole, increases with depth from 17.6% in Zone 1 to 30.5% in Zone 4. The northern bog lemming, *Synaptomys borealis,* is known from the Recent fauna of the Gaspé by a single individual trapped on Mt. Jacques Cartier, and now in the Canadian National Museum of Natural History. Nevertheless, it was one of the more common small mammals at all levels of the deposit, ranging from 6.8% to 14.85%—a total of 505 individuals from all levels. The masked shrew, *Sorex cinereus,* more common in boreal regions, increases relatively with depth. It comprises 7.5% of the Recent fauna, but 13.6% to 20.45% in the lowest levels of the deposit. Relative numbers of the woodland jumping mouse, *Napaeozapus insignis,* however, show a reduction in numbers with depth. It comprises 10.19% of the Recent fauna, but decreases with depth from 4.3% to 1.94% in the cave deposit, again suggesting increasingly harsh conditions correlated with stratigraphic depth. This is also suggested by the appearance of *Dicrostonyx hudsonius* and *Lepus arcticus*, true tundra forms, in Zone 4.

Upon the initial discovery of the cave, skeletons of moose (five adults and two calves), one caribou calf, five black bears, 11 porcupines, and two wolverines were found scattered about the cave floor. Many of their bones showed signs of rodent gnawing and at least one had been heavily gnawed by a large carnivore, probably bear or wolverine, so that all animals were not killed by the initial fall, but survived to die of hunger and exposure. Many of the bones were broken and the breakage patterns of some of the large bones will be studied by Dr. Gary Haynes, Department of Anthropology, Smithsonian Institution. He is trying to differentiate breakage patterns associated with early man as opposed to those that can occur under natural conditions. There is no evidence of man at the cave, so that the cave collection will serve as an uncontaminated sample of natural breakage patterns. Some of the moose bones have acquired an unusual polish with rounded surfaces on broken edges that an archaeologist would have called man-made. Obviously some natural mechanism is at work here that requires further study.

Based upon tooth wear and replacement, four of the entrapped moose were from three to four years old (at least two were bulls bearing antlers). One was a yearling about 1.5 years of age, and two calves were between 4–5 months of age. Assuming that most moose calve in June, these fell in in September or October. The age of the single caribou calf suggests this as well.

None of this large collection of vertebrate remains, unique in eastern North America north of the Wisconsinan terminal moraine, has been studied in detail. Research efforts to date have been concentrated primarily on sorting, identifying, and cataloguing this large collection. The evidence to date for biologic change with depth is strong and further excavation will almost certainly lead to increasing our knowledge of the late Pleistocene/early Holocene history of the area.

APPENDIX II

Radiocarbon dates pertinent to talus history of the St-Elzéar cave.

Lab no.	Material	Excavation levels	Age BP
QU-714	Bone fragments of moose (*Alces alces*)	St-Elzéar 1	410 ± 120
QU-715	Wood	St-Elzéar 2	Modern
QU-717	Bone fragments of moose (*Alces alces*)	St-Elzéar 4	4,400 ± 130
QU-745	Bone fragments of moose (*Alces alces*)	St-Elzéar 24	4,390 ± 120
QU-978	Wood	78-1	940 ± 120
GX-7016	Bone fragments of moose (*Alces alces*)	Surface of talus	5,110 ± 150 (^{13}C corrected)

APPENDIX III

^{230}Th/^{234}U ages (modified from Roberge and Gascoyne, 1978).

Lab no.	Material	Age Ka × 10^3	+1 σ Ka × 10^3	−1 σ Ka × 10^3
77023-1	Stalagmite	223.9	30.7	25.0
77027-1	Stalagmite	204.8	13.2	12.0
77039-1	Stalagmite	152.8	20.8	18.0
77026-1	Stalagmite	135.9	7.6	7.2
77040-2	Stalagmite	135.3	9.6	8.9
77040-1	Stalagmite	129.9	6.2	5.9
77029-1	Stalagmite	102.5	7.6	7.1

MEADOWCROFT ROCKSHELTER AND THE PLEISTOCENE/ HOLOCENE TRANSITION IN SOUTHWESTERN PENNSYLVANIA

J. M. Adovasio, J. Donahue, R. C. Carlisle, K. Cushman,
R. Stuckenrath, and P. Wiegman

ABSTRACT

Meadowcroft Rockshelter (36WH297) is a deeply stratified multicomponent site in Washington Co., southwestern Pennsylvania. The 11 well-defined stratigraphic units identified at the site span at least 16,000 and probably 19,000 years of intermittent occupation by human groups representing all the major cultural stages/periods now recognized in the prehistory of northeastern North America. Throughout the temporal sequence, the site served as a locus for hunting, collecting, and food processing activities which involved the seasonal exploitation of the Cross Creek Valley and contiguous uplands. Presently, Meadowcroft Rockshelter represents the earliest well-dated evidence of human beings in the New World and demonstrates the longest occupation sequence in the Western Hemisphere.

All of the extant ecofactual information including macrofaunal, microfaunal, macrofloral, and microfloral remains as well as various categories of geological/geomorphological data suggest that from ca. 9,300 or 9,000 B.C. (ca. 11,250 or 10,950 B.P.) to the present, the environment of Cross Creek was essentially modern in aspect. It also appears that the Late Pleistocene environment in that part of Cross Creek immediately adjacent to the rockshelter was not radically different from that of today. Given these conditions, it appears that the Pleistocene/Holocene transition in this portion of Pennsylvania was a low amplitude event of relatively short duration.

INTRODUCTION

John Guilday and Paul Parmalee (1982:163) observed in commenting on the archaeofauna from Meadowcroft Rockshelter that ". . . no other North American archaeological site has yielded the remains of so many vertebrate species." Indeed, the analysis and interpretation of the nearly one million element vertebrate faunal assemblage from Meadowcroft was not only the most extensive scrutiny of archaeologically recovered faunal remains that John Guilday (and his co-workers) had ever undertaken; it was also the last large archaeofaunal assemblage he examined. Given the size and diversity of this assemblage, the analysis phase occupied much of John's time as well as that of those by whom he was assisted from mid-1975 to 1980 when the exhaustive report (Guilday et al., 1980) on this material was completed for ultimate incorporation into the final Meadowcroft synthetic volume (Adovasio et al., n.d. *b*).

Clearly, John expended incredible time and patience on examining and recording the details of the Meadowcroft fauna and in assessing the potential of that data base for furthering our understanding of the prehistory and paleoecology of southwestern Pennsylvania. Unfortunately for all of us, John will not be able to witness the ultimate fruition of his countless hours of observation and attention to detail or to participate directly, and for the credit he so richly deserves, in the synthesis of all the Meadowcroft data sets.

At times, particularly at the beginning of our association, John was uncomfortable with what the Meadowcroft faunal materials seemed to reflect. They sometimes did not square with the "big picture" or with the ecological interpretations and reconstructions to which he subscribed and which were based on his personal examination of faunal remains from many other sites. Nevertheless, John unwaveringly adhered to a cardinal rule of science—to follow where the data lead and to keep an open mind. In the final analysis, this approach is John's greatest legacy to a new generation of faunal analysts, and at the end, it allowed him to attain the synthesis that comes when the logical conclusions of hard scientific observation fly in the face of what is established, what is perceived to be true or simply what one would like to believe is so.

One of the critical facets of the Meadowcroft ecofactual record tantalizingly touched upon, but by no means resolved, by the analysis of the vertebrate faunal assemblage alone is a subject that was dear to John's scholarly heart—the nature of the Pleistocene-Holocene "transition" not only in Pennsylvania, but in eastern North America as a whole. The interpretations that follow represent our current understanding of the Meadowcroft/Cross Creek data base as it applies to that theme. Of course, these interpretations are subject to revision before the final Meadowcroft/Cross Creek publication (Adovasio et al., n.d. *b*) appears.

GENERAL SETTING

LOCATION AND GENERAL GEOLOGY

Meadowcroft Rockshelter (36WH297) is a stratified, multi-component site 48.27 air km (78.84 km via road) southwest of Pittsburgh, Pennsylvania, and 4.02 surface km northwest of Avella in Washington Co., Pennsylvania (Fig. 1). The site is on the north bank of Cross Creek, a small tributary of the Ohio River which lies some 12.16 km to the west. The exact location of the site is 40°17'12"N; 80°29'0"W (U.S.G.S. Avella, Pennsylvania, 7.5' Quadrangle).

Meadowcroft Rockshelter is oriented approximately east to west and has a southern exposure. It is ca. 15.06 m above Cross Creek and 259.9 m above sea level. The area protected by the extant overhang is ca. 65 m², and the overhang itself is ca. 13 m above the modern surface of the site. In addition to water in Cross Creek, springs are abundant near the rockshelter. The prevailing wind is west to east across the mouth of the rockshelter. This nearly continuous ventilation quickly clears both smoke and insects.

Physiographically, Meadowcroft is on the Unglaciated Appalachian or Allegheny Plateau west of the Valley and Ridge Province of the Appalachian mountains, and northwest of the Appalachian Basin. The rockshelter is formed beneath a cliff of Morgantown-Connellsville sandstone, a thick fluvial or channel sandstone within the Casselman Formation (Flint, 1955) of Pennsylvanian age.

PHYSIOGRAPHY

Meadowcroft Rockshelter is in a maturely dissected topographic region. More than one-half of the 14,164.3 ha encompassed by the Cross Creek watershed are in valley slopes; upland and valley bottoms are in the minority. Maximum elevations in the Cross Creek drainage are generally above 396 m. At the divides on the east, elevations are above 426 m. Elevations at stream level are 310 m at Rea on the South Fork, 276 m at Avella, and 193 m normal pool level at the confluence with the Ohio River.

The main stem of Cross Creek flows for ca. 31.3 km. The maximum north-south watershed width is approximately 15 km. The prevailing stream pattern is dendritic, and numerous small creeks and runs supply the main stem of Cross Creek. The steep gradient headwaters and tributaries of Cross Creek have their sources in the hills of central Washington Co., Pennsylvania. Cross Creek and its major tributaries of North, Middle, and South Forks have an average gradient of 0.4%. Most of the gradient is in the upper watershed where the streams are small and of low volume. The drainage is northwest to west toward the West Virginia/Ohio border and the Ohio River.

The Cross Creek drainage pattern is very asymmetric; the northern tributaries are much shorter than those on the south. The drainage area south of Cross Creek is therefore much larger than that to the north. This condition is probably the result of a drainage pattern superimposed on a 3°–5° regional dip.

Adjacent to Cross Creek are four other major watersheds. Harmon Creek to the north parallels Cross Creek as does Buffalo Creek to the south. Both are quite similar to Cross Creek in structure, and both flow into the Ohio River. On the northeast is Raccoon Creek. This is a north-flowing stream which joins the Ohio River just west of the confluence with the Beaver River. To the east and southeast, the large Chartiers Creek watershed flows north and joins the Ohio River near its inception at Pittsburgh.

Present area topography developed during the latter Pleistocene when increased precipitation and runoff caused extensive downcutting. The area was unaffected directly by glacial ice as the mapped Wisconsinan Kent moraine (Woodfordian age) is some 50 km north of the site, whereas the later Lavery till is ca. 83 km to the north.

CONTEMPORARY CLIMATE

The present continental climate of the study area is characterized by a wide seasonal temperature range and by moderate precipitation that falls principally during the warmer parts of the year. The yearly temperature in Washington Co. averages 10.6°C with extremes ranging from −28.9°C during the winter months to 32.2°+C in July and August. The frost-free growing season averages 150 days. Precipitation averages approximately 1,016 mm per year with about 560 mm falling during the growing season. The amount of precipitation correlates closely with elevation and roughness of terrain, particularly on windward slopes. Temperatures tend to be lower in hilly areas than in more level places due to the effects of elevation and air drainage. Night temperatures are generally colder, and day temperatures are slightly higher on the valley bottoms than on hilltops.

CONTEMPORARY FLORA

Vegetative cover in the Cross Creek watershed today is a mosaic of second growth forest, abandoned fields, and pastures, agricultural tracts, and residential areas. Variation in these plant communities is promoted by the degree and duration of disturbances as well as by degree and exposure of slope. Human disturbance is now the principal determinant of vegetative cover.

Forest once covered essentially all of the watershed. Several forest types were common, including mixed oak on the hilltops and steep, south-facing slopes. Mixed mesophytic forest covered the more mesic north-facing slopes and headwater coves. Riverine vegetation spread along the alluvial flood plains.

Woodland now covers approximately 40% of the area. Little of this is old growth, and only a few stands approach the typical composition of virgin forest. Most of the area was heavily deforested during the early and middle 1800s when the land was first settled and cleared for cultivation. Areas allowed to re-establish forest have been logged repeatedly with resulting secondary forests.

Where woodlands have matured for a long period, they are beginning to resemble original stands (Wiegman, 1977). Mixed oak forests are common on the drier ridgetop sites and south-facing slopes, and they are typical on thin, stoney soils of mod-

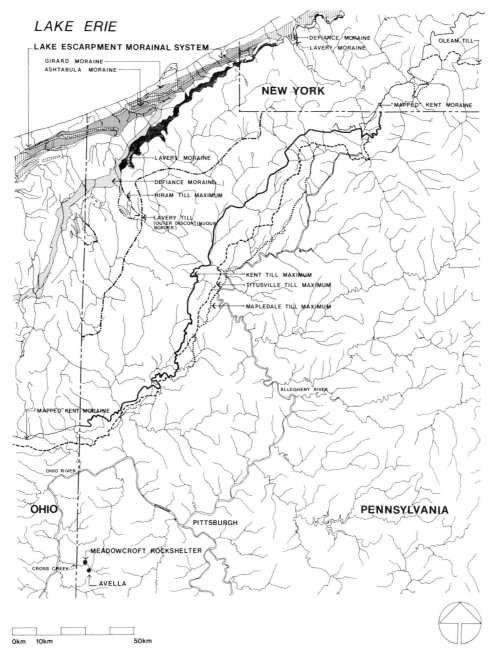

Fig. 1.—General view of the Upper Ohio Valley and contiguous regions showing maximum southward extension of the Wisconsinan glacial advance (after Johnson, 1981).

erate fertility. Dominant species are white oak (*Quercus alba*) and red oak (*Quercus rubra*). Associated species include red maple (*Acer rubrum*), sugar maple (*Acer saccharum*), hickories (*Carya* sp.), black birch (*Betula lenta*), American beech (*Fagus grandifolia*), tulip tree (*Liriodendron tulipifera*), chestnut oak (*Quercus prinus*), black oak (*Quercus velutina*), and scarlet oak (*Quercus coccinea*).

Vegetative layers below the canopy are sparsely populated. Typical shrubs include juneberry (*Amelanchier arborea*), maple-leaved viburnum (*Viburnum acerfolia*), flowering dogwood (*Cornus florida*), poison ivy (*Rhus radicans*), summer grape (*Vitis aestivalis*), and occasional small patches of mountain laurel (*Kalmia latifolia*). Individual plants are scattered or occur in small groups creating a rather open understory.

The herbaceous layer is equally poor in species diversity and population. During the spring, early saxifrage (*Saxifraga virginiensis*), rupanemone (*Anemonella thalictroides*), and starry campion (*Silene stellata*) are scattered over the rocky surface. Later dominants include shinleaf (*Pyrola elliptica*), spotted wintergreen (*Chimaphila maculata*), tick trefoil (*Desmodium* sp.), and black snakeroot (*Sanicula marilandica*). Grasses are also found in the dry oak woods and include linear leaf panicum (*Panicum linearifolium*), bushy panic grass (*Panicum dichotomum*), and Canada brome grass (*Bromus purgans*).

On moister north-facing slopes and headwater coves, mixed mesophytic forest is found. These stands occur on rich, well-drained, fertile soils with deep humus layers. Woodland dominants include sugar maple, American beech, tulip tree, white oak, red oak and white basswood (*Tilia heterophylla*). There are many other species in this forest which are common, but not dominant. These include red maple, yellow birch (*Betula alleghentiensis*), hickories, white ash (*Fraxinus americana*), black walnut (*Juglans nigra*), cucumber tree (*Magnolia acuminata*), black gum (*Nyssa sylvatica*), wild black cherry (*Prunus serotina*), and eastern hemlock (*Tsuga canadensis*).

The shrub layer is generally denser and more diverse than that of the mixed oak type. Prominant species include juneberry, flowering dogwood, witch-hazel (*Hamamelis virginiana*), spicebush (*Lindera benzoin*), pawpaw (*Asimina triloba*), Virginia creeper (*Parthenocissus quinquefolia*), wild plum (*Prunus americana*), and scattered occurrences of redbud (*Cercis canadensis*) on neutral or alkaline soils.

The herbaceous layer contains a myriad of species and is especially rich during the prevernal and vernal periods. Trillium is a prominent family with the large trillium (*Trillium grandiflorum*) the most common. Other species include numerous examples of violets (*Viola* sp.), blue phlox (*Phlox divaricata*), trout lily (*Erythronium americanum*), larkspur (*Delphinium tricorne*), spring beauty (*Claytonia virginica*), toothwarts (*Dicentra laciniata* and *Dicentra diphylla*), solomon's-seal (*Polygonatum biflorum*), and jack-in-the-pulpit (*Arisaema atrorubens*). After the canopy has closed there are fewer flowering species, but the general dense coverage remains.

The most prevalent forest type in the Cross Creek watershed today is mixed mesophytic. It most often occurs on steep slopes adjacent to larger streams. Here, the land is unsuitable for cultivation, and repeated logging provides a source of disturbance. In less accessible areas, the forest is well-developed maturing second growth.

Riverine forests grow on the flood plains of lower Cross Creek. Sycamore (*Platanus occidentalis*) and black willow (*Salix nigra*) are common in small groves and in bands adjacent to the stream. These woodlands blend with the mesophytic, and it is often difficult to draw distinct boundaries. Riverine forests share many species with mixed mesophytic forests including sugar maple, American beech, and white basswood. Associated species restricted to the flood plain are box elder (*Acer negundo*), occasional sweet buckeye (*Aesculus glabra*), black walnut, and silver maple (*Acer saccharinum*).

Shrub species under the forest canopy are also quite similar to those of the mixed mesophytic areas but are often quite dense. Species ordinarily restricted to this area are elder (*Sambucus canadensis*), ninebark (*Physocarpus opulifolius*), various willows (*Salix* sp.), hoptree (*Pletea trifolia*), bladdernut (*Staphylea trifolia*), silky dogwood (*Cornus amomum*), and smooth alder (*Alnus serrulata*). Other tree-covered areas are often the result of

significant vegetative alteration. Many abandoned fields now have extensive thickets of young black locust (*Robinia pseudoacacia*) or hawthorn (*Crataegus* sp.). Old fence rows are now lines of trees with black locust and black cherry as dominant species.

Strip mines are common throughout the Cross Creek watershed. Where reclamation has not been attempted, old spoil banks often have patches of black locust. More recent mines often have been regraded and planted with a variety of pines and spruces.

Abandoned fields and meadows are common in the Cross Creek watershed. They exist in various stages of succession depending on previous use and amount of time since abandonment. Woodland recovery occurs less rapidly in cultivated fields than in pastures. Moisture is a prominent controlling factor in woodland recovery; moister areas more rapidly achieve tree community development.

Old fields follow a general rejuvenation pattern. A variety of grasses and herbaceous plants first appear followed by thickets of woody shrubs, briars and young shade-intolerant trees such as red maple, black cherry, aspen (*Populus* sp.) and tulip tree. As the primary trees mature, shade-tolerant plants such as oak, sugar maple, beech, cucumber, hickories and gum begin growth. This final stage produces permanent forest cover.

Hackberry (*Celtis* sp.) was commonly recovered as a macrofossil at Meadowcroft (Cushman, 1982). Present distribution in the area apparently is less extensive than formerly. *Celtis occidentalis* and *Celtis tenuifolia* occur with two varieties (*Celtis occidentalis* var. *pumila*) and (*Celtis tenuifolia* var. *georgiana*). In the mixed mesophytic forest, *Celtis occidentalis* is an infrequent small tree. Only four hackberry specimens from Washington Co., Pennsylvania, are in the Carnegie Museum of Natural History Herbarium. Two *Celtis occidentalis* specimens were collected along Mingo Creek, near Riverview, Pennsylvania, in 1939 and 1942. Two others were collected from Little Pine Creek in 1927 and from Chartiers Creek west of Hendersonville in 1957. The most common plants in the immediate vicinity of Meadowcroft Rockshelter are listed in Table 1.

CONTEMPORARY FAUNA

The area near Meadowcroft Rockshelter supports a wide range of terrestrial and avian Carolinian fauna. This assemblage, however, is a pale reflection of that in the area's deciduous oak/ hickory forests after the end of the Wisconsinan glaciation. As late as the beginning of the 18th century, elk (*Cervus elaphus*), black bear (*Ursus americanus*), mountain lion (*Felis concolor*), wild cat (*Lynx rufus*), timber wolf (*Canis lupus*), fisher (*Martes pennanti*), otter (*Lutra canadensis*), beaver (*Castor canadensis*), wild turkey (*Meleagris gallopavo*), and passenger pigeon (*Ectopistes migratorius*) could be found in the hills of western Pennsylvania.

Hunting and rapid replacement of forest by farms and pastures soon altered the area's faunal population. Today, thanks chiefly to game conservation laws and secondary forest growth, the faunal assemblage has regained some of its former abundant character (Guilday, 1977). The white-tailed deer (*Odocoileus virginianus*), for instance, is once again found near the rockshelter but is the only large game species of the immediate vicinity. Presently, white-tailed deer, cottontail rabbit (*Sylvilagus floridanus*), gray and fox squirrel (*Sciurus carolinensis* and *S. niger*), ringneck pheasant (*Phasianus colchicus*), bobwhite quail (*Colinus virginianus*), ruffed grouse (*Bonasa umbellus*), muskrat (*Ondatra zibethicus*), and mink (*Mustela vison*) constitute the principal game

Table 1.—*Checklist of contemporary arboreal flora in the immediate vicinity of Meadowcroft Rockshelter.**

Taxa	Common name	Occurrence**
Acer saccharum	sugar maple	C
Acer rubrum	red maple	C
Fagus grandifolia	American beech	C
Quercus rubra	red oak	C
Quercus alba	white oak	C
Fraxinus americana	white ash	C
Ulmus americana	white elm	C
Ulmus rubra	slippery elm	C
Tsuga canadensis	eastern hemlock	C
Ostrya virginiana	hop-hornbeam	C
Platanus occidentalis	sycamore	C
Tilia americana	basswood	C
Sassafras albidum	white sassafras	C
Prunus serotina	wild black cherry	C
Betula nigra	red birch	C
Juniperus virginiana	red cedar	R
Carya tomentosa	mockernut hickory	R
Asimina triloba	pawpaw	U
Hamamelis virginiana	witch-hazel	U
Cercis canadensis	redbud	U

* From Carlisle et al., 1982:12, table 1.
** C = Common. R = Rare. U = Understory.

Table 2.—*Relative incidence of principal game species in the Cross Creek watershed as of December 1973.**

Species evaluated	Main stem Cross Creek**	South Fork Cross Creek**	North Fork Cross Creek**	Middle Fork Cross Creek**
Deer	L	L	M	L
Rabbit	M	M	M	M
Squirrel	L	L	M	L
Pheasant	S	S	S	S
Quail	L	L	L	L
Grouse	L	L	M	M
Muskrat	H	M	M	M
Raccoon	H	M	M	M
Mink	L	L	L	L
Fox	L	L	L	L

* After Grant, 1973.
** L = Low. M = Moderate. H = Heavy. S = Population dependent on stocking.

Table 3.—*Checklist of contemporary fish in the Cross Creek watershed.**

Taxa	Common name**
Cyprinidae	
Campostoma anomalum	Stoneroller
Clinostomus elongatus	Redside dace
Cyprinus carpio	Carp
Ericumba buccata	Silverjaw minnow
Nocomis micropogon	River chub
Notropis atherinoides	Emerald shiner
Notropis chrysocephalus	Striped shiner
Notropis stramineus	Sand shiner
Notropis rubellus	Rosyface shiner
Pimephales notatus	Bluntnose minnow
Pimephales promelas	Fathead minnow
Rhinichthys atratulus	Blacknose dace
Semotilus atromaculatus	Creek chub
Catostomidae	
Catostomus commersoni	White sucker
Hypentelium nigricans	Northern hog sucker
Centrarchidae	
Amplophites rupestris	Rock bass
Lepomis cyanellus	Green sunfish
Lepomis gibbosus	Pumpkinseed
Lepomis macrochirus	Bluegill
Micropterus dolomieui	Smallmouth bass
Micropterus salmoides	Largemouth bass
Percidae	
Etheostoma blennoides	Greenside darter
Etheostoma caeruleum	Rainbow darter
Etheostoma flabellare	Fantail darter
Etheostoma nigrum	Johnny darter
Cottidae	
Cottus bairdi	Mottled sculpin

* From Carlisle et al., 1982:15, table 3.
** Common names follow the list recommended by the American Fisheries Society Special Publication 6 (1970).

species of the area (Grant, 1973). The relative incidence of these species is plotted in Table 2.

Unlike many larger mammals, smaller mammals often survived the end of the Wisconsinan glaciation and the subsequent deforestation of the Historic period. During the 1973–1978 excavations, the field crews observed many representatives of this group including: gray squirrel, fox squirrel, gray fox (*Urocyon cinereoargenteus*), raccoon (*Procyon lotor*), and a large variety of rodents. Some mammals actually benefitted from extensive clearing and are now more widespread than they were aboriginally. These include: cottontail rabbit, woodchuck (*Marmota monax*), opossum (*Didelphis virginianus*), striped skunk (*Mephitis mephitis*), and red fox (*Vulpes vulpes*).

Although many animals found farther north or within the mountains of west-central Pennsylvania do not occur near Meadowcroft (for example, porcupine, *Erethizon dorsatum,* and snowshoe hare, *Lepus americanus*), the area does support an abundant and varied mammalian assemblage, which includes these additional species—chipmunk (*Tamias striatus*), southern flying squirrel (*Glaucomys volans*), meadow vole (*Microtus pennsylvanicus*), woodland or pine vole (*Microtus pinetorum*), southern bog lemming (*Synaptomys cooperi*), white-footed mouse (*Peromyscus leucopus*), meadow jumping mouse (*Zapus hudsonius*), short-tailed shrew (*Blarina brevicauda*), least shrew (*Cryptotis parva*), smoky shrew (*Sorex fumeus*), hairy-tailed mole (*Parascalops breweri*), least weasel (*Mustela nivalis*), long-tailed weasel (*Mustela frenata*) and several species of bats.

Domesticated animals such as the horse (*Equus caballus*), cow (*Bos taurus*), sheep (*Ovis aries*), pig (*Sus scrofa*), Norway rat (*Rattus norvegicus*), house mouse (*Mus musculus*), feral dog (*Canis familiaris*) and common cat (*Felis domesticus*) are, not unexpectedly, also found in the area today.

The avian fauna reflects continuing adjustment to changes in land utilization and vegetative cover. In general, forest-dwelling species have diminished in contrast to open-country forms since the onset of Euro-American settlement. As previously mentioned, the ruffed grouse and the introduced ringneck pheasant are important upland game birds. Man also has introduced the rock dove (*Columba livia*), English sparrow (*Passer domesticus*), and the starling (*Sturnus vulgaris*) to the inventory of birds which in aboriginal times included great numbers of migratory waterfowl, swans, ducks, and geese. The transient and resident avifauna now includes at least 30 species.

Terrestrial and riverine reptiles include black snake (*Coluber constrictor*), garter snake (*Thamnophis sirtalis*), snapping turtle (*Chelydra serpentina*), and box turtle (*Terrepene carolina*). Amphibians include various plethodontids, toads (*Bufo americanus*), tree frogs (*Rana clamitans*), and bullfrogs (*Rana catesbiana*).

Although not a large stream, Cross Creek is a direct tributary of the Ohio River, and many fish species were undoubtedly available aboriginally. The present riverine fauna is somewhat depauperate though not nearly so depleted as generally assumed. Polluted by raw sewage and mine effluvia, Cross Creek nevertheless supports a restricted fauna. Freshwater mussels are almost wholly absent, but a minimum of 26 fish species occur in small to moderate numbers (Table 3). Current field observations support the view of Cooper (1972) who noted general scarcity of game fish in the Cross Creek watershed.

EXCAVATION PROCEDURES

The excavation procedures employed at Meadowcroft Rockshelter are thoroughly detailed in other publications on the site (e.g., Adovasio et al., 1975, 1977a, 1977b, 1978a; Adovasio et al., 1978b, 1979–1980a, 1979–1980b; Adovasio, 1982). During the 466 working days of the 1973–1978 projects, approximately 60.5 m² of surface area inside the dripline and ca. 46.6 m² outside the dripline were excavated. Over 230 m³ of fill were removed. Nearly all of the excavation was conducted with trowels or smaller instruments, and excellent vertical and horizontal controls over the artifactual and "ecofactual" assemblage were able to be maintained.

As explained in many other Meadowcroft publications, 11 natural strata were distinguished during excavation. The earliest stratum is termed Stratum I; the most recent is Stratum XI. Two of the eleven strata (VI and X) occur only inside the dripline; the other strata are continuous across the site. In the interest of space, the reader is referred to Stuckenrath et al. (1982) for details on the 11 Meadowcroft strata. A composite profile of the site stratigraphy is presented in Fig. 2.

RADIOCARBON CHRONOLOGY

One hundred Meadowcroft samples were submitted for radiocarbon assay to the Radiation Biology Laboratory of the Smithsonian Institution. In all but two cases, the charcoal was derived from firepits, firefloors, or charcoal lenses within the deposits. The exceptions are portions of completely carbonized simple plaited basketry fragments (Adovasio et al., 1977a; Stile, 1982). To date, 70 of the samples have been processed. Twenty-two samples were too small to count. The results of the assays are presented in absolute stratigraphic order in Table 4, and the dates are plotted in Fig. 3.

The initial human occupation of the rockshelter is positively ascribable to the 15th millennium B.C.; the latest radiocarbon date on purely aboriginal materials is A.D. 1,265 ± 80 (685 B.P.). The deepest microstrata within Stratum IIa have produced two radiocarbon dates with limited cultural material in excess of 17,000 B.C. (18,950 B.P.) suggesting an even earlier initial occupation. Cross-dated lithics and ceramics from Strata VIII–XI indicate continued occupation or utilization of the site into the Historic period.

The radiocarbon sequence is remarkably consistent with the observed stratigraphy. Stratum IIa is the deepest and oldest culture-bearing depositional unit. For analysis and discussion purposes, Stratum IIa is subdivided into three subunits of unequal thickness labeled upper, middle, and lower Stratum IIa. Each of these subunits is bracketed by major roof spalling episodes, and each is well-dated by radiocarbon assay (Table 5). Upper Stratum IIa has a terminal date of 6,060 ± 110 B.C. (8,010 B.P.) from the uppermost living or occupation floor within this subunit and a date of 7,165 ± 115 B.C. (9,115 B.P.) from a slightly deeper occupational surface within the unit. At the base of upper Stratum IIa is a substantial roof spalling episode that marks the

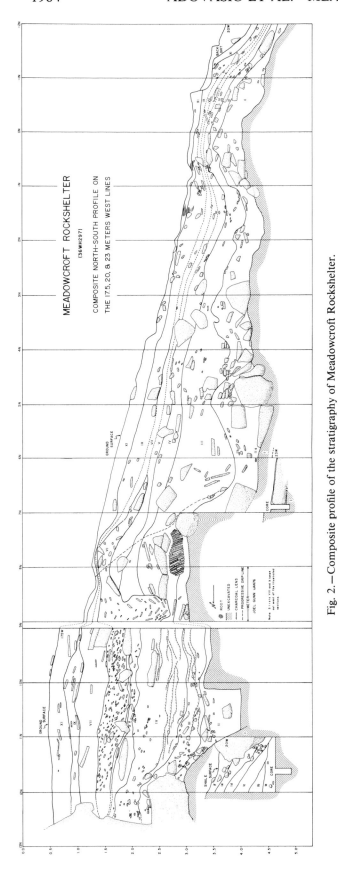

Fig. 2.—Composite profile of the stratigraphy of Meadowcroft Rockshelter.

boundary between this subunit and middle Stratum IIa. While the top of the roof spalling event that separates upper from middle Stratum IIa is undated, an assay of 9,350 ± 700 B.C. (11,300 B.P.) is available from directly beneath the roof spalling event at the top of middle Stratum IIa. Hence, for all intents and purposes, upper Stratum IIa dates ca. 9,000–6,000 B.C. (ca. 10,950–7,950 B.P.) and is of Holocene age.

Middle Stratum IIa, sealed from upper Stratum IIa by the roof spalling episode described above, is also initiated by a roof spalling episode. Directly beneath the latter roof spall is a date of 10,850 ± 870 B.C. (12,800 B.P.). Middle Stratum IIa is therefore bracketed by dates ranging ca. 11,000–9,000 B.C. (12,950–10,950 B.P.); it is of terminal Pleistocene age.

Lower Stratum IIa, which lies beneath the roof spalling episode that separates this subunit from middle Stratum IIa, has seven additional radiocarbon dates ranging from 17,650 ± 2,400 B.C. (19,600 B.P.) to 11,290 ± 1,010 B.C. (13,240 B.P.). The 18th millennium B.C. date constitutes the deepest date from the rockshelter that is associated with materials of indisputable human manufacture and also marks the onset of human utilization of this locality.

The maximum excavated depth of Stratum IIa varies between 70 cm and 90 cm in different portions of the rockshelter. At the interface of lower Stratum IIa and underlying Stratum I are several lenses of charcoal that have produced radiocarbon dates in the 20th and 29th millennia B.C. range. These dates are *not* associated with any cultural materials. Furthermore, they are separated from the deepest occupational floors within Stratum IIa by a considerable thickness of sterile deposits (Adovasio et al., 1980:588–589).

No questions have arisen about the post-10,000 B.C. (11,950 B.P.) Meadowcroft dates or their cultural associations. Some of the Stratum IIa dates, though collected and analyzed under identical conditions, have stimulated considerable discussion (for example, Haynes 1977, 1980; Mead, 1980; Dincauze, 1981). The discussion has centered on the suggestion that the radiocarbon dates from the oldest levels at the site, specifically from middle and lower Stratum IIa, are "too old" because of the injection of "dead" carbon in the form of coal particles or so-called organic solubles (Haynes, 1977, 1980). However, the evidence for particulate (Adovasio et al., 1978a) and nonparticulate (Adovasio et al., 1980, 1981) contamination of the Stratum IIa sam-

Table 4.—*Radiocarbon chronology from Meadowcroft Rockshelter as of August 1983.*[a]

Stratum (field designation)	Provenience/description	Lab designation	Date	Cultural period
XI (F-3)	Charcoal from firepit, middle one-third of stratum	SI-3013	A.D. 1,775 ± 50	Historic
X (F-25)	Charcoal from firepits	Samples too small to process	—	
IX (F-9)	Charcoal from firepit, upper one-third of stratum	SI-2363	A.D. 1,265 ± 80	Late Prehistoric
VIII (F-12)	Charcoal from firepit	SI-3023	A.D. 1,320 ± 100	
VII (F-13)	Charcoal from firepits, middle one-third of stratum	SI-2047 SI-3026	A.D. 1,025 ± 65 A.D. 660 ± 60	early Late Woodland
VI (F-63)	Charcoal from firepits and lenses	Samples too small to process	—	
V (F-14)	Charcoal from firepits, upper one-third of stratum	SI-3024 SI-3027 SI-3022 SI-2362 SI-2487	A.D. 285 ± 65 A.D. 160 ± 60 A.D. 70 ± 65 125 ± 125 B.C. 205 ± 65 B.C.	Middle Woodland
IV (F-16)	Charcoal from firepits, upper one-third of stratum	SI-2051 SI-1674 SI-2359 SI-3031	340 ± 90 B.C. 375 ± 75 B.C. 535 ± 350 B.C. 705 ± 120 B.C.	
	Charcoal from firefloor, middle one-third of stratum	SI-1665	865 ± 80 B.C.	
	Charcoal from firepit, middle one-third of stratum	SI-1668	870 ± 75 B.C.	
	Charcoal from firepits/firefloors, lowest one-third of stratum	SI-1660 SI-2049	910 ± 80 B.C. 1,100 ± 85 B.C.	
III (F-18)	Charcoal from firepits, upper one-third of stratum	SI-2066 SI-1664 SI-2053 SI-3030 SI-2046	980 ± 75 B.C. 1,115 ± 80 B.C. 1,140 ± 115 B.C. 1,150 ± 90 B.C. 1,165 ± 70 B.C.	Early Woodland
	Charcoal from firepit, middle one-third of stratum	SI-1679	1,305 ± 115 B.C.	
	Charcoal from firepits/firefloors, lowest one-third of stratum	Samples too small to process	—	
IIb (F-46 upper)	Charcoal from firepit, upper one-third of stratum[b]	SI-1681	1,260 ± 95 B.C.	
	Carbonized basketry fragment, upper one-third of stratum	SI-1680	1,820 ± 90 B.C.	
	Charcoal from firepits, middle one-third of stratum	SI-2063 SI-2058 SI-2054 SI-2356 SI-1685 SI-2358	2,000 ± 240 B.C. 2,020 ± 85 B.C. 2,055 ± 85 B.C. 2,430 ± 500 B.C. 2,870 ± 85 B.C. 4,340 ± 355 B.C.	Late Archaic

Table 4.—*Continued.*

Stratum (field designation)	Provenience/description	Lab designation	Date	Cultural period
	Charcoal from firefloor, lowest one-third of stratum	SI-2055	4,720 ± 140 B.C.	
	Charcoal from firepit, lowest one-third of stratum	SI-2056	3,350 ± 130 B.C.	Middle Archaic
	Charcoal from firepits/firefloors, lowest one-third of stratum	Samples too small to process	—	
IIa (F-46 lower)	Charcoal from firepits, upper one-third of stratum[c]	SI-2064 SI-2061	6,060 ± 110 B.C. 7,125 ± 115 B.C.[d]	Early Archaic
	Charcoal from firepit/firefloor, middle one-third of stratum	SI-2491	9,350 ± 700 B.C.	
	Charcoal from firepits, lowest one-third of stratum	SI-2489 SI-2065 SI-2488 SI-1872 SI-1686 SI-2354	10,850 ± 870 B.C. 11,290 ± 1,010 B.C.[e] 11,320 ± 340 B.C. 12,975 ± 620 B.C.[f] 13,170 ± 165 B.C. 14,225 ± 975 B.C.	Paleo-Indian
	Charcoal concentration, lowest level within stratum	SI-2062	17,150 ± 810 B.C.[g]	Paleo-Indian?
	Carbonized fragment of cut bark-like material, possible basketry fragment, lowest level within stratum	SI-2060	17,650 ± 2,400 B.C.	
	Charcoal concentration, base of stratum and directly above the Stratum I/IIa interface	DIC-2187	19,120 ± 475 B.C.	
I/IIa interface	Charcoal from lenses at the interface	SI-2121 SI-1687	19,430 ± 800 B.C.[h] 28,760 ± 1,140 B.C.[h]	No cultural associations
I (F-85) (Omega Unit)	Birmingham Shale	—	—	

  [a] All dates are uncorrected and are listed in absolute stratigraphic order.
  [b] Provenience incorrectly listed as "upper ½" in Adovasio et al., 1975:16, table 3.
  [c] Provenience incorrectly listed as "middle ⅓" in Adovasio et al., 1975:16, table 3.
  [d] This date has previously been listed incorrectly as 7,165 ± 115 B.C. in various publications (e.g., Adovasio et al., 1977a:33, table 7; Stuckenrath et al., 1982:80, table 2; 83, table 3, among others).
  [e] Date listed as 11,300 ± 1,000 B.C. in Adovasio et al., 1975:16, table 3 after rounding.
  [f] Date listed as 12,900 ± 200 B.C. in Adovasio et al., 1975:16, table 3.
  [g] Date incorrectly listed as 17,150 ± 801 B.C. in Stuckenrath et al., 1982:83, table 3.
  [h] Dates from the Strata I/IIa interface were originally and tentatively ascribed to the "upper (?) ⅓" of Stratum I in Adovasio et al., 1975:16, table 3.

ples is unconvincing. Of importance for understanding this point, the last remaining charcoal sample from lowest Stratum IIa was submitted to Dicarb Radioisotope Company. The sample came from a locus immediately below dates SI-2062 at 17,150 ± 810 B.C. (19,100 B.P.) and SI-2060, the cut bark basketry at 17,650 ± 2,400 B.C. (19,600 B.P.) but above sample SI-2121, the charcoal with no cultural associations at 19,430 ± 800 B.C. (21,380 B.P.). This charcoal sample was submitted to Dicarb with virtually no "background" information other than the fact that it was from somewhere in Cross Creek and "was probably older than 9,000 years." The sample was pretreated with sodium hydroxide and

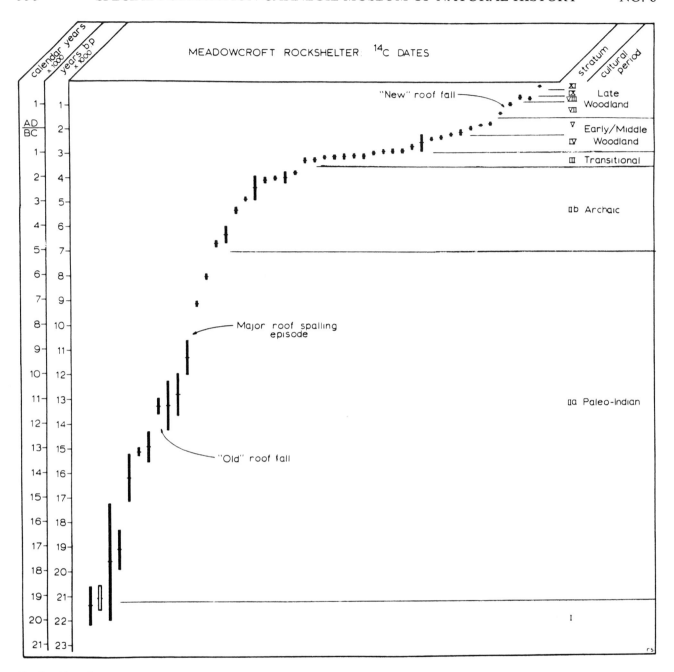

Fig. 3.—Plot of the Meadowcroft Rockshelter radiocarbon dates showing one standard deviation (from Stuckenrath et al., 1982:81, fig. 2).

hydrochloric acid, that is, not using the nitration pretreatment employed by the Smithsonian laboratory. Liquid scintillation counting of benzene rather than methane gas assay (which is standard practice at the Smithsonian) was used. Despite the different approaches, DIC-2187 provided a date in stratigraphic order of 19,120 ± 475 B.C. (21,070 B.P.). This appears in Fig. 3 as the hollow rectangle in the lower left of the plot. There was no evidence for any contamination of the sample (Irene Stehli, 1982, personal communication).

Table 5.—*Radiocarbon chronology from Strata I and IIa at Meadowcroft Rockshelter as of August 1983.*[a]

Stratum (field designation)	Provenience/description	Laboratory designation	Date	Cultural period
IIa (F46)	Charcoal from firepits, upper one-third of stratum[b]	SI-2064	6,060 ± 110 B.C.	Early Archaic
		SI-2061	7,125 ± 115 B.C.[c]	
	Charcoal from firepit/firefloor middle one-third of stratum	SI-2491	9,350 ± 700 B.C.	
	Charcoal from firepits, lowest one-third of stratum	SI-2489	10,850 ± 870 B.C.	Paleo-Indian
		SI-2065	11,290 ± 1,010 B.C.[d]	
		SI-2488	11,320 ± 340 B.C.	
		SI-1872	12,975 ± 620 B.C.[e]	
		SI-1686	13,170 ± 165 B.C.	
		SI-2354	14,225 ± 975 B.C.	
	Charcoal concentration, lowest level within stratum	SI-2062	17,150 ± 810 B.C.[f]	Paleo-Indian?
	Carbonized fragment of cut bark-like material, possible basketry fragment, lowest level within stratum	SI-2060	17,650 ± 2,400 B.C.	
	Charcoal concentration, base of stratum and directly above the Stratum I/IIa interface	DIC-2187	19,120 ± 475 B.C.	
I/IIa Interface	Charcoal from lenses at the interface	SI-2121	19,430 ± 800 B.C.[g]	No cultural associations
		SI-1687	28,760 ± 1,140 B.C.[g]	
I (F85) (Omega Unit)	No samples		No dates	No cultural associations

[a] All dates are uncorrected and are listed in absolute stratigraphic order.
[b] Provenience incorrectly listed as "middle ⅓" in Adovasio et al., 1975:16, table 3.
[c] This date has previously been listed incorrectly as 7,165 ± 115 B.C. in various publications (for example, Adovasio et al., 1977a:33, table 7; Stuckenrath et al., 1982:80, table 2; 83, table 3, among others).
[d] Date listed as 11,300 ± 1,000 B.C. in Adovasio et al., 1975:16, table 3 after rounding.
[e] Date incorrectly listed as 12,900 ± 200 B.C. in Adovasio et al., 1975:16, table 3.
[f] Date incorrectly listed as 17,150 ± 801 B.C. in Stuckenrath et al., 1982:83, table 3.
[g] Dates from the Strata I/IIa interface were originally and tentatively ascribed to the "upper(?) ⅓" of Stratum I in Adovasio et al., 1975:16, table 3.

THE DATA BASE

Several of the many discrete data sets from Meadowcroft Rockshelter are important both for paleoenvironmental reconstruction at and near the site and for the Mid-Atlantic region as a whole. They are also important for elucidating the nature of the Pleistocene/Holocene transition both in the general area and at the rockshelter. These data sets include vertebrate and invertebrate faunal as well as floral remains buttressed by detailed geological, geochemical, and sedimentological information. Extensive discussion of all the Meadowcroft data to date is provided in the appropriate chapters of Carlisle and Adovasio (1982).

The diverse Meadowcroft data sets range from poor to very good in the scope of information that they offer for understanding the paleoenvironment and the process of paleoenvironmental change at the rockshelter throughout the Holocene. Several data sets, however, are substantially incomplete for the critical Pleistocene/Holocene interface. Unfortunately, these include two of the potentially most revealing and diagnostic components, the faunal and floral remains.

FAUNAL REMAINS

Vertebrate faunal remains constitute the single most commonly encountered set of inclusions in the rockshelter. Over 115,166 bones or fragments were individually examined of the total of nearly one million recovered bones. Remains of at least 5,634 individual vertebrates, representing 151 taxa were identified: 149 to the species level—66 birds, 44 mammals, 26 reptiles, 8 fish, and 5 amphibian species are present (see, Adovasio et al., 1979–1980b: 108–109). Over 90% of the remains are from disintegrated digestive pellets regurgitated by raptorial birds, primarily owls, that formerly roosted on the cliff face of the rockshelter during the time that the deposits were building. Southern flying squirrel, the extinct passenger pigeon, and toad (*Glaucomys volans, Ectopistes migratorius, Bufo* sp.) account for 68% of all the identified vertebrates. Some idea of raptor activity at the site can be gathered from the fact that 44% of all the individual vertebrates from the site constitute but a single species, the mouse-sized southern flying squirrel—at least 2,503 individuals. Approximately 7% of the bone collection is probably attributable to Indian activity.

Bone preservation at all levels within the present dripline of the rockshelter is good, but remains of large and medium-sized vertebrates have been reduced to bone fragments or isolated teeth. In all, 23% of all fragments are charred, a figure that increases to 93% outside the dripline. No charring patterns were noted. All species were involved in what appears to have been the result of a random, accidental burning of fragments in the aboriginal hearth substrata; they were not necessarily burned as a result of aboriginal food preparation techniques.

The faunal reconstruction of the bone-bearing sequence at Meadowcroft advanced by Guilday and Parmalee (1982:170) is that of a temperate Carolinian biota typical of southwestern Pennsylvania until the 19th century.

In addition to the overall composition of the Meadowcroft fauna there is intraspecific evidence for temperate conditions as well. Late Pleistocene specimens of southern flying squirrel from New Paris Sinkhole No. 4, Pennsylvania, are larger on average than Recent Pennsylvania material, whereas southern bog lemming specimens are smaller (Guilday et al., 1964). They were deposited under boreal conditions, and the size characteristics are believed to have been caused by physiological adaptation. Measurements of southern flying squirrel humeri and southern bog lemming first lower molars were compared at different strata in the Meadowcroft deposit down to and including Stratum IIa to see if there was any indication of time-related size changes that might suggest climatic change. In both cases the results were negative, and the sample parameters agree with those of Recent material at all levels inside the dripline, suggesting temperate conditions throughout.

Of the 21 species of non-aquatic vertebrates that reached their lowest stratigraphic level inside the dripline (dated at ca. 9,350 B.C.; 11,300 B.P.), the presence of box turtle, timber rattlesnake, bobwhite, turkey, eastern mole, and pine or woodland vole implies that temperate Carolinian conditions prevailed in this part of the Upper Ohio Valley as early as 11,300 years ago.

The interpretation of the meager faunal remains in older levels from outside the dripline is not as clear. With the exception of the charred base of a white-tailed deer antler, all material is highly fragmented. Only four vertebrate species could be identified from associations dated at 9,350 B.C. (11,300 B.P.) or older in strata outside the dripline—white-tailed deer, eastern chipmunk, southern flying squirrel, and passenger pigeon. Charred fragments of toad, colubrid snake and deer mouse (*Peromyscus* sp.) are also present. Deer remains were not identified below the 14,225 ± 975 B.C. (16,175 B.P.) level in lower Stratum IIa. Passenger pigeon was found at 11,290 ± 1,010 B.C. to 13,170 ± 165 B.C. (13,240 to 15,120 B.P.) levels in lower Stratum IIa. One charred postcranial element of southern flying squirrel does occur at 17,150 ± 810 B.C. (19,100 B.P.) or older. The chipmunk remains may be suspect, as the bones are uncharred and the animal is an avid burrower, but white-tailed deer, southern flying squirrel, and passenger pigeon suggest, although they do not necessarily *demand,* a temperate setting. These species do, or did, range north rarely and marginally into Canadian zone situations, so that their ecological interpretation in these levels would be equivocal in the absence of the accompanying botanical record.

Unfortunately, the invertebrate faunal record from Meadowcroft Rockshelter is also more-or-less mute for the older levels of the site (pre-9,350 B.C.) as well. Although not common, mollusk remains are represented throughout the long occupational sequence (Lord, 1982). A minimum of 35 terrestrial and three aquatic gastropod species as well as 11 naiad (freshwater mussel) species have been identified. All recovered species are present in the area today, although several mussel species have been exterminated from the segment of Cross Creek adjacent to the rockshelter.

The gastropod fauna indicates that the local environment has remained essentially stable for at least the past 11,000 years before which time mollusks are poorly represented in the site's deposits. The vegetative and climatic regime reflected by the extant molluscan assemblage is characterized by a stable woodland, predominantly of oak and hickory—quite moist and shady but not extremely dense.

Floral Remains

Floral remains are the second most abundant class of material recovered from Meadowcroft Rockshelter. Included are moderately large sections of tree trunks and limbs, with and without bark, minute seeds and seed coats, fruits, charcoal, and small amounts of pollen. Vegetal remains are represented in all occupation levels, including Stratum IIa, and therefore span approximately 16,000 years of the site's history (Volman, 1981:4).

Paleoenvironmental reconstructions incorporating botanical data are incomplete for much of the Northeast. This is due to inadequate sampling of archaeological and natural deposits and/or to poor site deposition conditions (for example, abrasive soil texture, excessive water or extreme pH values). Consequently, few sites in the Northeast have produced quantities of well-preserved botanical remains.

The comments on Meadowcroft faunal preservation are also applicable in large measure to the floral assemblage. Over 97% of it is from upper Stratum IIa and above and is therefore of Holocene or modern aspect. Portions of this larger assemblage have been discussed elsewhere (Adovasio et al., 1977a, 1979–1980b; Adovasio and Johnson, 1981). A modest amount of floral material, usually charred, comes from middle and lower Stratum IIa and includes deciduous forest elements; this is in keeping with the Holocene character of the fauna.

The availability of both macrofloral elements and pollen from Meadowcroft is ideal and provides information on local and regional vegetation. Most macrobotanical remains are not susceptible to long-distance transport. Their presence in archaeological sites frequently indicates that the plant source was at or near the place of deposition. Taxa not represented in the cultural assemblage of a site were not necessarily absent from the local flora, however. In contrast, airborne pollen can originate some distance from the place of deposition, and most (but not all) pollen is a regional vegetation marker (Volman, 1981:4).

Approximately 10% of all macrobotanical materials from Meadowcroft were analyzed. This sample consists of more than 30,000 separate plant pieces. The remaining 90% of the material was examined to insure that the studied remains included adequate samples of all elements and taxa preserved at the site. Twelve complete sample columns also were taken. The sediment in these columns was processed in a hydrogen peroxide solution to separate the heavy (larger seeds and nutshell) from the light (smaller seeds and charcoal) fractions. All floral remains from these columns have been analyzed (Volman, 1981:13).

Among specimens of wood and charcoal, the chronological range and taxa diversity suggest that the vegetation at and near Meadowcroft was mixed conifer-hardwood from the time of initial human occupation onward. The two genera from the oldest plant-containing sediments, ca. 14,500 years ago, are *Quercus* sp. and *Carya* sp. Both genera remain in evidence to the present day (Volman, 1981:90). Two other genera abundant in the fossil record (*Celtis* sp. and *Prunus* sp.) are no longer extant with any frequency and decrease in numbers from the earliest through more recent strata. *Celtis* sp. could have been a relatively common species in the prehistoric forests around Meadowcroft. Present data suggest that this plant source may have provided food to both the human and animal populations of the area.

Pinus sp. appears in the prehistoric macrobotanical record about 11,000 years ago and remains to the present. It is particularly abundant in the period from about 3,000 years ago to the Historic period when, apparently, it was decimated by logging. *Pinus* sp. is a pioneer species intolerant of shade. Its presence in relative abundance may indicate that the forest surrounding Meadowcroft from ca. 10,000 years ago to the present was in the process of regeneration following some disturbance such as fire, storm damage, or land clearance (Volman, 1981: 91). *Juglans* sp., although a very shade tolerant species (Harlow and Harrar, 1969), persists from about 9,000 years ago to recent times in comparatively well-represented amounts. It is likely that *Juglans* sp. retained a dominant or co-dominant position in the forest canopy during its occupance in the Meadowcroft area.

The only genera represented through time with greater frequency than *Juglans* sp. are *Quercus* sp. and *Tsuga* sp., both of which are tolerant of a variety of light and moisture conditions. *Quercus* sp. is found at every time period at Meadowcroft and probably was (as it is presently) a major component of the mixed conifer-hardwood forest in the vicinity of the site. *Tsuga* sp. suggests the onset of possibly cooler, moister conditions, but its tolerance of marginal conditions renders this interpretation conjectural. Co-representation with *Pinus* sp. in comparatively large amounts at ca. A.D. 1025 and A.D. 1775 leads one to believe that the forest canopy may then have been partly or mainly *Pinus* sp. with *Tsuga* sp. a potential understory component.

Carya sp., *Ulmus* sp., *Fraxinus* sp., and *Fagus* sp. parallel each other in comparative amounts and frequency of representation through time. These genera are mesophytic and tolerant, although *Fagus* sp. is not usually found on dry soils (Harlow and Harrar, 1969). The greatest amount of *Fagus* sp. recovered from the fossil evidence corresponds with the greatest amount of *Tsuga* sp., which also prefers moist habitats. The periods ca. A.D. 1025 and A.D. 1775 at Meadowcroft may thus have been slightly wetter than previous times, based on the presence and apparent relative abundance of these two genera (Volman, 1981:93).

Nutshells of *Quercus* sp. and *Carya* sp. date to at least 9,000 years ago at Meadowcroft, and *Juglans* sp. nutshells are known from 6,000 year old levels. This supports the interpretation of the wood and charcoal data that mixed conifer-hardwood forests surrounded Meadowcroft at least since this time. Walnut trees probably were abundant for a long time in the prehistoric forest near the site. This is also reflected to some extent in the wood and charcoal remains as *Juglans* sp. occurs in small quantities at various times but is present in increased amounts in Strata VIII–XI.

Juglans sp. nutshell occurs in every time period represented at Meadowcroft, but its wood and charcoal are absent from the earliest strata. The greatest amounts of *Juglans* sp. nutshell parallel the largest number of *Juglans* sp. wood and charcoal specimens in most of the recent strata. *Carya* sp. nutshell occurs in greater relative frequency in the fossil record than does *Carya* sp. wood or charcoal.

Quercus sp. wood and charcoal occurs from the earliest levels at Meadowcroft, although acorn shell does not. The lack of acorns from the lowest levels probably can be attributed to a failure on the part of the aboriginal inhabitants of the site to collect them. Most of the recovered acorns are not carbon-ized, and they increase in quantity from Stratum V upward.

Fruits and seeds other than nutshell from Meadowcroft increase through time in numbers and diversity; however, the numbers recovered per stratum are low. The relative increase in weedy annuals since about 1,000 B.C., especially *Amaranthus* sp., suggests some increase in land clearance and/or disturbance. Pollen data from the same period corroborate this observation.

Fruit and seed remains from earlier time periods are, unfortunately, very scant. However, *Nyssa* sp. stones from at least 13,000 years ago argue that the environment in the vicinity of Meadowcroft then was at least as warm as that today, if not slightly warmer. Meadowcroft lies at the extreme margin of this plant's northernmost modern range (Volman, 1981:95; Fowells, 1965).

Although by no means common in the lowest occupational levels at the site, the extant Meadowcroft floral remains do extend to levels older than 9,350 B.C., the horizon below which faunal data are meager. Vegetation in the proximity of the site appears to have been mixed conifer-hardwood throughout the ca. 16,000 years represented by the floral remains. The presence in earliest time periods of *Quercus* sp., *Juglans* sp. and *Carya* sp. argues against boreal or tundra conditions near the site during late glacial times. Significantly, all species present in the fossil record are also present in the extant vegetation of the area.

GEOLOGICAL EVIDENCE

The faunal and floral data from Meadowcroft are limited in those strata that span the Pleistocene/Holocene boundary. Geological data for this critical time period, however, are abundant, continuous and not subject to interpretational distortions arising from sample size alone (Beynon, 1981; Beynon and Donahue, 1982). Much of this geological information has been presented in earlier publications (for example, Beynon and Donahue, 1982) but seldom discussed in the context of elucidating the Pleistocene/Holocene boundary in the study area. It is therefore useful to review here salient facts about the geology of sandstone rockshelters in general and of Meadowcroft Rockshelter in particular.

Detailed geoarchaeological examination of a number of sandstone rockshelters in eastern North America including Meadowcroft (Adovasio et al., n.d. *a*; Carlisle and Adovasio, 1982; Vento et al., 1980; Adovasio, 1982 (compiler); Yedlowski et al., 1982) has demonstrated the presence of a more-or-less "standardized" geological sequence in the formation of such sites and in the geological mechanisms for the accumulation of sediments. The rockshelters are developed within medium-bedded to massively bedded sandstones typically underlain by shales. Rockshelter development proceeds by fluvial downcutting through the sandstone unit(s) and then via undercutting of the less resistant shale thereby forming a re-entrant. In the case of Mead-

LONGITUDINAL PROFILE OF CROSS CREEK

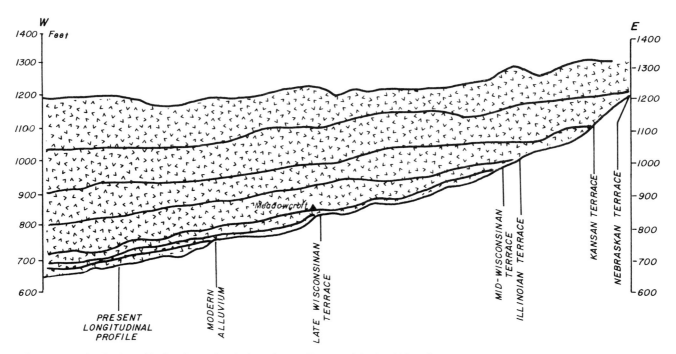

Fig. 4.—Longitudinal profile for Cross Creek from its confluence with the Ohio River on the west to its headwaters on the east. Uppermost line shows the elevation of uplands in the region, indicating that Cross Creek had ca. 51.8 m of relief in Nebraskan time. Five discontinuous terraces, Nebraskan, Kansan, Illinoian, early Middle Wisconsinan, and Late Wisconsinan in age, are indicated (after Beynon, 1981:210, fig. 43).

owcroft Rockshelter, the presence of a series of Pleistocene terraces along the Ohio River and Cross Creek can be used to establish the timing and mechanism of the initial development of the rockshelter (Benyon, 1981; Benyon and Donahue, 1982; Ray, 1974).

Progressive downcutting of Cross Creek can be documented by study of the terraces that developed along the Cross Creek drainage and by correlating them with terraces on the Ohio River (Fig. 4). As indicated in Fig. 4, Cross Creek in pre-Pleistocene time was a tributary of the north-flowing Monongahela-Beaver drainage system. The valley was wide, gently sloping and had not yet eroded into the Morgantown-Connellsville Sandstone in which the rockshelter ultimately was developed. Initial erosion of the sandstone occurred during the Yarmouth Interglacial. However, it was not until the Sangamon Interglacial that shale undercutting beneath the sandstone and the initial development of the rockshelter re-entrant occurred (Fig. 5).

Fluvial sedimentation during early Middle Wisconsinan time was responsible for deposition of a silty clay at the Stratum I/IIa interface. When Cross Creek resumed active downcutting in Middle Wisconsinan time, the rockshelter was enlarged but was *not again* affected by fluvial erosion and deposition. The Late Wisconsinan terrace of Cross Creek is ca. 3 m to 6 m below the top of Stratum I at Meadowcroft. Thus, at any point after middle Late Wisconsinan time (ca. 21,300 B.P.) the site was available for potential occupation and open to colluvial sedimentation.

All of the Meadowcroft sediments that post-date the deposition of the Stratum I/IIa interface silty clay are essentially of colluvial origin. Detailed field and laboratory examination of textural properties, mineralogy, and sedimentary structures of this colluvial pile indicate that it is derived via gradual but continuous erosion from a relatively limited number of sources. These include grain-by-grain attrition from the rockshelter ceiling and walls, rain-

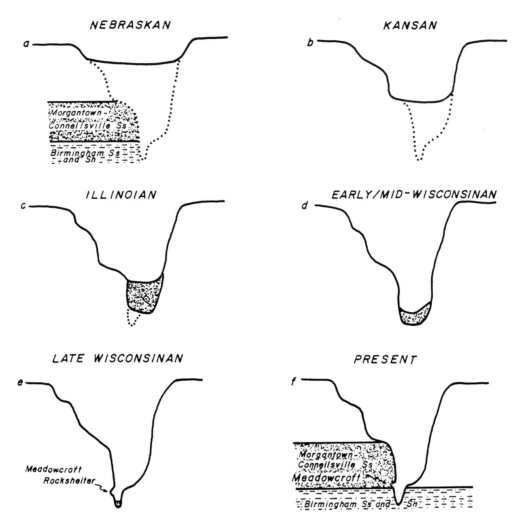

Fig. 5.—Sequential cross sections of Cross Creek at Meadowcroft Rockshelter. Progressive deposition and downcutting through Pleistocene terraces are illustrated (after Beynon, 1981:205, fig. 42; 213, fig. 44; 216, fig. 45; 218, fig. 47; 221, fig. 49; 225, fig. 51).

induced sheetwash from upland surfaces above the rockshelter, and rock fall (Beynon and Donahue, 1982:42–48). None of the other principal sediment source mechanisms for sandstone rockshelters in eastern North America (that is, stream action, aeolian deposition, or cultural activity) appear to have been of any importance for the accumulation of the relatively thick, generally poorly sorted Meadowcroft sediment pile.

Culturally derived sediments and modified sediments in the form of archaeological features are, of course, present in the Meadowcroft sequence from near basal Stratum IIa and proceeding upward through Stratum XI; however, this source constitutes a minor contribution to the total sediment pile.

When Stratum IIa cultural deposition began, Cross Creek was well below the basal occupational floor of the rockshelter. This fact, and information from intensive examination of the Stratum IIa sediments, indicate that flooding, stream deposition, and reworking of cultural deposits did not occur.

Neither dune nor loess sequences exist in the Meadowcroft sedimentological record. Dune fields are obviated by the humidity of the area, and loess deposits are not abundant anywhere in this part of Pennsylvania. Paleosols have not been discerned at Meadowcroft, and given the continuous nature of sedimentation at the site, it is not possible that they could have formed. In sum, once the Meadowcroft colluvial pile began to accumulate above the Stratum I/IIa interface, sedimentation quite unrelated to Cross Creek was continuous, proceeding to the present without interruption or erosional episodes (Beynon and Donahue, 1982:43).

Differences in strata composition at Meadowcroft are not due to climatic fluctuations such as those posited for the Pleistocene/Holocene boundary; rather, these differences arise from changes in the overall configuration of the rockshelter itself. The physical appearance of the rockshelter therefore effectively molds and conditions the character and composition of the sediment pile at any point in time as it has throughout the geological history of the site above the Stratum I/IIa interface.

Attempts to understand the interplay between natural and cultural factors in the development of Meadowcroft as a physical entity require recognition of the fact that once emplaced, colluvial sediments within the rockshelter were never reworked by water action. The readily identified presence of a distinct paleo-dripline within the colluvial sediments from lowest Stratum IIa upward through the sediment pile to the present floor of the site traces the gradual northward migration of the cliff (that is, the progressive retreat of the overhang) through time. It also, and, perhaps more graphically than any other indicator, testifies against any major reworking of the sediment pile, a process that would have eroded and erased the dripline. The position of the dripline is documented by fluctuations in calcium carbonate levels as well as by scanning electron microscopy of quartz grains from both inside and outside the rockshelter overhang. Discussions of this aspect of the Meadowcroft investigations are available in Adovasio et al. (1977a), Beynon (1981), Beynon and Donahue (1982), and Adovasio et al. (1983).

Detailed examination of continuous sediment column samples from various loci at Meadowcroft (for example, Adovasio et al., 1977a:14, fig. 7) as well as long-term scrutiny of modern attrition and sheetwash processes and products have permitted the systematic discrimination of sediments by size category and source within each stratum. The results of these examinations permit very detailed reconstruction of the entire depositional history of the site (Adovasio et al., 1977a; Beynon and Donahue, 1982; Adovasio et al., 1983). A summary of that part of the overall sequence bracketing the Pleistocene/Holocene transition is germane at this juncture for the additional interpretative potency it lends to the somewhat more restricted floral and faunal evidence.

The late Pleistocene and early Holocene at Meadowcroft are encompassed *within* Stratum IIa. As noted above, Stratum IIa is a relatively thick unit which varies from 70 cm to 90 cm in maximum excavated depth. It is divided into three subunits of unequal thickness each of which is bracketed by well-dated roof falls. Lower Stratum IIa dates from ca. 19,430 ± 800 B.C. (21,380 B.P.) to 10,850 ± 870 B.C. (12,800 B.P.) and is late Pleistocene in age. Middle Stratum IIa, it has been observed, dates from ca. 11,000–9,000 B.C. (12,950–10,950 B.P.) and is terminal Pleistocene in age. Upper Stratum IIa dates to the early Holocene, ca. 9,000–6,000 B.C. (ca. 10,950–7,950 B.P.).

During this long period, all three sediment sources contributed to the sediment pile at Meadowcroft, though in very unequal measure. In the entire lower Stratum IIa sequence, sedimentation was principally grain-by-grain attrition; rock fall episodes occurred only when the rockshelter roof became sufficiently unstable to spall off larger blocks. Modern attrition samples collected and analyzed for a continuous five year period suggest that cold and increased moisture *may* slightly increase the release of individual sand grains, but they certainly do not drastically accelerate the sedimentation rate from this source. During the deposition of lower Stratum IIa, there were no entrants in the overhang, and sheetwash could not contribute to the sediment pile.

Middle Stratum IIa also accumulated principally via grain-by-grain attrition punctuated by limited spalling, and this period reflects the first indications of the accumulation of limited amounts of sheetwash on the western margin of the site. The sheetwash was admitted through an entrant developed by a partial roof collapse termed the Old Roof Fall. Only a portion of the roof collapsed, but a large enough access route was created to elevate the sedimentation rate in that part of the rockshelter near the collapse.

Holocene-age upper Stratum IIa accumulated, as did middle and lower Stratum IIa, through a combination of attrition, limited rock spalling and the addition of some sheetwash to the western edge of the site. The sedimentation rates for upper and middle Stratum IIa are virtually identical, and both are only slightly elevated on average over the rates for lower Stratum IIa.

It is clear that during most of Stratum IIa time at Meadowcroft, the principal sediment source was grain-by-grain attrition. The sedimentation rate remained essentially the same except in those parts

Table 6.—*Calculated rates of sedimentation for Strata IIa–XI at Meadowcroft Rockshelter.**

Stratum	Duration (in radiocarbon years)	Average thickness	Rate of sedimentation	Dominant and subordinate sources
VIII–XI	925 years	35 cm	38 cm/1,000 years	sheetwash
VII	365 years	40 cm (average value)	110 cm/1,000 years	sheetwash and rock fall
VI	375 years	100 cm	267 cm/1,000 years	rock fall and sheetwash
V	625 years	30 cm	48 cm/1,000 years	sheetwash and attrition
IV	760 years	55 cm (average value)	72 cm/1,000 years	sheetwash and attrition
III	200 years	50 cm	250 cm/1,000 years	attrition, sheetwash and rock fall
IIb	4,700 years	40 cm	8.5 cm/1,000 years	attrition and rock fall
IIa	13,000 years	90 cm	6.9 cm/1,000 years	attrition and rock fall
Total	20,950 years	440 cm		
Average sedimentation value			21.0 cm/1,000 years	

* After Stuckenrath et al., 1982:89, table 4.

of the site affected by the creation of the western entrant. Although variation in rock fall frequency and location in Stratum IIa time may well have been controlled to some extent by climatic factors, rock fall was certainly never the most important contributor to the sedimentary history of the site.

Sedimentation rate and source are guided by factors that key on the configuration of the overhang in a virtually direct, one-to-one correlation. Where the overhang remains intact, gradual sedimentation takes place via grain-by-grain attrition and roof spalling alone. When entrants develop in the roof, sedimentation rates increase proportionately due to the influx of sheetwash. Despite the entrant of middle Stratum IIa age, sedimentation rate at the site did not change appreciably at the Pleistocene/Holocene boundary, which is marked at Meadowcroft by the middle/upper Stratum IIa interface, ca. 9,000 ± 500 B.C.

Haynes (1982:97) has suggested, ". . . that there is a pronounced stratigraphic break in alluvial successions throughout the conterminous United States that marks . . . the end of late Pleistocene fauna, . . . a major change in stream regimen . . . and a significant change in climate as reflected in the plant record." No comparable break is documented at Meadowcroft nor, we believe, in any other *closed* site of this time horizon (for example, Fort Rock Cave, Oregon; Wilson Butte Cave, Idaho).

One other observation on the sedimentation rate(s) reflected in Stratum IIa at Meadowcroft is worth a comment here. Sedimentation rates for the various Meadowcroft strata were calculated using the present radiocarbon chronology and the average thickness of each depositional unit (Table 6). As Table 6 shows, in those strata where grain-by-grain attrition is the predominant or only sediment source, sedimentation rate is, as Dincauze (1981:4) has put it, "exquisitely slow." Conversely, in strata where sheetwash from one or both major roof fall entrants is operative, sedimentation is quite rapid.

To ascertain approximately how slow sedimentation can be in areas affected by attrition alone, samples from the 5 m square attrition trap on the roof of the excavations from 1974 through 1978 were again scrutinized. The trap was divided into five 1 m by 5 m strips, and the amount of sediment falling on each of these strips was assessed. The samples from each strip generally yielded total weights ranging from 0.5 g to 10 g per day. Average values measured approximately 2 g per day per strip. The sediment collected in the trap consisted of quartz grains with a mean diameter of about 0.15 mm and a density of 2.65 g/cm^3. Using this information, an approximate if rough calculation can be made for modern sedimentation rate. With a fall of 2.65 g for one day, a volume of 1 cm^3 accrues. Over a 5 m^2 area (that is, one of the 1 m by 5 m strips comprising the attrition trap), this results in a sediment thickness of 2.0 × 10^{-5} cm. This in turn equals a rate of 7.3 cm/1,000 years. As attrition sediments have porosities of 15% to 25%, the actual sedimentation rate would increase to 8.75 cm/1,000 years. Burial of these same sediments causes compaction and re-

sults in an apparent decrease in rate. The significant point here is that the modern rate is perfectly compatible with the measured rates in lower Stratum IIa, where grain-by-grain attrition was the dominant sediment source.

OVERVIEW

The biological evidence from Meadowcroft Rockshelter presents a straightforward picture of a temperate "modern" biota. There is no problem of interpretation or correlation with other regional sites, paleontological and archaeological, that date from less than ca. 9,350 B.C. (11,300 B.P.). The Meadowcroft fauna agrees in content with archaeological faunas from 43 other regional sites ranging in cultural ascription from Archaic to Late Prehistoric (Guilday et al., 1980). All of these sites present a typical "Carolinian" fauna that one would expect to occur in the area today in the absence of European colonization.

This temperate fauna can be traced at other sites as far back as 7,290 ± 1,000 B.C. (9,240 B.P.) at Hosterman's Pit, Centre Co., Pennsylvania (Guilday, 1967). Other regional sites of some antiquity that contain a Recent fauna include Sheep Rock Shelter, Huntingdon Co., Pennsylvania, 6,970 ± 320 B.C. (8,920 B.P.) (Michels and Smith, 1967) and New Paris Sinkhole No. 3, Bedford Co., Pennsylvania, 6,620 ± 145 B.C. (8,570 B.P.) (Guilday et al., 1964).

Correlating the Meadowcroft data from older levels at the site with evidence from other mid-Appalachian sites is not as simple, however. Several writers, notably Haynes (1980) and Mead (1980) have suggested that the faunal and floral data from middle and lower Stratum IIa are climatically/ecologically inappropriate for the time period indicated by the radiocarbon dates. As amply noted in earlier Meadowcroft publications, however, the entire faunal sample from levels older than 9,350 ± 700 B.C. (11,300 B.P.), that is, from middle Stratum IIa and below, is limited to 11 identifiable specimens from among a total of 278 fragments. Only white-tailed deer, eastern chipmunk, southern flying squirrel, deer mouse, passenger pigeon, colubrid snake, and toad have been positively identified. As a group, these taxa suggest temperate conditions provided that they co-occurred near the rockshelter and were not transported some distance to the site by humans or raptors. Because each species might have occurred marginally in a boreal context (judging from modern ranges) their climatic implications would be ambiguous were it not for the botanical evidence which suggests that black gum (*Nyssa* sp.) occurred as early as 13,000 B.P., oak (*Quercus* sp.) and hickory (*Carya* sp.) as early as 14,500 B.P., and walnut/butternut (*Juglans* sp.) as early as 16,000 years ago (Volman, 1981:90, 93–94). Once again, these data suggest temperate conditions at the rockshelter. The vertebrate and invertebrate faunal evidence (including specific and, in the case of southern bog lemming and southern flying squirrel, infraspecific data) in concert with the floral remains therefore present a mutually supportive picture of the local environment which Meadowcroft's human occupants experienced at the close of the Pleistocene. That picture in general, the data lead us to believe, was not one substantively different from that of the present day. Unfortunately, preservation of fauna and flora at Meadowcroft prior to approximately 11,000 years ago is not good, but the information at hand reflects no radical ecological reorganization *at the site itself.*

In this regard it is important to note three frequently overlooked facts about Meadowcroft, the full ecological implications of which are as yet far from clear:

1) The Cross Creek drainage lies relatively far south (ca. 50 km) of the mapped Kent moraine, the maximum southern extent of the Wisconsinan ice. For much of the later Pleistocene human history of the site (that is, after ca. 12,830 B.C. or so), the discontinuous and only relatively recently mapped Lavery till, ca. 83 km north of Meadowcroft, marks the southern edge of the ice.

2) Meadowcroft occurs in a topographical setting which in one recent year had 40 to 50 more frost-free days than did the higher elevations or drainages contiguous to it. Indeed, modern Cross Creek has a more "southerly" temperature regime than any other drainage in the area. If this is now the situation, it may well have been the case prehistorically. In any event, as far as indications of late Pleistocene/Holocene environmental change at Meadowcroft Rockshelter are concerned, the

biotic record is silent. The fact that the Cross Creek drainage trends generally east-west rather than north-south may also help to explain its enduring temperate ecology as may the fact that the rockshelter itself faces south, away from the glacial front. Rather than an "either/or" ecological situation, however, it is not unreasonable to assume that Meadowcroft represents a floral and faunal mosaic perhaps not atypical or unique among other sites away from the glacial front.

3) Consistent with the greater number of frost-free days at Meadowcroft is its lower elevation (259.9 m) when compared not only to some other areas in the Cross Creek drainage but to well-studied paleontological sites in Pennsylvania and elsewhere. Hosterman's Pit stands at 377.9 m (Guilday, 1967:231). Not surprisingly, the mammalian faunal assemblage associated with its 7,290 ± 1,000 B.C. (9,240 B.P.) radiocarbon date is of Recent aspect (Guilday et al., 1977:80–81, table 23). As noted previously, New Paris Sinkhole No. 3, Pennsylvania, also has Recent fauna associated with its 6,620 ± 145 B.C. (8,570 B.P.) date. Leaving aside considerations of locational differences in comparison to Meadowcroft, these sites do not illuminate the characters of a Pleistocene/Holocene transition fauna in Pennsylvania.

New Paris Sinkhole No. 4, only a few meters from New Paris No. 3, however, has a 9,350 ± 1,000 B.C. (11,300 B.P.) date (at the 4.6 m level) and a fauna dominated by "boreal small rodents and insectivores" (Guilday et al., 1977:80). This is during middle Stratum IIa times at Meadowcroft (see Tables 4 and 5). It is important to note that New Paris No. 4 stands at 465 m above sea level (Guilday et al., 1977:79), that is, ca. 205 m *above* Meadowcroft's elevation. Furthermore, New Paris No. 4 is very near the foot of the Allegheny Front (Johnson, 1981:89). Its predominantly (but not completely) boreal fauna is therefore not unanticipated. Clearly, the utility of the data from this site for reconstructing the contemporaneous environment at Meadowcroft (on the *Unglaciated* Allegheny Plateau and at much lower elevation to name but two of the most obvious differences) is circumscribed. With some small deviations, the New Paris No. 4 fauna is "identical" (Guilday et al., 1977:77) to that at poorly dated but probably late Wisconsinan/early Holocene Clark's Cave in Bath Co., Virginia, which stands at 448 m, or 188 m above Meadowcroft's

elevation (Guilday et al., 1977:6, fig. 2). Though farther south than Meadowcroft, Clark's Cave was a short distance from the Allegheny Front.

Baker Bluff Cave in Sullivan Co., Tennessee, at 450 m elevation (Guilday et al., 1978:5) dates to as early as 17,150 ± 850 B.C. (19,100 B.P.) when deposition may have begun during the recovery phase of the Connersville Interstadial (Guilday et al., 1978:55). The lower fauna from this site suggest more temperate deciduous/coniferous conditions that thereafter disintegrated to boreal coniferous parkland (Guilday et al., 1978:77).

For matters of ecological reconstruction, it is an important fact that New Paris No. 4, Clark's Cave, and Baker Bluff Cave all contain fauna with *mixed* boreal, mid-continental, and eastern temperate elements though a general north to south, boreal to temperate cline is supported (Guilday et al., 1978:57). Pollen data (Davis, 1969a, 1969b; Miller, 1973) suggest that essentially modern mixed coniferous/deciduous forest characterized the Glaciated Allegheny Plateau—north of Meadowcroft and at much higher elevation—by ca. 7,000 B.C. (8,950 B.P.) (Johnson, 1981:92).

The importance of data from these and other sites for interpreting the Meadowcroft information cannot be underestimated. By the same token, however, climatic reconstruction for these sites, all of which are at much higher elevations (again, to name only one important variable) than Meadowcroft, cannot be used *directly* or vicariously to mold or condition our understanding of the possibly localized temperate climatic, vegetative, and faunal montage that characterized that site. Indeed, Meadowcroft's temperate setting may have been one of its primary attractions to aboriginal populations.

The geological data from Meadowcroft indicate that whatever the exact linkage between climatic variation and sedimentation at the site, the end of the Pleistocene and the onset of the Holocene is not signaled by any remarkable or, indeed, any *perceptible* interlude in the site's depositional history. Sedimentation rates at the Pleistocene/Holocene boundary apparently were not essentially different from those occurring at the site today.

The Meadowcroft geological, floral, and faunal data summarized above clearly lend support to the hypothesis that the Pleistocene/Holocene transition was not accompanied by any dramatic event or series of events witnessed in the extant ecofactual or depositional record at this exceptional site for which so many lines of evidence exist. These data sets benefit from the comprehensiveness of their collection, processing, and evaluation and imply that late Pleistocene conditions in areas south of the Wisconsinan glacial front were neither as uniform nor as inexorably harsh as some previous reconstructions for the Northeast and Mid-Atlantic areas might suggest. In the immediate vicinity of Meadowcroft, at least, there is very little to support the notion that late Pleistocene climatic conditions or the faunal/floral correlates of those conditions differed appreciably from circumstances that characterized the area in the early Holocene or, indeed, today. This does not mean that late Pleistocene/early Holocene ecological associations were the same in all details or that conditions were identical to those at present. It

seems, rather, that the local or microclimatic parameters and the indigenous biota were simply not *substantially* different. In sum, the authors share John Guilday's succinct characterization and interpretation of the Pleistocene/Holocene transition at Meadowcroft. Specifically, any environmental changes that occurred during the long span of time represented in the Meadowcroft faunal sequence took place within a mast forest context and apparently were of such a low order that the biota was not seriously disturbed at the site. The great majority of the faunal sequence post-dates the Wisconsinan/Holocene environmental change from boreal to temperate conditions, and it suggests that the period of transition may have been earlier, very rapid and relatively short in this part of western Pennsylvania.

ACKNOWLEDGMENTS

The excavations at Meadowcroft Rockshelter and throughout the Cross Creek drainage as well as the analysis of recovered materials were conducted under the auspices of the former Archaeological Research Program (now the Cultural Resource Management Program) of the Department of Anthropology, University of Pittsburgh. The incipient 1973 field project in addition to the 1977 and 1978 field seasons were directed by J. M. Adovasio. The 1974–1976 field projects were co-directed by J. M. Adovasio and J. D. Gunn. Analysis of all recovered materials is under the ultimate direction of J. M. Adovasio. Computer studies of the Meadowcroft/Cross Creek data base are supervised by J. D. Gunn at the University of Texas, San Antonio, and by R. Drennan at the University of Pittsburgh.

Generous financial and logistic support for the 1973–1978 excavations and analyses was provided by the University of Pittsburgh, the Meadowcroft Foundation, the National Geographic Society, the National Science Foundation, the Alcoa Foundation, the Buhl Foundation, the Leon Falk Family Trust, and Messrs. John and Edward Boyle of Oil City, Pennsylvania.

Radiocarbon assays were supplied via the Radiation Biology Laboratory, Smithsonian Institution, and in one instance by Dicarb Radioisotope Company. Line drawings for the figures used in this contribution are by K. Adkins, R. L. Andrews, and R. Stuckenrath. The manuscript version of this contribution was typed by G. LoAlbo Placone, Department of Anthropology, University of Pittsburgh.

LITERATURE CITED

ADOVASIO, J. M. 1982. Multidisciplinary research in the Northeast: one view from Meadowcroft Rockshelter. Pennsylvania Archaeologist, 52(3–4):57–68.

——— (compiler). 1982. The prehistory of the Paintsville Reservoir, Johnson and Morgan counties, Kentucky (by J. M. Adovasio, R. C. Carlisle, W. C. Johnson, P. T. Fitzgibbons, J. D. Applegarth, J. Donahue, R. Drennan and J. L. Yedlowski). Ethnology Monogr., 6:1–1074.

ADOVASIO, J. M., R. C. CARLISLE, K. CUSHMAN, J. DONAHUE, J. E. GUILDAY, W. C. JOHNSON, K. LORD, P. W. PARMALEE, R. STUCKENRATH, and P. WIEGMAN. n.d. a. Paleoenvironmental Reconstruction at Meadowcroft Rockshelter, Washington County, Pennsylvania. *In* Environment and extinctions: man in late glacial North America (J. I. Mead and D. J. Meltzer, eds.), Peopling of the Americas Series, Center for the Study of Early Man, Univ. Maine, Orono, in preparation.

ADOVASIO, J. M., J. DONAHUE, R. STUCKENRATH, and J. D. GUNN. 1981. The Meadowcroft Papers: a response to Dincauze. Quart. Rev. Archaeol., 2:14–15.

ADOVASIO, J. M., J. DONAHUE, K. CUSHMAN, R. C. CARLISLE, R. STUCKENRATH, J. GUNN, and W. C. JOHNSON. 1983. Evidence from Meadowcroft Rockshelter. Pp. 163–190, *in* Early man in the New World (R. Shutler, Jr., ed.), Sage Press, Beverly Hills, California, 223 pp.

ADOVASIO, J. M., J. D. GUNN, J. DONAHUE, and R. STUCKENRATH. 1975. Excavations at Meadowcroft Rockshelter, 1973–1974: a progress report. Pennsylvania Archaeologist, 45(3):1–30.

———. 1977a. Meadowcroft Rockshelter: retrospect 1976. Pennsylvania Archaeologist, 47(2–3):1–93.

———. 1977b. Progress Report on the Meadowcroft Rockshelter—A 16,000 year chronical. Pp. 37–159, *in* Amerinds and their paleoenvironments in northeastern North America (W. S. Newman and B. Salwen, eds.), Ann. New York Acad. Sci., 288:1–570.

———. 1978a. Meadowcroft Rockshelter, 1977: an overview. Amer. Antiq., 43:632–651.

ADOVASIO, J. M., J. D. GUNN, J. DONAHUE, R. STUCKENRATH, J. GUILDAY, and K. LORD. 1978b. Meadowcroft Rockshelter.

Pp. 140–180, *in* Early man in America from a circum-Pacific perspective (A. L. Bryan, ed.), Occas. Papers Dept. Anthropology, Univ. Alberta, 1:1–327.

———. 1979–1980*a*. Meadowcroft Rockshelter—retrospect 1977 (Part 1). North American Archaeologist, 1(1):3–44.

———. 1979–1980*b*. Meadowcroft Rockshelter—retrospect 1977 (Part 2). North American Archaeologist, 1(2):99–137.

ADOVASIO, J. M., J. D. GUNN, J. DONAHUE, R. STUCKENRATH, J. GUILDAY, and K. VOLMAN. 1980. Yes Virginia, it really is that old: a reply to Haynes and Mead. Amer. Antiq., 45:588–595.

ADOVASIO, J. M., and W. C. JOHNSON. 1981. The appearance of cultigens in the Upper Ohio Valley: a view from Meadowcroft Rockshelter. Pennsylvania Archaeologist, 51(1–2):63–80.

ADOVASIO, J. M. et al. (order of other authors not yet determined). n.d. *b*. The archaeology of Meadowcroft Rockshelter and the Cross Creek drainage. University of Pittsburgh Press, Pittsburgh, in preparation.

BEYNON, D. E. 1981. The geoarchaeology of Meadowcroft Rockshelter. Unpublished Ph.D. dissert., Dept. Anthropology, Univ. Pittsburgh, 283 pp.

BEYNON, D. E., and J. DONAHUE. 1982. The geology and geomorphology of Meadowcroft Rockshelter and the Cross Creek drainage. Pp. 31–52, *in* Meadowcroft: collected papers on the archaeology of Meadowcroft Rockshelter and the Cross Creek drainage (R. C. Carlisle and J. M. Adovasio, eds.), Dept. Anthropology, Univ. Pittsburgh, 270 pp.

CARLISLE, R. C., and J. M. ADOVASIO (eds.). 1982. Meadowcroft: collected papers on the archaeology of Meadowcroft Rockshelter and the Cross Creek drainage. Dept. Anthropology, Univ. Pittsburgh, 270 pp.

CARLISLE, R. C., J. M. ADOVASIO, J. DONAHUE, P. WIEGMAN, and J. E. GUILDAY. 1982. An introduction to the Meadowcroft/Cross Creek Archaeological Project, 1973–1982. Pp. 1–30, *in* Meadowcroft: collected papers on the archaeology of Meadowcroft Rockshelter and the Cross Creek drainage (R. C. Carlisle and J. M. Adovasio, eds.), Dept. Anthropology, Univ. Pittsburgh, 270 pp.

COOPER, E. L. 1972. Biological survey of headwater portion of Cross Creek watershed in Washington County, Pennsylvania. State College, Pennsylvania.

CUSHMAN, K. A. 1982. Floral remains from Meadowcroft Rockshelter, Washington County, southwestern Pennsylvania. Pp. 207–220, *in* Meadowcroft: collected papers on the archaeology of Meadowcroft Rockshelter and the Cross Creek drainage (R. C. Carlisle and J. M. Adovasio, eds.), Dept. Anthropology, Univ. Pittsburgh, 270 pp.

DAVIS, M. B. 1969*a*. Palynology and environmental history during the Quaternary Period. Amer. Scientist, 57:317–332.

———. 1969*b*. Climatic changes in southern Connecticut recorded by pollen deposition at Rogers Lake. Ecology, 50:409–442.

DINCAUZE, D. F. 1981. The Meadowcroft papers. Quart. Rev. Archaeol., 2:3–4.

FLINT, N. K. 1955. Geology and mineral resources of Somerset County, Pennsylvania. Pennsylvania Geological Survey County Report C56A, 267 pp.

FOWELLS, H. A. 1965. Silvics of forest trees of the United States. United States Dept. Agriculture Handbook, 271:1–762.

GRANT, K. E. 1973. Work plan for the Cross Creek watershed. Manuscript on file, Soil Conservation Service, Harrisburg.

GUILDAY, J. E. 1967. The climatic significance of the Hoster-man's Pit Local Fauna, Centre County, Pennsylvania. Amer. Antiq., 32:231–232.

———. 1977. Fauna. *In* Meadowcroft Rockshelter: retrospect 1976, by J. M. Adovasio, J. D. Gunn, J. Donahue, and R. Stuckenrath. Pennsylvania Archaeologist, 47(2–3):7–9.

GUILDAY, J. E., H. W. HAMILTON, E. ANDERSON, and P. W. PARMALEE. 1978. The Baker Bluff Cave deposit, Tennessee, and the late Pleistocene faunal gradient. Bull. Carnegie Mus. Nat. Hist., 11:1–67.

GUILDAY, J. E., P. S. MARTIN, and A. D. McCRADY. 1964. New Paris No. 4: A Pleistocene Cave deposit in Bedford County, Pennsylvania. Bull. Nat. Speleol. Soc., 26:121–194.

GUILDAY, J. E., and P. W. PARMALEE. 1982. Vertebrate Faunal Remains from Meadowcroft Rockshelter, Washington County, Pennsylvania: Summary and Interpretation. Pp. 163–174, *in* Meadowcroft: collected papers on the Archaeology of Meadowcroft Rockshelter and the Cross Creek drainage (R. C. Carlisle and J. M. Adovasio, eds.), Dept. Anthropology, Univ. Pittsburgh, 270 pp.

GUILDAY, J. E., P. W. PARMALEE, and H. W. HAMILTON. 1977. The Clark's Cave bone deposit and the late Pleistocene paleoecology of the central Appalachian Mountains of Virginia. Bull. Carnegie Mus. Nat. Hist., 2:1–87.

GUILDAY, J. E., P. W. PARMALEE, and R. C. WILSON. 1980. Vertebrate faunal remains from Meadowcroft Rockshelter (36WH297), Washington County, Pennsylvania. Manuscript on file, Dept. Anthropology, Univ. Pittsburgh, 140 pp.

HARLOW, W. M., and E. S. HARRAR. 1969. Textbook of dendrology. McGraw-Hill Book Co., New York, 512 pp.

HAYNES, C. V. 1977. When and from where did man arrive in northeastern North America: a discussion. Pp. 165–166, *in* Amerinds and their paleoenvironments in northeastern North America (W. S. Newman and B. Salwen, eds.), Ann. New York Acad. Sci., 288:1–570.

———. 1980. Paleoindian charcoal from Meadowcroft Rockshelter: is contamination a problem? Amer. Antiq., 45:582–587.

———. 1982. Pleistocene-Holocene boundary in the United States; alluvial stratigraphy and geochronology. American Quaternary Association Program and Abstracts, Seventh biennial conference, Univ. Washington, Seattle.

JOHNSON, W. C. 1981. Archaeological review activities in Survey Region IV, northwestern Pennsylvania: year end report of the Regional Archaeologist for the period September 1979 through August 1980. A report Prepared for the Pennsylvania Historical and Museum Commission by the Cultural Resource Management Program, Univ. Pittsburgh, Pittsburgh, Pennsylvania, Under the Supervision of J. M. Adovasio, in Accordance with the Provisions of Service Purchase Contract 645987.

LORD, K. 1982. Invertebrate faunal remains from Meadowcroft Rockshelter, Washington County, southwestern Pennsylvania. Pp. 186–206, *in* Meadowcroft: collected papers on the archaeology of Meadowcroft Rockshelter and the Cross Creek drainage (R. C. Carlisle and J. M. Adovasio, eds.), Dept. of Anthropology, Univ. of Pittsburgh, 270 pp.

MEAD, J. I. 1980. Is it really that old? A comment about the Meadowcroft "Overview." Amer. Antiq., 45:579–582.

MICHELS, J. W., and I. F. SMITH (eds.). 1967. A preliminary report of archaeological investigations at the Sheep Rock Shelter Site, Huntingdon, Pennsylvania. Pennsylvania State Univ., State College, Pennsylvania, 1:1–426; 2:427–943.

MILLER, N. G. 1973. Late-glacial and post-glacial vegetation change in southwestern New York State. Bull. New York State Mus. Sci. Service, 420:1–102.

RAY, L. L. 1974. Geomorphology and Quaternary geology of the glaciated Ohio River Valley—a reconnaissance study. U.S. Geological Survey Professional Papers, 826:1–74.

STILE, T. 1982. Perishable artifacts from Meadowcroft Rockshelter, Washington County, southwestern Pennsylvania. Pp. 130–141, in Meadowcroft: collected papers on the archaeology of Meadowcroft Rockshelter and the Cross Creek drainage (R. C. Carlisle and J. M. Adovasio, eds.), Dept. Anthropology, Univ. Pittsburgh, 270 pp.

STUCKENRATH, R., J. M. ADOVASIO, J. DONAHUE, and R. C. CARLISLE. 1982. The stratigraphy, cultural features and chronology at Meadowcroft Rockshelter, Washington County, southwestern Pennsylvania. Pp. 69–90, in Meadowcroft: collected papers on the archaeology of Meadowcroft Rockshelter and the Cross Creek drainage (R. C. Carlisle and J. M. Adovasio, eds.), Dept. Anthropology, Univ. Pittsburgh, 270 pp.

VENTO, F. J., J. M. ADOVASIO, and J. DONAHUE. 1980. Excavations at Dameron Rockshelter (15JO23A), Johnson County, Kentucky. Ethnology Monogr., 4:1–235.

VOLMAN, K. C. 1981. Paleoenvironmental Implications of Botanical Data from Meadowcroft Rockshelter, Pennsylvania. Unpublished Ph.D. dissert., Graduate College, Texas A&M Univ., 225 pp.

WEIGMAN, P. G. 1977. Flora. In Meadowcroft Rockshelter: retrospect 1976, by J. M. Adovasio, J. D. Gunn, J. Donahue, and R. Stuckenrath. Pennsylvania Archaeologist, 47(2–3): 5–7.

YEDLOWSKI, J. L., J. M. ADOVASIO, J. DONAHUE, R. C. CARLISLE, K. CUSHMAN, H. B. ROLLINS, and J. H. SCHWARTZ. 1982. Archaeological data recovery at Three Rockshelters in the Tombigee River Multi-Resource District, Alabama and Mississippi. An Interim Report Prepared for the U.S. Department of the Interior, National Park Service, Mid-Atlantic Region Under the Supervision of J. M. Adovasio, Principal Investigator, Under Contract Number C-54028(80).

Address (Adovasio): Professor and Chairman of Anthropology; Director, Cultural Resource Management Program, Department of Anthropology, University of Pittsburgh, Pittsburgh, Pennsylvania 15260.

Address (Donahue): Professor of Geology and Planetary Sciences and Anthropology, University of Pittsburgh, Pittsburgh, Pennsylvania 15260.

Address (Carlisle): Research Assistant Instructor, Assistant to the Chairman, Department of Anthropology and Editor, Cultural Resource Management Program, University of Pittsburgh, Pittsburgh, Pennsylvania 15260.

Address (Cushman): 3262 Cripple Creek Trail, Boulder, Colorado 80303.

Address (Stuckenrath): Radiation Biology Laboratory, Smithsonian Institution, Washington, D.C. 20560.

Address (Wiegman): Western Pennsylvania Conservancy, 316 Fourth Avenue, Pittsburgh, Pennsylvania 15222.

HISTORICAL BIOGEOGRAPHY OF FLORIDA PLEISTOCENE MAMMALS

S. David Webb and Kenneth T. Wilkins

ABSTRACT

The Pleistocene history of land mammal distributions in Florida can be drawn from that state's rich fossil record. The present essay interprets the geographic affinities of mammalian species from five rich Quaternary faunas as well as the modern fauna of Florida. Brief geographic accounts of the various mammal species are presented. The results are tabulated and interpreted according to a simple model which recognizes three routes by which land mammals could enter the Florida peninsula. The northern route accounted for between 4% and 10% of successive Quarternary faunas and is at its highest in the modern mammalian fauna. The Gulf Coastal Corridor has been far more important, transmitting between 17% and 34% of successive mammalian faunas, but is at its lowest in the modern fauna. The northern species apparently arrived more frequently within interglacial intervals (10.9% as compared with 5.9% of glacial faunas), whereas the Gulf Coastal Corridor species entered more abundantly during glacial intervals when the corridor was much wider (27.6% of glacial faunas as compared with 22.2% of interglacial faunas). Many Pleistocene species that occurred in Florida and around the Gulf Coastal Plain also extended northward up the Atlantic Coastal Plain to South Carolina. The classic view of Florida as an "ice-age winter resort" for northern or temperate taxa during glacial intervals is incorrect. Throughout the Pleistocene, about two-thirds of Florida's mammal fauna was also widespread in the southeastern United States and beyond. Endemic Florida species were extremely rare.

INTRODUCTION

Two circumstances render Florida a favorable place to study the historical biogeography of mammals. First, the state has produced a rich record of Pleistocene mammals. And secondly its location as a peninsula in the southeastern corner of the United States limits the potential routes for terrestrial faunal connections. Thanks especially to John Guilday's work to the north and to that of Claude Hibbard, Ernest Lundelius, and others to the west, it is possible to determine the probable geographic history of many Florida Pleistocene mammals.

In this essay, we propose a simple model of how Florida's Pleistocene environments may have affected its terrestrial biota during alternating glacial and interglacial intervals. We then recount the known distributional histories of Florida mammal groups. And finally we attempt crudely to quantify these diverse histories in terms of a few common dispersal patterns. In this manner, our simple model approximates the complex history of Pleistocene mammal distributions in Florida.

THE MODEL

In the Pleistocene, as at present, Florida was a south-facing cul-de-sac with its toe in the tropics. The northern opening into the peninsula extends 350 kms from the mouth of the Apalachicola River on the west to the north-flowing St. Johns River on the east at a latitude of about 30°. As indicated in Fig. 1, non-volant biota could enter the peninsula by any one of three more or less distinct routes. First, they could enter from the west along the Gulf Coastal Plain, crossing the Apalachicola, Ochlockonee and Suwannee rivers. Secondly, they could enter from the north by any of several relatively unimpeded avenues. It should be noted that although these northern approaches are often referred to as "highlands," that is a peculiarly Floridian perspective and the nearest truly "Appalachian" environments lie at least 240 kms farther north across southern Georgia. Or thirdly, terrestrial biota could enter the peninsula from the northeast along the Atlantic Coastal Plain.

The greatest environmental changes that affected the peninsula during the Pleistocene epoch resulted from sea-level changes of worldwide (eustatic) scope. During glacial time-intervals the sea fell some 140 m with respect to the present elevation of the peninsula. During interglacial intervals, the sea rose as little as 3 m and as much as 40 m above its present level, but progressively less in successive interglacial times. The piezometric surface (or regional water table level) tracked these sea-level cycles in a general

PLEISTOCENE CORRIDORS INTO
THE FLORIDA PENINSULA

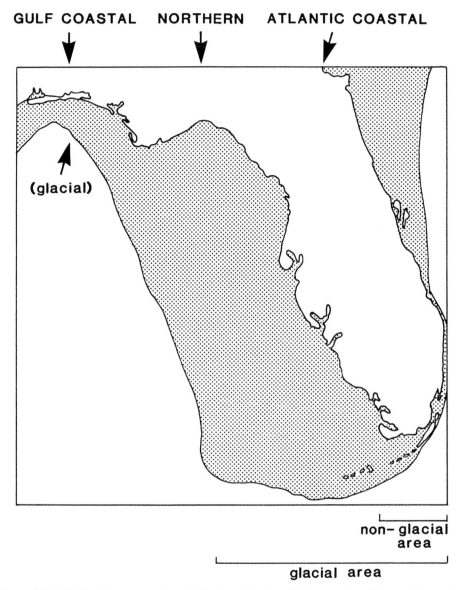

Fig. 1.—A geographic model indicating three more or less distinct corridors by which non-volant biota could enter the Florida peninsula. During non-glacial intervals the peninsula had approximately its present outline, whereas during glacial intervals an expanded Gulf Coastal Corridor (stippled area) emerged from the Gulf of Mexico.

way, so that the available surface water surely declined as sea-level dropped, especially in the central peninsula where permeable Paleogene limestone beds predominate. The most remarkable feature of the Florida landscape during glacial intervals was the major addition (essentially doubling) of lowland area as the shallow bottom of the Gulf of Mexico was exposed (Fig. 1).

In our simple model we do not attempt to calculate the redistribution of particular habitats during glacial and interglacial intervals. Palynological studies indicate that the peninsula tended to become more xeric during the last full glacial and presumably during earlier glacial intervals as well (for example, Watts, 1971). On the other hand, these same studies encourage a cautious attitude toward extrapolating widespread changes from a limited number of local samples. Delcourt and Delcourt (1983) recognized "minimal change in the composition of terrestrial plant communities and environmental conditions over the past 20,000 years . . . between 29°30′ and 33°N. latitude," due to a persisting maritime tropical airmass over Florida and adjacent areas. In this study we have not invoked any Pleistocene climatic or ecological shifts in the Florida peninsula, even though we acknowledge that subhumid habitats probably expanded proportionally over the enlarged peninsula during glacial epochs.

FAUNAL SUCCESSION

In Table 1 we present a sampling of the Pleistocene faunal succession in Florida. Many other local faunas could be listed; a number of others appear in Webb (1974) and Kurten and Anderson (1980); but we have selected five of the richest samples. They span the entire Pleistocene and are rather clearly assignable to glacial or interglacial intervals. We have also included the Recent fauna because it is both well-sampled and representative of a nonglacial interval. We will summarize the distributions of all of these species in an effort to test our simple biogeographic model. First, however, we offer a few specific comments on these selected faunas.

The Inglis local faunal was collected from a fissure in the Inglis Limestone at present sea level about 1 km inland from the Gulf of Mexico (Webb, 1974). Although the entire vertebrate fauna has not been described, the rich samples of unusual taxa have inspired the following special studies: Frazier, 1981, on *Erethizon*; Berta, 1981, on *Chasmaporthetes,* and, in press, on *Smilodon*; Wilkins, 1984, on *Geomys*; and Meylan, 1982, on the herpetofauna. The fauna correlates rather closely with the Curtis Ranch local fauna in Arizona which has been shown by magnetostratigraphic and radioisotopic studies to be 1.9 million years old (Johnson et al., 1975). Because the Inglis site lies at or just below present sea level but has no marine fossils or sediments it is correlated with a glacial interval, presumably the Nebraskan (Webb, 1974; Kurtén and Anderson, 1980).

The Coleman IIA fauna was obtained from alternating beds of sands and clays filling a sinkhole in the late Eocene Ocala Limestone in Sumter Co., Florida (Martin, 1974a). The site is 27 m above present sea level. Martin's analysis of this fauna indicates a latest Irvingtonian age and suggests that it was deposited during a glacial period (probably Kansan) when sea level was at least as low as at present. This fauna appears to correlate most closely with the Cumberland Cave fauna of Maryland.

Two Rancholabrean faunas are used in this paper to represent the Sangamonian interglacial interval. Deposits at both Arredondo IIA (Alachua Co.) and Reddick IA (Marion Co.) accumulated in fissures in the Eocene Ocala Limestone (Kurten and Anderson, 1980). These central peninsula sites lie at about 25 m above present sea level. The mammalian faunas of these sites have not yet been comprehensively treated. Recent curation of the extensive microfauna has yielded a more diverse faunal list (Table 1) for the Arredondo IIA site than indicated in Webb (1974). Gut and Ray (1963) described the geological context of the Reddick IA deposits and listed 53 mammalian taxa.

Weigel (1962) produced a list of vertebrates from the Vero deposits in Indian River Co. Recent curation of unsorted Vero material collected by Weigel has resulted in addition of three taxa to his list— *Mustela vison, Felix atrox,* and *Platygonus compressus*. These deposits gained notoriety due to association of extinct vertebrate fossils with human bones and artifacts (Sellards, 1917). Later Wisconsinan (Rancholabrean) and Holocene fossils (in strata 2 and 3, respectively) occur in sediments filling a former stream channel. Bone-bearing layers are underlain by a Pleistocene marine shell marl of the Anastasia Formation (Sellards, 1916). Deposition occurred during times of reduced sea level in the last glacial interval. Other coastal Florida sites contemporaneous with Vero include Melbourne and Seminole Field (Kurten, 1965).

The modern mammalian fauna of Florida includes approximately 53 native terrestrial species (Stevenson, 1976; Hamilton and Whitaker, 1979; Hall, 1981). Although not yet documented from Florida, the swamp rabbit (*Sylvilagus aquaticus*) is

Table 1.—*List of mammals from five Pleistocene sites and from the modern fauna of Florida. Geographic affinities denoted by letter: N, Northern; V, Varied; T, Tropical; W, Western; E, Endemic.*

Taxon	Inglis IA	Coleman IIA	Arredondo IIA	Reddick IA	Vero	Modern	Affinity
Didelphis virginiana		X	X	X	X	X	T
Scalopus aquaticus	X	X	X	X	X	X	V
Cryptotis parva	X	X	X	X	X	X	V
Sorex longirostris						X	V
Blarina carolinensis	X	X	X	X	X	X	V
Desmodus stocki	X		X	X			T
Myotis austroriparius		X	X	X		X	V
Myotis grisescens						X	N
Myotis keenii						X	N
Myotis sodalis						X	N
Myotis sp.	X						V
Pipistrellus subflavus	X	X				X	V
Eptesicus fuscus					X	X	V
Lasiurus borealis				X		X	N
Lasiurus seminolus					X	X	T
Lasiurus cinereus						X	N
Lasiurus intermedius			X	X	X	X	T
Nycticeius humeralis					X	X	V
Plecotus rafinesquii						X	V
Tadarida brasiliensis				X	X	X	T
Eumops glaucinus					X	X	T
Dasypus bellus	X	X	X	X	X		T
Dasypus novemcinctus						X	T
Kraglievichia paranense	X						T
Holmesina septentrionalis		X		X	X		T
Glyptotherium arizonae	X						T
Megalonyx leptostomus	X						T
Megalonyx sp.			X	X	X		T
Eremtherium mirabile	X						T
Glossotherium chapadmalensis	X						T
Glossotherium harlani				X	X		T
Tamias striatus						X	N
Tamias aristus			X				N
Sciurus carolinensis		X	X		X	X	V
Sciurus niger						X	V
Sciurus sp.	X		X	X			V
Glaucomys volans	X	X	X	X		X	V
Geomys propinetis	X						W
Geomys pinetis		X	X	X	X	X	V
Thomomys sp.			X				N
Castor canadensis						X	N
Erethizon kleini	X						T
Erethizon dorsatum		X					N
Hydrochoerus holmesi	X	X		X	X		T
Pitymys cf. mcnowni (=aratai)		X					N
Pitymys pinetorum			X	X	X	X	N
Microtus pennsylvanicus			X			X	N
Neofiber alleni		X	X	X	X	X	V
Ondatra cf. idahoensis	X						N
Synaptomys australis			X	X	X		N
Oryzomys palustris			X	X	X	X	T
Reithrodontomys humulis	X	X	X	X	X	X	V
Peromyscus polionotus			X	X	X	X	V
Peromyscus gossypinus			X	X	X	X	V
Peromyscus floridanus		X	X	X		X	E
Peromyscus sp.	X	X	X	X	X		V
Ochrotomys nuttalli		X	X	X		X	V

Table 1.—*Continued.*

Taxon	Inglis IA	Coleman IIA	Arredondo IIA	Reddick IA	Vero	Modern	Affinity
Sigmodon curtisi	X						W
Sigmodon bakeri		X					E
Sigmodon hispidus			X	X	X	X	T
Neotoma n. sp.	X						V
Neotoma floridana			X	X	X	X	V
Lepus alleni	X	X					W
Sylvilagus floridanus	X	X				X	V
Sylvilagus sp.			X	X	X	X	V
Urocyon sp.	X	X					V
Urocyon cinereoargenteus			X	X	X	X	V
Vulpes vulpes					X	X	N
Canis edwardi	X						V
Canis lupus		X					V
Canis latrans				X	X	X	W
Canis dirus (=*ayersi*)			X	X	X		V
Canis rufus					X		V
Tremarctos floridanus	X			X	X		T
Ursus americanus				X	X	X	V
Arctodus pristinus	X	X					V
Procyon n. sp.	X						V
Procyon lotor		X		X	X	X	V
Mustela frenata			X			X	V
Mustela vison					X	X	V
Lutra canadensis					X	X	V
Spilogale putorius	X	X		X	X	X	V
Mephitis mephitis		X		X		X	V
Conepatus leuconotus	X	X		X			T
Trigonictis macrodon	X						V
Felis cf. *inexpectata*	X						V
Felis onca		X	X	X	X		V
Felis concolor				X		X	V
Felis pardalis				X			T
Felis atrox					X		V
Felis rufus	X	X	X	X	X	X	V
Smilodon gracilis	X						V
Smilodon floridanus					X		V
Homotherium serum	X			X			V
Chasmaporthests ossifragus	X						V
Mammuthus sp.		X		X	X		V
Mammut americanum	X			X	X		V
Tapirus sp.	X						V
Tapirus veroensis			X	X	X		V
Equus sp.	X	X		X	X		V
Platygonus bicalcaratus	X						V
Platygonus cumberlandensis		X					V
Platygonus compressus			X	X	X		V
Mylohyus sp.		X	X		X		V
Mylohyus fossilis (=*nasutus*)				X			V
Hemiauchenia macrocephala	X		X	X	X		V
Hemiauchenia small species	X						T
Palaeolama mirifica		X	X	X	X		T
Odocoileus virginianus	X	X	X	X	X	X	V
Capromeryx arizonensis	X						W
Ovibovinae sp. (=?*Euceratherium* sp.)	X						N
Bison antiquus			X	X	X		V
Bison bison						X	V

thought to have recently invaded the panhandle region. The total fauna might be further increased through inclusion of the star-nosed mole (*Condylura cristata*) whose southernmost known occurrence is the Okefenokee Swamp in southern Georgia. Trapping in extreme northern Florida in the southern extension of this swamp will likely yield *Condylura.* The modern fauna is interpreted here as representative of an interglacial period of relatively high sea level.

PLEISTOCENE MAMMAL HISTORY

We attempted to determine the geographic affinities of each species in Table 1. We inferred these affinities from the known geographic range of each species or its sister-group during the same or immediately preceding mammal age. We recognized the following three distinctive distributional categories: northern (N), western (W), and tropical (T). In practice we were unable to distinguish between species that were distributed northward into Appalachian sites and those that were distributed northeastward along the Atlantic Coastal Plain. We also recognized two other distributional categories, namely endemic (E), and various (V), although these did not help sharpen our routes by which Pleistocene land mammals reached Florida. Before compiling the results of this tabulation, we may briefly recount the apparent Pleistocene history of Florida's major land mammal groups.

MARSUPIALS

Didelphis appeared during the Irvingtonian, but not in the early Inglis IA local fauna, and was a constant, often abundant presence thereafter in Florida. The reported occurrence in the late Blancan at Santa Fe I (Kurten and Anderson, 1980) is probably incorrect, based on Rancholabrean material mixed in the river site with Blancan specimens. Presumably it came from Neotropical origins by way of the Gulf Coast. Guilday (1958) noted its late or post-Pleistocene spread northward.

BATS

Fourteen species comprise Florida's modern bat fauna. By at least the Rancholabrean (for example, Reddick) Florida Pleistocene bat faunas displayed contributions from both the north (*Myotis* sp., *Lasiurus borealis*, *Lasiurus cinereus*) and from the Neotropics (*Desmodus*, *Lasiurus seminolus*, *Lasiurus intermedius*, *Eumops*, and *Mormoops*). The other bat species in Table 1 are rather widespread in North America and thus of uncertain geographic affinities.

The only bat present in the Rancholabrean of Florida which is not present in the state today is the vampire bat, *Desmodus stocki,* from Reddick and Arredondo (Gut, 1959; Jones, 1958; Hutchinson, 1967). Affinities of vampire bats lie within the Neotropical Realm where all three extant vampire species now occur. Their invasion of Florida presumably coincided with colonization by the Neotropical megafauna, whose blood surely comprised their diet. The earliest record for *Desmodus* (in Florida and elsewhere) is Inglis IA. *Desmodus* probably entered Florida along the Gulf Coast corridor during a glacial interval such as during Inglis IA or perhaps earlier times. The genus persisted in the more inland sites during the Sangamonian Interglacials, but is unknown from the Florida Wisconsinan despite the Wisconsinan presence of a megafauna seemingly adequate for the vampire's diet.

INSECTIVORES

None of Florida's insectivore species have narrow enough geographic ranges to provide useful results for this survey. The apparent late Pleistocene arrival of *Sorex longirostris* at Haile XIB is noteworthy, but does not clearly indicate any particular geographic route (Webb, 1974).

SCIURIDS

The distributions of tree squirrels (for example, *Sciurus* species, *Glaucomys volans*) occurring in the Quaternary of Florida are so widespread as to be inconclusive as regards geographic affinities. However, both western and northern connections are indicated by a ground squirrel and two chipmunks. The eastern chipmunk (*Tamias striatus*) now occurs restrictedly in Okaloosa Co. in the Florida panhandle in relictual woodland habitats characteristic of the Appalachian Mountains and other boreal environs (Jones, 1978). Similar northern affinities may reasonably be accorded to the Arredondo II fauna on the basis of presence of a mandible (UF 57101) tentatively referred to *T. aristus,* the extinct "noblest" chipmunk, previously known only from its Sangamonian type locality (Ladds, in northern Georgia) and distinguished from *T. striatus* by its 30% larger size (Ray, 1965*a*).

Florida's western connection is evinced by the occurrence of a ground squirrel, *Spermophilus* sp., in the Haile XIVA fauna (Alachua Co., Florida). This fauna is thought to have been deposited in the early Wisconsinan, a glacial interval when lowered sea levels opened the Gulf Coast Corridor as a west to east dispersal avenue. Careful study of this material suggests it to be distinct from *Spermophilus spilosoma, S. tridecemlineatus,* and *S. mexicanus* (Martin, 1974*b*).

POCKET GOPHERS

Two genera of pocket gophers occurred in Florida during the Pleistocene. The record of *Geomys* and more-or-less continuous, whereas that for *Thomomys* is spotty and of shorter duration. Florida fossil deposits record nearly two million years of evolution of the genus *Geomys.* A new species, described by Wilkins (1984), characterized the early Irvingtonian Inglis IA and mid-Irvingtonian Haile XVIA sites. This more primitive species presumably evolved into the extant *G. pinetis* by late Irvingtonian times (Coleman IIA).

The route by which *Geomys* entered Florida is uncertain, but either or both of two scenarios seem plausible. The first has pocket gophers travelling along the Gulf Coast corridor at times of lower sea level (such as characterized the time of Inglis IA deposition) when sandy marine terraces were emergent. Throughout the Cenozoic, the Mississippi River has surely been a significant barrier to many terrestrial species. Dispersal of pocket gophers across the Mississippi and other such rivers (for example, Apalachicola River in the Florida panhandle) was likely passive, via changes in river courses rather than active. Honeycutt and Schmidly (1979) discussed the significance of oxbowing in pocket gopher dispersal and gene flow. Crossing of rivers flowing over and beneath karst topography (for example Suwannee and other rivers of the Florida peninsula) probably occurred in several ways. Pocket gophers could swim through very shallow waters during moderate seasonal dry periods or even travel across the dry river beds on the ground surface during extreme droughts. The preceding scenarios characterize the moister interglacial periods when higher pi-

ezometric surfaces led to surface flow of rivers. Conversely, substantial lowering of water table levels during glacial intervals resulted in considerable underground movement of rivers through karst limestone and, thereby, abandonment of surficial river beds in many areas. At such times, pocket gopher dispersal was surely unimpeded by these dried river beds.

Alternatively, *Geomys* could have reached Florida via a midwestern route running through the Interior Lower Plateau situated west of the Appalachian Mountains and east of the Mississippi River. Parmalee and Klippel (1981) summarized late Wisconsinan and early Holocene records of *G.* cf. *bursarius* in this gap between extant *G. bursarius* in Illinois and Indiana and *G. pinetis* in Alabama and Georgia. As Guilday et al. (1971) suggested, it is indeed likely that *Geomys* pocket gophers were continuously distributed from the Midwestern plains into the southeastern coastal plains during the late Pleistocene (and perhaps at other earlier times). However, their contention that differentiation of the two extant species occurred since the Wisconsinan glaciation is unlikely in light of recent work (Wilkins, 1984) wherein the chronological range of *G. pinetis* was extended to the late Irvingtonian (late Kansan to early Illinoian Coleman IIA). Furthermore, Heaney and Timm (1983) argued that *G. pinetis* possesses several primitive features, thereby indicating an early (pre-late Irvingtonian) split from the *G. bursarius* species group. Because of the difficulty in distinguishing *G. pinetis* and *G. bursarius* on the basis of postcranial elements, specific identification of gopher material from late Pleistocene and early Holocene sites in the present hiatus region (as well as a better understanding of the evolutionary relationships and history of these species) must await dental and craniometric analyses.

Until recently, the only published record of Pleistocene *Thomomys* pocket gophers in the southeastern United States was Simpson's (1928) description of *T. orientalis* from the Sangamonian Sabertooth Cave deposit in west-central peninsular Florida (Citrus Co.). Wilkins (in press) reported an additional Sangamonian locality (Rock Springs, Orange Co.) for *Thomomys* sp. in central peninsular Florida. Re-examination of microfauna at the Florida State Museum Vertebrate Paleontology Collection revealed one *Thomomys* sp. mandible from the late Irvingtonian Coleman IIA site in Sumter Co., Florida. A fourth Florida Pleistocene record for *Thomomys* sp. is Williston IIIB (Levy Co.), a Sangamonian site collected in 1973 by Robert A. Martin. Hence, the temporal range of *Thomomys* sp. in Florida extends from the late Irvingtonian into the middle Rancholabrean.

Thomomys could possibly have reached Florida by either the Gulf Coastal or Midwestern routes proposed above for *Geomys* pocket gophers. Another alternative, however, is dispersal southward along the Appalachian Mountains. The presence of fossil *Thomomys potomacensis* in Maryland (Cumberland Cave, middle Irvingtonian) and West Virginia (lower levels of Trout Cave, late Irvingtonian) at times slightly antedating the earliest Florida record (late Irvingtonian Coleman IIA) suggests the Appalachians as the most likely of the three possible dispersal corridors.

MICROTINES

Representation of microtines in Florida's modern fauna (*Pitymys pinetorum, Microtus pennsylvanicus,* and *Neofiber alleni*) is less diverse than found in the state's Rancholabrean deposits (five species). Although *Microtus pennsylvanicus* was relatively widespread in Florida during the late Pleistocene, it has only recently been added to the list of living Florida mammals. An isolated population located in the coastal saltmarshes of Levy Co. apparently represents a relict population which has endured since the late Pleistocene moderation of the Florida climate (Woods et al., 1982). The depauperate microtine fauna at the early Irvingtonian Inglis IA site suggests that four of the five Florida species may not have reached the peninsula until the late Irvingtonian. In fact, Reddick comprises the earliest known record anywhere for *Pitymys pinetorum.* The record of all five microtine genera in Florida is essentially continuous from the late Irvingtonian through the Rancholabrean. This includes *Ondatra* as well, which is represented by the extant *O. zibethicus* in several Sangamonian and Wisconsinan deposits (Webb, 1974) although this temporal distribution is not evident in Table 1. *Synaptomys* last appears in Florida in the mid-Holocene Devil's Den site dated about 8,000 y.b.p. (Martin and Webb, 1974) and *Ondatra* continued until some 3,000 years ago.

Neofiber has been known in Florida more or less continuously at least since Irvingtonian time. During the early and medial Pleistocene these round-tailed muskrats were by no means Florida endemics, but ranged far to the west and through the southern Great Plains. Frazier (1977) attributed their very

late Pleistocene retreat to a relictual distribution in Florida to the advent of prolonged freezing winter temperatures during the last (and most severe) glacial epoch.

Voles, lemmings, and other microtines are generally presumed to have boreal affinities and to indicate cooler and moister climatic regimes. Repenning (1983) summarized the biogeography of the genus *Pitymys* in North America and traced their invasion of the continent via Beringia. Dispersal of microtines into Florida was undoubtedly from the north, although whether it was via the Appalachians or along lowland corridors east or west thereof is uncertain.

HISTRICOGNATH RODENTS

The Inglis IA fauna reveals already in the earliest Pleistocene the presence of the two hystricognath families that continuously occupied Florida up to the Recent, namely the porcupines (Erethizontidae) and the capybaras (Hydrochoeridae).

The distinctive small species of *Erethizon, E. kleini,* is unique to the Irvingtonian of Florida, but before Rancholabrean times gave way to the extant species, *E. dorsatum.* The final latest Pleistocene records occur at Seminole Field and possibly New Port Richey (Frazier, 1981). A pre-Pleistocene species of hydrochoerid, *Neochoerus dichroplax* described by Ahearn and Lance (1980) is supplanted by a more progressive species of *Neochoerus* and by the smaller living genus *Hydrochoerus.* Both of these large hydrochoerid rodents survived along the Gulf Coastal Plain in the southeastern United States until the latest Pleistocene. The demise of *Erethizon* in the southeast has been attributed by Ray and Lipps (1970) to the activity of man.

Both of these families of hystricognath rodents surely moved north from South America in the late Pliocene. Late Blancan records of hydrochoerids are known from low temperate latitudes in Mexico, Arizona, and Florida (Ahearn and Lance, 1980). Similar aged records of *Erethizon* are scattered from Mexico to Idaho (Frazier, 1981). Their most probable avenue into Florida is from the south or the west via the Gulf Coastal Corridor. The apparent elimination of hydrochoerids from the Great Basin by late Irvingtonian time may be related to increasing aridity in that region. By contrast two genera of hydrochoerids survived, in apparent sympatry, through the late Pleistocene in Florida and the southeastern coastal plain.

HARES AND RABBITS

An abundance of *Sylvilagus floridanus* is a regular feature of nearly all Pleistocene faunas in Florida, and the geographic distribution of that species is exceedingly broad. More notable are the abundant records of a large hare referable to *Lepus alleni* in the Inglis IA and Coleman IIA local faunas. These "jackrabbits" strongly indicate a broad coastal scrub or savanna connection between Florida and their western ranges. It is perhaps not surprising that these faunas represent early glacial intervals.

EDENTATES

The history of how edentate species entered Florida is relatively easy to deduce. Surely they originated ultimately from South America. Their most probable route is by circling the Gulf of Mexico through tropical and subtropical latitudes. In the Inglis local fauna six species represent six different families (three pilosan and three cingulate), including the earliest record of any megatheriid in North America. The other five taxa do not represent oldest records but apparently reached Florida during the late Blancan (late Pliocene) age, where they are first known in the Sante Fe River IA local fauna of probable glacial age (Webb, 1974; Robertson, 1976).

A seventh edentate genus apparently reached Florida during the late Irvingtonian. McDonald (1984) reported *Nothrotheriops* from the Coleman IIA local fauna, and explained its immigration by postulating a corridor of scrub-like vegetation along the Gulf of Mexico to semiarid western sites from which it was previously known (Lull, 1929). There is little doubt that the Coleman IIA local fauna represents a low-water glacial interval, possibly the Kansan (Webb, 1974; Kurtén and Anderson, 1980).

Since the Irvingtonian, the only additional edentate to reach Florida was the modern *Dasypus novemcinctus* which was accidentally introduced by human meddling, but has since spread eastward around the northern Gulf of Mexico and into Florida on its own. Clearly this was accomplished very recently during the present interglacial—a time when Gulf Coast Corridor is relatively narrowed. Its northward distribution has been limited by winter cold and its westward distribution by available moisture (Humphrey, 1974).

CARNIVORES

In general, the Florida records of Pleistocene carnivoran species were parts of widespread distributions. For example, Webb (1974) showed that the large sample of *Smilodon californicus* from the Rancho La Brea tar pits in California could not be distinguished from the sample of *Smilodon floridanus* in Rancholabrean sites in Florida. And Kurtén (1966) found a close resemblance between the large sample of *Tremarctos floridanus* from Florida and the smaller sample of spectacled bears from southern California. In view of such geographic continuity, it is difficult to determine direction of dispersal. The fact that *Tremarctos* is confined to low latitudes throughout its known (Pleistocene) record suggests that it either originated in Florida or extended its range there from a Neotropical source.

The most interesting geographic pattern with respect to Florida's Pleistocene carnivorans is their ultimate retreat to the American tropics. Hog-nosed skunks, jaguarundis, jaguars, ocelots, spectacled bears, and possibly margays, are all Florida Pleistocene species, now confined to tropical latitudes or Neotropical ranges. The northernmost limits of these taxa vary somewhat, but spectacled bears and jaguars have been recorded from the late Pleistocene in Tennessee (Guilday and Irving, 1967; Guilday and McGinnis, 1972). The ocelot, jaguarundi, and possibly margay, however, appear to have had ranges consistently restricted to latitudes below about 30° (Gillette, 1976; Ray et al., 1963; Ray, 1964).

In summary, it is difficult to determine the direction of immigration of Florida's Pleistocene carnivores, but the present geographic affinities of many are toward the New World tropics, and a few were essentially tropical.

UNGULATES

Most of Florida's Pleistocene ungulates are also distributed widely across North America. As with some of the carnivorans, however, a number share special ties to the present Neotropical Fauna. These species include llamas, tapirs, peccaries, certain deer, and certain proboscideans. Among Pleistocene llamas, for example, the commoner genus, *Hemiauchenia* is widespread from temperate North America through much of South America, whereas the rarer genus, *Palaeolama* is confined in North America to the Coastal Plain of South Carolina, Georgia, Florida, and Texas and also occurs in southern California (Webb, 1974). Simpson (1929) recognized the South American cervid genus *Blastocerus* in the late Pleistocene of Florida, although this remains to be confirmed by additional discoveries. An instructive case is the distribution of *Cuvieronius,* a progressive gomphotheriid, distinguished by its tusk-less, brevirostrine mandible. In the medial Pleistocene it ranged north into the southern Great Plains, but by late Pleistocene it occurred only along the coastal plain from South Carolina through Florida to Texas and is otherwise confined to Mesoamerica and South America. Several ungulate taxa including *Tapirus, Palaeolama,* and *Cuveronius* had essentially Neotropical and Coastal Plain distributions by the late Pleistocene.

RESULTS

Summary of Table 1 provides an estimate of proportional geographic affinities of the six Pleistocene and modern faunas (Table 2). From 61% to 68% of the taxa from each fauna have Pleistocene and/or modern ranges not conclusively indicating routes or directions of movement into Florida. Endemism (that is, occurrence only in Florida during the entire time interval under consideration) of these faunas is low, ranging from zero to two species or up to 5.1%. *Peromyscus floridanus* is endemic in four faunas; this species and *Sigmodon bakeri* are the only endemic mammals in the Coleman IIA fauna of latest Irvingtonian age.

The remaining one-third of the species in well-sampled Quaternary faunas has proved amenable to our interpretations of faunal affinities. We have recognized these species as invaders either from the west (having tropical or western affinities) or from the north. Species with western ties comprise negligible portions of these faunas except for Inglis IA which boasts four western taxa (8.5% of the fauna). Neotropical species comprise a much larger part of all-aged Florida faunas (between 15 and 28% of the successive faunas in Table 1). Both western and Neotropical species presumably entered Florida via the Gulf Coastal corridor, they therefore are summed under the heading "Gulf Coastal Corridor" in Table 2. Species with northern geographic affinities were seldom as numerous as the Neotropical contingent except in the present fauna which is notably high in northern taxa and notably low in Neotropical species. Nevertheless, the northern group is regularly represented during the Pleistocene and accounts for between 4% and 10% of each fauna.

DISCUSSION

Our results largely contradict the traditional view of Florida as a southern cul-de-sac into which northern fauna retreated during glacial intervals. In fact, we can find no example that justifies Colbert's (1942) labelling of Florida as an "Ice Age Winter Resort." Presumably such an example would show temperate

Table 2.—*Summary of geographic affinities of mammalian species in six Quaternary faunas in Florida. Paired table entries are numbers of species and corresponding percentages of each fauna.*

Geographic affinities	Inglia IA (glacial)	Coleman IIA (glacial)	Arredondo II (interglacial)	Reddick (interglacial)	Vero (glacial)	Modern (interglacial)
Tropical	12	7	8	15	14	8
	25.5%	17.9%	20.0%	27.8%	26.4%	15.1%
Western	4	1	—	1	1	1
	8.5%	2.6%		1.9%	1.9%	1.9%
Northern	2	3	4	2	3	10
	4.3%	7.7%	10.0%	3.7%	5.7%	18.9%
Varied	29	29	27	35	35	33
	61.7%	66.7%	67.5%	64.7%	66.0%	62.2%
Endemic	—	2	1	1	—	1
		5.1%	2.5%	1.9%		1.9%
Total taxa	47	39	40	54	53	53
Gulf Coastal Corridor (Tropical + Western)	16	8	8	16	15	9
	34.0%	20.5%	20.0%	29.7%	28.3%	17.0%

species surviving glacial intervals in Florida and then returning northward during interglacials. *Ondatra* comes close to fitting this pattern, but is not restricted to Florida at any time. John Guilday, more than a decade ago, pointed out an important distinction between the strong boreal influences reaching Appalachia during glacial intervals and the relatively limited northern influence that reached Florida (Guilday, 1971).

According to our data there is a moderate pattern of northern species distributing southward into Florida during the Pleistocene. The best examples are found, as might be expected, among microtine rodents. All five microtine species that reached Florida during the Pleistocene appear in medial or late Irvingtonian time. When it is fully studied the medial Irvingtonian site, Haile XVI A, will shed important light on the schedule of microtine first appearances. It is noteworthy that *Ondatra* and *Synaptomys* have retreated northward during Recent time and *Microtus* hangs on only by a minute relictual population. *Erethizon* has also retreated northward from Florida, following northern conifer belts. On the other hand, several very late Pleistocene arrivals in Florida such as *Vulpes vulpes, Sciurus niger,* and *Sorex longirostris* probably have northern origins. The number of truly boreal forms is very limited compared to even the low end of the gradient studied by Guilday et al. (1978). According to our tabulations, the average proportion of Florida land-mammal species having northern affinities is higher during interglacial epochs (10.9%) than during glacials (5.9%). This may be an artifact, based especially on the strong influence of northern groups in the Recent fauna.

Much the strongest pattern of land mammal distribution in Florida's Pleistocene record is the *Gulf Coastal Corridor* (Table 2). This corridor transmitted immigrant species from source areas that apparently lay both southward in the Neotropical Realm and westward at latitudes around 30° North. We have not always been able clearly to distinguish the western from the tropical patterns, and retain the impression that the two patterns are in many taxa interrelated. The sloth genus *Nothrotheriops,* for example, could be counted as western on the basis of the most familiar late Pleistocene records in western United States, but it could equally well be considered Neotropical on the basis of its ultimate origin in early Pleistocene time. Although the western pattern appears weaker than the tropical pattern in Pleistocene mammals, a very strong west-

LAND MAMMAL DISTRIBUTIONS INTO THE FLORIDA PENINSULA

A. Glacial Period

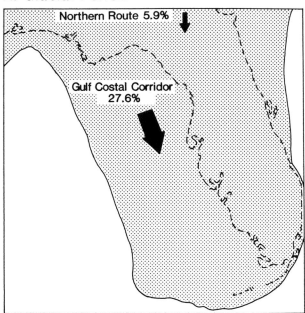

B. Interglacial Period

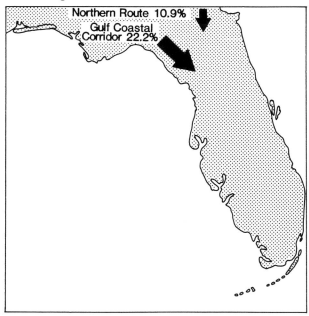

Fig. 2.—Preliminary calculations of apparent land-mammal dispersals into the Florida Peninsula (arrow widths proportional to percentage of Florida species) A) during glacial periods, B) during interglacial periods. The percentage of newly appearing taxa from each corridor is averaged over several glacial or interglacial intervals (see text).

ern pattern is evident in terrestrial lower vertebrates. The Inglis squamate reptile fauna, recently described by Meylan (1982), consists of 31 species, of which 15 are xeric-adapted forms with western counterparts, another three have circum-Gulf distributions, and two, *Rhineura* and *Ophisaurus,* are Florida relicts with former western ranges. Together western and Neotropical taxa entered Florida via the Gulf Coastal Corridor.

The strongest surge of new immigrants from the west and south appears to have arrived in the early Pleistocene. Possibly the first glacially lowered sea levels provided the strongest impetus for range extensions along the greatly broadened Gulf Coastal Corridor. The fact that the isthmian connection between South and North America had just formed also augumented the wave of new immigrants entering the southern end of the corridor, but this explains much more about the mammal faunal influx than it does about the herpetofaunal influx (Webb, 1976; cf. Meylan, 1982).

The actual record of Pleistocene mammals in the Gulf Coastal Plain westward and southward from Florida is sparse except in the latest Pleistocene, principally because the Mississippi Embayment persistently buried itself under a thick mass of sediments. The Ingleside fauna from Galveston, Texas, provides the richest late Pleistocene sample from the western Gulf of Mexico and is almost identical to Florida faunas of comparable age (Lundelius, 1972). This is the best direct evidence of the westward continuity of the Gulf Coastal Corridor with its maximum exposure during a glacial interval. The optimum conditions for burial of fossil faunas along the coastal plain probably developed during marine transgressive phases toward the end of each glacial interval when backfilling of stream and karst systems was most prevalent. While it is difficult to test this idea in early and medial Pleistocene chronologies, it does seem to be confirmed by the prevalence of late Wisconsinan sites during the last deglacial hemicycle.

The least important corridor for introducing land mammals into peninsular Florida is the Atlantic Coastal Plain. Indeed we were unable with certainty to distinguish any species that might have used that corridor rather than a more inland northern route. Thus the distinction we made a priori between these two potential routes in our Fig. 1 did not materialize in practice. On the other hand, this route exhibited great continuity with the Gulf Coastal Corridor—virtually all Pleistocene land mammals that entered Florida from the west or Neotropics also extended their ranges at least a short distance northward along the Atlantic Coastal Plain. In recent times expansion of the nine-banded armadillo's range into Georgia exemplifies this pattern. In the Pleistocene such Neotropical groups as glyptodonts, chlamytheres, armadillos, several groups of sloths, and capybaras, reached the coastal plain of South Carolina (Ray, 1965b; Roth and Laerm, 1980). Likewise a number of carnivorans and ungulates that had extended their ranges into the Neotropical region via Florida and the Gulf Coastal Plain maintained a foothold along the Atlantic Coastal Plain into late Pleistocene time; examples include *Tremarctos, Tapirus, Mylohyus,* and *Palaeolama.* The giant tortoises of the genus *Geochelone* also followed this route. Florida thus played a pivotal role in linking the Atlantic Coastal Plain with the Gulf Coastal Plain.

The role of the Gulf Coastal Corridor was potentially much greater during glacial intervals when sea levels were lower and the extent of the coastal plain vastly increased. In general this effect seems to be corroborated by our limited data set in Table 2; for the mean proportion of western and Neotropical species during glacial intervals (27.6%) exceeds that for interglacial (22.2%) (Fig. 2). The modern fauna has by far the lowest influence from Neotropical and western taxa. The early glacial interval represented by the Inglis I A local fauna, evidently played the most profound role in establishing a major Neotropical and western faunal contingent in the southeastern coastal plain region.

ACKNOWLEDGMENTS

First, we wish to acknowledge John Guilday's abiding influence both as a meticulous scientist and as a profoundly sensitive person. Among the others whose insights we have shared over the years we thank particularly Clayton Ray, Björn Kurtén, Ernest Lundelius, Elaine Anderson, Robert Martin, and Greg Mc-Donald. Our Florida Pleistocene work has been supported by grants from the National Geographic Society and the National Science Foundation.

Contribution no. 228 from the Vertebrate Paleontology Laboratory, Florida State Museum.

LITERATURE CITED

AHEARN, M. E., and J. F. LANCE. 1980. A new species of *Neochoerus* (Rodentia: Hydrochoeridae) from the Blancan (late Pleistocene) of North America. Proc. Biol. Soc. Washington, 93:435–442.

BERTA, A. 1981. The Plio-Pleistocene hyaena *Chasmaporthetes ossifragus* from Florida J. Vert. Paleo., 1:341–356.

———. In press. The sabrecat, *Smilodon gracilis* from Florida and a discussion of its relationships (Felidae; Mammalia). Florida State Mus. Bull. Nat. Sci.

COLBERT, E. H. 1942. Ice age winter resort. Nat. Hist., 50:16–21.

DELCOURT, P. A., and H. R. DELCOURT. 1983. Late Quaternary vegetational dynamics and community stability reconsidered. Quaternary Res., 19:265–271.

FRAZIER, M. K. 1977. New records of *Neofiber leonardi* (Rodentia: Cricetidae) and the paleoecology of the genus. J. Mamm., 58:368–373.

———. 1981. A revision of the fossil Erethizontidae of North America. Bull. Florida St. Mus., Biol. Sci., 27:1–76.

GILLETTE, D. G. 1976. A new species of small cat from the late Quaternary of southeastern United States. J. Mamm., 57:664–476.

GUILDAY, J. E. 1958. The prehistoric distribution of the opossum. J. Mamm., 39:39–43.

———. 1971. The Pleistocene history of the Appalachian mammal fauna. Pp. 233–262, *in* The distributional history of the biota of the southern Appalachians. Part III: Vertebrates (P. C. Holt, ed.), Virginia Polytech. Inst., Research Div. Monogr., 4:1–306.

GUILDAY, J. E., H. W. HAMILTON, E. ANDERSON, and P. W. PARMALEE. 1978. The Baker Bluff Cave Deposit, Tennessee, and the late Pleistocene Faunal Gradient. Bull. Carnegie Mus. Nat. Hist., 11:1–67.

GUILDAY, J. F., H. W. HAMILTON, and A. D. McCRADY. 1971. The Welsh Cave peccaries (*Platygonus*) and associated fauna, Kentucky Pleistocene. Ann. Carnegie Mus., 43:249–319.

GUILDAY, J. E., and D. C. IRVING. 1967. Extinct Florida spectacled bear *Tremarctos floridanus* (Gidley) from central Tennessee. Bull. Natl. Speleol. Soc., 29:149–162.

GUILDAY, J. E., and H. McGINNIS. 1972. Jaguar (*Panthera onca*) remains from Big Bone Cave, Tennessee and east central North America. Bull. Nat. Speleol. Soc., 34:1–14.

GUT, H. J. 1959. A Pleistocene vampire bat from Florida. J. Mamm., 40:534–538.

GUT, H. J., and C. E. RAY. 1963. The Pleistocene vertebrate fauna of Reddick, Florida. Quar. J. Florida Acad. Sci., 26:315–328.

HALL, E. R. 1981. The Mammals of North America. 2 vols., 2nd ed. 1181 pp. J. Wiley and Sons, New York.

HAMILTON, W. J., and J. O. WHITAKER. 1979. Mammals of the eastern United States. Cornell Univ. Press, Ithaca and London, second ed. 346 pp.

HEANEY, L. R., and R. M. TIMM. 1983. Relationships of pocket gophers of the genus *Geomys* from the central and northern Great Plains. Univ. Kans. Mus. Nat. Hist., Misc. Publ., 74:1–59.

HONEYCUTT, R. L., and D. J. SCHMIDLY. 1979. Chromosomal and morphological variation in the plains pocket gopher *Geomys bursarius*, in Texas and adjacent states. Occas. Papers Mus., Texas Tech Univ., 58:1–54.

HUMPHREY, S. R. 1974. Zoogeogaphy of the nine-banded armadillo (*Dasypus novemcinctus*) in the United States. Bioscience, 24:457–462.

HUTCHINSON, J. H. 1967. A Pleistocene vampire bat (*Desmodus stocki*) from Potter Creek Cave, Shasta County, California. Contrib. Mus. Paleont., Univ. California, 3:1–6.

JOHNSON, N. M., N. D. OPDYKE, and E. H. LINDSAY. 1975. Magnetic polarity stratigraphy of Pliocene-Pleistocene terrestrial deposits and vertebrate faunas. San Pedro Valley, Arizona. Bull. Geol. Soc. Amer., 86:5–12.

JONES, C. 1978. Eastern chipmunk, *Tamias striatus* (Linnaeus). Pp. 35–36, *in* Mammals, Vol. 1, Rare and Endangered Biota of Florida (J. N. Layne, ed.), Univ. Presses Florida, Gainesville, 52 pp.

JONES, J. K., JR. 1958. Pleistocene bats from San Josectio Cave, Nuevo Leon, Mexico. Univ. Kansas Publ., Mus. Nat. Hist., 9:389–396.

KURTÉN, B. 1965. The Pleistocene Felidae of Florida. Bull. Florida State Mus., Biol. Sci., 9:215–273.

———. 1966. Pleistocene bears of North America: 1. Genus *Tremarctos*, spectacled bears. Acta Zool. Fennica, 115:1–120.

KURTÉN, B., and E. ANDERSON. 1980. Pleistocene mammals of North America. Columbia Univ. Press, New York, 442 pp.

LULL, R. S. 1929. A remarkable ground sloth. Mem. Yale Peabody Mus., 3(2):1–39.

LUNDELIUS, E. L., JR. 1972. Fossil vertebrates from the late Pleistocene Ingleside Fauna, San Patricio County, Texas. Bur. Econ. Geol., Univ. Texas, Rept. Invest., 77:1–74.

MARTIN, R. A. 1974a. Fossil mammals from the Coleman IIA fauna, Sumter Co. Pp. 35–99, *in* Pleistocene mammals of Florida (S. D. Webb, ed.), Univ. Florida Presses, Gainesville, 270 pp.

———. 1974b. Fossil vertebrates from the Haile XIVA fauna, Alachua Co., Pp. 100–113, *in* Pleistocene mammals of Florida (S. D. Webb, ed.), Univ. Florida Presses, Gainesville, 270 pp.

MARTIN, R. A., and S. D. WEBB. 1974. Late Pleistocene mammals from the Devil's Den Fauna, Levy County. Pp. 114–145 *in* Pleistocene mammals of Florida (S. D. Webb, ed.), Univ. Florida Presses, Gainesville, 270 pp.

McDONALD, H. G. 1984. An Irvingtonian record of *Nothrotheriops* in Florida. *In* The biology of edentates (Edentata: Nothrotheriinae) (G. Montgomery, ed.), Smithsonian Inst. Press, in press.

MEYLAN, P. 1982. The squamate reptiles of the Inglis IA Fauna. Bull. Florida State Mus., Biol. Sci., 27:1–85.

PARMALEE, P. W., and W. E. KLIPPEL. 1981. A late Pleistocene population of the pocket gopher, *Geomys* cf. *bursarius* in the Nashville Basin, Tennessee. J. Mamm.,60:381–835.

RAY, C. E. 1964. The jaguarundi in the Quaternary of Florida. J. Mamm., 45:330–332.

———. 1965a. A new chipmunk, *Tamias aristus*, from the Pleistocene of Georgia. J. Paleontol., 39:1016–1022.

———. 1965b. A glyptodont from South Carolina. Charleston Mus. Leaflet, 27:1–12.

RAY, C. E., and L. LIPPS. 1970. Southerly distribution of porcupine in eastern United States during late Quaternary time. Bull. Georgia Acad. Sci., 28(2):24.

RAY, C. E., S. J. OLSEN, and H. J. GUT. 1963. Three mammals

new to the Pleistocene fauna of Florida, and a reconsideration of five earlier records. J. Mamm., 44:373–395.

REPENNING, C. A. 1983. *Pitymys meadensis* Hibbard from the valley of Mexico and the classification of North American species of *Pitymys* (Rodentia: Cricetidae). J. Vert. Paleo., 2: 471–482.

ROBERTSON, J. S. 1976. Latest Pliocene mammals from Haile XVA, Alachua County, Florida. Bull. Florida State Mus., Biol. Sci., 20:111–186.

ROTH, J. A., and J. LAERM. 1980. A late Pleistocene vertebrate assemblage from Edisto Island, South Carolina. Brimleyana, 3:1–29.

SELLARDS, E. H. 1916. Human remains and associated fossils from the Pleistocene of Florida. Florida Geol. Surv., 8th Ann. Rept., pp. 121–160.

———. 1917. On the association of human remains and extinct vertebrates at Vero, Florida. J. Geol., 25:2–4.

SIMPSON, G. G. 1928. Pleistocene mammals from a cave in Citrus County, Florida. Amer. Mus. Novitates, 328:1–16.

———. 1929. Pleistocene mammalian fauna of the Seminole Field, Pinellas County, Florida. Bull. Amer. Mus. Nat. Hist., 56:561–599.

STEVENSON, H. M. 1976. Vertebrates of Florida. Univ. Presses Florida, Gainesville, 607 pp.

WATTS, W. A. 1971. Postglacial and interglacial vegetation history of southern Georgia and central Florida. Ecology, 52: 676–690.

WEBB, S. D. 1974. Pleistocene mammals of Florida. Univ. Florida Presses, Gainesville, 270 pp.

———. 1976. Mammalian faunal dynamics of the Great American Interchange. Paleobiology, 2:220–234.

WEIGEL, R. D. 1962. Fossil vertebrates of Vero, Florida. Florida Geol. Surv., Spec. Publ., 10:1–59.

WILKINS, K. E. 1984. Evolutionary trends in Florida Pleistocene pocket gophers (genus *Geomys*) with description of a new species. J. Vert. Paleont., 3:166–181.

———. in press. Pleistocene mammals from the Rock Springs local fauna, central Florida. Brimleyana.

WOODS, C. A., W. POST, and C. W. KILPATRICK. 1982. *Microtus pennsylvanicus* (Rodentia: Muridae) in Florida: A Pleistocene relict in a coastal saltmarsh. Bull. Florida State Mus., Biol. Sci., 28:25–52.

Address: Florida State Museum, University of Florida, Gainesville, Florida 32611.

TWO IRVINGTONIAN (MEDIAL PLEISTOCENE) VERTEBRATE FAUNAS FROM NORTH-CENTRAL KANSAS

RALPH ESHELMAN AND MICHAEL HAGER

ABSTRACT

Two Irvingtonian localities, the Courtland Canal and Hall Ash local faunas, are reported from Jewell County, north-central Kansas. The Courtland Canal local fauna consists of eight aquatic and eight terrestrial mollusc, five fish, three amphibian, nine reptilian, seven bird, and 17 mammalian species including the rare *Soergelia mayfieldi*. The paleoclimatological and paleogeomorphological interpretations of the locality suggest a depositional environment of a slow-moving stream or larger river slough with a wooded border and relatively dry sandy prairie nearby. The Courtland Canal local fauna is regarded as middle Irvingtonian and comparable to early Kansan and late Aftonian age High Plain faunas such as the Sappa, Mullen I, and Rock Creek.

The Hall Ash local fauna was recovered from beneath the Hartford Ash (dated at 0.74 mybp) and consists of six aquatic and sixteen terrestrial mollusc, four amphibian, one reptilian, and 11 mammalian species. The paleoclimatological and paleogeomorphological interpretations of the site suggest a depositional environment of a prairie pond with prairie areas nearby. The Hall Ash local fauna is regarded as middle-late Irvingtonian and comparable to but possibly slightly older than late Kansan age High Plains faunas such as the Vera and Cudahy.

Collectively, both faunas suggest a paleoclimate of milder winters, cooler summers, and more effective moisture. We know of no environmental regime where these presently disharmonious species could live sympatrically today.

INTRODUCTION

The Courtland Canal local fauna was discovered while prospecting exposures along the Courtland Irrigation Canal on 22 May 1973, during field work on the White Rock local fauna (Eshelman, 1975). The locality contained abundant fossil bone protruding from a paleo-channel fill deposit exposed along an irrigation canal. A *Canis* jaw, elements of horse, proboscidian, camel, antelope, bovid, numerous birds, herps, fish, and molluscs were lying on the outcrop surface. Approximately 400 pounds of matrix were collected and washed for possible microfauna. Although the initial microsamples were not promising, complete jaws of *Ondatra* and *Blarina* were later recovered.

Preliminary study of the material suggested a Kansan age. It wasn't until the summer of 1978 that a field crew was organized and serious collection of the locality began. Between 29 May and 5 June, nearly 5 tons of matrix were washed and picked from the Courtland Canal site and nearly 2 tons from a newly found locality, the Hall Ash site. The latter site was discovered while prospecting an abandoned ash pit within sight of the Courtland Canal locality. The matrix was washed using the method of Hibbard (1949A). Some of the clayey matrix at the Courtland Canal site required three cycles of drying and washing before adequate breakdown of the matrix was suitable for picking.

GEOLOGY

The Courtland Canal site is exposed in a paleo-channel stream fill bisected by the Courtland Irrigation Canal on Myren Intermill's property in the E½, SE¼, Sec. 8, T1S, R6W, Webber, 7½ min. quad. topographic map, Jewell Co., Kansas, 1969, at an elevation of approximately 1,630 ft (Fig. 1). The section from bottom to top (Figs. 2 and 3) consists of interbedded and interfingered silts (Unit A), coarse limonitic sand (Unit B), a dark organic gley paleosol (Unit C), and a mottled reddish sandy-clay and clayey sand (Unit D). Unit D may be equivalent to the Loveland Formation. The gley paleosol has a dip 8° west and a strike of N 52°E. The majority of the specimens were collected from Unit A in light tan silts containing caliche nodules on the northeast side of the canal over a horizontal distance of approximately 20 m. All of the microfauna was collected from Unit A within a 1 square m between 1.6 and 2.3 m above the road at approximately the level where a skull of *Soergelia* was excavated. On the other side of the canal, a crushed proboscidian skull and antilocaprid and peccary material were recovered. Specimens not recovered from Unit A were discovered in spoil piles on either side of the canal in matrix believed to be equivalent to Unit A.

The Hall Ash site is located in the northeast corner of an abandoned ash pit, 0.9 km southeast of the Courtland Canal locality on the Harry Hall property in the E½, SW¼, Sec. 9, T1S, R6W, Scandia N.W. 7½ min. quad. topographic map, Jewell Co., Kansas, 1969, at an elevation of approximately 1,620 ft (Fig. 1). The section consists of a lower greenish gray silt (Unit A), possibly equivalent to the Sappa Formation, a bedded ash mixed with silts (Unit B) identified by fission-track as 0.7 million years old, suggesting the Hartford Ash and finally, an overlying red-molted clay with pebbles (Unit C) possibly equivalent to the Loveland

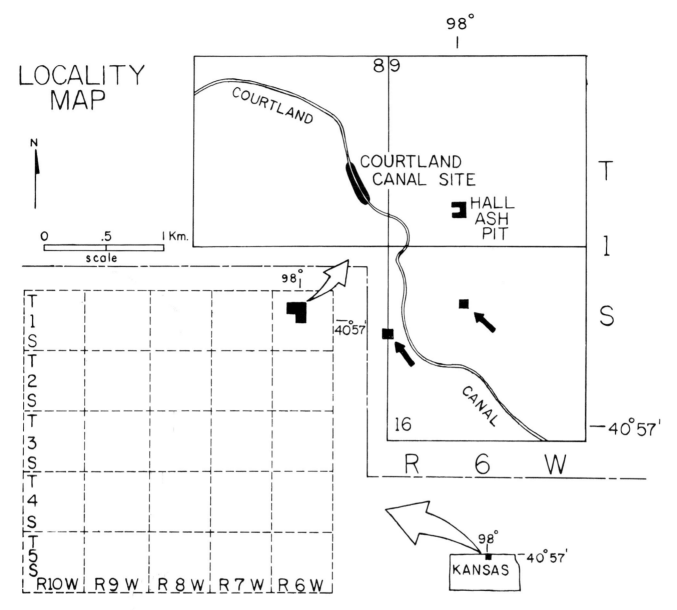

Fig. 1.—Locality map of Jewell Co., north-central Kansas showing locations of Courtland Canal and Hall Ash sites. Arrows point to test pit and Webber Road localities.

Formation. See Figs. 4 and 5 for section drawing and photo of site. Unit C lies above the ash over the entire quarry site. The ash (Unit B) is only exposed on the south wall of the quarry. The ash is believed to have accumulated in a paleovalley or some form of depression represented by Unit A. This exact relationship could not be substantiated at the ash site due to quarrying activity and extensive spoil heaps covering much of the quarry; but this is the relationship shown along Webber Road discussed below. All the fossils were collected from the northeast corner of the quarry in undisturbed sediments of Unit A. Apparently the ash has pinched out here and the silts of Unit A represent the edge of this assumed basin. In the southeast corner of the quarry the ash reaches a thickness of 3.96 m plus another probed 1.8 m

below the quarry floor. The bottom of Unit B was not reached. Along Webber Road, located between and south of the Courtland Canal and Hall Ash localities, at NW¼, SW¼, Sec. 16, T1S, R6W, Webber 7½ min. quad. topographic map, Jewell Co., Kansas, 1969, this same ash is exposed (see Fig. 1). Here the relationships of Unit A and B are clearly visible. The ash is exposed over an 18 m interval along the road. See Figs. 6 and 7 for geological section and photo of site. The above same geological relationships are also visible in what appears to be a test pit 0.8 km south of the Hall Ash site. Based on these two exposures we interpret the same stratigraphic sequence for the Hall Ash site even though the lower relationship is not visible anywhere in the pit. The Hall Ash fauna is therefore interpreted to be pre-ash in age. The Hall

Fig. 2.—(A) Overall view of Courtland Canal locality, figure standing at spot where *Soergelia* skull was collected. (B) Detail view of Courtland Canal locality, contact of Unit A and B at level of raised hand.

Ash locality is discussed by Fishel and Leonard (1955:108 and plate 10A, we believe the locality is more accurately located in Sec. 9, not Sec. 16) and referred to as Pearlette volcanic ash in the Sappa member of the Meade Formation. Zeller (1968) and Eshelman (1975) reviewed the Pleistocene geology of the region; in summary, the deposits are generally referred to as Kansan age sediments of the Sappa Formation.

AGE DETERMINATION OF THE HALL ASH

An ash sample was collected from the Hall Ash site and submitted to John Boellstorff for a fission-track age determination based on glass shards. These samples were tested providing a mean age of 0.706 ± 0.017 (see Miller and Eshelman, in press, table 3 for fission-track analysis of site). This date compares best to the Hartford Ash (0.74 ± 0.09) rather than to the more prevalent Pearlette Ash restricted (0.61 ± 0.04) (Boellstorff, 1978: table 1–1). The type Hartford Ash is located at Minnehaha Co., South Dakota. In addition, the Hartford Ash has been identified at the Little Sioux or County Line Section of Harrison Co., Iowa, where Gerry Paulson recovered an unpublished mammalian fauna (see Guilday and Parmalee, 1972).

AGE AND CORRELATION

The presence of *Ondatra annectens* and *Synaptomys* (*Mictomys*) cf. *S. meltoni* from both the Courtland Canal and Hall Ash local faunas suggest an Irvingtonian age for both faunas (see also the taxonomic discussion of *Ondatra*). Recovery of cf. *Allophaiomys* and *Titanotylopus* from the Courtland Canal local fauna suggests a somewhat earlier age than the Hall Ash local fauna where only *Microtus paroperarius* is present. The presence of the Hartford Ash (0.74 mybp) immediately above the Hall Ash local fauna also supports an Irvingtonian age.

GEOLOGIC SECTION AT INTERMILL SITE, COURTLAND CANAL SITE

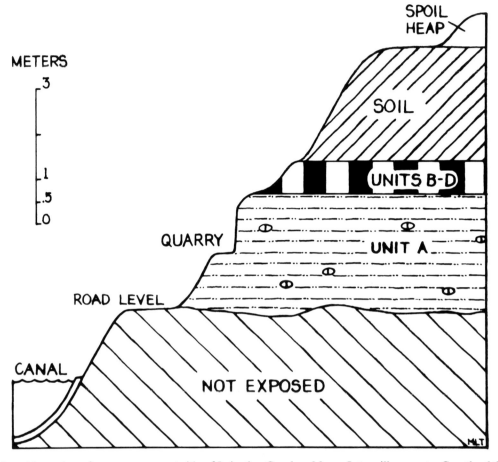

Fig. 3.—Stratigraphic section of exposures on east side of Irrigation Canal on Myren Intermill property, Courtland Canal locality.

The Little Sioux local fauna from Harrison Co., Iowa, worked by Gerald Paulson but unpublished (see Guilday and Parmalee, 1972) was also recovered from beneath the Hartford Ash and possesses a very similar fauna (personal communications, Gerald Paulson, 11 January 1980, and Miller and Eshelman, in press). Below the ash, Paulson recovered *Synaptomys (Mictomys) meltoni, Microtus paroperarius, Tamias, Sciurus,* and a *Sorex* like *S. dispar* but slightly larger, and a snail fauna predominated by *Vertigo,* a cool climatic indicator. Above the ash, Paulson recovered *Geomys,* a large

Blarina, Synaptomys (Mictomys) sp., and again *M. paroperarius,* but the snail fauna exhibited a more southern distribution.

The presence of *Microtus paroperarius* at both the Hall Ash and County Line sites, beneath the 0.7 mybp Hartford Ash, concurs with Van der Meulen's (1978:143) suggestion that *M. paroperarius* migrated into North America prior to 0.7 million years ago. Van der Meulen (1978:135) notes that the presence of *Pitymys meadensis* in the Cumberland Cave local fauna regarded as Middle Irvingtonian may mean *M. paroperarius* immigrated before *P. mead-*

Fig. 4.—(A) Overall view of Hall Ash locality looking northwest with Courtland Canal locality behind trees in left background at arrow. (B) Detail view of Hall Ash locality looking southeast, figure standing where Hall Ash local fauna collected.

ensis into North America. Repenning (1983:480) notes *Allophaiomys* and early *Microtus* are present in early Irvingtonian faunas of North America dated at about 1.9 mybp, but that *Pitymys* immigrated less than 1.2 mybp. Assuming that there are no ecological or sampling explanations, the absence of *P.*

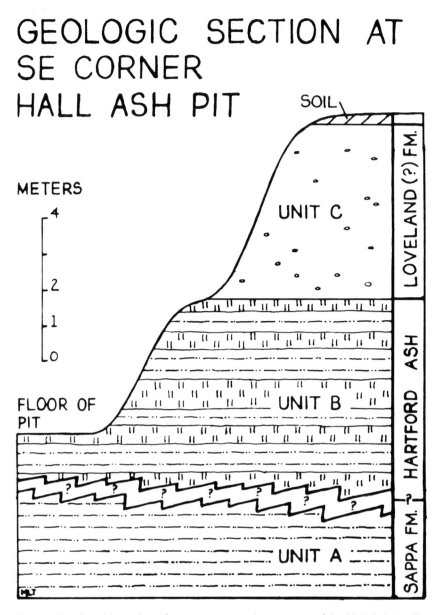

Fig. 5.—Stratigraphic section of exposures at southeast corner of the Hall Ash locality.

meadensis from the Hall Ash local fauna would therefore suggest a pre-Cudahy, pre-Cumberland Cave age.

The Sappa local fauna of Harlan Co., Nebraska (Schultz and Martin, 1970) is somewhat of an enigma as it was recovered from beneath the "S" type Pearlette ash dated at 1.2 mybp, but it compares most favorably with the younger late Kansan Age faunas such as the Cudahy (Kurten and Anderson, 1980:32; Eshelman and Hibbard, 1981:325) from beneath the "O" type Pearlette ash dated at 0.6

mybp. At least part of this disparity is that ash dates above faunas five only minimum age dates, and the amount of time between faunal disposition and ash deposition cannot be determined. In fact, Cudahy, Hall Ash, and Sappa may all be nearer or older than 1.2 mybp, but not necessarily directly correlative. The Sappa is still under study and future work may help to clarify this relationship.

In summary, the Courtland Canal local fauna is regarded as middle Irvingtonian and comparable to early Kansan and late Aftonian age High Plain fau-

Fig. 6.—Detail of exposures at Webber Road Ash locality showing Hartford Ash (Unit B, white layer) sandwiched between Unit A at bottom and Unit C at top.

nas such as the Sappa, Mullen I, and Rock Creek, respectively, whereas the Hall Ash local fauna is regarded as middle-late Irvingtonian and compa-

rable but possibly slightly older than the late Kansan age Vera and Cudahy local faunas.

TAXONOMY

Leslie Fay of Michigan State University prepared pollen samples collected from Units A and C from both the Courtland Canal and Hall Ash sites. Unfortunately, the samples were practically barren, providing only one elm pollen grain from Unit C of the Hall Ash site, and one each of elm, grass, and pine from the Courtland Canal site. Fay (letter to Hager, 18 January 1979) states, "The few grains recovered were battered, indicating some transport." Oxidation of the sediments is probably responsible for the lack of preserved pollen. These results are not surprising as pollen is rare from pre-Illinoian vertebrate localities from the Pleistocene of the High Plains (Kapp, 1970).

The molluscan faunule of the Hall Ash site (Miller and Eshelman, in press) consists of 17 terrestrial and six aquatic species, all collected from the Hall Ash site. The faunule is unusual in that it contains the first fossil occurrence of *Allogona profunda* from the Pleistocene of Kansas and probable undescribed species of *Gastrocopta* cf. *G.* near *armifera*.

Individuals of terrestrial taxa outnumber aquatic individuals 366 to 14. Miller also identified the molluscs from the Courtland Canal site. Eight aquatic and eight terrestrial species were identified.

Gerald Smith, of the Museum of Paleontology, University of Michigan, made preliminary identification of the fish remains. The herpetofauna has been completed by Rogers (1982) who reports all species have been identified from other Pleistocene localities and only the land tortoise *Geochelone* is extinct. David Steadman, Department of Vertebrate Zoology, Smithsonian Institution, made preliminary identification of the birds. A total of 24 mammalian taxa were identified by the senior author. The Courtland Canal local fauna is represented by a total of 41 vertebrate and 16 invertebrate taxa and the Hall Ash local fauna by 16 vertebrate and 24 invertebrate taxa. A complete list of all taxa known to date, from both the Courtland Canal and Hall Ash localities, is found in Table 1.

GEOLOGIC SECTION AT WEBBER ROAD ASH LOCALITY

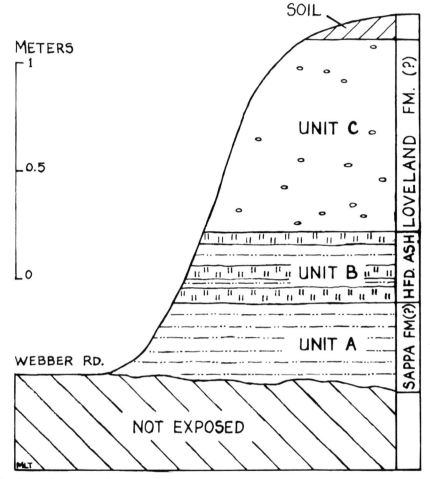

Fig. 7.—Stratigraphic section of exposures along east side of Webber Road at the Webber Road Ash locality.

COURTLAND CANAL LOCAL FAUNA

Class Mammalia
Order Insectivora
Family Soricidae
***Sorex* sp.**

Material.—Courtland Canal locality: USNM 304243, left posterior mandible fragment with ascending ramus.

Remarks.—The specimen is that of a large *Sorex*. Much of the detail is lost due to wear and breakage.

In general shape, the specimen resembles *Sorex (Neosorex) megapalustris* Paulson from the Cudahy local fauna. Comparison with *S. cudahyensis* Hibbard is difficult due to the lack of teeth. It differs from *Microsorex pratensis* in that the ascending ramus is more pointed and elongated and the form of the mandibular foramen is deeper and more excavated.

Table 1.—*Fossils from the Courtland Canal and Hall Ash localities.*

Taxa	Court-land Canal	Hall Ash
Mollusca		
Aquatic		
Ferrissia cf. *fragilis*		+
Ferrissia rivularis	+	
Gyraulus parvus	+	+
Helisoma anceps	+	
Helisoma trivolvis	+	
Physa sp. (broken)	+	
Physa gyrina		+
Pisidium compressum	+	
Planorbula campestris		+
Sphaerium sp.		+
Sphaerium simile	+	
Stagnicola caperata		+
Valvata tricarinata	+	
Terrestrial		
Carychium exiguum		+
Allogona profunda		+
Stenotrema leai		+
Zonitoides nitidus		+
Euconulus fulvus		+
Discus cronkhitei	+	+
Cionella lubrica		+
Helicodiscus parallelus	+	+
Nesovitrea electrina		+
Succinea ovalis		+
Hawaiia minuscula	+	+
Zonitoides arboreus	+	+
Gastrocopta cf. *G. armifera*		+
Gastrocopta armifera	+	
Vallonia gracilicosta	+	+
Vallonia pulchella	+	
Pupilla sinistra		+
cf. *Succinea*		+
Deroceras aenigma	+	
Class Osteichthyes		
Order Cypriniformes		
?*Minytrema* sp.	+	
Notropes spp.	+	
Semotilus cf. *atromaculatus*	+	
Order Siluriformes		
Ictalurus melas	+	
Order Perciformes		
Lepomis cf. *L. humilis*	+	
Class Amphibia		
Order Urodela		
Ambystoma tigrinum		+
Ambystoma maculatum		+
Order Anura		
Rana pipiens	+	+
Rana catesbeiana	+	
Bufo hemiophrys	+	+

Table 1.—*Continued.*

Taxa	Court-land Canal	Hall Ash
Class Reptilia		
Order Chelonia		
Geochelone sp.		+
Chrysemys picta	+	
Chrysemys scripta	+	
Order Squamata		
Heterodon nasicus	+	
Nerodia sipedon	+	
Thamnophis proximus	+	
Thamnophis radix	+	
Thamnophis sirtalis	+	
Columber constrictor	+	
Agkistrodon contortrix	+	
Class Aves		
Order Anseriformes		
Aserineae	+	
cf. *Anas crecca*	+	
size of *Anas discors*	+	
size of *A. clypeata*	+	
Order Galliformes		
cf. *Tympanuchus*	+	
Order Charadriiformes		
Scolopacidae		
size of *Capella gallinago*	+	
Order Passeriformes		
cf. Muscicapidae (Turdinae)	+	
Class Mammalia		
Order Insectivora		
Sorex aff. *S. cinereus*		+
Sorex sp.	+	
Microsorex cf. *M. pratensis*		+
Blarina sp.	+	
Order Edentata		
?*Glossotherium harlani*	+	
Order Carnivora		
Lutra cf. *L. canadensis*	+	
Canis latrans	+	
Order Rodentia		
Spermophilus cf. *S. richardsonii*	+	+
Spermophilus (*Spermophilus*) sp.	+	+
Cynomys sp.		+
Geomys sp.	+	+
Castoroides cf. *C. ohioensis*	+	
Phenacomys cf. *P. intermedius*		+
cf. *Allophaiomys* sp.	+	
Microtus paroperarius		+
Ondatra annectens	+	+
Synaptomys (*Mictomys*) cf. *S. meltoni*	+	+
Zapus sandersi		+
Order Perissodactyla		
Equus sp.	+	

Table 1.—*Continued.*

Taxa	Court-land Canal	Hall Ash
Order Artiodactyla		
Platygonus cf. *P. vetus*	+	
Titanotylopus sp.	+	
Antilocapridae gen. et sp. unident.	+	
Soergelia mayfieldi	+	
Order Proboscidea		
Gomphotheriidae gen. et sp. unident.	+	

Blarina sp.

Material.—Courtland Canal locality: USNM 304244, right mandible with incisor and P_4–M_3.

Measurement.—Length from tip of incisor to tip of condyle, 12.97 mm; length from tip of incisor to posterior border of M_3, 8.68 mm; M_1–M_3 length, 4.61 mm; M_1–M_2 length, 3.58 mm; P_4 1.38 by 1.07 mm; M_1, 2.13 by 1.37 mm; M_2, 1.63 by 1.17 mm; M_3, 1.22 by 0.91 mm (note the incisor appears to be slightly extended from the alveolus).

Remarks.—The specimen is smaller than all medial Pleistocene specimens; M_1–M_2 length of Cudahy specimen 3.8 mm, and width of M_1 of Cudahy specimen (Paulson, 1961:134) 1.4 mm; M_1–M_3 length of three Kanopolis specimens (Hibbard et al., 1978:16) 4.8 mm, 5.0 mm, and 5.1 mm. Hibbard (1956:163) mentions a *Blarina* from the late Blancan age Dixon local fauna as a large short-tailed shrew. Eshelman (1975:25) also reports a large *Blarina* from the similar age White Rock local fauna, having an M_1 length of 2.5 mm and width of 1.5 mm. Martin (1973:53) lists a new *Blarina* species (unnamed) from the Nebraskan? age Java local fauna. Martin gives no discussion or measurements for this taxon.

The external cingulum on M_1–M_2 is more pronounced than in Recent specimens. This character is also noted by Hibbard et al. (1978:16) for a dentary (UM 61263) recovered from a Cudahy local fauna equivalent (UM-K3-71) in Kansas. The lingual cingulum on M_2 does not appear to be better developed than Recent specimens as Paulson (1961:134) noted for a specimen (UM 36802) from the Cudahy local fauna.

Graham and Semken (1976:437) suggest the phena represented in living populations of *Blarina* are similar to those recovered in the late Kansan age Cumberland Cave local fauna. Whether the Courtland Canal specimen can be attributed to one of these phena, an extinct species as Martin (1973) suggests, or a combination of one of the above coupled with

its positive Bergman's response, cannot be determined with the material in hand.

Order Rodentia
Family Sciuridae

Spermophilus cf. *S. richardsonii* (Sabine)

Material.—Courtland Canal locality: USNM 304232, right M_1 or M_2.

Measurements.—3.32 mm (estimated) by 2.96 mm.

Remarks.—The isolated molar is broken along the anterior surface and is badly worn. It compares best to Recent specimens of *Spermophilus richardsonii*. Also recorded from the Cudahy (Paulson, 1961:134), this is the earliest record for *S. richardsonii*. Richardson's ground squirrel presently ranges north and west of the Courtland Canal locality in South Dakota, Wyoming, and Colorado (Hall and Kelson, 1959:339). Later age Illinoian and Rancholabrean fossil localities are also south of the present range (Kurten and Anderson, 1980:214).

Family Geomyidae
Geomys sp.

Material.—Courtland Canal locality: USNM 304245, right dentary with I; USNM 304246, right dentary with I; USNM 304268, left dentary with I, P_4–M_1; USNM 304247, six upper incisor fragments, five lower molars, two upper molars, P_4, P^4, M^3.

Remarks.—Paulson (1961:137) reported *Geomys tobinensis* from the Kansan age Cudahy fauna of southwest Kansas. The material reported here is too fragmentary for specific determination. No material of *Thomomys* is recognized, although Paulson (1961:137) and Hibbard et al. (1978:22) have recovered specimens from the Cudahy and Kanopolis faunas, respectively.

Family Castoridae
Castoroides cf. *C. ohioensis* Foster

Material.—Courtland Canal locality: USNM 304237, four incisor fragments; USNM 304238, proximal head of femur; USNM 304239, proximal half of ulna.

Measurements.—Maximum width of incisor 25.6 mm.

Remarks.—The incisor fragments may represent one or more incisors. Each exhibits enamel crinkling or crenulation. This character plus large size differentiates the specimens from *Paradipoides stovali*, Rinker and Hibbard (1952:99), and *Procastorides sweeti* Barbour and Schultz (1937:7), which are smaller and have no enamel crinkling (see also Woodburne, 1961:82, table III). The Courtland Canal specimens compare well to other *Castoroides* material in the USNM collections. *Castoroides ohioensis* is also reported from the Yarmouth age Kanopolis local fauna of Kansas (Hibbard et al., 1978:23) and several Illinoian faunas from the Great Plains (Hibbard, 1970:table 6).

Family Cricetidae
Subfamily Microtinae

cf. *Allophaiomys* sp.

Material.—Courtland Canal locality: USNM 304254, right mandible fragment with M_1 and M_2, left mandible fragment with M_2 and M_3, right mandible fragment with incisor fragment, two right M_1s, three left M_1s, right M_2, left M_2, two left M_3s, four right M^1s, four left M^1s, right M^2, two left M^2s.

Remarks.—In general, the material matches the descriptions given by Hibbard (1944:729) and Paulson (1961:148) for *Pitymys llanensis* which Hibbard recovered from the Cudahy local fauna. The M_2s in the left and right mandible fragments show a slight variation in that the first and second triangles are confluent and not closed as in all other material observed. The anterior loop or cap of M_1 is a simple crescent-shape which generally corresponds to Van der Meulen's (1978:107) morphotype 1a. This is the same configuration for *Pitymys* from the Wathena and Kentuck faunas and is more primitive than the trefoiled loop of the conard Fissure and Cudahy specimens (Van der Meulen, 1978:118). The *Pitymys llanansis* M_1s from the Conard Fissure virtually all belong type 1b. The mean length measurements of the first lower molar corresponds best to the Kentuck population (Table 2) but the range of variation overlaps all populations noted by Van der Meulen (1978:table 2). The Courtland Canal sample is also more elongate than *Pitymys* reported by Hibbard et al. (1978:table 3) for the younger Kanopolis local fauna of Yarmouth age. Geographical as well as temporal differences certainly account for some of this variation.

The following evolutionary lineage for *Pitymys* in North America is drawn from Zakrzewski in Hibbard et al. (1978:29), Van der Meulen (1978:118), and Repenning (1983:476): *Allophaiomys* sp. > *Allophaiomys guildayi* > *Pitymys llanensis* > *P. cumberlandensis* > *P. ochrogaster*. Martin (1975) correlates the Java fauna of South Dakota with the Kentuck and assigns both microtids to *Allophaiomys* cf. *A. pliocaenicus* Kormos. Van der Meulen (1978) believes the Kentuck and Wathena specimens possibly represent a new species of *Allophaio-*

Table 2.—*Molar measurements (mm) of* Allophaiomys *sp. from Courtland Canal locality.*

Tooth	Length			Width		
	N	Mean	Observed range	N	Mean	Observed range
M_1	6	2.99	2.85–3.24	4	1.27	1.20–1.33
M_2	4	1.86	1.73–1.99	4	1.17	1.12–1.24
M_3	3	1.58	1.47–1.78	3	0.88	0.83–0.92
M^1	8	2.64	2.29–2.97	10	1.44	1.26–1.57
M^2	3	1.83	1.81–1.85	3	1.14	1.10–1.22

mys. Whether the Courtland Canal material is referable to *Allophaiomys* or *Pitymys* and a possible new species cannot be determined without better comparative material. Eshelman has followed Repenning (1983) purely for convenience and cannot satisfactorily defend either generic placement without extensive further study. *Allophaiomys* is also reported from the Nash local fauna (Eshelman and Hibbard, 1981:323) and Fyllan Cave, Texas (Kurten and Anderson, 1980:258).

Ondatra annectens (Brown)

Material.—Courtland Canal locality: USNM 304252, adult left mandible with M_1–M_3; USNM 304253, right M^1, left M^2, left M_3.

Measurements.—Left mandible, M_1, 5.92 by 2.44 mm; M_2, 3.58 by 2.47 mm; M_3, 3.10 by 2.08 mm; right M^1, 4.40 by 2.00 mm; left M^2, 3.36 by 2.26 mm; left M_3, 3.37 by 2.14 mm.

Remarks.—The dentine tract height of M_1 (see Eshelman, 1975:39 for terminology) is 1.8 mm; the tooth is damaged near the position of measurement. This tract height corresponds to an intermediate position between Kansan and Illinoian age *Ondatra* measurements as plotted by Nelson and Semken (1970:fig. 2). The length versus width measurement falls within the scatter of the Cudahy fauna as plotted by Nelson and Semken (1970:fig. 1).

Nelson and Semken (1970) were able to demonstrate that the M_1 length/width ratio could be used to distinguish between warm and cool faunas. This study was substantiated by corresponding ratio differences between Recent northern and southern specimens and fossil glacial and interglacial specimens. The M_1 ratio from the Courtland Canal site (2.43) corresponds to the cooler glacial faunas of Nelson and Semken's graph (1970:fig. 7). Therefore, a glacial age of the Courtland Canal site is suggested by this method. *Ondatra annectens* has also been identified from Cumberland Cave, Java, Trout Cave, and Vera (Kurten and Anderson, 1980:266).

Table 3.—*Molar measurements (mm) for* Synaptomys *(Mictomys) cf.* S. meltoni *from the Courtland Canal and Hall Ash site.*

Tooth	Length			Width		
	N	Mean	Observed range	N	Mean	Observed range
M^1	5	3.05	2.79–3.38	5	1.32	1.23–1.45
M^3	2	1.90	1.82–1.98	2	0.92	0.88–0.96
M_1	6	2.77	2.59–2.94	6	1.35	1.26–1.47
M_2	2	1.94	1.85–2.02	2	1.14	1.08–1.19
M_3	2	1.61	1.46–1.75	2	1.03	0.93–1.12

Synaptomys (Mictomys) cf. S. meltoni Paulson

Material.—Courtland Canal locality: USNM 304248, right mandible with I, M_1, and M_2 fragment; USNM 304249, left mandible with M_2 and M_3; USNM 304250, four right M^1s, two left M^1s, right M^2, left M^3, right M^3, four right M_1s, one left M_1, one right M_2.

Remarks.—The enamel of the molars is differentiated into thick and thin portions as noted by Paulson (1961:142). There is no distinct, cement-filled labial re-entrant on the M_3s as is typical in *Synaptomys borealis* (Richardson). The first triangle of the M_1 has a convex rather than a concave posterior wall.

Both mandibles are damaged around the basal capsular process, but the process terminates near the posterior margin of the M_3. This character is noted by Hibbard (1952:10) and Paulson (1961:144) as being characteristic of *S. kansasensis* from the Kansan age Kentuck fauna. However, the incisor measurement (1.01 mm) and molar measurements (Table 3) are most closely similar to those of *S. meltoni* reported by Paulson (1961:143) from the Kansan age Cudahy fauna.

This suggests that the incisor basal capsular process position is too variable to be used reliably as a specific character. If two distinct species are present, *S. meltoni* is characterized by smaller, comparatively more square, first lower molars (less elongate and wider), whereas *S. kansasensis* is characterized by slightly larger, more rectangular, first lower molars (more elongate and narrower). Additional material may explain at least some of this variation as due to clinal change such as recognized by Hibbard (1963:211) and Eshelman (1975:37) for the subgenus *Synaptomys*.

One of the right upper molars has a re-entrant angle along the face of the anterior loop. This condition has also been reported by Hibbard (1944:738) in a specimen of ?*Pitymys* from the Tobin local fauna, except the re-entrant also contains cement.

A Recent specimen, *Pitymys nemoralis* from Douglas Co., Kansas, also processes a re-entrant on the anterior loop without cement (Hibbard, 1944:738). *Synaptomys meltoni* is also reported from the Sappa local fauna (Schultz and Martin, 1970).

Order Edentata
Family Mylodontidae

?*Glossotherium harlani* (Owen)

Material.—Courtland Canal locality: USNM 304251, badly fragmented left tibia.

Remarks.—The tibia is large in comparison to Rancholabrean age specimens noted by Stock (1925: table 88) for specimens from Rancho La Brea, California, and even larger than those given by Lundelius (1972:table 14) for specimens from Ingleside, Texas. This greater size is inconsistent with the general evolutionary trend of increased size and stoutness of build (Kurten and Anderson 1980:143). The greatest thickness of the distal extremity is 110 mm; the greatest width of distal extremity, 141 mm; the least width of shaft, 111 mm; and least thickness of shaft, 54 mm.

Family Canidae

Canis cf. C. latrans Say

Material.—Courtland Canal locality: USNM 304242, left mandible with M_1 and M_2.
Measurements.—M_1, 24.0 by 9.4 mm; M_2, 10.3 by 7.1 mm; depth of mandible at anterior edge of M_1, 19.4 mm; distance from anterior edge of P_1 to posterior edge of M_3, 83.4 mm; minimum depth of mandible between P_3 and P_4, 16.2 mm.

Remarks.—The mandible is approximately the size of that of a large coyote. However, the ascending ramus on the anterior border is more rounded, the vertical border of the mandible more rounded, and the posterior crest of the ramus is not as hooked as in Recent *Canis latrans* examined in the USNM collections. The width of the M_1 is broader than two late Blancan Borchers specimens (9.0 mm) reported by Getz (1960:363) and the late Pleistocene type of *C. l. harriscooki* (8.0 mm) reported by Slaughter (1961:503). All of the above parameters fall within the range of variation reported by Nowak (1979: appendix C) for Recent male populations or late Pleistocene specimens of *C. latrans*.

Lutra canadensis (Schreber)

Material.—Courtland Canal locality: USNM 304229, proximal half of right humerus.

Remarks.—The specimen compares identically with Recent humeri in the USNM collections, as well as with fossil humeri from the Irvingtonian Cumberland Cave local fauna assigned to *Lutra parvicuspis* Gidley and Gazin. Other than tooth characters, Gidley and Gazin (1933:349) note only that this extinct otter is somewhat larger than *Lutra canadensis*. Hall (1936:75) synonymized *L. parvicuspis* with *L. canadensis lataxina*. Gidley and Gazin (1938:42) again argued that *L. parvicuspis* is a distinct species. Kurten and Anderson (1980:158) refer *L. parvicuspis* to the extant species.

Order Perissodactyla
Family Equidae

Equus sp.

Material.—Courtland Canal locality: USNM 304223, left upper M¹? fragment; USNM 304221, right lower molar fragment; USNM 304222, left lower third premolar fragment?; USNM 304224, two upper incisors; USNM 304225, distal two-thirds of left metatarsal; USNM 304226, distal end of metapodial; USNM 304227 proximal two-thirds of phalanx; and USNM 304228, left ramus fragment with roots of DP₂.

Remarks.—The upper incisors have cups as does *Equus niobrarensis* Hay and *Equus laurentius* Hay, both from the Hay Springs local fauna. The right lower molar and left third lower premolar fragment are the size of specimens identified as *Equus* sp. from Hay Springs in the USNM collections. Both metatarsal fragments and the phalanx are the size of *Equus excelsus* specimens from Niobrara River in the USNM collection. All of the material compares well to *Equus* specimens from the early middle Pleistocene of the High Plains.

The right lower molar fragment, USNM 304221 is heavily worn and measures 23.0 by 15.8 mm. The metatarsal fragment, USNM 304225, has a greater length and width measurement of 52.5 by 39.0 mm on the distal end. The remaining material is too scrappy for measurement.

Order Artiodactyla
Family Tayassuidae

Platygonus cf. *P. vetus* Leidy

Material.—Courtland Canal locality: USNM 304240, left mandibular fragment with DP₂–P₄; USNM 304241, left P², left P⁴? fragment, right P⁴?, fragments of at least three upper molars and two canine fragments.
Measurements.—DP₂, 8.8 by 5.8 mm; DP₃, 10.4 by 7.5 mm; DP₄, 17.8 by 10.2 mm; LP², 12.4 by 11.5 mm; LP⁴?, 12.6 by 15.0 mm.

Remarks.—All of the upper teeth were recovered from within a one square meter area and are believed to be representative of one individual. The

upper premolars and deciduous lower premolars compare favorably in size to teeth of *Platygonus vetus* from Cumberland Cave, Maryland (medial Irvingtonian), in the USNM collections. A minimum of two individuals are indicated. *Platygonus* is also reported from the following High Plains, Kansan age faunas—Gilliland local fauna (Hibbard and Dalquest, 1966:31), Holloman local fauna (Hay and Cook, 1930), and Cudahy fauna (Paulson, 1961:151). Kurten and Anderson (1980:298) have synonymized *P. cumberlandensis* and *P. vetus*.

Family Camelidae

Titanotylopus sp.

Material.—Courtland Canal locality: USNM 304229, left M₁ fragment; USNM 304217, left metacarpal; USNM 304218, metatarsal proximal end fragment.

Remarks.—The metacarpal measures 439 mm in length, proximal transverse width, 83 mm; proximal anteroposterior width 54 mm; mid-length transverse width, 47 mm; and mid-length anteroposterior width, 44 mm. The length of the metacarpal is approximately 50 mm longer than *Camelops* specimens reported by Webb (1965:table 11).

Skinner and Hibbard et al. (1972:114) report a metacarpal length of 400.00 (F:AM 24900) from the Sand Draw local fauna referred to *Titanotylopus spatulus* (Cope). Meade (1945) reports a metacarpal length for four specimens of *T. spatulus* from the Blanco fauna from 395 to 497, proximal width 85 to 103 and distal width ranging from 104.5 to 148 mm. Hibbard and Riggs (1949) report the following measurements for *T. spatulus* from the Keefe Canyon local fauna for two metacarpals: length, 448, 455; width proximal end, 103 and 106; and width distal end, 136 and 140 mm. Barbour and Schultz (1934) report a metacarpal length of 425 for *T. fricki* synonymized with *T. spatulus* from the Broadwater local fauna. To our knowledge, no metacarpals are known that have been assigned to *T. nebraskensis* (Barbour and Schultz). In addition, an incisor, USNM 304219, may belong to this taxon.

Family Antilocapridae

Genus and species indeterminate

Material.—Courtland Canal locality: USNM 304255, left maxillary fragment with M¹ and M².
Measurements.—M¹, 12.3 by 9.3 mm; M², 14.7 by 10.2 mm.

Remarks.—The molars compare favorably in size and morphology with the extant *Antilocapra americanus* (Ord) and the extinct southwestern form

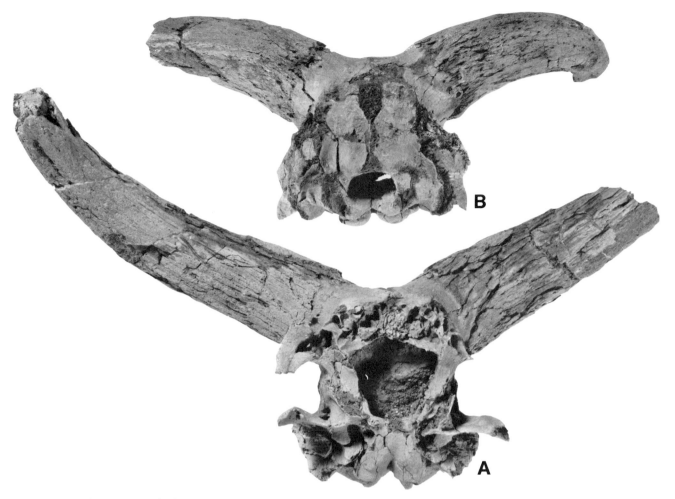

Fig. 8.—*Soergelia* cf. *S. mayfieldi*. (A) Dorsal-ventral view of occipital region and horn cores. (B) Ventral-anterior view of cranium posterior and horn cores.

Stockoceros onuarosagris (Roosevelt and Burden), (Colbert and Chaffee, 1939:14). The molars are approximately 25% smaller than *S. conklingi* (Stock). *Tetrameryx*? *knoxensis* (Hibbard and Dalquest, 1960), known only from horn cores, and *Capromeryx* sp. were recovered from the Late Kansan Gilliland local fauna of Knox County, Texas. Paulson (1961:151) reports an unidentified antilocaprid from the Cudahy fauna.

Family Bovidae

Soergelia cf. *S. mayfieldi* (Troxell)

Material.—Courtland Canal locality: USNM 304216, posterior one-third of cranium, one-half of the left horn core, most of the right horn core and most of the occipital region (Fig. 8).

Description.—The horn cores are flattened slightly more dorsal ventrally than anteroposteriorly as reported by Troxell (1915) for

the type specimen. The burrs on the horn core base do not appear as pronounced as in the type specimen. The horn cores curve up and outward from the cranium and then are directed forward nearly horizontally. This differs from Kurten and Anderson's (1980:332) description of *Soergelia mayfieldi* where the horn cores curve outward and downward and then somewhat forward. The Courtland Canal horn cores turn prominently forward. Harrington (letter to Eshelman, 5 January 1974) makes the following observations on the Courtland Canal specimen: The region of the frontals, the horn core burrs and the angle of rise of the horn cores all compare similarly to the European *Soergelia*. The striking difference is the greater horn core length in the Courtland Canal specimen. The horn cores appear to rise forward over a greater distance from the burrs before their downward deflection. In European *Soergelia* the horn cores usually are directed slightly back prior to the forward curve and slope downward more abruptly. Finally, the horn cores taper down much more rapidly in European specimens. The foramen magnum agrees with the type *Soergelia mayfieldi* in that the opening is on the dorsal surface of the occiput.

Table 4.—*Cranial measurements (mm) of* Soergelia mayfieldi *from Rock Creek, Courtland Canal, and* Soergelia elisabethae *from Basel, Germany.*

Measurement[1]	Soergelia mayfieldi		Soergelia elisabethae	
	Courtland Canal	Rock Creek[2]	D 376[3]	D 459[3]
Width between bases of horn cores	65	75		
Greatest diameter of left horn core burr	92	85	75.6	77.3
Least diameter of left horn core burr	78	74	65.9	59.8
Greatest diameter of right horn core burr	93	—	—	80.3
Least diameter of right horn core burr	78	—	—	62.8
Diameter of left horn core burr	285	—	—	232
Diameter of right horn core burr	274	—	239	229
Width of basi-occipital	58	55	—	—
Right horn core length on upper curve, burr to existing tip[4]	330	—	182	165
Right horn core length on lower curve, burr to existing tip[4]	305	—	—	—
Right horn core length, from tip to upper base	268	—	—	—
Distance between skull mid-line to beginning of horn core burr	40	—	—	40
Distance from mid-line of skull to tip of right horn core	304	—	194	198
Greatest width of auditory opening	154	—	—	—
Width of condyles	89	—	—	—
Depth, occipital crest to top of foramen magnum	67	—	—	—
Depth, occipital crest to lower border of foramen magnum	94	—	—	—
Greatest width of foramen magnum opening	40	—	—	34.4
Greatest height of foramen magnum opening	24	—	—	27.9
Angle of posterior divergence of horn core from cranium	79°	—	—	—
Angle of proximal horn core rise	21°	—	—	—

[1] Measurements after Skinner and Kaisen, 1947:fig. 1.
[2] Rock Creek, Texas, Troxell, 1915.
[3] Measurements by Leo Carson Davis from specimens in the Naturhistorisches Museum, 4051 Basel, Angnstinergasee 2, Germany; specimen D376 type specimen Süssenborn; D459 cast of original destroyed during WWII from Kapellenberg bei Rastenberg.
[4] Right horn core missing; estimated 30 mm of tip.

Measurements from Table 4 clearly indicate the horn cores of the Courtland Canal specimen are larger than the Rock Creek specimens. Sexual dimorphism most certainly played a role in this primitive musk oxen and may suggest that the Kansas specimen was a male.

Remarks.—The Courtland Canal specimen is probably the most complete cranium of *Soergelia* yet discovered in North America (Nelson and Neas, 1980:227). Harrington was the first to propose the synonymy of *Preptoceras* under *Soergelia*. Remains of *Soergelia* are rare in North America. Aside from the two specimens already mentioned, a third occurrence of uncertain age is reported by Harrington (1977) from the Old Crow River locality 11A of the Yukon. Harrington (1980:46) suggests *Soergelia* is a biostratigraphic indicator of Kansan age deposits in North America. The Courtland Canal specimen does not dispute this claim, although Kurten and Anderson (1980) suggest that the Rock Creek local fauna may be late Aftonian and that *Soergelia* spans a large part of the Irvingtonian.

Order Proboscidea
Family Gomphotheriidae

Genus and species indeterminate

Material.—Courtland Canal locality: USNM 304230, molar fragment.

Remarks.—The fragment is too scrappy for identification. Mr. Myren Intermill (personal communication, July 1973) states that when the Canal was dug, several large "elephant" bones were recovered and taken to the Kansas State University at Manhattan. During the field season of 1973, a badly crushed gomphothere skull was discovered on the west bank. During the 1978 field season the skull could not be relocated and is presumed eroded into the canal.

HALL ASH LOCAL FAUNA

Class Mammalia
Order Insectivora
Family Soricidae

Microsorex cf. *M. pratensis*

1944. *Microsorex pratensis* Hibbard, Bull. Geol. Soc. Amer., 55:722.
1980. *Sorex (Microsorex) pratensis* Diersing, J. Mamm., 61:76.
1980. *Microsorex pratensis* Kurten and Anderson, 1980.

Material.—Hall Ash locality: USNM 304256, right jaw fragment with P_4–M_3 associated incisor in jaw fragment.

Measurements.—The specimen measures 9.05 mm from tip of incisor to tip of condyle (9.2 mm in the Cudahy specimen); M_1–M_3 length is 3.18 mm (3.2 mm for two Cudahy specimens); length is 2.42 mm; P_4, 1.32 by 0.46 mm; M_1, 1.40 by 0.85 mm; M_2, 1.20 by 0.84 mm; M_3, 0.82 by 0.65 mm (Note—the specimen has subsequently been broken and the P_4 lost).

Remarks.—Viewed dorsally, the lingual side of the ramus posterior to the M_3 is arched abruptly labiad. This character serves to separate this form from those of the other small shrews, *Sorex cinereus* Kerr and *S. cudahyensis* Hibbard of the Cudahy fauna (Paulson, 1961:134). The mental foramen is positioned just posterior to the metaconid of the M_1. The ramus is heavier than *S. cinereus* where the mental foramen is directly below the metaconid of M_1. The internal temporal fossa is more squared ventrally and is larger as in *S. cinereus*. The incisor has one tubercle, whereas Paulson (1961:133) states that *S. pratensis* from the Cudahy local fauna has two well-defined tubercles, and the suggestion of a third. The specimen is about the size of *Sorex (Microsorex) hoyi* but the dentition is slightly heavier. The variation noted above in the number of incisor tubercles and the shape of the internal temporal fossa preclude positive identification.

Sorex aff. *cinereus* Kerr

Material.—Hall Ash locality: USNM 304257, right mandibular fragment with M_1–M_3.

Measurements.—M_1–M_3 length, 3.09 mm; M_1–M_2 length, 2.29 mm; M_1, 1.31 by 0.64 mm; M_2, 1.12 by 0.62 mm; M_3, 0.87 by 0.47 mm.

Remarks.—The specimen is the size of *Sorex cinereus,* but the dentition is narrower. Paulson (1961: 128) reports the presence of *S. cinereus* from the Cudahy local fauna, but states that the fossils are generally more robust than normal from Recent specimens, but overlap does occur. Hibbard (1956), Skinner and Hibbard et al. (1972), and Eshelman (1975:24) all report a small but slightly more robust

Sorex from the Dixon, Sand Draw, and White Rock local faunas, respectively. Eshelman (1975) states an M_2, V31973 from the Dixon local fauna measures 1.1 by 1.7 mm; this is obviously a typographical error and should read 1.1 by 0.7 mm. The Hall Ash site specimen does not appear to belong to this form, but is closely related to *S. cinereus*. The specimen is smaller than *S. cudahyensis* Hibbard. The mental foramen is just anterior to the metaconial on M_1, whereas it is typically more posterior in *S. cinereus* (Hibbard, 1944:719).

Three isolated ascending rami fragments (USNM 304258) appear to belong to this same taxon, but no positive assignment is attempted.

Order Rodentia
Family Sciuridae

Spermophilus cf. *S. richardsonii* (Sabine)

Material.—Hall Ash locality: USNM 304235, left M^1 or M^2.
Measurements.—2.36 by 3.64 mm.

Remarks.—The molar has a comparatively narrower basin between the paracone and metacone than is found in *Spermophilus franklini*; in this character the molar compares best to *S. richardsonii*. A lower molar (USNM 304267) may belong to this taxon, but no assignment is attempted.

Spermophilus (*Spermophilus*) sp. Cuvier

Material.—Hall Ash locality: USNM 304233, left M^1 or M^2; USNM 304234, left M_1 or M_2.
Measurements.—USNM 304233, 2.14 by 3.48 mm; USNM 304234, 2.19 by 3.22 mm.

Remarks.—The narrow triangular nature of the upper molar suggests that these teeth belong to the subgenus *Spermophilus*. In comparison to Recent specimens, the best fit is with *Spermophilus beldingi*. *Spermophilus richardsonii*, also within this subgenus, is known from the Cudahy fauna (Paulson, 1961:134); the upper molars, however, are more broadly triangular in this species than in the fossil teeth. *Spermophilus beldingi* is extant far to the west in the montane basin region of Idaho and Nevada, whereas *S. richardsonii* ranges nearer to the north and west in South Dakota, Wyoming, and Colorado (Hall and Kelson, 1959). The teeth do not appear to be those of *S. franklini* or *S. richardsonii*.

Cynomys sp.

Material.—Hall Ash locality: USNM 304236, left P_4.
Measurement.—3.35 by 3.87 mm.

Remarks.—The tooth, in early stage of wear, compares best with *Cynomys ludovicianus* (Ord). The specimen is larger than the late Blancan *C. hibbardi* Eshelman or *C.* cf. *C. vetus* Hibbard (Eshelman, 1975) and the extant of *C. gunnisoni* (Baird). *Cynomys niobrarius* Hay, *C. leucurus* Merriam, and *C. ludovicianus* are nearer in size.

Family Geomyidae

Geomys sp.

Material.—Hall Ash locality: USNM 304262, P₄, M³, isolated molar.

Remarks.—See remarks for this taxon in Courtland Canal local fauna.

Family Cricetidae
Subfamily Microtinae

Phenacomys cf. *P. intermedius* Merriam

Material.—Hall Ash locality: USNM 305261, right M₁.

Remarks.—The tooth is rooted, lacks cement, and falls within the occlusal variation noted by Guilday and Parmalee (1972:171) and Howell (1926) for *Phenacomys.* The internal re-entrant angles are deeper than the external ones, but are not as typically pronounced as in the late Pleistocene and living forms. In this character, the tooth in question is more similar to *Pliophenacomys primaevus* Hibbard (1938). The tooth measures 2.29 by 1.03 mm. The mean of M₁ length measurements given by Hibbard (in Skinner and Hibbard et al., 1972:103) for six specimens of *Pliophenacomys primarvus* is 2.86 mm. Guilday et al. (1964:table 20), however, found a range of variation from 1.8 to 3.3 mm with a mean of 2.67 mm for *Phenacomys* cf. *intermedius* from a late Pleistocene fauna.

Hibbard (1944:739) reported a M² questionably referred to *Phenacomys* from the Kansan Age Wilson Valley faunule. This report was confirmed in Hibbard's Cudahy faunal lists of 1949:1421 and 1970:419. *Phenacomys* sp. is also reported from Gerald R. Paulson's unpublished Kansan Age Little Sioux faunule, Lincoln Co., Iowa (Guilday and Parmalee, 1972:fig. 1) as well as from Cumberland and Trout Caves. Steward (1978:45) reports *P. intermedius* from the late Pleistocene Trapshoot local fauna of Rooks County, Kansas. With statistically significant numbers and more complete material, a new Kansan age species may be warranted.

Microtus paroperarius Hibbard

Material.—Hall Ash locality: USNM 304231, left jaw with M₁–M₂, two right jaws with M₁, two edentulous left jaws, six right

Table 5.—*Molar measurements (mm) of* Microtus paroperarius Hibbard *from the Hall Ash locality.*

Tooth	Length			Width		
	N	Mean	Observed range	N	Mean	Observed range
M₁	9	2.80	2.63–2.97	9	1.04	0.97–1.09
M₂	8	1.51	1.34–1.57	8	0.93	0.83–0.98
M¹	8	2.13	1.96–2.35	13	1.17	1.07–1.29
M²	10	1.67	1.40–1.80	10	1.04	0.95–1.14
M³	4	1.87	1.74–1.97	4	0.94	0.83–0.99

M₁s, six left M₂s, seven right M₂s, two left M₂s, twelve right M¹s, two left M¹s, seven right M²s, four left M²s, two right M³s, two left M³s.

Description.—Paulson (1961:144) indicates that in the Cudahy fauna approximately 20% of the M₁s of *M. paroperarius* have a closed fifth triangle. None of the Hall Ash specimens show this apparently variable character. Paulson (1961:144) also states that the "fourth labial re-entrant is normally present and developed enough to contain cement in more than half of the specimens." Of the Hall Ash specimens, approximately twenty percent exhibit cement in the fourth labial re-entrant.

Remarks.—Paulson (1961:145) states that a specimen of *M. paroperarius* recovered from beneath the "Pearlette ash" (type O) in Valley Co., Nebraska, is 11% smaller than the average specimen from the type locality. Additional specimens from Russell and Lincoln Co., Kansas (Hibbard, 1944:737, 740), are intermediate in size. Paulson (1961:145) suggests that these size differences are geographical (negative Bergmann's Response) and not species differences. The Hall Ash specimens (see Table 5) are also smaller than the Cudahy specimens [M₁ length: 2.80 mm (2.63–2.97 mm) versus 2.96 mm (2.6–3.4 mm) and M₁ width 1.04 mm (0.97–1.09 mm) versus 1.07 mm (1.0–1.2 mm)], supporting Paulson's conclusion. *M. paroperarius* is also identified from the Conard Fissure, Mullen I, and Vera local faunas.

Ondatra annecteus (Brown)

Material.—Hall Ash locality: USNM 304264, left M₁ posterior fragment.

Remarks.—See remarks for this taxon in Courtland Canal local fauna.

Synaptomys (Mictomys) cf. *S. meltoni* Paulson

Material.—Hall Ash locality: USNM 304263, right M₃.

Remarks.—See remarks for this taxon in Courtland Canal local fauna.

Family Dipodidae

Zapus sandersi Hibbard

Material.—Hall Ash locality: USNM 304259, right maxillary

fragment with P³–M²; USNM 304260, right mandibular fragment with M₁–M₃.

Measurements.—USNM 304259, M¹, 1.43 by 1.01 mm; M², 1.24 by 0.97 mm; USNM 304260, M₁, 1.50 by 0.94; M₂, 1.39 by 0.95; M₃, 0.79 by 0.70 mm.

Remarks.—Both specimens are in stage three of wear (Klingener, 1963) and probably represent the same individual. The anteroconid is not broader posteriorly as typical material of similar age assigned to *Zapus s. sandersi,* but is more similar in this character to the earlier subspecies form *Z. s. rexroadensis* (Klingener, 1963:258). Therefore, no subspecific assignment is made here and the validity of such an assignment is questioned.

The lower first molar exhibits a posterior anteroconid notch as seen in the Cudahy fauna (Paulson, 1961:fig. 4B and C). The second lower molar is wider than in the Recent forms as noted by Paulson (1961). The measurements agree with those given by Paulson (1961) and are smaller than those noted by Eshelman (1975:44) from the late Blancan age White Rock fauna. The greatest occlusal length of M₁–M₃ (3.68) is greater than that given for *Z. sandersi* from the Yarmouth age Kanopolis local fauna (3.43 mm) (Hibbard et al., 1978). *Zapus sandersi* is also known from the Sanders, Sappa, and Java local faunas.

PALEOECOLOGY

The molluscan fauna from the Hall Ash site (Miller and Eshelman, in press) exhibits a predominance of extant species with northern (39%) and eastern (26%) affinities. Combined with the total absence of snails with southern distributions, a climate of cooler summer temperatures than now characterize the area is inferred (Miller et al., 1980).

Miller (personal communication, 19 May 1983) states the following:

"The Courtland Canal molluscan assemblage is evenly distributed between aquatic and terrestrial taxa, although 73% of the individuals belong to aquatic species. The aquatic molluscs include *Ferrissia rivularis, Sphaerium simile, Valvata tricarinata,* and *Helisoma anceps* species which require permanent water habitats. *F. rivularis* is typically found in streams and rivers with some type of hard substrate. *S. simile* seems to prefer bodies of water with little or no current action. None of the *S. simile* show any sign of abrasion and one individual retains the articulated valves. The molluscs suggest that the depositional environment was either a small, slow moving stream or slough of a larger river. The terrestrial snails include species that could live in a grassland prairie situation with scattered trees or shrubs to provide shelter and some leaf humus."

The amphibians and reptiles from both faunas were published by Rogers (1982) who states:

"The Courtland Canal Fauna represents a permanent prairie pond (*Thamnophis radix, T. sirtalis*) with shallow, quiet water with a muddy bottom (*Chrysemys scripta*) bordered by a wooded hillside (*Agkistrodon contortrix*) and not far removed from a relatively dry, sandy prairie area (*Heterodon nasicus*). The Hall Ash Fauna represents a prairie pond (*Thamnophis radix*) bordered by an area of moist ground (*Ambystoma maculatus*). Prairie areas would have been nearby (*Geochelone*)."

Three species of ducks, one goose, and a sandpiper-like shore bird, all suggestive of a standing water habitat, were recovered from the Courtland Canal site. In addition an open country grouse from the same locality indicates a nearby grassland.

The Hall Ash site contains no fish or aquatic turtles. Individuals of terrestrial snails outnumbered aquatic individuals 276 to 13. Yet the Courtland Canal site contains significant fish and aquatic turtle remains, as well as clams, *Lutra* and *Castoroides,* all absent from the Hall Ash locality. The muskrat *Ondatra* and frogs are found in both localities, whereas salamanders are found only in the Hall Ash site.

Geomorphologically and sedimentologically, the Courtland Canal locality appears to represent a paleo-stream channel fill, whereas the Hall Ash site suggests a shallow depression, such as a prairie pond. The faunal remains support such an interpretation and suggest further that the Hall Ash locality may have had intermittent water and/or a moist grassy habitat. None of the fauna except *Agkistrodon* suggest wooded areas, but instead a grassy, low meadow environment near or adjacent to standing water.

Cool or cold climate indicators such as *Zapus, Synaptomys meltoni, Phenacomys* (see Kurten and Anderson, 1980:258), and *Ondatra annectans* (see taxonomic discussion) support the mollusc suggestions of cooler temperatures. The lack of lizards and cricetids and the abundance of microtids also agree with this interpretation. The only disharmonious taxon is *Geochelone* represented by one carapace fragment recovered on the surface of the Hall Ash locality. The burrowing habits of the extinct small *Geochelone* are unknown, but as Preston (1971) points out, the small species of *Geochelone* often occurs in association with the large, non-burrowing species (see also Holman, 1972:65). Living members of the genus *Geochelone* are confined to open sub-tropical or tropical regions where they feed on succulent plants (Brattstrom, 1961). Whether they could burrow to escape freezing weather or not, the winters must have been milder than exist in north-central Kansas today (Hibbard, 1960). It is possible that this specimen is intrusive or that the tortoise is a late "hold out" from the Yarmouthian. This interpretation would also suggest a younger age for the Hall Ash local fauna than for the Courtland Canal local fauna.

Rogers (1982:177) in interpreting the paleoecology of the amphibians and reptiles from the two faunas states:

"Species in the Courtland Canal and Hall Ash faunas that do not occur in Jewell Co., Kansas, today support the idea of a greenhouse effect on the Kansan paleoclimate of Jewell Co., Kansas. A comparison of species ranges with isolines of average solar radiation and vegetation maps indicates that *Chrysemys scripta, Agkistrodon contortrix, Ambystoma maculatus,* and *Bufo hemiophrys* are species that occur in areas with more water or more cloud cover than is found in Jewell Co., Kansas, today. With the increased moisture and moderated temperatures that would result from the greenhouse effect, this "northern" species and these "southern" species could exist together."

As numerous authors have hypothesized (Graham, 1976; Hibbard, 1960; Eshelman, 1975; Lundelius, 1976), more equable climates, with milder winters and cooler summers and more effective moisture are necessary for these seemingly disharmonious and allopatric species to have lived sympatrically in the past. Unless the physiologies of these animals have changed, there is no environmental regime known today which can serve as an analog for these past faunal associations (Lundelius, 1974).

ACKNOWLEDGMENTS

The authors would like to thank Messrs. Myren Intermill and Harry Hall for allowing us to collect on their property. The State of Kansas graciously allowed the field crew to camp and wash matrix at Lake Lovewell State Park.

The following persons are acknowledged for their expertise, for without their assistance this paper would be significantly incomplete: John Boellstorff, research geologist, Nebraska Geological Survey, Lincoln, Nebraska, fission-track ash dating; Leslie Fay, The Museum, Michigan State University, pollen analysis; Barry Miller, Department of Geology, Kent State University, Kent, Ohio, molluscs; Gerald Smith and Nancy Neff, Museum of Paleontology, University of Michigan, Ann Arbor, Michigan, fishes; Karl Rogers, Division of Scientific and Technological Studies, Adams State College, Alamosa, Colorado, amphibians and reptiles; David Steadman, Department of Vertebrate Zoology, Smithsonian Institution, Washington, D.C., birds; Jessica Harrison, geologist, Department of Exhibits, National Museum of Natural History, Smithsonian Institution, Washington, D.C., camels; C. R. Harrington, National Museum of Natural Sciences, Ottawa, Canada, *Soergelia*; and L. Carson Davis, University of Southern Arkansas, *Soergelia* measurements from Europe.

Augustana College, Rock Island, Illinois, provided Hager with a faculty research grant that provided funds for the field crew during the summer of 1978. All other expenses were borne by the investigators of the project. Michael Carleton, Division of Mammals, and Clayton Ray, Division of Paleobiology, National Museum of Natural History, Smithsonian Institution, Washington, D.C., kindly allowed access to their collections and use of research facilities.

Finally, the authors would like to acknowledge the field assistance of John Herman and Evelyne Eshelman.

LITERATURE CITED

Barbour, E. H., and C. B. Schultz. 1934. A new giant camel *Gigantrocamelus fricki*. Bull. Univ. Nebraska State Mus., 2:17–27.

———. 1937. An early Pleistocene fauna from Nebraska. Amer. Mus. Novitates, 942:1–10.

Boellstorff, J. 1978. North American Pleistocene stages reconsidered in light of probable Pliocene-Pleistocene continental glaciation. Science, 202:305–307.

Brattstrom, B. H. 1961. Some new fossil tortoises from western North America with remarks on zoogeography and paleoecology of tortoises. Paleont., 35:543–560.

Colbert, E. H., and R. G. Chaffee. 1939. A study of "Tetrameryx" and associated fossils from Papago Springs Cave, Sonoita, Arizona. Amer. Mus. Novitates, 1034:1–21.

Diersing, V. E. 1980. Systematics and evolution of pygmy shrews (Subgenus *Microsorex*) of North America. J. Mamm. 61:76–101.

Eshelman, R. E. 1975. Geology and paleontology of the early Pleistocene (late Blancan) White Rock Fauna from north-central Kansas. Pp. 1–60, *in* Studies on Cenozoic paleontology and stratigraphy, Claude W. Hibbard Memorial Vol. 4, Univ. Michigan Papers Paleont., 13.

Eshelman, R. E., and C. W. Hibbard. 1981. Nash Local Fauna (Pleistocene: Aftonian) of Meade County, Kansas. Contrib. Mus. Paleont., Univ. Michigan, 25:317–326.

Fishel, V. C., and A. R. Leonard. 1955. Geology and groundwater resources of Jewell County, Kansas. Bull. Kansas Geol. Surv., 115:1–152.

Getz, L. L. 1960. Middle Pleistocene carnivores from southwestern Kansas. J. Mamm., 41:361–365.

Gidley, J. W., and C. L. Gazin. 1933. New mammalia in the Pleistocene fauna from Cumberland Cave. J. Mamm., 14:343–347.

———. 1938. The Pleistocene vertebrate fauna from Cumberland Cave, Maryland. Bull. U.S. Nat. Mus., 171:1–99.

Graham, R. W. 1976. Late Wisconsin mammalian faunas and

environmental gradients of the eastern United States. Paleobiology, 2:343–350.

GRAHAM, R. W., and H. A. SEMKEN. 1976. Paleoecological significance of the short-tailed shrews (*Blarina*), with a systematic description of *Blarina ozarkensis*. J. Mamm., 57: 433–49.

GUILDAY, J. E., P. S. MARTIN, and A. D. MCCRADY. 1964. New Paris No. 4: A Pleistocene cave deposit in Bedford County, Pennsylvania. Nat. Speleo. Soc. Bull., 26:121–194.

GUILDAY, J. E., and P. W. PARMALEE. 1972. Quaternary Periglacial records of voles of the genus *Phenacomys* Merriam (Cricetidae: Rodentia). Quaternary Research, 2:170–175.

HALL, E. R. 1936. Mustelid mammals from the Pleistocene of North America. Carnegie Inst. Washington Publ., 473:41–119.

HALL, E. R., and K. R. KELSON. 1959. Mammals of North America. The Ronald Press, New York, 1:xxxi+1–546+90; 2:viii+547–1083+90.

HARRINGTON, C. R. 1977. Pleistocene mammals of the Yukon Territory. Unpublished Ph.D. dissert., Univ. Alberta, Edmonton, 1060 pp.

———. 1980. Faunal exchanges between Siberia and North America: Evidence from Quaternary land mammal remains in Siberia, Alaska, and Yukon Territory. Canadian J. Anthrop., 1:45–49.

HAY, O. P., and H. J. COOK. 1930. Fossil vertebrates collected near, or in association with, human artifacts at localities near Colorado, Texas; Frederick, Oklahoma; and Folsom, New Mexico. Proc. Colorado Mus. Nat. Hist., 9:4–40.

HIBBARD, C. W. 1938. An upper Pliocene fauna from Meade County, Kansas. Trans. Kansas Acad. Sci., 40:239–265.

———. 1944. Stratigraphy and vertebrate paleontology of Pleistocene deposits of southwestern Kansas. Bull. Geol. Soc. Amer., 55:707–745.

———. 1949. Pleistocene vertebrate paleontology in North America. Bull. Geol. Soc. Amer., 60:1417–1428.

———. 1949A. Techniques of collecting microvertebrate fossils. Contr. Mus. Paleont., Univ. Michigan, 8:7–19.

———. 1952. Vertebrate fossils from Late Cenozoic deposits of Central Kansas. Univ. Kansas, Paleontol. Contrib., 2:1–14.

———. 1956. Vertebrate fossils from Meade Formation of southwestern Kansas. Papers Michigan Acad. Sci., Arts, Letters, 41:145–203.

———. 1960. An interpretation of Pliocene and Pleistocene climates in North America. Michigan Acad. Sci. Ann. Rept., 62:5–30.

———. 1963. A late Illinoian fauna from Kansas and its climatic significance. Papers Michigan Acad. Sci., Arts, Letters, 48:187–221.

———. 1970. Pleistocene mammalian local faunas from the Great Plains and Central Lowland provinces of the United States. Pp. 395–433, *in* Pleistocene and Recent environments of the central Great Plains (W. Dort, Jr., and J. K. Jones, Jr., eds.), Univ. Press Kansas, Lawrence, 433 pp.

HIBBARD, C. W., and W. W. DALQUEST. 1960. A new antilocaprid from the Pleistocene of Knox County, Texas. J. Mamm., 41:20–23.

———. 1966. Fossils from the Seymour Formation of Knox and Baylor counties, Texas, and their bearing on the Late Kansan climate of that region. Contrib. Mus. Paleont., Univ. Michigan, 21:1–66.

HIBBARD, C. W., and E. S. RIGGS. 1949. Upper Pliocene vertebrates from Keefe Canyon, Meade County, Kansas. Bull. Geol. Soc. Amer., 60:829–860.

HIBBARD, C. W., R. J. ZAKRZEWSKI, R. E. ESHELMAN, G. EDMUND, C. D. GRIGGS, and C. GRIGGS. 1978. Mammals from the Kanopolis Local Fauna, Pleistocene (Yarmouth) of Ellsworth County, Kansas. Contrib. Mus. Paleont., Univ. Michigan, 25:11–44.

HOLMAN, J. A. 1972. Herpetofauna of the Kanopolis local fauna (Pleistocene: Yarmouth) of Kansas. Michigan Academician, 5(1):87–98.

HOWELL, A. B. 1926. Voles of the genus *Phenacomys*. N. Amer. Fauna, 48:1–66.

KAPP, P. O. 1970. Pollen analysis of pre-Wisconsin sediment from the Great Plains. Pp. 143–155, *in* Pleistocene and Recent environments of the Central Great Plains (W. Dort, Jr., and J. K. Jones, Jr., eds.), Univ. Press Kansas, Lawrence, 433 pp.

KLINGENER, D. 1963. Dental evolution of *Zapus*. J. Mamm., 44:248–260.

KURTÉN, B., and E. ANDERSON. 1980. Pleistocene mammals of North America. Columbia Univ. Press, New York, 442 pp.

LUNDELIUS, E. L., JR. 1972. Fossil vertebrates from the late Pleistocene Ingleside fauna, San Patricio County, Texas. Rept. Investigations Bur. Econ. Geol., Univ. Texas, 77:1–74.

———. 1974. The last fifteen years of faunal change in North America. Pp. 141–160, *in* History and prehistory of the Lubbock Lake Site (C. C. Black, ed.), Museum J., 15:1–160.

———. 1976. Vertebrate paleontology of the Pleistocene: an overview. Geoscience and Man, 13:45–59.

MARTIN, R. A. 1973. The Java local fauna, Pleistocene of South Dakota: A preliminary report. Bull. New Jersey Acad. Sci., 18:48–56.

———. 1975. *Allophaiomys* Kormas from the Pleistocene of North America. Pp. 97–100, *in* Studies on Cenozoic paleontology and stratigraphy, Claude W. Hibbard Memorial Vol. 3, Univ. Michigan Papers Paleont., 12:1–143.

MEADE, G. E. 1945. The Blanco fauna. Univ. Texas Publ., 4401: 509–556.

MILLER, B. B., R. E. ESHELMAN, and M. W. HAGER. 1980. A molluscan faunule collected beneath the Hartford(?) Ash (0.71 MYBP), Jewell County, Kansas. Abstract, Geol. Soc. Amer., north-central section, 14th Ann. meeting, Indiana Univ., Bloomington, 12:54.

MILLER, B. B., and R. E. ESHELMAN. In press. Pleistocene Molluscs From Jewell County, Kansas Associated With The Hartford Ash (0.706 MYBP). J. Paleont.

NELSON, M. E., and J. NEAS. 1980. Pleistocene musk oxen from Kansas. Trans. Kansas Acad. Sci., 83:215–229.

NELSON, R. S., and H. A. SEMKEN. 1970. Paleoecological and stratigraphic significance of the muskrat in Pleistocene deposits. Bull. Geol. Soc. Amer., 81:3733–3738.

NOWAK, R. M. 1979. North American Quaternary *Canis*. Monogr. Mus. Nat. Hist., Univ. Kansas, 6:1–154.

PAULSON, G. R. 1961. The mammals of the Cudahy fauna. Papers Michigan Acad. Sci., Arts, Letters, 46:127–153.

PRESTON, R. E. 1971. Pleistocene Turtles from the Arkalon Local Fauna of Southwestern Kansas. J. Herpetol., 5:208–211.

REPENNING, C. A. 1983. *Pitymys meadensis* Hibbard from the Valley of Mexico and the classification of North American species of *Pitymys* (Rodentia: Cricetidae). J. Vert. Paleo., 2:471–482.

RINKER, G. C., and C. W. HIBBARD. 1952. A new beaver and associated vertebrates, from the Pleistocene of Oklahoma. J. Mamm., 33:98–101.

ROGERS, K. L. 1982. Herpetofaunas of the Courtland Canal and Hall Ash Local Faunas (Pleistocene: Early Kansan) of Jewell County, Kansas. J. Herpetol., 16:174–177.

SCHULTZ, C. B., and L. D. MARTIN. 1970. Quaternary mammalian sequence in the Central Great Plains. Pp. 341–353, *in* Pleistocene and Recent Environments of the central Great Plains (W. Dort, Jr., and J. K. Jones, Jr., eds.), Univ. Press Kansas, Lawrence, 433 pp.

SKINNER, M. F., C. W. HIBBARD, ET AL. 1972. Early Pleistocene pre-glacial and glacial rocks and faunas of north-central Nebraska. Bull. Amer. Mus. Nat. Hist., 148:1–148.

SKINNER, M. F., and O. C. KAISEN. 1947. The fossil *Bison* of Alaska and preliminary revision of the genus. Bull. Amer. Mus. Nat. Hist., 89:1–256.

SLAUGHTER, B. H. 1961. A new coyote in the Late Pleistocene of Texas. J. Mamm., 42:503–509.

STEWARD, J. C. 1978. Mammals of the Trapshoot local fauna, Late Pleistocene of Rooks County, Kansas. Proc. Nebraska Acad. Sci., 88th Ann. meeting, 14–15 April, pp. 45–46.

STOCK, C. 1925. Cenozoic gravigrade edentates of western North America, with special reference to the Pleistocene Megalonychidae and Mylodontidae of Rancho La Brea. Carnegie Inst. Washington Publ., 331:1–206.

TROXELL, E. L. 1915. A fossil ruminant from Rock Creek, Texas, *Preptoceras mayfieldi* sp. nov. Amer. J. Sci., 40:479–482.

VAN DER MEULEN, A. J. 1978. *Microtus* and *Pitymys* (Arvicolidae) from Cumberland Cave, Maryland, with a comparison of some new and old world species. Ann. Carnegie Mus., 47:101–145.

WEBB, S. D. 1965. The osteology of *Camelops*. Bull. Los Angeles Co. Mus. Sci., 1:1–54.

WOODBURNE, M. O. 1961. Upper Pliocene geology and vertebrate paleontology of part of the Meade Basin, Kansas. Papers Michigan Acad. Sci., Arts, Letters, 46:61–101.

ZELLER, D. E. 1968. The stratigraphic succession. Bull. State Geol. Surv. Kansas, 189:1–81.

Address (Eshelman): Calvert Marine Museum, Solomons, Maryland 20688 and Research Associate, Department Paleobiology, Smithsonian Institution, Washington, D.C. 20560.

Address (Hager): Museum of the Rockies, Montana State University, Bozeman, Montana 59719.

PALEOECOLOGY OF A LATE WISCONSINAN/HOLOCENE MICROMAMMAL SEQUENCE IN PECCARY CAVE, NORTHWESTERN ARKANSAS

Holmes A. Semken, Jr.

ABSTRACT

Thirty-five taxa of rodents and insectivores, represented by 1,942 individuals, were recovered from seven levels within three fossiliferous stratigraphic units in Peccary Cave, northwestern Arkansas. Radiocarbon dates indicated that the fossils accumulated between 16,700 and 2,290 years ago. The two lower units also contained fossil remains typical of the late Wisconsinan megafauna; those in the uppermost unit were predominately modern species.

Statistical analysis of the associated micromammals reveals that neither Late Wisconsinan nor Holocene climates were static, that faunal turnover at the end of the Pleistocene was not catastrophic and that the modern community is a depauperate residuum developed by individualistic species response to climatic change. First, circa 16,700 years ago, the local community was dominated by individuals characteristic of a cool steppe with coniferous forest patches, (2) this gave way to a mixed-forest parkland association, (3) four dry boreal forest mammals then disappeared, and individuals representing both meadow and deciduous forest species increased in numbers to equal coniferous ecotypes in the parkland, (4) next, two additional boreal species were lost and deciduous parkland species became dominant, (5) megavertebrate extinction followed and was contemporaneous with a dramatic increase in both deciduous forest and temperate prairie elements, (6) finally, deciduous forest species prevailed but local enclaves for both prairie and boreal ecotypes remained. The modern closed deciduous forest must post-date 2,290 years B.P.

INTRODUCTION

INTERPRETATION OF PLEISTOCENE LOCAL FAUNAS

Since the recognition of multiple Pleistocene glaciations, the prediction of synchronous cooling in temperate regions has been based partially on the discovery of fossils of both arctic and boreal animals in mid-latitudes, for example, *Ovibos,* the barren ground muskox (Kitts, 1953). Conversely, deposits with remains of subtropical taxa, for example, *Neofiber,* the water rat (Blair, 1958), north of their present range were referred to interglacial stages. This concept is especially well-grounded when all taxa in a local fauna are representative of a discrete region (area of sympatry, Stephens, 1960) either to the north or south of their occurrence as fossils. A local fauna of this nature is regarded as harmonious and can be "explained by a comparatively simple environmental shift" (Hibbard and Taylor, 1960:37). This biome migration concept was designated the "Cliseral Shift Hypothesis" by Graham (1979). Harmonious local faunas, for example, the Robert (Schultz, 1967), Cragin Quarry (Hibbard and Taylor, 1960), and Elkader (Woodman, 1982) support the contention that existing North America biomes have antiquity and probably fluctuated north and south in response to glacial movements (Blair, 1958).

However, harmonious glacial-age local faunas are rare and most others, for example, Natural Chimneys (Guilday, 1962), contain several species that now are allopatric. Some, like the Sandahl (Semken, 1966), have a harmonious mammalian component but other classes, for example, molluscs (Miller, 1970), are disharmonious. These disharmonious local faunas, which include Peccary Cave, do not have a modern analog and, as a result are more difficult to interpret. Four hypotheses have been proposed. (1) The specimens were mixed as a result of reworking and thus constitute an "assemblage" rather than a local fauna. The Kentuck "assemblage" of McPherson Co., Kansas, contains both *Mictomys* (northern bog lemming) and *Sigmodon* (cotton rat), which now are separated from each other by at least 625 mi (1,000 km) (Hall, 1981), and it was explained in this manner (Hibbard, 1952; Semken, 1966). (2) Guilday (1962) considered slow fossil accumulation over a period of biome migration (cliseral shift) to explain the joint occurrence of boreal taxa (*Phenacomys* cf. *ungava, Sorex arcticus, Synaptomys borealis,* and others) with temperate forms (*Cryptotis parva, Neotoma floridana, Tamias striatus,* and others) in the Natural Chimneys local fauna. (3) Alternatively, Guilday (1962) suggested that the Natural Chimneys association may be a product of owls "grazing" on life zones developed on Appalachian topography during glacial advance. Patterned ground on Appalachian summits supported the previous existence of "tundra" (Guilday et al., 1964). Under this model a temperate forest and its associated fauna were restricted to valley floors with Canadian and boreal life zones occupying intermediate elevations. This interpretation is viable for montane regions but is difficult to invoke in areas with low relief (for example, Schultze Cave, Dalquest et al., 1969). (4) Disharmonious local faunas may represent climates unlike any known today. This non-uniformitarian interpretation was considered initially by both Braun (1955) and Hibbard (1960) but rarely was taken seriously until the 1970's.

Dalquest (1965) discussed the disharmonious Howard Ranch local fauna (16,775 ± 565 years B.P.) where topographic relief is insufficient for the life zone concept. He invoked Hibbard's (1955) equability model in which muted seasonal extremes and a reduction in the number of severe cold fronts were utilized to account for the disharmonious occurrence of presently allopatric taxa. Hibbard (1955) reasoned that the northern geographic limit of animals with southern distributions is largely controlled by the coldest days in winter and that the southern margin of animals with northern affinities is determined by the hottest days in summer. Reduction of seasonal extremes, caused by warmer winters and cooler summers, would permit many presently allopatric species to coexist.

Hibbard (1960) expanded the concept of equability to explain climatic conditions associated with many Pleistocene local faunas, the Illinoian climate of Meade Co., Kansas, compared to that of today "being generally cooler; the difference mostly or entirely in the summers." Hibbard (1960:23) further noted that "climatic zoning as known at the present with extreme low winter temperatures of the Interior is in large part due to the strong continentality of the climate." Slaughter (1965) subsequently suggested that megafaunal extinction at the end of the Wisconsinan was explained best by increased temperature extremes during the birthing season from a more continental post-glacial climate. Dalquest et al. (1969) proposed that a simple reduction in the number of cold fronts ("northers") on the Texas plains, because of a reduced climatic gradient, would result in overall warmer winters during glacial advance. Increased climatic extremes associated with the onset of Holocene climate would limit reproduction of species which require cool summers to now boreal regions, and thus reduce species diversity in temperate latitudes. The modern temperate fauna of the United States was regarded by Semken

(1974) and Martin and Webb (1974:138) as "impoverished," created by the northward withdrawal of animals now considered as either boreal or arctic in affinity. Graham (1976) emphasized that relative frequencies of boreal, deciduous, and steppe species were more balanced in late Wisconsinan than modern faunas of the eastern United States because the boreal species had immigrated into regions occupied by temperate animals in response to cooler summers. Cooler summers did not materially affect temperate species in mid-latitudes so these animals did not participate in a cliseral shift to the south. Thus, many late Pleistocene faunas are not only disharmonious but are characterized by increased species diversity. Because each boreal species responded individualistically (Graham, 1976) to glacial climates, some dispersed further to the south than others. This created a gradient with decreasing numbers of boreal species in lower latitudes. Both Graham (1979) and Guilday et al. (1978) attribute this increase in species diversity to more equable climates. This change in faunal composition permitted both Graham (1979) and Martin and Neuner (1978) to independently delineate similar Wisconsinan faunal complexes or provinces, none of which are directly analogous to any modern biome.

Most Wisconsinan local faunas, with the exceptions mentioned above, are disharmonious. However, the equable nature of associated climates is not uniformly applicable. A cluster of local faunas recently described from Iowa [Craigmile and Waubonsie (Rhodes, in press), Brayton (Dulian, 1975), Eagle Point (Rosenberg, 1983) and Elkader (Woodman, 1982)] and Wisconsin [Moscow (Foley, in press)] all suggest that Wisconsinan winters were substantially colder in this region. Cooler summers, which are compatible with the equability model, also were predicted.

The late Wisconsinan/Holocene micromammal sequence from Peccary Cave in northwestern Arkansas, which contains up to 35 species in a single level, also is characterized by high species diversity (Table 1). It not only lies between the equable local faunas of the southern United States and the "cold steppe" (Rhodes, in press) local faunas of the upper Mississippi valley, but also provides insight into the nature of changing climatic patterns associated with deglaciation.

LOCATION AND PREVIOUS INVESTIGATIONS

A diverse vertebrate fauna, the Peccary Cave local fauna, was collected by the Department of Geology,

University of Arkansas, from Peccary Cave, on Ben's Branch of Cave Creek, SW ¼, SE ¼, Sec. 22, R. 19W., T. 15N., Newton Co., Arkansas. Excavations were conducted from 1967 to 1969 under the auspices of National Science Foundation Grant GB6762; James H. Quinn and Charles R. Mc-Gimsy, principal investigators. Initial excavations by Jack McCutheon, cave discoverer, and L. Carson Davis were summarized by Davis (1969). Semken (1969) provided a short list of small mammals identified in the early stages of faunal analysis. Additional material was collected by the University of Iowa in the summers of 1971 and 1972. Quinn (1972) elaborated on the regional geomorphology, cave sedimentation, and the large mammals discovered during excavation. A typical late Pleistocene megamammal faunal list consisting of *Symbos* sp., *Cervalces* sp., *Tapirus terrestus, Platygonus* sp., *Dasypus bellus, Canis dirus, Sangamona* sp., *Mammut* sp., *Mammuthus* sp., *Odocoileus virginianus, O.* aff. *hemionus, Homo* sp., and possibly *Equus* was recorded from the cave. The stratigraphic position of these specimens was not published, but radiocarbon dates (Quinn, 1972) range from 2,230 ± 180 (I-4828); 2,980 ± 180 (I-3477); 4,290 ± 110 (I-3894) and 4,050 ± 190 (I-5392) years B.P. on charcoal; 9,510 ± 140 years B.P. (I-4830) on *Mesodon* shell; to 16,700 ± 250 years B.P. (I-5262) on bone collagen; and indicate that a late-glacial through postglacial sedimentary sequence was deposited in the cave. Davis (1973) interpreted the Peccary Cave herpetofauna in relation to this transition and predicted a warming then drying trend from the biostratigraphic sequence.

STRATIGRAPHY

Davis (1973), one of the site excavators, defined four lithologic units in Peccary Cave. Unit D, the oldest, is an unfossiliferous yellow sand and clay disconformably underlying unit C, a fossiliferous light reddish-brown (2.5 YR 4/6) clay grading upward into dark reddish-brown (2.5 YR 3/6-4) clay. Unit B, a fossiliferous dark red-brown (5 YR 3/3) clay, is conformable with Unit C but disconformable with overlying Unit A, a fossiliferous friable light-brown (5 YR 4/4) matrix with masses of mammal feces. The duration of either hiatus is not known.

The radiocarbon dates, 16,700 to 2,230 years B.P., can be supported on subjective evidence. The large mammal component has not been completely examined, but it includes at least the eight extinct taxa listed above (Quinn, 1972). Both demonstrate that some of the fill is Pleistocene in age. An open sinkhole with a developing talus cone (Davis, 1973), occupation by owls, woodrat nests, and Archaic Tradition artifacts indicate that sediment has continued to accumulate through the Holocene in the cavern.

The present source of sediment, primarily from the sinkhole, does not explain the bulk of fossiliferous matrix comprising lithologic units A, B, and C. Quinn (1972) reasoned that these essentially horizontal but channelled sediments, all of which contain fish and crayfish remains, were best explained by periodic flooding from Ben's Branch into a horizontal entrance. Subsequent excavation exposed this entrance and a charcoal date (I-4828) from the uppermost alluvium closing it suggests that major sedimentation ceased about 2,230 years B.P. Thus, these cave deposits clearly transcend the Pleistocene/Holocene boundary.

METHODS

EXCAVATION

Twenty-three trenches of varying lengths, depending on configuration of the cave, were excavated by square and level during the grant period (Davis, 1973, fig. 1). All were excavated in one foot intervals except those in Trenches 4, 8, 13, and 18 where smaller breakdown clast size permitted six inch divisions. All matrix was washed in the manner described by Hibbard (1949) and McKenna (1962) and the concentrate was picked for fossils (Davis, 1973).

NUMERICAL EVALUATION

Trenches 8 and 13 were selected for analysis of insectivore and rodent remains because of their smaller excavation intervals (six inches), high bone concentration, trench juxtaposition, and presence of the younger stratigraphic units (A and B). The discon-

formity between Units A and B, not visible until the profile dried, bisected level 3 in Trenches 8 and 13. Unfortunately, the sample was excavated by arbitrary level prior to delineation of the disconformity and that sample can not be separated into its natural stratigraphic components.

Trench 15, excavated in one foot intervals, was added as it also contained a high concentration of bone, was stratigraphically below Trenches 8 and 13 (Unit C), and test samples appeared to have a greater proportion of boreal taxa than that found in Trenches 8 and 13. The contact between Units B and C is conformable, and lower Unit B may be represented in uppermost level 1 of Trench 15 (T15-1).

The minimum number of individuals for each taxon was established for each square, usually on the maximum number of right or left first lower molars. A *Chi*-square analysis demonstrated that there was no significant difference between adjoining squares of each level in Trenches 8 and 13, and that the combined sample could be utilized for this analysis. Bias due to undulating stratigraphic contacts was minimal except in level 3 (T8/13-3) where statistical tests were neutral and the Unit A–B disconformity was developed.

Tests for faunal trends within the stratigraphic sequence (Tables 1 and 2) utilize the Freeman-Tukey Deviate, an exploratory statistical technique similar to *Chi*-square. The Freeman-Tukey Deviate, however, provides a method of examining the direction and significance of the relationship between actual and expected counts on a cell-by-cell basis. The analysis is conducted in the following manner: (1) tabulate minimum numbers of individuals (actual count, AC) for each taxon per level in a row and column format (Table 1); (2) total both columns and rows; (3) calculate the percent abundance (PA) of each taxon (row total of Table 1) in proportion to the grand total of individuals identified; (4)

compute the expected number (EN) of individuals in each cell assuming that the species were uniformly distributed in all levels in proportion to their percent abundance in the entire sample; this is done by multiplying each column total (CT) by the percent abundance (PA) of a given taxon for the entire sample (EN = PA × CT); (5) the Freeman-Tukey Deviate is obtained by the following equation: F-T D = $\sqrt{4AC + 2} - \sqrt{4EN + 1}$. The constants are added to insure a minimal value for each cell. The resulting deviate is a measure of the significance of the difference between the actual value and that expected. The sign notes the direction of the difference, a minus indicates that fewer specimens than expected were recovered, a plus indicates that more than expected numbers were present. A measure of 0.5 is the lowest significant deviate. The greater the deviate, the greater the disparity, and hence the significance, between the actual and expected values.

SYSTEMATIC DISCUSSIONS

INTRODUCTION

Specimens used in this analysis, those from Trenches 8, 13, and 15, are cataloged into the Paleontological Repository, Museum of Natural History, University of Iowa (SUI) by agreement with the University of Arkansas. Postcranial elements from these trenches and all other vertebrate remains, collected under NSF Grant GB6762, the bulk of the Peccary Cave sample, is or will be deposited at the University of Arkansas (UA) at Fayetteville. Collections from Trench 24, excavated in 1971 and 1972 with University of Iowa funds, also are housed in Iowa City (Megivern, 1982). All specimens of a given taxon are cataloged by trench. Thus, all *Sorex arcticus* from Trench 8 (T8) bear one number, those from Trench 13 (T13) another, but all are reposited by level and square in capsules with detailed provenience. Measured specimens are designated by either alphabetical or numerical subscripts.

The traditional listing of each species by taxonomic hierarchy has been condensed. However, the diagnostic characters identifying each species are noted in the following paragraphs because some identifications are more subjective than implied by a taxonomic list (Table 1).

MOLES

Both *Scalopus aquaticus* (eastern mole) and *Condylura cristata* (star-nosed mole) are present in Peccary Cave. *Scalopus*, characterized by robust proportions, shallow re-entrant angles between the trigonid and talonid, and two prominent lingual cusps on the lower molars, is readily separated from the diminutive *Condylura* with its deep re-entrants and

additional lobate median cusp. The *Parascalops* (hairy-tailed mole) lower molar, not identified in Peccary Cave, exhibits five lingual cusps, deep re-entrants, and both a smaller and more delicate configuration.

SHREWS

Eight species of shrews are present in Peccary Cave. Soricid mandibles greatly outnumber upper dental elements in Peccary Cave and the key developed by Guilday (1962) was the most useful taxonomic guide. The few rostra sufficiently complete to be identified by Junge and Hoffmann's (1981) criteria were assigned to comparable taxa. The pygmy shrew (*Sorex hoyi*, Junge and Hoffmann, 1981) was separated from the masked shrew (*S. cinereus* including *S. haydeni*) by the characteristically reduced entoconid in the former; *S. arcticus* (arctic shrew) exhibited both a *Sorex*-type articular condyle and a post-mandibular foramen; and *S. palustris* (water shrew) was distinguished on a deep hypoconid valley, large size, and absence of a post-mandibular foramen. *Cryptotis parva* (least shrew) displayed both a distinct concavity between the articular facets of the condyle and a reduced m3 talonid. *Blarina* (short-tailed shrew) was more massive than any other Peccary Cave shrew. A bivariate plot (Fig. 1) of the molar row length against median body depth suggests that there is character displacement by size in the Peccary Cave soricids. Thus, most fragmentary soricid remains also can be separated by size.

The *Blarina* cluster from T15-1 on the soricid plot (Fig. 1) is almost as large as that representing the five smaller species in the Peccary Cave sample.

Table 1.—*Actual (below diagonal) and expected (above diagonal) counts of rodents and insectivores from Peccary Cave Trenches 8 + 13 and 15. The columns become younger from right to left.*

Each cell is given as **expected / actual**.

TAXON	TRENCH 8+13: 1	2	3	4	5	TRENCH 15: 1	2	% TOT.
Scalopus aquaticus	2.0/2	1.8/2	1.9/3	1.9/5	0.9/3	11.9/9	8.6/5	1.5/29
Condylura cristata	0.4/–	0.4/1	0.4/–	0.4/–	0.2/–	2.4/2	1.7/2	0.3/5
Cryptotis parva	0.4/–	0.4/–	0.4/–	0.4/–	0.2/–	2.4/5	1.7/1	0.3/6
Sorex cinereus	2.3/–	2.1/–	2.1/–	2.1/2	1.0/1	13.5/20	9.8/9	1.7/32
Sorex palustris	0.4/1	0.4/–	0.4/–	0.4/–	0.2/1	2.4/1	1.7/1	0.3/5
Sorex arcticus	1.1/–	1.0/1	1.0/–	1.0/–	0.5/–	6.4/9	4.6/5	0.8/15
Sorex hoyi	0.3/–	0.2/–	0.3/–	0.2/1	0.1/–	1.6/2	1.2/1	0.2/4
Blarina brevicauda	5.3/2	4.8/1	4.9/3	4.8/3	2.2/2	31.1/39	22.5/25	3.9/75
Blarina hylophaga	6.0/8	5.6/8	5.6/18	5.6/15	2.5/3	35.1/27	25.3/7	4.4/86
Blarina carolinensis	1.8/2	1.6/3	1.6/1	1.6/1	0.7/–	10.4/15	7.5/3	1.3/25
Tamias striatus	1.6/6	1.4/3	1.4/3	1.4/–	0.7/–	9.2/9	6.6/1	1.1/22
Eutamias minimus	0.1/–	0.1/–	0.1/–	0.1/–	0.1/–	0.8/1	0.6/1	0.1/2
Marmota monax	1.2/5	1.1/3	1.1/2	1.1/2	0.7/1	7.2/2	5.1/2	0.9/17
S. tridecemlineatus	0.6/–	0.6/–	0.6/–	0.6/–	0.3/–	4.0/3	2.7/5	0.5/9
Sciurus cf. niger	1.6/4	1.5/5	1.5/3	1.5/4	0.7/1	9.6/4	6.9/2	1.2/23
Tamiasciurus hudsonicus	0.6/–	0.5/–	0.5/–	0.5/–	0.2/–	3.2/4	2.3/4	0.4/8
Glaucomys volans	0.6/2	0.5/1	0.5/–	0.5/1	0.2/–	3.2/1	2.3/2	0.4/7
Geomys bursarius	8.4/5	7.6/8	7.8/6	7.7/8	3.5/4	49.4/57	35.7/32	6.2/120
Castor canadensis	0.1/1	0.1/–	0.1/–	0.1/–	0.1/–	0.8/1	0.6/–	0.1/2
Erethizon dorsatum	0.6/2	0.5/1	0.5/–	0.5/–	0.2/1	3.2/2	2.3/2	0.4/8
Peromyscus cf. leucopus	2.9/6	2.7/8	2.7/8	2.7/3	1.2/2	16.7/8	12.1/6	2.1/41
Peromyscus cf. maniculatus	2.2/4	2.0/2	2.0/5	2.0/5	0.9/3	12.8/9	9.3/3	1.6/31
Peromyscus sp.	16.2/41	14.6/32	15.0/30	14.8/18	6.8/8	94.8/91	68.5/11	11.9/231
Onychomys leucogaster	0.1/–	0.1/1	0.1/–	0.1/–	0.1/–	0.8/–	0.6/–	0.1/1
Neotoma floridana	19.4/21	17.6/22	18.0/16	17.7/13	8.2/11	114.0/120	82.4/74	14.3/277
Oryzomys palustris	0.3/1	0.2/–	0.3/–	0.2/1	0.1/–	1.6/1	1.2/–	0.2/3
Ondatra zibethicus	0.4/–	0.4/–	0.4/1	0.4/1	0.2/–	2.4/2	1.7/1	0.3/5
Microtus pinetorum	9.5/13	8.7/8	8.8/11	8.7/8	4.0/1	55.8/72	40.3/23	7.0/136
Microtus ochrogaster	6.5/3	5.8/6	6.0/6	5.9/7	2.7/2	37.5/40	27.3/27	4.7/91
Microtus pennsylvanicus	26.7/3	24.1/4	24.7/5	24.3/17	11.2/7	156.2/158	112.9/185	19.6/379
Microtus xanthognathus	0.3/–	0.2/–	0.3/–	0.2/–	0.1/–	1.6/2	1.2/3	0.3/5
Synaptomys cooperi	1.2/1	1.1/1	1.1/2	1.1/6	0.7/3	7.2/3	5.1/2	0.9/18
Synaptomys borealis	0.6/1	0.6/–	0.6/–	0.6/–	0.3/1	4.0/3	2.7/2	0.5/7
Phenacomys intermedius	1.1/–	1.0/–	1.0/–	1.0/–	0.5/–	6.4/5	4.6/11	0.8/16
Clethrionomys gapperi	13.7/2	12.4/2	12.7/2	12.5/4	5.8/2	80.5/70	58.2/116	10.1/196
Zapus hudsonius	0.4/1	0.4/–	0.4/–	0.4/–	0.2/–	2.4/1	1.7/2	0.3/5
Total	137	124	125	125	57	798	576	1942

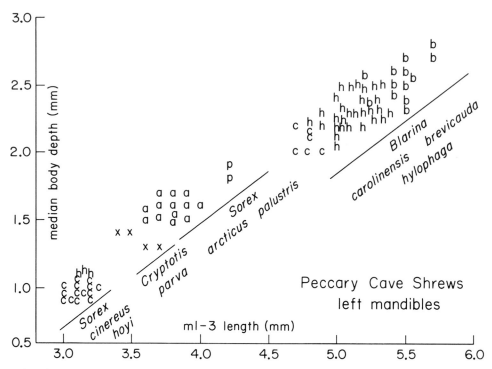

Fig. 1.—Bivariate plot of median body depth versus molar row length of Peccary Cave soricid left mandibles. The species of *Blarina* are designated by the linear index of Graham and Semken (1976).

Size variation within the *Blarina* sample also is much greater than that for any single modern subspecies (Graham and Semken, 1976). This implies either that *Blarina* was more variable circa 12,000 years ago or that more than one species of *Blarina* is present. Statistical inspection of a previously examined sample reveals a trimodel distribution and application of a linear index (Graham and Semken, 1976, Fig. 3) separated the Peccary Cave specimens into three groups. The largest group (Graham and Semken, 1976, Fig. 3) was comparable in size to *B. b. brevicauda*, the intermediate was similar to *B. b. kirtlandi* (and *B. b. hylophaga*) and the smallest equaled *B. b. carolinensis*. These phena must have been behaving as species to be sympatric near the cave (Graham and Semken, 1976).

Two *Blarina* phena previously regarded as subspecies were demonstrated to be species by Genoways and Choate (1972) who captured two "subspecies" without morphological intergradation in the same trap line in Nebraska. Subsequently, the sympatric distribution of *B. brevicauda* and *B. hylophaga* was mapped across southern Iowa and northern Missouri (Jones, 1982). Similarly, *B. brevicauda* and *B. carolinensis* are sympatric in Georgia and South

Carolina and *B. hylophaga* and *B. carolinensis* co-occur in Texas, Louisiana, and Arkansas (Jones, 1982). Thus, large, intermediate, and small species of *Blarina* currently are recognized in eastern North America. Although *B. hylophaga* is the size of *B. b. kirtlandi*, the two are not the same taxon. Karyologic data indicates that *B. b. kirtlandi* is a small subspecies of *B. brevicauda*. Morphologically, however, the *B. b. kirtlandi* cluster (Jones, 1982, fig. 3) shows greater separation from that of *B. brevicauda* than from either cluster of the karyologically distinct taxa—*B. hylophaga* or *carolinensis*.

B. hylophaga presently resides in the Peccary Cave vicinity, *B. carolinensis* is recorded within 75 mi (120 km) to the east and *B. brevicauda* has been collected approximately 250 mi (400 km) to the northeast of the cave. For this reason the intermediate-sized phenon of *Blarina* from Peccary Cave, originally regarded as *B. kirtlandi* by Graham and Semken (1976), is here redesignated *B. hylophaga* to be compatible with both recent biogeography and assignment of similarily sized specimens in other Middle Mississippi Valley local faunas (Jones, 1982).

Jones (1982) pooled suites of cranial characters taken from modern *Blarina* into five categories representative of modern taxa and searched 30 Rancholabrean local faunas for analogies. She did not detect three phena of *Blarina* in any single site, but recognized the presence of two at such geographically diverse sites as Crankshaft Cave, Missouri; Kline Cave, Texas; Baker Bluff Cave, Tennessee; New Paris #4, Pennsylvania; and Ladds, Georgia among others. Klippel and Parmalee (1982*a*) identified three phena of *Blarina* in Stratum 5 (Wisconsinan) of Cheek Bend Cave and noted that these phena were parapatric into the Holocene of central Tennessee. Therefore, as at Peccary Cave, multiple *Blarina* taxa are present in many other Wisconsinan local faunas.

SQUIRRELS

Seven species of squirrels are present in the Peccary Cave local fauna (Table 1). Remains of the woodchuck (*Marmota monax*), which weighs between 2.3 and 4.6 kg when alive, are readily separated by size from those of the next largest identified sciurid, the 0.45 to 1.4 kg fox squirrel (*Sciurus* cf. *niger*). All species of *Sciurus* are long-snouted and the diastema is thus longer than that of the short-faced red squirrel (*Tamiasciurus hudsonicus*). *Sciurus*, also is distinctly larger than *Tamiasciurus*, lacks accessory cuspids present in *Tamiasciurus*, and has a mandibular length clearly greater than mandibular depth. The fox squirrel (*S. niger*) and grey squirrel (*S. carolinensis*) both exhibit east/west clines with *S. carolinensis* increasing and *S. niger* decreasing in size toward the west. They approach the same size at their common western extreme (Purdue, 1980). Thus, size will not necessarily separate the two on the plains. Clinal shifts during the Holocene, demonstrated by Purdue (1980), complicate size as a taxonomic character. However, *S. carolinensis* usually exhibits both a tiny peg-like P3, absent in *S. niger*, and a two-rooted p4. Fox squirrels generally have three-rooted p4's. Neither *Sciurus* maxillary has a P3 alveolus and four of the six p4's are three-rooted. Thus, *S. niger* characters dominate the *Sciurus* sample, but *S. carolinensis* also might be present.

The three small forest squirrels are separated by the measurements of complete mandibles (Fig. 2). The alveolar row lengths of the eastern chipmunk (*Tamias striatus*) and the southern flying squirrel (*Glaucomys volans*) are almost identical in the Peccary Cave sample. For its size, however, *T. striatus* has a longer diastema than *G. volans*. The least chipmunk (*Eutamias minimus*) has a significantly smaller alveolar length than either *T. striatus* or *G. volans*. None of the Peccary Cave *Eutamias* retain a complete diastema, so a sample from Shield Trap, Montana (Geppert, 1984), was plotted for comparison. The two complete molar rows (diastema estimated) from Peccary Cave assigned to *Eutamias* (Fig. 2) best compare to the Shield Trap sample. Reasonably distinctive dental morphologies confirm this diagnosis. The six *Glaucomys* specimens from Peccary Cave have an alveolar row length that ranges from 6.2 to 6.8 mm, which compares to *G. volans* (6.2 to 7.3 mm), and are smaller than Late Pleistocene *G. sabrinus* (7.0 to 8.7 mm) in the eastern United States (Guilday et al., 1977). The alveolar row length of the 21 Peccary Cave *Tamias* (5.9 to 6.9 mm) compares well to that of 17 recent and 114 late Holocene specimens (5.8 to 6.9 mm) from Pennsylvania (Guilday et al., 1978). Because *T. striatus* exhibits a negative Bergman's response (Guilday et al., 1978), the small size of the Peccary Cave *Tamias* for its latitude apparently reflects a response to a colder climate.

The thirteen-lined ground squirrel (*Spermophilus tridecemlineatus*), known from complete mandibles and maxillae, is distinct from the other chipmunk-sized squirrels because it has rhombohedral instead of square molars and high anterior lophids.

GOPHERS

Geomys bursarius (plains pocket gopher) is a conspicuous component throughout Peccary Cave. It was identified by the presence of two-grooved upper incisors, rootless ovoid molars, equal p4 lophid widths, and parallel anterior and posterior faces on the p4 re-entrants.

MICE

Taxonomy of fossil *Peromyscus* is difficult because of the degree of inter- and intraspecific dental variation. Hooper (1957) and Guilday and Handley (1967) attempted to identify molars of *Peromyscus* by the degree of development of both accessory styles (-ids) and lophs (-ids). The relative abundance of these structures is fairly constant within each species but up to seven species (Guilday and Handley, 1967) share identical values. The mean percentages between other species differ by less than 5%, this is less than the intraspecific variance. For this reason *Peromyscus*, a common component of most late Pleistocene local faunas, frequently is listed as *Peromyscus* sp. Two common and widespread modern species, *P. leucopus* (white-footed mouse) and *P.*

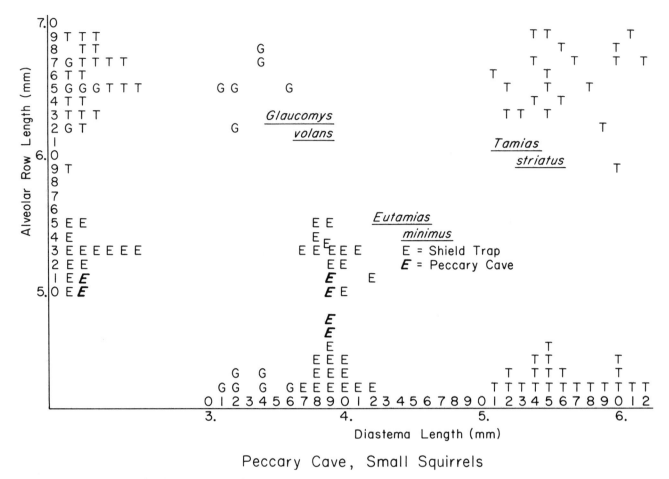

Fig. 2.—Bivariate plot of alveolar row versus diastema length of fossil *Glaucomys*, *Tamias*, and *Eutamias* from Peccary Cave. The control sample for *Eutamias* was taken from Shield Trap (Geppert, 1984).

maniculatus (deer mouse) have reasonably distinct ml's. Fifty percent of the former and 22% of the latter exhibit mesostylids and mesolophids. Guilday and Handley (1967) also observed that in young specimens, *P. leucopus* typically exhibited a well developed, symmetrical anteroconid with a deep anterior re-entrant and that the *P. maniculatus* anteroconid was structureless and characterized by a reduced buccal surface giving it an asymmetrical configuration. Peccary Cave ml's with wholly consistent characters were assigned to species. Those with either excessive wear or mixed character combinations were assigned to *Peromyscus* sp. Both the *P. leucopus* and *P. maniculatus* groups may be arbitrary. Five species of *Peromyscus* presently inhabit northwestern Arkansas (*P. leucopus, P. maniculatus, P. boylii* and *P. [Ochrotomys] nuttalli).* Any or all might be present in the fossil *Peromyscus* sample but none of the distinctive *Ochrotomys* mo-

lars were found. Neither could any specimens be attributed to *Reithrodontomys*.

The *Onychomys leucogaster* (northern grasshopper mouse) maxillary was identified by a long alveolar row (4.6 mm) and the sharp (90°) inflection between the anterior and dorsal borders of the zygomatic plate. Teeth of the rice rat (*Oryzomys palustris*) were separated from other Peccary Cave cricetids by a first upper and lower molar length greater than 1.8 mm, a rudimentary accessory root on both first molars, an anterocone (id) width nearly equal to the mean tooth width, and low cusps aligned opposite to (rather than diagonal to) each other. The posterior alveolus of the first upper molar of *Oryzomys* also is distinct because it is round, half the width of the tooth row, and offset to the labial side of the maxillary (Satorius-Fox, 1982).

The upper and lower molars of *Neotoma* (woodrat) are readily distinguished from those of other cricetids by their rooted, cementless, arvicoline-like form with four elements. The Peccary Cave *Neotoma* ml's typically exhibited the flattened lingual triangle of the anterior loop that separates *N. floridana* (eastern woodrat) from other species of *Neotoma* (Hibbard and Taylor, 1960; Patton, 1963; Semken, 1966). The Peccary Cave ml's are larger than *N. f. osagensis,* the subspecies presently living in northwestern Arkansas, and have accessory cusps in the re-entrants of m2. These cusps are lacking in the 40 examined *N. f. osagensis* but are found on both the first and second molars of *N. ozarkensis* (Brown, 1908), an Irvingtonian woodrat from the nearby Conard Fissure. The large size of the Peccary Cave *Neotoma* may relate to "giantism" characteristic of many late Wisconsinan species.

VOLES

The muskrat (*Ondatra zibethicus*) was identified by a ml length in excess of 7 mm, rooted molars with reticulate cement and high dentine tracts. Both the red-backed vole (*Clethrionomys gapperi*) and the heather vole (*Phenacomys intermedius*) also have rooted molars but the ml's are less than 4 mm in length. *Phenacomys* was separated from *Clethrionomys* by the absence of cement, tiny labial triangles, and closed first and second triangles in the former. The *Clethrionomys* ml has cement, equal labial and lingual triangles, and a confluent first and second triangle.

The "rootless" arvicolines also are common in Peccary Cave. The bog lemming (*Synaptomys*) lower molars, with lingual re-entrants extending across the tooth, could readily be divided into those with labial triangles (*S. cooperi*) and those without (*S. borealis*). This distinction can not be applied to upper molars because combinations of these characters are present in both species. *S. cooperi* presently exhibits a negative Bergman's response and Guilday et al. (1978) record the clinal nature of Pleistocene samples, predicted by Hibbard (1963), in the Appalachians. They also note that there is a paucity of sites between the largest Pleistocene specimens in Tennessee and a larger morph, described as *S. australis,* in Florida and suggested that intermediate forms might be discovered. *S. australis*-sized specimens have been collected with *S. cooperi*-sized fossils only in the Ladds assemblage, Georgia (Ray, 1967; Kurtén and Anderson, 1980), a site where the fauna may be mixed. The 16 Peccary Cave *Synaptomys* ml's (mean = 2.73, OR 2.5–2.9 mm) are

smaller than those of *S. australis* (mean = 3.5, OR 3.3–3.9 mm) but are slightly larger than the largest known eastern Pleistocene *S. cooperi* sample from Robinson Cave (mean = 2.65, OR 2.3–2.9), Tennessee (Guilday et al., 1978, Table 12). Thus, the intermediate-sized Peccary Cave sample along with those of the 14,800 year B.P. Waubonsie local fauna, Iowa (mean = 3.0, OR 2.6–3.3 mm), recorded by Rhodes (in press) demonstrate that *S. australis* is not as distinct in size as previously thought and that the cline extends to the west as well as to the south (Rhodes, in press).

The distinction between *Microtus (Pitymys) pinetorum* (woodland vole) and *M. (Pedomys) ochrogaster* (prairie vole) is difficult using the dental battery and the two frequently are not differentiated (Guilday et al., 1978). There are, however, at least four proposed techniques for separating these taxa. Patton (1963) devised a method utilizing the m3; specimens with nearly closed second and third triangles are *Pedomys,* open ones are *Pitymys.* This configuration appears diagnostic most (70%) of the time, but it is not generally applicable to the Peccary Cave sample because of the large numbers of isolated ml's from which the minimum number of individuals was obtained. The technique does confirm the diagnosis for mandibles with complete tooth rows. Johnson (1972), working with both the Glenwood local fauna, Iowa, and a recent sample, devised a technique using the width of the "isthmus" separating the fifth triangle of the ml from the anterior loop; widths greater than 0.2 mm being *Pedomys* and those smaller, *Pitymys.* Martin and Webb (1974) noted that in *Pitymys* the sixth re-entrant angle of the ml is deep and that the anterior border of the fourth triangle slopes posteriorly. In *Pedomys* the re-entrant is shallow and the anterior border of the fourth triangle is at a right angle to the axis of the tooth. Van der Meulen (1978) pointed out that only the medial part of the wide, shallow re-entrants characteristic of *Pedomys* are directed anteriorly (morphotype 1) and that the entire narrow, deep re-entrants of *Pitymys* are directed anteriorly (morphotype 4). He also observed that the enamel is the same thickness on both sides of the triangle in *Pitymys* (morphotype 4) but thin on the posterior border of *Pedomys* triangles (morphotype 1). All of these techniques have merit and some are related, for example, the deep re-entrants of *Pitymys* noted by both Martin and Webb (1974) and Van der Meulen (1978) will produce the narrow "isthmus" of Johnson (1972). All combinations of these characters have been observed in recent skulls of *Pitymys* from North Carolina and *Pedomys* from Kansas and are gradational between the two taxa; however, these distinctions are reasonable generalizations. The Peccary Cave sample was divided using the technique of Johnson (1972) because all specimens could be objectively assigned, knowing that some, especially those approaching 0.2 mm in width, are incorrectly diagnosed. Generally, the distinctive features noted by the other three methods were consistent with the assignment.

Microtus xanthognathus (yellow-cheeked vole) initially was recorded from Peccary Cave by Hallberg et al. (1974). Their sample, based on five M3's over 2.6 mm in length with cementless third labial re-entrants (Guilday and Bender, 1960), has since been augmented by two more M3's and three left and two right ml's from Trench 15. All of the latter exceeded 3.55 mm in length and clearly are separated from Peccary Cave specimens of *M. pennsylvanicus,* which do not exceed 3.2 mm. *M. xanthognathus* counts (Table 1) are based on M3's.

The Peccary Cave meadow vole (*Microtus pennsylvanicus*) sample includes many ml's with five or more closed triangles. This triangle count is characteristic of a variety of species of *Microtus* (Guilday, 1982), but distinctive M2's and M3's of *M. pennsylvanicus* are present in proportion to the ml's. At one time, any *Microtus* ml collected in central North America with five or more triangles was assigned to *M. pennsylvanicus.* Davis (1975) first challenged this assignment in the late Pleistocene Jones local fauna, Meade Co., Kansas. He counted 67 ml's with five closed triangles (*M. pennsylvanicus* type) and six ml's with three closed triangles (*M. ochrogaster* type). However, only 12 of 44 M2's were of the distinctive *M. pennsylvanicus* five element type. The remainder were of the four element *M. ochrogaster* type. Thus, the majority of four element M2's, Davis reasoned, were associated with five element ml's and had to belong to a *Microtus* other than *M. pennsylvanicus.* This animal was dubbed the "phantom" microtine because no specimen could definitely be assigned to it. Subsequently, Stewart (1978) identified the skull of *M. montanus,* which has both a five element ml and a four element M2, from the late Pleistocene Trapshoot local fauna in central Kansas. Guilday (1982) noted that the ml of the meadow vole in the eastern United States also is inseparable from that of the yellow-nosed vole (*M. chrotorrhinus*) which also has a four element

M2. The yellow-nosed vole does have distinctive M3's, with two lingual cementum-filled re-entrants on the posterior loop. Usually the meadow vole only has one.

The number of five versus four element *Microtus* M2's in Peccary Cave are in proportion to the number of ml's with five (*M. pennsylvanicus* type) and three (*M. ochrogaster* type) closed triangles. Only one *M. chrotorrhinus* type M3 is present (T15) in Peccary Cave. This pattern is recorded by Guilday (1982) in up to 5% of any modern *M. pennsylvanicus* sample and therefore is an expected variant in a large fossil *M. pennsylvanicus* sample. Few, if any, of the ml's assigned to *M. pennsylvanicus* could belong to either *M. montanus* or *M. chrotorrhinus.* It is conceded that either *M. montanus* or *M. chrotorrhinus* may be a rare species in the Peccary Cave fauna, but *M. pennsylvanicus* must account for most of the five closed triangle ml's.

Five *Microtus* ml's with a dental pattern characteristic of the root vole (*M. oeconomus*) are known from Peccary Cave. Each is characterized by four closed triangles but the anterior loops of two are more complicated than observed in two recent *M. oeconomus* ml's. Simple loops, present in the other three specimens, also are found in the Peccary Cave *M. pennsylvanicus* sample. Moreover, Semken (1966) examined 398 recent *M. pennsylvanicus* ml's and found either four closed triangles or aberrant dentitions in 1% of the specimens. Thus, five of 379 *M. pennsylvanicus* specimens (1.3%) with an *M. oeconomus* pattern are expected in a meadow vole sample of this size. The tundra dwelling root vole has not been identified in other late Wisconsinan local faunas from temperate North America. However, its presence south of Canada in Pleistocene deposits can not be dismissed because the tundra-dwelling collared lemming (*Dicrostonyx*) has been identified in 17 sites in the contiguous United States. Woodman (1982) also recorded both *Spermophilus parryi* (arctic ground squirrel) and *M.* cf. *miurus* (singing vole), both tundra species, as well as *Dicrostonyx* in the 20,500 year B.P. Elkader local fauna of northeastern Iowa. Thus, tundra components definitely were present at low latitudes and *M. oeconomus* might be expected in Wisconsinan deposits. As the Peccary Cave four closed triangle ml sample could represent aberrant individuals, they are cataloged as *M. pennsylvanicus, oeconomus* pattern. All are from Trench 15, levels 1 and 2, where the other *Microtus* molars are most abundant.

PALEOECOLOGY

MOLES
Biogeography

Scalopus aquaticus (eastern mole) presently is found in the Peccary Cave region, but *Condylura cristata* (star-nosed mole) ranges no closer to Peccary Cave than 550 mi (880 km). Where *Scalopus* and *Condylura* are sympatric, the star-nosed mole occupies low, wet ground near open water and the eastern mole is found in moist but well-drained sandy loam (Banfield, 1974).

Biostratigraphy

The eastern and star-nosed moles comprise 1.8% of the Peccary Cave small mammals (MNI). This relative abundance is similar to that noted in both Baker Bluff Cave, 1.6%, and Clark's Cave, 0.9%, (Guilday et al., 1977, 1978). In the Wisconsinan Appalachian sites, for example, Baker Bluff Cave, no one species exceeds 50% of the talpid sample. Conversely, Holocene sites, for example, Sheep Rock Shelter (Guilday and Parmalee, 1965) are dominated (>85%) by one species. This ratio also is reflected in recent owl pellet accumulations (Guilday et al., 1977). Both *Scalopus* and *Condylura* are common in the two lowest levels, T15-1 & 2 (Table 1). Only one *Condylura* specimen is recorded in the upper five levels (T8/13). The mucky ground necessary for the star-nosed mole, which primarily feeds on aquatic insects, apparently was severely reduced above the two lowest levels in response to the transition to post-glacial climate. If mole distribution is an index, only the two lower levels would be Pleistocene. A similar stratigraphic dicotomy is found in New Paris #4 (Guilday et al., 1964) where *Condylura* is found below the 5.5-m level and *Parascalops breweri* (hairy-tailed mole) is found above.

SHREWS
Biogeography

Blarina brevicauda (northern short-tailed shrew) today has the northernmost distribution of the three *Blarina* identified in Peccary Cave. It ranges into both the eastern deciduous and mixed conifer forests and extends west into the gallery forests of the Great Plains. Forest cover of some nature in necessary for its survival. *B. carolinensis* (southern short-tailed shrew) is an Austroriparian form that is found in the southeastern portion of the United States. Either *B. hylophaga* (western short-tailed

shrew [west]) or *B. kirtlandi* (Kirtland's short-tailed shrew [east]) separate the two. Thus, the intermediate-sized forms lie in a "temperate" zone between the cooler adapted *B. brevicauda* and warmer adapted *B. carolinensis*. Long-tailed shrews (*Sorex*) prefer mesic environments but most also occupy dry, open associations (Banfield, 1974). *Sorex palustris* (water shrew), which now lives at least 700 mi (1,120 km) north of the cave, is an exception and it requires an aquatic environment. The immigration of *S. arcticus* (arctic shrew) into northwestern Arkansas represents a southern dispersal of 600 mi (960 km), that of *S. hoyi* (pygmy shrew) is 510 mi (820 km), and *S. cinereus* (masked shrew) is 350 mi (560 km) south of its modern range. Peccary Cave contains the southwestern most known fossils for both *S. hoyi* and *S. arcticus*. *S. cinereus* is common in the Pleistocene of Texas and fossils of *S. palustris* have been recovered from a deposit in northcentral Texas. *Cryptotis parva* (least shrew), presently a resident of northwestern Arkansas, has a center of distribution in the southeastern United States where it prefers open grass-covered areas, marshes, and glades (Banfield, 1974). It is commonly associated with boreal species in the Pleistocene of the eastern United States and Guilday et al. (1978:27) noted that the presence of this temperate form seems anomalous to the boreal context indicated by its associates. However, the northern limit of *Cryptotis* is within 50 mi (80 km) of and may be narrowly sympatric with the southern limits of the boreal shrews in east-central Wisconsin. It, along with *B. carolinensis,* which is dramatically disjunct from the area of sympatry (Fig. 3), does indicate that rainfall was sufficient for tree growth but insufficient to generate a closed forest.

Biostratigraphy

Although all of the soricids (7% of the local fauna) are associated on one or more of the seven tested Peccary Cave levels, they are not randomly distributed and clearly are most abundant in two lowest levels (Table 1; Fig. 4). This may be directly related to the higher diversity of prey if the comparison of the number of mammals (mean = 33) in the two lowest levels (T15-1 & 2) to that (mean = 21) in the five upper levels (T8/13) is an index. *Cryptotis* has been identified only in the lowest two levels of the cave (T15), *Sorex cinereus* is found in the lower

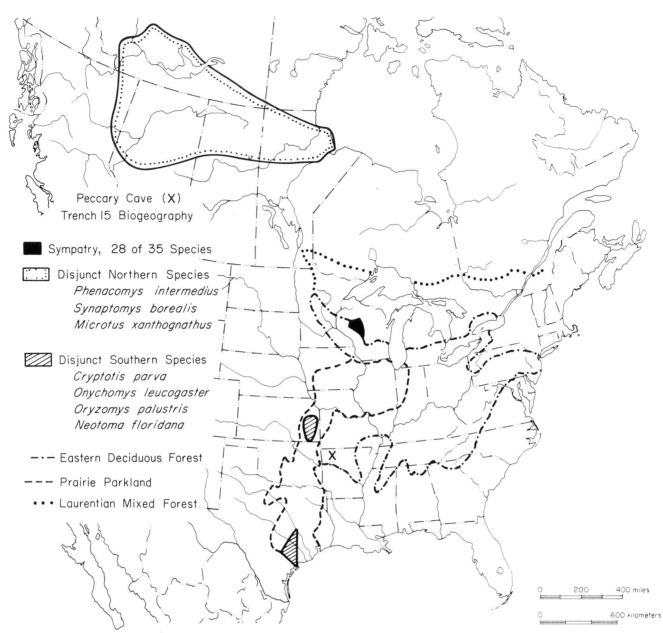

Fig. 3.—Area of sympatry for the Trench 15, level 1 or 2 faunules, Peccary Cave local fauna, Newton Co., Arkansas. Ecoregions after Bailey (1981).

four levels (T15, T8/13-4 & 5), and the other *Sorex* are sporadically represented but present throughout the section. *Blarina brevicauda* is most abundant at the bottom of the section but is dramatically reduced in numbers in the upper levels (Fig. 4). *B. hylophaga* is common in the lower levels, diminishes, and then becomes dominant in the middle portion of the sampled units. *B. carolinensis,* also most common in the lower portion of the section, is absent in T8/

13-5 and gradually increases upward into level two (T8/13-2). Its upper mode lies above that of the intermediate-sized *B. hylophaga.* This gradual replacement of one *Blarina* phenon by another through time is better exhibited and statistically confirmed in Table 2. *B. brevicauda* is found in substantially greater than expected numbers low in the section, *B. hylophaga* dominates in the middle but is under represented both earlier and later, and *B. caroli-*

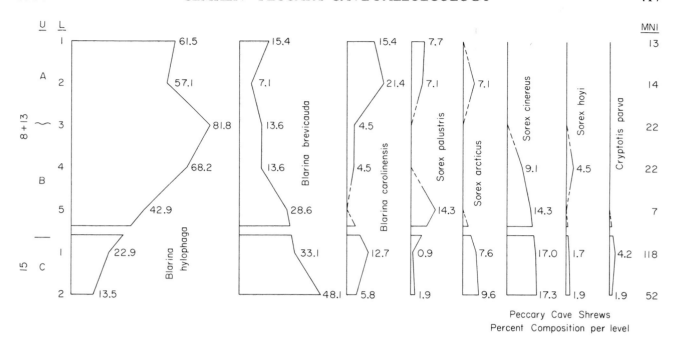

Fig. 4.—Relative abundance of soricids (%) in each level of Trenches 8 + 13 and 15, Peccary Cave local fauna, Newton Co., Arkansas.

nensis has greater than expected abundance both low and high in the column. The replacement pattern of one phenon of *Blarina* by another demonstrates both that the morphotypes are behaving as species and that the soricids are adjusting individualistically as proposed by King and Graham (1981) rather than as members of an "organismic" community.

The change in relative dominance of *Blarina* in Peccary Cave from large to small species through time suggests that temperatures generally increased during deposition of the examined portion of the section. The conditions could have been either relatively mesic if the intermediate-sized phenon is *kirtlandi* or, almost certainly, xeric if *hylophaga* is present. General summer warming also is implied by the upward reduction in all species of *Sorex*, each of which now lives only to the north of Arkansas. It may be significant that *S. palustris*, a semiaquatic, is absent in the portion of the column that is dominated by the xeric adapted *B. hylophaga*.

SQUIRRELS
Biogeography

The thirteen-lined ground squirrel (*Spermophilus tridecemlineatus*), now restricted to the prairies of central North America, has been recovered east of its present range in nine Wisconsinan sites in the Appalachian region (Semken, 1983, fig. 12-9). Suitable habitat also was present in the upper Mississippi/middle Missouri River region because it is in the Craigmile and Waubonsie local faunas of southwestern Iowa (Rhodes, 1982), Brayton local fauna of west-central Iowa (Dulian, 1975), Moscow Fissure of southwestern Wisconsin (Foley, in press), Crankshaft Cave in eastern Missouri (Parmalee et al., 1969), Bat Cave (Hawksley et al., 1973) and Brynjulfson Caves (Parmalee and Oesch, 1972) in central Missouri. It clearly was distributed widely during the Wisconsinan and indicates that open ground was more widespread in the eastern United States during late-glacial time than at present (Guilday et al., 1978:55). It also occurs in the Peccary Cave local fauna (Table 1, Fig. 5).

As in the Appalachian sites, the thirteen-lined ground squirrel in Peccary Cave is associated with a variety of forest squirrels including the fox squirrel (*Sciurus* cf. *niger*), woodchuck (*Marmota monax*), red squirrel (*Tamiasciurus hudsonicus*), southern flying squirrel (*Glaucomys volans*), eastern chipmunk (*Tamias striatus*), and least chipmunk (*Eutamias minimus*). Except for the absence of the

Table 2.—*Freeman-Tukey deviates for rodents and insectivores from Peccary Cave Trenches 8 + 13 and 15. The signs (+ or −) indicate the direction of the actual from the expected count. Any deviate over 0.5 is increasingly significant.*

		TRENCH 8 + 13						TRENCH 15	
TAXON		1	2	3	4	5		1	2
Scalopus aquaticus	D	+0.2	+0.3	+0.8	+1.8	+1.6		−0.8	−1.3
Condylura cristata	B	−0.2	+0.8	−0.2	−0.2	+0.1		−0.1	+0.4
Cryptotis parva	D	−0.2	−0.2	−0.2	−0.2	+0.1		+1.4	−0.3
Sorex cinereus	B	−1.8	−1.7	−1.7	+0.1	+0.2		+1.6	−0.2
Sorex palustris	B	+0.8	+0.8	−0.2	−0.2	+1.1		−0.8	−0.3
Sorex arcticus	B	−0.9	+0.2	−0.8	−0.8	−0.3		+1.0	+0.3
Sorex hoyi	B	−0.1	−0.1	−0.1	+1.1	+0.2		+0.4	+0.1
Blarina brevicauda	D	−1.6	−2.0	−0.8	−0.8	+0.1		+1.3	+0.6
Blarina hylophaga	D	+0.8	+1.0	+3.8	+3.0	+0.4		−1.4	−4.6
Blarina carolinensis	D	+0.3	+1.0	−0.3	−0.3	−0.5		+1.3	−1.8
Tamias striatus	D	+2.4	+1.2	+1.2	−1.2	−0.5		+0.1	−2.8
Eutamias minimus	B	+0.2	+0.2	+0.2	+0.2	+0.2		+0.4	+0.6
Marmota monax	D	+2.3	+1.4	+0.8	+0.8	+0.5		−2.3	−1.5
S. tridecemlineatus	S	−0.4	+0.6	−0.4	−0.4	−0.1		−0.4	+1.3
Sciurus cf. niger	D	+1.5	+2.0	+1.1	+1.6	+0.5		−2.0	−2.2
Tamiasciurus hudsonicus	B	−0.4	−0.3	−0.3	−0.3	−0.1		+0.5	+1.1
Glaucomys volans	D	+1.3	+0.7	−0.3	+0.7	−0.1		−0.6	−0.1
Geomys bursarius	S	−1.2	+0.2	−0.6	+0.2	+0.4		+1.1	−0.6
Castor canadensis		+1.3	+0.2	+0.2	+0.2	+0.2		+0.4	−0.4
Erethizon dorsatum	B	+1.3	+0.7	−0.3	−0.3	+1.1		−0.6	−0.1
Peromyscus cf. leucopus	D	+1.6	+2.4	+2.4	+0.3	+0.8		−2.4	−1.9
Peromyscus cf. maniculatus		+1.1	+0.2	+1.7	+1.7	+1.6		−1.1	−2.4
Peromyscus sp.		+4.8	+3.6	+3.2	+0.8	+0.5		−0.4	−9.8
Onychomys leucogaster	S	+0.2	+1.3	+0.2	+0.2	+0.2		+0.2	+0.2
Neotoma floridana	D	+0.4	+1.0	−0.4	−1.1	+1.0		+0.6	−0.9
Oryzomys palustris	D	+1.0	+0.1	−0.7	+1.1	+0.2		−0.5	−1.0
Ondatra zibethicus		−0.2	−0.2	+0.8	+0.8	−0.1		−0.1	−0.3
Microtus pinetorum	D	+1.1	−0.2	+0.7	−0.2	−1.7		+2.1	−3.1
Microtus ochrogaster	S	−1.5	+0.2	+0.1	+0.5	−0.3		+0.4	−0.1
Microtus pennsylvanicus	B	−6.7	−5.6	−5.3	−1.6	−1.3		+0.2	+6.0
Microtus xanthognathus	B	−0.1	+0.1	−0.1	+0.1	+0.2		−0.3	+1.3
Synaptomys cooperi	B	+0.1	+0.1	+0.8	+2.8	+1.8		−1.8	−1.5
Synaptomys borealis	B	+0.6	−0.4	+0.6	−0.4	+1.0		+0.1	+0.3
Phenacomys intermedius	B	−0.9	−0.8	−0.8	−0.8	−0.3		−0.5	+2.4
Clethrionomys gapperi	B	−4.3	−5.7	−4.1	−3.0	−1.8		−1.2	+6.2
Zapus hudsonius	B	+0.8	+0.8	−0.2	−0.2	−0.1		−0.8	+0.4

$$F-TD = \sqrt{4 \cdot count + 2} - \sqrt{4 \cdot expected\ count + 1}$$

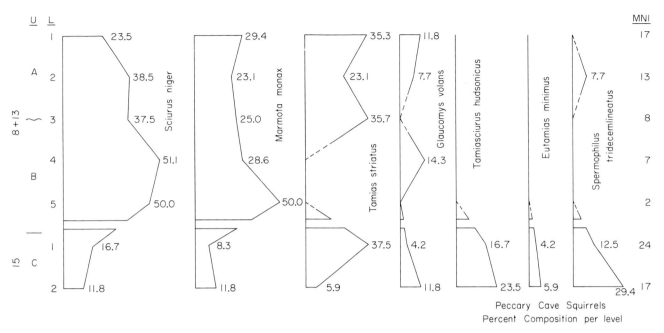

Fig. 5.—Relative abundance of sciurids (%) in each level of Trenches 8 + 13 and 15, Peccary Cave local fauna, Newton Co., Arkansas.

northern flying squirrel (*Glaucomys sabrinus*), the species composition of the Peccary Cave sciurids (Table 1) is identical to that of the Appalachian local faunas (Guilday et al., 1978).

Peccary Cave records of both the least chipmunk (*Eutamias minimus*) and the red squirrel (*Tamiasciurus hudsonicus*) extend the Wisconsinan range of these animals into northwestern Arkansas, over 550 mi (880 km) south of their present southern nonmontane limit in Wisconsin. *Eutamias* occupies a variety of habitats but boreal parkland is preferred and closed forests are avoided. *Tamiasciurus* also is characteristic of boreal parkland but it will occupy both closed conifer and mixed conifer/hardwood forests. The other squirrels in Peccary Cave are denizens of the eastern deciduous forest but they do range into the southern mixed conifer/hardwood forest. All of the Peccary Cave sciurids coexist today in the central Wisconsin area of sympatry (Fig. 3).

Biostratigraphy

Overall, squirrels comprise 4.5% of the Peccary Cave local fauna, a misleading figure because there is a substantial change in relative abundance up-section (Fig. 5). Squirrels comprised 2.8% of the local fauna in the lowest level (T15-2) and progressively increased through 3.0, 3.5, 5.6, 6.3, 10.6, to 12.5% in the uppermost level (T8/13-1). Excluding *Spermophilus*, the percentages gradually increase from 1.9% in T15-2 to 12.5% in T8/13-1. This indicates increasing forest density during deposition. The significance of this progression is shown by the Freeman-Tukey deviates (Table 2). The eastern chipmunk (*Tamias striatus*), fox squirrel (*Sciurus* cf. *niger*), woodchuck (*Marmota monax*), and southern flying squirrel (*Glaucomys volans*), all with centers of distribution in the eastern deciduous forest are under represented at the bottom of the section (T15-2) and over represented at the top (T8/13-1). Conversely, the thirteen-lined ground squirrel (*Spermophilus tridecemlineatus*), a prairie ecotype, is present in much greater than expected numbers in the lowest level (T15-2) and is under represented above. The only forest squirrels that are present in greater than expected numbers low in the section are the red squirrel *Tamiasciurus hudsonicus*) and least chipmunk (*Eutamias minimus*), both of which center in the boreal forest. Thus, the sciurids suggest that a boreal parkland, with a deciduous forest component, was in the vicinity of Peccary Cave during deposition of T15-2. The steppe aspect of this diminished by T15-1 and conifers became much less important by the beginning of T8/13-5 time. Mature deciduous forest elements were progressively more important during each depositional phase and resembled modern climax forests at the top sampled interval. The reduction in relative abundance (Fig. 5) of the fox squirrel, which prefers open forest, in level T8/13-1 probably reflects closure of the deciduous forest.

GOPHERS
Biogeography

The plains pocket gopher (*Geomys bursarius*) presently is found in sandy soils as close as 40 mi (64 km) to the southwest of Peccary Cave. It is excluded from the Ozark Highlands because of both shallow soil and dense forest growth. The gopher, which feeds on the roots, stems and tubers overlying its burrow, is restricted to regions sufficiently open for lush surface growth. It is widely distributed in prairies but also is common in parkland situations. The discovery of *Geomys* as far east as Welsh Cave, central Kentucky (Guilday et al., 1971), and Cheek Bend Cave in central Tennessee (Parmalee and Klippel, 1981*a*), where a closed canopy forest now exists, indicates that open ground extended much further east during late Wisconsinan time than it does at present. The presence of the closely related but "isolated" population of *Geomys pinetis* (southeastern pocket gopher) in southern Georgia, Alabama, and northern Florida prompted Guilday et al. (1971) to suggest that the pocket gopher and hence open ground was ubiquitous in the southeastern United States during the late Pleistocene. The occurrence of the thirteen-lined ground squirrel (*Spermophilus tridecemlineatus*) in late glacial deposits of Pennsylvania, Virginia, Kentucky, Tennessee, and Florida (Semken, 1983) supports this conclusion. This eastern and southern range extension of *Geomys* suggests that less precipitation than that of today characterized the late Pleistocene of the southeastern and south-central United States. This does not imply either aridity or extensive prairies. The number of tree squirrels and boreal species from each of these local faunas preclude this possibility. Precipitation, while reduced, must have been more effective. A widely distributed parkland is compatible with each local fauna. The Peccary Cave local fauna differs from those in Iowa and Wisconsin in that *Thomomys talpoides* (northern plains pocket gopher), which also extended its range well to the east during the Wisconsinan (Rhodes, in press; Foley, in press; Rosenberg, 1983) is not present.

Biostratigraphy

Geomys bursarius is the fourth most common species (6% of the local fauna) and is present in all levels in Peccary Cave (Table 1). Initially it is under represented in the section (T15-2) but the situation is reversed in T15-1 (Table 2). It is present in near expected numbers in all of the upper levels but is rare in T8/13-3 and especially T8/13-1. Its low abundance in the upper levels is attributed to forest closure. This is supported by increased numbers of forest species, especially sciurids, in this part of the section. However, the predominance of grassland species in T15-2 precludes its scarcity in this level being attributed to a closed forest. Level T15-2 also is the most "boreal" of all and probably is best described in terms of a cool steppe, a condition not attractive to *Geomys*.

MICE
Biogeography

The new world rats and mice, represented by four genera and 584 individuals (30% of the total fauna), comprise the second largest family identified in Peccary Cave. Collectively, the white-footed mice (*Peromyscus*) and woodrats (*Neotoma*) accounted for 580 of these individuals. The rice rat (*Oryzomys*) and grasshopper mouse (*Onychomys*), known from three and one individuals respectively, comprise the remainder.

When *Peromyscus leucopus* (white-footed mouse) and *P. maniculatus* (deer mouse) are sympatric, the former occupies woodlands and the latter more open brushy areas. The deer mouse will reside in any dryland habitat while the white-footed mouse will select more mesic situations. Both feed on seeds, nuts, and insects and indicate deciduous vegetation. The northern grasshopper mouse (*Onychomys leucogaster*), relatively rare in its modern range, is a carnivore that is restricted to the Great Plains. Where short-tailed shrews are absent, grasshopper mice become the major small carnivore. The single specimen of *Onychomys*, which occurs within 110 mi (176 km) of Peccary Cave, indicates a more open environment in Arkansas at that time.

Neotoma floridana (eastern woodrat) is the second most common mammal, after the meadow vole (*Microtus pennsylvanicus*) in Peccary Cave. The eastern woodrat presently is common in the Ozarks where bluffs, caves, and fissures, their preferred habitat, are well developed. They also frequent wooded terrain and build nests against fallen logs, in hollow trees, or in burrows (Sealander, 1979). At the present time the eastern woodrat lives no further north than central Missouri in the Mississippi Valley but fossils have been recovered from post-altithermal cave fills (Eshelman, 1971) and deposits near bluffs (Fay,

1980) in Iowa. These "outliers" of *Neotoma* have disappeared and suggest that factors other than deglaciation are responsible for its present distribution. Western species of the genus range north to the Yukon and the eastern woodrat is found as far north as both South Dakota and New York. Therefore, the restricted northern limit of this southeastern form in the Mississippi valley at present probably is related to something other than cold winters.

Meadow jumping mice (*Zapus hudsonius*) now range approximately 100 mi (160 km) to the north of Peccary Cave. These mice, which feed on seeds, insects, and fruits, prefer meadow and edge zone habitats for nests and feeding.

The rice rat (*Oryzomys palustris*) also is rare in Peccary Cave, but it occurs sporadically throughout the column (Table 1). This suggests that this semi-aquatic, deciduous forest biome inhabitant previously, as now, was only locally present in northwestern Arkansas (Sealander, 1979). It indicates the presence of wet meadows with dense vegetation bordering lentic environments at the time of deposition.

Biostratigraphy

The relative abundance of *Peromyscus* changes substantially in the section (Table 2). Collectively, *Peromyscus,* a browser, is strongly under represented in the lowest levels (T15-1 & 2) and over represented in the upper levels (T8/13-all) of the cave. *Peromyscus,* along with the sciurids (Fig. 5, Table 2), becomes dominant upsection at the expense of the arvicolines. Both *P.* cf. *leucopus* (white-footed mouse), which prefers dry deciduous woodlands and *P.* cf. *maniculatus* (deer mouse), which characteristically inhabits more open brushy areas are rare in T15. *P.* cf. *maniculatus* then becomes abundant in the lower levels of T8/13. While it is still common in the upper levels of T8/13, *P.* cf. *leucopus* replaces it as the most common of the two. It is difficult to interpret the pattern precisely because both of these mice occupy a variety of habitats, but a transition from open terrain in T15-2 through a scrub forest (T8/13-5, 4, & 3) to a dry, closed deciduous forest (T8/13-1 & 2) is apparent. While the overall complexion of the T8/13-2 faunule clearly is that of a forest association, it is significant that the grasshopper mouse (*Onychomys leucogaster*), thirteen-lined ground squirrel (*Spermophilus tridecemlineatus*), and prairie pocket gopher (*Geomys bursarius*), all prairie ecotypes, exhibit increases in abundance in this level. This may be related to altithermal conditions recorded elsewhere in the Ozarks (Semken, 1983).

Neotoma reflects at least 10.5% of the fauna in any level and may represent almost 20% of the individuals (T8/13-5). Its actual, with respect to expected, numbers (Table 2) fluctuate greatly and show no apparent pattern. This suggests that local rather than regional influences primarily affected the population. The presence of the rice rat (*Oryzomys palustris*) in Peccary Cave is compatible with the meadow habitat suggested by the numbers of voles but its southeastern distribution is incongruous with the boreal center of distribution of most of the associated voles and shrews. The occurrence of the rice rat demonstrates, along with *Blarina carolinensis, B. hylophaga,* and *Neotoma floridana,* that temperate species did not vacate Arkansas when the boreal fauna immigrated. These four species have two areas of sympatry (Fig. 3) on the prairie-forest border of the southern plains. Both are strictly disharmonious with the common sympatry for the squirrels, voles, and insectivores, less *Blarina,* (Fig. 3) but substantiate a parkland environment for the late Wisconsinan around the cave. Both *Peromyscus leucopus* and *P. maniculatus* are found in the cricetid sympatries. *Reithrodontomys* (harvest mouse), a common component of Holocene local faunas, is not identified in the Peccary Cave sample. It appears to have been rare during the Pleistocene east of the Great Plains.

VOLES

Biogeography

The most abundant mammal in the Peccary Cave local fauna (Table 1) is the meadow vole (*Microtus pennsylvanicus*). It is recorded as both more numerous and widespread in eastern and central North America (Zakrzewski, in press) during the Pleistocene when it ranged south to central Texas (Lundelius, 1967), Louisiana, and Florida (Martin, 1968). This animal, which avoids closed forests (Banfield, 1974), indicates that expanses of suitable grass cover (for example, meadows, mesic prairies, or forest edge situations) were more widespread than at present. Martin (1968) predicts that both 5°F cooler July temperatures and increased moisture toward the southwest were necessary for the meadow vole to colonize the Gulf coastal plain. The meadow vole digs shallow burrows in the summer but resides in spherical grass nests constructed under snow in winter. It is active all winter, as are all of the Peccary Cave voles, and consumes cached food or green grass bases growing under the snow (Banfield, 1974). Its modern southern distribution roughly parallels isograds on snow cover maps (Visher, 1954) and it may reflect increased Wisconsinan snow cover in Arkansas.

Hall (1981) mapped the northern bog lemming (*Synaptomys borealis*), which lives in the boreal forest, and the southern bog lemming (*S. cooperi*), a mixed forest form, as being nearly allopatric today. Although they now occur together only in portions of southeastern Manitoba, eastern Quebec, and northern New Brunswick (Banfield, 1974), the two were widely sympatric and co-occurred as far south as Peccary Cave, Arkansas, and Baker Bluff Cave, Tennessee, during the Pleistocene. *S. cooperi*-like forms (*S. australis*) ranged south into both Florida and Texas (Lundelius, 1967) during late Wisconsinan time. During late glacial time in the northeastern United States, the northern bog lemming was both common and more abundant than the southern bog lemming (Guilday et al., 1978). This ratio is reversed to the south in both Peccary and Baker Bluff caves and *S. borealis* was not present further south in the Cheek Bend local fauna, Tennessee (Klippel and Parmalee, 1982*b*), or in any central Texas locality (Lundelius, 1967). Both species ranged hundreds of kilometers further south during the Wisconsinan but neither abandoned their present unglaciated northern range. The isolating mechanisms geographically separating the two at present were either reorganized or different during the Pleistocene. Both animals, as the name implies, were associated with bogs and meadows of the Arkansas Pleistocene.

The heather vole (*Phenacomys intermedius*), now restricted to the boreal forest of Canada in eastern North America, was nearly ubiquitous east of the Rocky Mountains during the late Wisconsinan. Guilday and Parmalee (1972) recorded 16 extralimital Pleistocene localities with *Phenacomys,* one of which (Peccary Cave) was approximately 800 mi (1,280 km) south of its present limits. Since then specimens have been recovered in Cheek Bend Cave, Tennessee, which also is 800 mi to the south, by Parmalee and Klippel (1982*b*) and 600 mi (960 km) south of its modern range in northcentral Nebraska (Voorhies, 1981). This distribution was generally harmonious with the Pleistocene range of the northern bog lemming. Dry, open forest with a coniferous component and an understory of heaths (dwarf birch, willow, and blueberry) is the prime habitat of the heather vole. This environment must have been more widespread during late glacial time in order for this browsing arvicoline to colonize now temperate regions of the eastern United States.

The red-backed vole, *Clethrionomys gapperi,* rep-

resented by 196 individuals, is the third most common species in Peccary Cave (10% of the entire fauna, 23% of all voles), as it is in most other late Wisconsinan local faunas. *Clethrionomys* comprises 17.6% of all voles in Baker Bluff Cave, 10% in Carrier Quarry Cave, and 14% in Robinson Cave (Guilday et al., 1978), all of which arc in eastern Tennessee at approximately the same latitude as Peccary Cave. Klippel and Parmalee (1982*b*) record the red-backed vole as representing 12% of all voles in Pleistocene portions of the Cheek Bend local fauna. It also is in the Pleistocene of Texas (Lundelius, 1967) but is not recorded by Webb (1974) in the Pleistocene of Florida. *Clethrionomys,* which now is restricted to the uppermost reaches of the Mississippi River in central North America, is a common vole in the mixed conifer/deciduous forest, especially where the floor is littered with logs and stumps. It also frequents areas with a brushy understory in forest edge situations, and aspen bluff/shrubby coulees on the prairie (Banfield, 1974). It feeds primarily on the petioles of broad leafed forbs and shrubs in summer and on heaths which remain green under the snow in winter. Unlike *Phenacomys, Clethrionomys* will graze to supplement its diet.

Microtus xanthognathus, the yellow-cheeked vole, presently is restricted to the northern boreal forest and bordering tundra west of Hudson Bay. It primarily is sylvan but does construct runways from its burrow to sphagnum bogs (Banfield, 1974). This taxon, which feeds on horsetails and lichens, is widely distributed in the late Wisconsinan of the United States. It is known from Peccary Cave, Arkansas (Hallberg et al., 1974), and Baker Bluff Cave, Tennessee (Guilday et al., 1978). Both sites are at least 1,500 mi (2,400 km) south of its modern range. It is common in sites north of these two localities in both the upper Mississippi Valley (Rhodes, in press; Foley, in press; Woodman, 1982; Rosenberg, 1983) as well as three Appalachian sites reviewed by Guilday et al. (1978). The presence of this vole in Arkansas reinforces the interpretation of boreal forest in the late Pleistocene of northwestern Arkansas.

The prairie vole, *Microtus ochrogaster,* does not inhabit the immediate vicinity of Peccary Cave today but it has been collected 50 mi (80 km) to the northwest. *M. ochrogaster* is restricted to dry grasslands where there is both sufficient grass to cover its runways and soil for burrows (Banfield, 1974). It will locate in shrubby terrain but never enters wooded regions. Arkansas populations of the prairie vole presently are restricted to prairie enclaves, for example, Grand Prairie, but it is predicted to have

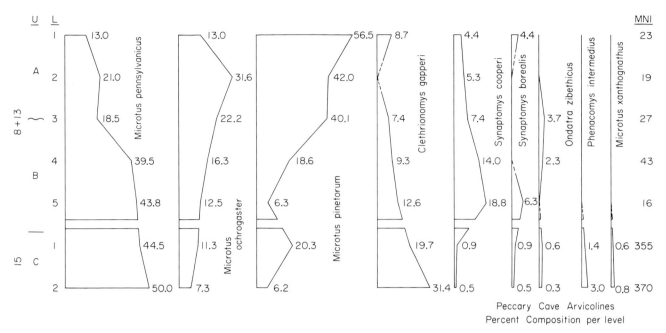

Fig. 6.—Relative abundance of arvicolines (%) in each level of Trenches 8 + 13 and 15, Peccary Cave local fauna, Newton Co., Arkansas.

been more widely distributed in Arkansas in the past (Sealander, 1979). The prairie vole's declining abundance upsection in Peccary Cave (Table 2; Fig. 6) and its absence in the vicinity today suggest that Sealander (1979) is correct.

The woodland vole, *Microtus pinetorum,* is the only arvicoline found in the Peccary Cave local fauna that still inhabits the region. As the name implies, these voles live in woodlands, especially dry, mature deciduous forests, where they construct burrows in litter on the forest floor. They feed (from the underside) on bulbs and tubers as well as green leaves, nuts, and fruits which fall to the ground. These semi-fossorial forms rarely enter fields and never do so when prairie voles are present. They are found throughout the Peccary Cave section, represent approximately 16% of all voles, and clearly were not forced out of the region by late Wisconsinan climates. Their continued presence demonstrates that dry deciduous woodland survived in the area when the boreal elements immigrated into northwestern Arkansas during the Wisconsinan. The woodland vole also is common in all of the known late Pleistocene faunas from the Appalachians.

Biostratigraphy

Arvicoline rodents, represented by nine species and 852 individuals, comprise 44% of the local fauna and are the most abundant animals in Peccary Cave. This relative abundance, which is in concordance with that at New Paris #4 (41%), Clark's Cave (47%) and Baker Bluff Cave (47%) in the Appalachians (Guilday et al., 1978), but greater than that at Brynjulfson #1 (28%) in the Ozarks (Parmalee and Oesch, 1972). Almost half of the Peccary arvicolines, 379 individuals, are meadow voles (*Microtus pennsylvanicus*) which now range south only to northern Missouri, approximately 300 mi (480 km) to the north. It is followed, in decreasing order of abundance (Table 1, Fig. 6), by the red-backed vole (*Clethrionomys gapperi*), woodland vole (*M. pinetorum*), prairie vole (*M. ochrogaster*), southern bog lemming (*Synaptomys cooperi*), heather vole (*Phenacomys intermedius*), northern bog lemming (*S. borealis*), muskrat (*Ondatra zibethicus*), and yellow-cheeked vole (*M. xanthognathus*).

The meadow vole (*Microtus pennsylvanicus*) comprises 50% of the T15-2 voles (Fig. 6). It progressively declines to 13% in relative abundance in the uppermost level (T8/13-1). The red-backed vole, *Clethrionomys gapperi,* initially representing 31% of the voles, exhibits a corresponding decline upsection. Collectively, the patterns (Table 2; Fig. 6) of these two species indicate a progressive diminuation in importance of both boreal meadow and boreal forest associations. The dominance of boreal habitats in the lowest levels (T15-1 & 2) is supported by both the heather vole, *Phenacomys intermedius,* and the yellow-cheeked vole, *M. xanthognathus,* which are sympatric today only in northern Canada (Fig. 3). These two species have not been identified above the lowest two levels, are less abundant in T15-1 than in T15-2, and suggest that the environment transgressed from an open, dry mixed forest (Canadian) parkland with an understory of heaths to a parkland like that near the confluence of the boreal, deciduous forest, and prairie biomes (Fig. 3). Reduction in relative abundance or loss of these species above T15-1 indicates that the mixed parkland gradually became more temperate (deciduous) in aspect through deposition of T8/13-3.

The decline in abundance or loss of the four above species was accompanied by an immediate but short-lived (T8/13-5) increase in both *Synaptomys borealis* and *S. cooperi. S. borealis,* now confined to the boreal forest, was rare but continued to survive in the cave area later than the other strictly boreal species. This probably was because of locally favorable conditions. Its southern counterpart, *S. cooperi,* now characteristic of the Canadian (Laurentian) mixed forest, continued as a major but progressively less abundant element upsection.

These boreal voles were replaced in a near reciprocal pattern by voles of the temperate prairie, *Microtus ochrogaster,* and deciduous forest, *M. pinetorum.* Thus, there was a gradual replacement of boreal meadow by prairie species among the grazing voles and a transition from predominately conifer (for example, *M. xanthognathus*), through mixed (for example, *Synaptomys cooperi*), to predominately deciduous forest (for example, *M. pinetorum*) species among the browsing voles in the parkland. The closed deciduous forest of the Ozark Uplands today apparently was not completely established until after 2,290 years B.P. when major deposition ceased in the cave.

SUMMARY AND CONCLUSIONS

PALEOECOLOGY

Peccary Cave microvertebrates are not randomly distributed throughout the section. Arvicolines constitute 64.2% of the total faunule in the lowest level (T15-2) but decline to 16.9% in the uppermost level (T8/13-1). Squirrels, primarily deciduous forest species, conversely expand from 2.8% in T15-2 to 12.5% in T8/13-1. The distribution of micromammals in the Peccary Cave sequence indicates both that Wisconsinan climates were not monolithic chronologically and that faunal turnover at the end of the Pleistocene in northwestern Arkansas was not catastrophic, but typified by both individualistic and progressive responses. First, faunal composition changed from one dominated by individuals characteristic of cool steppe with dominantly coniferous mixed forest inliers to one of a mixed forest parkland. This was followed in order by: (2) extirpation of four dry boreal forest species, (3) an increase in both meadow and forest species to balanced representation between deciduous and boreal coniferous ecotypes in the parkland, (4) loss of two more boreal species and dominance of deciduous parkland ecotypes, (5) megavertebrate extinction and a dramatic increase in both deciduous forest and temperate prairie elements, (6) development of the eastern deciduous forest with local enclaves of prairie and boreal conifer stands, and finally (7) gradual closure of the deciduous forest to a density approaching modern conditions. A completely modern faunal analogue is not present in the section and must post-date 2,290 years B.P.

Eight taxa are over-represented significantly in the lowest level (T15-2). These are *Blarina brevicauda, Eutamias minimus, Spermophilus tridecemlineatus, Tamiasciurus hudsonicus, Microtus pennsylvanicus, M. xanthognathus, Phenacomys intermedius,* and *Clethrionomys gapperi.* All, except *B. brevicauda* are boreal and, of these, only *T. hudsonicus* will occupy closed forests. Each is typical of either open parkland or meadow/prairie associations. Fourteen taxa are clearly under-represented in T15-2: *Scalopus aquaticus, B. hylophaga, B. carolinensis, Tamias striatus, Marmota monax, Sciurus* cf. *niger, Geomys bursarius, Peromyscus* cf. *leucopus, P.* cf. *maniculatus, P.* sp., *Neotoma floridana, Oryzomys palustris, M. pinetorum,* and *Synaptomys cooperi.* With the exception of *G. bursarius* (steppe) and *S. cooperi* (boreal but in mixed forest) all have distributions centered in the deciduous forest. Thus, the T15-2 faunule represents a cool, generally dry steppe, but one unlike that recorded by Rhodes (in press) in southwestern Iowa. The mixed forest aspect of the parkland was better developed in Arkansas than in Iowa. High species diversity, including a minor but diverse deciduous forest community within the T15-2 faunule and the disharmonious sympatry demonstrate that this parkland is not analogous to any existing boreal ecosystem. The flora probably resembled that presently occurring in the tension zone (Curtis, 1959) of central Wisconsin (Fig. 3), the area of sympatry, but grassland rather than forest dominated the landscape. Boreal conifers, understory heaths, and moss were more important than deciduous elements. Precipitation was low enough to permit extensive development of an open parkland but was sufficiently effective to maintain meadows (*Synaptomys*), marshes (*Sorex palustris*), and ponds (*Ondatra zibethicus*) as well as forest groves (*Sciurus* and *Tamiasciurus*) in favorable topographic locations. Summer temperatures must have averaged at least 12°F cooler than those in Arkansas today (Fig. 3) to permit immigration of the boreal component. Winter temperatures may have been similar to those presently in Central Wisconsin or slightly warmer. They were not significantly colder than those in Wisconsin now because many of the temperate deciduous forest species living in Arkansas today (for example, *Neotoma* and *Blarina carolinensis*) also resided in the area during the Wisconsinan.

The most striking change in relative abundance of species (Table 2) occurred between levels T15-2 and T15-1. Eleven species reversed their position as either under (−) or over (+) represented by at least 1.5 Freeman-Tukey units (Table 2). *Cryptotis parva*, *Sorex cinereus*, *Blarina carolinensis*, *Spermophilus tridecemlineatus*, *Geomys bursarius*, *Neotoma floridana*, and *Microtus pinetorum*, all of which now range in temperate regions were present in greater than expected numbers in T15-1. *M. xanthognathus*, *Phenacomys intermedius*, *Clethrionomys gapperi*, and *Zapus hudsonius*, all boreal species, became substantially reduced. At the same time 10 species moved at least 0.5 Freeman-Tukey units toward equivalence between their actual and expected abundance (*Scalopus aquaticus*, *B. hylophaga*, *Tamias striatus*, *Glaucomys volans*, *Peromyscus* cf. *maniculatus*, *P.* sp., *Oryzomys palustris*, *Tamiasciurus hudsonicus*, *Erethizon dorsatum*, and *M. pennsylvanicus*). The direction of faunal turnover is clear; deciduous forest forms either changed from under- to over-representation or became less "deficient" in the sample; boreal and grassland ecotypes, except *Geomys bursarius*, were of reduced importance. The T15-1 faunule suggests that the parkland became more heavily wooded, brush occurred with heaths in the understory, and deciduous

trees occupied a greater proportion of the landscape than previously (T15-2). This implies warmer summers, more effective precipitation (first *Oryzomys*) and less severe winters than during T15-2 time. The T15-1 faunule also has the highest species diversity (35 species) of any level and the most equal balance between boreal (43%), steppe (9%), and deciduous forest (37%) species. It represents an "equable" fauna as described in the *Introduction* and suggests that equable Wisconsinan climates apparently are more characteristic of late Wisconsinan time.

There possibly is a hiatus between T15-1 and T8/13-5, but a veneer of Unit B deposits at the top of T15-1 and a less dramatic change in relative abundance than between the lower two levels suggest that the time lag is minor. T8/13-5 is characterized by the loss of four boreal species (*Eutamias minimus*, *Tamiasciurus hudsonicus*, *Microtus xanthognathus*, and *Phenacomys intermedius*). With the exception of both species of *Synaptomys*, *Erethizon dorsatum* and *Sorex palustris*, all other boreal species are found in reduced numbers. Because *Erethizon* survived in Arkansas until historic time, its boreal classification is an artifact of history and it should not be considered a "boreal" indicator. The other three taxa are associated with wetlands and their increase is in accord with predicted moisture increases shown by loss of the four dry conifer parkland indicators. Deciduous forest species increase dramatically in T8/13-5 with substantial expansion in proportions of *Scalopus aquaticus*, *Marmota monax*, *Sciurus* cf. *niger*, *Peromyscus* cf. *leucopus*, *P.* cf. *maniculatus*, and *Neotoma floridana*. Only *M. pinetorum* and *Glaucomys volans* of the deciduous forest group decrease in relative numbers. This may relate to restructuring of the forest from predominantly coniferous to deciduous and increased distance between surviving deciduous trees. The *P.* cf. *maniculatus* peak in T8/13-5 suggests an increase of brush in the understory. This is supported by reduction of three taxa—two prairie species, *Geomys bursarius* and *Spermophilus tridecemlineatus* and the grazing *Microtus pennsylvanicus*, which tolerates cool, temperate grasslands. Wetlands also occupied a greater area and may have restricted available habitat for grazing rodents. The climate of T8/13-5 time undoubtedly had still warmer summers and increasingly effective rainfall compared to that of the underlying T15-1. Winters probably were warmer.

The number of individuals associated with boreal environments continued to decrease in T8/13-4. Only *Sorex hoyi* and *Synaptomys cooperi* of this group increased in relative abundance but the former along with *S. cinereus* last appeared in this level. Deciduous forest forms (for example, *Glaucomys* and *Sciurus*) continued to become more abundant but boreal conifers, indicated by *Clethrionomys,* still were locally present. The increase in both the prairie vole (*M. ochrogaster*) and woodland vole (*M. pinetorum*) denotes overall temperate conditions. Continued increases in summer temperature and precipitation permitted the development of a continuous deciduous forest in the area; conifers became restricted to small enclaves.

Because the A/B disconformity bisects level T8/13-3, it undoubtedly contains a mixture of specimens from both stratigraphic units. Nevertheless the last extinct megavertebrates are associated with this level and are most likely Unit B specimens. There is no evidence of any major climatic change in the T8/13-3 micromammal assemblage other than a sharp reduction in the relative abundance of *Microtus pennsylvanicus.*

Level T8/13-2, dated 4,290 years B.P. is completely within Unit A. The faunule basically represents that of a temperate deciduous forest. Both *Microtus ochrogaster* and *M. pinetorum,* temperate species, become more abundant than *M. pennsylvanicus* for the first time. T8/13-2 is unique because *Onychomys leucogaster,* a steppe indicator, is confined to this level. The two other steppe species recorded in Peccary Cave, *Spermophilus tridecemlineatus,* and *Geomys bursarius,* also exhibit a mode in this level. These animals are not sufficiently numerous to suggest major prairie encroachment but evidence for either balds or upland prairie enclaves is better developed here than in the other T8/13 levels. Because T8/13-2 immediately post-dates the Altithermal, the prairie indicators may represent a remnant of more widespread prairie present during the Unit A/B hiatus. Evidence of prairie encroachment to the east during the Altithermal also is suggested by fossil mammals recovered from both the Cherokee Sewer site in Iowa (Semken, 1980) and Rodgers Shelter in Missouri (Wood and McMillan, 1976).

The T8/13-1 faunule has the most modern aspect of the seven examined faunules and represents a nearly closed deciduous forest. Boreal indicators still are present but the seven species only account for 8% of the individuals in the faunule. They undoubtedly were relictual but their presence in the post-Altithermal of Arkansas suggests that the Ozarks were a refugium for boreal species late into the Holocene. Individuals of the two steppe species, *Microtus ochrogaster* and *Geomys bursarius,* both presently extirpated, comprise 6% of the fauna. The remainder are either widespread species, primarily *Peromyscus,* or deciduous forest taxa. The woodland vole (*M. pinetorum*), eastern chipmunk (*Tamias striatus*), woodchuck (*Marmota monax*), western short-tailed shrew (*Blarina hylophaga*) and fox squirrel (*Sciurus niger*) are most common. The modern deciduous forest of the area, which does not host either steppe or boreal species, developed sometime after 2,290 year B.P.

THE PLEISTOCENE/HOLOCENE BOUNDARY IN PECCARY CAVE

If the traditional biostratigraphic concept of megavertebrate extinction is used, the Pleistocene/Holocene boundary in the Peccary Cave sequence, is best placed at the disconformity separating Units A and B (T8/13-3). Extinct megavertebrates are common in both Units B and C but are rare in the overlying Unit A. However, Davis (1973) notes that all large mammal remains are sparse in Unit A and believes that the paucity is partly attributable to closure of the horizontal entrance by alluviation. Four post-5,000 year B.P. radiocarbon dates, one (I-3894) from T8/13-2, support middle to late Holocene age for Unit A. If these dates from Unit A are correct and megavertebrate extinction is chronologically precise, Units B and C are Pleistocene and Unit A is Holocene. The Unit A/B disconformity then represents an early to mid-Holocene hiatus. The sparse remains of extinct megavertebrates in Unit A are reworked if this division is correct. Quinn (1972:95–96), however, believes that some extinct species became "entombed at the very end of the Altithermal" in Peccary Cave and that the Ozarks provided a post-10,000 year B.P. refugium for them.

Micromammal relative abundance shows no obvious discontinuities (Figs. 4–6) separating the Pleistocene from the Holocene. Biostratigraphic boundaries usually are based on (1) last appearances, (2) first appearances and/or (3) coincident changes in relative abundance between units. Taken in order: (1) Eight of 33 species recorded in the lowest Peccary Cave level (T15-2) are not known from the uppermost level (T8/13-1). Five of these (*Microtus xanthognathus, Phenacomys intermedius, Eutamias minimus, Tamiasciurus hudsonicus,* and *Cryptotis parva*) disappear in the interval between levels T15-1 and T8/13-5. All, except *C. parva,* are boreal and their loss is expected. The interim between T8/13-

4 and 3 is a second datum marked by the last appearance of *Sorex hoyi* and *S. cinereus,* also boreal species. *Ondatra zibethicus* is not recorded above T8/13-3 and finally, two boreal species, *S. arcticus* and *Condylura cristata* (Table 1) disappear above T8/13-2. Major loss of the boreal components occurs between T15-1 and T8/13-5, but extirpation is a continuing process throughout the section. (2) Only four species in the Peccary Cave section are not present in T15-2. *Oryzomys palustris* and *Castor canadensis* are first recorded in T15-1. *Onchomys leucogaster* appears in T8/13-2 (Table 1) and *Notiosorex crawfordi* (desert shrew), which occupies northwestern Arkansas today, is recorded in other, presumably younger, Unit A sediments of other trenches. Thus, appearance of new taxa is limited and indicates that the modern fauna of Arkansas is largely an impoverished residue of a more diverse Wisconsinan community. (3) Ten species (Figs. 4–6) achieve maximum relative abundance in T15-2, two in T15-1, four in T8/13-5, three in T8/13-4, two in both T8/13-3 and 2 and one in T8/13-1. The differences in relative abundance are much greater between T15-1 and 2 than between any other two levels. The difference across the Unit A/B disconformity is much less than expected for a 5,000 year hiatus separating Pleistocene and Holocene deposits. Faunal turnover in the Peccary Cave section (Figs. 4–6, Table 2) is highly individualistic and generally reflects a progressive transition between late Wisconsinan (T15-2) and post-Altithermal (T8/13-1) time. This confirms gradual environmental change, following an abrupt change between T15-2 and 1, interpreted above for the Peccary Cave section.

To examine the boundary question from another perspective, the faunal counts (Table 1) for sciurids (Fig. 7) and arvicolines (Fig. 8) were normalized (to 130 individuals per level to minimize sample size differences) and replotted to reflect species density (percent distribution throughout the entire fauna rather than an individual level). While these graphs appear similar to the relative abundance per level plots (Figs. 4–6), they weigh the distributions independently and graphically reflect the Freeman-Tukey deviates (Table 2). The distribution of *Microtus pinetorum* is illustrative. This taxon gradually comprises a greater proportion of each faunule upsection (Fig. 6) but its absolute abundance (density) is fairly stable (Fig. 8) through time. The arvicolines as a family were almost decimated (Fig. 8) after deposition of T15-1. However, the proportions of

each species changed gradually (Fig. 6), both across the break and within the radically thinned ranks of voles (Fig. 8) in the upper five levels. This suggests that the specific habitat of each vole, except for the extirpated yellow-cheeked and heather voles, remained in the Peccary Cave vicinity but occupied substantially less area. The animals in these reduced niches then responded in a patterned manner (T8/13) to the factor(s) which created the most abrupt biostratigraphic change, presumably the new climatic regime discussed in the previous section. The sharpest inflection for the sciurids appears between T15-1 and T8/13-5 immediately after, and perhaps in response to, the major reduction in the number of voles. A gradual increase in deciduous forest squirrels upsection is almost reciprocal to the demise of the arvicolines.

The position of the Pleistocene/Holocene boundary in the Peccary Cave section is difficult to establish, perhaps because of its transitional nature. Based on the loss of megamammals and radiocarbon dates, the most logical boundary is in the disconformity that bisects T8/13-3. This position can be supported faunally by the last appearance of both *Sorex cinereus* and *S. hoyi,* an abrupt decline in the abundance of *Microtus pennsylvanicus,* an increase in numbers of *Tamias striatus,* the first appearance of *Onychomys leucogaster,* and an increase in numbers of two other steppe species (*M. ochrogaster* and *Spermophilus tridecemlineatus*). The majority of last appearances, however, are concentrated in the interval between T15-1 and T8/13-5 where five species, including four boreal forms disappear. T8/13-5 also is marked by a sharp increase in the importance of sciurids. Although the major change in density from dominance of voles (boreal) to sciurids (deciduous) occurs between T15-1 and T8/13-5, major faunal reorganization occurred earlier, between T15-2 and T15-1. The greatest change in relative abundance of species, confirmed by dramatic differences in sign of the Freeman-Tukey deviates, is between these two levels and, as noted in the paleoecology section, the only "reversal" in paleoecological trend is recorded here. The first appearance of two species (*Oryzomys palustris* and *Castor canadensis*) also occurs at this time.

Thus, in Peccary Cave, community reorganization was the first event to mark the Pleistocene/Holocene transition, maximum replacement of voles by sciurids followed, and finally the Wisconsinan megavertebrates disappeared. Modern closed deciduous forest conditions post-date T8/13-1. The exact

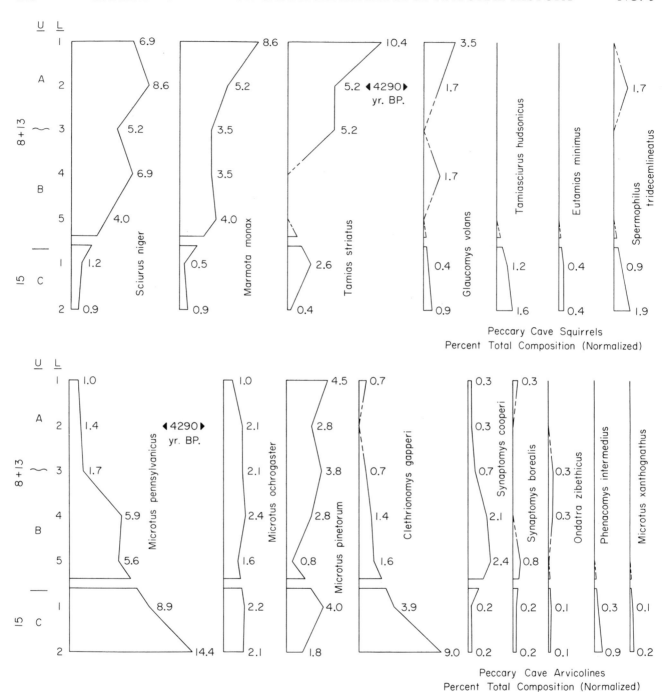

Fig. 7 (above).—Sciurid density (%), normalized to 130 MNI, in Trenches 8 + 13 and 15, Peccary Cave local fauna, Newton Co., Arkansas.
Fig. 8 (below).—Arvicoline density (%), normalized to 130 MNI, in Trenches 8 + 13 and 15, Peccary Cave local fauna, Newton Co., Arkansas.

position of the boundary then becomes a matter of definition. Climatic change probably triggered the first event. This initiated a series of individualistic responses marked by failure of most boreal small mammals to reproduce in the area, their replace-

ment by deciduous forest species and ultimately in megavertebrate extinction. The small mammals continued to adjust well into late Holocene (post-4,000 years B.P.) time. This scenario supports progressive but ultimately radical climatic change as

the prime motivator for Wisconsinan megavertebrate extinction as proposed by Graham and Lundelius (in press). The Peccary Cave chronofauna,
along with that recorded by Guilday et al. (1978) at
Baker Bluff Cave, Graham (1976) at Friesenhahn
Cave, Klippel and Parmalee (1982a) at Cheek Bend,
and Guilday et al. (1964) at New Paris, also demonstrates that Wisconsinan climates were not monolithic but characterized temporally as well as geographically, by biologic, ecologic, and climatic
variability.

ACKNOWLEDGMENTS

The author wishes to thank the late James H. Quinn, University of Arkansas, for bringing the Peccary Cave local fauna to his
attention; Jack McCutcheon, property owner, both for the discovery and recognition of the significance of the fossils; and to
L. Carson Davis, Southern Arkansas University who, along with
McCutcheon, excavated, waterscreened, and processed the vertebrate remains.

For professional assistance I thank: Michael D. Carleton, University of Michigan; L. Carson Davis, Southern Arkansas University; Russell W. Graham, Illinois State Museum; the late John
E. Guilday, CMNH; Robert S. Hoffmann, University of Kansas;

Emmet T. Hooper, University of Michigan; Ernest L. Lundelius,
University of Texas; Larry D. Martin, University of Kansas;
Richard S. Rhodes II, University of Iowa; Gerald R. Smith,
University of Michigan; and Michael R. Voorhies, University of
Nebraska. Financial assistance for excavation was provided by
NSF Grant GB6762. The manuscript was prepared on the word
processor by Rebecca R. Bush; figures were drafted by Joyce E.
Chrisinger.

I want to especially thank Richard S. Rhodes II for his diligent
work on the manuscript and both he and L. Carson Davis for
their continued support of this project.

LITERATURE CITED

BAILEY, R. G. 1981. Ecoregions of North America (map). USDI
Fish and Wildlife Service, Washington, D.C.

BANFIELD, A. W. F. 1974. The mammals of Canada. Univ.
Toronto Press, Toronto, xxv + 438 pp.

BLAIR, W. F. 1958. Distributional patterns of vertebrates in the
southern United States in relation to past and present environments. Pp. 433–568, in Zoogeography (C. L. Hubbs,
ed.), American Assoc. Adv. Sci., Publ., 51.

BRAUN, E. L. 1955. Phytogeography of unglaciated eastern
United States and its interpretation. Botanical Rev., 21:297–
375.

BROWN, E. L. 1908. The Conard Fissure, a Pleistocene bone
deposit in northern Arkansas: with a description of two new
genera and twenty new species of mammals. Mem. Amer.
Mus. Nat. Hist., 9:157–208.

CURTIS, J. T. 1959. The vegetation of Wisconsin. Univ. Wisconsin Press, Madison, 657 pp.

DALQUEST, W. W. 1965. New Pleistocene formation and local
fauna from Hardeman County, Texas. J. Paleontol., 39:63–
79.

DALQUEST, W. W., E. ROTH, and F. JUDD. 1969. The mammal
fauna from Schulze Cave, Edwards County, Texas. Bull.
Florida St. Mus., 13:206–276.

DAVIS, L. C. 1969. The biostratigraphy of Peccary Cave, Newton County, Arkansas. Proc. Arkansas Acad. Sci., 23:192–
196.

———. 1973. The herpetofauna of Peccary Cave, Newton
County, Arkansas. Unpublished Masters thesis, Univ. Arkansas, Fayetteville, 85 pp.

———. 1975. Late Pleistocene geology and paleoecology of the
Spring Valley Basin, Meade County, Kansas. Unpublished
Ph.D. dissert., Univ. Iowa, Iowa City, 170 pp.

DULIAN, J. J. 1975. Paleoecology of the Brayton local biota,
late Wisconsinan of southwestern Iowa. Unpublished Masters thesis, Univ. Iowa, Iowa City, 50 pp.

ESHELMAN, R. E. 1971. The paleoecology of Willard Cave, Delaware County, Iowa. Unpublished Masters thesis, Univ. Iowa,
Iowa City, 72 pp.

FAY, L. P. 1980. Mammals of the Garrett Farm and Pleasant
Ridge local biotas (Holocene), Mills County, Iowa. Pp. 74–
75, in Amer. Quat. Assoc. Sixth Biennial Meeting Abs.,
Orono, Maine.

FOLEY, R. L. In press. Late Pleistocene (Woodfordian) vertebrates from the Driftless Area of southwestern Wisconsin,
the Moscow local fauna. Illinois State Mus. Repts. Invest.,
39.

GENOWAYS, H. H., and J. R. CHOATE. 1972. A multivariate
analysis of systematic relationships among populations of
the short-tailed shrew (genus Blarina) in Nebraska. Syst.
Zool., 21:106–116.

GEPPERT, T. J. 1984. Small mammals of Shield Trap, East Pryor
Mountains, Montana. Unpublished Masters thesis, Univ.
Iowa, Iowa City, 45 pp.

GRAHAM, R. W. 1976. Late Wisconsinan mammalian faunas
and environmental gradients of the eastern United States.
Paleobiology, 2:343–350.

———. 1979. Paleoclimates and late Pleistocene faunal provinces in North America. Pp. 49–69, in Pre-Llano Cultures
of the Americas: problems and paradoxes (R. L. Humphrey
and D. Stanford, eds.), Washington Anthro. Soc., Washington, D.C., 150 pp.

GRAHAM, R. W., and E. L. LUNDELIUS, JR. In press. Coevolu-
tionary disequilibrium and Pleistocene extinction. In Quaternary extinctions (P. S. Martin and R. G. Klein, eds.), Univ.
Arizona Press, Tucson.

GRAHAM, R. W., AND H. A. SEMKEN, JR. 1976. Paleoecological
significance of the short-tailed shrew (Blarina), with a systematic discussion of Blarina ozarkensis. J. Mamm., 57:
443–449.

GUILDAY, J. E. 1962. The Pleistocene local fauna of the Natural

Chimneys, Augusta County, Virginia. Ann. Carnegie Mus., 36:87–122.

———. 1982. Dental variation in *Microtus xanthognathus, M. chrotorrhinus,* and *M. pennsylvanicus* (Rodentia: Mammalia). Ann. Carnegie Mus., 51:211–230.

GUILDAY, J. E., and M. S. BENDER. 1960. Late Pleistocene records of the yellow-cheeked vole, *Microtus xanthognathus* (Leach). Ann. Carnegie Mus., 35:315–330.

GUILDAY, J. E., H. W. HAMILTON, E. ANDERSON, and P. W. PARMALEE. 1978. The Baker Bluff Cave deposit, Tennessee, and the late Pleistocene faunal gradient. Bull. Carnegie Mus. Nat. Hist., 11:1–67.

GUILDAY, J. E., H. W. HAMILTON, and A. D. MCCRADY. 1971. The Welsh Cave peccaries (*Platygonus*) and associated fauna, Kentucky Pleistocene. Ann. Carnegie Mus., 43:249–320.

GUILDAY, J. E., and C. O. HANDLEY, JR. 1967. A new *Peromyscus* (Rodentia: Cricetidae) from the Pleistocene of Maryland. Ann. Carnegie Mus., 39:91–103.

GUILDAY, J. E., P. S. MARTIN, and A. D. MCCRADY. 1964. New Paris No. 4: A Pleistocene cave deposit in Bedford County, Pennsylvania. Bull. Nat. Speleo. Soc., 26:121–194.

GUILDAY, J. E., and P. W. PARMALEE. 1965. Animal remains from the Sheep Rock Shelter (36 Hu 1), Huntingdon County, Pennsylvania. Pennsylvania Arch., 35:34–49.

———. 1972. Quaternary periglacial records of voles of the genus *Phenacomys* Merriam (Cricetidae: Rodentia). Quaternary Res., 2:170–175.

GUILDAY, J. E., P. W. PARMALEE, and H. W. HAMILTON. 1977. The Clark's Cave bone deposit and the late Pleistocene paleoecology of the central Appalachian Mountains of Virginia. Bull. Carnegie Mus. Nat. Hist., 2:1–87.

HALL, E. R. 1981. The mammals of North America. John Wiley and Sons, New York, 1:xv + 1–600, +90; 2:vi + 601–1181 +90.

HALLBERG, G. R., H. A. SEMKEN, JR., and L. C. DAVIS. 1974. Quaternary records of *Microtus xanthognathus* (Leach), the yellow-cheeked vole, from northwestern Arkansas and southwestern Iowa. J. Mamm., 55:640–645.

HAWKSLEY, O., J. F. REYNOLDS, and R. L. FOLEY. 1973. Pleistocene vertebrate fauna of Bat Cave, Pulaski County, Missouri. Bull. Nat. Speleo. Soc., 35:61–87.

HIBBARD, C. W. 1949. Techniques of collecting microvertebrate fossils. Contrib. Mus. Paleont., Univ. Michigan, 8:7–19.

———. 1952. Vertebrate fossils from Late Cenozoic deposits of Central Kansas. Univ. Kansas, Paleont. Contrib., Vertebrata, 2:1–14.

———. 1955. The Jinglebob interglacial (Sangamon?) fauna from Kansas and its climatic significance. Contrib. Mus. Paleont., Univ. Michigan, 12:179–228.

———. 1960. An interpretation of Pliocene and Pleistocene climates in North America. Ann. Rept. Michigan Acad. Sci., Arts and Letters, 62:5–30.

———. 1963. A late Illinoian fauna from Kansas and its climatic significance. Papers Michigan Acad. Sci., Arts, Letters, 48:181–221.

HIBBARD, C. W., and D. W. TAYLOR. 1960. Two late Pleistocene faunas from southwestern Kansas. Contrib. Mus. Paleont., Univ. Michigan, 16:1–223.

HOOPER, E. T. 1957. Dental patterns in mice of the genus *Peromyscus.* Misc. Publ. Mus. Zool., Univ. Michigan, 99:1–59.

JOHNSON, P. C. 1972. Mammal remains associated with Nebraska phase earth lodges in Mills County, Iowa. Unpublished Masters thesis, Univ. Iowa, Iowa City, 71 pp.

JONES, C. A. 1982. Paleontology and systematic relationships of short-tailed shrews (Genus *Blarina*). Unpublished Masters thesis, Fort Hays State Univ., Hays, Kansas, 210 pp.

JUNGE, J. A., and R. S. HOFFMANN. 1981. An annotated key to the long-tailed shrews (Genus *Sorex*) of the United States and Canada, with notes on middle American *Sorex.* Occas. Papers Mus. Nat. Hist., Univ. Kansas, 94:1–48.

KING, F. B., and R. W. GRAHAM. 1981. Effects of ecological and paleoecological patterns on subsistence and paleoenvironmental reconstructions. Amer. Antiq., 46:128–142.

KITTS, D. B. 1953. A Pleistocene musk ox from New York and the distribution of the musk-oxen. Amer. Mus. Novitates, 1607:1–8.

KLIPPEL, W. E., and P. W. PARMALEE. 1982a. Diachronic variation in insectivores from Cheek Bend Cave and environmental change in the Midsouth. Paleobiology, 8:447–458.

———. 1982b. The paleontology of Cheek Bend Cave, Maury County, Tennessee. Phase II report to the Tennessee Valley Authority, Knoxville.

KURTÉN, B., and E. ANDERSON. 1980. Pleistocene mammals of North America. Columbia Univ. Press, New York, 442 pp.

LUNDELIUS, E. L. 1967. Late-Pleistocene and Holocene faunal history of central Texas. Pp. 287–319, *in* Pleistocene extinctions (P. S. Martin and H. E. Wright, Jr., eds.), Yale Univ. Press, New Haven, 453 pp.

MARTIN, L. D., and A. M. NEUNER. 1978. The end of the Pleistocene in North America. Trans. Nebraska Acad. Sci., 6:117–126.

MARTIN, R. A. 1968. Late Pleistocene distribution of *Microtus pennsylvanicus.* J. Mamm., 49:265–271.

MARTIN, R. A., and S. D. WEBB. 1974. Late Pleistocene mammals from the Devil's Den Fauna, Levy County. Pp. 114–145, *in* Pleistocene mammals of Florida (S. David Webb, ed.), Univ. Florida Press, Gainesville, x + 270 pp.

MCKENNA, M. C. 1962. Collecting small fossils by washing and screening. Curator, 3:221–235.

MEGIVERN, K. J. 1982. Paleoclimatic significance of the Pleistocene Insectivora and Rodentia of Trench 24, Peccary Cave, Newton County, Arkansas. Unpublished Masters thesis, Univ. Iowa, Iowa City, 56 pp.

MILLER, B. B. 1970. The Sandahl molluscan fauna (Illinoian) from McPherson County, Kansas. Ohio J. Sci., 70:39–50.

PARMALEE, P. W., and W. E. KLIPPEL. 1981a. A late Pleistocene population of the pocket gopher, *Geomys* cf. *bursarius,* in the Nashville Basin, Tennessee. J. Mamm., 62:831–835.

———. 1981b. A late Pleistocene record of the heather vole (*Phenacomys intermedius*) in the Nashville Basin, Tennessee. J. Tennessee Acad. Sci., 56:127–129.

PARMALEE, P. W., and R. D. OESCH. 1972. Pleistocene and Recent faunas from the Brynjulfson Caves, Missouri. Illinois State Museum, Rept. Investigation, 25:52 pp.

PARMALEE, P. W., R. D. OESCH, and J. E. GUILDAY. 1969. Pleistocene and Recent vertebrate faunas from Crankshaft Cave, Missouri. Illinois State Museum, Rept. Investigation, 14:1–37.

PATTON, T. H. 1963. Fossil vertebrates from Miller's Cave, Llano County, Texas. Bull. Texas Memorial Mus., 7:1–41.

PURDUE, J. R. 1980. Clinal variation of some mammals during the Holocene in Missouri. Quaternary Res., 13:242–258.

QUINN, J. H. 1972. Extinct mammals in Arkansas and related C^{14} dates circa 3,000 years ago. 24th International Geol. Cong., 12:89–96.

RAY, C. E. 1967. Pleistocene mammals from Ladds, Bartow County, Georgia. Bull. Georgia Acad. Sci., 25:120–150.

RHODES, R. S., II. In press. Paleoecology and regional paleoclimatic implications of the Farmdalion Craigmile and Woodfordian Waubonsie mammalian local faunas, southwestern Iowa. Illinois State Mus. Repts. Invest., 40.

ROSENBERG, R. S. 1983. The paleoecology of the late Wisconsinan Eagle Point local fauna, Clinton County, Iowa. Unpublished Masters thesis, Univ. Iowa, Iowa City, 70 pp.

SATORIUS-FOX, M. R. 1982. Paleoecological analysis of micromammals from the Schmidt Site, a Central Plains Tradition village in Howard County, Nebraska. Univ. Nebraska Div. Archaeological Res. Technical Rpt., 83-13, 87 pp.

SCHULTZ, G. E. 1967. Four superimposed Late-Pleistocene vertebrate faunas from southwest Kansas. Pp. 321–336, in Pleistocene extinctions (P. S. Martin and H. E. Wright, Jr., eds.), Yale Univ. Press, New Haven, 453 pp.

SEALANDER, J. A. 1979. A guide to the Arkansas mammals. River Road Press, Conway, Arkansas, 313 pp.

SEMKEN, H. A., JR. 1966. Stratigraphy and paleontology of the McPherson *Equus* beds (Sandahl local fauna), McPherson Co., Kansas. Contrib. Mus. Paleont., Univ. Michigan, 20: 121–178.

———. 1969. Paleontological implications of micromammals from Peccary Cave, Newton County, Arkansas. Geol. Soc. America, South Central Sec. Abs., p. 27.

———. 1974. Micromammal distribution and migration during the Holocene. Amer. Quaternary Assoc. Abs., Madison, 3:25.

———. 1980. Holocene climatic reconstructions derived from the three micromammal bearing cultural horizons of the Cherokee Sewer Site, northwestern Iowa. Pp. 67–99, in The Cherokee excavations (D. C. Anderson and H. A. Semken, Jr., eds.), Academic Press, New York, 277 pp.

———. 1983. The Holocene mammalian record of the eastern and central United States. Pp. 182–207, in Late-Quaternary environments of the United States: The Holocene (H. E. Wright, Jr., ed.), Univ. Minnesota Press, Minneapolis, 277 pp.

SLAUGHTER, B. H. 1965. Animal ranges as a clue to late Pleistocene extinction. Pp. 155–167, in Pleistocene extinctions (P. S. Martin and H. E. Wright, Jr., eds.), Yale Univ. Press, New Haven, 453 pp.

STEPHENS, J. J. 1960. Stratigraphy and paleontology of a late Pleistocene basin, Harper County, Oklahoma. Bull. Geol. Soc. Amer., 71:1675–1702.

STEWART, J. D. 1978. Mammals of the trapshoot local fauna, Late Pleistocene of Rooks County, Kansas. Proc. Nebraska Acad. Sci., Abs. for 88th Annual Meeting.

VAN DER MEULEN, A. J. 1978. *Microtus* and *Pitymys* (Arvicolidae) from Cumberland Cave, Maryland, with a comparison of some new and old world species. Ann. Carnegie Mus., 47:101–145.

VISHER, S. S. 1954. Climatic atlas of the United States. Harvard Univ. Press, Cambridge, 403 pp.

VOORHIES, M. R. 1981. A fossil record of the porcupine (*Erethizon dorsatum*) from the Great Plains. J. Mamm., 62: 835–837.

WEBB, S. D. (ed.). 1974. Pleistocene mammals of Florida. Univ. Florida Press, Gainesville, x + 270 pp.

WOOD, W. R., and R. B. McMILLAN. 1976. Prehistoric man and his environments. Academic Press, New York, 271 pp.

WOODMAN, N. 1982. A subarctic fauna from the late Wisconsinan Elkader site, Clayton County, Iowa. Unpublished Masters thesis, Univ. Iowa, Iowa City, 56 pp.

ZAKRZEWSKI, R. J. In press. The fossil record of *Microtus* in North America. *In* The Biology of New World *Microtus* (R. A. Tamarin, ed.), Spec. Publ. Amer. Soc. Mamm.

Address: Department of Geology, University of Iowa, Iowa City, Iowa 52242.

LATE PLEISTOCENE AND EARLY RECENT MAMMALS FROM FOWLKES CAVE, SOUTHERN CULBERSON COUNTY, TEXAS

WALTER W. DALQUEST AND FREDERICK B. STANGL, JR.

ABSTRACT

Stratified deposits from Fowlkes Cave in the Chihuahuan Desert of Trans-Pecos Texas yielded abundant fossil remains of late Pleistocene vertebrates from a 20 cm thick layer sandwiched between seemingly barren layers. The mammalian fossils include both species adapted to boreal habitat (*Sorex vagrans, S. palustris, Eutamias cinericollis, Microtus mexicanus*) and forms typical of arid desert habitat (*Dipodomys merriami, D. spectabilis, Peromyscus eremicus, Onychomys leucogaster, O. torridus*), all from a single stratum that must have accumulated in a relatively short time. Similar associations have been reported from caves in southern New Mexico but were thought to have resulted from postmortum mixing of materials. We suggest that, because the Guadalupe Mountains, 100 km to the north, are known to have been glaciated, the improbable association found at Fowlkes Cave resulted from glaciation. Meltwater streams permitted the existence of cool meadows and thickets while, back from the oasis-like streamsides, desert conditions existed and desert mammals lived. The now barren desert ranges, the Delaware and Apache mountains, formed an avenue by which some alpine mammals emigrated from the Guadalupe Mountains to the Davis Mountains where they are now isolated.

INTRODUCTION

Culberson Co. (Fig. 1) is large (3,787 sq mi), and lies in extreme western, or Trans-Pecos Texas. Most of the county'consists of desert flats, scrub, grassland, and rugged hills and mesas, typical of the Chihuahuan Desert. In the northwestern corner of the county the Guadalupe Mountains, a southward extension of the southern Rocky Mountains, rise to 8,700 ft elevation and support a montane flora and fauna. The Recent mammals of the Guadalupe Mountains have interested mammalogists since Bailey (1905) collected specimens there in 1901. For more recent accounts see Genoways et al. (1979). The Recent mammals of the desert lowlands of Culberson Co. are less well-known but were treated by Davis and Robertson (1944) and by Schmidly (1977). Numerous other accounts deal, at least in part, with the mammals of Culberson Co. and are cited in the bibliographies of the four papers mentioned.

The geology of Culberson Co. is complicated but much of the exposed surface rock consists of Permian limestones in which caves are relatively common. Similar cave-forming limestones occur in adjacent New Mexico and extend nearly across the entire southern border of that state. Carlsbad Cavern, for example, lies in this Permian limestone series. For a summary of some 25 early Recent and late Pleistocene mammalian faunas from caves in Trans-Pecos Texas and southern New Mexico (see Harris, 1974).

The cave faunas reported to date, from the Guadalupe Mountains and southern New Mexico, occur in a band approximately 100 km from north to south and 400 km from east to west, and many lie at relatively high elevations, where modern conditions are more mesic than those typical of the Chihuahuan Desert.

In 1979, Mr. J. M. Fowlkes of Pecos, Texas, told us of a sinkhole-type cave on his large ranch (70 sections, 44,800 acres) north of Kent, Texas. This cave lies 100 km south of the caves in the Guadalupe Mountains and southern New Mexico, where Pleistocene and early Recent mammalian faunas have been described. It seemed possible that fossils from Fowlkes Cave might contribute important information concerning an area south of the previously known caves, at lower elevation, and in what is now arid, typical Chihuahuan Desert. With the cooperation of Mr. Fowlkes, we were able to obtain a quantity of the black, silty surficial deposits in the cave, and of the upper part of the underlying late Pleistocene sediments. Both samples yielded abundant bones of mammals, birds, and reptiles. The fossils of lower vertebrates from the Pleistocene deposit will be reported elsewhere by other workers. Only the early Recent and Pleistocene mammals are dealt with here.

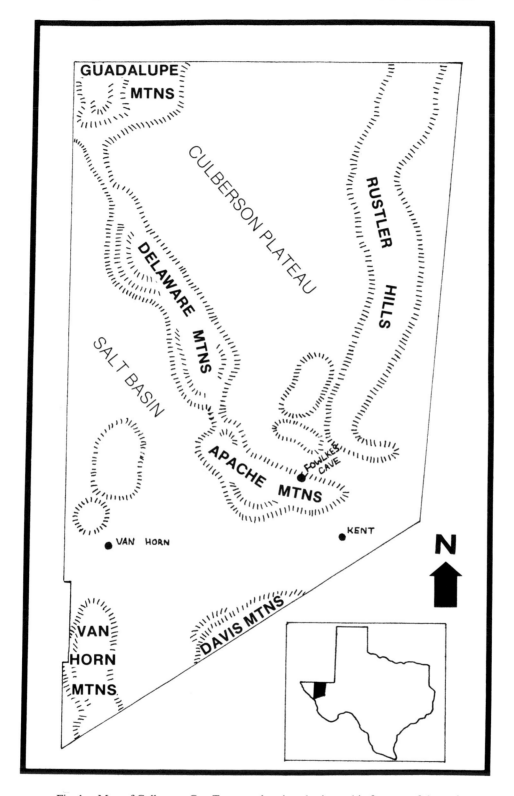

Fig. 1.—Map of Culberson Co., Texas, and major physiographic features of the region.

DESCRIPTION OF FOWLKES CAVE AND ITS SURROUNDINGS

Fowlkes Cave lies approximately 10 km north of Kent (a gasoline station and general store plus a few buildings north of Interstate Highway 10). The exact location of the cave is not detailed at the request of the landowner. It lies near the edge of the southern termination of the barren, limestone hills termed the Apache Mountains, which are a southern extension of the Delaware Mountains. The Delaware Mountains, in turn, constitute a southern extension of the Guadalupe Mountains. Just to the south are the Davis Mountains, where montane conditions exist and where the flora and, in part, the fauna, is similar to that of the Guadalupe Mountains. Zoogeographically, Fowlkes Cave is especially important in that its location is in the desert lowlands between the Guadalupe Mountains to the north and Davis Mountains to the south, and almost on the flanks of the high ridge between these mountain ranges. Montane mammals that occur in the Davis Mountains today presumably reached the fir-oak-pine woodlands of the Davis Mountains from the Guadalupe Mountains in the Pleistocene, when ecological conditions between the two mountain ranges were suitable for the existence of these species (for speculation see Schmidly, 1977:17–18). The chain of hills (Delaware Mountains–Apache Mountains) would be a logical pathway between the two montane highlands, and Fowlkes Cave might preserve a sample of the mammalian emigrants. Study to date, reported here, suggests that this is true.

Fowlkes Cave is a sinkhole (Fig. 2) whose entrance is located on the slopes of a limestone hill (Fig. 3). From the cave mouth to the aluvial fan below measures approximately 90 m. The slope, from cave mouth to aluvial fan, is approximately 45 degrees. The slopes of the entire hill is so steep that only grasses and a few desert shrubs such as cacti and juniper bushes, with roots anchored in crevices in the limestone, manage to survive on the hill. Below the hill the aluvial fans support a rich flora of yuccas, cacti, and other desert shrubs. The aluvial fans grade into creosote bush flats (Fig. 4). Several kilometers away are dry washes and fringing vegetation that might, in a cooler climatic cycle, have supported riparian marshes and permanent running water.

The cave entrance measures approximately 2 m². The opening drops 3 m to a shallow step-like shelf. Beneath this is a sheer drop of 6 m to a platform 1 m wide, followed by another sheer drop of 5 m to the debris-filled upper chamber. From this point one can scramble 2 m down the slopes of a debris cone to the lower chamber. Both upper and lower chambers are on the same axis, the upper extending northwest and the lower southeast. The upper chamber is filled to a considerable but unknown depth by dust, trash, cave-swallow droppings and nests, and limestone fragments that have spalled from the cave walls. Its dimensions, from a point below the entrance shaft, are approximately 22 by 12.5 m. A small pit in the back leads to other chambers and passages that were explored in part but do not play a part in this work. The lower chamber measures approximately 25.5 by 8 m from a point below the entrance shaft, but part of the chamber is nearly filled to the level of the upper chamber by debris that collected beneath the entrance shaft. Stalactites, stalagmites, partial cave pearls, cave coral, and other intricate shapes and struc-

Fig. 2.—Sinkhole entrance to Fowlkes Cave.

Fig. 3.—Sinkhole entrance (arrow) to Fowlkes Cave, as seen from alluvial fan. Distance from foreground figure to cave is approximately 90 m.

tures of carbonate line the walls and ceilings. Although bats have occupied the cave in the past, only cave swallows seem to have lived there in late years.

Mr. Fowlkes stated that in past years the cave was mined for bat guano, and evidence of this work, in the form of wooden planks and scrap metal, litter the slope near the cave, while ropes, cables, and ladders were found in the cave itself. The mining, however, seems to have been very limited and confined to digging in the upper chamber. Probably the product was swallow guano rather than bat guano. Partly for this reason, we excavated only in the lower chamber. The sediments here were undisturbed, save for one hole approximately a 1 m² that penetrated the surficial black silts near the middle of the chamber floor. A yard-square piece of ½-inch mesh hail screen lay here, apparently left by unsuccessful artifact hunters in past years.

Fig. 4.—View from Fowlkes Cave entrance, to the northeast, of alluvial fans and desert flats. Dark areas of vegetation in distance fringe currently dry watercourses.

MATERIALS AND METHODS

A rectangle 10 by 20 ft was laid off in the lower chamber, along the southern wall of the cave and with the eastern end extending into the debris cone at the base of the shaft. Matrix was actually gathered only from the western 10 ft length of the area, however, though exploratory pits were made at several places in the eastern end. The black surficial silt, less than 1 ft deep at the western extremity but more than 16 inches deep at the eastern end, closer to the debris cone, was collected. The actual weight of the material was probably near 1,000 pounds. The silt was shoveled into buckets and, with pulleys and ropes, lifted out of the cave and passed down a steel cable to a truck on the alluvial fan below. There the buckets were loaded into burlap bags and the empty buckets pulled up to the cave mouth to be refilled.

When the surficial, black Recent materials had been removed, the yellow Pleistocene sediments were exposed. Unlike most cave deposits, these were stratified. There was very little mud or clay-like material in the Pleistocene deposits. A trench showed an upper layer (1) of approximately 35 cm thickness, composed of irregular limestone fragments and pebbles, bits of stalactitic limestone, cave pearls, and crystals of water-clear calcite. Average diameter of the rock fragments, exclusive of stalactite fragments, was between 15 and 25 mm. Beneath this was a layer (2) averaging approximately 20 cm thick, composed of smaller fragments of similar materials, between 5–15 mm in diameter. Layers 1 and 2 did not seem to have a distinct boundary separating them, but the nature of the materials made this difficult to determine. Beneath Layer 2 was a third layer (3) of perhaps 30 cm in thickness, composed of finer materials, the consistency of coarse sand. This

layer was set off distinctly from the lower surface of Layer 2. Beneath Layer 3, calcareous structures, partly "cave coral" from the cave walls and partly travertine structures formed locally, in the sediments, were interlocked so intricately that deeper digging was difficult and investigation has not been done to date.

An exploratory excavation was made in the Pleistocene sediments where the artifact hunters had dug through the black silts. However, though fossils were abundant, they were incorporated into successive, parallel layers of travertine rock, from which their removal was judged to be too difficult and expensive to attempt.

The coarse, gravel-like materials of Layer 1 were examined carefully in several places but no fossil bones were detected. Therefore, we attempted to dig away and discard all of Layer 1. We cannot be certain that we were successful at this, for the lower limits of this layer were not sharply demarked. Because fossils, if present at all, were very rare here, we doubt that contamination of Layer 2 occurred.

Layer 3 also seemed to be barren of fossils. Layer 2, the productive layer, was removed from the surface of Layer 3 with ease, for the two layers are sharply demarked. Sediments of Layer 3 were examined in many places but no fossils of any kind were found. Limb bones, vertebra, or jaws of even small mice should have been readily apparent in the sand-like materials, even in the light of a gasoline lantern.

In contrast to layers 1 and 3, Layer 2 contained abundant vertebrate fossils. Quite early in the excavation, the jaw of a diminutive antilocaprid, *Capromeryx,* was found in place near the middle of the layer. This extinct species was sufficient to establish the Pleistocene age of the layer, and deeper excavation was thought undesirable at that time. Thirty buckets filled with Pleistocene sediments, with estimated total weight in excess of 1,000 pounds, together with remains found in Recent silts, form the basis of this report.

The early Recent silts were washed and sorted in the customary fashion (Hibbard, 1949) and presented no problems. The half-ton of Pleistocene matrix was also washed but this resulted in the elimination of only about 200 pounds of the finely-divided material. However, the coarse, clean, washed matrix was reduced, with the aid of gold-pans, to approximately 50 pounds of concentrate containing almost all of the teeth, bones and bone fragments. This concentrate was sorted under jewelers' lenses to obtain the teeth and jaws used in this report.

SYSTEMATIC ACCOUNTS OF PLEISTOCENE SPECIES

Of the 42 taxa listed from the Pleistocene deposits of the Fowlkes Cave local fauna, the two ungulates (*Mylohyus* and *Capromeryx* cf. *C. furcifer*) are extinct. Of the 40 extant forms, a pocket mouse (*Perognathus*) can not be identified to species and a harvest mouse (*Reithrodontomys fulvescens*) is only tentatively identified. Thirty-eight species have been identified with reasonable certainty. In the accounts of species of the Pleistocene fauna that follow, the geographic range and present day habitat requirements of each species are given, if pertinent, and the basis of identification is explained. Numbers in parentheses in each account are those of the Midwestern State University Collection of Fossil Vertebrates and pertain to material mentioned in Table 1. Scientific and vernacular names, and systematic arrangement of species are taken from Jones et al. (1982).

Sorex vagrans Baird

Vagrant Shrew (11928, 11929)

The vagrant shrew does not occur in Texas today but is found in the mountains of south-central New Mexico (Hall, 1981), and ranges southward onto the Mexican Plateau. It occurred in Eddy Co., southern New Mexico, in post-Pleistocene times (Findley, 1965). In New Mexico it is found in greatest numbers in streamside thickets, marshes, and meadows. It occupies woodlands where the soil is soft and humid. In New Mexico, it is an alpine form that does not occur at low elevations.

Sorex palustris Richardson

Water Shrew (11947)

Water shrews range across much of Canada and mountainous areas of the United States, but today do not occur south of the mountains of western-central Arizona and northern New Mexico. In the late Pleistocene, the species lived on what is now the Mesquite Plains of the southern Texas Panhandle (Dalquest 1965). Typically, the water shrew lives along the margins of small, swift, cold-water streams in alpine habitat but is often found in marshes and wet meadows beside brooks and rills.

Notiosorex crawfordi (Coues)

Desert Shrew (11930)

This must have been an abundant species near Fowlkes Cave in the late Pleistocene, for it is represented by more than 100 fossil specimens. The desert shrew ranges from California to Arkansas and from southern Colorado to central Mexico. It has a broad tolerance to environmental conditions, and is found from dry deserts to tropical forests. Its preferred habitat seems to be arid brushland, and it is rarely trapped by collectors. However, one individual was captured under a rock in typical Chihuahuan desert habitat in Jeff Davis Co., about 15 km south of Fowlkes Cave.

Myotis lucifugus (LeConte)

Little Brown Bat (11931)

The only specimen is a lower jaw fragment but preservation of the fossil is good. Size and dentition match specimens of the living *Myotis lucifugus occultus* Hollister. Hall (1981) indicates that the only record of *Myotis lucifugus* from Texas is a specimen from Fort Hancock, a site well east of Fowlkes Cave, but the species ranges widely in New Mexico. It is typically a woodland bat, but wanders into other habitat in its nocturnal hunting activities. Presence near Fowlkes Cave today, as a transient, would not be unexpected.

Myotis velifer (J. A. Allen)

Cave Myotis (11933)

The cave myotis lived in Fowlkes Cave in the recent past, for many bones were discovered on the surface of the black silts of the cave floor and were scattered through the surficial silts. Fossil bones were common in the late Pleistocene sediments as well. Jaws from the Pleistocene part of the cave average

Table 1.—*Modern, early Recent, and Pleistocene Fauna of Fowlkes Cave. Note: Column 1 includes species known to live in the vicinity of the cave today. Column 2 lists species whose remains were found in the early Recent deposits of the cave. Column 3 indicates Pleistocene records from the cave.*

Species	1	2	3	Basis for Pleistocene record
Sorex vagrans			+	2 lower jaws with teeth
Sorex palustris			+	1 maxillary with teeth
Notiosorex crawfordi	+	+	+	2 rostra, 14 maxillaries, 114 lower jaws
Myotis lucifugus			+	1 lower jaw with M_1–M_3
Myotis velifer	+	+	+	1 rostrum, 2 maxillaries, 27 lower jaws
Pipistrellus hesperus	+			
Eptesicus fuscus		+	+	7 lower jaws
Antrozous pallidus	+	+		
Mormoops megalophyla	+			
Tadarida braseliensis	+			
Homo sapiens	+	+		
Lepus californicus	+	+	+	Several fragments of limb bones
Sylvilagus floridanus			+	1 lower jaw with teeth
Sylvilagus audubonii	+	+	+	2 lower jaws, premaxillary with incisors
Marmota flaviventris			+	5 cheek teeth
Eutamias cinericollis			+	Lower jaw fragment with M_2
Ammospermophilus interpres	+			
Spermophilus spilosoma			+	Lower jaw fragment with M_2; isolated M_1^1
Spermophilus variegatus	+	+	+	Isolated teeth
Cynomys ludovicianus			+	1 M^3
Thomomys bottae	+	+	+	5 skulls, 15 palates, 71 lower jaws
Pappogeomys castanops	+	+	+	4 upper incisors, 4 cheek teeth
Perognathus hispidus	+	+	+	1 skull, 116 lower jaws, many maxillaries
Perognathus sp.		+	+	3 lower jaws
Perognathus flavus	+	+	+	657 lower jaws, many maxillaries
Dipodomys ordii	+	+	+	107 lower jaws
Dipodomys merriami	+	+	+	14 lower jaws
Dipodomys spectabilis	+	+	+	77 lower jaws
Reithrodontomys megalotis	+	+	+	84 lower jaws
Reithrodontomys montanus	+	+	+	19 lower jaws
Reithrodontomys fulvescens		+	+	2 tentatively referred lower jaws
Peromyscus eremicus	+	+	+	55 lower jaws
Peromyscus maniculatus	+	+	+	16 lower jaws
Peromyscus leucopus	+	+	+	10 lower jaws
Peromyscus boylii			+	22 lower jaws
Peromyscus pectoralis	+	+	+	6 lower jaws
Peromyscus difficilis		+	+	3 lower jaws
Onychomys leucogaster			+	183 lower jaws
Onychomys torridus	+	+	+	6 lower jaws
Sigmodon hispidus	+	+	+	1 palate, 44 maxillaries, 189 lower jaws
Neotoma micropus	+	+	+	23 lower jaws
Neotoma albigula	+	+	+	33 lower jaws
Neotoma mexicana			+	1 lower jaw
Microtus mexicanus			+	2 lower jaws
Erethizon dorsatum			+	1 incisor fragment
Canis latrans	+	+		
Bassariscus astutus	+	+		
Felis rufus			+	Maxillary fragment with teeth
Dicotyles tajacu	+			
Mylohyus sp.			+	Tusk fragment
Capromeryx cf. C. furcifer			+	Lower jaw
Odocoileus hemionus	+			

larger than Recent specimens and probably represent *Myotis velifer magnimolaris* Choate and Hall, a race now found in Kansas, Oklahoma, and north-central Texas (Dalquest and Stangl, in press).

The cave myotis ranges from Kansas southward over much of Mexico, with a distinct preference for arid habitat. Its presence in Fowlkes Cave is not surprising. There are numerous records of the species from late Pleistocene caves in Texas.

Eptesicus fuscus (Beauvois)
Big Brown Bat (11934)

The big brown bat is widespread, ranging from the tropical jungles of southern Mexico to southern Canada and from coast to coast. We did not capture living specimens at Fowlkes Cave, but it may occur there, at least at times. *Eptesicus fuscus* is not usually considered a desert bat but we have specimens taken in typical Chihuahuan desert habitat in Presidio Co., Texas. There are numerous records of the big brown bat from the Pleistocene of Texas.

Lepus californicus Gray
Black-tailed Jackrabbit (11935)

Black-tailed jackrabbits range widely over much of western United States and Mexico, and are common about Fowlkes Cave today. However, the only fossils recovered in the cave include the proximal end of a femur, distal half of a tibia, distal end of a humerus, and part of an inominate. The sutures of the bones are closed but they are all a bit smaller than bones of modern jackrabbits from north-central Texas.

The scarcity of jackrabbit remains in the Pleistocene sediments of Fowlkes Cave probably results from the habits of the species. They prefer level areas of ground, with a cover of grass or scattered clumps of brush. They would avoid the steep, barren hillside where the mouth of Fowlkes Cave is located. Adults are too large to be brought into the cave by barn owls. *Lepus californicus* has been recorded from almost all late Pleistocene cave deposits in Texas where the microvertebrate fossils have been studied.

Sylvilagus floridanus (Allen)
Eastern Cottontail (11936)

The lower jaw fragment contains the M_2 and the alveoli of the other cheek teeth. The M_2 is 2.2 mm long, the trigonid is 2.8 mm wide, and the talonid is 2.15 mm wide. Alveolar length P_2–M_3 is 13.8 mm. These measurements are too great for *Sylvilagus*

audubonii and are like those of the few available specimens of *S. f. robustus* from the Davis Mountains, to the south of Fowlkes Cave. Some authors consider the Davis Mountains cottontail to belong to a distinct species, *S. robustus* Bailey.

The Davis Mountains cottontail lives in the pine-oak woodland of the Davis Mountains and the Guadalupe Mountains. It ranges into the oak-juniper belt where there is ample cover, but does not enter the desert habitat like that surrounding Fowlkes Cave today. The presence of the single specimen shows that the range of the eastern cottontail extended completely across the gap between the ranges of the present populations in the late Pleistocene. The single fossil also indicates existence of areas of thicker brush or woodland in the late Pleistocene than occur near Fowlkes Cave today.

Sylvilagus audubonii (Baird)
Desert Cottontail (11937–11939)

In addition to the material listed in Table 1, there are numerous isolated cheek teeth and some limb bones that probably belong to this species. Many of the teeth are of immature animals. The desert cottontail is a small species but adult individuals are probably too large to be readily overcome by barn owls, and the fossils of adult cottontails are probably of animals that fell into the cave trap.

Eutamias cinericollis (J. A. Allen)
Gray-collared Chipmunk (11941)

Chipmunk is represented by a lower jaw fragment with the M_2. The specimen is referred to the species that lives today in the Guadalupe Mountains, but the reference is tentative. Other species occur in the mountains of New Mexico.

Eutamias cinericollis inhabits montane habitat and does not descend to the desert. It probably ranged into the Fowlkes Cave area in the late Pleistocene along cool, ice-fed streams, along with the vagrant shrew, water shrew, voles, and other montane species. There are records from the late Pleistocene cave deposits in the Guadalupe Mountains and southern New Mexico.

Marmota flaviventris (Audubon and Bachman)
Yellow-bellied Marmot (11940)

The five cheek teeth are clearly of a single individual. They are P^4, M^1, M_1, and right and left M_2's, with roots closed but almost unworn.

Marmots today occur no closer to Fowlkes Cave

than the mountains of northern New Mexico, and their range extends northward almost to the arctic region. There are numerous late Pleistocene records from southern New Mexico and southward as far as Nuevo Leon, Mexico (Stock, 1942). The yellow-bellied marmot seems to be an obligatory inhabitant of rocky areas, cliffs and talus. In the western parts of the range of the species (Nevada, Oregon, etc.), marmots occur in desert habitat but they do not occur today in Chihuahuan Desert.

Spermophilus spilosoma Bennett
Spotted Ground Squirrel (11942)

A lower jaw fragment has the alveoli of P_4–M_3 but only M_2 present. There is also an isolated M^1, tentatively referred. Alveolar length of the tooth row is 7.8 mm, and the M_2 measures 1.9 by 1.9 mm. The isolated tooth measures 2.1 by 1.8 mm. The jaw ramus is not narrowed at the site of P_4, as is the similar-sized antelope ground squirrel, *Ammospermophilus*. The jaw and teeth differ from *Ammospermophilus* in other details also.

Spermophilus spilosoma lives in prairie and desert habitat from South Dakota to central Mexico. It shows definite preference to sandy soils, which are almost lacking about Fowlkes Cave today but must have been present in the late Pleistocene.

Spermophilus variegatus (Erxleben)
Rock Squirrel (11943)

One enamel cap of an upper molar had not erupted, and is definitely of the rock squirrel. However, there are three well-preserved upper and one lower cheek teeth, and several worn or broken teeth, that are only tentatively referred to this species. The enamel pattern and size of these teeth is like that of the rock squirrel, but the lingual halves of the crowns are markedly narrower than in *S. variegatus*. Numerous specimens of the modern forms from Trans-Pecos Texas and the Edwards Plateau were studied, and none had teeth like these fossils. In shape, the teeth resemble those of *S. columbianus* but are much larger than that northern species. *S. beecheyi* (Richardson) and *S. annulatus* Audubon and Bachman are large species but their teeth do not have the narrowed, triangular shape like that of the fossils. The fossils show several different stages of wear, and represent several different individual squirrels. They probably represent a slightly aberrant population of *S. variegatus*. If rock squirrels live or lived in an area, their remains may be expected in the cave deposits. Fossils are absent in the late Pleistocene

deposits of caves on the Edwards Plateau, though they are common in the early Recent surficial silts of the same caves. The species is recorded from the caves in the Guadalupe Mountains (Williams Cave, Ayer, 1937; Upper Sloth Cave, Logan and Black, 1979) but seems to be absent from the well-studied caves of southern New Mexico (Harris, 1974). Apparently rock squirrels lived in the Trans-Pecos area in the late Pleistocene but did not extend their range eastward to the Edwards Plateau and northward to southern New Mexico until the Recent.

Cynomys ludovicianus (Ord)
Black-tailed Prairie Dog (11944)

The prairie dog is represented in the cave fauna by a single worn M^3. This diurnal, large squirrel would not be prey for the barn owl, and the tooth must have arrived in the cave in some fashion other than most of the other fossils.

Prairie dogs lived in Culberson Co. until quite recently, but we found no evidence of their existence near Fowlkes Cave today. Doubtlessly they have been exterminated by man. There are records of the prairie dog from caves in southern New Mexico, and from Pratt Cave, in the Guadalupe Mountains (Lundelius, 1979), but fossils of this species are usually rare in caves. The rodents are active only by day, and are too large for the cave-inhabiting owls, like the barn owl, to carry. They live on level grasslands, usually far from the rugged terrain where caves occur.

Thomomys bottae (Eydoux and Gervais)
Botta's Pocket Gopher (11953–11956)

Remains of *Thomomys* are abundant in the cave and include nearly complete skulls, numerous palates, and many lower jaws. Isolated teeth are present by the thousands. Obviously the small pocket gopher was a favored prey of the barn owls in the late Pleistocene as it is today. The cave record shows that it was abundant in the past, both in the Pleistocene and early Recent, but it is a rare resident near the cave today. We trapped one specimen about 2 km from the cave mouth.

Schmidly (1977) notes that the common pocket gopher of the deeper soils of the Trans-Pecos is *Pappogeomys castanops*, whereas *Thomomys bottae* is found on thinner soils of hillsides and mountains, usually in the vicinity of lecheguilla plants. It has been suggested (Reichman and Baker, 1972) that *Pappogeomys* is replacing *Thomomys* in the Trans-Pecos. This seems to be true at Fowlkes Cave, where remains of *Thomomys* are abundant in late Pleistocene and early Recent sediments. We set traps near the cave, where gopher workings were observed near lecheguilla plants on thin soils of hillsides, and expected to capture *Thomomys*. Instead, the traps took only *Pappogeomys*. Our only *Thomomys* was trapped in deep soils, and a *Pappogeomys* was taken in a trap set 30 meters away. *Pappogeomys* remains are rare in the cave sediments, but the species today lives about the cave almost everywhere that gophers can exist. Changes in the environment caused by cattle or other activities of man may be responsible for the increasing scarcity of *Thomomys* in Culberson Co.

It is possible that the cave record may be deceptive with regard to *Pappogeomys*. The animals are quite large and their burrows are often deeper in the ground than those of *Thomomys*. Owls may be less able to capture the larger species. We are unable to explain the ability of the barn owl to capture *Thomomys* so easily, in view of the almost completely fossorial habits of the rodents. However, examination of regurgitated pellets of barn owls collected from areas where *Thomomys* live, almost invariably show *Thomomys* to be a common food item. Whatever feature of the habits of *Thomomys* that makes them susceptible to predation by barn owls may not operate for *Pappogeomys*.

Pappogeomys castanops (Baird)
Yellow-faced Pocket Gopher (11957)

Fossils are rare and include four upper incisors, one P^4 and three P_4's. In contrast, *Thomomys* is represented by nearly one hundred skulls and jaws. Although fragmentary, the fossils show definitely that *Pappogeomys* lived about Fowlkes Cave in the late Pleistocene. There are records from cave deposits in the Guadalupe Mountains and southern New Mexico as well, but most suggest relatively late Pleistocene or early Recent age. *Pappogeomys* are found from Kansas and Colorado southward deep into Mexico. It is not known from the Pleistocene cave deposits of the Edwards Plateau, but does occur there in early Recent cave sediments. Presumably, *Pappogeomys* invaded Texas in the late Pleistocene.

Genus *Perognathus*

Five species of pocket mice occur in the Trans-Pecos today, but one of these (*Perognathus nelsoni*) is not recorded from Culberson Co. It may, of course, have lived there in the past. We trapped specimens of the large species, *Perognathus hispidus*, and the diminutive species, *P. flavus*, less than 1 km from the cave. We did not find the other species (*P. nelsoni, P. intermedius,* and *P. penicillatus*), although special efforts were made to catch them.

Identification of fossil *Perognathus* jaws presents special problems. No characters of the enamel pattern seem reliable in separating species, and any enamel pattern is transient because surface patterns of the teeth wear down swiftly and change with age and wear. Relative size of P_4 versus M_3 may be helpful but is of little practical value for the Fowlkes Cave fossils, because most have lost the M_3. Transverse breadth of the lower incisor and alveolar length of the tooth row are the only effective means of identifying the fossils, and these are of limited value in separation of some species.

Alveolar length of the lower tooth row was determined in 25 adult, randomly-chosen specimens of the five species of pocket mice. Measurements were taken with an occular micrometer, with the jaw held at approximately a 40° angle, and viewed from the lingual side. Extremes of measurements were: *P. hispidus,* 4.0–4.75 mm; *P. nelsoni,* 3.35–3.70 mm; *P. intermedius,* 3.25–3.70 mm; *P. penicilliatus,* 3.10–3.90 mm; *P. flavus,* 2.65–3.30 mm.

The incisor breadth is readily measured with a caliper. In *P. hispidus* the breadth exceeds .65 mm; in the other species the breadth is less than .6 mm. *P. hispidus* is readily identified by the breadth of the incisor. In the other species, however, there is so much overlap that the character is of no value. Surprisingly, the little *P. flavus* sometimes has the breadth of the incisor greater than that of some *P. penicillatus*.

Perognathus hispidus lower jaws are therefore separated with confidence on the basis of incisor breadth and alveolar length of tooth row. The tiny *P. flavus* can not be identified with certainty, but more than 90% of the jaws of this species have the length of the tooth row less than 3.15 mm. More than 90% of the *P. penicillatus* examined have the tooth row more than 3.15 mm, as do all *P. nelsoni* and *P. intermedius*. Thus, lower jaws with the alveolar length of the tooth row 3.15 mm or less are referred to *P. flavus* with the reservation that a very few small (tooth row 3.1–3.2 mm) *P. penicillatus* might be included. No firm method of separating fossils of the three medium-sized species of pocket mice was found.

Alveolar length of the tooth row of *Perognathus* tends to be greater in aged animals, where the bone often retreats from the roots of the teeth when these are excessively worn. This is apparent only in jaws with greatly worn teeth. It was also noted that measurements of alveolar length of the tooth row tended to be slightly greater in jaws lacking P_2 and M_3. Perhaps the loss of teeth involves slight but imperceptible damage to the alveoli.

Perognathus hispidus Baird

Hispid Pocket Mouse (11945)

Remains of this species are abundant in the Pleistocene and early Recent deposits of the cave. It was clearly a favored food of the barn owls.

The hispid pocket mouse has a geographic range that extends from North Dakota southward deep into Mexico, and includes much of Texas. It is primarily a prairie or grassland mouse. Its abundance in the past may be due to the apparent absence of competing medium-sized species of pocket mice. *P. hispidus* is known from many late Pleistocene cave deposits, including those in the Guadalupe Mountains and southern New Mexico.

Perognathus flavus Baird

Silky Pocket Mouse (11947)

The silky pocket mouse is the most abundant species of mouse in the cave fauna, represented by 657 complete to fragmentary lower jaws and numerous palates and maxillary fragments.

Perognathus flavus occurs from Wyoming to southern Mexico and from western Arizona to the Gulf Coast of Texas, including all of the Trans-Pecos. It has a broad range of tolerance to environmental conditions and inhabits prairies and deserts, level ground and rocky, broken land. No particular significance can be placed on its occurrence in Fowlkes Cave. Wherever found, it seems to be a favored prey of barn owls.

Perognathus sp. (11946)

Three lower jaws are from medium-sized pocket mice. They may be of any of the following: *Perognathus intermedius, P. nelsoni,* or *P. penicillatus.*

Perognathus nelsoni and *P. intermedius* prefer stony soils and are rarely found far from rocks and rough, broken land. *P. nelsoni* seems not, at this time, to occur north of the Davis Mountains, whereas *P. intermedius* has been found only along the western border of Culberson Co. The present absence of these species in the vicinity of Fowlkes Cave and the areas to the north and east, may be significant in view of the scarcity of remains of medium-

sized pocket mice in the Pleistocene sediments of the cave.

Perognathus penicillatus occupies loose, sandy soils, and its range includes virtually the entire Trans-Pecos. No suitable habitat for the desert pocket mouse was found within several kilometers of the cave, and trapping in the sandy areas that we did find, yielded no specimens of *P. penicillatus.* Sandy soils certainly existed closer to the cave in the late Pleistocene, in view of the abundance of other sand-dwelling species (for example, *Dipodomys ordii, Onychomys leucogaster*) in the cave fauna. It is probable, on the basis of modern distribution, that the three fossil jaws listed belong to *P. penicillatus.* However, where *P. penicillatus* is found today, it is usually abundant, whereas *P. intermedius* and *P. nelsoni* are rarely abundant in any area. If *P. penicillatus* lived near Fowlkes Cave at all, it should have been represented by numerous fossil jaws. The extent of predation on this species by barn owls seems not to have been investigated in detail.

Genus *Dipodomys*

Three species of kangaroo rats occur near Fowlkes Cave today—the large *Dipodomys spectabilis,* medium-sized *D. ordii,* and small *D. merriami.* The three are readily separated by size of the lower jaw and incisor. In *D. spectabilis* the breadth of the incisor is greater than 1.0 mm, with no overlapping of the measurements of those of *D. ordii.* Incisor breadth of *D. ordii* is .8 to .95 mm; in *D. merriami* it is less than .8 mm. Although no actual overlap in measurements was found in specimens from the Trans-Pecos, it is possible that some slight overlap may occur.

Alveolar length of tooth row is greater than 5.6 mm in *Dipodomys spectabilis,* 4.8–5.5 mm in *D. ordii,* and less than 4.75 in *D. merriami.*

The breadth of the incisor is the best character to use in separation of fossil jaws. The incisor usually remains in the fossil jaws and is readily measured with a caliper. If the incisor is missing, the alveolar length of the tooth row is available. Unfortunately, the M_3 of the fossil *Dipodomys* is often missing, and usually the thin bone bordering the alveolus posteriorly is broken away as well. It is usually impossible to then estimate the position of the posterior border of the alveolus. *Dipodomys spectabilis* can then be identified by the large size of cheek teeth, and *D. merriami* can be separated from *D. ordii* by smaller molars.

Dipodomys ordii Woodhouse

Ord's Kangaroo Rat (11948)

Ord's kangaroo rat is found in desert areas of the west from southern Canada to southern Mexico, including all of the Trans-Pecos. It shows a strong preference for loose, sandy soils and rarely occurs on clay soils or stony ground. The abundance of this species at Fowlkes Cave in Pleistocene deposits suggests that loose, sandy soils must have existed near the cave in the late Pleistocene. No suitable habitat for the species was noted around the cave today. However, one specimen was taken in a trap less than 1 km from the cave, along with several Merriam kangaroo rats. It was not recognized as *D. ordii* until it was being prepared as a study specimen. A small population survives in the area, though suitable habitat is several kilometers away. Whether the specimen represents a remnant of the Pleistocene population or is a more recent invader can not be known.

Dipodomys spectabilis Merriam

Banner-tailed Kangaroo Rat (11950)

The banner-tailed kangaroo rat is found from northern New Mexico to central Mexico, and over much of the Trans-Pecos. It inhabits the high desert grasslands but occurs also on mesquite flats and aluvial fans. It does not seem to like the extensive creosote bush flats with their stony soils, or steep hillsides. It is moderately common about Fowlkes Cave, and a number of specimens were trapped near the cave.

Dipodomys merriami Mearns

Merriam's Kangaroo Rat (11949)

Merriam's kangaroo rat has a geographic range extending from northern Nevada to central Mexico and California to western Texas. Its preferred habitat in Texas is the creosote bush flats, but it has a relatively great range of tolerance to conditions. *D. merriami* and *D. spectabilis* are commonly sympatric. Wherever found, *D. merriami* is an indicator of true, arid, desert conditions. It is the common kangaroo rat near Fowlkes Cave today. Its relatively scarcity (14 jaws versus 107 *D. ordii* and 77 *D. spectabilis*) indicates a change in environment between the late Pleistocene and modern times.

Genus *Reithrodontomys*

Three species of harvest mice are sympatric in Trans-Pecos Texas today—*Reithrodontomys meg-alotis, R. montanus,* and *R. fulvescens. R. montanus* can readily be separated from the other two forms by the alveolar length of the tooth row—less than 3.0 mm in *R. montanus,* but greater than 3.0 mm in *R. megalotis* and *R. fulvescens.* Although *R. fulvescens* is larger and heavier-bodied than *R. megalotis,* there is some overlap in measurements of breadth of lower incisor, alveolar length of M_1–M_3, and length of M_1. In jaws retaining a M_3 in the proper stages of wear, *R. megalotis* exhibits an "C" pattern to the enamel, while the pattern in *R. fulvescens* appears as a "S."

Reithrodontomys megalotis (Baird)

Western Harvest Mouse (11912)

The geographic range of *Reithrodontomys megalotis* extends from southern Canada to southern Mexico and includes much of the United States west of the Mississippi. It is found throughout the Trans-Pecos, and is common about Fowlkes Cave today. Its habitat extends from hot desert to alpine meadows, and grassy plains and prairies to cliffs and talus slides. The species is not a reliable indicator of past climate. It has been found in the Pleistocene fauna of Pratt Cave, Culberson Co. (Lundelius, 1979).

Reithrodontomys montanus (Baird)

Great Plains Harvest Mouse (11911)

Reithrodontomys montanus is a grassland mammal, usually found where the grass is relatively short but dense enough to offer overhead cover. It is most often found in moderately arid climate where soils are well-drained. It occurs in the Trans-Pecos today, but usually is sparsely distributed and rare. We trapped one specimen near Fowlkes Cave. The 18 jaws recorded from Fowlkes Cave (versus 84 jaws of *R. megalotis*) suggests that *R. montanus* was relatively more abundant near the cave in the late Pleistocene, and that grasslands were present closer to the cave than at present. We know of no records of *R. montanus* from the caves in the Guadalupe Mountains or southern New Mexico.

Reithrodontomys cf. *fulvescens* J. A. Allen

Fulvous Harvest Mouse (11913)

Two lower jaws in the cave collection have the alveolar length of the lower tooth row and breadth of the lower incisor greater than the maximum expected in *Reithrodontomys megalotis* (3.25 and .56 mm respectively). These measurements are within the lower limits for *R. fulvescens.* Both, unfortunately, lack the diagnostic M_3. They are hesitantly referred to *R. fulvescens,* but may be unusually large late Pleistocene representatives of *R. megalotis.*

The geographic range of *Reithrodontomys fulvescens* extends from Kansas and Missouri southward through Mexico to Nicaragua. It is rare in the Trans-Pecos, seemingly confined to the eastern half of the area. It has not been recorded from Culberson Co. but has been taken in Reeves Co., 50 mi to the east (Schmidly 1977). The fulvous harvest mouse is a grassland species that prefers denser and taller grass than does *R. montanus.* The two species rarely occur together. There are no records of *R. fulvescens* from the Pleistocene cave deposits in the Guadalupe Mountains or southern New Mexico.

Genus *Peromyscus*

This genus presents special problems in identification in that seven species are known to live today in Culberson Co. (Schmidly, 1977) and another (*Peromyscus crinitus* Merriam) has been recorded from Dry Cave, Eddy Co., New Mexico (Harris, 1974), slightly more than 100 mi to the north. *P. maniculatus, P. leucopus, P. eremicus,* and *P. pectoralis* occur within 1 km or so of Fowlkes Cave, and *P. boylii, P. difficilis,* and *P. truei* occur at higher elevations, in the northern part of the county. Criteria for the identification of lower jaws of these eight species, based on Recent specimens, have been discussed elsewhere (Dalquest and Stangl, 1983). Only a brief summary is given here.

The incisor-base capsule is rated in three grades—Grade 0 (only a slight bulge when seen from directly above), Grade 1 (strong bulge when seen from directly above but not forming a blunt, finger-like projection), and Grade 2 (capsule so strongly developed as to form a blunt, finger-like projection). This character serves to separate *Peromyscus maniculatus, P. leucopus, P. crinitus,* and most *P. eremicus* from the other four forms concerned here.

The anteroconid of M_1 is also rated in three grades—Grade 0 (not divided), Grade 1 (narrowly divided), and Grade 2 (separated by a broad anterior valley). Other complications of the enamel pattern of M_1 found to be of value include the presence or absence of ectostylids, mesostylids, ectolophids, and mesolophids. Measurements of the alveolar lengths of M_1–M_3 serve to identify *Peromyscus difficilis* and aid in identification of other forms.

Measurements and characters overlap broadly in some species but, in combination, permit identification of nearly all jaws of some species, some individuals of other species, and tentative identification of still others.

Peromyscus crinitus is readily separated from the other concerned species. The incisor-base capsule is Grade 2, anteroconid not divided, lophids and stylids absent on M_1, and alveolar length of M_1-M_3 is 3.5 mm or less. The incisor-base capsule is uniquely narrow. It would be possible to confuse this species with lower jaws of large *Reithrodontomys* if the incisor-base capsule were severely damaged.

Only four specimens of *Peromyscus truei* were available from Texas. All had the incisor-base capsule Grade 0, two had divided anteroconids (Grade 2), one had an ectostylid, and another had an ectolophid. Alveolar length of the lower cheek-tooth row ranged from 3.85 to 4.20 mm. Specimens from Arizona, Utah, and westward usually had M_1's with deeply divided anteroconids and strongly developed stylids and lophids. The Texas mice appear to be more variable and to have more simple M_1's than those to the west. Four specimens are too few to characterize the species but it would appear that it might be confused with large individuals of *P. pectoralis* or *P. boylii,* or even small individuals of *P. difficilis.* No fossils are referred to *P. truei.*

All specimens of *Peromyscus* from Fowlkes Cave that were utilized were more or less fragmentary lower jaws. Another 22 jaws are moderately well-preserved but either are too worn or have critical features too damaged to be identified. These have been referred to *Peromyscus* sp. (11922).

Peromyscus difficilis has the incisor-base capsule Grade 0, anteroconid divided (90%), stylids and lophids usually present (86% and 82%), and the alveolar length of M_1–M_3 is 4.1 mm or greater. Only the very largest *P. pectoralis* might be mistaken for *P. difficilis,* having the M_1–M_3 length 4.1 mm.

Peromyscus boylii and *P. pectoralis* have the incisor-base capsule Grade 0 and the M_1–M_3 length almost the identical. However, lower jaws are readily separated because *P. boylii* has relatively few stylids (29%) and almost never has lophids in the major valleys of M_1. *P. pectoralis* usually has stylids (86%) and lophids (82%). The anteroconid of M_1 is divided in *P. boylii* in about 60% of the specimens studied, but in *P. pectoralis* the anteroconid is almost always divided, and is usually more deeply and strongly divided than in *P. boylii.* Thus, *P. boylii* and *P. pectoralis* can be separated with confidence in almost all instances.

Separation of these two species from *P. eremicus* is also a problem, for *P. eremicus* may have a Grade

0 incisor capsule. More than a third of the specimens examined (39%), resemble *P. boylii* and *P. pectoralis* in this respect. Because *P. eremicus* usually (82%) lacks the divided anteroconid, it differs from *P. pectoralis,* where the anteroconid is almost always divided. Moreover, the 18% of *P. eremicus* that do have the anteroconid divided, it is always Grade 1. In *P. pectoralis,* it is Grade 2 in 61% of the Recent specimens studied. Thus, *P. pectoralis* is separated from *P. eremicus* with confidence, although it is possible that a very few specimens might be wrongly identified by the listed characters.

There are no good characters that will reliably separate the lower M_1 of *P. boylii* from those *P. eremicus* that have Grade 0 incisor-base capsules. Although the alveolar length of the lower tooth row overlaps broadly in the two species (3.55–3.95 mm versus 3.65–4.20 mm), approximately 90% of the specimens of *P. eremicus* have the measurement 3.85 mm or less, whereas 79% of the *P. boylii* have the length 3.85 mm or greater. This is used in the present identifications.

Identification of those species with well-developed incisor-base capsules (*Peromyscus maniculatus, P. leucopus,* and *P. eremicus*) is more difficult. However, *P. leucopus* can be separated from the other two species in almost all instances. *P. leucopus* is larger, and the alveolar length of the lower tooth row, although it overlaps that of the other species broadly (*leucopus* 3.55–4.15 mm; *maniculatus* 3.42–4.00 mm; *eremicus* 3.55–3.95 mm), almost 90% of the *P. eremicus* have the tooth row less than 3.85 mm, and 96% of the *P. maniculatus* have it less than 3.85 mm, whereas 79% of the *P. leucopus* have the tooth row 3.95 mm or greater. Further, all *P. leucopus* have undivided anteroconids and lack lophids in the M_1, whereas 21% of the *P. maniculatus* and 5% of the *P. eremicus* have divided anteroconids and 29% of the *P. eremicus* possess lophids on the M_1's. Still further, almost all (96%) of the *P. leucopus* have the incisor-base capsule Grade 2, whereas 36% of the *P. maniculatus* and 39% of the *P. eremicus* have the incisor-base capsule Grade 1.

As a check, all fossil jaws identified as *Peromyscus leucopus* because they had Grade 2 incisor-base capsules and the alveolar length of the M_1–M_3 measured 3.95 or more, were checked for divided anteroconids and presence of lophids. None were found.

Separation of *Peromyscus maniculatus* from those *P. eremicus* that have Grade 1 incisor-base capsules is more difficult. Jaws with incisor-base capsules Grade 2 and tooth rows measuring less than 3.95

mm are considered to be *P. maniculatus,* though a very few small *P. leucopus* could be included or even some *P. eremicus.* There are average differences between Recent specimens of the two species (for example, 92% of *P. eremicus* have lophids present on M_1 versus only 7% in *P. maniculatus*), but we find no character that can be of reliable use. In Recent material, only 60% of the *P. eremicus* resemble *P. maniculatus* in having Grade 1 or Grade 2 incisor-base capsules. There are 19 jaws of this type, and 55 jaws referred to *P. eremicus* because they have Grade 0 incisor-base capsules but alveolar length of lower tooth row 3.85 mm or less. If only *P. eremicus* were included in the collection, these figures would be reverse, providing that the percentages were the same in the late Pleistocene as they are today.

There is evidence of the presence of *Peromyscus maniculatus* in the Fowlkes Cave collection, on the basis of specimens with Grade 2 incisor-base capsules but there are 19 jaws that might represent either *P. maniculatus* or *P. eremicus.*

With the foregoing reservations, it is thought that six species of *Peromyscus* are represented in the cave collection.

Peromyscus eremicus (Baird)
Cactus Mouse (11919)

Peromyscus eremicus ranges from southern Nevada and Utah southward through the deserts of Mexico, and from western Texas to California. It is found in suitable habitat throughout the Trans-Pecos. It inhabits the hot desert flats and rocky hillsides above the creosote bush flats, where cover in the form of cactus, yuccas, or other plants or rocks occur. It is commonly associated with the Merriam kangaroo rat and southern grasshopper mouse. The few records of this species from the late Pleistocene are open to question in view of the difficulty of identification.

Peromyscus maniculatus (Wagner)
Deer Mouse (11914)

This species is almost ubiquitous, ranging over most of temperate North America and penetrating into almost every habitat where small mammals survive. The present inhabitant of Culberson Co., including the Guadalupe Mountains (Schmidly 1977) is *P. m. blandus* Osgood, a pale race usually considered a desert mouse. Today it is abundant about Fowlkes Cave, inhabiting the desert flats and the lower margins of the aluvial fans, but rarely enters rocky land or hillsides where other species of *Peromyscus* occur. Because the population of the Guadalupe Mountains as well as that of the desert about Fowlkes Cave are both referable to *P. m. blandus,* that is probably the race that lived near the cave in the late Pleistocene.

Peromyscus leucopus (Rafinesque)
White-footed Mouse (11915)

Peromyscus leucopus is found over much of eastern United States, from southern Canada to southern Mexico, and westward to Arizona, including virtually all of Texas. The desert race, *P. l. tornillo* Mearns, prefers denser cover than most desert mammals, and is most common in thickets along streambeds and dry washes, but it enters rocky areas as well. There are few records of *P. leucopus* from the late Pleistocene of western Texas and southern New Mexico, probably due to difficulty in identification.

Peromyscus boylii (Baird)
Brush Mouse (11918)

Peromyscus boylii ranges from northern California and Utah to southern Mexico, and western Texas to California. In Trans-Pecos, Texas it is found mostly at higher elevations, in the pinyon-juniper belt and upward into the pine forests. In some localities, however, as in southern Brewster Co., it is found in the denser woody vegetation fringing canyons at low elevation, not far from the Rio Grande. The presence of rocky cliffs or talus in rather dense brushland or woods seems to be a requirement for the species.

There is no habitat at present near Fowlkes Cave suitable for *P. boylii,* but it is found in the Davis Mountains to the south and the Guadalupe Mountains to the north. Its presence in the cave fauna indicates heavier cover on the hills about the cave in the late Pleistocene. The only record for the late Pleistocene of southern New Mexico is that of Harris (1974), and this record is queried.

Peromyscus pectoralis Osgood
White-ankled Mouse (11917)

The comparative scarcity, six only, of jaws referable to *P. pectoralis* is surprising, for it occurs near the cave today. The species ranges from Texas and southeastern New Mexico southward into Mexico. Preferred habitat is moderately dense vegetative cover growing in and about cliffs, rocky talus, or rough, broken land. In the Trans-Pecos, Texas it is

found with *P. boylii* in the pinyon-juniper belt and downward into desert habitat on rocky hills. Near Fowlkes Cave it was trapped almost exclusively at the bases of juniper trees on rocky hills. There seem to be no records of this species from late Pleistocene deposits in Culberson Co. or southern New Mexico.

Peromyscus difficilis (J. A. Allen)
Rock Mouse (11916)

Peromyscus difficilis ranges from northern Colorado to southern Mexico, in high desert habitat. Past records from the Trans-Pecos are all from the flanks of mountain ranges at relatively high elevations. Preferred habitat is cliffs, talus, and rocky slopes. We did not find this species near Fowlkes Cave but suspect that it may yet survive on the higher limestone hills of the Apache Mountains. Its presence in the cave fauna suggests conditions essentially similar to those about the cave today. The rock mouse has been listed by Harris (1974) from Dry Cave, Eddy Co., New Mexico, and by Lundelius (1979) from Pratt Cave, northern Culberson Co.

Onychomys leucogaster (Wied-Neuwied)
Northern Grasshopper Mouse (11958)

The northern grasshopper mouse is readily separated from *Onychomys torridus* by strongly developed rather than poorly developed incisor-base capsule and greater length of lower tooth row and teeth (see Carleton and Eshelman, 1979). *Onychomys leucogaster* ranges from southern Canada to northern Mexico, through arid regions of the western United States. It has a strong preference for sand and sandy soils, and its abundance in the cave sediments indicates the presence of sandy soils nearby which are absent at the present time.

At present, *Onychomys leucogaster* does not occur near Fowlkes Cave but does live east of the Pecos River, to the east, and in the extreme western part of Trans-Pecos Texas. Its absence near the cave probably results from increased aridity in the post-Pleistocene. *O. leucogaster* has been reported from Pratt Cave, in northern Culberson Co., by Lundelius (1979) and from late Pleistocene cave sites in southern New Mexico (Harris, 1974).

Onychomys torridus (Coues)
Southern Grasshopper Mouse (11959)

Three of the six fossil jaws are somewhat damaged and certain identification is not possible. The other three are definitely *Onychomys torridus*. This long-tailed grasshopper mouse is the present inhabitant of the area about Fowlkes Cave and most of the Trans-Pecos. For identifying characters, see foregoing account. *O. torridus* ranges from central Nevada southward over the hotter deserts of the western United States and Mexico.

This species of grasshopper mouse is less confined to sandy habitat, and is found on the stony soils of the creosote bush flats and the varied cover of aluvial fans, but avoids hills and broken land. The comparative scarcity of the species in the cave fauna suggests that it was just invading the area from the south when the fossil-containing deposit was forming. Lundelius (1979) found both *O. torridus* and *O. leucogaster* in the early Recent deposits of Pratt Cave, in northern Culberson Co., and postulated that the relatively temperate McKittrick Canyon, where Pratt Cave is located, may have served as a refuge for *O. leucogaster* into the Holocene. The fauna of Fowlkes Cave substantiates this, for the replacement of *O. leucogaster* by *O. torridus* in southern Culberson Co. was underway in the late Pleistocene.

Sigmodon hispidus Say and Ord

Hispid Cotton Rat (11923–11926)

Present-day distribution of the cotton rat is from Nebraska to Panama and Arizona to the Atlantic. Wherever it occurs, it is, in its proper habitat of dense grass and weeds, often abundant. It is an important item in the diet of the barn owl, and this doubtless accounts for its abundance in Fowlkes Cave. Cotton rats occur throughout the Trans-Pecos. Some were trapped in tall, dense grass and weeds along a dry wash less than 1 km from the cave. *Sigmodon*, probably *S. hispidus*, has been recorded from late Pleistocene cave deposits in Eddy and Grant cos., New Mexico (Harris, 1974), and the post-Pleistocene deposits in Pratt Cave, northern Culberson Co. (Lundelius, 1979).

A related species, *Sigmodon ochrognathus* Bailey, is found in the Davis Mountains, and specimens were taken some 30 mi south of Fowlkes Cave. In the Guadalupe Mountains of northern Culberson Co., *S. ochrognathus* does not occur, but *Microtus mexicanus*, a vole, is found instead. *M. mexicanus* is not found in the Davis Mountains. Both species occupy similar habitat—the scattered, sometimes extensive, areas of grass, weeds, and succulent vegetation at or above the pine-oak belt on the mountains. *S. ochrognathus* is the ecological vicar of *M. mexicanus*, the former a southern form with tropical

affinities, whereas the latter is a northern species derived from boreal faunas.

It seemed important to determine whether any of the *Sigmodon* material from Fowlkes Cave might represent *S. ochrognathus*, rather than *S. hispidus*. We could find no reliable characters that would separate jaws or teeth of the two species, and Martin (1979) reached identical conclusions. Lundelius (1979) referred all *Sigmodon* from Pratt Cave to *S. hispidus* on the basis of large size. However, although *S. hispidus* averages larger than *S. ochrognathus*, and some extremely large *S. hispidus* are larger than the largest *S. ochrognathus* available, there is a great deal of overlap in measurements between the two species. The largest *Sigmodon* specimens in the Fowlkes Cave collection can only be *S. hispidus* but some of the smaller fossils could be *S. ochrognathus*.

Because *Sigmodon hispidus* currently occurs near Fowlkes Cave, and *Microtus mexicanus*, a Pleistocene resident of the region, is not known to occur in sympatry with *S. ochrognathus*, all the *Sigmodon* material is referred to *S. hispidus*.

Genus Neotoma

Three species of woodrats (*Neotoma micropus, N. albigula,* and *N. mexicana*) occur in the Trans-Pecos today, and another *N. cinerea* (Ord), has been found in late Pleistocene and early Recent cave deposits in northern Culberson Co. and southern New Mexico. The four species may be separated as follows. *Neotoma micropus* is relatively large. The width of the second lophid of M_1 is greater than 1.94 mm (Dalquest et al., 1969; not a completely reliable character according to Lundelius, 1979) and there are no strong dentine tracts on the anteroexternal side of M_1 (Lundelius, 1979).

Neotoma albigula is a small woodrat. The second lophid of M_1 usually measures less than 1.94 mm and there are no dentine tracts on the anteroexternal side of M_1 (Lundelius, 1979). Although there is some overlap in measurements of this character, it probably serves in the majority of instances to separate *N. albigula* from *M. micropus*. Because a number of specimens have the breadth of the second lophid distinctly greater than 1.94 mm, and a number of others have the breadth distinctly less, it is quite certain that both *Neotoma micropus* and *N. albigula* are represented in the fauna.

Neotoma mexicana is the smallest woodrat of the four species here considered. None of the 20 *N. mexicana* specimens from the Davis Mountains in the Midwestern State University collection had the breadth of the second lophid of M_1 greater than 1.75 mm, whereas none of the specimens of *N. albigula* examined had the breadth of the lophid less than 1.90 mm. Moreover, in *N. mexicana* the M_1 has well-developed dentine tracts extending from one-quarter to one-third of the height from root to top of crown in the unworn tooth. These are absent or vestigial in *N. albigula* (Lundelius, 1979). This species can be separated from other Trans-Pecos woodrats with certainty.

Neotoma cinerea is a large, stout-bodied woodrat, as known from skulls from northern New Mexico. The breadth of the second lophid of M_1 is greater than 1.94 mm, and there are well-developed dentine tracts on the anteroexternal side of M_1 (Lundelius, 1979). Although *N. cinerea* has been reported from Pleistocene and Post-Pleistocene cave deposits in northern Culberson Co. (Logan and Black, 1979; Lundelius, 1979) and southern New Mexico (Harris, 1974), this species was absent in Fowlkes Cave.

There are 30 maxillary fragments, eight lower jaws lacking M_1, and many isolated teeth (11930) that have been identified only as *Neotoma* sp.

Neotoma micropus Baird
Southern Plains Woodrat (11927)

The geographic range of *Neotoma micropus* extends from Kansas southward into northeastern Mexico, and from the western two-thirds of Texas westward to Colorado. It is found throughout the Trans-Pecos in suitable habitat such as desert flats and the more gentle slopes of aluvial fans. It seems, in the Trans-Pecos, to avoid the cliffs and talus where *N. albigula* lives. It does require cover of cactus, trees, and yuccas, where it constructs stick nests on the surface of the ground. It is usually common and prominent wherever found. It has been recorded from late Pleistocene and early Recent deposits of northern Culberson Co. and southern New Mexico (Harris, 1974; Logan and Black, 1979; Lundelius, 1979).

Both *Neotoma micropus* and *N. albigula* occur about Fowlkes Cave today, and *N. micropus* is by far the most abundant of the two, probably because its habitat, the desert lowlands, is so extensive compared to the rocky cliffs and talus required by *N. albigula*. In the collection of fossils, it is *N. albigula* that is most common. Today, *N. albigula* occurs in rocky areas about the cave, and even has lived in

the cave itself, whereas *N. micropus* is found no closer than several hundred meters from the cave on the aluvial fans below. The relative numbers of fossils of the two species does not indicate that *N. albigula* was more abundant in the late Pleistocene but rather that *N. micropus* is a large, heavy rat that must be near the upper limit of size of prey that a barn owl can overcome and carry. This is supported by the fact that only four of the *N. micropus* specimens are of adults, and the others are all more or less immature. In contrast, the bulk of the *N. albigula* fossils are of adults.

Neotoma albigula Hartley
White-throated Woodrat (11928)

The white-throated woodrat ranges from southern Utah and Colorado southward into central Mexico and from central Texas to California. It occurs throughout the Trans-Pecos. Favored habitat is crevices in rocky outcrops, cliffs, and talus slides, and less commonly in vegetation, such as patches of dense cactus. There are records of the species from late Pleistocene caves in southern New Mexico (Harris, 1974) and Pleistocene and early Recent caves in northern Culberson Co. (Logan and Black, 1979; Lundelius, 1979). *Neotoma albigula* and *N. micropus* are often sympatric in the Trans-Pecos, but, as mentioned in the account of *N. micropus,* occupy mutually exclusive habitat. The presence of *N. albigula* in the cave fauna only indicates arid conditions and the presence of rocky habitat nearby.

Neotoma mexicana Baird
Mexican Woodrat (11929)

This is rare in the cave fauna, represented by a single lower jaw. *Neotoma mexicana* is primarily a southern, even tropical, woodrat, but in the northern parts of its range it is found only at higher elevations in mountains. It ranges from the mountains of northern Colorado southward to Guatemala and from the Trans-Pecos to Arizona. It occurs in the Davis Mountains and the Guadalupe Mountains, in rocky habitat and talus slides. The bulk of its range in Texas is in the pine-oak belt, but specimens were trapped in the pinyon-juniper belt 33 km south of Fowlkes Cave, and it doubtless occurs even closer. It certainly does not occur at present in the desert habitat near the cave.

Microtus mexicanus (Sassure)

Mexican Vole (11932)

The Mexican vole is found in alpine habitat from southern Colorado southward through mountains of Arizona, New Mexico, and western Texas, and the highlands of Mexico almost to the Guatemala border. It enters Texas only in the Guadalupe Mountains, where it is locally common in grassy areas and in succulent vegetation at elevations of 7,800 ft or greater (Schmidly, 1977).

The two jaws from Fowlkes Cave are well-preserved and identification is confident. The presence of the fossils is enough to establish the presence of cool grasslands near the cave in the late Pleistocene.

Erethizon dorsatum (Linnaeus)

Porcupine (11933)

Material includes one incisor fragment. The nature and angle of the occlusal surface and cross-sectional shape show that a porcupine rather than a beaver or marmot is involved. Porcupines range from northern Mexico into Canada and virtually over the entire United States. The species is common in the Trans-Pecos, and doubtless lives near Fowlkes Cave today though we found no specimens. The tooth probably washed into the cave or was brought in by woodrats.

Felis rufus Schreber

Bobcat (11935)

Bobcats range over most of temperate North America southward to southern Mexico. They are found throughout the Trans-Pecos. There are records from late Pleistocene cave deposits in southern New Mexico (Harris, 1974). The specimen from Fowlkes Cave is of a kitten, and is the only evidence of carnivore in the Pleistocene fauna.

Mylohyus sp.

Long-nosed Peccary (11936)

Two fragments of tusk were found separately, and when fitted together formed part of the base of a peccary tusk. The root canal was almost closed when the animal died. Wrinklings on the surface of the fragment closely resemble those on isolated tusks of *Prosthenops* from the early Hemphillian Ocote local fauna of Mexico, and *Prosthenops* is the presumed ancestor of *Mylohyus*. No tusks of *Mylohyus* are available for comparison. The cross-sectional shape of the tusk base is different from that of tusks of *Platygonus* or the modern *Tayassu*.

Capromeryx cf. *furcifer* (Matthew)

Extinct Antilocaprid (11937)

The lower jaw was discovered in place, during excavation, in approximately the center of the Layer 2 sediments, and was sufficient to establish the Pleistocene age of this layer. The teeth are the proper size for *Capromeryx furcifer* but, without the diagnostic horn core, identification to species is only tentative. *C. furcifer* was presumably a grassland species, and its presence in the cave suggests the presence nearby of prairies that are not found in the near vicinity today.

The specimen consists of a ramus broken across just in front of the posterior alveolus of P_2 and just back of M_3. Much of the ventral border of the jaw is missing also. P_2 is lost but P_3–M_3 are present and well-preserved. The posterior part of the root of M_3 is exposed, showing the height of the third lobe to be 30.8 mm. The fossil is of a young-adult animal. Alveolar length of P_3–M_3 is 49.1 mm. Antilocaprid teeth, especially the molars, are wedge-shaped and increase in length at the occlusal and alveolar levels with increased age.

STRATIFICATION OF DEPOSITS

Of the 42 taxa listed from Fowlkes Cave, two are extinct, and a pocket mouse and the fulvous harvest mouse are only tentatively identified. Thirty-eight extant species, however, are identified, and the distribution and habitat requirements of these species are well known. In view of our knowledge of the distribution and habitat requirements of these 38 species today, the association of some of the species found at Fowlkes Cave seems absurd. Nowhere today would one find *Dipodomys merriami*, *Onychomys torridus*, and *Peromyscus eremicus*, forms typical of hot deserts, living in proximity to such alpine species as *Sorex vagrans*, *Sorex palustris*, and *Microtus mexicanus*. The latter forms would be expected along a trout stream among mountain peaks or far to the north.

However, the fossils were all found in a single 20-cm (8-inch) thick layer of sediments sandwiched between thicker layers that appeared to be barren of fossils. The stratigraphic evidence indicates that these species did live in association.

Harris (1970, 1974) encountered a fauna in Dry Cave, New Mexico, composed of species very similar to those making up the Fowlkes Cave local fauna. Because he did not himself collect the fossils first obtained from Dry Cave, and some of the collections made later could have been subject to postmortum mixing, Harris assumed that the fauna of Dry Cave represented a mixture composed of cold-climate elements, derived from a period of glacial advance, and a warm-climate element derived from an interstadial time.

The Fowlkes Cave fauna seems not to have been subject to postmortum mixing. There were tens of thousands of bones and bone fragments in Layer 2, sandwiched, between layers of sediment that were barren of fossils. Had mixing occurred, there would have been fossils in the layers above and below Layer 2, and there seemed not to have been. The stratification in the Fowlkes Cave sediments is unusual. Most Pleistocene cave deposits consist of unstratified red clay rather than stratified layers of limestone pebbles and fragments.

No record was kept of the level within Layer 2 at which bones were imbedded. Fossils did seem most abundant in the lowest part of the layer, and larger bones, such as limb bones of large rodents, lay at various angles to each other, although most seemed to lie parallel to the surface of Layer 2. It seems impossible that there was appreciable stratification within Layer 2, and if there was, it would have been next to impossible to detect. It seems equally unlikely that appreciable postmortum mixing might have occurred after Layer 3 was deposited and before Layer 1 was formed. In short, we think Layer 2 must have been deposited during one relatively brief period of time.

The stratification of the Fowlkes Cave deposits may have resulted from climatic conditions. The finely-divided, sand-like, materials of Layer 3 may have been formed by more effective shattering of limestone and spalling of the cave lining in the cold of a glacial advance. The coarser but still fine materials of Layer 2 may represent less-fine fragmentation at the terminal part of the glacial advance, and the coarse materials of Layer 1 might be the result of erosion during still-warmer climate immediately following the end of the Pleistocene climate. The Recent, of course, is represented by the black, surficial silts.

We doubt that there was standing or running water in Fowlkes Cave that might have stratified sediments. The cave chambers containing the fossils are many meters above the level of the aluvial fans below, and fractures in the limestone would have drained water away swifty. Today, except for occasional drips, the cave is dry. Snail fossils are only moderately common in the cave sediments, but a collection was made. Identification by Dr. Richard W. Fullington, Dallas Museum of Natural History, included: *Glyphyalini identata paucilirata* (Morelet), *Hawiia minuscula* (Binney), *Pupoides albilabris* (Adams), *Helicodiscus singleyanus* (Pilsbry), *Succinea* sp., *Gastrocopta pellicuda hordeacella* (Pfeiffer), *Gastrocopta pentodon* (Say), and *Gastrocopta armifera* (Say).

Dr. Fullington notes (personal communication) that all of the listed species are land snails and all occur in Culberson Co. today, although *Gastrocopta indentata, G. armifera,* and *Helicodiscus singleyanus* live only in isolated colonies high in the Guadalupe Mountains. Dr. Fullington finds it interesting that none of the xeric or western species of *Oreohilix, Holospira,* and *Ashmunella* are represented.

Absence of standing water in Fowlkes Cave seems highly probable, both because the cave would have been well-drained, and because none of the snails in the fauna are aquatic species. The only feature that could conceivably be responsible for the stratification in the cave is climate. It is assumed that colder climate caused greater fragmentation of limestone debris and the calcite lining materials in the cave than occurred during warmer cycles.

TAPHONOMY

Bones of mammals found in caves probably reached their resting places by four main routes: (1) carcasses were brought into the cave as prey of predatory birds or mammals; (2) the animals flew, walked, fell or climbed into the cave; (3) mammals died nearby and their bones washed or fell into the cave; (4) bones were brought into the cave as nesting material by woodrats.

Fowlkes Cave would seem to be an ideal pitfall trap, for the opening lies level with the surrounding rock. However, it seems not to have been an effective trap for mammals. We did find the mummified bodies of two snakes on the surface debris. These had fallen into the cave, perhaps in attempting to catch cave swallows. However, there were no bones of deer or other large mammals present on the surface debris, unlike most caves of the sinkhole type. Further, the Pleistocene fauna is strikingly deficient in ungulates and carnivores (1 *Capromeryx* jaw, 1 peccary tooth, 1 bobcat jaw fragment), and bones of adult rabbits are few. Ungulates, carnivores, and rabbits are the forms expected to be trapped in a pitfall cave.

Although the entrance of Fowlkes Cave is large today, it is possible that it was much smaller in the late Pleistocene. However, even the early Recent sediments yielded relatively few remains thought to have been trapped in the cave. It is more probable that the steep limestone hillside at the site of the cave was a place where few ungulates, rabbits, or carnivores would travel. Jackrabbits, for example, rarely leave level or gently sloping ground, and the few jackrabbit bones found in the Pleistocene sediments in the cave may have belonged to a single individual. Also, the bare rock surrounding the cave entrance may have made the entrance so obvious as to warn approaching animals of their danger. Lastly, just within the opening to the vertical shaft is a ledge that may have prevented many animals from falling deeper into the cave trap.

Whatever the reasons, Fowlkes Cave seems not to have sampled the large mammals that doubtless lived in the vicinity in the past. The few bones of larger animals present (antilocaprid jaw, peccary tusk, porcupine tooth, and perhaps even the marmot teeth) probably were washed into the cave after they died nearby, or were brought into the cave by woodrats.

In modern times, some mammals, such as bats, doubtless reside in the cave by day. The mummified body of a *Myotis velifer* and numerous scattered bones of the same species were found on a bed of bat guano in a side chamber, and the articulated skeleton of a ringtail (*Bassariscus astutus*) was found nearby. Remains of woodrats were commonly noted on the surface debris. Ringtails and woodrats, as well as *Peromyscus,* are agile climbers and probably enter and leave the cave at will, and did so in the past as well. Only the most skillful climbing animals would be able to scale the vertical walls of the entrance shaft.

The bulk of the fossils from Fowlkes Cave were brought to the cave by barn owls. These owls, after capturing and devouring prey, retire to sheltered and protected sites, such as caves or old buildings, to roost. When the flesh and most connective tissues of their meal has been digested, the remaining material, consisting of bones, teeth and hair, is regurgitated and falls to the ground beneath the roost.

Hibbard (1941) noted that owls might have been the source of fossils he recovered from Pleistocene terrace deposits in Kansas, and others (for example, Dalquest et al., 1969; Carleton and Eshelman, 1979) have attributed the fossils of small mammals found in Pleistocene caves to barn owls. Evidence suggests that barn owls brought almost all of the Fowlkes Cave fossils into the cave. Although no barn owls were seen in the cave while we were working there, they resided in the cave in the recent past for the mummified body of one was recovered from the surface debris. Doubtless they lived there in the late Pleistocene as well. The mammal fossils recovered, with the exceptions noted, are all of small forms, that could be overcome and eaten by the owls. In the case of the southern plains woodrat and cottontail rabbits, the bulk of the specimens are of immature individuals. Note that the owls need not carry the complete carcasses of mammals to the cave, although they might do so as food for nestlings, but only eat their prey where captured and return to the cave with the bones in their stomachs. The bones would later be deposited in pellets on the cave floor.

When the fossil-bearing sediments were screenwashed (see Materials and Methods), small clusters of bones were often noted when the matrix began to separate in the water. These clusters of bones were the size of and almost certainly were ancient owl pellets. There was no cement holding the bones together, and apparently they remained together only because the bones were so interlocked (as in fresh owl pellets). Most disintegrated immediately, but by moving swiftly, a few were recovered with only the outer-most bones falling away. These have been preserved (11943, 11944) as fossil owl pellets. Thus, there is quite strong proof that the bulk of the small mammal fossils in Fowlkes Cave were brought to the cave by owls.

The identification of the source of the fossils as barn owl pellet-derived is important in understanding the composition of the Fowlkes Cave local fauna. Owls are certainly, to a large extent, opportunistic hunters and take whatever prey is available. Thus, one might expect prey species that live in rocky habitat, aluvial fans, and on steep hillsides, to be most common in the fauna, because steep hills and the aluvial fans are closest to the cave. The creosote bush flats are farther away, though great areas of such habitat are within 1 km or so of the cave. Species that inhabit riparian habitat, however, were probably 10 or more km distant from the cave, even in the late Pleistocene (see Fig. 4). Thus species that lived in streamside meadows and marshes (wandering shrew, water shrew, vole) would be expected to be rare in the cave fauna, as they are.

It is assumed that barn owls prey most heavily on the small mammals that are most abundant nearest their roosts. The owls seem to have preference for certain kinds of prey, at least in any given locality (see Goyer et al., 1981), or certain species of mammals may simply be easier for the owls to capture because they tend to move about in the open rather than under cover. Owls hunting from perches find the rocky hillsides and aluvial fans offer such perches in abundance, whereas the creosote bush flats are almost devoid of suitable perches. It is most likely, however, that the owls fly no farther than they must to obtain their prey. In times of relative scarcity of desert rodents, the Pleistocene owls may have flown to the distant streamside meadows to feed on shrews and voles, but rarely would waste the time and energy to fly so far unless it were necessary. We are informed (personal communication) by Dr. Keith A. Arnold, Texas A&M University, that barn owls are known to regularly fly 10 mi or more to hunt, and to hunt by hawk-like quartering in areas barren of perches.

PALEOECOLOGY

Thirty-eight of the 42 taxa of mammals identified fall into three groups: (1) Chihuahuan desert species, that live near the cave today; (2) alpine types, that today are found in the Canadian or Hudsonian lifezones of the Davis, Guadalupe, or southern Rocky mountains; and (3) species found in both desert and alpine habitat in Texas and southern New Mexico (Table 2). The latter forms are of less value in interpretation of paleoecology and paleoclimate at the cave site.

The placement of some of these species might be argued. For example, *Peromyscus boylii* does occur in sites in the desert where conditions are suitable, but these may be relict populations. *Peromyscus difficilis* is, over much of its range in Mexico, a mouse of talus slides, cliffs, and rocky sites on the desert. In Texas and New Mexico, however, records are all from high elevations.

The desert-species composition from Fowlkes Cave shows a definite difference from the modern fauna. *Pappogeomys castanops* is rare as a fossil, vastly outnumbered by *Thomomys bottae*. The reverse is true in the modern fauna. It is possible that the scarcity of *Pappogeomys* is related to the ability of barn owls to capture this gopher. *Dipodomys ordii,* rare in the modern fauna near the cave, outnumbers *D. merriami* (107 to 14) in the fossil fauna. *D. merriami* is abundant about the cave at present. The abundance of *D. ordii* is thought to indicate the presence of sandy soils nearby in the late Pleistocene. *Onychomys leucogaster,* not found in most of the Trans-Pecos today, is the common grasshopper mouse in the fossil fauna, outnumbering *O. torridus* 183 to six. Again, presence of extensive sandy soils in the past is here indicated.

Of the alpine species listed, some would require a somewhat more temperate climate, with more green vegetation growing throughout the year, and presence of more junipers or other trees, before they could exist on the rocky hillsides and aluvial fans near the cave. Probably only a moderate increase in rainfall, more equitably distributed throughout the year, would permit these species to live in the vicinity of the cave today. We would include here *Myotis lucifugus, Sylvilagus floridanus, Eutamias cincericollis, Marmota flaviventris, Peromyscus boylii, P. difficilis,* and *Neotoma mexicana.* Other than *Marmota flaviventris,* these species are known to live in the Davis or Guadalupe mountains, at no great distance from Fowlkes Cave.

Marmota flaviventris now occurs no farther south than the mountains of northern New Mexico, southern Utah, and southern California (Hall, 1981). There are late Pleistocene records from sites far south of its present range, including: Arizona (Meade and Phillips, 1981), southern New Mexico (Harris, 1974), Trans-Pecos Texas (Logan and Black, 1979), and Nuevo Leon, Mexico (Stock, 1942). Lundelius (1979) reported marmot remains found in association with bone dated by carbon-14 at slightly more than 2,000 years B.P., showing that marmots existed in the Trans-Pecos until quite late in the Recent age. Even

Table 2.—*Paleoecological affinities of mammalian species identified in Fowlkes Cave local fauna.*

Species	Chihuahuan desert	Alpine habitat	Generalized
Sorex vagrans		+	
Sorex palustris		+	
Notiosorex crawfordi	+		
Myotis lucifugus		+	
Myotis velifer	+		
Eptesicus fuscus			+
Sylvilagus floridanus		+	
Sylvilagus audubonii	+		
Lepus californicus	+		
Eutamias cinericollis		+	
Marmota flaviventris		+	
Spermophilus spilosoma	+		
Spermophilus variegatus			+
Cynomys ludovicianus	+		
Thomomys bottae			+
Pappogeomys castanops	+		
Perognathus hispidus	+		
Perognathus flavus	+		
Dipodomys ordii	+		
Dipodomys merriami	+		
Dipodomys spectabilis	+		
Reithrodontomys montanus	+		
Reithrodontomys megalotis			+
Peromyscus maniculatus			+
Peromyscus leucopus			+
Peromyscus eremicus	+		
Peromyscus boylii		+	
Peromyscus pectoralis			+
Peromyscus difficilis		+	
Onychomys leucogaster	+		
Onychomys torridus	+		
Sigmodon hispidus			+
Neotoma albigula	+		
Neotoma micropus			+
Neotoma mexicana		+	
Microtus mexicanus		+	
Erethizon dorsatum			+
Felis rufus			+
Totals	17	10	11

a conservative outline of geographic range constructed from these records indicates a vastly greater range in the late Pleistocene, including all of Arizona and New Mexico, Trans-Pecos Texas, much of Chihuahua, and parts of Coahuila, Durango, Zacatecas, Nuevo Leon, and San Luis Potosi.

The extermination of marmots over so great an area must have been primarily due to the increased aridity in post-Pleistocene times. Harris (1970) stated "If the mountains of Arizona and southern New Mexico are excepted, marmots occur in the west where there is sufficient winter and early spring precipitation to support green plants during the critical spring-early summer drought period, regardless of elevation. At present, approximately 2 inches of winter precipitation seems to be sufficient in the southern Rockies." Harris feels that marmots were eliminated from Arizona and southern New Mexico (and by implication from Texas and Mexico as well) by droughts of moderate extent but at critical periods in the late winter and early spring.

Increasing aridity of the early Recent is doubtless also responsible for the local elimination of the eastern cottontail (*Sylvilagus floridanus robustus*), chipmunk, brush mouse, rock mouse, and Mexican woodrat, as well as the marmot. Some of these species live today on the slopes of the Davis Mountains, at elevations higher than those at Fowlkes Cave, and where green or woody vegetation is more common. Moisture, not elevation or temperature, appears to be the limiting factor.

Ecologically, the most unusual members of the Fowlkes Cave local fauna are the vagrant shrew, water shrew, and Mexican vole. These forms demand mesic conditions, and the water shrew seems to be confined to the vicinity of cool streamlets of clear, swiftly-moving water. Water shrews do enter marshy areas and damp meadows when these are not far removed from running water. Wandering shrews prefer damp meadows and marshes but enter woodlands if the soil is soft and preferably damp. The Mexican vole is a mouse of grassy meadows, preferably of short, dense grass, but enters areas of taller grass, weedy areas and even the margins of dense brushland if the shrubs are low. The habitat demands of these species can scarcely have changed in the past 10,000 years, and since they are present in the cave, habitat suitable for them must have occurred nearby, within the (10-mi) maximum hunting-range of the barn owl. Because these species are rare in the fauna, this habitat must have been near the extreme range of the owls. Because the habitat was certainly riparian, cold, running water and fringing damp meadows and willow thickets must have been present where the dry washes now exist.

Glaciation of the Guadalupe Mountains occurred, and the ice must have been melting when the Fowlkes Cave fauna lived in the vicinity. Only this will account for the presence of *Sorex* and voles in the fauna. Swift streams of cool water might also have brought sandy terraces into existence, which have since been eroded away. The Ord kangaroo rat and short-tailed grasshopper mouse are sand-dwelling species, abundant in the Pleistocene fauna. Cool, swift streams with fringing meadows do not contradict the presence of Chihuahuan Desert habitat back from the streams. Where permanent-water streams flow today in deserts, oasis-like conditions commonly exist. In some places in the Great Basin, such sites are refugia for mesic-habitat mammals such as voles. Nowhere today are found, so far as we are aware, cool streams through Chihuahuan Desert where wandering shrew, water shrew, or Mexican vole are brought into near contact with desert mammals, but somewhat equivalent conditions do exist.

The reconstructed ecological conditions near Fowlkes Cave during the period when the fossils were deposited are as follows. Cool streams with fringing borders of willows and meadowlands existed where there are now only dry washes, 10 km or more from the cave. The hillsides and aluvial fans supported somewhat more vegetation, including more junipers and perhaps some other trees. The creosote bush flats existed as today, but soils were sandy on terraces closer to the streams. Some of the typical desert mammals, such as Merriam kangaroo rat and long-tailed grasshopper mouse, were just entering the area. Precipitation need not have been much greater but was perhaps more equitably distributed. Summer temperatures were probably little lower than those today and winters could have been but little cooler. The Merriam kangaroo rat and long-tailed grasshopper mouse remain active in winter, and apparently cannot tolerate extended periods of freezing conditions. The Ord kangaroo rat, at least, apparently becomes dormant in freezing weather, and is thus able to survive in the colder deserts of the north. The scarcity of the Merriam kangaroo rat and long-tailed grasshopper mouse suggest that cooler climate of the late Pleistocene was just giving way to the hot, desert climate of modern conditions.

ACKNOWLEDGMENTS

We are deeply indebted to J. M. Fowlkes for allowing us to work in Fowlkes Cave, to collect Recent mammals on his ranch, and for numerous other kindnesses. Edward Matelski is due our thanks for his aid in collecting the fossil-bearing sediments, and for his efforts in washing and sorting of the materials collected. Guy Nelson and Gregory Zolnerowich also aided in the work involved in collecting and transporting the fossil matrix. Dr. Robert J. Baker permitted us the use of specimens in The Museum, Texas Tech University, and Dr. David J. Schmidly of Texas A&M University made certain pertinent specimens available. We wish to thank Dr. Norman Horner, Midwestern State University, for taking the photographs. (Craig Hood and Gregory Zolnerowich reviewed an earlier draft of this manuscript. Shirley Neunaber typed the manuscript in its final form.

LITERATURE CITED

AYER, M. Y. 1937. The archeological and faunal material from Williams Cave, Guadalupe Mountains, Texas. Proc. Acad. Nat. Sci. Philadelphia, 88:599–618.

BAILEY, V. 1905. Biological survey of Texas. N. Amer. Fauna, 25:1–222.

CARLETON, M. D., and R. E. ESCHELMAN. 1979. A synopsis of fossil grasshopper mice, genus Onychomys, and their relationships to Recent species. Mus. Zool., Univ. Michigan, Papers on Paleo., Hibbard Memm. Vol., 7:1–63.

DALQUEST, W. W. 1965. A new Pleistocene formation and local fauna from Hardeman County, Texas. J. Paleo., 39:63–79.

DALQUEST, W. W., E. ROTH, and F. JUDD. 1969. The mammalian fauna of Schulze Cave, Edwards County, Texas. Bull. Florida State Mus., 13:205–276.

DALQUEST, W. W., and F. B. STANGL, JR. 1983. Identification of seven species of Peromyscus from Trans-Pecos Texas by characters of the lower jaws. Occas. Papers Mus., Texas Tech Univ., 9:1–12.

———. In press. The taxonomic status of Myotis magnimolaris Choate and Hall.

DAVIS, W. B., and J. L. ROBERTSON, JR. 1944. The mammals of Culberson County, Texas. J. Mamm., 25:254–273.

FINDLEY, J. S. 1965. Shrews from Hermit Cave, Guadalupe Mountains, New Mexico. J. Mamm., 46:206–210.

GENOWAYS, H. H., R. J. BAKER, and J. E. CORNELY. 1979. Mammals of the Guadalupe Mountains National Park, Texas. Pp. 271–332, in Biological investigations in the Guadalupe Mountains National Park, Texas (H. H. Genoways and R. J. Baker, eds.), Nat. Park Serv., Proc. Trans. Ser., 4:xvii + 1–422.

GOYER, N., A. L. BARR, and A. R. P. JOURNET. 1981. Barn Owl pellet analysis in northwestern Harris County, Texas. Southwestern Nat., 26:202–204.

HALL, E. R. 1981. The mammals of North America. John Wiley and Sons, New York.

HARRIS, A. H. 1970. The Dry Cave mammalian fauna and late pluvial conditions in southeastern New Mexico. Texas J. Sci., 22:3–27.

———. 1974. Wisconsin age environments in the northern Chihuahuan Desert: evidence from the higher vertebrates. Pp. 23–52, in Transactions of the symposium on the biological resources of the Chihuahuan Desert Region (R. H. Waver and D. H. Riskind, eds.), Nat. Park. Serv., Trans. Proc., 3:xxii + 1–658.

HIBBARD, C. W. 1941. The Borchers Fauna, a new Pleistocene interglacial fauna from Meade County, Kansas. Bull. Kansas St. Geol. Surv., Univ. Kansas Publ., 38:197–220.

———. 1949. Techniques of collecting microvertebrate fossils. Contrib. Mus. Paleo., Univ. Michigan, 8:7–19.

JONES, J. K., JR., D. C. CARTER, H. H. GENOWAYS, R. S. HOFFMANN, and D. W. RICE. 1982. Revised checklist of North American mammals north of Mexico. Occas. Papers Mus., Texas Tech Univ., 80:1–22.

LOGAN, L. E., and C. C. BLACK. 1979. The Quaternary vertebrate fauna of Upper Sloth Cave, Guadalupe Mountains National Park, Texas. Pp. 141–158, in Biological investigations in the Guadalupe Mountains National Park, Texas (H. H. Genoways and R. J. Baker, eds.), Nat. Park Serv., Proc. Trans. Ser., 4:xvii + 1–422.

LUNDELIUS, E. L. 1979. Post-Pleistocene mammals from Pratt Cave and their environmental significance. Pp. 239–258, in Biological investigations in the Guadalupe Mountains National Park, Texas (H. H. Genoways, and R. J. Baker, eds.), Nat. Park Serv., Proc. Trans. Ser., 4:xvii + 1–422.

MARTIN, R. A. 1979. Fossil history of the rodent genus Sigmodon. Evol. Monogr., 2:1–36.

MEADE, J. I., and A. M. PHILLIPS. 1981. The late Pleistocene and Holocene fauna and flora of Vulture Cave, Grand Canyon, Arizona. Southwestern Nat., 26:257–288.

REICHMAN, O. J., and R. J. BAKER. 1972. Distribution and movements of two species of pocket gophers (Geomyiidae) in an area of sympatry in the Davis Mountains, Texas. J. Mamm., 53:21–33.

SCHMIDLY, D. J. 1977. The mammals of Trans-Pecos Texas. Texas A&M Univ. Press, College Station, 225 pp.

STOCK, C. 1942. The cave of San Josecito, Nuevo Leon, Mexico. Bull. Geol. Soc. Amer., 52:1822.

Address: Department of Biology, Midwestern State University, Wichita Falls, Texas 76308.

Present address (Stangl): Department of Biological Sciences, Texas Tech University, Lubbock, Texas 79408.

A LATE PLEISTOCENE MAMMALIAN FAUNA FROM CUEVA QUEBRADA, VAL VERDE COUNTY, TEXAS

Ernest L. Lundelius, Jr.

ABSTRACT

Deposits from Cueva Quebrada, Val Verde County, Texas, contain a mammalian fauna consisting of chiropterans, *Spilogale* sp., *Mephitis mephitis, Canis* sp., *Urocyon cinereoargenteus, Bassariscus* cf. *astutus, Arctodus simus, Ammospermophilus interpres, Thomomys bottae, Pappageomys castanops, Perognathus* sp., *Baiomys taylori, Onychomys leucogaster, Peromyscus* sp., *Neotoma* sp., *Sylvilagus* sp., *Lepus* cf. *californicus, Equus* cf. *scotti, Equus francisci*, cf. *Camelops, Navajoceros fricki, Stockoceros* sp., and *Bison* sp.

Much of the bone is charred and extensively broken. Only five specimens show damage that can be readily attributed to carnivore activity. Both spiral and rectilinear fractures are present. Although few cut marks were seen on the bones, the general breakage pattern is consistent with human activity.

The fauna has only one extant species, *Baiomys taylori*, that is not found in the Amistad area today. It suggests a more humid climate at the time of accumulation of the deposits.

The composition of the fauna indicates a late Pleistocene age which is confirmed by three radiocarbon dates of 12,280 ± 170 B.P., 13,920 ± 210 B.P., and 14,300 ± 220 B.P.

INTRODUCTION

Cueva Quebrada is located on the north side of a small tributary canyon to the Rio Grande River about 2 mi north of the mouth of the Pecos River, 18 mi (28 km) west of Comstock, Val Verde County, Texas, at 29°44′N by 101°24′W (Figs. 1, 2). It is located immediately east of a large shelter known as Conejo Shelter, the archaeology of which was studied by Alexander (1974). The excavations in Cueva Quebrada were done under the supervision of Alexander in connection with the Texas Archaeological Salvage Project's (TASP) contract with the National Park Service in the Amistad Reservoir. It is recorded in the TASP files as 41 VV 162A.

The archaeological material found at this site was negligible, but a considerable amount of faunal material of Pleistocene age was recovered from the lower units. This material provides new information on the Pleistocene fauna of southwest Texas. The condition of the bones and their distribution in the cave raises questions as to what were the agent(s) responsible for bringing the bones into the cave and what caused the burning.

ACKNOWLEDGMENTS

I thank the following people for assistance in this study: Ms. Melissa Winans prepared the computer generated maps and scatter diagrams (Figs. 8–11), and made available her measurements on the metapodials of *Equus* samples from San Josecito Cave, Channing, Texas, and Rock Creek, Texas; Dr. Dee Ann Story and Dr. Solveig Turpin helped with finding and interpreting the available records of the excavation; Dr. Arthur Harris made available his original measurements of *Equus* metapodials from Dry Cave, New Mexico; Mr. Elton Prewitt provided information on the stratigraphy of the deposits. The figures of the teeth and bones were drawn by Mrs. Pam Westerby. My wife Judith Lundelius edited the manuscript. Financial assistance was provided by the Geology Foundation of the University of Texas at Austin.

MATERIALS AND METHODS

The cave was excavated as an archaeological site according to standard archaeological methods. Provenience data on some of the bones was kept in terms of precise horizontal grid coordinates and depth below a datum. Other bone material was recorded only from a given square and stratigraphic unit. All specimens were scored for degree of burning—burned was defined as having all the organic material burned which left the bone light blue-gray or white in color and often deformed; charred was defined as blackened but with organic material still present; unburned bones showed no evidence of scorching. All specimens with horizontal grid and vertical control were used to construct maps with the aid of a computer mapping routine to determine any patterns of occurrence of burning and/or taxonomic units.

The materials listed for each taxon are not necessarily all specimens of that taxon, but are those that are most useful for identification. The detailed provenience of each specimen is also

456

Fig. 1.—Photo of Cueva Quebrada and Conejo Shelter. Cueva Quebrada is the small opening to the right and slightly below the larger Conejo Shelter.

omitted. A complete list with provenience is available from the Vertebrate Paleontology Laboratory, Texas Memorial Museum, University of Texas, Austin.

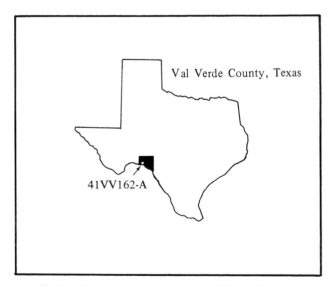

Fig. 2.—Map showing the location of Cueva Quebrada.

Measurements were taken with dial calipers and a microscope reticle. Abbreviations are: TMM—Texas Memorial Museum, M Recent Skeletal collection of Vertebrate Paleontology Laboratory, TNHC—Texas Natural History Collection of Texas Memorial Museum, TAMU—Texas A&M University. The material from Cueva Quebrada is catalogued under TMM 41238.

STRATIGRAPHY

The cave is small, measuring 30 ft by 15 ft with the long axis oriented north-south (Fig. 3). A number of stratigraphic units were recognized (Fig. 4) but notes with descriptions of these units are not available. Layer I at the top was composed of white limestone dust. It was completely devoid of artefacts and bone. Layers II through IV were also composed of limestone dust but were various shades of gray and brown presumably related to the amount of finely divided charcoal mixed with the limestone dust. Layer V was a light yellow color. The stratigraphic layers were approximately horizontal but

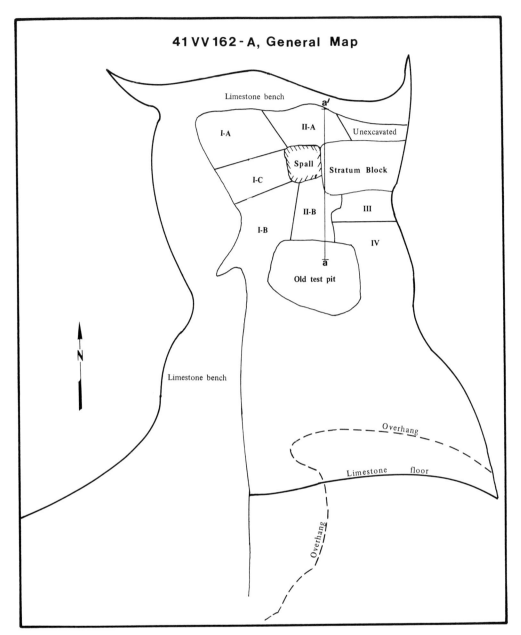

41 VV 162 - A, General Map

Fig. 3.—Map of Cueva Quebrada showing the excavation areas. A-A′ is the line of section shown in Fig. 4. One inch equals 6.3 ft.

the surfaces dividing them were frequently uneven. The cross section (Fig. 4) shows that the surface of

Layers III and IV were eroded in places. All layers have limestone spalls of various sizes.

AGE

The presence of several extinct taxa in the deposits of Cueva Quebrada indicates a late Pleistocene age. In addition there are three radiocarbon dates (Valastro et al., 1977, 1979). They are as follows:

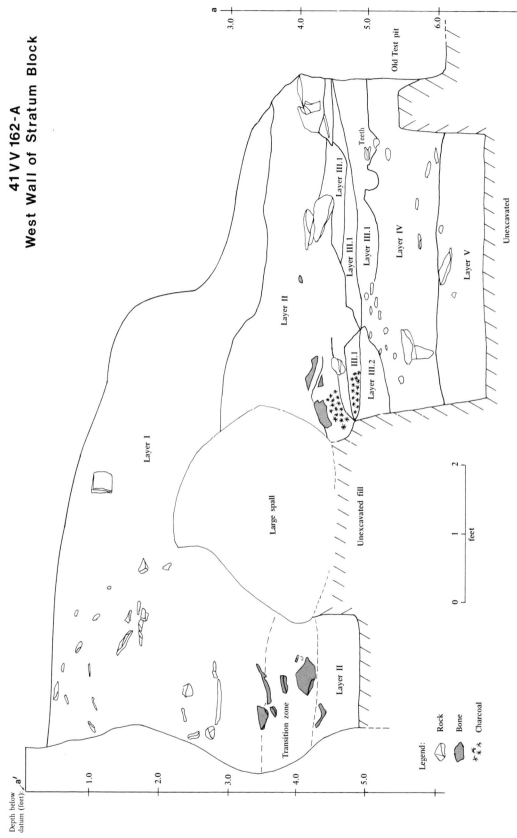

Fig. 4.—North-south section along line A-A' of deposits in Cueva Quebrada.

TX-879 12,280 ± 170 on charcoal from units I and II, 1.04–1.19 m below datum.

TX-880 13,920 ± 210 on one piece of wood from Unit IC, top of dark layer, 1.37 m below datum, immediately above a broken femur of *Equus scotti* (TMM 41238-317).

TX-881 14,300 ± 220 on one piece of wood .18 m north of TMM 41238-317, 1.37 m below datum.

These dates are consistent with the late Pleistocene age indicated by thc fauna. Two of the dates, TX-880 and TX-881, overlap at ±1.96 standard deviations and are not significantly different. The third, TX-879, does not overlap the other two at 1.96 standard deviations and is slightly younger. All the dated samples are from levels that contain abundant bone and would seem to be good indicators of the age of the bone.

TAPHONOMY

The fauna of Cueva Quebrada is not extensive, but the condition of the bones and the sparse evidence of human involvement raise questions as to the agent or agents responsible for bringing the bones into the cave. Much of the bone is burned, some so thoroughly as to have lost all organic material and to have been plastically distorted as is seen in much of the *Bison* bone from Bonfire Shelter in the same region (Dibble and Lorrain, 1968). Burning, especially in caves, is usually thought of as indicating human activity. As pointed out by Dibble and Lorraine (1968) large accumulations of organic material undergoing decomposition may generate enough heat to start combustion. In contrast to the situation at Bonfire Shelter, where there is evidence that large numbers of *Bison* were driven over the cliff and fell into the cave, the maximum number of large animals represented by the bones from Cueva Quebrada is approximately nine, which would not seem to provide enough organic material to initiate spontaneous combustion. There is also no evidence that all these animals were introduced into the cave at the same time. On the contrary, the general lack of articulation and the distribution of burned and charred bone in the deposits suggests a number of separate episodes of burning.

An examination of the maps showing the horizontal distribution of burned, charred and unburned bone for which there was precise data shows little concentration of burned or charred bones. There is a slight indication of increased burning and charring of bones of *Equus* in the central part of the cave. This is consistent with the observation of E. Prewitt (personal communication) that there were no hearths or fire pits.

The possibility that a packrat nest was ignited either by natural or human agency should not be overlooked. These nests are sometimes very large (up to 2 m in height and diameter) and represent a considerable amount of fuel (Van Devender, personal communication).

The agent or agents responsible for bringing the bones into the cave cannot be identified certainly. The bones of the small animals, such as rodents, probably were brought in by owls. They are generally unbroken, which is characteristic of bones derived from owl pellets. A number of the bones of the larger animals such as horse are not broken or the fragments were not widely dispersed and it is possible to reassemble them. The skull of *Arctodus* was burned and then fractured but was found intact. The thoracic and lumbar vertebrae of *Arctodus* were articulated, a foot of *Equus scotti* was articulated, and the phalanges of a foot of *Stockoceros* were articulated.

Many of the breaks in the long bones show a spiral fracture pattern indicating breakage as fresh bones. Some of the horse metapodials show rectilinear fractures. All of the skull material of horse, bison, and antelope is extensively broken.

The postmortem alteration of bones by humans and animals has recently been considered by a number of people (Bonnichsen, 1973, 1979; Brain, 1981; Haynes, 1980, 1983). Both experimental and field studies have identified criteria that indicate carnivore damage to bones. When these criteria are applied to the bones from Cueva Quebrada, a few seem to show features indicating carnivore damage. Three bones show what appear to be tooth marks. A horse first phalanx (TMM 41238-322) has conical depressions on both the anterior and posterior faces that have the shape and size of the canine of a wolf (Fig. 5A). Fragments of the surface bone that were driven into the depression are still present. The proximal end of a femur of *E. scotti* (TMM 41238-290) shows a similar depression on the posterior surface just below the head (Fig. 5B). These depressions are similar to puncture marks made by a wolf on the femur of a bison figured by Haynes (1983: fig. 1). There are also grooves that are U shaped in cross section on the ball of the femur that resemble those produced by carnivore teeth. If *Arctodus* was a carnivore as Kurtén (1967) suggests, it is possible that the depressions on the horse phalanx and femur were made by the carnassials of that animal. The protoconid of the M_1 is approximately the right size and shape.

Two horse tibias (TMM 41238-152, 412) show damage resembling that reported by Haynes (1980, 1983) to the tibias of moose by wolves (Fig. 5). In one (TMM 41238-152), the cnemial crest and the anterior part of the proximal articular surfaces are removed. The other (TMM 41238-412) lacks the proximal end, and the posterior part of the distal end is removed. No grooves that might have been made by carnivore teeth are visible on these two bones.

The distal end of a humerus of *Stockoceros* has a large hole in the epicondylar area and a deep conical depression on the inside of the medial wall of the epicondylar fossa. This latter depression seems to have been made by a more slender tooth than the one that made the depressions in the other two bones. This suggests that more than one carnivore was in-

volved in chewing the bones. Although *Arctodus* is the only carnivore represented by skeletal material in the cave deposits that is large enough to have made the large depressions, the late Pleistocene fauna included large canids (*Canis lupus* and *Canis dirus*) which were widespread in North America and were almost certainly members of the fauna of the Amistad region. There is no guarantee that any of these predators were responsible for bringing the animals into the cave, but they could have scavenged the carcasses found there.

All the bones were examined for cut marks, particularly near the ends, that might indicate human butchering activity. Only one was found. The proximal end of the horse femur (TMM 41238-290) has three linear V shaped grooves that resemble grooves on bones known to have been made by tools (Fig. 5C). Thus the breakage pattern, the surface alterations and the burning, suggest that both man and animals were responsible for the presence and condition of the bones in the cave.

ENVIRONMENTAL INTERPRETATION

The Pleistocene fauna from Cueva Quebrada has only a few extant taxa that are useful in a paleoecological analysis. The relatively poor representation of the smaller animals of the fauna has made species identification difficult for many taxa. Most of the extant taxa have a wide distribution in the southwest today and are still living in Val Verde County. One exception is *Baiomys taylori* which is now found in

more mesic areas to the east in Texas and at higher elevations to the south in Mexico.

The extinct fauna, with two species of horse, an antilocaprid, and *Bison,* suggests prevalence of open country, grasslands or savannah, on the uplands. The canyons, as now, probably supported brushy vegetation and provided habitats for many non-grassland and savannah species.

SPECIES ACCOUNTS

Class Mammalia
Order Chiroptera

Material.—An incomplete rostrum (TMM 41238-476); a number of isolated teeth.

Discussion.—The bat material is inadequate for reliable identification.

Order Carnivora
Family Mustelidae

Spilogale sp.

Material.—Right edentulous juvenile mandibular ramus (TMM 41238-571).

Description.—The alveoli for all the post incisor teeth are present. The alveoli for the M_1 show clearly that this tooth had two small lateral roots under the

midpoint of the tooth. This distinguishes it from comparable sized species of *Mustela* in which the M_1 has only one very small median root on the labial side. The horizontal ramus is very shallow and the surface of the bone is spongy indicating that it is from a young animal. The distinction between *S. gracilis* and *S. putorius* cannot be made on the basis of the material reported here.

Discussion.—The spotted skunk *Spilogale gracilis* is widely distributed in Trans-Pecos Texas where it is usually found in rocky and brushy conditions (Schmidly, 1977).

Mephitis mephitis (Schreber)

Material.—Left mandibular ramus with M_1, roots of P_4 alveolus for M_2 (TMM 41238-178); left mandibular ramus with roots of P_4, M_1 (TMM 41238-406).

Table 1.—*Dental and mandibular measurements of fossil and Recent specimens of* Urocyon cinereoargenteus *and* Vulpes macrotis.

Specimen	Length P$_4$	Length M$_1$	Anterior width M$_1$	Posterior width M$_1$	Length talonid M$_1$	Length M$_2$	Depth mandible at midlength M$_1$
			Urocyon cinereoargenteus				
Cueva Quebrada							
TMM 41238-252	7.2	—	—	—	—	—	12.7
TMM 41238-366	—	11.5	5.5	5.3	4.5	6.1	12.4
Recent							
M-1662	7.5	11.6	4.7	4.8	4.7	6.2	11.3
M-998	7.1	13.1	4.9	5.0	4.7	6.3	10.9
M-2063	7.7	11.4	4.9	4.4	4.2	6.6	11.6
M-3453	6.9	12.6	4.8	5.1	5.0	6.7	10.7
M-4822	7.3	12.3	5.3	5.3	4.8	7.4	12.1
			Vulpes macrotis				
M-1200	7.7	12.2	4.5	4.2	3.8	—	11.3
M-1412	7.7	11.8	4.7	4.1	3.6	5.1	10.5
M-1269	7.7	11.7	4.5	4.1	3.3	5.6	10.6
M-1910	8.4	12.6	4.8	4.5	3.4	5.6	11.8

Description.—The M$_1$ of TMM 41238-178 is deeply worn especially on the lingual side where the wear extends below the enamel. The trigonid is longer than the talonid and the talonid is relatively narrow as in *Mephitis* and in contrast to *Conepatus* in which the talonid is longer than the trigonid and is very much wider than the trigonid. Specimen TMM 41238-406 has the crown of M$_1$ and P$_4$ broken away. It is assigned to *M. mephitis* on the basis of size and the presence of well developed lateral roots under the center of M$_1$. *Mustela vison*, which has an M$_1$ slightly smaller than that of *M. mephitis*, has a very small central root only on the labial side.

Discussion.—*Mephitis mephitis* is present today in the vicinity of Cueva Quebrada. It occupies a wide variety of habitats in Trans-Pecos Texas but is least common in rough rocky terrain (Schmidly, 1977).

Family Canidae

Canis sp.

Material.—One first phalanx (TMM 41238-201).

Description.—The phalanx is similar in size and proportions to those of living specimens of *Canis latrans* which is common in Val Verde County today.

Urocyon cinereoargenteus (Schreber)

Material.—Part of a left horizontal ramus with part of P$_2$, P$_4$, alveoli for P$_1$, P$_3$ and M$_1$ (TMM 41238-252); part of a right horizontal ramus with M$_1$, M$_2$ alveoli for P$_4$ and M$_3$ (TMM 41238-366).

Description.—The specimens from Cueva Quebrada are similar to *U. cinereoargenteus* in both size and morphology. The cuspule directly posterior to the main cusp of the P$_4$ is relatively larger than that of *Vulpes macrotis*. The talonid of the M$_1$ is broad and has equidimensional entoconid and hypoconid, whereas in *V. macrotis* the hypoconid is larger than the entoconid and the talonid is relatively narrower and shorter (Table 1). The size of the P$_4$ and the depth of the mandible are closer to the corresponding dimensions of *U. cinereoargenteus* than to *Vulpes vulpes*, which is larger, or to *V. macrotis* which is smaller.

Discussion.—*Urocyon cinereoargenteus* is widespread in Texas today including the Trans-Pecos region where it is usually found in rocky areas in association with pinyon-juniper forests (Schmidly, 1977).

Family Procyonidae

Bassariscus cf. *astutus* (Lichtenstein)

Material.—Left calcaneum (TMM 41238-560).

Description.—The calcaneum of *Bassariscus* is distinguishable from those of similar sized animals such as *Mephitis*, *Conepatus*, *Spilogale*, and *Mustela vison* on the basis of both size and morphology. It is smaller than the corresponding bone in *Mephitis* and *Conepatus* and is larger than that of *Spilogale*. The proximal articular facet extends posteriorly over the dorsal surface of the calcaneal tuber, which is not the case in any of the skunks. The calcaneum of *Mustela vison* is the same size as that of *Bassariscus* and also has a proximal articular facet that extends posteriorly onto the dorsal surface of the calcaneal tuber but the form of the dorsal articular facet differs. The part of the dorsal articular facet that lies on the dorsal surface of the calcaneal tuber is concave in *Bassariscus* and convex in *Mustela vison* and the anterior portion of the facet is not abruptly turned ventrally as in *Bassariscus astutus*. The calcanei of these two animals can be further distinguished by the relatively longer pre-proximal articular facet length in *B. astutus*. This is approximately one of the half total length of the calcaneum in *B. astutus* and one third the total calcaneum length in *M. vison*. In all these characters the specimen from Cueva Quebrada resembles *B. astutus*.

Discussion.—*Bassariscus astutus* is widely distributed in Texas today and is present in the vicinity of Cueva Quebrada. Dalquest et al. (1969) have noted the general absence of *B. astutus* from Pleistocene faunas of central Texas. The one exception is a record from the Pleistocene Red Fill unit of Longhorn Cavern, Burnet County, Texas (Semken, 1961). These authors have suggested that *B. astutus* has moved eastward into central Texas in post-Pleistocene time. It is known from a number of Pleistocene faunas in the southwest such as Rancho La Brea, California (Akersten et al., 1979), Smith Creek Cave, Nevada (Miller, 1979), Rampart Cave, Arizona (Meade, 1981), Upper Sloth Cave in the Guadelupe Mountains of West Texas (Logan and Black, 1979); Burnet Cave and Dry Cave in eastern New Mexico (Harris, 1977). Except for the Longhorn Cavern record, the Cueva Quebrada occurrence is the easternmost Pleistocene record for this species.

According to Schmidly (1977) the most important factor influencing the distribution of *B. astutus* is the presence of rocky areas such as cliffs and canyons. These situations were surely present on the Edwards Plateau during the Pleistocene as they are today and yet *Bassariscus* was apparently absent.

Family Ursidae

Arctodus simus (Cope)

Material.—Skull (TMM 41238-72); ten thoracic and lumbar vertebrae (TMM 41238-249); right femur (TMM 41238-347).

Description.—The *Arctodus* material from Cueva Quebrada is being described and compared with other *Arctodus* material by Kurtén, Lundelius, and Johnson and the reader is referred to that paper for more detailed information. The skull is fairly com-

plete but has been burned and is distorted (Fig. 6). The teeth have been essentially destroyed by the burning and no characters can be discerned on them. Some idea of tooth size can be obtained from the alveoli in the skull. The skull is proportionately broader across the canines and orbits than in *Ursus* as noted by Kurtén (1967). This skull is morphologically similar to other known skulls but is somewhat small. In most dimensions it is within the range reported by Kurtén (1967) but others are below the observed range (Table 2). Some of the dimensions probably have been affected by the distortion caused by the burning but there seems to be no doubt that the skull is at the small end of the size range of this species. Kurtén (1967) presents evidence for considerable sexual dimorphism in *Arctodus simus* with the females being noticeably smaller than the males. On this basis the small size of the Cueva Quebrada specimen probably indicates that it is a female.

The femur is long and slender with a wide distal end. Although it is longer than the femur of *Ursus americanus* it is not relatively more massive as might be expected. Kurtén (1967) has noted this characteristic of the femur of *Arctodus*. The femur is slightly small in comparison to other North American *Arctodus simus* material. An examination of Table 3 shows that the Cueva Quebrada femur is slightly shorter than other femora reported by Kurtén (1967).

Discussion.—Remains of this large extinct bear have been found over a large part of North America (Kurtén and Anderson, 1980) but are not common. Kurtén (1967) has suggested that this large bear was predominantly carnivorous on the basis of its skull shape, primarily its short, broad rostrum which is similar to that of the large cats. If this interpretation of the mode of life is correct, *Arctodus* could have been an important agent in bringing other animals into the cave (see the section on taphonomy).

Order Rodentia
Family Sciuridae

Ammospermophilus interpres (Merriam)

Material.—Left upper molar (M¹ or M²) (TMM 41238-730); right M³ (TMM 41238-731).

Description.—The upper molar is broadly triangular. The anteroloph does not join the protocone by turning abruptly posteriorly as in *Spermophilus spilosoma, S. tridecemlineatus,* and *S. mexicanus.* The metaconule shows no tendency to fuse with the metacone and the metaloph does not join the pro-

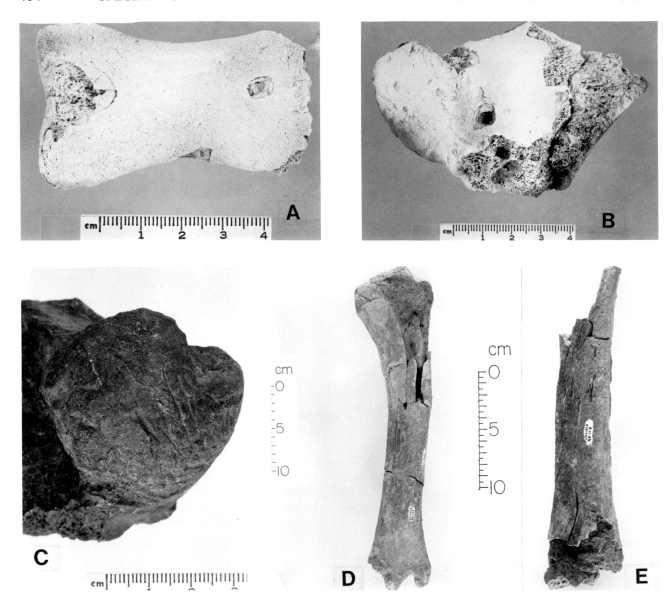

Fig. 5.—Bones showing carnivore and possible human damage. A) Anterior surface of first phalanx of *Equus francisci* (TMM 41238-322) showing perforations caused by carnivores; B) proximal end of left femur of *E. scotti* (TMM 41238-290) showing a perforation caused by a carnivore tooth; C) ball of left femur of *E. scotti* (TMM 41238-290) showing possible cut marks from a tool and grooves from carnivore gnawing; D) right tibia of *Equus scotti* (TMM 41238-152) with cnemial crest missing; E) right tibia of *Equus scotti* (TMM 41238-412) showing damage to distal end.

tocone. The mesostyle was apparently small and has been largely removed by wear.

The tooth is larger than the molars of *Eutamias cinereicollis, E. quadrivittatus,* and *Tamias striatus,* and smaller than the molars of the three species of *Spermophilus* mentioned above. It is the same size as the M¹ and M² of three specimens of *Ammosper-*

mophilus interpres from Brewster County, Texas (Table 4). The M³ is rounded in occlusal view, wear has removed many of the features of the crown. The anteroloph is close to the metaloph. The upper M¹ or M² is more similar to the upper molars of this species than to any other sciurid and is tentatively assigned to it. The M³ is the same size as those of

Table 2.—*Skull measurements of* Arctodus simus *from Cueva Quebrada and other North American specimens*

Measurements	TMM 41238-72	Observed range of North American Arctodus simus[1]
Basal length	343	330–440
Condylobasal	366	352–463
Extreme length	371	373–521
Palatal length	207	185–260
Zygomatic width	246	222–364
Rostral width at canines	101 est.	100–136
Width over M²	118	110–150
Interorbital width	112	117–153
Width over postorbital processes	161	147–205
Width over postorbital constriction	107 est.	98–107
Postorbital height	113	128–187
Width of nasal opening	69	76–107

[1] From Kurtén, 1967, Table 5.

Table 4.—*Measurements of upper molars of fossil and Recent specimens of* Ammospermophilus interpres.

Specimens	Tooth	Length (mm)	Width (mm)
Cueva Quebrada			
TMM 41238-730	M¹ or M²	1.74	2.03
TMM 41238-731	M³	1.85	1.87
Recent			
TNHC 4002	M¹	1.63	1.74
	M²	1.63	2.03
	M³	1.72	1.90
TNHC 4005	M¹	1.74	2.00
	M²	1.79	2.05
	M³	1.87	1.96
TNHC 4003	M¹	1.70	2.03
	M²	1.74	2.07
	M³	1.85	1.98

A. interpres, but the extensive wear makes a detailed comparison impossible. Its assignment to *A. interpres* is uncertain.

Discussion.—*Ammospermophilus interpres* is currently found in the area of Cueva Quebrada in Val Verde County, where it is an inhabitant of canyons with bare rocks and cliffs (Schmidly, 1977).

Family Geomyidae

Thomomys bottae (Eydoux and Gervais)

Material.—Right mandibular ramus with incisor, M₂ and alveoli for P₄, M₁, and M₃ (TMM 41238-439); left P₄ (TMM 41238-722); left P⁴ (TMM 41238-732).

Description.—The anterior lobe of the P₄ is longer than wide and has dentine tracts on both sides. The posterior lobe of the P₄ is oval with dentine tracts on both sides. The M₂ is strongly narrowed lingually

Table 3.—*Measurements of femora of* Arctodus simus *from Cueva Quebrada and other North American specimens.*

Measurements	TMM 41238-248	Observed range of North American Arctodus simus[1]
Greatest length	508	513–678
Greatest proximal width	123	122–165
Caput diameter	63	61–77
Least transverse width of shaft	42	41–62
Greatest distal width over epicondyles	104	108–137

[1] From Kurtén, 1967, Table 18.

to produce a tear drop shape. The dentine tract on the labial side is much wider than the one on the lingual side and is the site of a shallow groove on this side of the tooth. The P⁴ is bilobed with the anterior lobe more flattened than in a specimen of *T. bottae* from Dona Ana County, New Mexico, and the posterior lobe is less constricted labially.

Discussion.—This species occurs today in Val Verde County where it lives under a wide variety of environmental conditions (Davis, 1960; Schmidly, 1977).

Pappageomys castanops (Baird)

Material.—Right mandibular ramus with incisor and P₄ (TMM 41238-353); right mandibular ramus with incisor (TMM 41238-405); left mandibular ramus with alveoli for incisor, P₄, M₁₋₃, probably a juvenile (TMM 41238-566); upper incisor (TMM 41238-616); right edentulous mandibular ramus with alveoli for incisor and P₄, M₁₋₃ (TMM 41238-553); right mandibular ramus with incisor, alveoli for P₄, M₁ (TMM 41238-552); four upper incisors (TMM 41238-407).

Description.—The P₄ has dentine tracts on the antero-external and the antero-internal faces of the anterior lobe and on both the internal and external edges of the posterior lobe. The upper incisors are unisulcate.

Discussion.—*Pappageomys castanops* occurs throughout Trans-Pecos Texas (Schmidly, 1977) and is recorded in the vicinity of Langtry (Hall and Kelson, 1959). This species is reported by Schmidly (1977) to occupy areas of sandy loam with a minimum depth of 6 to 8 inches.

Table 5.—*Measurements of premolars of fossil and Recent specimens of* Perognathus.

Specimens	P⁴		P₄		
	Length (mm)	Posterior width (mm)	Length (mm)	Anterior width (mm)	Posterior width (mm)
Cueva Quebrada					
TMM 41238-675	.73	.85	.91	.73	.88
TMM 41238-738	—	—	.60	.54	.65
Perognathus flavus					
TMM M-2666	.685	.93	—	—	—
TMM M-887	.65	.91	—	—	—
TMM M-6059	.65	.85	—	—	—
TMM M-6060	.57	.85	—	—	—
TMM M-6061	.65	.86	—	—	—
TMM M-6062	.685	1.02	—	—	—
TMM M-6063	.73	.89	—	—	—
TMM M-2666	.72	.91	—	—	—
TMM M-6064	—	—	.52	.51	.57
TMM M-6065	—	—	.51	.49	.60
TMM M-6066	—	—	.54	.47	.59
TNHC 6114	.55	.85	.59	.54	.64
TNHC 6115 worn	.73	.86	.55	.46	.60
TNHC 407	.72	.96	.60	.52	.60
Perognathus penicillatus					
TNHC 4039	.85	1.02	.76	.55	.73
TNHC 4040	.75	1.09	.73	.57	.75
TMM M-1994	.98	1.06	.76	.62	.86
TMM M-1993	.91	1.14	.73	.65	.81
Perognathus intermedius					
TNHC 2558	.83	1.14	.67	.65	.80
TMM M-3730	.82	1.06	.80	.57	.78
Perognathus nelsoni					
TNHC 3256 worn	1.02	1.06	.76	.62	.73
TNHC 3257	.89	1.03	.68	.60	.73
TNHC 3261	.89	1.14	.70	.68	.78

Table 6.—*Measurements of teeth of* Onychomys leucogaster *from Cueva Quebrada.*

Specimens	Tooth	Length	Anterior width	Posterior width
TMM 41238-741	M₁	1.81	.95	1.18
TMM 41238-744	M₂	1.55	1.16	1.18
TMM 41238-783	M₂	1.65	1.32	1.17
TMM 41238-757	M¹	2.04	1.25	1.29
TMM 41238-758	M¹	1.85	1.05	1.25

Family Heteromyidae

Perognathus sp.

Material.—Left maxilla with P⁴ (TMM 41238-675); right mandible fragment with P₄ (TMM 41238-738); left mandible fragment with P₄ (TMM 41238-737); two molars (TMM 41238-711, 728).

Description.—The P⁴ consists of three major cusps, one anterior and two posterior that form a transverse loph with wear. In size and morphology the P⁴ is similar to that tooth in *P. flavus* (Table 5). It differs from the P⁴ of *P. penicillata, P. intermedius* and *P. nelsoni* in being smaller (Table 5).

The two P₄'s are somewhat different in size but both have four cusps. One, TMM 41238-737, is comparable in size to the P₄ of *P. penicillatus, P. nelsoni* and *P. intermedius.* One other, TMM 41238-738, is comparable in size to the P₄ of *P. flavus* (Table 5). The material is inadequate to allow a positive specific assignment. The different size of the P₄'s suggests that two species might be represented.

Discussion.—Six species of *Perognathus* are found today in Trans-Pecos Texas. Two of them, *P. flavus* and *P. nelsoni,* are known to occur in Val Verde County and two more, *P. hispidus* and *P. penicillatus* might have occurred there in the past in view of their present distributions. The three that fall into the size category of the Cueva Quebrada specimens, *P. flavus, P. nelsoni,* and *P. penicillatus* occupy a variety of environments in this region (Schmidly, 1977).

Family Cricetidae

Baiomys taylori (Thomas)

Material.—Right M₁ (TMM 41238-776).

Description.—The M₁ is similar to a series of M₁'s of recent *B. taylori* from San Patricio County, Texas, in size and morphology. This is one of the smallest of the North American cricetid rodents. The size (length 1.36 mm, maximum width .82 mm) is slightly larger than 4 specimens from San Patricio County, Texas (length 1.19 to 1.24 mm, maximum width .76 to .81 mm). The tooth is simple with no accessory cuspules and the major cusps are relatively high and inclined forward.

Discussion.—This species is not found in the Val Verde County area today. Its primary distribution in Texas is the Gulf Coastal Plain and south Texas. The two records closest to Val Verde County are Boerne, Kendall County, Texas, and six miles southwest of Geronimo, Coahuila, Mexico (Hall and Kelson, 1959:660). The area of Texas where it occurs today has a more humid climate than does Val Verde County. Its presence in the Pleistocene deposits of Cueva Quebrada implics more effective moisture for that area in the past.

Onychomys cf. leucogaster

Material.—Left M^1 (TMM 41238-757); right M^1 (TMM 41238-758); right M_2 (TMM 41238-744); left M_1 (TMM 41238-771).

Description.—The teeth all have the simple pattern with high cusps and open valleys that are characteristic of this genus. The assignment of the *Onychomys* material from Cueva Quebrada to *O. cf. leucogaster* is based on size. The dimensions of the teeth from Cueva Quebrada are closer in value to those given by Carleton and Eshelman (1979) for *O. leucogaster* than for *O. torridus* (Table 6).

Discussion.—*O. leucogaster* is recorded from northeastern Terrell County and from the Edwards Plateau to the east (Hall and Kelson, 1959; Schmidly, 1977).

Peromyscus sp.

Material.—Nine left M^1's (TMM 41238-718, 740, 742, 743, 759, 772 through 775); one right M^1 (TMM 41238-777); one right M_1 (TMM 41238-760); two right M_2's (TMM 41238-778, 779).

Discussion.—The specific identification of individual *Peromyscus* teeth is difficult and unreliable. The Cueva Quebrada sample shows some variation in the size and complexity of the M^1's which may indicate the presence of more than one species.

Neotoma sp.

Material.—Left mandible fragment with incisor (TMM 41238-554); right edentulous mandible (TMM 41238-533); two left M^1's (TMM 41238-724, 746); right M^1 (TMM 41238-750); two left M^2's (TMM 41238-725, 745); left M^3 (TMM 41238-739); right M^3 (TMM 41238-751); five left M_1's (TMM 41238-726, 747, 753, 727, 748); left M_2 (TMM 41238-723).

Description.—Three species of *Neotoma* occur today in Trans-Pecos Texas, *N. albigula*, *N. micropus*, and *N. mexicanus* (Schmidly, 1977). In addition *N. floridana* occurs in eastern Texas up to the edge of the Edwards Plateau with an isolated record near Rock Springs in Edwards County (Goldman, 1910) and *N. cinerea* occurs in the mountains of northern New Mexico (Hall and Kelson, 1959). *N. cinerea* has been reported in a mid-Holocene fauna from the southern Guadelupe Mountains of Hudspeth County, Texas (Lundelius, 1979). Both of these latter species might be expected to have occurred farther south during the Pleistocene when the climate was cooler and/or wetter than present.

Neotoma cinerea and *N. mexicana* can be readily distinguished from the other three species on the basis of the presence of a dentine tract on the antero-external side of the M_1. The differentiation of the

Table 7.—*Measurements of teeth of* Neotoma *sp. from Cueva Quebrada.*

Teeth and specimens	Length at base	Width at base	Width of second loph
M^1			
TMM 41238-724	3.28	2.33	—
TMM 41238-746	3.42	2.38	—
M^2			
TMM 41238-745	2.64	2.05	—
TMM 41238-725	2.68	2.23	—
TMM 41238-735	2.54	2.34	—
M^3			
TMM 41238-739	1.84	1.83	—
TMM 41238-751	1.95	1.70	—
M_1			
TMM 41238-748	3.06	1.78	1.78
TMM 41238-753	3.01	1.92	1.92
TMM 41238-747	3.50	1.91	1.91
TMM 41238-726	3.28	1.89	1.89
M_2			
TMM 41238-723	2.70	1.97	—

other three species on the basis of individual teeth and mandibles is difficult. Dalquest et al. (1969) found no qualitative dental characters that separate *N. floridana* from *N. micropus* and the teeth do not differ in size. Dalquest et al. (1969) reported that *N. albigula* and *N. micropus* can be separated on the basis of the width of the second loph of the M_1. They report that in all Texas specimens of *N. albigula* they measured, the width of the second loph of the M_1 was less than 1.94 mm and in *N. micropus* this measure was greater than 1.94 mm. The samples they studied showed no overlap. M. Winans (pers. comm.) has found that in samples of 72 *N. micropus* and 68 *N. albigula* from Texas this measurement showed extensive overlap and indicates that this character is not reliable when large samples from varied geographic areas are compared. This is also true when specimens of both species from Trans-Pecos Texas are compared.

Four M_1's are available from Cueva Quebrada. None have the dentine tract on the antero-labial side which rules out their assignment to either *N. cinerea* or *N. mexicana*. The dimensions and qualitative characters of the *Neotoma* teeth from Cueva Quebrada do not allow a specific identification (Table 7).

Discussion.—*N. micropus* is known today as a living animal in western Val Verde County; *N. albigula* is recorded 7.5 mi (11.5 km) to the west of Val Verde County (Schmidly, 1977) and the nearest known occurrence of *N. floridana* is 80 mi (123 km) to the east. *N. floridana* might be expected to have extended its range farther west during the Pleistocene but more material will be needed to demonstrate its presence.

Order Lagomorpha

Sylvilagus sp.

Material.—Two upper molars (TMM 41238-754, 755); two right P_3's (TMM 41238-756, 789); left P^2 (TMM 41238-790); fragment of left maxillary (TMM 41238-618); fragment of left mandible (TMM 41238-572); proximal ends of two left humeri (TMM 41238-559, 398); ventral ends of two right scapulae (TMM 41238-585, 562); two partial pelves (TMM 41238-473, 403); distal ends of two right femora (TMM 41238-622, 648); proximal ends of two right tibias (TMM 41238-455, 623); distal end of one right tibia (TMM 41238-626); three right calcanei (TMM 41238-240, 480, 627).

Description.—The material assigned to *Sylvilagus* is unmistakable in both size and morphology as to its generic assignment but is totally inadequate to permit specific identification.

Discussion.—Two species of *Sylvilagus*, *S. floridanus* and *S. audubonii* occur today in Val Verde County.

Lepus cf. *californicus* (Gray)

Material.—Left mandibular ramus with P_2, P_3 alveolus for M_1 (TMM 41238-499); partial right edentulous right maxilla (TMM 41238-590); partial right ilium (TMM 41238-386).

Description.—The material assigned to *Lepus californicus* is within the size range of modern specimens of that species. The measurements of the dentition are P_3 L. 3.76 mm, W. 3.30 mm; P_4 L. 3.00 mm, W. 3.49 mm. The morphology of the P_3 is also like that of living specimens of *L. californicus*. *Lepus townsendi* has been reported from Pleistocene faunas from Schultze Cave in Edwards County, Texas, 80 km (50 mi) to the east of Cueva Quebrada (Dalquest et al., 1969), Dry Cave Eddy County, New Mexico, 419 km (262 mi) to the north of Cueva Quebrada (Harris, 1970) and Burnet Cave in the Guadelupe Mountains, 395 km (247 mi) north of Cueva Quebrada. A comparison of the Cueva Quebrada specimen with two specimens of *L. townsendi* from Minnesota (TMM M-3347) shows that the latter species has fewer crenulations of the enamel in the re-entrant angles of the P_3. Recent specimens of *L. californicus* from Texas have more crenulated enamel in the re-entrants of the P_3. The Cueva Quebrada material is assigned to *L. californicus* on this basis.

Discussion.—*L. californicus* is found in the area

of Cueva Quebrada today. Pleistocene records of *L. townsendi* mentioned above are well to the south of its southernmost occurrence today and raise the question as to the extent of its southward extension at that time. Dry Cave is located at an altitude of 1,300 m (4,200 ft) which is 900 m higher than Cueva Quebrada as well as being farther north. However Schultze Cave is only 300 m higher than Cueva Quebrada and is no farther north. If the specimen from Schultz Cave is *L. townsendi* there is a real possibility that it also reached Val Verde County during late Pleistocene time. Additional material from more localities will be necessary to settle this question.

Order Perissodactyla

Family Equidae

The taxonomy of the Pleistocene horses is currently confused. Although it is usually possible to assign horse material from one locality to one or more groups it is difficult to decide on the name or names that should be applied. This is the case with the horses from Cueva Quebrada. There are clearly two horses represented, a large form and a small one, and most of the material can be assigned to one or the other on the basis of size.

Equus cf. *scotti* Gidley

Material.—Three right M^2's (TMM 41238-64, 68, 103); fragments of upper molars (TMM 41238-433); left upper premolar (TMM 41238-99); right upper premolar (TMM 41238-105); two right upper molars (TMM 41238-62, 60); two upper molars (TMM 41238-61, 191); right horizontal ramus with P_3–M_3 (TMM 41238-2); right P_2 (TMM 41238-74); left M_3 (TMM 41238-102); three left lower premolars (TMM 41238-65, 83, 66); two right lower premolars (TMM 41238-104, 278); four right lower molars (TMM 41238-70, 67, 255, 253); two left lower molars (TMM 41238-197, 258); one unworn lower cheek tooth (TMM 41238-196); distal ends of two right humeri (TMM 41238-225, 194); right radius (TMM 41238-195); distal end of right femur (TMM 41238-37); proximal end of left femur (TMM 41238-290); right tibia (TMM 41238-152); left tibia (TMM 41238-159); distal ends of two left tibias (TMM 41238-47, 23); distal end of one right tibia (TMM 41238-153); right metacarpal (TMM 41238-48); left metacarpal (TMM 41238-41); two right metatarsals (TMM 41238-88, 42); three right calcanei (TMM 41238-139, 96, 138); two right astragali (TMM 41238-79, 158); one left astragalus (TMM 41238-205); left articulated pes with all tarsals and proximal end of metatarsal (TMM 41238-142 through 151); fifteen first phalanges (TMM 41238-322, 90, 303, 305, 109, 14, 160, 304, 202, 301, 45, 193, 17, 302, 169); six second phalanges (TMM 41238-15, 174, 44, 166, 175, 192); three third phalanges (TMM 41238-316, 461, 256).

Description.—The material assigned to this taxon is from a heavily built horse about the size of a riding

Table 8.—*Measurements of teeth of* Equus scotti *from Cueva Quebrada.*

Specimens and teeth	Measurements		
	Upper teeth		
	Length along ectoloph	Width normal to para-mesostyle	Width normal to meso-metastyle
TMM 41238-62 Pm	32.0	28.2 est.	30.6
TMM 41238-105 Pm	30.7	29.9	29.0
TMM 41238-99 Pm	32.1	28.3	31.1
TMM 41238-60 M	27.1	28.3	27.2
TMM 41238-61 M	30.9	25.3	26.2 est.
TMM 41238-191 M	31.4	27.6	28.8
	Lower teeth		
	Length	Anterior width	Posterior width
TMM 41238-64 P_2	44.4	—	—
TMM 41238-68 P_2	37.9	14.3	16.4
TMM 41238-74 P_2	34.1	13.3	15.8
TMM 41238-102 M_3	31.8	13.7	13.2
TMM 41238-65 Pm or M	31.0	17.2	18.5
TMM 41238-66 Pm or M	31.4	19.0	—
TMM 41238-253 Pm or M	33.7	15.4	14.0 est.
TMM 41238-197 Pm or M	27.7	15.5	14.9
TMM 41238-70 Pm or M	31.5	17.1	15.2
TMM 41238-83 Pm or M	28.2	—	18.6
TMM 41238-255 Pm or M	28.9	14.7	15.0
TMM 41238-104 Pm or M	31.7	17.6	18.0
TMM 41238-278 Pm or M	—	16.5	—
TMM 41238-67 Pm or M	33.7	15.1	14.0
TMM 41238-2 P_3	30.3	15.5	17.7
TMM 41238-2 P_4	27.0	18.3 est.	18.0
TMM 41238-2 M_1	27.1	16.7	—
TMM 41238-2 M_2	26.8	15.6 est.	15.0
TMM 41238-2 M_3	—	—	14.0

horse. The upper premolars and molars are large, very close to the dimensions given by Dalquest (1964) for specimens assigned by him to *E. scotti* from three localities in north Texas including the type locality in Briscoe County, Texas. The enamel pattern is comparable in complexity (Fig. 7). The three specimens figured by Dalquest show some variation in the fossettes and the development of the pli caballin. The latter tends to be better developed on the premolars in the three specimens illustrated by Dalquest (1964) and the same is true of the Cueva Quebrada material.

The lower premolars and molars are also large and high crowned when unworn (Table 8). All the P_3–M_3's have open U-shaped linguaflexids with flattened metastylids (Fig. 7). In the one specimen in which the position of each tooth can be established (TMM 41238-2), the ectoflexids of P_3 and P_4 are blunt and do not enter the metaconid-metastylid isthmus. The ectoflexids of M_1 and M_2 are narrower than those of P_3 and P_4 and extend into the metaconid-metastylid isthmus. The ectoflexid of M_3 extends to the level of, but not into, the isthmus.

The postcranial material that can be assigned to this taxon with some confidence consists of limb and foot bones. Scatter diagrams show the metapodials match those from the type locality of *E. scotti* in Briscoe County, Texas, in both length and width (Figs. 8, 9). The horse phalanges from Cueva Quebrada fall into two size groups. The larger size group agrees with phalanges of *E. scotti* and are assigned to that taxon (Tables 9–12).

Discussion.—There is little agreement on the taxonomy of the large late Pleistocene horses. In the southern Great Plains region these horses have been assigned to a number of species. Stock and Bode (1937) assigned large horse teeth from Blackwater Draw, New Mexico, to *E. excelsus*. Quinn (1957) recognized *E. caballus caballus* from two localities (Blackwater Draw, New Mexico and Lubbock, Texas), *E. caballus* from Blackwater Draw and *E. midlandensis* from three localities (Blackwater Draw, J. O. Baggett Ranch in Ector County, Texas, and Scharbaur Ranch in Midland County, Texas). Lundelius (1972) assigned the large horse material from the gray sand unit at Blackwater Draw to *E. scotti* on the basis of morphological similarity. Harris and Porter (1980) assigned much of the larger horse material from the Dry Cave, New Mexico fauna and the Blackwater Draw material to *E. niobrarensis* and differentiated it from *E. scotti* on the basis of its smaller size.

However, a scatter diagram of length versus transverse width of the distal articular surface of metatarsals of large horses from various late Pleistocene localities shows a clustering of specimens from Blackwater Draw, Knox County, Texas, and Cueva Quebrada (Fig. 9). This cluster includes the type of *E. scotti* and other specimens from the type locality in Briscoe County, Texas. Part of the Dry Cave sample of metatarsals assigned to *E. niobrarensis* by Harris and Porter (1980) plot with the specimens from Rock Creek, Blackwater Draw, Scharbaur Ranch, and Cueva Quebrada and are considered to be the same taxon. One Dry Cave specimen (UTEP 31-64) assigned to *E. scotti* by Harris and Porter (1980) on the basis of its size, is somewhat longer (304 mm). Although it plots outside the cluster of the other large horses, it would not extend the range beyond expectation if the total ranges of the other

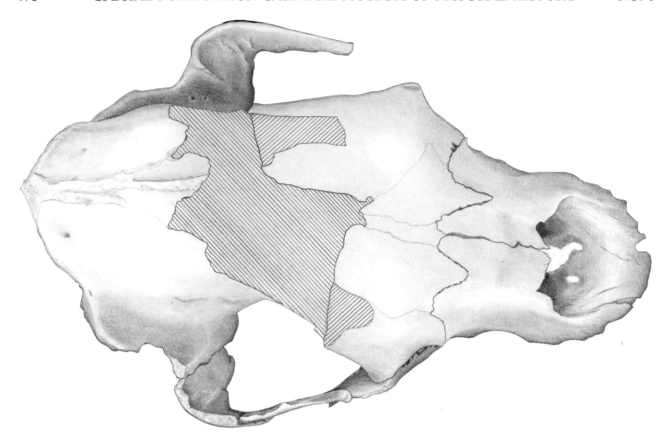

Fig. 6.—Dorsal view of skull of *Arctodus simus* (TMM 41238-72) from Cueva Quebrada. ×.5.

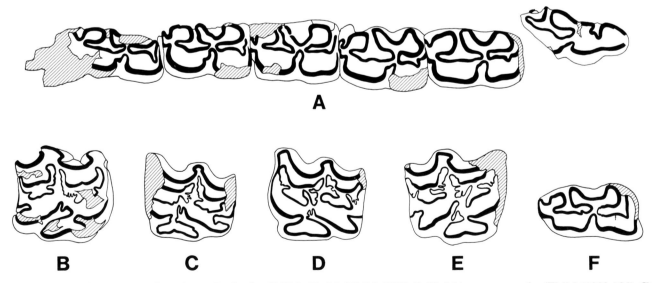

A

B C D E F

Fig. 7.—Teeth of *Equus scotti* from Cueva Quebrada. A) Right P$_2$–M$_3$ (TMM 41238-2); B) right upper premolar (TMM 41238-105); C) left upper molar (TMM 41238-61); D) left upper molar (TMM 41238-191); E) right upper molar (TMM 41238-62); F) right lower molar (TMM 41238-67). ×.75.

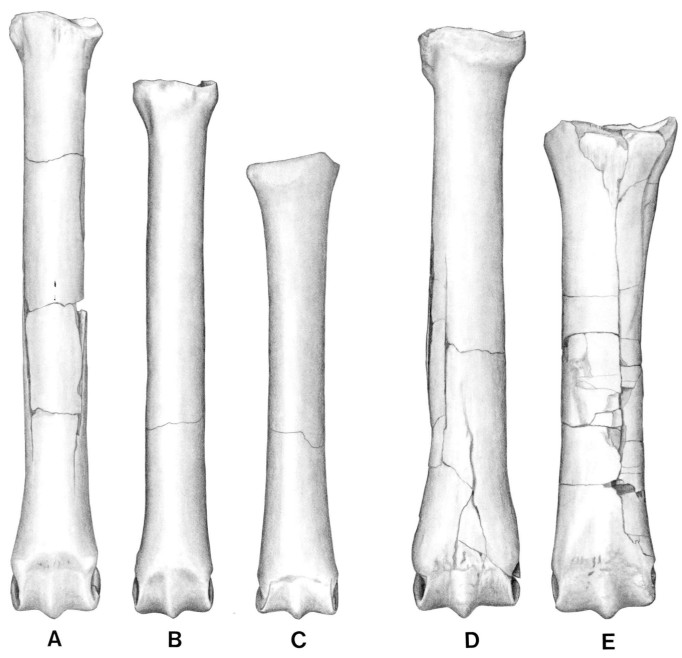

A B C D E

Fig. 8.—Metapodials of equids from Cueva Quebrada. A) Right third metatarsal of *Equus francisci* (TMM 41238-46); B) right third metatarsal of *Equus francisci* (TMM 41238-227); C) left third metacarpal of *Equus francisci* (TMM 41238-629); D) right third metatarsal of *Equus scotti* (TMM 41238-42); E) right third metacarpal of *Equus scotti* (TMM 41238-48). ×.55.

samples shown in Fig. 8 are considered. The plot of the distal width versus trochlear diameter shows more heterogeneity (Fig. 10). The Cueva Quebrada specimens are closely associated with *E. scotti*.

An examination of the scatter diagram of the length versus distal width and distal width versus trochlear diameter of the metacarpals leads to the same conclusion, although there seems to be more variation in the distal width (Figs. 11, 12) and less clear cut differentiation of the Dry Cave sample. The large

Table 9.—*Measurements of major limb bones of* Equus scotti *from Cueva Quebrada.*

Bones and specimens	Length	Proximal width	Distal width
Femur			
TMM 41238-37	—	—	102
Tibia			
TMM 41238-47	—	—	80
TMM 41238-23	—	—	83 (min.)
Humerus			
TMM 41238-255	—	—	86
TMM 41238-194	—	—	85
Radius			
TMM 41238-195	353	88	84

metacarpals from Cueva Quebrada are associated with those of *E. scotti* from Rock Creek and the larger metacarpals from Dry Cave. The teeth of the large horse from Cueva Quebrada show no significant differences in either enamel pattern or size from those of *E. scotti* from Briscoe County, Texas (Table 8, Fig. 7).

It is not clear which names should be applied to the large late Pleistocene horses and a thorough investigation of this problem is beyond the scope of this study. *E. caballus* is inappropriate as it was applied by Linnaeus (1758) to the domestic horse. The status of the other species is uncertain. The oldest name is *E. excelsus* but the type consists of a maxilla with P^4–M^3. The type of *E. scotti* is a skull, jaws, and part of a skeleton. Topotype material con-

sisting of several individuals is available. The type of *E. midlandensis* consists of left and right mandibular rami, P^1, P^3–M^1, a metatarsal III, and phalanges of fore and hind foot believed to come from one individual.

All the material assigned to the various species listed above, with the possible exception of *E. excelsus,* appear to form a relatively homogeneous group in size and morphology in so far as they are presently known. The best known type material is that of *E. scotti* and the Cueva Quebrada large horse is assigned to that taxon. If future work demonstrates that *E. excelsus* and *E. scotti* are synonymous, *E. excelsus* will become the proper name.

The type material of *E. scotti* is from the Irvingtonian Rock Creek local fauna in Briscoe County, Texas, which is about .75 million years old. It is realized that the application of this name to material of late Rancholabrean age implies the continuity of one species over this period of time. This is well within the range of species longevity given by Kurtén (1968) for Pleistocene perissodactyls of Europe.

Equus francisci Hay

Material.—Left maxillary fragment with P^4–M^3 (TMM 41238-157); right upper premolar (TMM 41238-639); right upper molar (TMM 41238-100); left upper cheek tooth (TMM 41238-220); right M^3 (TMM 41238-485); right lower cheek tooth (TMM 41238-75); right ilium (TMM 41238-232); right humerus (TMM 41238-234); right pelvis (TMM 41238-231); two left radii (TMM 41238-226, 131); distal part of right radius (TMM 41238-285); distal end of left radius (TMM 41238-313); two right femora (TMM 41238-39, 229); distal half of left femur (TMM 41238-38); distal ends of two right tibias (TMM 41238-19, 342); distal ends of

Table 10.—*Measurements of metacarpals of* Equus scotti *from Cueva Quebrada.*

Measurements	TMM 41238-48	TMM 41238-41	TMM 41238-13	TMM 41238-78
Length	237	237	231	—
Transverse width of proximal end	58.9	56.2	47.7	—
Antero-posterior width of proximal end	—	36.8	35.6 (min.)	33.6 (min.)
Transverse diameter of posterior articular facet for the magnum	32.5	33.6	37.0	31.1
Transverse width of facet for magnum	43.9	41.9	46.2	40.7
Transverse width of shaft at midpoint	38.6	38.5	36.7	36.1
Antero-posterior diameter of shaft at midpoint	28.8	29.0	27.4	27.7
Transverse width of distal end above the articular facet	48.8	51.2	49.2	—
Transverse width of distal articular surface	48.5	51.8	52.4 est.	—
Antero-posterior diameter of inner distal articular facet	31.8	35.7	33.1	—
Antero-posterior diameter of central ridge of distal articular surface	39.9	41.5	38.3	—

Table 11.—*Measurements of metatarsals of* Equus scotti *from Cueva Quebrada.*

Measurements	TMM 41238-88	TMM 41238-42
Length	291	289
Transverse width of proximal end	59.5	53.0
Antero-posterior width of proximal end	52.7	51.2
Transverse diameter of posterior articular facet	29.3	28.3
Transverse width of facet for ectocuneiform	51.5	48.9
Transverse width of shaft at midpoint	40.2	37.1
Antero-posterior diameter of shaft at midpoint	35.5	33.4
Transverse width of distal end above the articular surface	—	51.0
Transverse width of distal articular surface	—	50.7
Antero-posterior diameter of inner distal articular facet	39.4 est.	32.0
Antero-posterior diameter of central ridge of distal articular surface	—	38.0

Table 12.—*Measurements of phalanges of* Equus scotti *from Cueva Quebrada.*

Phalanges and specimens	Length	Proximal width	Distal width	Mid width
First phalanx				
TMM 41238-90	—	—	46.4	30.5
TMM 41238-301	—	—	43.0	35.7
TMM 41238-322	—	—	44.7	30.1
TMM 41238-45	80.4	58.4	47.9	36.8
TMM 42138-14	81.7	54.3	47.8	35.8
TMM 42138-17	83.0	—	—	35.5
TMM 42138-160	81.1	51.5	45.8	35.6
TMM 42138-202	—	—	—	34.7
TMM 42138-193	74.9	58.5	45.3	33.3
TMM 41238-305	78.2	59.1	45.1	36.6
TMM 41238-302	78.8	58.4	—	36.9
TMM 41238-304	79.5	58.5	46.2	36.2
TMM 41238-169	—	56.9	—	—
Second phalanx				
TMM 41238-175	36.7	54.6	—	—
TMM 41238-44	38.7	55.3	51.1	49.0
TMM 41238-166	38.3	47.2	—	41.0
TMM 41238-15	38.6	55.1	50.6	47.6
TMM 41238-192	39.3	54.6	48.7	44.6
TMM 41238-174	40.3	57.5	51.0	47.3

two left tibias (TMM 41238-207, 208); two right metatarsals (TMM 41238-46, 227); one left metatarsal (TMM 41238-1); two left metacarpals (TMM 41238-154, 629); distal ends of two metatarsals (TMM 41238-228, 660); proximal end of right metatarsal (TMM 41238-87); distal ends of two metacarpals (TMM 41238-40, 49); one left and two right calcanei (TMM 41238-118, 457, 165); two right astragali (TMM 41238-52, 312); seven first phalanges (TMM 41238-16, 4, 6, 161, 630, 692, 3); distal end of first phalanx (TMM 41238-425); six second phalanges (TMM 41238-299, 176, 424, 168, 438, 213); one third phalanx (TMM 41238–315).

Description.—The upper dentition is much less complex than that of *E. scotti* (Fig. 13). The fossettes of the available teeth have virtually no secondary plications. The pli caballin is small on both premolars and molars. The protocones of the upper teeth are short (except for TMM 41238-220) and most have grooves on their lingual faces. In size and qualitative characters the upper teeth from Cueva Quebrada are similar to the dentition of the type of *E. francisci* (TAMU 2518) reported by Lundelius and Stevens (1970).

Only one lower cheek tooth can be confidently assigned to this taxon (TMM 41238-75). It is smaller than most of the lower teeth assigned to *E. scotti.* Several teeth of the latter species are very close to the same length (TMM 41238-2, M_1 and M_2, TMM 41238-83, 255, 258) but are noticeably wider (Ta-

bles 8, 13). The Cueva Quebrada specimen is somewhat longer than any of the lower teeth of the type specimen of *E. francisci* but is close to the same width of that specimen (Lundelius and Stevens, 1970: table 1). Some of the length difference may be the result of the difference in the stage of wear of the two specimens. The Cueva Quebrada specimen is less deeply worn. Its antero-posterior length at the midpoint of its crown height is 23.7 mm which is close to the widths of the lowers of the type of *E. francisci.* The ectoflexid just reaches the metaconid-metastylid isthmus and a small pli caballinid is present. Both these characters differ from the type of *E. francisci* in which the ectoflexid does not reach the metaconid-metastylid isthmus and no pli caballinid is present. The metaconid-metastylid groove is open but is V-shaped as in the type of *E. francisci.*

The limb bones are smaller and more gracile than those of *E. scotti* and the metapodials are long and slender (Tables 14–16). A scatter diagram of the length versus distal width of the metatarsals shows that two of the three specimens from Cueva Quebrada are grouped with the type of *E. francisci* and specimens of *E. quinni* (Slaughter et al., 1962) and a large sample of *Equus* sp. from Channing, Texas (Fig. 9). One metatarsal (TMM 41238-227) is closer

Table 13.—*Measurements of teeth of* Equus francisci *from Cueva Quebrada.*

	Measurements		
	Upper teeth		
Specimens and teeth	Length along ectoloph	Width normal to para-mesostyle	Width normal to meso-metastyle
TMM 41238-157 P⁴	23.1	24.9	25.2
TMM 41238-157 M¹	22.1	24.4	23.5
TMM 41238-157 M²	22.5	23.3	24.9
TMM 41238-157 M³	23.7	20.1	21.8
TMM 41238-639 Pm	24.3 (min.)	24.8	—
TMM 41238-100 M	21.7	23.2	24.3
TMM 41238-220 M	—	21.1	—
	Lower teeth		
	Length	Anterior width	Posterior width
TMM 41238-75 Pm or M	27.8	11.6	12.0

in length to metatarsals from the Slaton fauna assigned to *E. conversidens* by Dalquest (1967), the Cedazo fauna from Aguascalientes, Mexico assigned to *E. conversidens* by Mooser and Dalquest (1975), the more slender metatarsals from the Dry Cave fauna assigned to *E. conversidens* by Harris and Porter (1980) and a large sample from San Josecito Cave, Nuevo Leon, assigned to *E. conversidens* by Stock (1950, 1953). The Cueva Quebrada metatarsal is narrower distally than all of the others with

Table 14.—*Measurements of major limb bones of* Equus francisci *from Cueva Quebrada.*

Bones and specimens	Length	Proximal width	Distal width
Femur			
TMM 41238-39	310	—	80
TMM 41238-229	315	108	82
TMM 41238-38	—	—	83
Tibia			
TMM 41238-159	345	81 (min.)	62
TMM 41238-152	357	91 (min.)	—
TMM 41238-208	—	—	65
TMM 41238-207	—	—	59
TMM 41238-342	—	—	52
Humerus			
TMM 41238-234	232	80	66
Radius			
TMM 41238-226	315	—	65
TMM 41238-130	—	—	69
TMM 41238-180	—	—	67

Table 15.—*Measurements of metacarpals of* Equus francisci *from Cueva Quebrada.*

Measurements	TMM 41238-628	TMM 41238-154
Length	223	237
Transverse width of proximal end	44.4	—
Antero-posterior width of proximal end	29.3	—
Transverse diameter of posterior articular facet for the magnum	28.7	—
Transverse width of facet for magnum	33.3	—
Transverse width of shaft at midpoint	27.2	29.4
Antero-posterior diameter of shaft at midpoint	24.1	23.7
Transverse width of distal end above the articular facet	39.1	39.1
Transverse width of distal articular surface	39.7	—
Antero-posterior diameter of inner distal articular facet	27.7	27.2
Antero-posterior diameter of central ridge of distal articular surface	30.4	30.4

the exception of the Cedazo material which is shorter. The proportions of the distal end (trochlear diameter versus distal width) show much more homogeneity in the Cueva Quebrada metatarsals (Fig. 10) and plots with the San Josecito sample.

The two small metacarpals do not show the same amount of variation as the metatarsals. They are shorter than those from Channing, Texas, and more slender than those from San Josecito but they are similar to two from Dry Cave (Figs. 11, 12).

The phalanges assigned to this taxon are slender and show no evidence of heterogeneity (Table 17).

Discussion.—The taxonomy of the small Pleistocene horses is still uncertain. There is no agreement on the number of species or on the characters or the limits of variation that might characterize the different species. All of the small horse material from Cueva Quebrada indicates the presence of only one species with the possible exception of the somewhat short metatarsal (TMM 41238-227). If only one species is represented the three metatarsals represent the range of length for this species. This is not appreciably greater than the ranges shown by two other much larger samples of metatarsals of comparable sized horses, one from Channing, Texas, and one from San Josecito Cave, Nuevo Leon, Mexico (Fig. 9). The Cueva Quebrada sample would be inter-

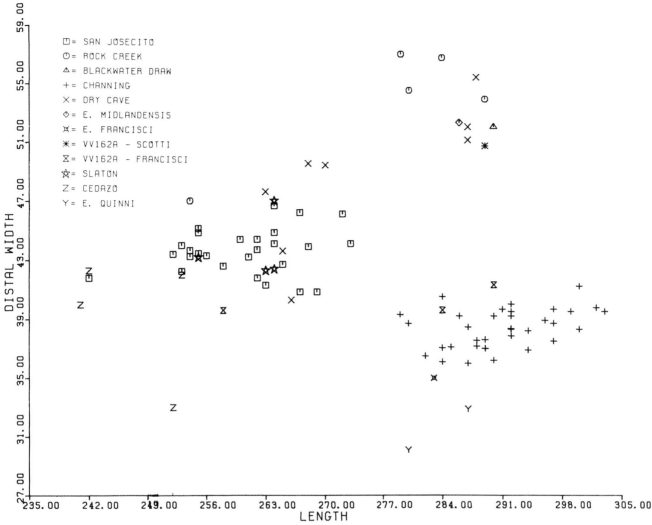

Fig. 9.—Scatter diagram of length versus distal width of metatarsals of Pleistocene *Equus*.

Table 16.—*Measurements of metatarsals of* Equus francisci *from Cueva Quebrada.*

Measurements	TMM 41238-227	TMM 41238-1	TMM 41238-46	TMM 41238-228	TMM 41238-87
Length	258	284	290	—	—
Transverse width of proximal end	41.8	—	45.8	—	44.6
Antero-posterior width of proximal end	39.8	—	—	—	—
Transverse diameter of posterior articular facet	19.2	—	—	—	—
Transverse width of facet for ectocuneiform	39.8	—	40.0	—	40.6
Transverse width of shaft at midpoint	28.6	27.8	31.3	29.2	—
Antero-posterior diameter of shaft at midpoint	28.4	27.7	28.7	30.4	—
Transverse width of distal end above the articular surface	38.9	38.5	40.9	41.7	—
Transverse width of distal articular surface	39.6	39.6	41.3	40.5	—
Antero-posterior diameter of inner distal articular facet	27.8	29.2	29.4	29.0	—
Antero-posterior diameter of central ridge of distal articular surface	30.0	31.3 est.	33.6 est.	31.4	—

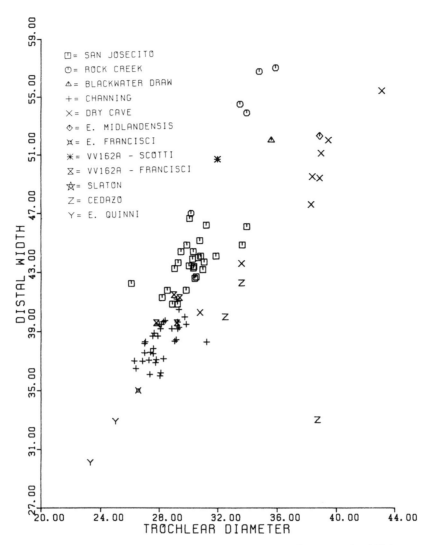

Fig. 10.—Scatter diagram of distal width versus trochlear diameter of metatarsals of Pleistocene *Equus*.

mediate in size. If two species are represented the other small horse material shows no clear evidence of this. A single species appears more probable and is adopted here.

Which name should be applied to the Cueva Quebrada material, is not clear. Lundelius and Stevens (1970) demonstrated that the type of *E. francisci* had long, slender metapodials. In their study of the Cedazo fauna from Aguascalientes, Mooser and Dalquest (1975) associated somewhat stockier metapodials with dentitions they believed were referable to *E. conversidens* and metapodials similar in proportions to those of *E. francisci* and Cueva

Quebrada specimens were associated with dentitions they believed could be assigned to *E. tau*. If these associations are correct, *E. francisci* would be a junior synonym of *E. tau*.

Order Artiodactyla
Family Camelidae

cf. *Camelops*

Material.—Distal end of a third or fourth metapodial (TMM 41238-326).

Description.—The form and size of the articular surface match those of *Camelops* from other late

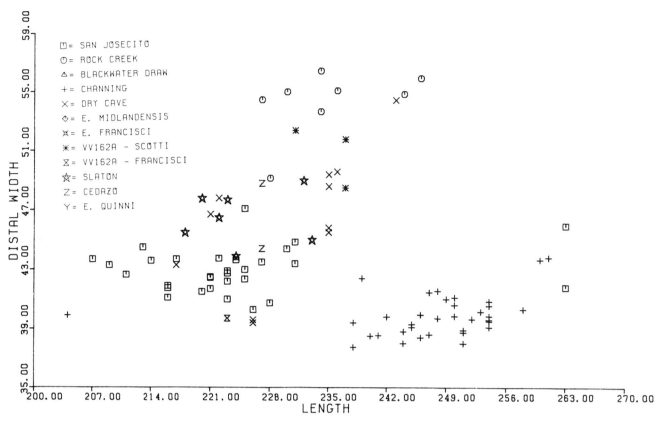

Fig. 11.—Scatter diagram of length versus distal width of metacarpals of Pleistocene *Equus.*

Pleistocene localities. The crest on the articular surface is confined to the ventral part. The width of the articular surface is 40.0 mm, the antero-posterior diameter is 42.3 mm. This is slightly below or at the lower end of the size range of *Camelops* given by Webb (1965). The differences are slight and are probably not significant in view of the small size of the sample available to Webb.

Discussion.—*Camelops* sp. is a widespread large camel in late Pleistocene faunas in the western part of North America (Kurtén and Anderson, 1980). Remains of a large camel, probably *Camelops,* are known from Bonfire Shelter in Val Verde County, Texas (Frank, 1968).

Family Cervidae

Navajoceros fricki Kurtén

Material.—Proximal part of a left femur (TMM 41238-362).

Description.—The femur is intermediate in size between those of *Odocoileus hemionus* and *Cervus canadensis.* The length of the Cueva Quebrada specimen cannot be determined but the length of the femur given by Kurtén (1975), of 328 mm would not be an impossible estimate. The proximal width is 89 mm.

Table 17.—*Measurements of phalanges of* Equus francisci *from Cueva Quebrada.*

Phalanges and specimens	Length	Proximal width	Distal width	Mid width
First phalanx				
TMM 41238-3	72.4	41.7	33.8	25.2
TMM 41238-16	76.3	40.5	35.1	26.3
TMM 41238-6	87.6	45.3	36.6	26.7
TMM 41238-4	81.7	48.2 est.	35.2	27.6
TMM 41238-161	78.1	41.4 est.	34.4	26.2
TMM 41238-630	73.1	43.6	33.7	26.5
TMM 41238-692	86.2	—	—	28.3
Second phalanx				
TMM 41238-438	34.6	42.2	39.4	35.2
TMM 41238-299	33.4	37.9	33.7	31.1
TMM 41238-213	34.8	41.0	37.0	32.5
TMM 41238-168	33.8	38.5	37.6	33.5
TMM 41238-176	26.8	36.1	34.1	30.6
TMM 41238-424	—	41.4	—	—

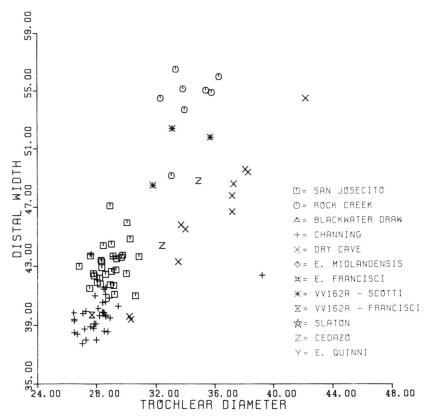

Fig. 12.—Scatter diagram of distal width versus trochlear diameter of metacarpals of Pleistocene *Equus*.

Discussion.—This large deer has been reported from a number of localities in the Rocky Mountains from Wyoming to Mexico (Kurtén and Anderson, 1980). The Cueva Quebrada record is the easternmost occurrence known in the United States. Although Val Verde County is located well east of the Rocky Mountains as such, the rocky, steep terrain associated with the canyons in the region of Cueva Quebrada apparently provided the proper environment.

Family Antilocapridae

Stockoceros sp.

Material.—Fragment of a right frontal with basal part of anterior horn core and base of posterior horn core (TMM 41238-462); right M^3 (TMM 41238-11); right mandibular fragments with P_4–M_2 (TMM 41238-9); right P_4 (TMM 41238-466); two left M_3's (TMM 41238-12, 10); lower incisor (TMM 41238-410); distal ends of one right and two left humeri (TMM 41238-216, 215, 513); one juvenile left radius with distal epiphysis gone (TMM 41238-172); distal part of juvenile right radius with epiph-

Fig. 13.—Teeth of *Equus francisci* from Cueva Quebrada. A) Left P^4–M^3 (TMM 41238-157); B) right upper premolar (TMM 41238-639); C) right lower molar or premolar (TMM 41238-75). ×.75.

Table 18.—*Measurements of teeth of* Stockoceros *sp. from Cueva Quebrada.*

Specimens	Teeth	Length	Anterior width	Posterior width	Width of fourth lobe
TMM 41238-11	M^3	17.3	9.0	7.8	—
TMM 41238-466	P_3	9.7	5.1	4.9	—
TMM 41238-9	P_4	9.5	5.0	5.0	—
TMM 41238-9	M_1	10.1	4.7	5.4	—
TMM 41238-9	M_2	11.5	6.2	6.3	—
TMM 41238-10	M_3	—	6.9	6.3	4.8
TMM 41238-12	M_3	19.4	6.9	6.8	4.7

Table 19.—*Measurements of major limb bones of* Stockoceros *sp. from Cueva Quebrada.*

Bones and specimens	Proximal width	Distal width
Humerus		
TMM 41238-215	—	30.6
TMM 41238-513	—	30.7
Radius		
TMM 41238-172	30.4	—
TMM 41238-658	29.3	—
Tibia		
TMM 41238-107	—	29.5
TMM 41238-21	—	29.5
TMM 41238-368	—	31.2
TMM 41238-642	48.5	—
TMM 41238-257	—	27.2
Metacarpal		
TMM 41238-101	22.6	—

ysis gone (TMM 41238-112); proximal part of right radius (TMM 41238-658); proximal end of left ulna (TMM 41238-173); part of shaft of a left femur (TMM 41238-171); shaft and distal epiphyses of right femur (TMM 41238-5); shaft and distal end of juvenile right tibia (TMM 41238-107); proximal part of one right and one left tibia (TMM 41238-286, 642); distal parts of right and left tibia (TMM 41238-257, 21); distal epiphysis of a left tibia (TMM 41238-368); two right and one left astragali (TMM 41238-369, 370, 611); proximal end of right metacarpal (TMM 41238-101); distal epiphysis of a metapodial (TMM 41238-217); distal end of a metapodial (TMM 41238-641); articulated pair of first, second and third, phalanges (TMM 41238-7); eight first phalanges (TMM 41238-222, 54, 111, 356, 245, 58, 110, 602); distal end of a first phalanx (TMM 41238-395); proximal end of first phalanx (TMM 41238-123); eleven second phalanges (TMM 41238-291, 508, 56, 55, 57, 113, 394, 472, 292, 294, 279).

Description.—The fragment of the frontal bone has the lower part of the anterior horn core and the base of the posterior horn core. The anterior edge of the anterior horn core is broken away so the entire cross section of the horn core cannot be determined, but it appears to have been oval as in *Stockoceros onusrosagris* (Skinner, 1942). The breakage of the base of the posterior horn core makes it impossible to determine the angle between the two horn cores but it appears to have been approximately that shown for *Stockoceros onusrosagris* by Skinner (1942:fig. 10). The prominent pit at the base of the horn cores at the junction with the supraorbital process stated by Skinner (1942) to be characteristic of *Stockoceros* is discernable on the Cueva Quebrada specimen but is not obvious because of the breakage of the supraorbital process. The supraorbital foramen is located immediately anterior to the core base.

The dentition of the Cueva Quebrada antilocaprid is similar to that of *Stockoceros onusrosagris* in both size and morphology (Table 18). The M^3 has a smaller metacone that is seen on the M^3 of *Antilocapra americana*, a point of similarity with *S. onusrosagris*, and a prominent heel is present on the external

side of the posterior end. This character is variable in both *A. americana* and *S. onusrosagris* (Skinner, 1942). The lower dentition is similar to that of *Stockoceros onusrosagris* from Papago Springs Cave. The P_4 consists of two lobes, a large anterior one and a much smaller posterior one. The posterior fold is closed by the union of the hypoconid and entoconid to form a posterior median fossette. The anterior median fold is closed by the junction of the paraconid and metaconid to form a median fossette in the anterior lobe. The form of the two P_4's from Cueva Quebrada is not as molariform as those figured by Skinner (1942:fig. 16). They are more like those figured by Colbert and Chaffee (1939:fig. 7). The M_3's are three lobed teeth with weakly developed posterior ridges. In neither of these specimens does the posterior ridge approach the stage of a fourth lobe shown in some specimens from Papago Springs Cave (Skinner, 1942). The size of the dentition is close to that of the Papago Springs Cave sample and is smaller than that of *A. americana* (Table 18). The postcranial material is too incomplete to provide measurements that would allow comparison with those of *S. onusrosagris* from Papago Springs Cave given by Roosevelt and Burden (1934), Colbert and Chaffee (1939), and Skinner (1942). Comparison with a skeleton of *A. americana* shows the Cueva Quebrada specimens to be somewhat smaller. Measurements of some postcranial elements are given in Table 19, 20.

Discussion.—Antilocaprids of the genus *Stocko-*

Table 20.—*Measurements of phalanges of* Stockoceros *sp. from Cueva Quebrada.*

Phalanges and specimens	Length	Depth large dist. art. surface	Proximal width	Proximal depth
First phalanges				
TMM 41238-58	—	12.5	—	—
TMM 41238-54	40.4	10.1	11.2	15.1 (min.)
TMM 41238-356	40.1	12.6	12.0 (min.)	15.9
TMM 41238-111	39.9 (min.)	12.3	—	—
TMM 41238-245	39.1	11.5	11.5	14.9
TMM 41238-222	39.9	12.2	11.9	15.7
TMM 41238-7	44.1	11.3	11.7 (min.)	16.8
TMM 41238-7	44.4	12.2	12.3	16.9
TMM 41238-110	—	12.7	13.1	16.0
TMM 41238-123	—	—	13.2	16.8
TMM 41238-602	—	10.7	—	—
Second phalanges				
TMM 41238-291	25.1	12.4	—	—
TMM 41238-508	26.9	12.3	11.6	14.9
TMM 41238-279	26.2	11.6	10.8	14.0
TMM 41238-56	27.1	10.1	10.5	—
TMM 41238-55	25.4	11.8	10.9	14.2
TMM 41238-57	24.3	11.0	10.3	13.4 (min.)
TMM 41238-394	24.2	12.4	10.7	14.2
TMM 41238-472	27.0	11.6	11.4	14.7
TMM 41238-292	26.4	12.8	11.1	14.1
TMM 41238-294	25.0	11.1	10.1	13.4

ceros are known from a number of late Pleistocene localities in the southwestern United States and Mexico. Two species *S. conklingi* and *S. onusrosagris* have been named but the differences between them are slight and Kurtén and Anderson (1980) have suggested that detailed quantitative studies on all samples may clarify their relationship.

Family Bovidae

Bison sp.

Material.—Two partial horn cores (TMM 41238-22); distal end of a metapodial (TMM 41238-447); first phalanx (TMM 41238-432).

Description.—The two horn cores, probably from the same individual are too large to be from a modern bison. They are the right size for either *B. antiquus* or *B. occidentalis,* but the material is too incomplete to allow a specific identification.

Discussion.—Bison remains are not common in Pleistocene deposits in Trans-Pecos Texas, but are common farther north on the Great Plains. Bison remains are abundant in deposits in Bonfire Shelter in the same area (Dibble and Lorrain, 1968). It can be shown there that on at least three occasions Paleo Indians drove a herd of bison over the cliff and into the shelter.

LITERATURE CITED

AKERSTEN, W. A., R. L. REYNOLDS, and A. E. TEJADA-FLORES. 1979. New mammalian records from the late Pleistocene of Rancho La Brea. Bull. Southern California Acad. Sci., 78: 141–143.

ALEXANDER, R. K. 1974. The archaeology of Conejo Shelter: A study of cultural stability at an archaic rockshelter site in southwestern Texas. Unpublished dissert., Univ. Texas, Austin, 340 pp.

BONNICHSEN, R. 1973. Some operational aspects of human and animal bone alteration. Pp. 9–24, *in* Mammalian osteology: North America (B. M. Gilbert, ed.), Missouri Archaeol. Soc., Spec. Publ., Columbia, 337 pp.

———. 1979. Pleistocene bone technology in the Beringian refugium. Nat. Mus. Man Merc. Ser., Archaeol. Surv. Canada Paper, 89:1–297.

BRAIN, C. K. 1981. The hunters or the hunted? An introduction to African cave taphonomy. Univ. Chicago Press, Chicago, 365 pp.

CARLETON, M. D., and R. E. ESHELMAN. 1979. A synopsis of fossil grasshopper mice, genus *Onychomys,* and their relationships to Recent species. Univ. Michigan, Papers Paleont., 21:1–63.

COLBERT, E. H., and R. G. CHAFFEE. 1939. A study of *Tetrameryx* and associated fossils from Papago Springs Cave, Sonoita County, Arizona. Amer. Mus. Novit. 1034:1–21.

DALQUEST, W. W. 1964. *Equus scotti* from a high terrace near Childress, Texas. Texas J. Sci., 16:350–358.

———. 1967. Mammals of the Pleistocene Slaton local fauna of Texas. Southwestern Nat., 12:1–30.

DALQUEST, W. W., E. ROTH, and F. JUDD. 1969. The mammal fauna of Schulze Cave, Edwards County, Texas. Bull. Florida State Mus., Biol. Sci., 13:205–276.

DAVIS, W. B. 1960. The mammals of Texas. Bull. Tex. Game and Fish Comm., 27:1–252.

DIBBLE, D. S., and D. LORRAIN. 1968. Bonfire shelter: a stratified bison kill site, Val Verde County, Texas. Misc. Papers, Texas Memorial Mus., 1:9–138.

FRANK, R. M. 1968. Identification of miscellaneous faunal remains from Bonfire Shelter. Pp. 133–134 Appendix, *in* Bonfire Shelter: a stratified bison kill site, Val Verde County, Texas (D. S. Dibble and D. Lorrain), Misc. Papers, Texas Memorial Mus., 1:9–138.

GOLDMAN, E. A. 1910. Revision of the wood rats of the genus *Neotoma.* N. Amer. Fauna, 34:1–124.

HALL, E. R., and K. R. KELSON. 1959. Mammals of North America. The Ronald Press Co., New York, 1:xxx + 1–546 + 79; 2:viii + 547–1083 + 79.

HARRIS, A. H. 1970. The Dry Cave mammalian fauna and late pluvial conditions in southeastern New Mexico. Texas J. Sci., 22:3–27.

———. 1977. Wisconsin age environments in the northern Chihuahuan Desert: Evidence from the higher vertebrates. Pp. 23–52, *in* Transactions of the Symposium on the Biological Resources of the Chihuahuan Desert Region, United States and Mexico (R. H. Wauer and D. H. Riskind, eds.), National Park Service. Proc. Trans. Ser., 3:ixxii + 1–658.

HARRIS, A. H., and L. S. W. PORTER. 1980. Late Pleistocene horses of Dry Cave, Eddy County, New Mexico. J. Mamm., 61:46–65.

HAYNES, G. 1980. Evidence of carnivore gnawing on Pleistocene and Recent mammalian bones. Paleobiology, 6:341–351.

———. 1983. A guide for differentiating mammalian carnivore taxa responsible for gnaw damage to herbivore limb bones. Paleobiology, 9:164–172.

KURTÉN, B. 1967. Pleistocene bears of North America. 2. Genus *Arctodus*, short-faced bears. Acta. Zool. Fennica, 117:1–60.

———. 1968. Pleistocene mammals of Europe. Aldine Publ. Co., Chicago, 317 pp.

———. 1975. A new Pleistocene genus of American mountain deer. J. Mamm., 56:507–508.

KURTÉN, B., and E. ANDERSON. 1980. Pleistocene mammals of North America. Columbia Univ. Press, New York, 442 pp.

LINNAEUS, C. 1758. Systema Naturae. 10th ed. Stockholm, v. 1, 824 pp.

LOGAN, L. E., and C. C. BLACK. 1979. The Quaternary vertebrate fauna of Upper Sloth Cave, Guadelupe Mountains National Park, Texas. Pp. 141–158, *in* Biological investigations in the Guadelupe Mountains National Park, Texas (H. H. Genoways and R. J. Baker, eds.), National Park Service Proc. Trans. Ser., 4:xvii + 1–442.

LUNDELIUS, E. L., JR. 1972. Vertebrate remains from the Gray Sand. Pp. 148–163, *in* Blackwater Locality No. 1, A stratified early man site in eastern New Mexico (J. J. Hester, ed.), Publ. Fort Burgwin Research Center, 8:1–239.

———. 1979. Post-Pleistocene mammals from Pratt Cave and their environmental significance. Pp. 239–257, *in* Biological Investigations in the Guadelupe Mountains National Park, Texas (H. H. Genoways and R. J. Baker, eds.), National Park Service Proc. Trans. Ser., 4:xvi + 1–442.

LUNDELIUS, E. L., JR., and M. S. STEVENS. 1970. *Equus francisci* Hay, a small stilt legged horse, Middle Pleistocene of Texas. J. Paleo., 44:148–153.

MEADE, J. J. 1981. The last 30,000 years of faunal history within the Grand Canyon, Arizona. Quat. Res., 15:311–326.

MILLER, S. J. 1979. The archaeological fauna of Smith Creek Canyon. Pp. 273–329, *in* The Archaeology of Smith Creek Canyon, eastern Nevada (D. R. Tuohy and D. L. Randall, eds.), Nevada State Mus. Anthro. Papers, 17:273–329.

MOOSER, O., and W. W. DALQUEST. 1975. Pleistocene mammals from Aguascalientes, central Mexico. J. Mamm., 56: 781–820.

QUINN, J. H. 1957. Pleistocene Equidae of Texas. Bur. Eco. Geol., Univ. Texas, Rept. Invest., 33:5–51.

ROOSEVELT, Q., and J. W. BURDEN. 1934. A new species of antilocaprine, *Tetrameryx onusrosagris,* from a Pleistocene cave deposit in southern Arizona. Amer. Mus. Novit., 754: 1–4.

SCHMIDLY, D. J. 1977. The mammals of Trans-Pecos Texas. Texas A&M University Press, College Station, 225 pp.

SEMKEN, H. A. 1961. Fossil vertebrates from Longhorn Cavern, Burnet County, Texas. Texas J. Sci., 13:290–310.

SKINNER, M. F., 1942. The fauna of Papago Springs Cave, Arizona and a study of *Stockoceros* with three new antilocaprines from Nebraska and Arizona. Bull. Amer. Mus. Nat. Hist., 80:143–220.

SLAUGHTER, B., W. W. CROOK, JR., R. K. HARRIS, D. C. ALLEN, and M. SEIFERT. 1962. The Hill-Shuler local faunas of the Upper Trinity River, Dallas and Denton Counties, Texas. Univ. Texas Bur. Econ. Geol., Rept. Invest., 48:1–75.

STOCK, C. 1950. 25,000-year-old horse; the skeleton of an ice age horse makes a return trip to Mexico. Eng. Sci. Monthly, 14:16–17.

———. 1953. El caballo pleistoceno (*Equus conversidens leoni*) de la cueva de San Josecito, Aramberri, Nuevo Leon. Mem. Congr. Cient. Mexico, 3:170–171.

STOCK, C., and F. D. BODE. 1937. The occurrence of flints and extinct animals in pluvial deposits near Clovis, New Mexico, Part III—Geology and vertebrate paleontology of the late Quaternary near Clovis, New Mexico. Proc. Acad. Nat. Sci. Philadelphia, 88:219–241.

VALASTRO, S., E. M. DAVIS, JR., and A. VARELA. 1977. University of Texas at Austin, Radiocarbon Dates XI. Radiocarbon, 19:280–325.

———. 1979. University of Texas at Austin, Radiocarbon Dates XIII. Radiocarbon, 21:257–273.

WEBB, S. D. 1965. The osteology of *Camelops*. Bull. Los Angeles County Mus., Sci., 1:1–54.

Address: Department of Geological Sciences, University of Texas at Austin, Austin, Texas 78712.

ALASKAN MEGABUCKS, MEGABULLS, AND MEGARAMS: THE ISSUE OF PLEISTOCENE GIGANTISM

R. Dale Guthrie

ABSTRACT

Pleistocene large mammals from Alaska appear to be larger than their living counterparts. This seems to be the case for cervids (*Rangifer, Cervus,* and *Alces*) and bovids (*Ovis, Ovibos,* and *Bison*). The circumstances of this larger body size are examined and it is concluded that the primary reason behind such large bodied, large horned or antlered, individuals was two-fold (1) they belonged to populations which were kept understocked by high winter mortality and (2) experienced a long peak in nutrient availability during the growth season. The length of this seasonal peak in nutrient availability is seen as critical to studies of body size changes. Wild sheep are used as an exemplary case and are examined in greater detail than other species.

Large non-ruminant ungulates seem to respond differently than ruminants to the changing conditions from mid-Pleistocene, to late Pleistocene, through the Holocene. This may be due to their more conservative life histories. Pleistocene ground squirrels (*Spermophilus*) in Alaska seem to have responded differently than ungulates in their body size changes. Their special life history features, particularly hibernation and reproduction may account for this difference.

INTRODUCTION: THE ISSUE OF CAUSES UNDERLYING BODY SIZE CHANGES

We are accustomed to thinking in typological symbolic images when reconstructing past creatures. For example, "how did a wooly mammoth look?" Yet we realize that there was considerable variation at any one time and particularly through time. How a wooly mammoth looked depends on which wooly mammoth one picks and from what time period. My main emphasis in this paper is that the resources available to wooly mammoths and their ilk influenced these differences. We see subtle versions of these influences today even in our own species. French women born before World War II, are two inches (50 mm) shorter than their counterparts born in the 1960's (Zeldin, 1983).

Because we can easily study the differences among contemporary populations of wild mammals, they can help us picture how these interactions of available resources and body size must work through time. The large black bull buffalo, *Syncerus caffer,* of East Africa is twice the size of its small subspecific effeminately colored reddish-brown counterpart in West Africa. The horn size differences are even more exaggerated. Likewise, the small Florida white-tailed deer, *Odocoileus virgianus,* are almost unrecognizable as conspecific to the subspecies which grows to immense proportions in the northern corn-belt states.

Wildlife managers understand the basic developmental and ecological reasons why these differences in individual "quality" exist (for example, Schmidt and Gilbert, 1978). Animals reared on overstocked ranges of poor forage quantity and qual-ity during the growth-seasons are considerably smaller than the same genotypes reared on high quality understocked ranges just as with French girls.

One could conclude the intuitively obvious, that if these differences persist for a long enough time the balances of natural selections will also be skewed and genetic changes will ultimately emerge. Poor growth-season conditions select for smaller animals and the opposite for good growth conditions. For large mammals, at least, there is a wealth of data and experience which supports these conclusions. Reconstructing what is meant by poor and good growing seasons as one goes back through time, however, is no easy task, and what one might intuitively consider poor "quality" growing conditions may not be so to ungulates. This is particularly true in the cold north.

It is difficult to imagine that glacial conditions, so severe as to rid Alaska of virtually all trees, would result in a richer growth season for ungulates than the warmer wetter conditions of the postglacial, but that indeed seems to be the case.

Unglaciated Alaska was an arm of Asia during most of the Pleistocene, and was a peculiarly different place than the now ubiquitous damp spruce woodlands and wet tundra (Hopkins et al., 1982). A herbaceous mammoth steppe extended from Europe across northern Asia to Alaska. The harsh climatic conditions of increased aridity and coolness along with different seasonal patterns of temperature and moisture created this xeric herbaceous vegetation that supported an amazing array of large mammal species. In other studies (for example, Guthrie, 1982) I have discussed the paleoclimatic significance of these Alaskan large mammal faunas, and the possible conditions leading to their local and regional extinctions (Guthrie, 1984). In this study I wish to look to another dimension, that of body size changes.

SEASONAL RESOURCE AVAILABILITY AND BODY SIZE

By way of brief introduction to some selection dynamics of body size, let me say that mammals and many other organisms have an intrinsic growth rate potential which tends to level-off or stop at a predetermined size. Those organisms exhibiting this "plateaued" pattern of growth are called "determinate." All ungulates are relatively determinate in growth pattern, with sexes showing similar growth rates (males often slightly higher). Female growth usually reaches a plateau—and slows to a stop—earlier than the male; this creates a marked sexual dimorphism among adults. The earlier plateau of the females is usually due to an adaptation in which growth energy is normally diverted to reproduction.

The various evolutionary and developmental forces which affect growth rate and the point at which determinate growth is reached (body size) are complex and quite contextual. I have argued, however, that the most important factor is the quality and quantity of resources during the growth season (Guthrie, 1984).

There are a number of diverse selection forces which would increase body size—social advantages, physiological benefits, predatory or antipredatory benefits, and others. However, there are costs to being bigger in a competitive world. The benefits of larger body size are counterbalanced by absolute limitations of essential basic resources. The larger body size variants may experience increased debilitation and mortality from insufficient nutrients or energy, exposure to predation from spending more time searching for food, and other diverse, contextual stresses in this same vein. When a higher quality and quantity of food becomes available, animals usually respond first by moving closer to their maximum genetic size potential (see Bell, 1971; Hanley, 1980, for a discussion). Then, given sufficient time, that genetic potential is extended upward, that is, there is selection for greater body size.

Thus, the character of the growth season is of maximum importance. However, the time spent outside the growth season is important as well, because this is the time of major adult mortality which, in turn, influences the degree of competition for the succeeding growth season's resources. Reductions in the available quality and quantity of resources, either through scarcity or through competition, would select for a reduced growth rate and/or postpone growth to subsequent growth seasons. Increased quality of available resources would select for increased growth rate. We thus end up with several potential ways to increase body size in relation to changing resources (Fig. 1).

There are ultimately three evolutionary avenues to increased body size. I will argue that one of these was the major element in ungulate gigantism during the Pleistocene.

The first avenue is an *improvement* in food quality and a concomitant ability to use this "bonus" quality for a more rapid growth within any one growing season. The second often accompanies a *decline* in food quality (and reduced mortality) and compensates by retaining the ability to growth over more years; instead of stopping growth at, say, the end of the second year, it continues on to the fourth. The third evolutionary avenue to larger body size normally accompanies a *lengthening of the season* when growth resources are available. Thus an animal is able to grow for a longer time within each annual growth season. These different patterns are portrayed in Fig. 1 and discussed in more detail below.

Large Body Size Due to Higher Peak in Resource Quality

For herbivores, young vegetation is normally very palatable, that is, it has low concentrations of fiber, phytoliths, and antiherbivory compounds. It is also high in growth materials for the herbivore—soluable carbohydrates, phosphorus, nitrogen, potassium, and others—which are in easily digestable forms.

However, the maximum capability of ungulates to use growth nutrients is usually around or below the spring peak in vegetational quality, so that additional dietary quality during this time is not readily usable (that is, ungulates have intrinsic volume and assimilation limits to their capacities as well as metabolic limits). Plains, mountain, and tundra ungulates are generally mobile; they follow early plant

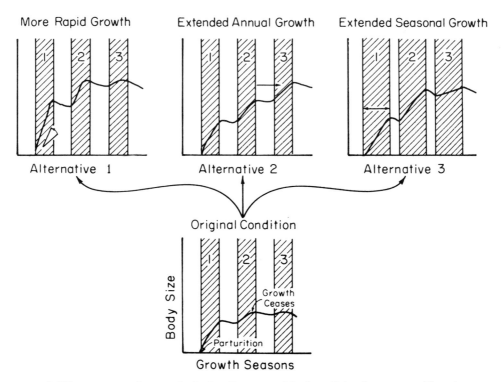

Fig. 1.—There are several different ways to increase body size. From an original condition (lower part of figure) one can either increase *growth rate* (Alternative 1), or the same rate can be maintained for *more growth seasons* (Alternative 2), or the *length of the annual growth season* can be extended (Alternative 3). Although growth rates were possibly somewhat greater, the third alternative seems to be the major factor in explaining the general phenomenon of Pleistocene gigantism. The zones numbered 1, 2, and 3 represent growth seasons.

growth topographically or geographically—thus extending the effective duration of the maximum consumption of quality resources. Ungulates can find peak food quality by pursuing age (phenological) gradients of the same plant species or taking advantage of a succession of different plants in the same location. Most ungulate growth occurs during this period when plants are at their highest quality. It is unlikely that the ungulate growth rate could be greatly enhanced during this time (that is, on supplemental higher quality resources ungulate growth *rates* cannot be greatly improved). I am speaking here about extant high quality, ungulate populations; there are some populations in every species for which growth rates are suboptimal even during this peak resource season because range is poor, or competition is great.

In summary, there are species in which a higher peak in food quality increases body size, but most ungulates benefit only modestly from supplemental higher quality food during the peak of the growth

season. Higher quality food for ungulates and the accompanying increased growth rates do not seem to be the only, or even the major factor in accounting for the general phenomenon of Pleistocene gigantism. I do not want to discount the importance of nutrient quality during the growth season in the determination of body size. It is of obvious importance and is a major element in my model. The question at hand, however, is whether the *peak* of nutrient quality was *higher* in the Pleistocene. There are several lines of evidence which suggest that a higher nutrient peak was not the only or even the major cause of Pleistocene gigantism. This evidence is discussed in more detail later.

Larger Body Size Achieved in Additional Annual Growing Seasons

Most species of vertebrates do use several annual growth seasons to reach full size; a single growth season is often not sufficient to attain optimum body size for the ecological position they occupy. Generally speaking, a reduction in juvenile and adult mortality (increased life expectancy) selects for prolonged growth. Stated another way, a more conservative, elongated life history usually results in allocations of resources into additional growth over a succession of years. The increased time investments in larger body size pay off in competitive ability, in greater stature and reproductive advantages, and in reduced predation. Because individual productivity can be spread over a longer period of time, litter size is normally reduced and greater care is given each offspring.

This phenomenon can be seen for example, among the elephants (proboscidians), which grow slowly over many years and in general have a conservative life history. The same is also true for other major groups which attain a large body size (for example, rhinos and cetaceans).

When growth is sustained in additional annual growth seasons an evolutionary shift to a reduced rate of growth almost always occurs. There are several ways of evaluating whether this was or was not the case for ungulates in the Pleistocene. First, there should be a striking change in the survivorship and growth curves, but (at least among the ruminants) such change is not apparent. There is no indication that mortality rates were lower in Pleistocene artiodactyls. In fact, several authors (for example, Geist, 1971) have argued that many extant large artiodactyls are shorter-lived than their smaller counterparts (large Marco Polo sheep, *Ovis ammon* versus the smaller bighorn sheep, *Ovis canadensis*). In comparisons of many closely related extant artiodactyl species, larger species or subspecies body size seems to be associated with higher adult mortality—quite contrary to what this second explanation of gigantism would predict. Large forms seem to be larger at the same individaul age as their smaller counterparts (Klein, 1965) suggesting that their larger size is not due to extending growth over more annual growth seasons. Also, larger species within genera or larger subspecies within species mature earlier (or as early as the smaller forms), suggesting that their large size is not a more conservative trait. If we could imagine many of the Pleistocene forms as chronospecies or chronosubspecies of extant ungulate populations, this same principle seems to have temporal as well as spatial dimension.

In summary, many species do attain large body size by continuing to grow for additional years and

this has been an important element in some gigantism in the past (proboscidians, for example), but it does not seem to account for Pleistocene gigantism or Holocene dwarfing.

Large Body Size Via Longer Growth Seasons Each Year

Species can become much larger in body size without greatly increasing growth rates or greatly changing survivorship or mortality curves. They simply need to be supplied with growth nutrients for a longer period during each annual growth season. The growth peak can be extended. Although supplementing ungulate diet during their normal growth peak seldom results in major growth changes, supplemental feeding before and after the normal growth peak greatly increases annual growth gains without affecting the growth rate. This indicates that ungulates have the developmental potential for prolonged seasonal growth if nutrients are available.

If this extension of the growth season occurred naturally it would theoretically: (1) allow animals to reach their genetic maximum potential body size and (2) select for animals with greater genetic size potentials.

I should emphasize that it is not just the length of season which is important but the conjunction of *high nutrient levels maintained over a longer period of time* that is required to maximize body size.

Additionally, heavy competition for survival resources through interspecific or intraspecific competition (outside the growth season) is required to hold population numbers low enough to insure range quality (and hence, ungulate quality) during subsequent growth seasons. This addendum is a necessary part of any alternative model of events; it is likely to have occurred during much of the Pleistocene. The intensifying winter or dry season which accompanied the cooling process, and perhaps also, interspecific competition during the non-growth seasons within the context of the greater species diversity, would tend to reduce population numbers.

I will focus my discussion of Pleistocene gigantism on Alaskan mountain sheep (*Ovis*) and, then to a lesser extent, on bison (*Bison*), elk (*Cervus*), caribou (*Rangifer*), moose (*Alces*), and muskoxen (*Ovibos*) and discuss why these same principles do not necessarily apply to such animals as mammoth (*Mammuthus*), horses (*Equus*), and ground squirrels (*Spermophilus*).

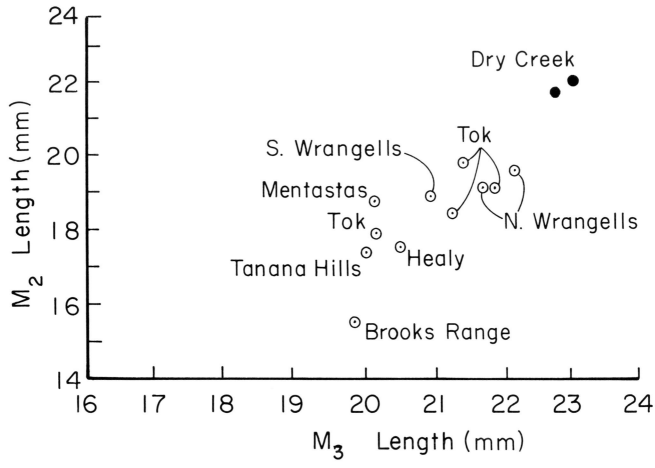

Fig. 2.—A comparison of body size in recent Dall sheep (*Ovis dalli*) using length of molars, illustrating the larger size of the well preserved Dry Creek specimens (dated at around 11,000 B.P.). All of the measurements taken on recent sheep are from mature "trophy class" rams from different areas in Alaska.

FOSSIL ALASKAN DALL SHEEP: IMPLICATIONS FOR PLEISTOCENE SEASONALITY

My first real awareness that Alaskan fossil sheep could be larger than their living counterparts came from studying an archaeological site. Fossil sheep teeth from the Dry Creek archaeological site (Powers et al., in press) have been plotted against those of some trophy ram skulls available to the author. Even with these few samples, it is apparent that the Dry Creek sheep were probably larger (Fig. 2).

Additionally, in Figs. 3 and 4 comparisons have been made of sheep skull characters and horn core dimensions from other fossil sheep collected in central Alaska. They also differ significantly from extant rams. These fossils came from the Fairbanks mining district and were removed over the years of peak mining activity earlier in this century; they now reside in the American Museum of Natural History. The silt sediments from which the fossils were collected are mainly Rancholabrean in age, predominantly Late Wisconsin Glacial (Guthrie, 1968). The recent sheep skulls were taken from all over the state—with a particular bias for large trophy rams—so the differences between fossil and recent sizes may even be greater than appear here.

Because the modern treeline in the north is correlated with the botanical length of the summer growing season (number of degree-days), we have

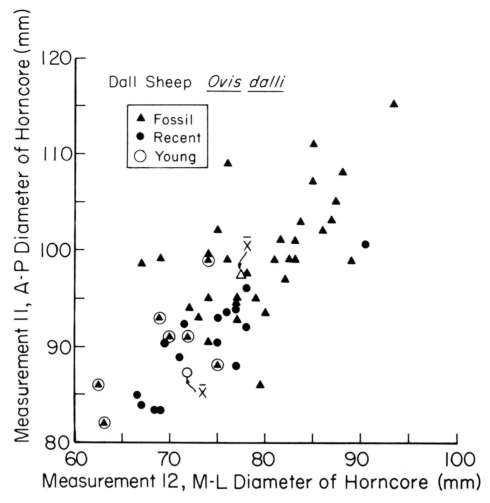

Fig. 3.—A comparison of horn size in fossil and recent Dall sheep (*Ovis dalli*) from Alaska. The fossils are mainly from the Fairbanks, Alaska, area, whereas the recent sheep were adult rams collected from numerous places throughout Alaska.

assumed that a predominantly herbaceous flora in the north and boreal forests in temperate latitudes could only be achieved by shortening the present growth season. Treeline in central Alaska now averages about 1,000 m above sea level. Pollen and macrofossil evidence indicates the absence, or virtual absence, of spruce (*Picea*) during the late Wisconsin Glacial in the same area (Matthews, 1974; Ager, 1975). This means a shift in treeline of at least some 870 m (Fairbanks, Alaska, is 133 m above sea level).

In fact, the reduction of trees in Alaska and the Yukon Territory during the Wisconsin Glacial (as in the case of the Holocene reduction of the boreal forests over what is now the Great Plains) is probably not a simple matter of length of summer growing season. It may be more closely controlled by

additional factors such as seasonal moisture changes, like the aridity which now controls the southern border of the boreal forest (La Roi, 1982).

If Alaskan Pleistocene plant communities were known in specific detail we could approach the matter of climate and seasonality directly. However pollen is difficult to identify to lower taxonomic levels for the non-woody plants—the very ones, I contend, we would have to know about. Macrofossil plant identifications have yet to be done at a scale which would be of much assistance. I also contend that physical indicators in the fossil record of mean temperature, that is, permafrost and ice wedge formation, may also be relatively uninformative about seasonality. The problem has to be approached more obliquely.

In the north most large mammals grow only dur-

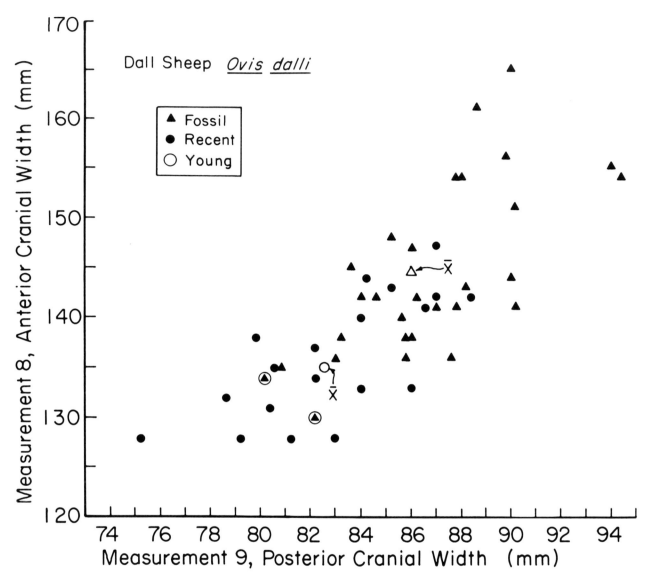

Fig. 4.—A comparison of body size in recent Dall sheep using skull measurements as an index. The recent sheep are adult rams collected from all over Alaska. The fossil sheep came mainly from the Fairbanks, Alaska, area.

ing a relatively short time of the year and experience growth dormancy during the remaining portion (Norden et al., 1968). During this latter period they do not have the potential for growth; even when given high quality food they continue to lose weight. Because of this winter dormancy, actual body size can therefore serve as a rough index to summer range—the duration and quality of summer growth resources—given, of course, that one does not use body size out-of-context of ecological settings, or compare forms which are not closely related.

To avoid as many problems as possible with the latter two issues my main analysis will use North American mountain sheep. There are two species in North America *Ovis canadensis* and *Ovis dalli*. Considerable information exists about their biology, although the biology of these species is far from being totally understood. The most important advantage of using sheep, however, is that we have fossils from many different restricted localities where there are still existing sheep populations. Sheep, unlike bison and wapiti, are not good colonizers (Geist,

1971), nor are they given to erratic movements or long migrations, so it is likely that the fossil sheep lived in areas near where their remains were found and that they are related to those sheep now living in the same area.

There are a number of variables which affect size and some of these are species-specific, such as the physical constraints on an insectivorous hibernating bat or a weasel which has to follow a mouse's tunnel. There are no obvious outside physical constraints of this nature on the American sheep. They have a broad thermal-neutral zone, so it is not likely that temperature is an important metabolic size regulator. Likewise, they do not depend directly on size to fend off or escape predators. However, size of body and horns are, as with many ungulates, very important socially (Geist, 1971). Size is also directly affected by nutrition, both ontogenetically and phylogenetically. Generally, one can argue convincingly for a balance between the social advantages of large size versus the metabolic expenses of rapid growth and early reproductive involvement (Geist, 1966); and that these forces determine the *optimum balance* of size "strategy" for any point in time.

The resource or nutritional determinant has several faces. One is sheer quantity of herbage available, the other is its quality, and still another is the span of time it is available. In conjunction one must consider also the severity of the winter bottleneck. I would like to propose the following hypotheses as part of a general theory accounting for Pleistocene gigantism in American mountain sheep:

1. Competition during the *peak* of the growth season is not the ultimate force determining body or horn size.
2. During the *peak* of the growth season, body and horn growth are not limited by *quality* of suitable herbage.
3. Body and horn size are also not limited during tle major part of the growth season by *quantity* of herbage.
4. The physiological-nutritional debts accumulated during the winter are not the major determinants of body size.
5. The colonization phenomenon (Geist, 1971) does not of itself explain body size trends in fossil North American sheep.
6. The major variable in body size, both past and present, in North American sheep is the duration of the peak of high quality resources—the length

of season during which the sheep *can* grow at a maximum rate.

As with any life history feature there are necessarily a number of interlacing variables. All of the six items referred to in the above hypotheses probably do affect body size, some more than others, in different situations and for different reasons. What I have tried to do conceptually is show that the sixth hypothesis is the principal component to the geographic and historical (chronoclinal) variations in body and horn size in North American sheep. I will next discuss the items individually.

1. *Competition during the peak of the growing season is not the ultimate force determining body or horn size.* Sheep numbers and body quality seem to be density dependent (Heimer and Smith, 1975), so intraspecific competition is certainly important in sheep biology. Competition is particularly significant for northern sheep on winter ranges which usually are quite discrete and restricted and it seems here to be the fundamental density regulator (Geist, 1971). Also first green-up often occurs in limited areas and it may be that on very poor early spring or autumn ranges intraspecific competition does play a role. Once into the full flush of spring and summer, however, it is unlikely that even a highly selective grazer like sheep meets with intraspecific competition (Whitten, 1975). The summertime dearth of feeding displacement behavior which is common among Dall sheep in winter also points to reduced competition during the growth season peak.

2. *During the peak of the growth season, body and horn growth are not limited by quality of suitable herbage.* Sheep select early seasonal growth stages of plants when available. These are more palatable and more digestable than later growth stages (Oleberg, 1956; Stoddart and Smith, 1955). Klein (1970) has reviewed the adaptations of alpine plants which enable them to grow rapidly and exhibit an unusually high nutrient content. Whitten (1975) showed that most plants in the spring and summer diets of Mt. McKinley Park sheep had protein percentages (dry weight) of 25–30%. Compare this to high quality domestic alfalfa at only 16% or rolled oats at only 11% protein (Church, 1972). There are upper limits to the amount of protein sheep can digest. As domestic sheep are about the same size as wild sheep and have about the same intrinsic growth rate we can make some very general comparisons. Protein percentages of over 11% are considered adequate for optimum domestic ram growth

(NRC, 1971; Church, 1972). Wallach (1972) recommends a maximum ration of 12 to 19% protein for wild sheep in captivity. It thus appears that sheep ranges provide more than adequate quality during the peak of the growing season. The extremely high protein quality of sheep spring and summer forage is matched by other nutrients (Whitten, 1975). Winter's (1980) comparisons of the nutrient content of plants available to a high quality sheep population with one of lower quality sheep population during the growth peak found no significant differences. Given that wild sheep are getting food of more than ample quality, can we be sure that they are getting it in sufficient quantity?

3. *Body and horn size are also not limited during the major part of the growth season by quantity of herbage.* This assertion is not easy to confirm directly as it is difficult to analyze the amount of forage consumed in the wild. I think there are ways, however, to indirectly illustrate that ample quantities of food are available.

Palmer (1944) reported that Dall sheep require 6 lbs forage dry weight per day. Hoefs (1974) calculating numbers of bites per day, and estimating quantity per bite, found wild sheep to exceed this amount. Arnold (1964) found that the maximum rate of biting is 100 bites per minute when vegetation is small and thinly distributed necessitating many bites. Horejsi (1976) found wild sheep on summer range to be below this amount (55 bites per minute in June). Hoefs (1974) observed a feeding rate of 45 to 55 bites per minute in dense emerging vegetation in May and June.

The length of time it takes to fill or "top-up" the rumen in summer could also be used as a crude index of quantity of food available. Viereck (1963) found that summer feeding periods averaged only 1.1 hrs; Whitten's (1975) average for the summer was 1.6 hr. On winter range Viereck (1963) found the feeding time averaged much longer (2.43 hr). Studies of domestic sheep have shown that the rate of food intake increases and the rate of grazing time decreases with improved pasture quality (Arnold, 1960a, 1960b, 1962). Also, Blaxter et al. (1961) observed that sheep appear to feed to a given point of distention of their digestive tracts. Thus the brevity of feeding periods is an indicator of pasture quantity, assuming at least a moderate rate of passage. Because the ruminant system is geared to increase transit time with increased food quality (rates of digestion), one must also temper the evaluation of feeding time with some estimate of time of passage.

The higher the food quality the more quickly it moves through the gut, allowing increased rates of consumption—so in many ways quality and quantity are potentially related. To be assured that the short feeding periods in the summer are indicative of ample food quantity, rather than a rumen which is filled with poor quality, slowly digesting food, one can look at comparative faecal pellet counts. Summer food has a rapid transit rate (Hoefs, 1974) as can be seen by increased faecal pellet production during spring and early summer (41 defaecations per 24 hrs in spring as opposed to 30 per 24 hrs in October—pellet number per defaecation being similar). This is true despite the fact the percentage of undigestible fiber consumed (the bulk of faecal material) is much higher in autumn.

Hoefs' observation that sheep have a high number of bites per hour in the first feeding periods of the day but decline later also suggests that quantity is not limiting during most of the summer.

4. *The accumulated physiological-nutritional debts during the winter are not the major determinant of body size.* The best way to test this hypothesis is to capture sheep from a low quality population in early spring when they are at their lowest physiological ebb, then provide them with ample high quality food, and see if they respond rapidly to the increase in growth resources. I did this with seven yearling Dall sheep (three male, four female) from Healy, Alaska. They were caught in late March and fed in captivity on a moderate to high quality dietary plane. They responded rapidly in growth that first spring, one of the males grew a 28 cm horn segment, whereas the average in the wild is about 18 (Nichols, 1978). The length of horn in the captive sheep was longer than any I have found in wild Dall sheep. The first autumn in captivity the yearling rams inseminated the yearling ewes and the ewes raised lambs the following spring. This again is evidence of their high quality, as yearling ewes do not commonly rear lambs in the wild.

It may be that in an extremely poor winter there could be some ungulate disability and lag-time in spring recovery. However, the traditional view of winter stress being the sole limiting factor, as in the management of deer quality, has been discredited. Klein (1965) showed that summer resources were the major factor.

Two of my captive sheep became severely emaciated when ill. They began to gain weight immediately after being cured. One two-year-old male who recovered in midwinter caught up with his peers in horn growth and body weight in early spring. This also is evidence for no or very little lag in recovery from winter debilitation.

Winter mortality is important in reducing spring competition for many animals—it leaves fewer individuals among which to divide the next summer's resources. I suspect winter severity is the causative agent in correlations between general population quality and density (Woodgerd, 1964), both between populations and within single populations through time (Heimer and Smith, 1975). I will argue later that in the production of large sheep, length of summer growing season is often secondarily associated with severe winters.

5. *The colonization phenomenon does not itself explain body size trends in fossil North American sheep.* Geist (1971) proposed that colonization results in larger animals and that intraspecific size variability is due to this phenomenon. McDonald (1981) expands on this theory in his interpretation of large body size in bison. The expansion into new habitat usually does result in a size increase (probably both developmentally and genetically) for most species. As populations reach or exceed carrying capacity, however, body size is again reduced. There are numerous examples of this delayed size reduction in introduced or colonizing species, for example, an introduced bighorn sheep population (Woodgerd, 1964).

Geist proposed that as the glaciers receeded, the sheep population colonizing the newly exposed habitat became larger. One would expect then a chronoclinal gradient from small to large in the fossil record across the glacial-postglacial boundary. However, just the reverse is true as I have shown in a later section, which suggests some other cause.

As Geist observed, sheep species do exhibit a geographic gradient of increased size toward glaciated areas. However, Klein (personal communication) has argued that this is a secondary correlate of physiographic-phenological features which promote a long, nutrient rich growing season, being more likely to occur in areas susceptible to glaciation (that is, well-watered, high mountain pastures).

6. *The major variable of sheep body size, both past and present in North America is the duration of the high quality resource peak—the length of the growing season.* As I have already mentioned, there is a portion of the year during which wild sheep cannot grow because of an inherent growth dormancy. The other part of the year they *may* grow if resources beyond maintenance demands are avail-

able. For wild sheep populations, however, resources are such that the sheep *can* only grow in a portion of this annual available potential growth period (Fig. 1). From studies with Dall sheep reared in captivity, I have determined that the potential growth period begins early in the spring—well before new green growth is currently available to wild sheep, and that growth potential continues well after plants on the sheep's natural range have senesced. Although I have not been able to match the quality of resources available to wild sheep in summer alpine pastures, I have been able to extend the period of actual growth in the captive sheep by providing quality feed to support an early onset of growth in the spring and to sustain growth in the autumn and early winter. Because most of the captive sheep are still living I have no osteological measurements to compare with other populations. Hoefs (1974), however, has already shown a much more rapid growth rate and absolute weight in captive Dall sheep kept on nutritionally supplmented rations. I do have horn growth measurements; these seem to be as equally good an indicator of dietary quality as body size (Geist, 1971). The horns of my captive sheep are much larger than the horns of wild sheep of the same age and area (Guthrie, in preparation).

It seems reasonable to assume that if high quality resources are generally available in ample quantities during the peak growth season, then the remaining variable to body size is the *duration* of this nutritional peak.

One can visualize the "quality growth season" hypothesis in experimental terms. We know that northern ungulates can only grow rapidly during a short segment of the year. We also know that if resources are kept at a minimum maintenance level during a large portion of the potential growth season, ungulates will mature stunted in size. Now, we can imagine an experiment in which we gave a group of sheep unlimited quantities of the very highest quality food for two weeks during that growth season, but keep them on maintenance rations for the remainder. Because there are intrinsic limits to consumption, digestion, assimilation of food, and to growth rates, the two weeks' access to optimum resources would still not allow them to reach maximum genetic potential. However, if one continues to expand the period of free access to quality resources, adding a week at a time, the animals will eventually reach a point at which they will reach maximum genetic potential. From observing wild sheep reared in captivity we can see that maximum growth comes only when high quality resources are supplied throughout the entire annual period of potential growth. All of the hypotheses discussed seem to point to the support of this last hypothesis that the major variable in body and horn size development is the seasonal duration of high quality, high quantity resources.

Winter on northern sheep ranges is usually severe. Sheep exploit high windblown slopes that are relatively free of snow (Summerfield, 1974), but winter vegetation is sparse and of low quality. The thick fat deposits that sheep develop on energy rich forbs and grass in the autumn decline steadily on meager winter resources. Rams lose a considerable fraction of these resources during the rut in late November–early December (Geist, 1966). Ewes must carry the new lambs during late winter when their fat reserves are depleted and range resources are at the very lowest level.

As a consequence of this winter bottleneck, American mountain sheep have a relatively high mortality rate; the majority of lambs are dead by the end of their second winter (Murie, 1944; Murphy, 1974). However, once past this second winter, sheep have an unusually rectangular survivorship curve which drops off only in old age. Most two-year-old sheep will live for another six to eight years. This high adult survivorship in the face of severe winters is due primarily to two things: (1) the use of escape terrain—mountainous slopes where sheep can usually out-run predators uphill, and (2) the highly nutritious summer growth resource that alpine pastures provide for growth, maintenance, and fat reserves.

Klein (1970) has reviewed the factors responsible for the extraordinary quality of arctic and alpine plants. These cool-season plants are adapted to grow very rapidly in a short summer. This rapid growth produces leafy tissue with high levels of nitrogen and phosphorous and little undigestible fiber. The lack of shading by an upper story of trees means that more light reaches these low growth forms and the long days increase the length of the photosynthetic period, which in turn means less nighttime catabolic activity. Also Chapin (1977) has shown that northern herbs have the ability to extract phosphorous from the soil more rapidly than their southern counterparts. During their early seasonal growth stages, these plants are relatively undefended by antiherbivore compounds. In addition, biomass production on a 24-hr basis can be as high in the arctic as in many temperate climate communities (Bliss, 1962).

Klein (1965, 1970) presented a model of ungulate strategy, which capitalized on this brief, high quality plant growth by physically following its wave of appearance over the spring and summer landscape. Because of topographic and altitudinal variations, plant phenology will vary considerably throughout the summer. This local variation is further exaggerated by the relatively low sun angle in the north. On some low southern exposures new plant growth occurs shortly after the snow and may not disappear until late summer. Northern ungulates exploit these variations in their summer range and are able to find young plant growth for a considerable portion of the summer season. Whitten (1975) and Winters (1980) have documented this process with Dall sheep movements. Dall sheep tend to range higher and higher as the summer progresses, following the appearance of new plant growth up the slopes.

Large-bodied and large-horned sheep occur today on ranges with wide altitudinal spans where they can exploit a long season of high quality alpine growth. Other factors are also important in determining the length of the ungulate growth season in addition to sun angle, altitude, and aspect; most of these features can be summed as *climate*. Jet stream and cyclonic patterns can change the air temperature, cloud cover, and moisture regime to greatly lengthen or shorten the growth season for any one locality. Decreased snow, for example, can accelerate the onset of spring greenery. Aridity may be an important factor limiting plant growth on well drained, windy slopes with shallow soils; early summer showers may contribute significantly to plant quality. Cloud cover may act to retard or prolong the growth season depending on its occurrence in spring or autumn. Certainly air temperature is of primary importance in determining length of growth season.

In Alaska, Dall sheep horn size increases from west to east in both the Brooks Range and the Alaska Range (Heimer and Smith, 1975). The fact that the summer snowline altitiudes also follow the same gradients (Pewe, 1975) suggests more equable temperatures (and hence plant growth) at higher altitudes from west to east. In fact Pewe's mapped estimates of high altitude summer snowlines correspond roughly to the gradients of large-bodied sheep populations throughout Alaska.

It is likely that a number of factors can potentially

lengthen the peak of available growth resources and that these may vary in importance among living sheep populations. Any one factor may not necessarily be used to explain the causes of greater body size in a given population without empirically examining the circumstances in question.

With the above information as background I will now turn to the problem posed by the large size of Pleistocene sheep, how they differ from the living sheep and the possible causes of this change.

Today, the narrow altitudinal confinement of sheep pasture between the talus of the mountain crests and treeline was stretched in the Pleistocene. The absence of a woodland meant access to the early plant emergence of the lowlands, which would extend the length of the sheep's peak growth season. Sheep fossils, in now wooded hills around Fairbanks, many kilometers from their current distribution, testifies to the greater access of grassy foothills in the Pleistocene. The largest bodied sheep species today, *Ovis ammon*, makes considerable use of foothills (Skogland and Petocz, 1975) as part of pastures with tremendous altitudinal expanse. In these habitats, Marco Polo sheep (*Ovis ammon poli*) face severe winters, and few individuals have ever been known to live past nine years (Skogland and Petocz, 1975). The winters were undoubtedly harsh for Alaskan Pleistocene sheep as well, with winds (Guthrie, 1982) and exposure providing access to winter forage, while at the same time leaching the available nutrients. This balance of tight winter bottleneck keeping populations well below competitive margins for growth resources and a long growing peak would select for animals which allocate nutrients to rapid body growth and violent early participation in rut (Geist, 1966). This same selection balance-point seems to have existed throughout the Pleistocene geographic range of North American sheep.

Although sheep occurred only in the western part of the North American continent, their fossil record is comparatively complete. Alaska and the Yukon Territory have produced a number of specimens and there are some fossils from British Columbia. There are also Wisconsin-aged sheep fossils from Jaguar Cave in Idaho; Little Box Elder Cave, Catlaw Cave, and Winnemucca in Nevada; Rampart Cave and Stanton Cave in Arizona; Glendale, California; and near Lake Bonneville, Utah. These localities are reviewed by Harington (1978).

Stock and Stokes (1969) discussed the Wisconsin-age sheep which some authors give separate specific status—*Ovis catclawensis* because of their much larger body size (an almost American version of the giant *Ovis ammon* of Asia). Stock and Stokes provided a date of 15,030 ± 210 years B.P. on associated organic material from Dry Cave in Nevada. A review of these more southern sheep fossils reveals that they do average much larger than *O. canadensis,* but appear to be quite closely related, and in fact, are now considered by many to be a chronological subspecies, *O. candensis catclawensis.* Harris and Mundel (1974) found that these large sheep were widespread in the western United States and Mexico during the late Wisconsin. They propose that "environmental deterioration" near the end of the Wisconsin and into the Holcene resulted in selection for smaller sized sheep. Thus extant bighorn sheep, *O. canadensis,* are reduced or dwarfed representatives of the late Pleistocene sheep from the same areas.

This also seems to be the case with *O. dalli* and its fossil counterparts. Harington (1978) has concluded that the Wisconsin-aged sheep from the Yukon Territory are much larger than sheep living there today; measurments on fossil skeletal material average about 10 to 11% larger than the same measurements on extant sheep from the Yukon Territory. Harington suggests that the larger size of these fossil sheep can be better appreciated when one notes that hind leg length between living rams and ewes (exhibiting considerable sexual dimorphism) is only in the order of 6%. As pointed out earlier, I have recorded this same phenomenon in fossil and living sheep in the Yukon-Tanana Uplands in central Alaska. This additional evidence for postglacial size reduction in *Ovis* now allows one to conclude that such dwarfing was a general phenomenon throughout North America.

HOLOCENE SIZE REDUCTIONS IN *BISON* AND *CERVUS*

In addition to sheep, there are two other North American megafaunal grazers which did not become extinct at the end of the late Pleistocene, *Bison* and *Cervus*—both are present in Alaska well into the Holocene.

The decline in horn and body size of *Bison* sp. in

the early Holocene is probably the best documented case of rapid evolution in mammals. Harington's (1978) most recent date for large-horned superbison (*B. priscus*) in the Yukon Territory is just prior to 12,000 B.P. We have a date 10,600 B.P. on a very large bison mandible from the Dry Creek site (Powers et al., in press). After this time (11,000 to 10,000 B.P.) bison begin to decline rapidly in size toward the living species sizes in Great Plains archaeological sites beginning at 11,000 B.P. As is the case in Alaska, these Great Plains bison become progressively smaller until the present. Vereshchagin (1967) has discussed a parallel phenomenon in central Eurasia and Sher (1971) has reported the same decline in Siberia.

The mid-Pleistocene increase and subsequent reduction in body size which occurred in bison evolution characterizes many ungulate size trends throughout the Quaternary (Guthrie, 1980). The peak in bison size was reached in the mid-Pleistocene, though the chronology varied somewhat in different areas. Bison remained large bodied through to the early part of the late Pleistocene (Illinoian in North America) and then began a decline in size. This decline plateaued somewhat during the last glaciation and then accelerated in a rapid descent in body and horn size during the postglacial. The timing of this last decline also varied with geographic location.

The major postglacial decline in bison body size was a widespread event, occurring in northern and southern Europe, in northern Asia, and much of North America. I contend that the major cause of this decline in body size was an abbreviation of the annual optimum growth season. In the case of bison, the effects of this shorter growing season must have been exacerbated by increasing postglacial plant zonation (Guthrie, 1984) which reduced local vegetational diversity.

In the New World the postglacial winter bottleneck was opened by the spreading of bison's optimal short-grass, *Bouteloua-Buchloe,* habitat, so bison numbers increased concurrent with, and despite the declining body size.

There must have been a somewhat cyclical relationship initiated by climatic change—an abbreviated season for plant growth resulted in plant zonation; such zonation decreased local herbaceous diversity; which in turn, further shortened the growing season for most ungulates (Guthrie, 1984). Thus, the reductions in "ungulate-growth-days" were not a simple matter of a subtraction of the actual decrease in "degree days."

The effects of the shortened season were intensified for bison in the New World by the postglacial disappearance of bison's competitors, which further exaggerated the differences between good and poor bison habitat; due to the absence of regrowth from the tall grass species eaten by the more fiber-tolerant large caecalids, and the decline in whole-plant nutrient quality because the low quality parts (stems) were not removed by the (now extinct) mammoths and equids. The short-grass plains however became dominated by bison, but the simplicity of vegetation probably resulted in a decreased phenological succession of local plant maturation—shortening the peak availability of high quality grasses. These C^4 grasses tend to be lower in peak quality than their C^3 counterparts (Mattson, 1980).

The fossil record of elk (or wapiti), *Cervus,* is not as complete as that of *Bison,* but it is clear in the far north at least that *Cervus* underwent a postglacial decline in body size (Guthrie, 1966; Harington, 1978). This phenomenon is best seen in Europe where the extant *Cervus* (red deer) is much smaller than the Pleistocene forms (Altuna, 1972).

The large wapiti *Cervus elaphus* described by Shackleton and Hills (1977) dating at 9,670 ± 160 years B.P. illustrates the same principle seen in *Bison* and *Ovis.* Even during the glacial recession into the early Holocene this megafaunal grazer had not yet reached the reduced size of its modern representatives. For example, the *Cervus* specimens from the Star Carr Site dated at around 9,000 B.P. in England are larger than those from any of the current populations in the British Isles (Clark, 1954).

As with *Ovis* and *Bison,* the Pleistocene *Cervus* have disproportionally larger horns or antlers than their Holocene descendants. This is significant because social organs are an even more sensitive indicators of range quality than body size. In times of stress these are organs of "low growth priority" and are reduced first developmentally, the evolutionarily. The same processes operate for exceptionally high quality ranges, except in the reverse—the social organs grow larger.

The antlers of *Cervus* are informative in this regard, because they give us additional information about the Pleistocene growth season length. Antlers grow somewhat like woody plants, having their "meristemative" tissue on the ends of the tines. The calcification process takes place in a proximal-distal gradient. Elk or wapiti raised in captivity, where they are fed a highly nutritious diet from early spring to the time the velvet is shed, develop antlers shaped differently than antlers of wild wapiti; the tines are unusually long. This seems to be due to the maintenance of a high dietary plane for a longer time than in the wild. Although antlers from captive wapiti tend to be somewhat larger than their wild counterparts due to a better quality diet, it is these long tines that give them their unusual form.

This form is unusual in extant wild *Cervus,* but it is common in the large Pleistocene *Cervus* antlers, suggesting two things—that the range quality was excellent and that the high quality existed for a relatively long season each year.

Kurten (1968) has discussed this post-Pleistocene dwarfing for a number of species, including brown bears, *Ursus arctos.* Garutt (1964) has shown that it was characteristic of late glacial Siberian mammoth, *Mammuthus primigenius,* just prior to extinction.

Unfortunately, we know all too little about why

some extant populations, counterparts of species discussed above, are of higher quality than other populations. We can only say that dietary resources are accessible and are usually of high quality because of various combinations of situations. All ungulate populations usually have access to high quality food resources for some part of the growth season; however, it is the *duration* of the quality peak which seems to be important.

Bison, for example, use high quality cool-season grasses in the early spring, but then shift to more fibrous, but moderate quality warm-season grasses throughout their growth season, but the duration of the peak relates to numerous seasonal variables including plant phenology, moisture, and fire. Although the proximate variables may differ somewhat, much the same is also true for these other species.

HOLOCENE BODY SIZE REDUCTION IN *RANGIFER, OVIBOS,* AND *ALCES*

One would expect to find an increase in body size in reindeer (*Rangifer*), muskoxen (*Ovibos*), and moose (*Alces*), when they expanded in the Holocene to dominate the northern ungulate faunas. Presumably, caribou, muskoxen, and moose become more abundant with the rise of more moist conditions and the accompanying increase in shrubs and tundra vegetation. I found that, in fact, the Alaska Pleistocene specimens of these three species did average larger in body size (Figs. 5 to 11) than living Alaskan species. (The Alaskan subspecies are the largest forms of the extant species.)

In Alaska, summer forage for caribou, muskoxen, and moose is profuse and nutritious—yet, their Pleistocene ancestors were even larger than those of today. How can this be? We know of no floristic sources that would have greatly increased the quality of summer forage during the Pleistocene. The only obvious route to larger body size is to extend the period in which quality forage is available. The largest moose today are found on the Alaska Peninsula where summers are long and cool. The caribou on the Alaska Peninsula and the Aleutian Islands, with a similar seasonal pattern, are also among the largest of the subspecies.

Alaskan moose, muskoxen, and caribou, like bison on the Great Plains (Guthrie, 1980), experienced a postglacial increase in numbers. Interspecific competetion declined with the demise of most of the other megafauna as forage for moose, mus-

koxen, and caribou expanded. This widespread vegetational change resulted in a greater quantity of winter forage (willow shrubs for moose and muskoxen, and lichen for caribou, Guthrie, 1982). It thus allowed greater numbers of animals through the winter bottleneck, but new climatic regime did not provide more growth resources per animal in the summer.

Lengths and widths of M^1, M^2, and M^3 are compared between fossil moose from interior Alaska and several extant populations from the same area. The differences are all highly significant (Figs. 8, 9 and 10). The data are presented in Table 1 and 2. I have not as yet located sufficient moose metapodials, but Dr. David R. Klein of the Cooperative Wildlife Unit, University of Alaska, graciously let me use his data for caribou metatarsals. The fossils are significantly longer. There were two caribou metatarsals from Adak Island which exceeded the fossil metatarsals in length. The Adak caribou were transplanted from the central Alaskan Nelchina herd, and have become enormous on this virgin range with a long cool maritime growing season. One adult bull was weighed at 700 lbs (320 kg), twice the weight of healthy bulls from most other populations. Measurements were also taken on fossil Alaskan muskoxen (*Ovibos*) skulls to compare with wild animals living today in the Canadian High Arctic. The fossils are significantly larger (Table 2 and Fig. 11).

The moose, muskoxen, and caribou fossils used

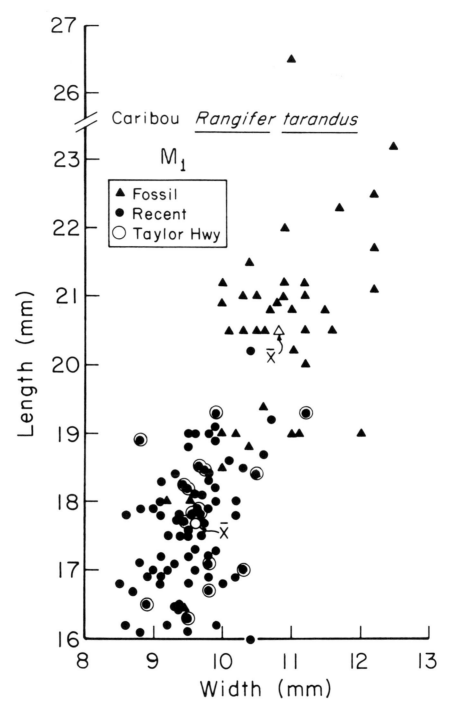

Fig. 5.—A comparison of fossil and recent caribou (*Rangifer*) from interior Alaska using the first lower molars. The fossil animals (triangles) and the recent specimens (circled dots) are from the same range of hills near Fairbanks, Alaska. The uncircled black dots represent recent caribou taken from all over Alaska.

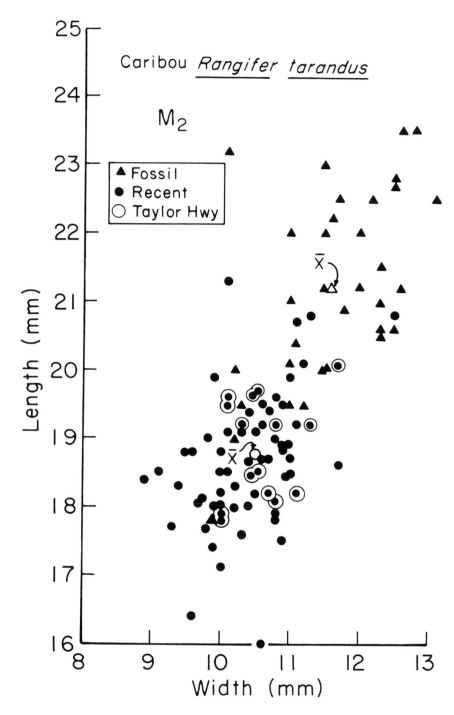

Fig. 6.—A comparison of fossil and recent caribou (*Rangifer*) from interior Alaska using second lower molars. All details the same as Fig. 5.

Table 1.—*Comparison of dental and leg bone measurements between fossil and Recent caribou (*Rangifer tarandus*).*

Tooth or bone	Length (mm)	Width (mm)	Significance	
M_1 (Fossil)	mean = 20.50 ± 1.63 n = 41	mean = 10.79 ± 0.75 n = 39	Student-t L = 12.51 W = 10.47	v. sig. v. sig.
M_1 (Recent)	mean = 17.66 ± 0.94 n = 88	mean = 9.58 ± 0.52 n = 87		
M_2 (Fossil)	mean = 21.21 ± 1.41 n = 33	mean = 11.62 ± 0.86 n = 33	Student-t L = 10.92 W = 8.51	v. sig. v. sig.
M_2 (Recent)	mean = 18.81 ± 0.89 n = 88	mean = 10.45 ± 0.59 n = 88		
M_3 (Fossil)	mean = 23.93 ± 1.93 n = 23	mean = 11.02 ± 1.11 n = 24	Student-t L = 7.23 W = 8.38	v. sig. v. sig.
M_3 (Recent)	mean = 21.56 ± 1.23 n = 88	mean = 9.66 ± 0.55 n = 88		
Metacarpals (Fossil)	mean = 19.61 ± 1.12 n = 71	mean = 43.86 ± 2.78 n = 70		
Metatarsals (Fossil)	mean = 27.51 ± 1.36 n = 53	mean = 43.13 ± 2.26 n = 52	Student-t L = 16.86	v. sig.
Metatarsals (Recent)	mean = 23.68 ± 1.48 n = 163	Extant *Rangifer* from Alaska (various populations)		

for comparison came from Fairbanks area and were collected in conjunction with the placer gold mining earlier in the century. They date mainly from late Pleistocene (late Rancholabrean). Pewe (1975) and I (Guthrie, 1968) discussed the collection in previous papers. The collections are now in the American Museum of Natural History.

DWARFED MONOGASTRICS OF THE MAMMOTH STEPPE

Although present during the last glacial, horses and hemionids (*Equus*), wooly mammoths (*Mammuthus*), and camelids (*Camelops*) became extinct at the beginning of the Holocene. Their extinction is informative because these species are somewhat of an exception to the Pleistocene gigantism trend. Although wooly mammoths are large by anyone's standards, they are not large for proboscidians. Relatives in the central and southern part of the continent were much larger (for example, *Mammuthus columbi*), as are the living elephants today. The hemionids and horses were also quite small throughout most of the late Pleistocene in Alaska—not much larger than Shetland ponies. The few bones and teeth of camels (*Camelops*) also do not seem to be as large as some of their relatives further south in North America, and they did not approach the size of the old world *Camelus*.

My assessment of this unusual reduced size among the mammoth steppe non-ruminant fauna (I will include camels in this category [Guthrie, 1984] with qualifications) is that they suffered most from the shorter growing season in the north.

Ruminants are capable of taking advantage of very high nutrient planes, these caecal-colon digesters are not. They are adapted to an extremely long season of medium-to-low quality nutrient levels (Guthrie, 1984). They fit in the second category in Fig. 1. That is, they are large because they prolong their growth over a number of years. A long growing season on

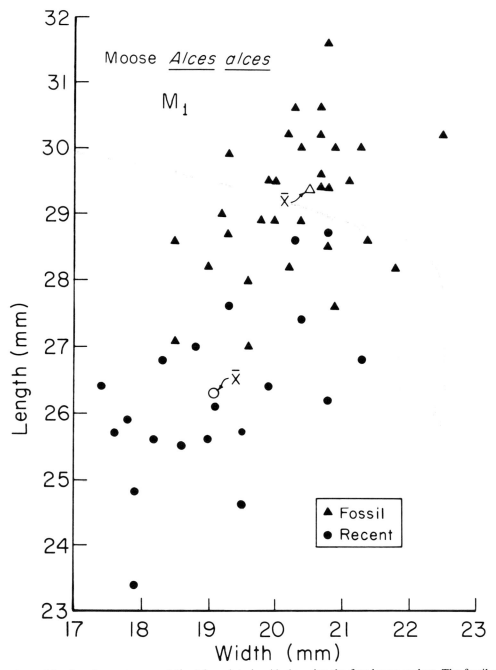

Fig. 8.—A comparison of fossil and recent moose (*Alces*) from interior Alaska using the first lower molars. The fossil moose teeth are taken from sediments of late Pleistocene age in the Fairbanks, Alaska, area.

gulates today are the ones which are adapted to eating the lower quality forage (more fibrous, fewer nutrients), that correlation can cloud the issue of

Pleistocene gigantism. The concept behind this body size-dietary fiber association is that a larger caecum or rumen is required to efficiently incubate roughage

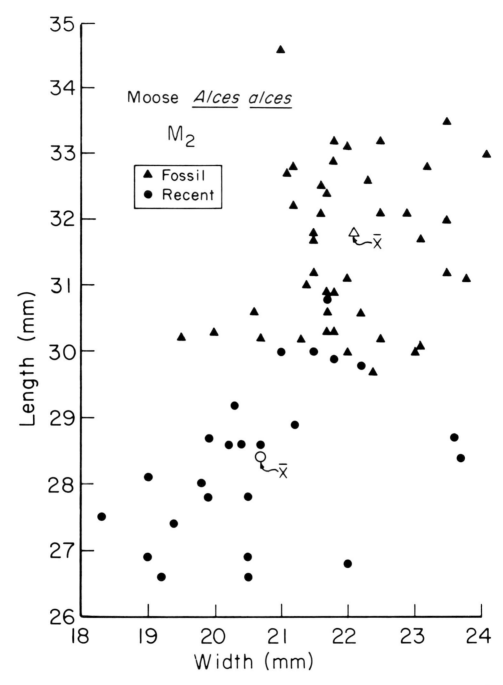

Fig. 9.—A comparison of fossil and recent moose (*Alces*) from interior Alaska using second lower molars. Specimens as Fig. 8.

and hence would select for large bodied animals (Janis, 1975). Ungulates thus fall on a gradient of two opposing foraging strategies (Bell, 1971; Wes-

toby, 1974, 1978; Hanley, 1980). High quality forage is sparsely distributed, and to take full advantage of it one has to be quite selective. Being selective

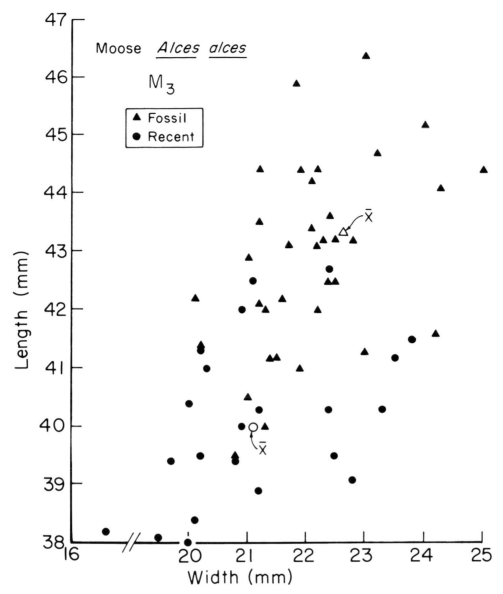

Fig. 10.—A comparison of fossil and recent moose (*Alces*) from interior Alaska using third lower molars. Specimens as Fig. 8.

requires much greater time and results in less quantity per given effort. These quantity limitations limit body size, so that more selective browsers such as American deer (*Odocoileus*) or roe deer (*Capreolus*) are relatively small bodied. The lower quality forage, however, is normally abundant and quickly obtained, thus less selective grazers can get adequate quantities with which to grow large, but seem to be, seasonally, on the margins of obtaining sufficient

quality. The wapiti or red deer (*Cervus*) are sympatric with the above small deer and are contrastingly larger bodied and forage less selectively.

Each strategy is caught by the opposite limits. For a red deer to grow larger, it must increase the forage quality by being more selective, but this would limit quantity. Roe deer in order to grow larger must increase forage quantity, by being less selective, but this would decrease quality. Limited in opposite

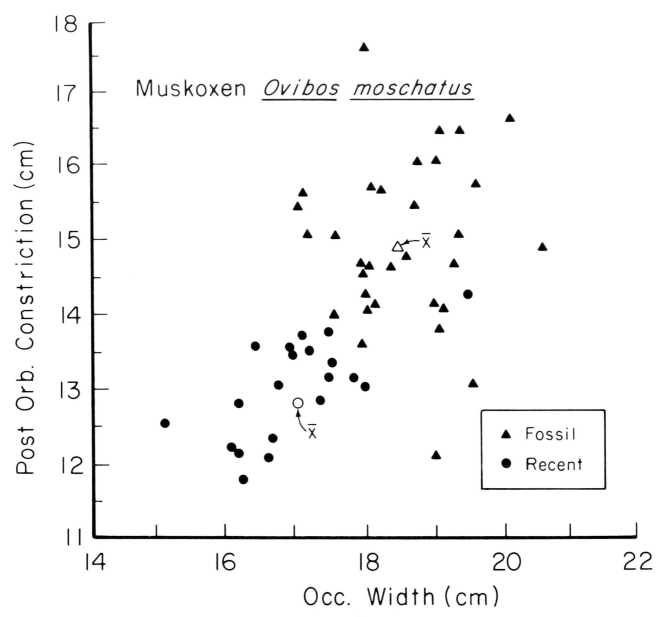

Fig. 11.—A comparison of fossil and recent muskoxen. The muskoxen fossils are all from Alaska. The measurements were from recent skulls taken on individuals from various places in the Canadian High Arctic.

ways, each species, occupying a different balance point in these strategies, can live together without much competition.

I would argue, however, that the train of this thinking provides very few clues to Pleistocene gigantism. The reason is that *both* quantity and quality can be increased. For example, roe deer and red deer are both quite small on the Iberian Peninsula (Altuna, 1972), yet both species are gigantic (over twice the size) in southern Siberia. Forage quantity and quality can vary absolutely up and down the scale for both large and small ungulates as they did for, say, small saiga and large bison in the Alaskan Pleistocene which were larger than their living relatives to the south or for, say, large mammoths or small horses in the Alaskan Pleistocene which were both smaller than their relatives to the south.

MINI-GROUND SQUIRRELS AMONG THE GIANT MEGAFAUNA

Cursory examination of Alaskan Pleistocene ground squirrels showed them to be smaller than their extant relatives for the treeless arctic. A detailed comparison of several osteoloical measurements confirmed that assessment. The mean length of adult male and female skulls from arctic Alaska is 61 mm and 58 mm, respectively, whereas the Pleistocene specimens from the Fairbanks area averaged 53 mm. All other measurements showed this same size difference. The fossils were more comparable to subspecies further south. Why with all this gigantism around them would not the Pleistocene ground squirrels also have been large?

One obvious difference is that they are a hibernator and hibernators seem to be governed by a little different balance of life history forces than non-hibernators. A long cold season of hibernation requires a greater lipid reserve (Mrosovsky, 1976; Wrazen and Wrazen, 1982). Greater body size means a greater absolute amount of reserve energy and greater likelihood of surviving the winter (Kiell and Millar, 1978).

There is, consequently, a body size gradient among ground squirrels from south to north with increasing length of hibernation, just as there is in analogous fashion with increasing altitude. In fact, that altitudinal analogy can help us unravel the puzzle of Pleistocene body size. For with increasing latitude and altitude, ground squirrels are not only larger (Levenson, 1979), they have lower reproductive rates in the form of smaller litters (Bronson, 1980; Murie et al., 1980; Murie and Harris, 1982). Having to reach near adult body size before the first winter and living at high altitudes or latitudes, (facing a shorter growing season) these squirrels must grow faster than their temperate latitude or lowland counterparts. Thus, there are altitudinal and latitudinal gradients of intrinsic growth rates (Kiell and Millar, 1978; Koeppe and Hoffman, 1981; Murie and Harris, 1982). Although forage quality in northern and mountainous areas is known to be rich (Klein, 1970) the ground squirrels obviously grow large, not simply because of a greater growth season peak or length but because they allocate fewer resources to reproduction in the form of reduced litters.

Thus body size among ground squirrels is a better reflection of the length of hibernation period (in context with other life history features) than it is a reflection of their growing season resources. Increased length of peak growing season seems to be translated into increased litter size (at the expense of body size), a quite different phenomenon than occurs in larger nonhibernators.

Thus the small Alaskan Pleistocene ground squirrels, rather than indicating a shorter growing season suggest a longer one with a shorter hibernation period than occurs in those same areas today. This is consistent with our generalizations from the megafauna.

PLEISTOCENE GIGANTISM FROM AN ALASKAN PERSPECTIVE

I have tried to present a theory accounting for the reductions in body size of Alaskan ungulates from the peak of the glacial phase to and through the postglacial. This conceptual model of seasonal shifts and its impact on evolution of life history strategies may be used to better understand Pleistocene gigantism in general. Although there were large mammals throughout the Tertiary, it is during the Quaternary (particularly the middle portions) that many groups experience an increased body size in many of their phylogenetic lines. Some of the equids became immense (for example, *Equus giganteus*) and so did the bovids (for example, *Bison latifrons*), cervids (for example, *Cervalces, Megaceros*), proboscidians (for example, *Mammuthus columbi*) and many other groups. Additionally many had enormous tusks, horns, and antlers unparalleled in the Tertiary. Geist (1971) and Edwards (1967) discuss these in detail. The question is why? Why were large body size and large social organs selected during the

Pleistocene when we normally think of the cooler climate in terms of climatic deterioration? Generalizing from the Beringian situation I propose that the effect was global due to several interrelated features, all a product of that climatic shift:

1. A tightening of the winter or dry season bottleneck increased off-season mortality and hence *decreased competition* during the following growth season.
2. A more stressful season of plant growth fostered less conservative plants which experienced *marked seasonal pulses of nutrients* and as a consequence were more digestible to ungulates—grasses, forbs, and shrubs (all in complex mosaic communities).
3. The shortening growth seasons in the early and mid-Pleistocene still were not limiting to body size and were probably compensated for by an accompanying reduction in large mammal species diversity. However, at least in the north, the further abbreviated growth seasons in the late Pleistocene did become limiting. Thus, many groups, particularly the conservative large caecalids (proboscidians, equids, and rhinocerids) underwent body size reductions, and then became extinct at the beginning of the Holocene. Among the large mammals that did survive this revolutionary time, it is apparent that most experienced at least some body size reduction. From an ungulate's eye view the late Pleistocene afforded a longer growing season than did the Holocene.

I have argued elsewhere (Guthrie, 1974, 1980, 1982, 1984) that where there is intraspecific competition for growth resources (and there usually is among ungulates) heavy mortality in winter increases the quality and quantity of available growth herbage per individual during the following growth season (particularly during the critical beginning and end of that time). This is a well-known principle among wildlife managers who increase the hunting bag during the non-growth season to increase individual deer quality in the herd (Klein, 1965).

Increasing resources by decreasing competition not only increases the quality of available growth nutrients, it can also increase their duration over a longer growth season by increasing the quantity of the rarer nutritious early plants in the spring and late ones in the autumn—per *individual* animal.

The increased duration of growth resources pushes the individuals nearer their maximum size growth potential. If there are inherent advantages to being larger when growth nutrients are available (and there usually are these advantages for most ungulates), the selection balance shifts toward a larger body size optimum.

Larger individuals gain social stature sooner and hence experience reproductive privilege sooner, thus reproduce earlier. When the bottleneck of winter survival is tightened, large size becomes a decided fitness advantage as one cannot gamble on too conservative a growth, or reproductive strategy when the chances of dying during the winter have increased. Geist (1971) pointed out that the rapidly growing, large "daring," species and subspecies of mountain sheep (*Ovis*) are the ones with very short life expectancies. This is a general pattern in other artiodactyl species as well where the factors outlined above seem to be in play.

This selection regime affects social organs in the same manner as it does body size—except its effects are even more exaggerated. The relative size of an individual's social organs (antlers, horns, tusks, manes, and similar structures) are often the major clue to social rank and hence reproductive privilege. The resources that are devoted to fighting and display paraphernalia become optimal evolutionary investments when nutrients for growth abound *and* when the likelihood of dying before the next reproductive season has increased.

This combination of a long peak growth season in an environment which is kept understocked because of high competition during the winter bottleneck seems to have reached its zenith during the middle Pleistocene. The more open winter bottleneck of the Tertiary and early Pleistocene environment changed to a tighter bottleneck and a long peak ungulate growth season during the mid-Pleistocene. In the late Pleistocene the growth season shortened and this change continued through the Holocene.

The postglacial dwarfing can thus be seen as a continuation (albeit accelerated) of the general trend toward reduced body size in ungulates since the mid-Pleistocene. This was of course exaggerated in the north where shortened ungulate growth seasons had a more profound effect, but it can be seen occurring at mid-latitudes as well (*Bison latifrons* into *B. antiquus* and *B. antiquus* into *B. bison*).

It is possible that an increased seasonality during the mid-Pleistocene tended to favor plants higher in available nutrients (that is, which grow more rapidly and are less well defended by antiherbivory compounds and fiber and contain high percentages of nitrogen and phosphorus). These are grasses, forbs, and shrubs—the growth forms which ungulates broadly seem to have evolved to use as their dietary staples.

If the sequence of my argument is sound thus far, several questions implied by this train of thought now require investigation. The most prominent of these is the need to look beyond such generalizations as annual mean temperatures or July maximums, and their correlations with modern biota, in our work toward reconstruction of the past. Several authors (for example, Guilday et al., 1977; Whithead, 1973) have shown that there was no simple climatic displacement of the biomes we recognize as integral today. Changes more profound than mean annual shifts occurred; there had to be major changes in *seasonal* relationships (Guthrie, 1984). A number of paleoclimatologists have mentioned this likelihood. Hare (1976), for example, proposes that there was "probably less seasonal variation in the general pattern of circulation than today." Circumpolar wave number two today undergoes a large shift of longitude between summer and winter (Hare, 1976). This would not have happened in Wisconsin times, because it is "forced by large thermal differences between land and sea that develop today during the low-albedo summer and high-albedo late winter and spring seasons."

Whatever the origins of these different climatic states, we can conclude from the ungulate size strategies that the ecological norms of today were different then. Along with Guilday et al. (1977), we must conclude that there may be no good analogues in the biomes of today with those of the Pleistocene.

So for optimum large northern ungulate body growth, there would be an early spring moisture peak and the summers would be dry (somewhat arid, to keep the woody plants down) but with spotty showers to moisten the herbaceous growth. The strategies of these "mammoth steppe" megafaunas were mostly nomadic, judging by their plains counterparts today, so that they could migrate to the areas of high quality plant growth.

Autumns may be less critical to body size because in most northern ungulates somatic growth declines before fall and autumn is a time body metabolism switches to depositing fat for winter. Autumn somatic growth could be prolonged by clear skies, which result in warm days without moisture, but also clear skies would increase nighttime radiation loss (creating nighttime frosts). This may not be a detrimental element however as many northern

herbs are moderately frost resistant. Early snows, however, would shorten the growth season, so it is possible that they came later in the year than they do today.

The Pleistocene regulator of densities was undoubtably severe winters. A severe winter for an ungulate can involve deep snows which require increased energy expenditure (we know there were no deep snows, Guthrie, 1982, in Beringia) or poor over-wintering plant resources (which probably was the case). Standing dead cool-season plants are notoriously poor winter food (Peden et al., 1974). They are made even worse by exposure (lack of protective snow-cover). Winters must have been times of great reliance on body fat. Ruminants on such a diet can recycle nitrogen to break down carbohydrates, but do so inefficiently on forage at very low nutrient levels (Church, 1972). The high mortality (particularly of the early age classes) is ironically a boon to ungulate quality as it greatly decreases competition for limited spring growth resources.

There are anatomical indicators that the winters were in fact difficult for ungulates. The cementum layers on the teeth are laid down in an annual cycle of deposition, like a tree ring. In teeth from the large ungulates of the mammoth steppes the winter annulus is so constructed that it is strikingly apparent to the naked eye—without having to make histological sections as one has to do for more southern species. This annulus is produced by growth cessation in response to winter debilitation.

A weather summary for the glacial conditions which promoted Alaskan gigantism would seem to involve a long dry windy summer, hazy clear skies with little precipitation, early springs and late autumns, with little winter snow cover.

Isdo (1976) proposed the climatic "thermal blanket" theory for desert or extremely dusty areas. The atmospheric particles serve to maintain ground level temperatures higher than usual by reducing radiation loss. The implications of this theory of thermal blanketing may apply to glacial atmosphere. Glacial outwash fans produced prodigious quantities of rock flour in the silt-clay size range—a newly milled cafeteria of mineral nutrients. These fine particles were lifted by winds and distributed across the countryside as loess (Pewe, 1975). It is difficult to fully reconstruct the atmosphere of the mammoth steppe, but the thick loess deposits across northern Eurasia and Alaska, and south of the North American continental ice sheets which produced an enriched mineral mantle suggest a dense "dust bowl" haze throughout the growing season. In the far north the "thermal blanket" of dust would probably have its greatest effect during late summer and autumn because radiation loss to the cold night sky increases as the days get shorter.

Surface dust also serves to decrease late winter and spring albedos. In several windy regions of Alaska today surface areas stripped free of snow contribute dust which covers the remaining snow with a thin coat of silt; this in turn hastens early sublimation or melting and produces early greenery.

The Asian people who moved into Alaska at the end of the Pleistocene must have caught the last edge of these colorful sunsets and hazy summers; a major shift had begun toward the recent climate. They were among the last to see the remaining giants of the Beringian Pleistocene and experience the climatic vestiges responsible for those animals. It was a revolutionary time.

ACKNOWLEDGMENTS

I wish to thank the National Park Service and the National Geographic Society who jointly funded the North Alaska Range Project. Under that project, we excavated the Dry Creek Site and, while doing so, I was stimulated to ponder the reasons underlying the large body size of the Pleistocene Alaskan faunas and have the time to research those ideas. Also I wish to thank Gary Selinger who measured the fossil and recent ground squirrel bones for me as an extra credit course. Colleagues provide an essential sounding board and coffee-room critique for one's ideas and I would like to acknowledge their contribution. Dave Klein, Roger Powers, Bob White, John Bryant, and others have significantly affected my thinking on ecology and gigantism.

LITERATURE CITED

AGER, T. A. 1975. Late Quaternary environment history in the Tanana Valley, Alaska. Inst. Polar Studies, Ohio State Univ., Columbus, 54:1–230.

ALTUNA, J. 1972. Fauna de Mamiferos de los Yacimientos. Prehistoricos de Guipuzcoa Munibe, Ano 24, Fasc. 1–4.

ARNOLD, G. W. 1960a. Selective grazing by sheep of two forage species at different stages of growth. Australian J. Agric. Res., 11:1062–1133.

———. 1960b. The effect of quantity and quality of pasture available to sheep and their grazing behavior. Australian J. Agric. Res., 11:1034–1043.

———. 1962. The influence of several factors in determining the grazing behavior of border Leichester × merino sheep. J. British Grassland Soc., 17:41–51.

———. 1964. Factors within plant associations affecting the behavior and performance of grazing animals. Pp. 10–38, in Grazing in terrestrial and marine environments (D. J. Crisp, ed.), Blackwell, Oxford, 430 pp.

BELL, R. H. V. 1971. A grazing system in the Serengeti. Sci. Amer., 225:86–93.

BLAXTER, K. L., F. M. WAINMAN, and R. S. WILSON. 1961. The regulation of food intake by sheep. Anim. Prod., 3:51–61.

BLISS, L. C. 1962. Adaptations of arctic and alpine plants to environmental conditions. Arctic, 15:117–144.

BRONSON, M. 1980. Altitudinal variation in emergence time of golden-mantled ground squirrels (Spermophilus columbianus). J. Mamm., 61:124–126.

CHAPIN, F. S., III. 1977. Temperature compensation in phosphate absorption occurring over diverse time scales. Arctic and Alpine Res., 9:139–148.

CHURCH, D. C. 1972. Digestive physiology and nutrition of ruminants. Published by D. C. Church, Oregon State Univ., Corvalis, 3:1–351.

CLARK, J. D. G. 1954. Excavations at Star Carr. Cambridge Univ. Press, Cambridge, 200 pp.

EDWARDS, W. E. 1967. The late-Pleistocene extinction and diminution in size of many mammalian species. Pp. 141–154, in Pleistocene extinctions (P. S. Martin and H. E. Wright, Jr., ed.), Yale Univ. Press, New Haven, 453 pp.

GARUTT, V. E. 1964. Das Mammut (Mammuthus primigenius Blusmenb.). Neue Brehm-Buherei, Wittenberg-Lutherstadt, Siemsen, 283 pp.

GEIST, V. 1966. The evolutionary significance of mountain sheep horns. Evolution, 20:558–566.

———. 1971. Mountain sheep—A study of behavior and evolution. Univ. Chicago Press, Chicago, 383 pp.

GUILDAY, J. E., P. W. PARMALEE, and H. W. HAMILTON. 1977. The Clark's Cave bone deposit and the late Pleistocene paleoecology of the central Appalachian Mountains of Virginia. Bull. Carnegie Mus. Nat. Hist., 2:1–88.

GUTHRIE, R. D. 1966. The extinct wapiti of Alaska and Yukon Territory. Canadian J. Zool., 44:47–57.

———. 1968. Paleoecology of the large mammal community in interior Alaska during the late Pleistocene. Amer. Midland Nat., 79:346–363.

———. 1974. Environmental influences on body size, social organs, population parameters, and extinction of Beringian mammals. Khabarovsk Conference on Beringia. Soviet Academy of Science, Vladivostock, 80 pp.

———. 1980. Bison and man in North America. Canadian J. Anthro., 1:55–75.

———. 1982. Mammals of the mammoth steppe as paleoenvironmental indicators. Pp. 307–326, in Paleoecology of Beringia (D. M. Hopkins, J. V. Matthews, C. E. Schweger, S. B. Young, eds.), Academic Press, New York, 489 pp.

———. 1984. Mosaics, allelochemics and nutrients: an ecological theory of late Pleistocene megafaunal extinctions. In Pleistocene extinctions (P. S. Martin and R. G. Klein, eds.), Univ. Arizona Press, Tuscon, second edition.

HANLEY, T. A. 1980. Nutritional constraints on food and habitat selection by sympatric ungulates. Unpublished Ph.D. dissert., Univ. Washington, 176 pp.

HARE, F. K. 1976. Late Pleistocene and Holocene climates: some persistent problems. Quat. Res., 6:507–517.

HARINGTON, C. R. 1978. Quaternary vertebrate faunas of Canada and Alaska and their suggested chronological sequence. Syllogeus Series, 15:1–105.

HARRIS, A. H., AND P. MUNDEL. 1974. Size reduction in big horn sheep (Ovis canadensis) at the close of the Pleistocene. J. Mamm., 55:678–680.

HEIMER, W. E. and A. C. SMITH. 1975. Ram horn growth and population quality—Their significance to Dall sheep management in Alaska. Alaska Dept. Fish and Game, Wildlife Tech. Bull. 5:1–41.

HOEFS, M. E. G. 1974. Food selection by Dall's sheep Ovis dalli dalli. Pp. 759–786, in Behavior of ungulates and its relation to management (V. Geist and F. Walther, eds.), Internat. Union Conserv. of Nat. Res., Switzerland, n.s., 24:759–786.

HOPKINS, D. M., J. V. MATTHEWS, JR., C. E. SCHWEGER, and S. B. YOUNG. 1982. Paleoecology of Beringia. Academic Press, New York, 489 pp.

HOREJSI, B. 1976. Suckling and feeding behavior in relation to lamb survival in bighorn sheep. (Ovis canadensis canadensis Shaw). Unpublished Ph.D. thesis, Univ. Calgary, Calgary, 201 pp.

ISDO, S. B. 1976. A perspective on the role of climatic dust in climatic change theory abstracts. Amer. Quat. Assoc. (AMQUA), 4th biennial meeting, Tempe, Arizona, pp. 59–60.

JANIS, C. 1975. The evolutionary strategy of the Equidae and the origins of the rumen and caecal digestion. Evolution, 30: 757–774.

KIELL, D. J., and J. S. MILLAR. 1978. Growth of juvenile arctic ground squirrels (Spermophilus parryi) at McConnell River N.W.T. Canadian J. Zool., 56:1475–1478.

KLEIN, D. R. 1965. Ecology of deer range in Alaska. Ecol. Monog., 35:259–284.

———. 1970. Tundra ranges of the boreal forests. J. Range. Mgmt., 23:8–14.

KOEPPL, J. W., and R. S. HOFFMANN. 1981. Comparative growth rates (postuatae) of four species of ground squirrels. J. Mamm., 61:41–57.

KURTEN, B. J. 1968. Pleistocene mammals of Europe. Weidenfield and Nicolson, London, 317 pp.

LA ROI, G. H. 1982. From aspen parkland to the Arctic coast: A vegetation transect. (Abstracts) Annual Meeting, Society of Range Management, Calgary, Alberta, 1982.

LEVENSON, H. 1979. Sciuvid growth rates: some corrections and additions. J. Mamm., 60:232–234.

MATTHEWS, J. V. 1974. Wisconsin environment of interior Alaska; pollen and macrofossil analysis of a 27-meter core from Isabella Basin (Fairbanks, Alaska). Canadian J. Earth Sci., 11:828–841.

MATTSON, W. J. 1980. Herbivory in relation to plant nitrogen content. Ann. Rev. Ecol. Syst., 11:119–162.

McDONALD, J. N. 1981. The Northern American Bison: Their classification and evolution. Univ. of California Press, Berkeley, 316 pp.

MROSOVSKY, N. 1976. Lipid programmes and life strategies in hibernators. Amer. Zool., 16:685–697.

MURIE, A. 1944. The wolves of Mount McKinley. Fauna Ser., U.S. Fauna National Parks, 5:xix + 1–238.

MURIE, J. O., D. A. BOAG, and V. K. KIEVETT. 1980. Litter size in Columbian ground squirrels (Spermophilus columbianus). J. Mamm., 61:237–244.

MURIE, J. O., and M. A. HARRIS. 1982. Annual variation of spring emergence and breeding in Columbian ground squirrels (Spermophilus columbianus). J. Mamm., 63:431–439.

MURPHY, E. C. 1974. An age structure and a reevaluation of the population dynamics of Dall sheep (Ovis dalli dalli). Unpublished M.S. thesis, Univ. Alaska, Fairbanks, 113 pp.

NICHOLS, L., JR. 1978. Dall's sheep. Pp. 179–190, in Big game in North America (J. L. Schmidt and D. L. Gilbert, eds.), Stackpole Books, Harrisburg, 494 pp.

NORDEN, H. C., I. M. COWAN, and A. T. WOOD. 1968. Nutritional requirements of blacktailed deer in captivity. Pp. 89–96, in Comparative nutrition of wild animals (M. A. Crawford, ed.), Symp. Zool. Soc. London, 21.

NRC. 1971. Recommended nutrient allowances for sheep. Nat. Academy Sci., Washington, D.C., 107 pp.

OLEBERG, K. 1956. Factors affecting the nutritive value of range forage. J. Range. Mgmt., 9:220–225.

PALMER, L. J. 1944. Food requirements of some Alaskan game mammals. J. Mamm., 25:49–54.

PEDEN, D. G., G. M. VAN DYNE, R. W. RICE, and R. M. HANSEN. 1974. The trophic ecology of Bison bison in short grass prairie. J. App. Ecol., 11:489–498.

PEWE, T. L. 1975. Quaternary geology of Alaska. U.S. Geol. Soc. Prof. Paper, 835:1–212.

POWERS, W. R., R. D. GUTHRIE, and J. F. HOFFECKER. In press. Dry Creek: archeology and paleoecology of a late Pleistocene Alaskan hunting camp.

SCHMIDT, J. L., and D. L. GILBERT. 1978. Big game of North America: ecology and management. Stackpole Books, Harrisburg, 494 pp.

SHACKELTON, D. M., and L. V. HILLS. 1977. Post-glacial ungulates (Cervus and Bison) from Three Hills, Alberta. Canadian J. Earth Sci., 14:963–986.

SHER, A. V. 1971. Pleistocene mammals and stratigraphy of the far Northeast USSR and North America. Geol. Inst., USSR Academy of Sciences, Moscow, 310 pp. (Trans. Ameri. Geol. Inst., Published 1974, in International Geol. Review, 16:1–284).

SKOGLAND, T., and R. G. PETOCZ. 1975. Ecology and behavior of Marco Polo sheep (Ovis ammon poli) in Pamir during winter. Report from FAO/UNPD Project: Conservation and Utilization of Wildlife Resources, 23 pp.

STOCK, A. D., and W. L. STOKES. 1969. A re-evaluation of Pleistocene bighorn sheep from the Great Basin and their relationship to living members of the genus Ovis. J. Mamm., 54:805–807.

STODDART, L. A., and A. D. SMITH. 1955. Range Management. McGraw-Hill, New York, second ed., 433 pp.

SUMMERFIELD, B. C. 1974. Population dynamics and seasonal movement patterns of Dall sheep in the Atigun Canyon area Brooks Range Alaska. Unpublished Master Sci. thesis, Univ. Alaska, College, 212 pp.

VERESCHAGIN, N. K. 1967. The mammals of the Caucasus. Izadtel'stvo Akademu Nauk. (SSSR English Translation) Israel Program for Scientific Translations, Jerusalem, 89 pp.

VIERECK, L. A. 1963. Sheep and goat investigations: range survey. Unpublished report on project W-6-R-3, E, 2-a. Alaska Dept. Fish and Game, Juneau.

WALLACH, J. D. 1972. Nutrition and feeding of captive ruminants in zoos. Pp. 292–307, *in* Digestive physiology and nutrition of ruminants (D. C. Church, ed.), Corvalis, Oregon, 780 pp.

WESTOBY, M. 1974. An analysis of diet selection by large generalist herbivores. Amer. Nat., 108:290–304.

———. 1978. What are the biological bases of varied diets? Amer. Nat., 112:627–631.

WHITHEAD, D. R. 1973. Late-Wisconsin vegetational changes in unglaciated eastern North America. Quaternary Res., 3: 621–631.

WHITTEN, K. 1975. Habitat relationships and population dynamics of Dall sheep (*Ovis dalli dalli*) in Mt. McKinley National Park. Unpublished M.S. thesis, Univ. Alaska, Fairbanks, 117 pp.

WINTERS, J. 1980. Summer habitat and food utilization by Dall sheep and its relation to body and horn size. Unpublished M.S. thesis, Univ. Alaska, Fairbanks, 109 pp.

WOODGERD, W. 1964. Population dynamics of bighorn sheep on Wild-horse Island. J. Wildlife Mgmt., 28:381–391.

WRAZEN, J. A., and L. A. WRAZEN. 1982. Hoarding, body mass dynamics, and torpor as components of the survival strategy of the eastern chipmunk. J. Mamm., 63:63–72.

ZELDIN, T. 1983. The French. Collins, London, 542 pp.

Address: Institute of Arctic Biology, University of Alaska, Fairbanks, Alaska 99701.

QUATERNARY MARINE AND LAND MAMMALS AND THEIR PALEOENVIRONMENTAL IMPLICATIONS—SOME EXAMPLES FROM NORTHERN NORTH AMERICA

C. R. Harington

ABSTRACT

Workers on Quaternary mammals should try to obtain as much information as possible from fossil evidence available—especially from well-preserved specimens. Examples discussed here are drawn from marine (seal and walrus) and land mammal remains (mammoth, stag-moose, saiga antelope, and arctic ground squirrel) from northern North America.

It is particularly important that: specimens are correctly identified; stratigraphic and sedimentological situations of skeletons found in place are carefully assessed; associated plant and animal remains are evaluated for paleoenvironmental clues; direct and indirect geochronological data applying to the specimens are gathered—where possible testing the validity of such data by cross-checking; bone surfaces are closely examined so that signs of predator-prey relationships, scavenging, and disease may be detected; an effort is made to establish individual ages of the specimens (for example, by tooth-sectioning); in the case of unusually well-preserved specimens, stomach contents, droppings and soft parts are analyzed in order to gain information on feeding habits and other ecological parameters.

Above all, Quaternary paleobiologists should view these long-dead animals as realistically as possible, trying to see their place in the web of life—as mammalogists view living species. Finally, we must try to pass on significant paleoenvironmental information in an appealing way to the public—they should be aware of the great changes that can occur in landscapes and their biota over even relatively short periods of geological time.

INTRODUCTION

As a vertebrate paleontologist working on Quaternary mammals, I am concerned that as much evidence as possible on the habitat of these mammals is collected, and that it is properly analyzed. The closer we can come to discovering the natural habits and habitats of these mammals the better. Both marine and land mammals deserve study in this respect. It is particularly important to gather such information on articulated skeletons found in primary position, and from associated fossils in enclosing sediments. Indeed, the nature of the sediments themselves may help to explain the environment of deposition. Although whatever is discovered about the environment of deposition of a species is not necessarily typical of that species' habitat, it is important to accumulate such paleoenvironmental evidence in the hope that habitat preferences of certain Pleistocene mammals may be more precisely defined.

Sometimes, fossils of mammalian species with well-defined adaptations and habitat needs provide clues to the nature of past environments in an area—the basic assumption being that species represented by fossils had ecological requirements similar to those of the same or closely allied living species. Among the cases mentioned here, perhaps the ringed seal and saiga antelope best exemplify this kind of evidence (Fig. 1).

Occasionally, other fossils, such as plant macrofossils, pollen, insect remains, and mollusc shells directly associated with ancient skeletons shed light on habitats of extinct mammals. Examples used here are the Babine Lake mammoth and the stag-moose (*Cervalces scotti*) from Columbia, New Jersey.

Rarely, nearly complete animals, including stomach contents, droppings and other soft parts of ice age mammals are preserved, providing direct evidence of their environmental adaptations and eating habits. The Berezovka mammoth from frozen ground in Siberia is an outstanding international example (Augusta and Burian, 1963). The potential of this kind of evidence is seen in a 12,000-year-old arctic ground squirrel skeleton found in its nest in frozen ground near Dawson, Yukon.

I wish to emphasize that significant paleoenvironmental information should not be restricted to scientists. We must be concerned with passing on our information in an appealing way to the public: they should be aware of the great changes that can occur in landscapes and their biota over even relatively short periods of geological time.

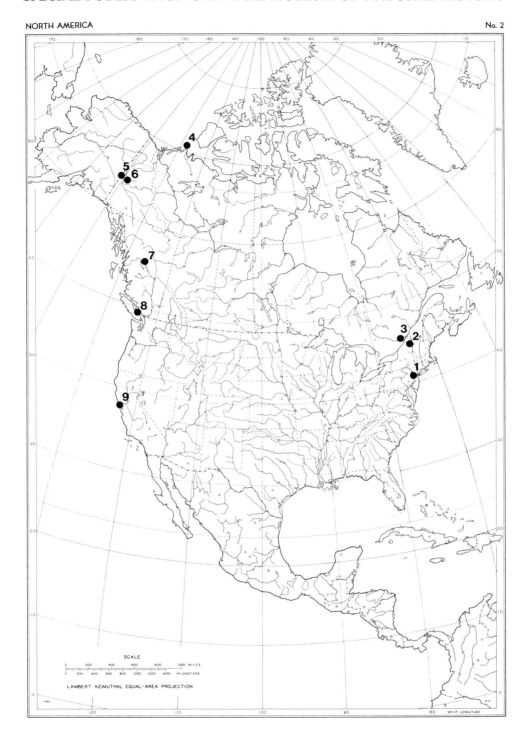

Fig. 1.—Localities of specimens mentioned in text. Key: 1) Columbia, New Jersey (stag-moose, *Cervalces scotti*); 2) Plattsburgh, New York (harbour seal, *Phoca vitulina*); 3) Hull, Québec (ringed seal, *Phoca hispida*); 4) Baillie Islands, Northwest Territories (saiga antelope, *Saiga tatarica*); 5) Glacier Creek, Yukon Territory (ground squirrel carcass, *Spermophilus parryi*); 6) Dominion Creek, Yukon Territory (ground squirrel nest and skeleton, *Spermophilus parryi*); 7) Babine Lake, British Columbia (Columbian mammoth, *Mammuthus* cf. *columbi*); 8) Qualicum Beach, British Columbia (walrus, *Odobenus rosmarus*); 9) San Francisco Bay, California (walrus, *Odobenus rosmarus*).

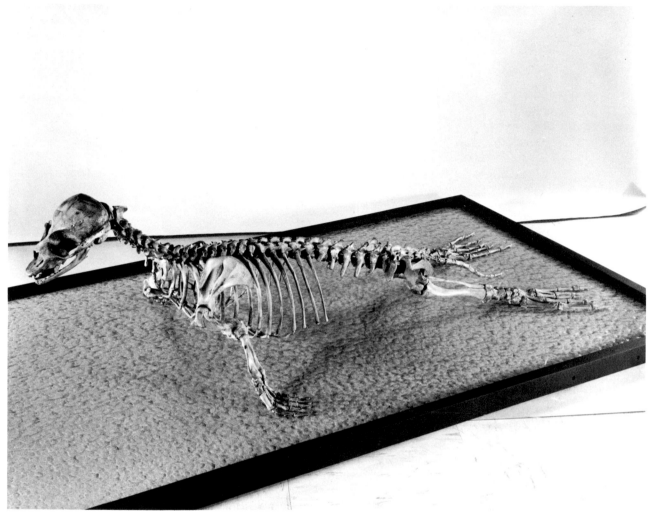

Fig. 2.—Left side of a ringed seal (*Phoca hispida*; NMC 6830) skeleton from Champlain Sea deposits at Hull, Quebec.

EXAMPLES

MARINE MAMMALS

First, I want to focus on marine mammals of the Champlain Sea (an inland sea that covered the St. Lawrence Lowlands from about 12,000 to 10,000 years ago). Although many species of marine mammals that once lived in this sea now occupy the Gulf of St. Lawrence, at least three, the bowhead whale (*Balaena mysticetus*), ringed seal (*Phoca hispida*), and bearded seal (*Erignathus barbatus*) are closely adapted to arctic waters (Mansfield, 1967).

A skeleton of a ringed seal (Fig. 2) found in place with *Hiatella arctica* shells in marine clay at Hull, Québec, was important in determining the coldness

of the Champlain Sea's early stages. Ironically, because the specimen was incorrectly identified [as probably a young harp seal (*Phoca groenlandica*), then as a young harbour seal (*Phoca vitulina*), eventually being displayed in the National Museum of Canada as a young harp seal], it was not until some 80 years after collection that its paleoenvironmental implications were realized! A detailed study of the specimen (Harington and Sergeant, 1972) showed that it was a ringed seal, and, despite its small size, an adult about seven years old. Ringed seals are now the commonest and most widespread arctic seals (Fig. 3). They are adapted to keep breathing holes

Fig. 3.—Ringed seal (*Phoca hispida*). Courtesy of New York Zoological Society, Carleton Ray.

open in sea ice during winter and to pupping on fast ice (sea ice attached to the coast) in spring. So, it is reasonable to conclude that fast ice clung to the northern coast of the sea. Perhaps the individual was so small for its age because of poor nourishment caused by early separation from its mother. Such cases occur on relatively straight coasts with unstable fast ice, and these conditions may have prevailed along the northern margin of the Champlain Sea. Finally, it is worth noting that corroboration of an arctic phase of the sea has since come from studies of marine ostracodes (Cronin, 1981). According to these studies, conditions during the *Hiatella arctica* phase (from 11,600 to about 11,000) were frigid to subfrigid, with bottom temperatures of 0 to 12°C, and normal marine salinities (18–35 PPT). Therefore, if I were asked to predict the geological age of the Hull specimen, I would be inclined to place it between 11,600 and approximately 11,000 B.P.

Again, the importance of correct identification of specimens is shown in the case of another Champlain Sea seal fossil found in 1901 at Plattsburgh, New York (Bishop, 1921; Hartnagel and Bishop, 1921). A tibial shaft of a juvenile, considered for many years to represent a hooded seal (*Cystophora cristata*), has been relocated and reidentified as part of a harbour seal (*Phoca vitulina*; NYSM 892-D/8569; Fig. 4) (Ray, 1983). T. M. Cronin identified the following benthic foraminifera from a clay sample with the bone—*Haynesia obiculare,* and *El-*

phidium cf. *albiumbiculatum,* suggesting that the clay was deposited during the *Hiatella arctica* or *Mya arenaria* (approximately 11,000–10,000 B.P.) phase of the Champlain Sea. Evidently, the harbour seal (Fig. 5) occupied shallow water along the western margin of the sea's southern arm. Such an environment corresponds to the known habitat preference of harbour seals today—and is quite different from that of the hooded seal (Mansfield, 1967).

As a museum scientist, I feel it is particularly important to transmit significant paleoenvironmental information in an appealing way to the public. If someone asked me what the Champlain Sea would have looked like about 11,000 years ago, I would try to convey the following impression based on fossil finds made during the past 130 years (Harington, 1980*a*; Mott et al., 1981), and using modern observations of the species concerned to achieve a more vivid result:

"Under the bright summer sun, blue-green water scattered with pans of ice extends in a broad sweep eastward. To the north, partly bounded by freshly exposed, rounded granitic bluffs lies the dazzling Laurentide ice sheet, streaked and smudged along its edge by gray heaps of morainic debris. Some fast ice still clings to the north shore, where several ringed seals bob in the cold sea off the floe-edge. They twist and turn, feeding on abundant shrimp-like crustaceans. Two small flocks of Common Eiders whistle over the wavelets stirred up by cold winds draining off the edge of the ice sheet. They curve toward a low, rocky island where females are nesting. Other eiders in the colony dive in shallow waters nearby for clam-like molluscs.

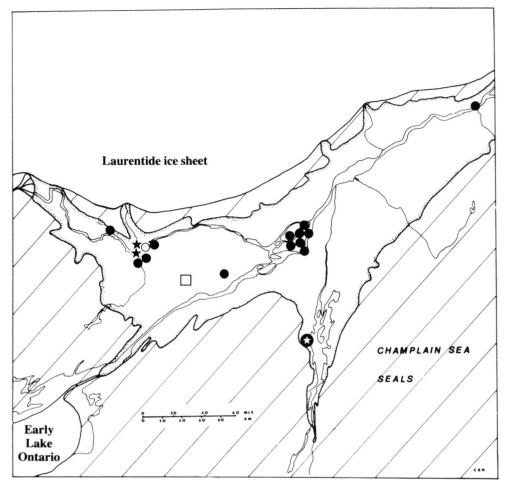

Fig. 4.—Localities of seal remains from Champlain Sea deposits. Stippling indicates the approximate maximum limit of Champlain Sea. The cluster of symbols at the centre marks Montréal, and at the left marks Ottawa. Key: ringed seal, *Phoca hispida* (black star); harp seal, *Phoca (Pagophilus) groenlandica* (white dot); bearded seal, *Erignathus barbatus* (white square); harbour seal, *Phoca vitulina* (white star); unidentified seal remains (black dot).

In deeper water several kilometres off the sea's northern coast, a massive black shape cruises at the surface, periodically blowing a v-shaped mist into the air. The migrating bowhead whale moves steadily westward toward the broad sloping sands of the Petawawa Delta, the sun making rainbows in its moist breath. The first white whales have arrived from the Gulf of Saint Lawrence. A pod of 20 appear from above like a corps de ballet. They slide smoothly through the cool greenish water, their gleaming white backs periodically breaking the surface sending chevron-like wakes behind. A few new-born calves, like small shadows, press close to the females. Their heart-shaped flukes beat rapidly to keep up. Hundreds of belugas already loll in shallow, river-warmed waters in the southern embayment of the sea, where Lake Champlain now lies.

On the sandy beaches of the south shore, small birds pecking fitfully with their sharp bills scuttle back and forth in harmony with the advancing and retreating waves. Dead starfish and stray patches of kelp are partly exposed on the beach. Just off-shore,

the sea is alive with spawning capelin. Their silvery bodies, in hundreds, rise and fall with the waves. Multitudes lie rotting in rows at the level of the last high tide, their stench overriding the salt tang of the onshore breeze. Through the glare of sun on the water, silhouettes of about a dozen harp seals can be seen leaping and cavorting. A solitary bearded seal, whiskers glistening white against its rotund gray body, hauls out on a nearby sandbar to rest.

On the northeastern margin of the sea, several of the Monteregian Islands (tops of the volcanic mounds of Mount Royal, Mont St. Hilaire, etc.) appear darkly below. Waves roll in strongly on their shores, churning up and scattering chalky, quadrangular valves of *Hiatella arctica*. Farther east near what is now Québec City, tundra stretches northward from a large coastal lagoon to the ice front some 13 kilometres away. The taller grasses and sedges near the edge of the lagoon quiver in the cold breeze. Dense mats of mouse-ear chickweed almost cover the gravel there. Patches of creamy mountain avens contrast with the waxy

Fig. 5.—Harbour seal (*Phoca vitulina*). Courtesy of New York Zoological Society, Carleton Ray.

green of shrub willows and the duller green of crimson-flowered mountain sorrel. A hare, sheltering behind a granite boulder, nibbles on succulent willow shoots, its ears delicately attuned to each new sound. Frightened, it zig-zags toward the end of the lagoon and pauses. There, brilliant flashes of yellow mark the tall flowering heads of groundsel, which dominates the surrounding moist, peaty ground. Southward, across the 60-kilometre-wide threshold to the Champlain Sea, lies more tundra. Only in the valleys far to the south do stunted spruce and birch trees mark the pioneering northern fringe of the Boreal Forest, in whose shade stalks a beady-eyed marten."

Many problems remain concerning this inland sea and its environment. Why have no walrus fossils been reported from Champlain Sea deposits when specimens are known from 15 localities (see for example Harington, 1977: fig. 3) in the eastern approaches to the sea? Important habitat requirements, such as accessible banks of abundant molluscs (on which walruses are known to feed), and suitable hauling-out grounds (particularly near the north shore of the sea), seem to have been present. Furthermore, walrus bones tend to preserve well, especially skulls.

Continuing on the topic of ice age walruses, but shifting to the Pacific coast, I wish to comment on a remarkable walrus skeleton (Fig. 6) found near Qualicum Beach, Vancouver Island. The skeleton was found in place in Dashwood glaciomarine clayey silt of early Wisconsin age (S. Hicock, personal communication, 1981). A radiocarbon date on bone

collagen of >40,000 B.P. (I-11617) supported the relatively early age of this specimen based on its stratigraphic position. Further, finding the specimen in the basal clay is in accord with features of this

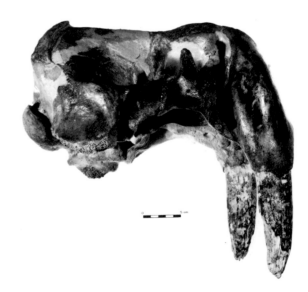

Fig. 6.—Right side of walrus (*Odobenus rosmarus*; NMC 38490) cranium. This is part of an individual skeleton found in early Wisconsin glaciomarine clay near Qualicum Beach, British Columbia.

unit as described by Fyles (1963) (that is, an epineritic to littoral marine clay containing dropstones probably derived from floating sea ice, and abundant well-preserved pelecypod shells). Mollusc assemblages collected from the unit are living today in the southern part of Bering Sea (Wagner, 1959), which approximates the southern limit of Pacific walrus (*Odobenus rosmarus divergens*) range. In other words, the stratigraphic, geochronological, and paleontological evidence seems to hang together. In the hope that more paleoenvironmental (particularly paleotemperature) data could be gained, I excavated a block of clay from directly beneath the skeleton, and passed it on to Sigrid Lichti-Federovich for analysis of marine microorganisms. Unfortunately only three species of marine diatoms (*Plagiogramma staurophorum, Amphora* sp., and *Grammatophora* sp.) were noted, and the paucity of specimens precluded a meaningful paleoenvironmental interpretation based on this evidence. Suffice to say that walruses lived far south of their present range during late Pleistocene glacial advances (for example, records from San Francisco Bay, Fig. 7; Kittyhawk, North Carolina; Montrouge near Paris; and Tokyo, Harington, 1975, and this paper).

LAND MAMMALS

Paleoenvironmental evidence derived from Quaternary land mammals is also of interest, and I wish to mention a few pertinent cases.

Remains of a partly articulated mammoth skeleton were exposed during stripping at a mining site on Babine Lake, central British Columbia. The bones lay in silty pond deposits in a bedrock depression, and were overlain by a thin layer of gravel and a thick layer of glacial till. Although no molar teeth were found, limb proportions indicated that the specimen was a large mammoth, like the Columbian mammoth (*Mammuthus* cf. *columbi*). Two radiocarbon dates of 42,900 ± 1860 B.P. (GSC-1657) and 43,800 ± 1830 B.P. (GSC-1687) on wood from the silty layer, and another of 34,000 ± 690 B.P. (GSC-1754) on mammoth bone suggested that, during this part of the Olympia Interglaciation, the vegetation near Babine Lake was similar to present shrub tundra just beyond the treeline in northern Canada (Fig. 8). Probably birch and willow shrubs bordered the pond and were scattered over the landscape. Open ground supported abundant grasses, *Artemisia* and other composites, various members of the rose family, pinks, willow-herb, and several other herbs. Although the exact contemporaneity of

the plant and mammoth remains is questionable, both types of fossils suggest the presence of grassland in the Babine Lake area from about 34,000 to 43,000 B.P. (Harington et al., 1974).

I am presently studying, in collaboration with Gary J. Sawyer of the Natural Science Museum, Blairstown, New Jersey, a virtually complete skeleton of an extinct stag-moose (*Cervalces scotti*; NSM 264) from a bog near Columbia, New Jersey (Figs. 9–10). This case is noteworthy because it is providing the first clear evidence on the habitat of this moose-like animal. The specimen was excavated from the base of a marl layer about 1 m thick, that was overlain by approximately 1 m of consolidated peat and about 0.3 m of rich black loam. A "leaf layer" containing abundant remains of *Potamogeton* and *Najas*, commonly found on pond or lake margins (J. V. Matthews, Jr., personal communication, 1979), occurring just below the bones yielded a radiocarbon date of 11,600 ± 170 B.P. (I-11335). A bone sample gave a date of 11,230 ± 160 B.P. (I-11286). Dominant species of gastropod molluscs in the marl surrounding the *Cervalces* bones are *Gyraulus parvus* and *Valvata tricarinata*, suggesting a large cold lake with abundant vegetation and a muddy bottom with some stones (M. F. I. Smith, personal communication, 1979), which seems to fit the paleobotanical evidence. Among the sphaeriids, *Pisidium variabile* and *P. nitidum* were most common, whereas *P. ferrugineum, P. casertanum, Musculium securis, Sphaerium rhomboideum*, and *S. simile* were rarer (G. L. Mackie, personal communication, 1979). The dominant species (*Pisidium variabile*) presently occurs in a variety of habitats, but *Sphaerium rhomboideum* may define the type of habitat more closely (an environment of deposition in a eutrophic lake less than 5 m deep, pH 7–9, dissolved oxygen >6 mg/l, total alkalinity 60–250 mg $CaCO_3$/l, free CO_2 <15 mg/l). Looking at a map showing stages in the retreat of the Laurentide ice sheet (Prest, 1969), I calculate that some of these moose-like animals lived about 600 km south of the southern margin of the ice approximately 11,000 years ago. *Cervalces scotti*, according to the above radiocarbon dates, was contemporaneous with the American mastodon (*Mammut americanum*) in this part of New England (Parris, 1983).

Turning from eastern North America to the extreme northwestern part of this continent, I wish to mention the apparent intrinsic value of saiga antelope (*Saiga tatarica*; Fig. 11) bones as paleoenvironmental indicators. Nine saiga specimens are

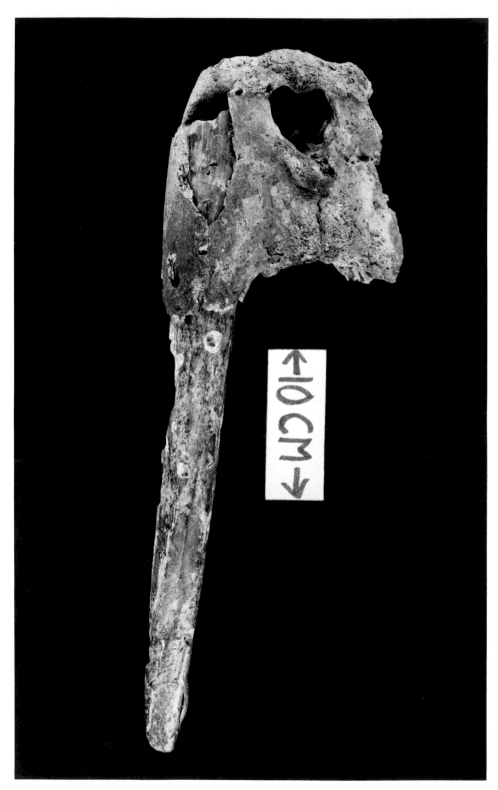

Fig. 7.– Front view of an anterior cranial fragment of a walrus (*Odobenus rosmarus*; California Academy of Sciences 16677) dredged from the bottom of San Francisco Bay, California. A radiocarbon date of 27,200 ± 950 B.P. (I-9994) was obtained on the tusk. A check date using the accelerator technique may be advisable.

Fig. 8.—Restoration of the Babine Lake mammoth (*Mammuthus* cf. *columbi*) as it may have appeared in its natural shrub tundra environment during the Olympia Interglaciation. Ink sketch by Charles Douglas.

Fig. 9.—Restoration of the extinct stag-moose (*Cervalces scotti*). Pictured under winter conditions in this sketch by R. Bruce Horsfall. Copyright 1937 by the American Philosophical Society.

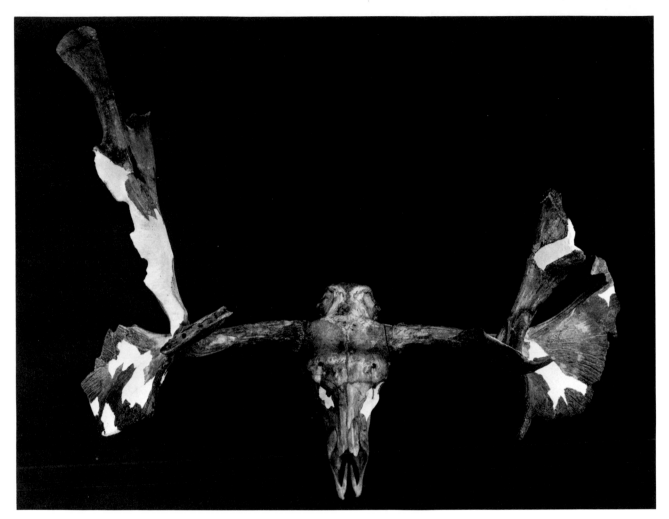

Fig. 10.—Top view of cranium with partial antlers of a stag-moose (*Cervalces scotti*; NSM 264) excavated with most of the remainder of the skeleton from a bog near Columbia, New Jersey. Lighter areas are restored. A bone sample from the specimen yielded a radiocarbon date of 11,230 ± 160 B.P. (I-11335).

known from late Pleistocene deposits in Eastern Beringia (six from central Alaska, two from northern Alaska—one dated at 37,000 ± 900 B.P. (GSC-3050)—and one from Baillie Islands, Northwest Territories; Fig. 12). Probably most are of Wisconsin age. Because living saigas of central Eurasia are particularly adapted to dry steppe-grasslands, I suggest that they crossed broad, steppe-like plains of the northern Bering Isthmus during glacial phases of the late Pleistocene (Fig. 13). Presumably the kind of northern steppe to which they had adapted once extended eastward, up the Yukon valley to central Alaska, and along the Arctic Coastal Plain to Baillie Islands in Canada. Saiga antelope remains appear

to be useful indicators suggesting the presence of steppe-like vegetation, generally low, flattish terrain, rather arid climatic conditions, and shallow snow cover in winter (Harington, 1980b, 1981).

Of potential paleoenvironmental interest are several nests of late Pleistocene arctic ground squirrels (*Spermophilus parryi*) that have been collected from frozen silts near Dawson, Yukon, during the past 15 years. One example from Dominion Creek contained nesting grasses, part of a seed cache, fecal pellets, and skeletal fragments of an individual ground squirrel. The nesting grasses gave a date of 12,200 ± 100 B.P. (GSC-2641). Part of the nest debris was analyzed for fossil pollen and revealed

Fig. 11.—Saiga antelope (*Saiga tatarica*). This lightly built animal with its lyre-shaped horns and expanded nasal region evidently occupied steppe-like terrain in northwestern North America during the late Pleistocene.

abundant fungal elements and spores as well as numerous slime mold and bryophyte spores. The angiosperm pollen, which composed only about 20% of the total assemblage, represented several grasses (Gramineae), cotton grass (*Eriophorum* sp.), willow (*Salix* sp.), Chenopodiaceae/Amaranthaceae, several species of Compositae, and much sage (*Artemisia* sp.) (D. M. Jarzen, personal communication, 1975). Pollen analysis of another nest from Hunker Creek near Dawson yielded 73.5% grasses (Gramineae), 14.6% sage (*Artemisia* sp.), 7.4% Chenopodiaceae, 2.6% willow (*Salix* sp.), 1.0% Compositae, 0.3% birch (*Betula* sp.), traces of *Lycopodium* sp., *Botrychium* sp., *Sphagnum* sp., and a great

abundance of fungal spores (C. McAtee, personal communication, 1979). As more work is completed on these nests (keeping in mind modern analogues, for example, Krog, 1954; Bee and Hall, 1956; Banfield, 1974), probably we will be able to tell exactly what grasses were chosen for nests, what plants and fungi were eaten, and what kinds of seeds were cached. It may be possible to recreate, on this basis, a microcosm of the paleoenvironment of this part of Eastern Beringia toward the close of the last glaciation. Indeed, these ground squirrels could be considered as small, furry botanists, industriously sampling vegetation within a limited range of their nests and storing it away in underground herbaria.

Fig. 12.—Left side of left horncore with attached portions of frontal and parietal bones of a saiga antelope (*Saiga tatarica*; NMC 12090) from Baillie Islands, Northwest Territories. It is the first recorded saiga specimen from Canada.

In 1981, M. Peschke, a placer miner on Glacier Creek, Yukon Territory, picked up a ball of silt with bits of hair projecting from it. I later found that it was a carcass of a ground squirrel (*Spermophilus parryi*; Fig. 14). An x-ray showed that a virtually complete skeleton lay inside the skin and hair. The animal had been curled up as if it had died during hibernation—perhaps more than 10,000 years ago (Fig. 15). A detailed study of this specimen may also yield valuable paleoenvironmental evidence.

CONCLUSION

Workers on Quaternary mammals should try to wring a maximum of information from the fossil evidence available—particularly from significant, well-preserved specimens. I suggest that valid goals are: (a) correct identification of the mammal—even fragments of some species, such as horncores of the saiga antelopes can have intrinsic value as paleoenvironmental indicators; (b) careful assessment of the stratigraphic and sedimentological situation of skeletons found in place—sedimentology often yields data on environment of deposition; (c) collection of associated plant and animal macrofossil remains, in addition to substantial samples of matrix enclosing the bones, in case they may yield valuable paleoenvironmental clues; (d) gathering direct (for example, radiocarbon dates on bone) and indirect geochronological data (for example, radiocarbon dates on associated mollusc shells or plants, dates on enclosing tephra, and other material), where possible making an effort to cross-check such data; (e) close examination of bone surfaces so that relationships between prey and predators or scavengers (including humans), or possible effects of disease may be detected; (f) to establish the age of the individual represented by noting the degree of tooth wear and suture fusion, and by sectioning teeth and other methods; (g) analyses of fecal droppings, stomach contents and ancient "tooth-jam"[1] in unusually well-preserved specimens, in order to gather data on eating habits of the species in question.

The most reliable paleoenvironmental evidence is that which is clearly associated with a specimen, is readily identifiable, abundant, undisturbed, based on various sources, and which, ultimately, is compatible. Above all, it is important for Quaternary paleontologists to view these long-dead animals as realistically as possible, trying to see their place in the web of life—as mammalogists would view living species.

---

[1] Remains of vegetation eaten by an animal, which may be preserved in "lakes" of ungulate teeth.

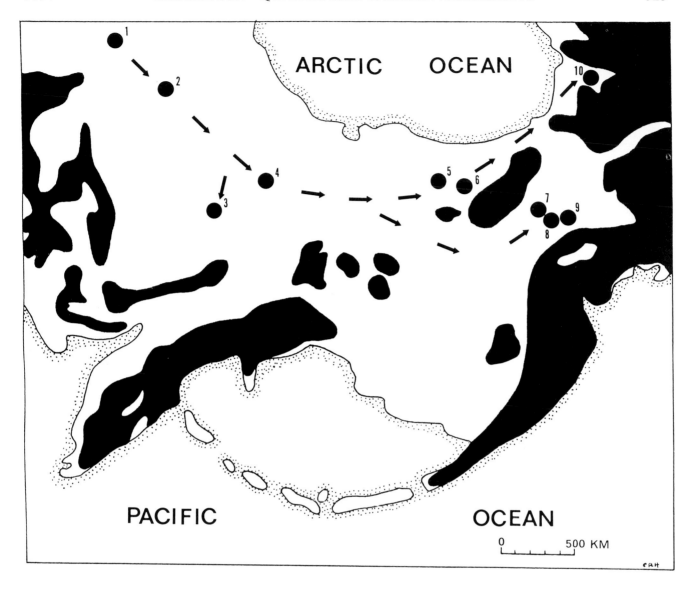

Fig. 13.—Generalized map of Beringia during the peak of the last glaciation showing localities of saiga (*Saiga tatarica*) fossils (black dots) (Harington, 1980*b*). Arrows indicate possible lowland routes used by saigas moving from western to eastern Beringia via the broad exposed plains of the Bering Isthmus (right of center). Glaciated areas are shown in black, ocean shorelines are stippled and land is white. Localities 1 to 4 are in northeastern Siberia, 5 to 9 are in Alaska, and 10 is in the Northwest Territories.

ACKNOWLEDGMENTS

Paleoenvironmental work usually requires a great deal of interdisciplinary cooperation. In this case, I would particularly like to thank: D. E. Sergeant, Graham Beard, S. Hicock, S. Lichti-Federovich, H. Tipper, R. J. Mott, J. G. Fyles, Gary J. Sawyer (who collected and is owner of the *Cervalces* specimen from near Columbia, New Jersey), J. V. Matthews, Jr., M. F. I. Smith, G. L. Mackie, the late Harold Schmidt, D. M. Jarzen, C. McAtee, M. Peschke, and D. Drummond.

Finally, I would like to pay tribute to the late John Guilday, who provided me with cheerful advice, help, and encouragement since I began work in the field of Quaternary mammals.

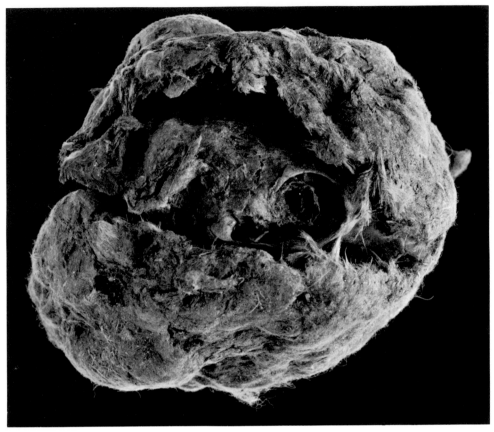

Fig. 14.—Carcass of a ground squirrel (*Spermophilus parryi*; NMC 37557)—perhaps more than 10,000 years old—from organic silt deposits on Glacier Creek, Yukon Territory. Note fur projecting through silt, and right side of back part of cranium exposed centrally.

LITERATURE CITED

AUGUSTA, J., and Z. BURIAN. 1963. A book of mammoths. Paul Hamlyn, London, 50 pp.

BANFIELD, A. W. F. 1974. The mammals of Canada. Univ. Toronto Press, Toronto, 438 pp.

BEE, J. W., and E. R. HALL. 1956. Mammals of northern Alaska on the Arctic Slope. Univ. Kansas Mus. Nat. Hist., Misc. Publ., 8:1–309.

BISHOP, S. C. 1921. Remains of a fossil phocid from Plattsburg, New York. J. Mamm., 2:170.

CRONIN, T. M. 1981. Paleoclimatic implications of late Pleistocene marine ostracodes from the St. Lawrence Lowlands. Micropaleontology, 27:384–418.

←

Fig. 15.—X-ray image of ground squirrel (*Spermophilus parryi*; NMC 37557) from Glacier Creek, Yukon Territory, suggesting that the animal had died curled up nose to tail in hibernating position.

Fyles, J. G. 1963. Surficial geology of Horne Lake and Parksville map-areas, Vancouver Island, British Columbia. Mem. Geol. Surv. Canada, 318:1–142.

Harington, C. R. 1975. A postglacial walrus from Bathurst Island, Northwest Territories. Canadian Field-Nat., 89:249–261.

———. 1977. Marine mammals in the Champlain Sea and the Great Lakes. Ann. New York Acad. Sci., 288:508–537.

———. 1980a. Whales and seals of the Champlain Sea. Trail & Landscape, 15(1):32–47.

———. 1980b. Radiocarbon dates on some Quaternary mammals and artifacts from northern North America. Arctic, 33:815–832.

———. 1981. Pleistocene saiga antelopes in North America and their paleoenvironmental implications. Pp. 193–225, in Quaternary paleoclimate (W. C. Mahaney, ed., Geo Books, Univ. East Anglia, Norwich, 464 pp.

Harington, C. R., and D. E. Sergeant. 1972. Pleistocene ringed seal skeleton from Champlain Sea deposits near Hull, Quebec—a reidentification. Canadian J. Earth Sci., 9:1039–1051.

Harington, C. R., H. W. Tipper, and R. J. Mott. 1974. Mammoth from Babine Lake, British Columbia. Canadian J. Earth Sci., 11:285–303.

Hartnagel, C. A., and S. C. Bishop. 1921. The mastodons, mammoths and other Pleistocene mammals of New York State. Bull. New York State Mus., 241–242:1–110.

Krog, J. 1954. Storing of food items in the winter nest of the Alaskan ground squirrel, Citellus undulatus. J. Mamm., 35:586.

Mansfield, A. W. 1967. Seals of arctic and eastern Canada. (2nd edition, revised). Bull. Fish. Res. Board Canada, 137:1–35.

Mott, R. J., T. W. Anderson, and J. V. Matthews, Jr. 1981. Late-glacial paleoenvironments of sites bordering the Champlain Sea based on pollen and macrofossil evidence. Pp. 129–171, Quaternary paleoclimate (W. C. Mahaney, ed.), Geo Books, Univ. East Anglia, Norwich, 464 pp.

Parris, D. C. 1983. New and revised records of Pleistocene mammals of New Jersey. The Mosasaur, 1:1–21.

Prest, V. K. 1969. Retreat of Wisconsin and Recent ice in North America. Geol. Surv. Canada Map, 1257A.

Ray, C. E. 1983. Hooded seal, Cystophora cristata: supposed fossil records in North America. J. Mamm., 64:509–512.

Wagner, F. J. E. 1959. Palaeoecology of the marine Pleistocene faunas of southwestern British Columbia. Bull. Geol. Surv. Canada, 52:1–67.

Address: National Museum of Natural Sciences, National Museums of Canada, Ottawa K1A 0M8, Canada.

PHYLETIC TRENDS AND EVOLUTIONARY RATES

Larry D. Martin

ABSTRACT

Single lineage phylogenies are compared to branching phylogenies. Such phylogenies are as theoretically sound and testable as are comparable branching phylogenies. Single lineage phylogenies provide unique data concerning evolutionary rates and processes. They have direct bearing on models of "punctuated equilibrium" and the importance of natural selection in evolution. Single lineage phylogenies also yield our most exact biostratigra-

phies and might be utilized to form a new scheme of absolute dating.

Examples of single lineage phylogenies are drawn from the saber-toothed cats and arvicolid rodents. These examples show long-term evolutionary trends which vary dramatically in their rates of change.

INTRODUCTION

Many workers have accepted two assumptions about the fossil record that seem to be mutually exclusive. One of these assumptions is that the fossil record is so good that major evolutionary changes, without gradual intermediaries, must be the result of unusual evolutionary processes (that is, macroevolution) rather than simply the result of missing sediments and their correlated geological time. The other assumption is that the fossil record is so bad that it is impossible to draw evolutionary lineages and hence the fossil record can give no useful clues to evolutionary history that cannot be better obtained from the study of modern biology (Patterson, 1981). The truth probably lies between these two viewpoints, and I will present evidence for phylogenetic lineages and argue that they do provide insight into evolutionary processes. On the other hand

these lineages seem to require no special evolutionary models and are gradual and overlapping when they can be examined in detail. Changes without intermediate forms in these examples seem to be the result of missing geological time. Many of the objections to lineage reconstruction are based on the idea that the recovery of a member of an ancestral population is an unlikely event. This objection may be valid for poorly known groups whose samples are separated by many millions of years, and lineage hypotheses in such situations are usually stated within the framework of higher taxonomic units. These hypotheses are useful as long as their real nature is understood. Highly refined lineages may sometimes be proposed. The recognition and use of such lineages is a promising area of vertebrate paleontology.

LINEAGE PHYLOGENIES

Lineage phylogenies imply that the direct ancestors of a taxon are contained within earlier taxa in the phylogeny. They do not imply that all species within that taxon (for instance a genus) must be ancestral to a subsequent taxon (another genus) anymore than that all subspecies within a species are in themselves directly ancestral to a later species. This is a general statement of ancestry, and the direct ancestor at the species level does not have to be identified. Statements of general ancestry may be falsified by showing that the common ancestor of two taxa (one of which is thought to be ancestral) must have preceded in time the development of the characters that are thought to unite them.

Special phylogenies may be proposed on a species

to species basis for segments of evolutionary lines. These phylogenies can be rigorously tested and defended. Objections to ancestor recognition are numerous in the cladistic literature, but it seems that many of these objections are due to an unnatural fear of paraphyletic groups (taxa that do not include all descendents of the common ancestor). This is a difficult policy to understand as all taxa must at some point have ancestors that are not included in the same taxon and whatever taxa those ancestors belong to must then be paraphyletic by definition. It would thus seem to be a waste of time to argue whether paraphyletic taxa are "natural" or not, as they exist in any classification containing ancestral forms. However, it would seem to be a fear of creating paraphyletic taxa that has been the motivation for arguments that phylogenies should always be shown on branching diagrams. Branching diagrams have also been defended as more "probable" and as testable while single lineages have often been treated as imaginary constructs without scientific merit.

Hennig demonstrated that only the presence of shared homologies at the level of their origin (synapomorphies) can unite taxa phylogenetically by testing between alternate hypotheses of the relative recency of common ancestry. These hypotheses must take the form of a three taxon statement of the sort: "taxa A and B share a common ancestor not shared by taxon C." The presence of complex and identical structures shared by "C" and "B," but absent in "A" and in all more remotely related groups would render this hypothesis unparsimonious under the criteria outlined below. This process of "outgroup" comparisons relies upon higher-level phylogenies previously accepted on the basis of other characters. It is by these comparisons that the "level of origin"

of homologies is determined. This procedure is truly reciprocally illuminative (Hull, 1967), although it may appear outwardly circular. Shared primitive characters (symplesiomorphies) and characters unique to only one taxon in the analysis (autapomorphies) have no relevance to three taxon tests (see further discussion by Wiley, 1976). A theory of relationships is considered corroborated only so long as it remains the most workable (parsimonious) interpretation of the biology of the organisms considered. When it does not meet these standards, it is said to be falsified.

Because primitive character states *must* temporally precede derived states, establishing the level of origin of homologies provides a relative temporal sequencing of the appearance of monophyletic groups (Fig. 1, A, II), as well as groupings of taxa that share more immediate common ancestors than other taxa in the analysis (Fig. 1, A, I). The statement of relationships is always a comparative hypothesis, and it is not possible to say that one taxon is *the* sister group of another.

TWO TAXON HYPOTHESES

The cladistic analysis of Hennig cannot provide a complete statement of phylogeny because it cannot test some types of phylogenetic relationships. Its application always results in a statement of comparative relationship that must be expressed in terms of three groups. Such an analysis cannot distinguish among three geometric relationships (Fig. 1, B: I, II, III) for any two taxa (B and C of Fig. 1, B). By convention, some workers have accepted type II in which common ancestors are hypothetical and evolution is assumed to be branching (see, for example, Tattersall and Eldredge, 1977).

I distinguish between simple phylogenies (I) and complex phylogenies ("cladograms," II, and III). A simple phylogeny has, as its basis a one-to-one relationship between ancestor and daughter taxa and thereby implies direct ancestry. A complex phylogeny is a combination of two or more simple phylogenies which share a common lineage until a point of cladogenesis (branching) occurs (Fig. 1, C). The point of cladogenesis is unique to a complex phylogeny and marks that point where parts of an evolving population become members of separate simple phylogenies. The time transected at a point of cladogenesis (insofar as it is a unique point) is only a single generation in length and is pragmatically un-

recognizable. Taxa are, in fact, defined on the basis of recognized evolutionary change along a segment of a simple phylogeny even when cladogenesis is known to have occurred. Evolution presents us with a continuum of simple phylogenies resolving themselves into complex phylogenies. Systematic analysis is applied to segments of this continuum. Because taxonomy involves the partitioning of a continuum, we must eventually separate parents from their offspring as we extend our analysis from the present into the past.

A hypothesis of an ancestor-descendant relationship between two taxa follows from a tested hypothesis of the relative recency of common ancestry (a three taxon hypothesis) based on shared derived character suites. A two taxon model may therefore be falsified by refutation of the common ancestor hypothesis upon which it is based. If, for example, it is demonstrated that "B" of Fig. 1, A shared a unique ancestor with "A," instead of with "C," none of the models of Fig. 1, B are correct.

The type II phylogeny of Fig. 1, B implies not only: 1) a cladogenetic event, and 2) a hypothetical taxon, but also that 3) this taxon has not been sampled and probably exists in some geographic region outside of the study area. Thus the local sequence

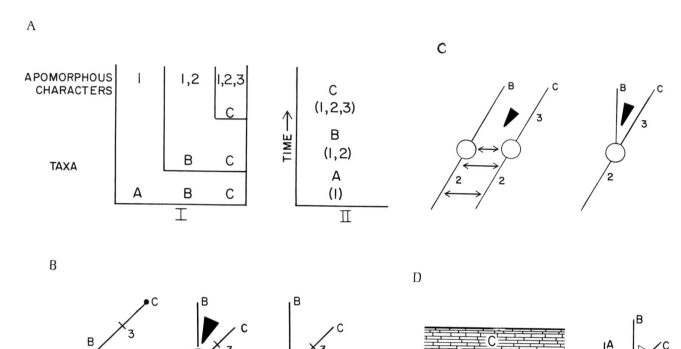

Fig. 1.—A) phylogenetic nesting arrangement (I) for three taxa (A, B, C) is provided by (II) temporal sequence of appearance of monophyletic groups; B) Alternate phylograms for taxa B and C of (A.), the black wedge implies the presence of a physical barrier and the open circle a hypothetical common ancestor. C) The cladogram as a combination of simple phylogenies which share a common lineage until branching occurs; D) A local sequence of taxa, (A, B, C) as explained by a branching sequence with two hypothetical common ancestors and two hypothetical barriers.

(Fig. 1, D) must be accounted for by 4) local extinction of B followed or caused by 5) immigration of descendants of the hypothetical taxon into the region of study. None of these assumptions are tested at the three taxon level. If taxa A, B, and C are related by derived characters as in Fig. 1, A and stratigraphically as in Fig. 1, D, a hypothesis of direct ancestry without a cladistic (branching) event is the most parsimonious interpetation of their geometry and should be maintained until branching is demonstrated.

The most basic assumption of a complex phylogram is the existence of a cladogenetic (vicariance) event. This is some aspect of the geographic or ecological environment that effectively divides up the population. This aspect of the environment I call a vicariator and symbolize by a wedge (Fig. 1, B–D). A cladogenetic event implies the existence of a vi-

cariator, even if its nature is not known. If two taxa (B and C, for example) are found to exist at the same time, and form a holophyletic assemblage, they must represent sister clades which shared a common ancestor until a cladogenetic event occurred. Their geometric relationships must be as in either model II or III of Fig. 1, B. A hypothesis of phyletic descent whereby taxon B evolves into taxon C without a cladistic event (Model 1) has thereby been refuted.

A cladogenetic event may also be demonstrated by the distribution of derived character states. Except for reversals in daughter taxa (which are assumed to be rare), an ancestral population must be primitive in all character states where it differs from a daughter taxon. Again, this is a two taxon statement, easily falsifiable in terms of relative parsimony because it is applicable to all morphological states by which the taxa differ. If taxon B of Fig. 1,

B is found to have derived characters unique to itself, type I and type II phylogenies are rejected. Both of these geometric arrangements are also rejected if taxon C can be shown to predate the evolutionary appearance of taxon B.

Falsification of a type I (simple) phylogeny corroborates a complex phylogeny and its assumptions. If a species is defined as a single evolutionary lineage, as was done by Wiley (1978), then the test of a type I phylogram is also a test of the hypothesis that a vertically stacked sequence of populations represents such a lineage. If a type I phylogeny has been refuted, type II and type III phylograms may be similarly tested until some geometric relationship is considered sufficiently corroborated.

Direct species-to-species lineages require close stratigraphic control, but they are clearly possible. Any time that we accept the continuance of one species through several geological horizons we have accepted all of the assumptions and probabilities that are required for species-to-species lineages. Such examples are actually common and have been used to support the idea of evolutionary "stasis" in punctuated equilibrium models. The American badger, *Taxidea taxus,* seems to be a good example, as the living species appears to extend back into the late Pliocene some 3.5 million years ago (Hibbard, 1970). Of course the complete anatomy of the Pliocene badger isn't known, and there could have been considerable anatomical change without our detecting it. In fact, we are always dealing with a subset of the total suite of characters, and the evolution we observe (or lack of it) must largely be a result of the particular set of characters we choose to observe. This is an important consideration when we talk about stasis versus change. We must realize that the great majority of characters in all organisms have been in stasis for tens of millions of years. For instance, red blood corpuscles in mammals are nearly uniform throughout the order and their basic features must have been established some time in the Mesozoic.

Size reduction in lineages of large mammals that survived the late Pleistocene extinction provide another set of examples of species lineages. There are many such examples, and in some the ice-age form has been placed in a separate species, whereas in others it is kept in the same species as its living descendants. The best documented lineage is that of the plains bison, *Bison bison bison,* whose descent from *B. antiquus* has been described by a number of workers (Schultz and Frankforter, 1946; Schultz and Martin, 1970a; Wilson, 1974). In this lineage the size reduction is gradual with very little change in proportions.

One of the important questions in any phylogenetic analysis is the direction of evolutionary change (polarity) in the characters used. Cladists tend to favor outgroup comparisons (comparison with character states in related groups) to determine polarity, but two other methods can also be applied to proposed lineages. The first of these is stratigraphic position. The youngest member of a lineage should be derived in all ways that it differs from the oldest member. Obviously stratigraphy in itself cannot be used to establish a lineage, as it only gives a *hypothesis of polarity.* This hypothesis can then be applied to all the available samples from stratigraphic positions lying between the two age extremes. In every case each successive stratigraphic stage should become more like the youngest sample in features that are changing. There is no reason to expect this result if a lineage is not present and if many successive stages are available; alternative explanations soon become absurdly complicated.

Functional analysis can also be used to formulate a hypothesis of polarity. If the function of a structure is well understood, then it should be possible to identify changes that improve the function; we would expect such changes to be more derived. It should also be possible to evaluate the importance of the structure to the organism and how likely a reversal in polarity might be.

EXAMPLES OF LINEAGES

I have previously applied Kurten's terminology for saber-toothed cats (dirk-toothed and scimitar-toothed) to all cats with the proper morphology (Martin, 1980), rather than to just the Smilodontini and Homotherini as he first applied them (Kurten, 1968). I also added a third category, the conical-toothed cats, which includes all the living felids. The dirk-toothed cats have the longest upper canines and seem in most respects to be the most highly modified.

The dirk-toothed adaptive type has developed independently at least four times (Fig. 2). The pattern

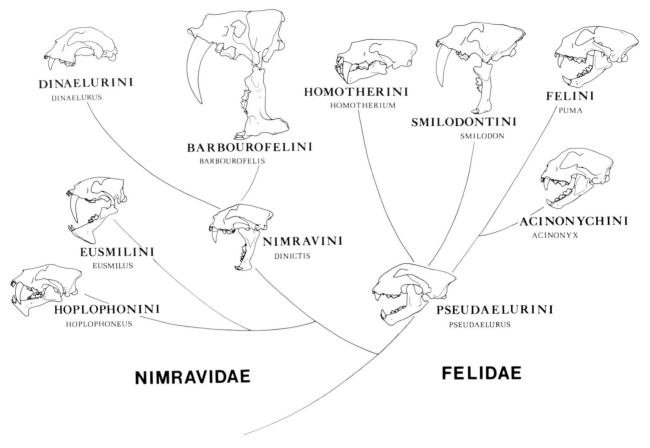

Fig. 2.—Suggested phylogeny of the cats (from Martin, 1980). The Smilodontini, Barbourofelini, Eusmilini, and Hoplphonini are dirk-toothed cats.

seems to be one of extinction followed by redevelopment of the adaptive type. In North America these redevelopments seem to be spaced at fairly regular intervals. The two best known lineages involve the Barbourofelini, a group of nimravid cats, and the Smilodontini, a group of felid cats.

The dirk-toothed cats of the tribe Barbourofelini are known from a number of sites whose relative stratigraphic positions are known. This makes it an excellent group to test some of the hypothesis of lineage relationships. The Barbourofelini may be united by the following derived features which separate them from all other cats: 1) upper canine flattened with prominent internal and external grooves; 2) enormous enlargement of the origin of the superficial masseter muscle; 3) serration of the cutting edges of all of the deciduous and permanent teeth, and 4) displacement of the lower cheek laterally on a distinct buttress of the lower jaw.

The oldest members of the tribe are found in Europe and the Barbourofelini first appeared in North America about eleven million years ago, with *Barbourofelis whitfordi* in the Clarendonian. The last form in the lineage is *B. fricki* which became extinct about seven million years ago. When we compare the earliest European form, where both the skull and lower jaws are known (*Sansanosmilus palmidens*), we find that there has been—enlargement of the canines and carnassials; reduction in the size of the P_3^3; change from an inclined to a vertical occipit; lowering of the glenoid fossae on the skull until they extend to the lower edge of the upper carnassials; development of a distinct area of overlap between P_4 and M_1, and reduction in the height of the coronoid process on the ramus. The direction of the polarity of all these characters is also confirmed by outgroup comparison with living felid cats. There is also a remarkable increase in size (Fig. 3).

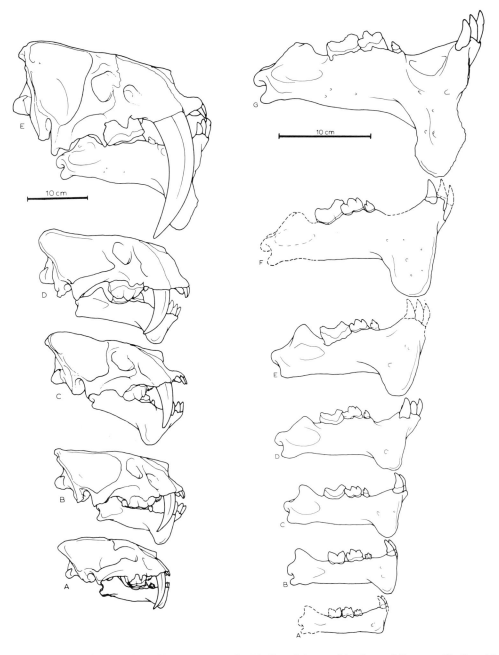

Fig. 3.—Stratigraphic sequence from oldest (A) to youngest for skull and jaws of barbourofelin cats. Skulls with lower jaws: A) *Sansanosmilus palmidens* (postorbital bar not developed yet); B) *Barbourofelis whitfordi* (earliest North American form); C) *B. morrisi*; D) *B. lovei*; E) *B. fricki*. Right rami: A) *Syrtosmilus syrtensis*; B) *Sansanosmilus palmidens*; C) *Barbourofelis whitfordi*; D) *B. morrisi*; E) *B. lovei*; F) *Barbourofelis* sp. from the Jack Swayze Quarry, Clark Co., Kansas; G) *B. fricki*.

The polarity of characters can also be tested functionally. Many of the specializations found in the skulls and lower jaws of saber-toothed cats are related to the elongation of the upper canines. The functional bite of the animal is the arc that lies be-

tween the tips of the upper and lower canines. The lower jaw has to rotate through a considerable distance before there is any separation of the canines. In the most specialized saber-toothed cat, *Barbourofelis fricki,* the total rotation must be slightly more than 115°. Features that facilitate this rotation include: 1) the vertical occipital region; 2) the lowering of the glenoid fossa on the skull; 3) the lowering of the coronoid process on the ramus, and 4) the enlargement of the paramastoid processes. Features that facilitate the carnassial bite include: 1) the enlargement of the upper carnassial; 2) the reduction of P_3^3 and 3) the overlapping of P_4 on M_1. We can thus see that the functional interpretation of these characters is in agreement with their stratigraphic distribution and with outgroup comparisons with other cats. When we look at the proposed lineage for *Barbourofelis* (Fig. 3), we can see that the changes in these features form a progressive series. We can also see that not all features change at the same rate, and that changes in a mechanical system stop after the system is perfected. For instance, a postorbital bar is achieved sometime between *Sansanosmilus* and *Barbourofelis,* and that region remains fairly constant after the development of the postorbital bar. The relationship of the P_4 to the M_1 is also fairly constant after an initial overlap is achieved.

We can make an additional check on our understanding of the evolutionary process in some other group of similar saber-toothed cats.

The Barbourofelini were very widespread during the Miocene, and are now known from France, Germany, Spain, Turkey, Mongolia, and a number of sites in the United States. They also appear to have occurred in the Miocene of Africa (Kurten, 1976; Ginsburg, 1978). The African material is not very good, and I accept it as belonging to the Barbourofelini on the basis of comparisons made by its describers. One of the African forms is Burdigalian in age and would probably be one of the oldest and most primitive barbourofelin known (Fig. 3), if it is correctly assigned. It would certainly be the smallest known species.

In order for the Barbourofelini to enter North America they must have had a northern Asiatic population around eleven million years ago. This population was probably continuous with the European population. The American populations seem to have been well separated from the European one soon after the dispersal into North America. This is a classical cladogenic event through dispersal with separate lineages of Barbourofelini evolving contemporaneously in North America and Eurasia. The European lineage shows very little increase in size (Fig. 4), but it does show the progressive reduction of P_3 and the overlap of P_4 and M_1 found in North American *Barbourofelis* lineage. It is clear that the

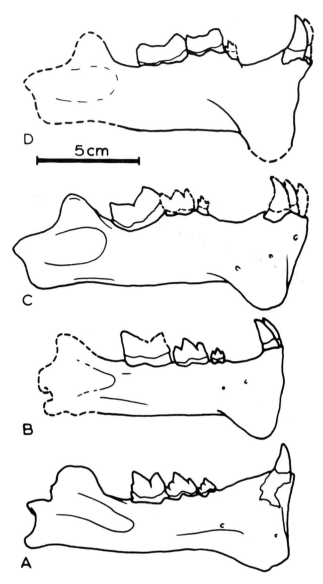

Fig. 4.—Stratigraphic sequence from oldest (A) to youngest (D) of barbourofelin cats in Eurasia. A) *Prosansanosmilus peregrinus,* Langenau 1 (Heizmann et al., 1980); B) *Sansanosmilus palmidens,* Sansan (Ginsburg, 1961); C) *Sansanosmilus jourdani vallesiensis* (Beaumont and Crusafont-Pairo, 1982); D) "*Sansanosmilus*" *piveteaui* (Ozansoy, 1965). Size does not increase very much but the P_3 becomes relatively smaller and loses a root, the carnassial becomes larger and it and P_4 tilt posteriorly.

overall rate of known evolutionary change was less in the European branch, and ultimately this resulted in less modification at the time that the European population became extinct.

We can make an additional check on our understanding of the evolutionary processes in the Barbourofelini by looking at the same sort of series in some other group of saber-toothed cats. If the functional progression has been correctly interpreted, and if the changes are the result of natural selection to improve those functions, we might expect to see the same pattern of changes in any group of animals that become functional dirk-toothed cats. This seems to actually be the case, and we can draw a parallel example from the dirk-toothed felids of the late Pliocene and Pleistocene, the Smilodontini. The genus *Megantereon* shares with *Smilodon* a short tail and shortened distal limb segments. The Smilodontini are felid cats and have the divided auditory bulla characteristic of that family. The Barbourofelini are nimravids (Fig. 2) and some workers consider their relationship to the true cats (felids) to be remote. The oldest known *Megantereon* seems to be the Pliocene, *M. hesperus* (Schultz and Martin, 1970*b*). The American lineage terminated in *Smilodon floridanus* about 10,000 years ago. The total duration of the lineage in North America is something on the order of 5,000,000 years. This lineage in North America is well represented by lower jaws (Fig. 5) but not by skulls of *Megantereon*. If we compare the skull of the European *M. megantereon* to *Smilodon floridanus,* we find that they differ in the younger form having enlarged the carnassial, reduced P_3^3, lowered the glenoid fossa, enlarged the paramastoid process, lowered the coronoid process on the ramus, overlapped M_1 with P_4 and increased in size dramatically. These changes are the same as those observed in *Barbourofelis,* and when we observe the stacked series of lower jaws (Fig. 5), we see that they develop in the same ordered, gradual progression seen in *Barbourofelis* (Fig. 3). There is one notable exception. The Barbourofelini have a progressive increase in the size of the dependent flange on the ramus. The Smilodontini have a progressive decrease in the relative size of the dependent flange, and in the Smilodontin lineage the occipit region actually becomes more inclined as the flange decreases in size.

The arvicolid rodents (voles and lemmings) provide a number of detailed species lineages (Martin, 1979). The arvicolids are grass-eating rodents, and show progressive hypsodonty in the evolution of their molars until most forms have ever-growing teeth. As the crowns of their molars become higher, the roots become lower and appear later ontogenetically. The function of the roots to anchor the tooth in the jaw is at the same time taken over by grooves in the enamel of the tooth crown that expose

the dentine covered by root cementum. The peridontal ligaments attach to the root cementum within these grooves, which are called dentine tracts. The dentine tracts take over the functions of the diminishing roots and the measurement of the height of the dentine tracts above the base of the tooth crown is a good measure of the hypsodonty of the molar. In evergrowing teeth the dentine tracts form continuous bands from the top to the bottom of the tooth crown.

Besides increasing the crown height of the molars, arvicolids increase the lifespan of their molars by increasing the resistance of their molars to abrasion. They do this by changing the arrangement of the apatite prisms in their molar enamel (Koenigswald, 1980), and the functional significance and directions of these changes have been worked out by Koenigswald.

The functional significance of the different arrangements of apatite prisms in the enamel of arvicolid molars seems to be related to the direction of chewing forces (Koenigswald, 1980). The primitive arrangement for arvicolids (the only type found in most of the early members of the arvicolid radiation and in some cricetids) is radial enamel. This arrangement when stressed by the directional forces of chewing permits fractures to transect the entire thickness of the enamel. Arvicolids with only radial enamel usually compensate for this by having very thick enamel. Lamellar or tangential enamel inhibit fractures from passing through the enamel and thereby preserve the enamel edges, extending the lifespan of the tooth.

The *Mimomys* lineage published by Koenigswald (1980) shows the general pattern of evolutionary change in enamel histology for an arvicolid rodent, and these changes are gradual and regular. They parallel very closely the other changes which are related to the rate of tooth wear and subsequent increases in hypsodonty. Such patterns argue very strongly for the effect of natural selection in these evolutionary patterns and provide little support for either stochastic or macroevolutionary models.

In North America, the muskrats (Ondatrini) have been the most extensively studied arvicolid rodents (Nelson and Semken, 1970; Martin, 1979; Zakrzewski, 1974), and the following species lineage is well accepted: *Pliopotamys minor, P. meadensis, Ondatra idahoensis, O. annectens, O. nebracensis, O. zibethicus.* Within this lineage we find the following evolutionary trends: 1) increase in size; 2) increase in hypsodonty and in dentine tract height;

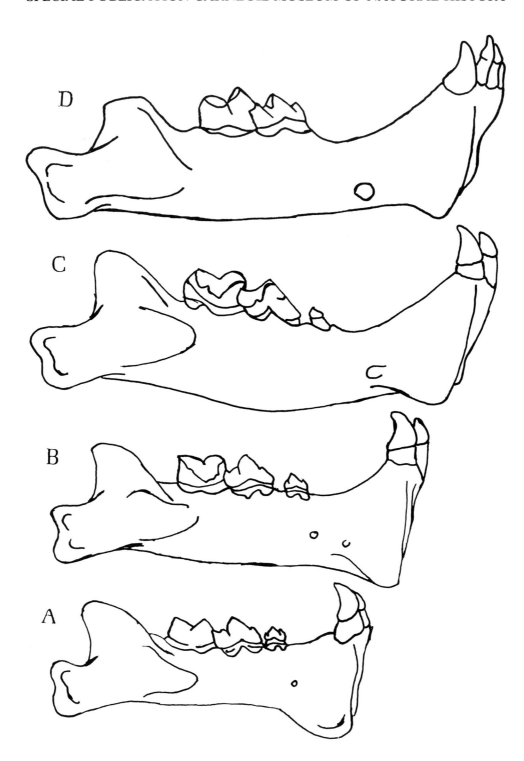

Fig. 5.—Stratigraphic sequence from oldest (A) to youngest (D) of lower jaws of Smilodontin cats in North America: A) *Megantereon hesperus*; B) *M. gracilis*; C) *Smilodon fatalis*; D) *Smilodon floridanus*.

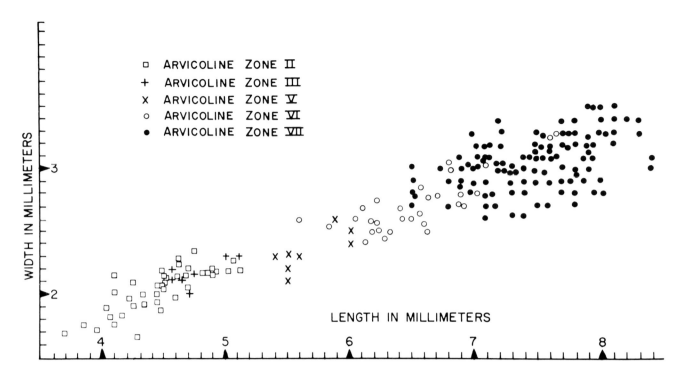

Fig. 6.—Measurements of lower first molars of muskrats showing increase in size from older (Arvicoline Zone II) to younger (Zone VII) arvicoline zones in North America (from Martin, 1979). Note that the size increase is overlapping except for that area where no sample was available (Zone IV).

3) increase in crown complexity on M_1, and 4) addition of cement to the re-entrant angles of the molar crowns. In Figs. 6 and 7 we can see that the increases of size and crown complexity are gradual and overlapping.

The arvicolid molars consist of a series of generally alternating triangles. In the M_1 of very primitive forms like *Prosomys*, there is an anterior loop followed by three alternating triangles. These correspond to the anteroconid, metaconid, protoconid, and hypoconid, plus the posterior cingulum in the cricetine M_1. Most of the evolutionary modification of the grinding surface of M_1 achieved by increasing the length and complexity of the anteroconid, and the elaboration of the crown pattern results from selection acting on crenulations of the anteroconid (anterior loop). In early arvicolids these crenulations occupy the position of alternating triangles in later forms. The new triangles are formed by enlargement of these crenulations coupled with enlargement of their re-entrants. The oldest known muskrat, *Plioptamys minor*, has a highly crenulated anterior loop followed by five distinct alternating triangles and a

posterior loop. The fourth and fifth triangles become better separated from the anterior loop in *P. meadensis* and *Ondatra idahoensis*. In *Ondatra annectens*, *O. nebracensis*, and *O. zibethicus*, the posterior portion of the anterior loop opens broadly into sixth and seventh triangles. Cement first appears as tiny deposits in the bases of the re-entrant angles of the molars of *O. idahoensis*, and eventually fills re-entrants in later forms. The muskrat chronocline published by Nelson and Semken (1970) includes four samples that are reasonably well dated: Hagerman, 3.5 m.y.; Borchers, 1.9 m.y.; Cudahy, 0.6 m.y.; and the Wisconsin samples, all of which should date less than 100,000 years before present. Borchers is lumped with Grandview which seems to be a slightly older fauna, and the curve could probably be improved in that area. Dentine tracts are extremely low or absent in the Hagerman sample, and I start the curve at zero at that point. There is a fairly slow steady increase in both size of M_1 and dentine tract height from the Hagerman to the Cudahy faunas, a time of nearly two million years. After the Cudahy faunas the last 600,000 years has been a time of very

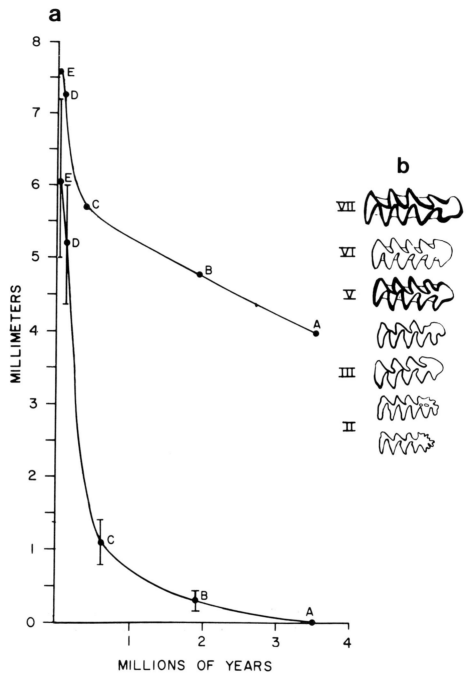

Fig. 7.—Graph *a* showing change in dentine tract height on M_1 (lower curve) and length of M_1 (upper curve) for muskrats (measurements adapted from Nelson and Semken, 1970); (A) Hagerman Local Fauna, (B) Borchers and Grandview local faunas, (C) Cudahy local faunas, (D) Wisconsinan faunas, and (E) modern muskrats. Part *b*; Left M_1's of muskrats arranged in stratigraphic sequence with oldest at the bottom and North American arvicoline zone listed on the left. From bottom to top: *Pliopotamys minor, P. meadensis, Ondatra idahoensis, O.* cf. *annectens, O. annectens, O. nebracensis, O. zibethecus* (from Martin, 1979).